COMPUTATIONAL FLUID MECHANICS AND HEAT TRANSFER

Series in Computational Methods in Mechanics and Thermal Sciences

W. J. Minkowycz and E. M. Sparrow, *Editors*

Anderson, Tannehill, and Pletcher, Computational Fluid Mechanics and Heat Transfer
Aziz and Na, Perturbation Methods in Heat Transfer
Baker, Finite Element Computational Fluid Mechanics
Jaluria and Torrance, Computational Heat Transfer
Patankar, Numerical Heat Transfer and Fluid Flow
Shih, Numerical Heat Transfer

PROCEEDINGS
Shih, Editor, Numerical Properties and Methodologies in Heat Transfer: Proceedings of the Second National Symposium

COMPUTATIONAL FLUID MECHANICS AND HEAT TRANSFER

Dale A. Anderson
Professor of Aerospace Engineering
Iowa State University

John C. Tannehill
Professor of Aerospace Engineering
Iowa State University

Richard H. Pletcher
Professor of Mechanical Engineering
Iowa State University

HEMISPHERE PUBLISHING CORPORATION, New York
A subsidiary of Harper & Row, Publishers, Inc.

Cambridge Philadelphia San Francisco Washington
London Mexico City São Paulo Singapore Sydney

COMPUTATIONAL FLUID MECHANICS AND HEAT TRANSFER

5 6 7 8 9 0 EBEB 8 9 8 7

This book was set in Press Roman by Hemisphere Publishing Corporation. The editor was Brenda M. Brienza; the production supervisor was Miriam Gonzalez; and the typesetters were A. Wayne Hutchins, Sandra F. Watts, and Peggy M. Rote. Edwards Brothers, Inc. was printer and binder.

Library of Congress Cataloging in Publication Data

Anderson, Dale A. (Dale Arden), date
Computational fluid mechanics and heat transfer.

(Series in computational methods in mechanics and thermal sciences)
Bibliography: p.
Includes index.
1. Fluid mechanics. 2. Heat–Transmission.
I. Tannehill, John C. II. Pletcher, Richard H.
III. Title. IV. Series.
QA901.A53 1984 532 83-18614
ISBN 0-89116-471-5

To our wives and children: Marleen, Greg, and Lisa Anderson

Marcia, Michelle, and John Tannehill

Carol, Douglas, Laura, and Cynthia Pletcher

To our wives and children, Madison, Greg, and Lisa Anderson;

Marion, Nicole, and [illegible];

Carol, Douglas, Leah, and Cynthia Fletcher

CONTENTS

PREFACE

This book is intended to serve as a text for introductory courses in computational fluid mechanics and heat transfer [or, synonymously, computational fluid dynamics (CFD)] for advanced undergraduates and/or first-year graduate students. The text has been developed from notes prepared for a two-course sequence taught at Iowa State University for more than a decade. No pretense is made that every facet of the subject is covered, but it is hoped that this book will serve as an introduction to this field for the novice. The major emphasis of the text is on finite-difference methods.

The material has been divided into two parts. The first part, consisting of Chapters 1-4, presents basic concepts and introduces the reader to the fundamentals of finite-difference methods. The second part of the book, consisting of Chapters 5-10, is devoted to applications involving the equations of fluid mechanics and heat transfer. Chapter 1 serves as an introduction, while a brief review of partial differential equations is given in Chapter 2. Finite-difference methods and the notions of stability, accuracy, and convergence are discussed in the third chapter.

Chapter 4 contains what is perhaps the most important information in the book. Numerous finite-difference methods are applied to linear and nonlinear model partial-differential equations. This provides a basis for understanding the results produced when different numerical methods are applied to the same problem with a known analytic solution.

Building on an assumed elementary background in fluid mechanics and heat transfer, Chapter 5 reviews the basic equations of these subjects, emphasizing forms most suitable for numerical formulations of problems. A section on turbulence modeling is included in this chapter. Methods for solving inviscid flows using both conservative and nonconservative forms are presented in Chapter 6. Techniques for solving the boundary-layer equations for both laminar and turbulent flows are discussed in Chapter 7. Chapter 8 deals with equations of a class known as the "parabolized" Navier-Stokes equations which are useful for flows not adequately modeled by the boundary-layer equations, but not requiring the use of the full Navier-Stokes equations. Parabolized schemes for both subsonic and supersonic flows over external surfaces and in confined regions are included in this chapter. Chapter 9 is devoted to methods for the complete Navier-Stokes equations, including the Reynolds averaged form. A brief introduction to methods for grid generation is presented in Chapter 10 to complete the text.

At Iowa State University, this material is taught to classes consisting primarily of aerospace and mechanical engineers although the classes often include students from other branches of engineering and earth sciences. It is our experience that Part 1 (Chapters 1-4) can be adequately covered in a one-semester, three-credit-hour course. Part 2 of the book contains more information than can be covered in great detail in most one-semester, three-credit-hour courses. This permits Part 2 to be used for courses with different objectives. Although we have found that the major thrust of each of Chapters 5 through 10 can be covered in one semester, it would also be possible to use only parts of this material for more specialized courses. Obvious modules would be Chapters 5, 6, and 10 for a course emphasizing inviscid flows or Chapters 5, 7-9, (and perhaps 10) for a course emphasizing viscous flows. Other combinations are clearly possible. If only one course can be offered in the subject, choices also exist. Part 1 of the text can be covered in detail in the single course or, alternatively, only selected material from Chapters 1-4 could be covered as well as some material on applications of particular interest from Part 2. The material in the text is reasonably broad and should be appropriate for courses having a variety of objectives.

For background, students should have at least one basic course in fluid dynamics, one course in ordinary differential equations, and some familiarity with partial differential equations. Of course, some programming experience is also assumed.

The philosophy used throughout the CFD course sequence at Iowa State and embodied in this text is to encourage students to construct their own computer programs. For this reason, "canned" programs for specific problems do not appear in the text. Use of such programs does not enhance basic understanding necessary for algorithm development. At the end of each chapter, numerous problems are listed that necessitate numerical implementation of the text material. It is assumed that students have access to a high-speed digital computer.

We wish to acknowledge the contributions of all of our students, both past and present. We are deeply indebted to F. Blottner, S. Chakravarthy, G. Christoph, J. Daywitt, T. Holst, M. Hussaini, J. Ievalts, D. Jespersen, O. Kwon, M. Malik, J. Rakich, M. Salas, V. Shankar, R. Warming, and many others for helpful suggestions for improving the text. We would like to thank Pat Fox and her associates for skillfully preparing the illustrations. A special thanks to Shirley Riney for typing and editing the manuscript. Her efforts were a constant source of encouragement. To our wives and children, we owe a debt of gratitude for all of the hours stolen from them. Their forbearance is greatly appreciated.

Finally, a few words about the order in which the authors' names appear. This text is a collective work by the three of us. There is no junior or senior author. The final order was determined by a coin flip. Despite the emphasis on finite-difference methods in the text, we resorted to a "Monte Carlo" method for this determination.

Dale A. Anderson
John C. Tannehill
Richard H. Pletcher

PART ONE

FUNDAMENTALS OF FINITE-DIFFERENCE METHODS

CHAPTER

ONE

INTRODUCTION

1-1 GENERAL REMARKS

The development of the high-speed digital computer has had a great impact on the way in which principles from the sciences of fluid mechanics and heat transfer are applied to problems of design in modern engineering practice. Problems can now be solved at very little cost in a few seconds of computer time which would have taken years to work out with the computational methods and computers available twenty years ago. The ready availability of previously unimaginable computing power has stimulated many changes. These were first noticeable in industry and research laboratories where the need to solve complex problems was the most urgent. More recently, changes brought about by the computer have been occurring in university classrooms where students are being exposed to the fundamentals which must be mastered in order to make the best use of modern computational tools. It is hoped that the present textbook will contribute to the organization and dissemination of some of this new information.

We have been witnessing the rise to importance of a new methodology for attacking the complex problems in fluid mechanics and heat transfer which has become known as computational fluid dynamics (CFD). In this computational (or numerical) approach, the equations (usually in partial differential form) which govern a process of interest are solved numerically. Some of the ideas are very old. The evolution of numerical methods, especially finite-difference methods for solving ordinary and partial differential equations began at about the turn of the century. The automatic digital computer was invented by Atanasoff in the late 1930's (see Gardner, 1982) and was used from nearly the beginning to solve problems in fluid dynamics. Still, these events alone did not revolutionize engineering practice. The explosion in computational activity did not begin until a third ingredient, general availability of high-speed digital computers, occurred in the 1960's.

Traditionally, both experimental and theoretical methods have been used to develop designs for equipment and vehicles involving fluid flow and heat transfer. With the advent of the digital computer, a third method, the numerical approach, has become available. Although experimentation continues to be important, especially when the flows involved are very complex, the trend is clearly toward greater reliance on computer based predictions in design.

This trend can be largely explained by economics (Chapman, 1979). Over the years, computer speed has increased much more rapidly than computer costs. The net effect has been a phenomenal decrease in the cost of performing a given calculation. This is illustrated in Fig. 1-1 where it is seen that the cost of performing a given calculation has been reduced by a factor of 10 every 8 years. (Compare this with the trend in the cost of peanut butter in the past 8 years.) This trend in the cost of computations is based on the use of the best computers available. It is true that not every user will have easy access to the most recent computers. When improvements in numerical algorithms are taken into account, the reduction in cost is even greater. Chapman (1979) cites an impressive example of this trend in computing efficiency. "A numerical simulation of the flow over an airfoil using the Reynolds averaged Navier-Stokes equations can be conducted on today's supercomputers in less than half an hour for less than $1000 cost in computer time. If just one such simulation had been attempted 20 years ago on computers of that time (e.g., IBM 704 Class) and with algorithms then known, the cost in computer time would have amounted to roughly $10 million, and the results for that single flow would not be available until 10 years from now, since the computation would have taken about 30 years to complete." This trend of lower costs for computations is expected to continue for some time into the future. On the other hand, the costs of performing experiments have been steadily increasing in recent years.

The suggestion here is not that computational methods will soon completely replace experimental testing as a means to gather information for design purposes. Rather, it is believed that computer methods are likely to be used more extensively in the future. In most fluid flow and heat transfer design situations it will still be neces-

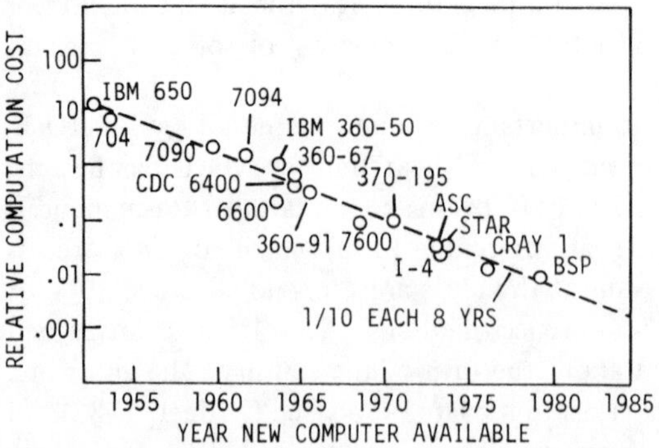

Figure 1-1 Trend of relative computation cost for a given flow and algorithm (Chapman, 1979).

sary to employ some experimental testing. However, computer studies can be used to reduce the range of conditions over which testing is required. The need for experiments will probably remain for quite some time in applications involving turbulent flow where it is presently not economically feasible to utilize computational models which are free of empiricism. In applications involving multiphase flows, boiling, or condensation, especially in complex geometries, the experimental method remains the primary source of design information. Progress is being made in computational models for these flows but the work remains in a relatively primitive state compared to the status of predictive methods for laminar single phase flows over aerodynamic bodies.

1-2 COMPARISON OF EXPERIMENTAL, THEORETICAL, AND NUMERICAL APPROACHES

As mentioned in the previous section, there are basically three approaches or methods which can be used to solve a problem in fluid mechanics and heat transfer. These methods are:

1. Experimental
2. Theoretical
3. Numerical (CFD)

The theoretical method is often referred to as an analytical approach while the terms numerical and computational are used interchangeably. In order to illustrate how these three methods would be used to solve a fluid flow problem, let us consider the classical problem of determining the pressure on the front surface of a circular cylinder in a uniform flow of air at a Mach number (M_∞) of 4 and a Reynolds number (based on the diameter of the cylinder) of 5×10^6.

In the experimental approach, a circular cylinder model would first need to be designed and constructed. This model must have provisions for measuring the wall pressures and it should be compatible with an existing wind tunnel facility. The wind tunnel facility must be capable of producing the required freestream conditions in the test section. The problem of matching flow conditions in a wind tunnel can often prove to be quite troublesome, particularly for tests involving scale models of large aircraft and space vehicles. Once the model has been completed and a wind tunnel selected, the actual testing can proceed. Since wind tunnels require large amounts of energy for their operation, the wind tunnel test time must be kept to a minimum. The efficient use of wind tunnel time has become increasingly important in recent years with the escalation of energy costs. After the measurements have been completed, wind tunnel correction factors can be applied to the raw data to produce the final wall pressure results. The experimental approach has the capability of producing the most realistic answers for many flow problems; however, the costs are becoming greater every day.

In the theoretical approach, simplifying assumptions are made in order to make the problem tractable. If possible, a closed-form solution is sought. For the present problem, a useful approximation is to assume a Newtonian (see Hayes and Probstein, 1966) flow of a perfect gas. With the Newtonian flow assumption, the shock layer (region between body and shock) is infinitesimally thin, and the bow shock lies adjacent to the surface of the body as seen in Fig. 1-2(a). Thus, the normal component of the velocity vector becomes zero after passing through the shock wave since it immediately impinges on the body surface. The normal momentum equation across a shock wave (see Chapter 5) can be written as

$$p_1 + \rho_1 u_1^2 = p_2 + \rho_2 u_2^2 \tag{1-1}$$

where p is the pressure, ρ is the density, u is the normal component of velocity, and the subscripts 1 and 2 refer to the conditions immediately upstream and downstream of the shock wave, respectively. For the present problem [see Fig. 1-2(b)], Eq. (1-1) becomes

$$p_\infty + \rho_\infty V_\infty^2 \sin^2 \sigma = p_{\text{wall}} + \rho_{\text{wall}} \cancelto{0}{u_{\text{wall}}^2} \tag{1-2}$$

or

$$p_{\text{wall}} = p_\infty \left(1 + \frac{\rho_\infty}{p_\infty} V_\infty^2 \sin^2 \sigma\right) \tag{1-3}$$

For a perfect gas, the speed of sound in the freestream is

$$a_\infty = \sqrt{\frac{\gamma p_\infty}{\rho_\infty}} \tag{1-4}$$

where γ is the ratio of specific heats. Using the definition of Mach number

$$M_\infty = \frac{V_\infty}{a_\infty} \tag{1-5}$$

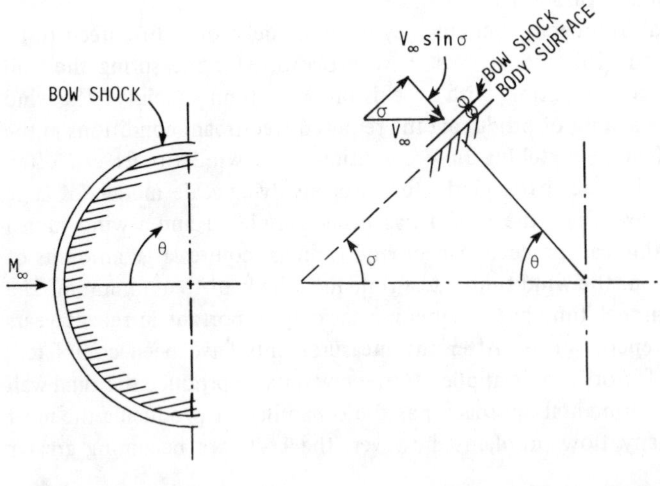

Figure 1-2 Theoretical approach. (a) Newtonian flow approximation. (b) Geometry at shock.

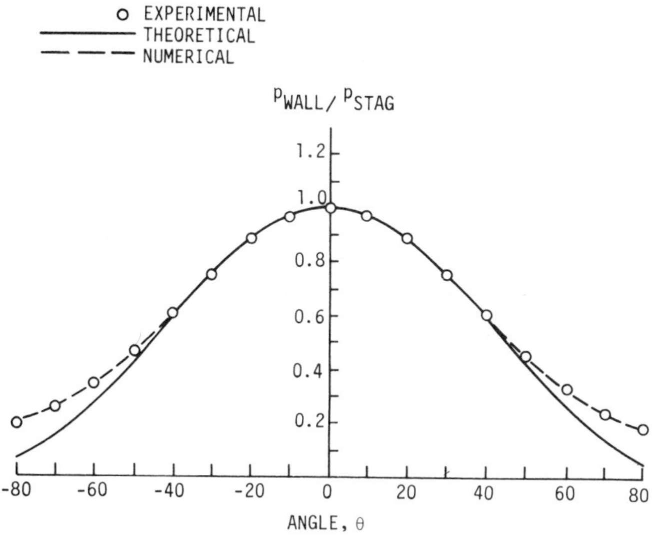

Figure 1-3 Surface pressure on circular cylinder.

and the trigonometric identity

$$\cos\theta = \sin\sigma \tag{1-6}$$

Equation (1-3) can be rewritten as

$$p_{\text{wall}} = p_\infty(1 + \gamma M_\infty^2 \cos^2\theta) \tag{1-7}$$

At the stagnation point, $\theta = 0°$ so that the wall pressure becomes

$$p_{\text{stag}} = p_\infty(1 + \gamma M_\infty^2) \tag{1-8}$$

After inserting the stagnation pressure into Eq. (1-7), the final form of the equation is

$$p_{\text{wall}} = p_\infty + (p_{\text{stag}} - p_\infty)\cos^2\theta \tag{1-9}$$

The accuracy of this theoretical approach can be greatly improved if, in place of Eq. (1-8), the stagnation pressure is computed from Rayleigh's pitot formula (Shapiro, 1953)

$$p_{\text{stag}} = p_\infty \left[\frac{(\gamma+1)M_\infty^2}{2}\right]^{\gamma/(\gamma-1)} \left[\frac{\gamma+1}{2\gamma M_\infty^2 - (\gamma-1)}\right]^{1/(\gamma-1)} \tag{1-10}$$

which assumes an isentropic compression between the shock and body along the stagnation streamline. The use of Eq. (1-9) in conjunction with Eq. (1-10) is referred to as the modified Newtonian theory. The wall pressures predicted by this theory are compared in Fig. 1-3 to the results obtained using the experimental approach (Beckwith and Gallagher, 1961). Note that the agreement with the experimental results is quite good up to about ±35°. The big advantage of the theoretical approach is that "clean," general information can be obtained, in many cases, from a simple formula as in the present example. This approach is quite useful in preliminary design work since reasonable answers can be obtained in a minimum amount of time.

In the numerical approach, a limited number of assumptions are made and a high-speed digital computer is used to solve the resulting governing fluid dynamic equations. For the present high Reynolds number problem, inviscid flow can be assumed since we are only interested in determining wall pressures on the forward portion of the cylinder. Hence, the Euler equations are the appropriate governing fluid dynamic equations. In order to solve these equations, the region between the bow shock and body must first be subdivided into a computational grid as seen in Fig. 1-4. The partial derivatives appearing in the unsteady Euler equations are replaced by appropriate finite differences at each grid point. The resulting equations are then integrated forward in time until a steady-state solution is obtained asymptotically after a sufficient number of time steps. The details of this approach will be discussed in forthcoming chapters. The results of this technique (Daywitt and Anderson, 1974) are shown in Fig. 1-3. Note the excellent agreement with experiment.

In comparing the methods we note that a computer simulation is free of some of the constraints imposed on the experimental method for obtaining information upon which to base a design. This represents a major advantage of the computational method which should be increasingly important in the future. The idea of experimental testing is to evaluate the performance of a relatively inexpensive small scale version of the prototype device. In performing such tests, it is not always possible to simulate the true operating conditions of the prototype. For example, it is very difficult to simulate the large Reynolds numbers of aircraft in flight, atmospheric re-entry conditions, or the severe operating conditions of some turbomachines in existing test facilities. This suggests that the computational method, which has no such restrictions,

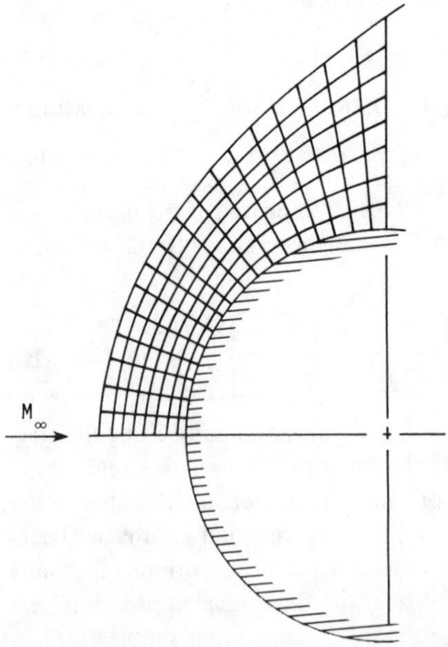

Figure 1-4 Computational grid.

Table 1-1 Comparison of approaches

Approach	Advantages	Disadvantages
Experimental	1. Capable of being most realistic	1. Equipment required 2. Scaling problems 3. Tunnel corrections 4. Measurement difficulties 5. Operating costs
Theoretical	1. Clean, general information which is usually in formula form	1. Restricted to simple geometry and physics 2. Usually restricted to linear problems
Numerical	1. No restriction to linearity 2. Complicated physics can be treated 3. Time evolution of flow can be obtained	1. Truncation errors 2. Boundary condition problems 3. Computer costs

has the potential of providing information not available by other means. On the other hand, computational methods also have limitations; among these are computer storage and speed. Other limitations arise due to our inability to understand and mathematically model certain complex phenomena. None of these limitations of the computational method are insurmountable in principle and current trends show reason for optimism about the role of the computational method in the future. As seen in Fig. 1-1, the relative cost of computing a given flowfield has decreased by almost three orders of magnitude during the past 20 years, and this trend is expected to continue in the near future. As a consequence, many believe that wind tunnels will someday play a secondary role to the computer, much in the same manner as ballistic ranges now perform secondary roles to computers in trajectory mechanics (Chapman, 1975).

Some of the advantages and disadvantages of the three approaches are summarized in Table 1-1. In closing, it should be mentioned that it is sometimes difficult to distinguish between the different methods. For example, when numerically computing turbulent flows, the eddy viscosity models which are frequently used are obtained from experiments. Likewise, many theoretical techniques which employ numerical calculations could be classified as numerical approaches.

1-3 HISTORICAL PERSPECTIVE

As one might expect, the history of CFD is closely tied to the development of the digital computer. Most problems were solved using methods that were either analytical or empirical in nature until the end of World War II. Prior to this time, there were a few pioneers using numerical methods to solve problems. Of course, the calculations were performed by hand and a single solution represented a monumental amount of work. Since that time, the digital computer has been developed and the routine calculations required in obtaining a numerical solution are carried out with ease.

The actual beginning of CFD or the development of methods crucial to CFD is a matter of conjecture. Most people attribute the first definitive work of importance to Richardson (1910) who introduced point iterative schemes for numerically solving Laplace's equation and the biharmonic equation in an address to the Royal Society of London. He actually carried out calculations for the stress distribution in a masonry dam. In addition, he clearly defined the difference between problems which must be solved by a relaxation scheme and those which we refer to as marching problems.

Richardson developed a relaxation technique for solving Laplace's equation. His scheme used data available from the previous iteration to update each value of the unknown. In 1918, Liebmann presented an improved version of Richardson's method. Liebmann's method used values of the dependent variable both at the new and old iteration level in each sweep through the computational grid. This simple procedure of updating the dependent variable immediately reduced the convergence times for solving Laplace's equation. Both the Richardson method and Liebmann's scheme are usually used in elementary heat transfer courses to demonstrate how apparently simple changes in a technique greatly improve efficiency.

Sometimes the beginning of modern numerical analysis is attributed to a famous paper by Courant, Friedrichs, and Lewy (1928). The acronym CFL, frequently seen in the literature, stands for these three authors. In this paper, uniqueness and existence questions were addressed for the numerical solutions of partial differential equations. Testimony to the importance of this paper is evidenced in its republication in 1967 in the *IBM Journal of Research and Development.* This paper is the original source for the CFL stability requirement for the numerical solution of hyperbolic partial differential equations.

In 1940, Southwell introduced a relaxation scheme which was extensively used in solving both structural and fluid dynamic problems where an improved relaxation scheme was required. His method was tailored for hand calculations in that point residuals were computed and these were scanned for the largest value. The point where the residual was largest was always relaxed as the next step in the technique. During the decades of the 1940's and 1950's, Southwell's methods were generally the first numerical techniques introduced to engineering students. Allen and Southwell (1955) applied Southwell's scheme to solve the incompressible, viscous flow over a cylinder. This solution was obtained by hand calculation and represents a substantial amount of work. Their calculation added to the existing viscous flow solutions which began to appear in the 1930's.

During World War II and immediately following, a great deal of research was done on the use of numerical methods for solving problems in fluid dynamics. It was during this time that Professor John von Neumann developed his method for evaluating the stability of numerical methods for solving time-marching problems. It is interesting that Professor von Neumann did not publish a comprehensive description of his methods. However, O'Brien, Hyman, and Kaplan (1950) later presented a detailed description of the von Neumann method. This paper is significant because it presents a practical way of evaluating stability that can be understood and used reliably by scientists and engineers. The von Neumann method is the most widely used technique in computational fluid dynamics for determining stability. Another of the important contributions appearing at about the same time was due to Peter Lax (1954). Lax developed a technique for computing fluid flows including shock waves which represent discontinuities in the flow variables. No special treatment was required for computing the shocks. This special feature developed by Lax was due to the use of the conservation-law form of the governing equations and is referred to as shock capturing.

At the same time, progress was being made on the development of methods for both elliptic and parabolic problems. Frankel (1950) presented the first version of the successive overrelaxation (SOR) scheme for solving Laplace's equation. This provided a significant improvement in the convergence rate. Peaceman and Rachford (1955) and Douglas and Rachford (1956) developed a new family of implicit methods for parabolic and elliptic equations in which sweep directions were alternated and the allowed step size was unrestricted. These methods are referred to as alternating direction implicit (ADI) schemes and are used extensively today.

Books treating various aspects of CFD began to appear in the late fifties and early sixties. The early book by Richtmyer (1957) and later Richtmyer and Morton (1967) provided a source of information for solving marching problems while Forsythe and Wasow (1960) emphasized methods for problems of the elliptic type. Research in CFD also continued at a rapid pace during the decade of the sixties. Early efforts at solving flows with shock waves used either the Lax approach or an artificial viscosity scheme introduced by von Neumann and Richtmyer (1950). Early work at Los Alamos included the development of schemes like the particle-in-cell method (PIC), which used the dissipative nature of the finite-difference scheme to smear the shock over several mesh intervals (Evans and Harlow, 1957). In 1960, Lax and Wendroff introduced a method for computing flows with shocks which was a second-order scheme that avoided the excessive smearing of the earlier approaches. The MacCormack (1969) version of this technique is one of the popular methods in use today for solving problems with shocks. Gary (1962) presented early work demonstrating techniques for fitting moving shocks, thus avoiding the smearing associated with the previous shock-capturing schemes. Moretti and Abbett (1966) and Moretti and Bleich (1968) applied shock-fitting procedures to multidimensional supersonic flow over various configurations. Richtmyer and Morton also described early shock-fitting schemes. Even today, we see either shock-capturing or shock-fitting methods used to solve problems with shock waves.

During the past 15 years, the progress made and the number of researchers working in computational fluid dynamics has expanded at an ever increasing rate. For this reason, it would be difficult to provide a short history of this time period and include all whose contributions are significant. However, the paper by Hall (1981) is recommended for a summary of events since 1950. In closing, three additional papers should be noted because they were published to inform the general scientific community about computational fluid dynamics. Macagno (1965) and Harlow and Fromm (1965) published papers in the French magazine, *La Houille Blanche*, and *Scientific American*, respectively. They explained the usefulness of numerical methods applied to fluid mechanics and heat transfer and provided a number of computed examples demonstrating results obtained. A more recent article by Levine (1982) appeared in *Scientific American* detailing the potential of computational methods. As in the earlier papers, this served as a vehicle for informing the scientific community of the status of computational fluid dynamics and the new generation of supercomputers.

CHAPTER

TWO

PARTIAL DIFFERENTIAL EQUATIONS

2-1 INTRODUCTION

Many important physical processes in nature are governed by partial differential equations (PDE's). For this reason it is important to understand the physical behavior of the model represented by the PDE. In addition, knowledge of the mathematical character, properties, and the solution of the governing equations is required. In this chapter we will discuss the physical significance and the mathematical behavior of the most common types of PDE's encountered in fluid mechanics and heat transfer. Examples are included to illustrate important properties of the solutions of these equations. In the last sections, we extend our discussion to systems of PDE's and present a number of model equations, many of which are used in Chapter 4 to demonstrate the application of various finite-difference methods.

2-2 PHYSICAL CLASSIFICATION

2-2.1 Equilibrium Problems

Equilibrium problems are problems in which a solution of a given partial differential equation is desired in a closed domain subject to a prescribed set of boundary conditions (see Fig. 2-1). Equilibrium problems are boundary value problems. Examples of such problems include steady-state temperature distributions, incompressible inviscid flows, and equilibrium stress distributions in solids.

Sometimes equilibrium problems are referred to as jury problems. This is an apt name since the solution of the partial differential equation at every point in the

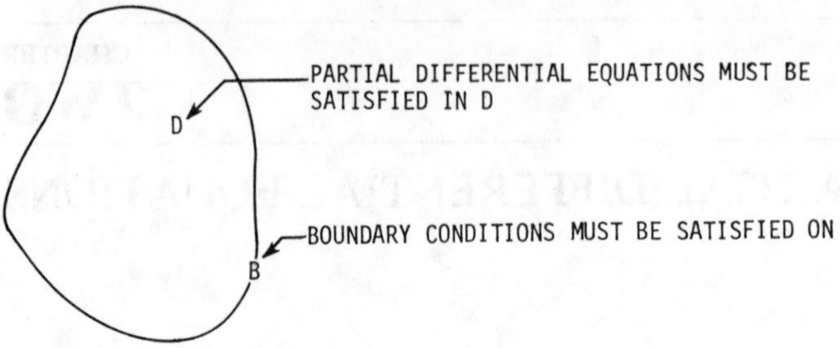

Figure 2-1 Domain for an equilibrium problem.

domain depends upon the prescribed boundary condition at every point on B. In this sense the boundary conditions are certainly the jury for the solution in D. Mathematically, equilibrium problems are governed by elliptic partial differential equations.

Example 2-1 The steady-state temperature distribution in a conducting medium is governed by Laplace's equation. A typical problem requiring the steady-state temperature distribution in a two-dimensional solid with the boundaries held at constant temperatures is defined by the equation

$$\nabla^2 T = \frac{\partial^2 T}{\partial x^2} + \frac{\partial^2 T}{\partial y^2} = 0 \qquad 0 \leqslant x \leqslant 1,\ 0 \leqslant y \leqslant 1 \tag{2-1}$$

with boundary conditions:

$$T(0, y) = 0$$
$$T(1, y) = 0$$
$$T(x, 0) = T_0$$
$$T(x, 1) = 0$$

The two-dimensional configuration is shown in Fig. 2-2.

Solution: One of the standard techniques used to solve a linear PDE is separation of variables (Greenspan, 1961). This technique assumes that the unknown temperature can be written as the product of a function of x and a function of y, i.e.,

$$T(x, y) = X(x)Y(y)$$

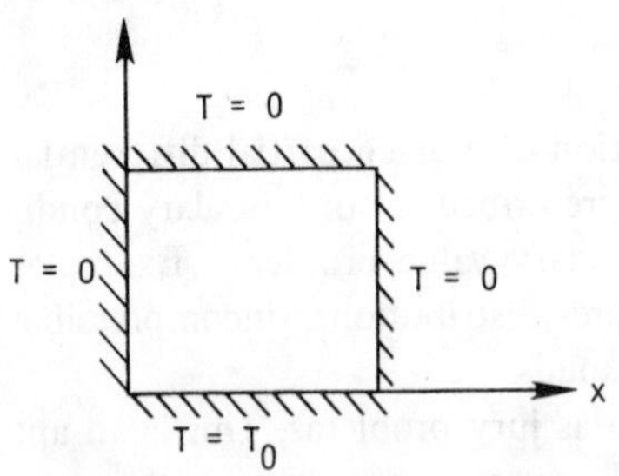

Figure 2-2 Unit square with fixed boundary temperatures.

After this form of the temperature is substituted into Laplace's equation, two ordinary differential equations are obtained. The resulting equations and homogeneous boundary conditions are:

$$X'' + \alpha^2 X = 0 \qquad Y'' - \alpha^2 Y = 0$$

$$X(0) = 0$$

$$X(1) = 0 \qquad Y(1) = 0 \tag{2-2}$$

The prime denotes differentiation and the factor α^2 arises from the separation process and must be determined as part of the solution to the problem. The solutions of the two differential equations given in Eq. (2-2) may be written

$$X(x) = A \sin(n\pi x) \qquad Y(y) = C \sinh[n\pi(y-1)]$$

The boundary conditions enter the solution in the following way:

1. $$T(0,y) = 0 \rightarrow X(0) = 0$$
$$T(x,1) = 0 \rightarrow Y(1) = 0$$

These two conditions determine the kinds of functions allowed in the expression for $T(x,y)$. The boundary condition $T(0,y) = 0$ is satisfied if the solution of the separated ordinary differential equation satisfies $X(0) = 0$. Since the solution in general contains sine and cosine terms, this boundary condition eliminates the cosine terms. A similar behavior is observed by satisfying $T(x, 1) = 0$ through $Y(1) = 0$ for the separated equation.

2. $$T(1,y) = 0 \rightarrow X(1) = 0$$

This condition identifies the eigenvalues, i.e., the particular values of α which generate eigenfunctions satisfying this required boundary condition. Since the solution of the first separated equation, Eq. (2-2), was

$$X(x) = A \sin(\alpha x)$$

a nontrivial solution for $X(x)$ exists which satisfies $X(1) = 0$ only if $\alpha = n\pi$ where $n = 1, 2, \ldots$.

3. $$T(x,0) = T_0$$

The prescribed temperature on the x axis determines the manner in which the eigenfunctions are combined to yield the correct solution to the problem. The solution of the present problem is written

$$T(x,y) = \sum_{n=1}^{\infty} A_n \sin(n\pi x) \sinh[n\pi(y-1)] \tag{2-3}$$

In this case functions of the form $\sin(n\pi x) \sinh[n\pi(y-1)]$ satisfy the PDE and three of the boundary conditions. In general, an infinite series composed of products of trigonometric sines and cosines and hyperbolic sines and cosines is required to satisfy the boundary conditions. For this problem, the fourth boundary condition along the lower boundary of the domain is given as:

$$T(x, 0) = T_0$$

We use this to determine the coefficients A_n of Eq. (2-3). Thus we find†

$$A_n = \frac{2T_0}{n\pi} \frac{[(-1)^n - 1]}{\sinh(n\pi)}$$

The solution $T(x, y)$ provides the steady temperature distribution in the solid. It is clear that the solution at any point interior to the domain of interest depends upon the specified conditions at all points on the boundary. This idea is fundamental to all equilibrium problems.

Example 2-2 The irrotational flow of an incompressible, inviscid fluid is governed by Laplace's equation. Determine the velocity distribution around the two-dimensional cylinder shown in Fig. 2-3 in an incompressible, inviscid fluid flow. The flow is governed by

$$\nabla^2 \phi = 0$$

where ϕ is defined as the velocity potential, i.e., $\nabla\phi = \mathbf{V}$ = velocity vector. The boundary condition on the surface of the cylinder is

$$\mathbf{V} \cdot \nabla F = 0 \tag{2-4}$$

where $F(r, \theta) = 0$ is the equation of the surface of the cylinder. In addition, the velocity must approach the freestream value as distance from the body becomes large, i.e., as $x, y \to \infty$

$$\nabla\phi = \mathbf{V}_\infty \tag{2-5}$$

Solution: This problem is solved by combining two elementary solutions of Laplace's equation which satisfy the boundary conditions. This superposition of two elementary solutions is an acceptable way of obtaining a third solution only because Laplace's equation is linear. For a linear PDE, any linear combination of

†See Prob. 2-1.

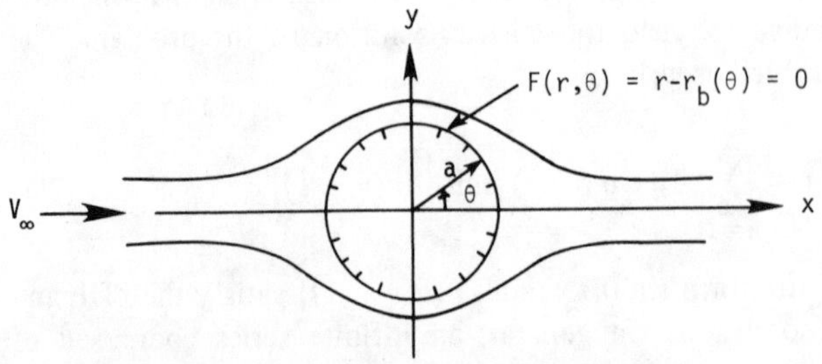

Figure 2-3 Two-dimensional flow around a cylinder.

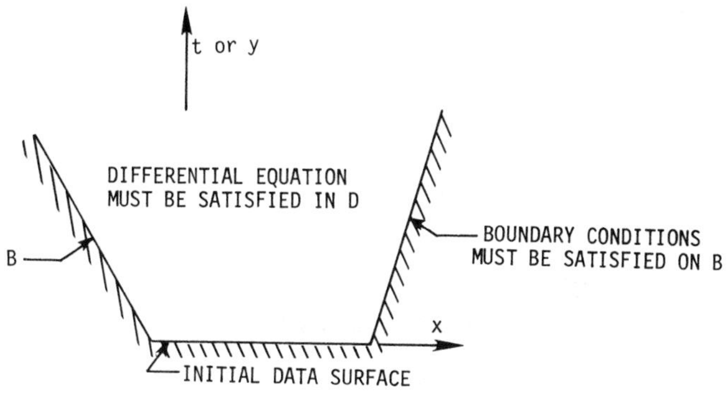

Figure 2-4 Domain for a marching problem.

solutions is also a solution (Churchill, 1941). In this case, the flow around a cylinder can be simulated by adding the velocity potential for a uniform flow to that for a doublet (Karamcheti, 1966). The resulting solution becomes

$$\phi = V_\infty x + \frac{K \cos \theta}{\sqrt{x^2 + y^2}} = V_\infty x + \frac{Kx}{x^2 + y^2} \tag{2-6}$$

where the first term is the uniform oncoming flow and the second term is a solution for a doublet of strength $2\pi K$.

2-2.2 Marching Problems

Marching or propagation problems are transient or transient-like problems where the solution of a partial differential equation is required on an open domain subject to a set of initial conditions and a set of boundary conditions. Figure 2-4 illustrates the domain and marching direction for this case. Problems in this category are initial value or initial boundary value problems. The solution must be computed by marching outward from the initial data surface while satisfying the boundary conditions. Mathematically these problems are governed by either hyperbolic or parabolic partial differential equations.

Example 2-3 Determine the transient temperature distribution in a one-dimensional solid (Fig. 2-5) with a thermal diffusivity α if the initial temperature in the solid is 0° and at all subsequent times, the temperature of the left side is held at 0° while the right side is held at T_0.

Solution: The governing differential equation is the one-dimensional heat equation

$$\frac{\partial T}{\partial t} = \alpha \frac{\partial^2 T}{\partial x^2} \tag{2-7}$$

with boundary conditions

$$T(0, t) = 0 \qquad T(1, t) = T_0 \tag{2-8}$$

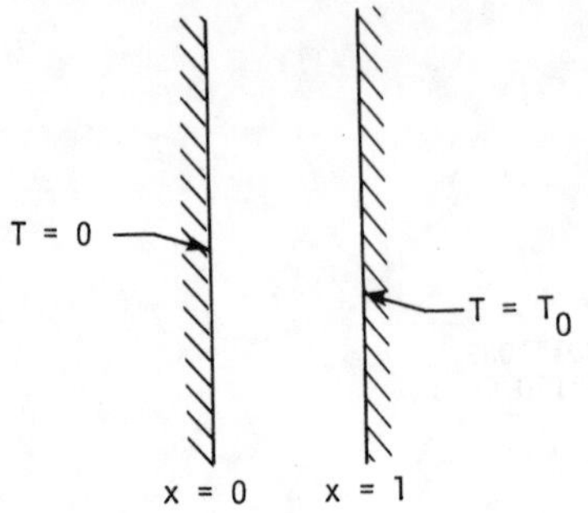

Figure 2-5 One-dimensional solid.

and initial condition

$$T(x, 0) = 0 \tag{2-9}$$

A solution to this problem is most easily obtained by introducing the variable $u = T - T_0 x$. This transforms the boundary conditions into a homogeneous set and permits the use of separation of variables to solve the heat equation

$$\frac{\partial u}{\partial t} = \alpha \frac{\partial^2 u}{\partial x^2}$$

with boundary conditions

$$u(0, t) = 0 \qquad u(1, t) = 0$$

and the initial distribution

$$u(x, 0) = -T_0 x$$

Separation of variables may be used and the solution is written in the form $u(x, t) = V(t)X(x)$. If we denote the separation constant by $-\beta^2$, it is necessary to solve the ordinary differential equations

$$V' + \alpha\beta^2 V = 0 \qquad X'' + \beta^2 X = 0$$
$$X(0) = X(1) = 0$$

with the initial distribution on u as noted above. The solution to this problem is

$$u(x, t) = \sum_{n=1}^{\infty} \frac{2T_0(-1)^n}{n\pi} e^{-n^2\pi^2\alpha t} \sin(n\pi x) = T - T_0 x$$

or

$$T = T_0 x + \sum_{n=1}^{\infty} \frac{2T_0(-1)^n}{n\pi} e^{-n^2\pi^2\alpha t} \sin(n\pi x) \tag{2-10}$$

Example 2-4 Find the displacement $y(x, t)$ of a string of length l stretched between $x = 0$ and $x = l$ if it is displaced initially into position $y(x, 0) = \sin \pi x/l$ and released from rest. Assume no external forces act on the string.

Solution: In this case the motion of the string is governed by the wave equation

$$\frac{\partial^2 y}{\partial t^2} = a^2 \frac{\partial^2 y}{\partial x^2} \tag{2-11}$$

where a is a positive constant. The boundary conditions are

$$y(0, t) = y(l, t) = 0 \tag{2-12}$$

and initial conditions

$$y(x, 0) = \sin \frac{\pi x}{l}, \quad \frac{\partial}{\partial t} y(x, t)|_{t=0} = 0 \tag{2-13}$$

The solution for this particular example is

$$y(x, t) = \sin \left(\pi \frac{x}{l}\right) \cos \left(a\pi \frac{t}{l}\right) \tag{2-14}$$

Solutions for problems of this type usually require an infinite series to correctly approximate the initial data. In this case, only one term of this series survives because the initial displacement requirement is exactly satisfied by one term.

The physical phenomena governed by the heat equation and the wave equation are different but both are classified as marching problems. The behavior of the solutions to these equations and methods used to obtain these solutions are also quite different. This will become clear as the mathematical character of these equations is studied.

Typical examples of marching problems include unsteady inviscid flow, steady supersonic inviscid flow, transient heat conduction and boundary-layer flow.

2-3 MATHEMATICAL CLASSIFICATION

A general second-order partial differential equation is the standard model used to present the mathematical classification of partial differential equations. Consider the PDE

$$a\phi_{xx} + b\phi_{xy} + c\phi_{yy} + d\phi_x + e\phi_y + f\phi = g(x, y) \tag{2-15}$$

where a, b, c, d, e, and f are functions of (x, y), i.e., we consider a linear equation. While this restriction is not essential, this form is convenient to use. Frequently consideration is given to quasi-linear equations which are defined as equations which are linear in the highest derivative. In terms of Eq. (2-15) this means that a, b, and c could be functions of x, y, ϕ, ϕ_x, and ϕ_y. For our discussion, however, we assume that Eq. (2-15) is linear and the coefficients depend only upon x and y.

Certain standard or canonical forms for the classes of PDE's can now be developed. There are three types of partial differential equations which Eq. (2-15) represents. These are the elliptic, parabolic, and hyperbolic types. This terminology used in classifying PDE's is by analogy with the general second-order equation in analytic geometry. The type of conic section depends upon the discriminant and the same terminology is used here. Specifically, we say the PDE is hyperbolic at a point (x_0, y_0) if

$$b^2 - 4ac > 0 \tag{2-16}$$

and the associated canonical form is

$$\phi_{\xi\xi} - \phi_{\eta\eta} = h_1(\phi_\xi, \phi_\eta, \phi, \xi, \eta) \tag{2-17}$$

The equation is parabolic at a point (x_0, y_0) if

$$b^2 - 4ac = 0 \tag{2-18}$$

with a canonical form

$$\phi_{\xi\xi} = h_2(\phi_\xi, \phi_\eta, \phi, \xi, \eta) \tag{2-19}$$

and elliptic at a point (x_0, y_0) when

$$b^2 - 4ac < 0 \tag{2-20}$$

where the canonical form is

$$\phi_{\xi\xi} + \phi_{\eta\eta} = h_3(\phi_\xi, \phi_\eta, \phi, \xi, \eta) \tag{2-21}$$

The wave equation, the heat equation, and Laplace's equation are examples of each of these types of equations. The *characteristic coordinate form* of the hyperbolic PDE given by

$$\phi_{\xi\eta} = h_4(\phi_\xi, \phi_\eta, \phi, \xi, \eta) \tag{2-22}$$

is also frequently encountered. This is usually a more convenient form of the equation to use when an analytic solution is constructed.

Suppose a transformation is performed on Eq. (2-15) of the form

$$(x, y) \to (\xi, \eta) \tag{2-23}$$

This means that x and y are transformed into new independent variables ξ and η. We also require that this transformation be nonsingular which provides that a one to one relationship exists between (x, y) and (ξ, η). We are assured of a nonsingular mapping provided that the Jacobian of the transformation

$$J = \frac{\partial(\xi, \eta)}{\partial(x, y)} = \xi_x \eta_y - \xi_y \eta_x \tag{2-24}$$

is nonzero (Taylor, 1955). To apply the transformation given by Eq. (2-23) to Eq. (2-15), each derivative is replaced by repeated application of the chain rule. For example

$$\begin{aligned} \frac{\partial \phi}{\partial x} &= \xi_x \frac{\partial \phi}{\partial \xi} + \eta_x \frac{\partial \phi}{\partial \eta} \\ \frac{\partial^2 \phi}{\partial x^2} &= \xi_x^2 \frac{\partial^2 \phi}{\partial \xi^2} + 2\xi_x \eta_x \frac{\partial^2 \phi}{\partial \xi \partial \eta} + \eta_x^2 \frac{\partial^2 \phi}{\partial \eta^2} + \xi_{xx} \frac{\partial \phi}{\partial \xi} + \eta_{xx} \frac{\partial \phi}{\partial \eta} \end{aligned} \tag{2-25}$$

Substitution into Eq. (2-15) yields

$$A\phi_{\xi\xi} + B\phi_{\xi\eta} + C\phi_{\eta\eta} + \cdots = g(\xi, \eta)$$

where $A = a\xi_x^2 + b\xi_x\xi_y + c\xi_y^2$

$B = 2a\xi_x\eta_x + b\xi_x\eta_y + b\xi_y\eta_x + 2c\xi_y\eta_y$

$C = a\eta_x^2 + b\eta_x\eta_y + c\eta_y^2$

An important result of applying this transformation is immediately clear. The discriminant of the transformed equation becomes

$$B^2 - 4AC = (b^2 - 4ac)(\xi_x\eta_y - \xi_y\eta_x)^2 \tag{2-26}$$

where

$$\xi_x\eta_y - \xi_y\eta_x = J = \frac{\partial(\xi, \eta)}{\partial(x, y)}$$

Therefore, any real nonsingular transformation does not change the type of partial differential equation.

2-3.1 Hyperbolic PDE's

The characteristic coordinate form for hyperbolic PDE's is that given in Eq. (2-22). To obtain this form, we must have (ξ, η) defined as roots of the equations $A = 0$ and $C = 0$, respectively. Thus

$$a\xi_x^2 + b\xi_x\xi_y + c\xi_y^2 = 0$$

or

$$a\left(\frac{\xi_x}{\xi_y}\right)^2 + b\frac{\xi_x}{\xi_y} + c = 0 \tag{2-27}$$

Consider surfaces $\xi(x, y) = \text{constant}$. Along these surfaces

$$d\xi = \frac{\partial\xi}{\partial x}dx + \frac{\partial\xi}{\partial y}dy = 0$$

or

$$\frac{\xi_x}{\xi_y} = -\frac{dy}{dx}$$

Substituting into Eq. (2-27) yields the characteristic differential equation

$$a\left(\frac{dy}{dx}\right)^2 - b\frac{dy}{dx} + c = 0$$

The roots of this equation are

$$\frac{dy}{dx} = \frac{b \pm \sqrt{b^2 - 4ac}}{2a} \tag{2-28}$$

The solutions of this pair of differential equations yield the surfaces $\xi(x, y) = \text{constant}$ and $\eta(x, y) = \text{constant}$. The solutions for ξ and η are called the characteristics of the

PDE [Eq. (2-15)]. The variables ξ and η are called characteristic coordinates and they define the coordinates used in Eq. (2-22). For this reason, Eq. (2-22) is called the characteristic coordinate form for hyperbolic equations.

By employing a change of dependent variables from ϕ to Φ, an interesting reduction of the PDE Eq. (2-15) to the canonical form for hyperbolic equations

$$\Phi_{\xi\xi} - \Phi_{\eta\eta} = h(\Phi, \xi, \eta) \tag{2-29}$$

can be made if the coefficients a, b, c, d, and e are assumed to be constants (Gary, 1969). This reduction is accomplished by letting λ_1 and λ_2 denote the roots of the characteristic differential equation $a\lambda^2 - b\lambda + c = 0$. The transformation (mapping)

$$\begin{aligned} y - \lambda_1 x &= \xi + \eta \\ y - \lambda_2 x &= \xi - \eta \end{aligned} \tag{2-30}$$

is applied to eliminate the cross-derivative term. The first derivative terms are eliminated by a change of dependent variable given by

$$\phi = \Phi e^{-\alpha\xi} e^{-\beta\eta} \tag{2-30a}$$

where α and β must be suitably chosen (see Prob. 2-5). This mapping reduces the general equation, Eq. (2-15), to the required form. The similarity with the simple wave equation

$$\phi_{tt} - \phi_{xx} = 0 \tag{2-31}$$

suggests an example showing further reduction to characteristic coordinates. In this case $a = 1, b = 0$, and $c = -1$ and Eq. (2-28) becomes

$$\frac{dt}{dx} = \pm 1$$

Integration of this differential equation yields

$$\begin{aligned} x + t &= \xi \\ x - t &= \eta \end{aligned}$$

Using this mapping we obtain

$$\begin{aligned} \phi_{tt} &= \phi_{\xi\xi} - 2\phi_{\xi\eta} + \phi_{\eta\eta} \\ \phi_{xx} &= \phi_{\xi\xi} + 2\phi_{\xi\eta} + \phi_{\eta\eta} \end{aligned}$$

thus

$$\phi_{\xi\eta} = 0$$

This is the standard form for the wave equation written in characteristic coordinates. An example demonstrating the utility of this form of the second-order wave equation is instructive.

Example 2-5 Solve the second-order wave equation

$$u_{tt} = c^2 u_{xx} \tag{2-32}$$

on the interval

$$-\infty < x < +\infty$$

with initial data

$$u(x, 0) = f(x)$$
$$u_t(x, 0) = g(x)$$

Solution: The transformation to characteristic coordinates permits simple integration of the wave equation

$$u_{\xi\eta} = 0$$

where $\xi = x + ct, \eta = x - ct$.

We integrate to obtain the solution

$$u(x, t) = F_1(x + ct) + F_2(x - ct) \tag{2-33}$$

This is called the D'Alembert (Wylie, 1951) solution of the wave equation. The particular forms for F_1 and F_2 are determined from the initial data.

$$u(x, 0) = f(x) = F_1(x) + F_2(x)$$
$$u_t(x, 0) = g(x) = cF_1'(x) - cF_2'(x)$$

This results in a solution of the form

$$u(x, t) = \frac{f(x + ct) + f(x - ct)}{2} + \frac{1}{2c} \int_{x-ct}^{x+ct} g(\tau)\, d\tau \tag{2-34}$$

A distinctive property of hyperbolic partial differential equations can be deduced from the solution of Eq. (2-32) and the geometry of the physical domain of interest. Figure 2-6 shows the characteristics which pass through the point (x_0, t_0). The right running characteristic has a slope $+(1/c)$ while the left running one has slope $-(1/c)$. The solution $u(x, t)$ at (x_0, t_0) depends only upon the initial data contained in the interval

$$x_0 - ct_0 \leqslant x \leqslant x_0 + ct_0$$

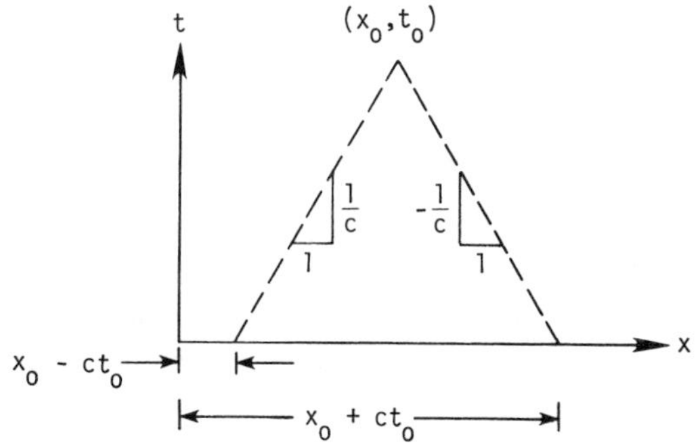

Figure 2-6 Characteristics for the wave equation.

The first term of the solution given by Eq. (2-34) represents propagation of the initial data along the characteristics while the second term represents the effect of the data within the closed interval at $t = 0$.

A fundamental property of hyperbolic PDE's is the limited domain of dependence exhibited in Example 2-5. This domain of dependence is bounded by the characteristics which pass through the point (x_0, t_0). Clearly, the solution $u(x_0, t_0)$ depends only upon information in the interval bounded by these characteristics. This means that any disturbance which occurs outside of this interval can never influence the solution at (x_0, t_0). This behavior is common to all hyperbolic equations and is nicely demonstrated through the solution of the second-order wave equation. The basis for the name initial value or marching problem is clear. Initial conditions are specified and the solution is marched outward in time or in a time-like direction.

The term pure initial value problem is frequently encountered in the study of hyperbolic PDE's. Example 2-5 is a pure initial value problem, i.e., there are no boundary conditions which must be applied at $x =$ constant. The solution at (x_0, t_0) depends only upon initial data.

In the classification of partial differential equations, many well-known names are associated with the specific problem types. The most well-known problem in the hyperbolic class is the Cauchy problem. This problem requires that one obtain a solution u to a hyperbolic PDE with initial data specified along a curve C. A very important theorem in mathematics assures us that a solution to the Cauchy problem exists. This is the Cauchy-Kowalewsky theorem. This theorem asserts that if the initial data are analytic in the neighborhood of x_0, y_0 and the function u_{xx} (applied to our second-order wave equation of Example 2-5) is analytic there, a unique analytic solution for u exists in the neighborhood of x_0, y_0.

Some discussion is warranted regarding the type of problem specification which is allowed for hyperbolic equations. For our second-order wave equation, initial conditions are required on the unknown function and its first derivatives along some curve C. It is important to observe that the curve C must not coincide with a characteristic of the differential equation. If an attempt is made to solve an initial value problem with characteristic initial data, a unique solution cannot be obtained (see Example 2-6). As will be discussed further in Section 2-4, the problem is said to be "ill-posed."

Example 2-6 Solve the second-order wave equation in characteristic coordinates.

$$u_{\xi\eta} = 0$$

subject to initial data

$$u(0, \eta) = \phi(\eta) \qquad u_\xi(0, \eta) = \psi(\eta)$$

Solution: The characteristics of the governing PDE are defined by $\xi =$ constant and $\eta =$ constant. In this case the initial data are prescribed along a characteristic.

Suppose we attempt to write a Taylor series expansion in ξ to obtain a solution for u in the neighborhood of the initial data surface $\xi = 0$. Our solution must be in the form

$$u(\xi, \eta) = u(0, \eta) + \xi u_\xi(0, \eta) + \frac{\xi^2}{2} u_{\xi\xi}(0, \eta) + \cdots$$

From the given initial data $u(0, \eta)$ and $u_\xi(0, \eta)$ are known. It remains to determine $u_{\xi\xi}(0, \eta)$.

The governing differential equation requires

$$u_{\xi\eta}(0, \eta) = 0$$

However, we already have the condition that

$$u_{\xi\eta}(0, \eta) = \psi'(\eta) = 0$$

Therefore

$$\psi(\eta) = \text{constant} = c_1$$

We may also write

$$\frac{\partial u_{\xi\eta}}{\partial \xi} = \frac{\partial u_{\xi\xi}}{\partial \eta} = 0$$

Integration of this equation yields

$$u_{\xi\xi} = f(\xi)$$

In view of the given initial data, we conclude that

$$u_{\xi\xi}(0, \eta) = \text{constant} = c_2$$

and

$$u(\xi, \eta) = \phi(\eta) + \xi c_1 + \frac{\xi^2}{2} c_2$$

or

$$u(\xi, \eta) = \phi(\eta) + g(\xi)$$

We are unable to uniquely determine the function $g(\xi)$ when the initial data are given along the characteristic $\xi = 0$.

Proper specification of initial data or boundary conditions is very important in solving a PDE. Hadamard (1952) provided insight in noting that a well-posed problem is one in which the solution depends continuously upon the initial data. The concept of the well-posed problem is equally appropriate for elliptic and parabolic PDE's. An example for an elliptic problem is presented later in Section 2-4.

2-3.2 Parabolic PDE's

A study of the solution of a simple hyperbolic PDE provided insight on the behavior of the solution of that type of equation. In a similar manner, we will now study the solution to parabolic equations. Referring to Eq. (2-15), the parabolic case occurs when

$$b^2 - 4ac = 0$$

For this case the characteristic differential equation is given by

$$\frac{dy}{dx} = \frac{b}{2a} \tag{2-35}$$

The canonical form for the parabolic case is given by

$$\phi_{\xi\xi} = g(\phi_\xi, \phi_\eta, \phi, \xi, \eta) \tag{2-36}$$

This form may be obtained by identifying ξ and η as

$$\eta = y - \lambda_1 x$$

$$\xi = y - \lambda_2 x$$

where λ_1 is given by Eq. (2-35). In view of Eq. (2-35), we obtain only one characteristic. We must choose λ_2 to insure linear independence of ξ and η. This requires that the Jacobian

$$\frac{\partial(\xi, \eta)}{\partial(x, y)} = f(\lambda_1, \lambda_2) \neq 0 \tag{2-37}$$

When λ_2 is selected satisfying this requirement and the transformation to (ξ, η) coordinates is completed, the canonical form given by Eq. (2-36) is obtained.

Parabolic partial differential equations are associated with diffusion processes. The solutions of parabolic equations clearly show this behavior. While the partial differential equations controlling diffusion are marching problems, i.e., we solve them starting at some initial data plane and march foward in time or in a time-like direction, they do not exhibit the limited zones of influence that hyperbolic equations have. In contrast, the solution of a parabolic equation at time t_1 depends upon the entire physical domain $(t \leqslant t_1)$ including any side boundary conditions. To illustrate further, Example 2-3 required that we solve the heat equation for transient conduction in a one-dimensional solid. The initial temperature distribution was specified as were the temperatures at the boundaries. Figure 2-7 illustrates the domain of dependence for this parabolic problem at t_1.

This shows that the solution at $t = t_1$ depends upon everything which occurred in the physical domain at all earlier times. The solution given by Eq. (2-10) also exhibits this behavior. Another example illustrating the behavior of a solution of a parabolic equation is of value.

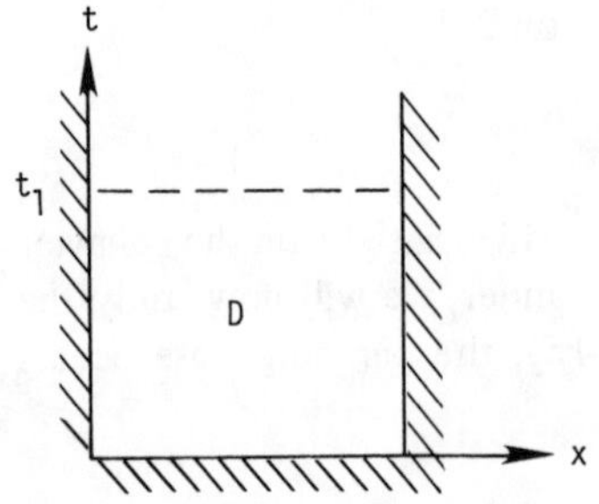

Figure 2-7 Domain of dependence for a simple parabolic problem.

Example 2-7 The unsteady motion due to the impulsive acceleration of a flat plate in a viscous, incompressible fluid is known as the Rayleigh problem and may be solved exactly. If the flow is two-dimensional, only the velocity component parallel to the plate will be nonzero. Let y be the coordinate normal to the plate and x be the coordinate along the plate. The equation which governs the velocity distribution is

$$\frac{\partial u}{\partial t} = \nu \frac{\partial^2 u}{\partial y^2} \tag{2-38}$$

where ν is the kinematic viscosity. The time derivative term is the local acceleration of the fluid while the right-hand side is the resisting force provided by the shear stress in the fluid ($\tau = \nu\rho\, \partial u/\partial y$). This equation is subject to the boundary conditions

$$u(0, y) = 0$$
$$u(t, 0) = U \qquad t > 0$$
$$u(t, \infty) = 0$$

Solution: The solution of this problem provides the velocity distribution on a two-dimensional flat plate impulsively accelerated to a velocity U from rest. An interesting method frequently used in solving parabolic equations is to seek a similarity solution (Hansen, 1964). In finding a similarity solution, we introduce a change in variables which results in reducing the number of independent variables in the original PDE (Churchill, 1974). In this case we attempt to reduce the partial differential equation in (y, t) to an ordinary differential equation in a new independent variable η. For this problem let

$$f(\eta) = \frac{u}{U}$$

and

$$\eta = \frac{y}{2\sqrt{\nu t}}$$

The governing differential equation becomes

$$\frac{d^2 f}{d\eta^2} + 2\eta \frac{df}{d\eta} = 0$$

with boundary conditions

$$f(0) = 1$$
$$f(\infty) = 0$$

This ordinary differential equation may be solved directly to yield the solution

$$u = U\left(1 - \frac{2}{\sqrt{\pi}} \int_0^{\eta} e^{-\eta^2}\, d\eta\right) \tag{2-39}$$

Using the definition of the error function

$$\text{erf}\,(\eta) = \frac{2}{\sqrt{\pi}} \int_0^\eta e^{-\eta^2}\, d\eta \tag{2-40}$$

the solution becomes

$$u = U[1 - \text{erf}\,(\eta)]$$

This shows that the layer of fluid which is influenced by the moving plate increases in thickness with time. In fact, the layer of fluid has thickness proportional to $\sqrt{\nu t}$. This indicates that the growth of this layer is controlled by the kinematic viscosity ν and the velocity change in the layer is induced by diffusion of the plate velocity into the initially undisturbed fluid. We see that this is a diffusion process as is one-dimensional transient heat conduction.

2-3.3 Elliptic PDE's

The third type of partial differential equation is elliptic. As we previously noted, jury problems are governed by elliptic PDE's. If Eq. (2-15) is elliptic, the discriminant is negative, i.e.,

$$b^2 - 4ac < 0 \tag{2-41}$$

and the characteristic differential equation has no real roots. For this case, the roots are given by

$$\lambda_{1,2} = \frac{b \pm i\sqrt{4ac - b^2}}{2a}$$

and the characteristics are both complex.

The canonical form for the elliptic equation can be obtained by a coordinate transformation. In this case the transformation to characteristic coordinates is complex. Let

$$y - \lambda_1 x = \xi + i\eta \qquad y - \lambda_2 x = \xi - i\eta \tag{2-42}$$

and in addition, let

$$\phi = \phi_1 e^{-\alpha\xi - \beta\eta} \tag{2-43}$$

If Eq. (2-15) is transformed using Eqs. (2-42) and (2-43), a new expression is obtained and may be written

$$\phi_{1_{\xi\xi}} + \phi_{1_{\eta\eta}} + g\phi_1 = f(\xi, \eta) \tag{2-44}$$

While this equation is similar to Eq. (2-21), the functional form is simplified and ϕ_1 appears explicitly.

The dependence of the solution upon the boundary conditions for elliptic PDE's has been previously discussed and demonstrated in Example 2-1. However, another example is presented here to reinforce this basic idea.

Example 2-8 Given Laplace's equation on the unit disk

$$\nabla^2 u = 0 \qquad 0 \leqslant r < 1 \qquad -\pi \leqslant \theta \leqslant \pi$$

subject to boundary conditions

$$\frac{\partial u}{\partial r}(1, \theta) = f(\theta) \qquad -\pi \leqslant \theta \leqslant \pi$$

what is the solution $u(r, \theta)$?

Solution: This problem can be solved by assuming a solution of the form

$$u(r, \theta) = \frac{a_0}{2} + \sum_{n=1}^{\infty} r^n (a_n \cos n\theta + b_n \sin n\theta)$$

The correct expressions for a_n and b_n can be developed using standard techniques (Garabedian, 1964). For this example, the expressions for a_n and b_n depend upon the boundary conditions at all points on the unit disk. This dependence on the boundary conditions should be expected for all elliptic problems. The important point of this example is that a solution of this problem exists only if

$$\int f(\theta)\, dl = 0$$

over the boundary of the unit disk (Zachmanoglou and Thoe, 1976). This may be demonstrated by applying Green's theorem to the unit disk. In this problem, the boundary conditions are not arbitrarily chosen but must satisfy the integral constraint shown above.

2-4 THE WELL-POSED PROBLEM

The previous section discussed the mathematical character of the different partial differential equations. The examples illustrated the dependence of the solution of a particular problem upon the initial data and boundary conditions. In our discussion of hyperbolic partial differential equations, it was noted that a unique solution to a hyperbolic partial differential equation cannot be obtained if the initial data are given on a characteristic. Similar examples showing improper use of boundary conditions can be constructed for elliptic and parabolic equations.

The difficulty encountered in solving our hyperbolic equation subject to characteristic initial data has to do with the question of whether or not the problem is "well-posed." In order for a problem involving a partial differential equation to be well-posed, the solution to the problem must exist and be unique, and the solution must depend continuously upon the initial or boundary data. Example 2-6 led to a uniqueness question. Hadamard (1952) has constructed a simple example which demonstrates the problem of continuous dependence on boundary data.

Example 2-9 A solution of Laplace's equation

$$u_{xx} + u_{yy} = 0 \qquad -\infty < x < \infty \qquad y \geqslant 0$$

is desired subject to the boundary conditions ($y = 0$)

$$u(x, 0) = 0$$

$$u_y(x, 0) = \frac{1}{n} \sin(nx) \qquad n > 0$$

Solution: Using separation of variables we obtain

$$u = \frac{1}{n^2} \sin(nx) \sinh(ny)$$

If our problem is well-posed, we expect the solution to depend continuously upon the boundary conditions. For the data given, we must have

$$u_y(x, 0) = \frac{1}{n} \sin(nx)$$

We see that u_y becomes small for large values of n. The solution behaves in a different fashion for large n. As n becomes large u approaches e^{ny}/n^2 and u grows without bound even for small y. But, $u(x, 0) = 0$ so that continuity with the initial data is lost. Thus we have an ill-posed problem. This is evident from our earlier discussions. Since Laplace's equation is elliptic, the solution depends upon conditions on the entire boundary of the closed domain. The problem given in this example requires the solution of an elliptic differential equation on an open domain. Boundary conditions were given only on the $y = 0$ line.

Problems requiring the solution of Laplace's equation subject to different types of boundary conditions are identified with specific names. The first of these is the *Dirichlet problem* (Fig. 2-8). In this problem, a solution of Laplace's equation is required on a closed domain subject to boundary conditions which require the solution to take on prescribed values on the boundary. The *Neumann problem* also requires the solution of Laplace's equation in D. However, the normal derivative of u is specified on B rather than the function u. If s is the arc length along B, then

$$\nabla^2 u = 0 \quad \text{in } D$$

$$\frac{\partial u}{\partial n} = g(s) \quad \text{on } B$$

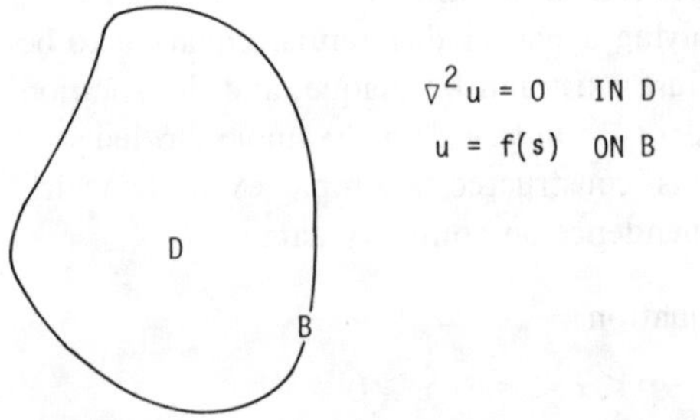

Figure 2-8 Dirichlet problem.

The specification of the Dirichlet and Neumann problems leads one to speculate about the existence of a boundary value problem requiring specification of a combination of the function u and its normal derivative on the boundary. This is called the mixed or third boundary value problem (Zachmanoglou and Thoe, 1976). This is also referred to as *Robin's problem.* Mathematically this problem may be written as

$$\nabla^2 u = 0$$

in D and

$$a_1(s)\frac{\partial u}{\partial n} + a_2(s)u = h(s)$$

on B. The assignment of the names Dirichlet, Neumann, and Robin to the three boundary value problems noted here is generally used to define types of boundary or initial data specified for any partial differential equation. For example, if the comment "Dirichlet boundary data" is used, it is understood that the unknown, u, is prescribed on the boundary in question. This is accepted regardless of the type of differential equation.

2-5 SYSTEMS OF EQUATIONS

In applying numerical methods to physical problems, systems of equations are frequently encountered. It is the exceptional case where a physical process is governed by a single equation. In those cases where the process is governed by a higher-order PDE, the PDE can usually be converted to a system of first-order equations. This can be most easily demonstrated by two simple examples.

The wave equation [Eq. (2-32)] can be written as a system of two first-order equations. Let

$$v = \frac{\partial u}{\partial t} \qquad w = c\frac{\partial u}{\partial x}$$

then we may write

$$\begin{aligned} \frac{\partial v}{\partial t} &= c\frac{\partial w}{\partial x} \\ \frac{\partial w}{\partial t} &= c\frac{\partial v}{\partial x} \end{aligned} \tag{2-45}$$

If we introduce u as one of the variables in place of either w or v, then u can be seen to satisfy the second-order wave equation.

Many physical processes are governed by Laplace's equation [Eq. (2-1)]. As in the previous example, Laplace's equation can be replaced by a system of first-order equations. In this case, let u and v represent the unknown dependent variables. We require that

$$\frac{\partial u}{\partial x} = +\frac{\partial v}{\partial y}$$
$$\frac{\partial u}{\partial y} = -\frac{\partial v}{\partial x} \tag{2-46}$$

These are the famous Cauchy-Riemann equations (Churchill, 1960). These equations are extensively used in conformally mapping one region onto another.*

The equations most frequently encountered in CFD may be written as first-order systems. We must be able to classify systems of first-order equations in order to correctly treat them. Consider the linear system of equations

$$\frac{\partial \mathbf{u}}{\partial t} + [A]\frac{\partial \mathbf{u}}{\partial x} + [B]\frac{\partial \mathbf{u}}{\partial y} + \mathbf{r} = 0 \tag{2-47}$$

We assume for simplicity that the coefficient matrices $[A]$ and $[B]$ are functions of t, x, y and we restrict our attention to two space dimensions. The dependent variable $\mathbf{u}$ is a column vector of unknowns and $\mathbf{r}$ depends upon $\mathbf{u}, x, y$.

According to Zachmanoglou and Thoe (1976) there are two cases that can be definitely identified for first-order systems. The system given in Eq. (2-47) is said to be hyperbolic at a point in (x, t) if the eigenvalues of $[A]$ are all real and distinct. Richtmyer and Morton (1967) define a system to be hyperbolic if the eigenvalues are all real and $[A]$ can be written as $[T][\lambda][T]^{-1}$ where $[\lambda]$ is a diagonal matrix of eigenvalues of $[A]$ and $[T]^{-1}$ is the matrix of left eigenvectors. The same can be said of the behavior of the system in (y, t) with respect to the eigenvalues of the B matrix.

This point can be illustrated by writing the system of equations given in Eq. (2-45) as

$$\frac{\partial \mathbf{u}}{\partial t} + [A]\frac{\partial \mathbf{u}}{\partial x} = 0 \tag{2-48}$$

where

$$\mathbf{u} = \begin{bmatrix} v \\ w \end{bmatrix}$$

$$[A] = \begin{bmatrix} 0 & -c \\ -c & 0 \end{bmatrix}$$

The eigenvalues, λ, of the $[A]$ matrix are found by setting

$$\det |[A] - \lambda[I]| = 0$$

*It should be noted that some differences exist in solving Laplace's equation and the Cauchy-Riemann equations. A solution of the Cauchy-Riemann equations is a solution of Laplace's equation but the converse is not necessarily true.

thus

$$\begin{vmatrix} -\lambda & -c \\ -c & -\lambda \end{vmatrix} = 0$$

or

$$\lambda^2 - c^2 = 0$$

The roots of this characteristic equation are

$$\lambda_1 = +c$$

$$\lambda_2 = -c$$

These are the characteristic differential equations for the wave equation, i.e.,

$$\left(\frac{dx}{dt}\right)_1 = +c$$

$$\left(\frac{dx}{dt}\right)_2 = -c$$

The system of equations in this example is hyperbolic and we see that the eigenvalues of the $[A]$ matrix represent the characteristic differential equations of the wave equation.

The second case which can be identified for the system given in Eq. (2-47) is elliptic. Equation (2-47) is said to be elliptic at a point in (x, t) if the eigenvalues of $[A]$ are all complex. An example illustrating this behavior is given by the Cauchy-Riemann equations.

Example 2-10 The system given in Eq. (2-46) may be written

$$\frac{\partial \mathbf{w}}{\partial x} + [A] \frac{\partial \mathbf{w}}{\partial y} = 0$$

where

$$\mathbf{w} = \begin{bmatrix} u \\ v \end{bmatrix}$$

and

$$[A] = \begin{bmatrix} 0 & -1 \\ 1 & 0 \end{bmatrix}$$

The eigenvalues of $[A]$ are

$$\lambda_1 = +i$$

$$\lambda_2 = -i$$

Since both eigenvalues of $[A]$ are complex, we identify the system as elliptic. Again, this is consistent with the behavior we are familiar with in Laplace's equation.

The first-order system represented by Eq. (2-47) can exhibit hyperbolic behavior in (x, t) space and elliptic behavior in (y, t) space depending upon the eigenvalue structure of the A and B matrices. This is a result of evaluating the behavior of the PDE by examining the eigenvalues in (x, t) or (y, t) independently.

Hellwig (1977) has presented classification for systems of equations of the form

$$a_1 \frac{\partial u}{\partial x} + b_1 \frac{\partial v}{\partial x} + a_2 \frac{\partial u}{\partial y} + b_2 \frac{\partial v}{\partial y} = f_1$$

$$\hat{a}_1 \frac{\partial u}{\partial x} + \hat{b}_1 \frac{\partial v}{\partial x} + \hat{a}_2 \frac{\partial u}{\partial y} + \hat{b}_2 \frac{\partial v}{\partial y} = f_2$$

This system may be written as a matrix system of the form

$$[A] \frac{\partial \mathbf{w}}{\partial x} + [C] \frac{\partial \mathbf{w}}{\partial y} = \mathbf{F} \tag{2-49}$$

where

$$\mathbf{w} = \begin{bmatrix} u \\ v \end{bmatrix} \qquad \mathbf{F} = \begin{bmatrix} f_1 \\ f_2 \end{bmatrix}$$

and

$$[A] = \begin{bmatrix} a_1 & b_1 \\ \hat{a}_1 & \hat{b}_1 \end{bmatrix} \qquad [C] = \begin{bmatrix} a_2 & b_2 \\ \hat{a}_2 & \hat{b}_2 \end{bmatrix}$$

Let

$$B = \begin{vmatrix} a_1 & b_2 \\ \hat{a}_1 & \hat{b}_2 \end{vmatrix} + \begin{vmatrix} a_2 & b_1 \\ \hat{a}_2 & \hat{b}_1 \end{vmatrix}$$

and

$$D = B^2 - 4|A||C|$$

where $|A|$ means the determinant of the $[A]$ matrix. The system given by Eq. (2-49) is elliptic if $D < 0$, hyperbolic if $D > 0$, and parabolic if $D = 0$.

Several questions now appear regarding behavior of systems of equations with coefficient matrices where the roots of the characteristic equations contain both real and complex parts. In those cases, the system is mixed and may exhibit hyperbolic, parabolic, and elliptic behavior. The physical system under study usually provides information which is very useful in understanding the physical behavior represented by the governing PDE. Experience gained in solving mixed problems provides the best guidance in their correct treatment.

The classification of systems of second-order PDE's is very complex. It is difficult to determine the mathematical behavior of these systems except for simple cases. For example, the system of equations given by

$$\mathbf{u}_t = [A]\mathbf{u}_{xx}$$

is parabolic if all the eigenvalues of $[A]$ are real. The same uncertainties present in classifying mixed systems of first-order equations are also encountered in the classification of second-order systems.

2-6 OTHER DIFFERENTIAL EQUATIONS OF INTEREST

Our discussion in this chapter has centered on the second-order equations given by the wave equation, the heat equation and Laplace's equation. In addition, systems of first-order equations were examined. A number of other very important equations should be mentioned since they govern common physical phenomena or they are used as simple models for more complex problems. In many cases, exact analytical solutions for these equations exist.

1. The first-order, linear wave equation

$$\frac{\partial u}{\partial t} + c\,\frac{\partial u}{\partial x} = 0 \tag{2-50}$$

This equation governs propagation of a wave moving to the right at a constant speed c. This is called the advection equation in meteorology.

2. The inviscid Burgers equation

$$\frac{\partial u}{\partial t} + u\,\frac{\partial u}{\partial x} = 0 \tag{2-51}$$

This is also called the nonlinear first-order wave equation. This equation governs propagation of nonlinear waves for the simple one-dimensional case.

3. Burgers equation

$$\frac{\partial u}{\partial t} + u\,\frac{\partial u}{\partial x} = \nu\,\frac{\partial^2 u}{\partial x^2} \tag{2-52}$$

This is the nonlinear equation [Eq. (2-51)] with diffusion added. This particular form is very similar to the equations governing fluid flow and can be used as a simple nonlinear model for numerical experiments.

4. The Tricomi equation

$$y\,\frac{\partial^2 u}{\partial x^2} + \frac{\partial^2 u}{\partial y^2} = 0 \tag{2-53}$$

This equation governs problems of the mixed type such as inviscid transonic flows. The properties of the Tricomi equation include a change from elliptic to hyperbolic character depending upon the sign of y.

5. Poisson's equation

$$\frac{\partial^2 u}{\partial x^2} + \frac{\partial^2 u}{\partial y^2} = f(x,y) \tag{2-54}$$

This elliptic equation governs the temperature distribution in a solid with heat sources described by the function $f(x, y)$. Poisson's equation also determines the electric field in a region containing a charge density $f(x,y)$.

6. The advection-diffusion equation

$$\frac{\partial \xi}{\partial t} + u\frac{\partial \xi}{\partial x} = \alpha \frac{\partial^2 \xi}{\partial x^2} \tag{2-55}$$

This particular expression represents the advection of a quantity ξ in a region with velocity u. The quantity α is a diffusion or viscosity coefficient.

7. The Korteweg-deVries equation

$$\frac{\partial u}{\partial t} + u\frac{\partial u}{\partial x} + \frac{\partial^3 u}{\partial x^3} = 0 \tag{2-56}$$

The motion of nonlinear dispersive waves is governed by this equation.

8. The Helmholtz equation

$$\frac{\partial^2 u}{\partial x^2} + \frac{\partial^2 u}{\partial y^2} + k^2 u = 0 \tag{2-57}$$

This equation governs the motion of time-dependent harmonic waves where k is a frequency parameter. Applications include the propagation of acoustic waves.

Many of the equations cited here will be used to demonstrate the application of finite-difference methods in later chapters. While the list of equations is not exhaustive, examples of the various types of PDE's are included.

PROBLEMS

2-1 The solution of Laplace's equation for Example 2-1 is given in Eq. (2-3). Show that the expression for the Fourier coefficients A_n is correct as given in the example. Hint: Multiply Eq. (2-3) by $\sin(m\pi x)$ and integrate over the interval $0 \leqslant x \leqslant 1$ to obtain your answer after using the boundary condition $T(x, 0) = T_0$.

2-2 Show that the velocity field represented by the potential function in Eq. (2-6) satisfies the surface boundary condition given in Eq. (2-4).

2-3 Demonstrate that Eq. (2-14) is the solution of the wave equation as required in Example 2-4. Use the separation of variables technique.

2-4 Show that the type of partial differential equation is unchanged when any nonsingular, real transformation is used.

2-5 Derive the canonical form for hyperbolic equations [Eq. (2-29)] by applying the transformations given in Eq. (2-30) and Eq. (2-30*a*) to Eq. (2-15). Determine the correct form for α and β.

2-6 Show that the canonical form for parabolic equations given in Eq. (2-36) is correct.

2-7 Show that a solution to Example 2-8 exists only if

$$\int f(\theta)\, dl = 0$$

on the unit circle.

2-8 Classify the following PDE's

$$\frac{\partial^2 u}{\partial t^2} + \frac{\partial^2 u}{\partial x^2} + \frac{\partial u}{\partial x} = -e^{-kt}$$

$$\frac{\partial^2 u}{\partial x^2} - \frac{\partial^2 u}{\partial x\,\partial y} + \frac{\partial u}{\partial y} = 4$$

2-9 Classify the behavior of the following system of PDE's in (t, x) and (t, y) space

$$\frac{\partial u}{\partial t} + \frac{\partial v}{\partial x} - \frac{\partial u}{\partial y} = 0$$

$$\frac{\partial v}{\partial t} - \frac{\partial u}{\partial x} + \frac{\partial v}{\partial y} = 0$$

2-10 (*a*) Write the Fourier cosine series for the function

$$f(x) = \sin(x) \qquad 0 < x < \pi$$

(*b*) Write the Fourier cosine series for the function

$$f(x) = \cos(x) \qquad 0 < x < \pi$$

2-11 Find the characteristics of each of the following PDE's

(*a*)
$$\frac{\partial^2 u}{\partial x^2} + 3\frac{\partial^2 u}{\partial x\,\partial y} + 2\frac{\partial^2 u}{\partial y^2} = 0$$

(*b*)
$$\frac{\partial^2 u}{\partial x^2} - 2\frac{\partial^2 u}{\partial x\,\partial y} + \frac{\partial^2 u}{\partial y^2} = 0$$

2-12 Transform the PDE's given in Prob. 2-11 into canonical form.

2-13 Obtain the canonical form for the following elliptic PDE's

(*a*)
$$\frac{\partial^2 u}{\partial x^2} + \frac{\partial^2 u}{\partial x\,\partial y} + \frac{\partial^2 u}{\partial y^2} = 0$$

(*b*)
$$\frac{\partial^2 u}{\partial x^2} - 2\frac{\partial^2 u}{\partial x\,\partial y} + 5\frac{\partial^2 u}{\partial y^2} + \frac{\partial u}{\partial y} = 0$$

2-14 Transform the following parabolic PDE's to canonical form

(*a*)
$$\frac{\partial^2 u}{\partial x^2} - 6\frac{\partial^2 u}{\partial x\,\partial y} + 9\frac{\partial^2 u}{\partial y^2} + \frac{\partial u}{\partial x} - e^{xy} = 1$$

(*b*)
$$\frac{\partial^2 u}{\partial x^2} + 2\frac{\partial^2 u}{\partial x\,\partial y} + \frac{\partial^2 u}{\partial y^2} + 7\frac{\partial u}{\partial x} - 8\frac{\partial u}{\partial y} = 0$$

2-15 Find the solution of the wave equation

$$\frac{\partial^2 u}{\partial x^2} - \frac{\partial^2 u}{\partial y^2} = 0 \qquad y \geqslant 0$$

with initial data

$$u(x, 0) = 1$$

$$u_y(x, 0) = 0$$

2-16 Solve Laplace's equation

$$\nabla^2 u = 0 \qquad 0 \leqslant x \leqslant \pi \qquad 0 \leqslant y \leqslant \pi$$

subject to boundary conditions

$$u(x, 0) = \sin x + 2 \sin 2x$$
$$u(\pi, y) = 0$$
$$u(x, \pi) = 0$$
$$u(0, y) = 0$$

2-17 Repeat Prob. 2-16 with

$$u(x, 0) = -\pi^2 x^2 + 2\pi x^3 - x^4$$

2-18 Determine the solution of the heat equation

$$\frac{\partial u}{\partial t} = \frac{\partial^2 u}{\partial x^2} \qquad 0 \leqslant x \leqslant 1$$

with boundary conditions

$$u(t, 0) = 0$$
$$u(t, 1) = 0$$

with an initial distribution

$$u(0, x) = \sin (2\pi x)$$

2-19 Repeat Prob. 2-18 if the initial distribution is given by

$$u(0, x) = 1 - \cos (4\pi x)$$

CHAPTER

THREE

BASICS OF FINITE-DIFFERENCE METHODS

3-1 INTRODUCTION

In this chapter, basic concepts and techniques needed in the formulation of finite-difference representations are developed. In the finite-difference approach, the continuous problem domain is "discretized" so that the dependent variables are considered to exist only at discrete points. Derivatives are approximated by differences resulting in an algebraic representation of the partial differential equation (PDE). Thus, a problem involving calculus has been transformed into an algebraic problem.

The nature of the resulting algebraic system depends on the character of the problem posed by the original PDE (or system of PDE's). Equilibrium problems usually result in a system of algebraic equations which must be solved simultaneously throughout the problem domain in conjunction with specified boundary values. Marching problems result in algebraic equations which usually can be solved one at a time (although it is often convenient to solve them several at a time). Several considerations determine whether the solution so obtained will be a good approximation to the exact solution of the original PDE. Among these considerations are truncation error, consistency, and stability, all of which will be discussed in the present chapter.

3-2 FINITE DIFFERENCES

One of the first steps to be taken in establishing a finite-difference procedure for solving a PDE is to replace the continuous problem domain by a finite-difference mesh or grid. As an example, suppose that we wish to solve a PDE for which $u(x,y)$ is the dependent variable in the square domain $0 \leqslant x \leqslant 1$, $0 \leqslant y \leqslant 1$. We establish a grid on the domain by replacing $u(x, y)$ by $u(i\,\Delta x, j\,\Delta y)$. Points can be located according to values of i and j so difference equations are usually written in terms of the general point (i, j) and its neighbors. This labeling is illustrated in Fig. 3-1. Thus, if we think of $u_{i,j}$ as $u(x_0, y_0)$ then

$$u_{i+1,j} = u(x_0 + \Delta x, y_0) \qquad u_{i-1,j} = u(x_0 - \Delta x, y_0)$$

$$u_{i,j+1} = u(x_0, y_0 + \Delta y) \qquad u_{i,j-1} = u(x_0, y_0 - \Delta y)$$

Often in the treatment of marching problems, the variation of the marching coordinate is indicated by a superscript, such as u_j^{n+1}, rather than a subscript. Many different finite-difference representations are possible for any given PDE and it's usually impossible to establish a "best" form on an absolute basis. First of all, the accuracy of a difference scheme may depend on the exact form of the equation and problem being solved and secondly, our selection of a "best" scheme will be influenced by the aspect of the procedure which we are trying to optimize, i.e., accuracy, economy, programming simplicity, etc.

The idea of a finite-difference representation for a derivative can be introduced by recalling the definition of the derivative for the function $u(x, y)$ at $x = x_0$, $y = y_0$.

$$\frac{\partial u}{\partial x} = \lim_{\Delta x \to 0} \frac{u(x_0 + \Delta x, y_0) - u(x_0, y_0)}{\Delta x} \tag{3-1}$$

Here, if u is continuous, it is expected that $[u(x_0 + \Delta x, y_0) - u(x_0, y_0)]/\Delta x$ will be a "reasonable" approximation to $\partial u/\partial x$ for a "sufficiently" small but finite Δx. In fact, the mean value theorem assures us that the difference representation is exact for some point within the Δx interval. The difference approximation can be put on a more formal basis through the use of either a Taylor-series expansion or Taylor's formula with a remainder. Developing a Taylor-series expansion for $u(x_0 + \Delta x, y_0)$ about (x_0, y_0) gives:

$$u(x_0 + \Delta x, y_0) = u(x_0, y_0) + \left.\frac{\partial u}{\partial x}\right)_0 \Delta x + \left.\frac{\partial^2 u}{\partial x^2}\right)_0 \frac{(\Delta x)^2}{2!} + \cdots$$

$$+ \left.\frac{\partial^{n-1} u}{\partial x^{n-1}}\right)_0 \frac{(\Delta x)^{n-1}}{(n-1)!} + \left.\frac{\partial^n u}{\partial x^n}\right)_\xi \frac{(\Delta x)^n}{n!} \qquad x_0 \leqslant \xi \leqslant (x_0 + \Delta x) \tag{3-2}$$

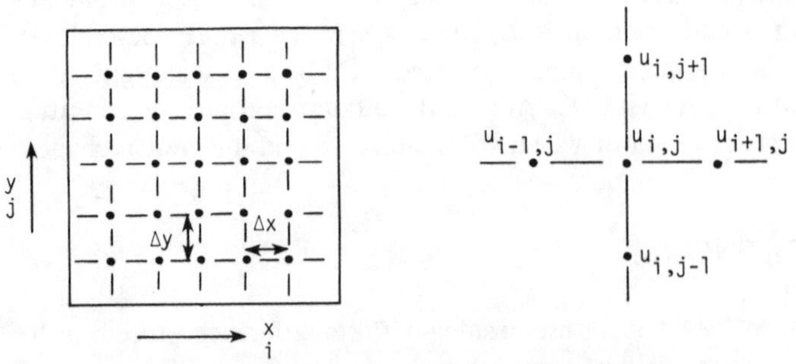

Figure 3-1 A typical finite-difference grid.

where the last term can be identified as the remainder. Thus, we can form the "forward" difference by rearranging Eq. (3-2)

$$\left.\frac{\partial u}{\partial x}\right)_{x_0, y_0} = \frac{u(x_0 + \Delta x, y_0) - u(x_0, y_0)}{\Delta x} - \left.\frac{\partial^2 u}{\partial x^2}\right)_0 \frac{\Delta x}{2!} - \cdots \tag{3-3}$$

Switching now to the i, j notation for brevity, we consider

$$\left.\frac{\partial u}{\partial x}\right)_{i,j} = \frac{u_{i+1,j} - u_{i,j}}{\Delta x} + \text{Truncation error} \tag{3-4}$$

where $(u_{i+1,j} - u_{i,j})/\Delta x$ is obviously the finite-difference representation for $\partial u/\partial x)_{i,j}$. The truncation error is the difference between the partial derivative and its finite-difference representation. We can characterize the limiting behavior of the truncation error (T.E.) by using the order of (O) notation whereby we write,

$$\left.\frac{\partial u}{\partial x}\right)_{i,j} = \frac{u_{i+1,j} - u_{i,j}}{\Delta x} + O(\Delta x)$$

where $O(\Delta x)$ has a precise mathematical meaning. Here, when the truncation error is written as $O(\Delta x)$ we mean $|\text{T.E.}| \leqslant K|\Delta x|$ for $\Delta x \to 0$ (sufficiently small Δx). K is a positive real constant. As a practical matter, the order of the truncation error in this case is found to be Δx raised to the largest power which is common to all terms in the truncation error.

To give a more general definition of the O notation, when we say $f(x) = O[\phi(x)]$, we mean that there exists a constant K, independent of x such that $|f(x)| \leqslant K|\phi(x)|$ for all x in S where f and ϕ are real or complex functions defined in S. We often restrict S by $x \to \infty$ (sufficiently large x) or as is most common in finite-difference applications, $x \to 0$ (sufficiently small x). More details on the O notation can be found in Whittaker and Watson (1927).

Note that $O(\Delta x)$ tells us nothing about the exact size of the T.E. but rather how it behaves as Δx tends toward zero. If another difference expression had a T.E. = $O[(\Delta x)^2]$, we might expect or hope that the T.E. of the second representation would be smaller than the first for a convenient Δx but we could only be *sure* that this would be true if we refined the mesh "sufficiently," and "sufficiently" is a quantity which is hard to estimate.

An infinite number of difference representations can be found for $\partial u/\partial x)_{i,j}$. For example we could expand "backwards"

$$u(x_0 - \Delta x, y_0) = u(x_0, y_0) - \left.\frac{\partial u}{\partial x}\right)_0 \Delta x + \left.\frac{\partial^2 u}{\partial x^2}\right)_0 \frac{(\Delta x)^2}{2} - \left.\frac{\partial^3 u}{\partial x^3}\right)_0 \frac{(\Delta x)^3}{6} + \cdots \tag{3-5}$$

and obtain the "backward" difference representation,

$$\left.\frac{\partial u}{\partial x}\right)_{i,j} = \frac{u_{i,j} - u_{i-1,j}}{\Delta x} + O(\Delta x) \tag{3-6}$$

We can subtract Eq. (3-5) from Eq. (3-2), rearrange, and obtain the "central" difference

$$\left.\frac{\partial u}{\partial x}\right)_{i,j} = \frac{u_{i+1,j} - u_{i-1,j}}{2\,\Delta x} + O(\Delta x)^2 \tag{3-7}$$

We can also add Eq. (3-2) and Eq. (3-5) and rearrange to obtain an approximation to the second derivative

$$\left.\frac{\partial^2 u}{\partial x^2}\right)_{i,j} = \frac{u_{i+1,j} - 2u_{i,j} + u_{i-1,j}}{(\Delta x)^2} + O(\Delta x)^2 \tag{3-8}$$

It should be emphasized that these are only a few examples of the possible ways in which first and second derivatives can be approximated.

It is convenient to utilize difference operators to represent finite differences when particular forms are used repetitively. Here we define the first forward difference of $u_{i,j}$ with respect to x at the point i, j as

$$\Delta_x u_{i,j} = u_{i+1,j} - u_{i,j} \tag{3-9}$$

Thus, we can express the forward finite-difference approximation for the first partial derivative as

$$\left.\frac{\partial u}{\partial x}\right)_{i,j} = \frac{u_{i+1,j} - u_{i,j}}{\Delta x} + O(\Delta x) = \frac{\Delta_x u_{i,j}}{\Delta x} + O(\Delta x) \tag{3-10}$$

Similarly, derivatives with respect to other variables such as y can be represented by

$$\frac{\Delta_y u_{i,j}}{\Delta y} = \frac{u_{i,j+1} - u_{i,j}}{\Delta y}$$

The first backward difference of $u_{i,j}$ with respect to x at i, j is denoted by

$$\nabla_x u_{i,j} = u_{i,j} - u_{i-1,j} \tag{3-11}$$

It follows that the first backward-difference approximation to the first derivative can be written as

$$\left.\frac{\partial u}{\partial x}\right)_{i,j} = \frac{u_{i,j} - u_{i-1,j}}{\Delta x} + O(\Delta x) = \frac{\nabla_x u_{i,j}}{\Delta x} + O(\Delta x) \tag{3-12}$$

The central-difference operators $\bar{\delta}$, δ and δ^2 will be defined as

$$\bar{\delta}_x u_{i,j} = u_{i+1,j} - u_{i-1,j} \tag{3-13}$$

$$\delta_x u_{i,j} = u_{i+1/2,j} - u_{i-1/2,j} \tag{3-14}$$

$$\delta_x^2 u_{i,j} = \delta_x(\delta_x u_{i,j}) = u_{i+1,j} - 2u_{i,j} + u_{i-1,j} \tag{3-15}$$

and an averaging operator, μ as

$$\mu_x u_{i,j} = \frac{u_{i+1/2,j} + u_{i-1/2,j}}{2} \tag{3-16}$$

It is convenient to have specific operators for certain common central differences although two of them can be easily expressed in terms of first-difference operators

$$\bar{\delta}_x u_{i,j} = \Delta_x u_{i,j} + \nabla_x u_{i,j} \tag{3-17}$$

$$\delta_x^2 u_{i,j} = \Delta_x u_{i,j} - \nabla_x u_{i,j} = \Delta_x \nabla_x u_{i,j} \tag{3-18}$$

Using the newly defined operators, the central-difference representation for the first partial derivative can be written as

$$\left.\frac{\partial u}{\partial x}\right)_{i,j} = \frac{u_{i+1,j} - u_{i-1,j}}{2\,\Delta x} + O(\Delta x)^2 = \frac{\bar{\delta}_x u_{i,j}}{2\,\Delta x} + O(\Delta x)^2 \tag{3-19}$$

and the central-difference representation of the second derivative as

$$\left.\frac{\partial^2 u}{\partial x^2}\right)_{i,j} = \frac{u_{i+1,j} - 2u_{i,j} + u_{i-1,j}}{(\Delta x)^2} + O(\Delta x)^2 = \frac{\delta_x^2 u_{i,j}}{(\Delta x)^2} + O(\Delta x)^2 \tag{3-20}$$

Higher-order forward- and backward-difference operators are defined as

$$\Delta_x^n u_{i,j} = \Delta_x(\Delta_x^{n-1} u_{i,j}) \tag{3-21}$$

and

$$\nabla_x^n u_{i,j} = \nabla_x(\nabla_x^{n-1} u_{i,j}) \tag{3-22}$$

As an example, a forward second-derivative approximation is given by:

$$\frac{\Delta_x^2 u_{i,j}}{(\Delta x)^2} = \frac{\Delta_x(u_{i+1,j} - u_{i,j})}{(\Delta x)^2} = \frac{u_{i+2,j} - u_{i+1,j} - u_{i+1,j} + u_{i,j}}{(\Delta x)^2}$$

$$= \frac{u_{i+2,j} - 2u_{i+1,j} + u_{i,j}}{(\Delta x)^2} = \left.\frac{\partial^2 u}{\partial x^2}\right)_{i,j} + O(\Delta x) \tag{3-23}$$

We can show that forward- and backward-difference approximations to derivatives of any order can be obtained from

$$\left.\frac{\partial^n u}{\partial x^n}\right)_{i,j} = \frac{\Delta_x^n u_{i,j}}{(\Delta x)^n} + O(\Delta x) \tag{3-24}$$

and

$$\left.\frac{\partial^n u}{\partial x^n}\right)_{i,j} = \frac{\nabla_x^n u_{i,j}}{(\Delta x)^n} + O(\Delta x) \tag{3-25}$$

Central-difference representations of derivatives of orders greater than the second can be expressed in terms of Δ and ∇ or δ. A more complete development on the use of difference operators can be found in many textbooks on numerical analysis such as Hildebrand (1956).

Most of the PDE's arising in fluid mechanics and heat transfer involve only first- and second-partial derivatives, and generally we strive to represent these derivatives using values at only two or three grid points. Within these restrictions, the most frequently used first-derivative approximations on a grid for which $\Delta x = h =$ constant are

$$\left.\frac{\partial u}{\partial x}\right)_{i,j} = \frac{u_{i+1,j} - u_{i,j}}{h} + O(h) \tag{3-26}$$

$$\left.\frac{\partial u}{\partial x}\right)_{i,j} = \frac{u_{i,j} - u_{i-1,j}}{h} + O(h) \tag{3-27}$$

$$\left.\frac{\partial u}{\partial x}\right)_{i,j} = \frac{u_{i+1,j} - u_{i-1,j}}{2h} + O(h^2) \tag{3-28}$$

$$\left.\frac{\partial u}{\partial x}\right)_{i,j} = \frac{-3u_{i,j} + 4u_{i+1,j} - u_{i+2,j}}{2h} + O(h^2) \tag{3-29}$$

$$\left.\frac{\partial u}{\partial x}\right)_{i,j} = \frac{3u_{i,j} - 4u_{i-1,j} + u_{i-2,j}}{2h} + O(h^2) \tag{3-30}$$

$$\left.\frac{\partial u}{\partial x}\right)_{i,j} = \frac{1}{2h}\left(\frac{\bar{\delta}_x u_{i,j}}{1 + \delta_x^2/6}\right) + O(h^4) \tag{3-31}$$

The most common three-point second-derivative approximations for a uniform grid, $\Delta x = h =$ constant, are

$$\left.\frac{\partial^2 u}{\partial x^2}\right)_{i,j} = \frac{u_{i,j} - 2u_{i+1,j} + u_{i+2,j}}{h^2} + O(h) \tag{3-32}$$

$$\left.\frac{\partial^2 u}{\partial x^2}\right)_{i,j} = \frac{u_{i,j} - 2u_{i-1,j} + u_{i-2,j}}{h^2} + O(h) \tag{3-33}$$

$$\left.\frac{\partial^2 u}{\partial x^2}\right)_{i,j} = \frac{u_{i+1,j} - 2u_{i,j} + u_{i-1,j}}{h^2} + O(h^2) \tag{3-34}$$

$$\left.\frac{\partial^2 u}{\partial x^2}\right)_{i,j} = \frac{\delta_x^2 u_{i,j}}{h^2(1 + \delta_x^2/12)} + O(h^4) \tag{3-35}$$

The compact, three-point schemes given by Eqs. (3-31) and (3-35) having fourth-order truncation errors deserve a further word of explanation. (See also Orszag and Israeli, 1974.) Letting $\partial u/\partial x)_{i,j} = v_{i,j}$, Eq. (3-31) is to be interpreted as

$$\left(1 + \frac{\delta_x^2}{6}\right) v_{i,j} = \frac{\bar{\delta}_x u_{i,j}}{2h}$$

or

$$\frac{1}{6}\left(v_{i+1,j} + 4v_{i,j} + v_{i-1,j}\right) = \frac{\bar{\delta}_x u_{i,j}}{2h} \tag{3-36}$$

which provides an *implicit* formula for the derivative of interest, $v_{i,j}$. The $v_{i,j}$ can be determined from the $u_{i,j}$ by solving a tridiagonal system of simultaneous algebraic equations, which can usually be accomplished quite efficiently. Tridiagonal systems commonly occur in connection with the use of implicit difference schemes for second-order PDE's arising from marching problems and will be defined and discussed in some detail in Chapter 4. For now it is sufficient to think of a tridiagonal system as the arrangement of unknowns which would occur if each difference equation in a system only involved a single unknown variable evaluated at 3 adjacent grid locations. The interpretation of Eq. (3-35) proceeds in a similar manner providing an *implicit* representation of $\partial^2 u/\partial x^2)_{i,j}$. Some difference approximations for derivatives which involve more than three grid points are given in Table 3-1. For completeness, a few common difference representations for mixed partial derivatives are presented in Table 3-2. These will prove useful for schemes to be discussed in later chapters. The mixed derivative approximations in Table 3-2 can be verified by using the Taylor-series expansion for two variables

Table 3-1 Difference approximations using more than three points

Derivative	Finite-difference representation	Equation
$\left.\frac{\partial^3 u}{\partial x^3}\right)_{i,j} =$	$\frac{u_{i+2,j} - 2u_{i+1,j} + 2u_{i-1,j} - u_{i-2,j}}{2h^3} + O(h^2)$	(3-38)
$\left.\frac{\partial^4 u}{\partial x^4}\right)_{i,j} =$	$\frac{u_{i+2,j} - 4u_{i+1,j} + 6u_{i,j} - 4u_{i-1,j} + u_{i-2,j}}{h^4} + O(h^2)$	(3-39)
$\left.\frac{\partial^2 u}{\partial x^2}\right)_{i,j} =$	$\frac{-u_{i+3,j} + 4u_{i+2,j} - 5u_{i+1,j} + 2u_{i,j}}{h^2} + O(h^2)$	(3-40)
$\left.\frac{\partial^3 u}{\partial x^3}\right)_{i,j} =$	$\frac{-3u_{i+4,j} + 14u_{i+3,j} - 24u_{i+2,j} + 18u_{i+1,j} - 5u_{i,j}}{2h^3} + O(h^2)$	(3-41)
$\left.\frac{\partial^2 u}{\partial x^2}\right)_{i,j} =$	$\frac{2u_{i,j} - 5u_{i-1,j} + 4u_{i-2,j} - u_{i-3,j}}{h^2} + O(h^2)$	(3-42)
$\left.\frac{\partial^3 u}{\partial x^3}\right)_{i,j} =$	$\frac{5u_{i,j} - 18u_{i-1,j} + 24u_{i-2,j} - 14u_{i-3,j} + 3u_{i-4,j}}{2h^3} + O(h^2)$	(3-43)
$\left.\frac{\partial u}{\partial x}\right)_{i,j} =$	$\frac{-u_{i+2,j} + 8u_{i+1,j} - 8u_{i-1,j} + u_{i-2,j}}{12h} + O(h^4)$	(3-44)
$\left.\frac{\partial^2 u}{\partial x^2}\right)_{i,j} =$	$\frac{-u_{i+2,j} + 16u_{i+1,j} - 30u_{i,j} + 16u_{i-1,j} - u_{i-2,j}}{12h^2} + O(h^4)$	(3-45)

Table 3-2 Difference approximations for mixed partial derivatives

Derivative	Finite-difference representation	Equation
$\left.\frac{\partial^2 u}{\partial x\,\partial y}\right)_{i,j} =$	$\frac{1}{\Delta x}\left(\frac{u_{i+1,j} - u_{i+1,j-1}}{\Delta y} - \frac{u_{i,j} - u_{i,j-1}}{\Delta y}\right) + O(\Delta x, \Delta y)$	(3-46)
$\left.\frac{\partial^2 u}{\partial x\,\partial y}\right)_{i,j} =$	$\frac{1}{\Delta x}\left(\frac{u_{i,j+1} - u_{i,j}}{\Delta y} - \frac{u_{i-1,j+1} - u_{i-1,j}}{\Delta y}\right) + O(\Delta x, \Delta y)$	(3-47)
$\left.\frac{\partial^2 u}{\partial x\,\partial y}\right)_{i,j} =$	$\frac{1}{\Delta x}\left(\frac{u_{i,j} - u_{i,j-1}}{\Delta y} - \frac{u_{i-1,j} - u_{i-1,j-1}}{\Delta y}\right) + O(\Delta x, \Delta y)$	(3-48)
$\left.\frac{\partial^2 u}{\partial x\,\partial y}\right)_{i,j} =$	$\frac{1}{\Delta x}\left(\frac{u_{i+1,j+1} - u_{i+1,j}}{\Delta y} - \frac{u_{i,j+1} - u_{i,j}}{\Delta y}\right) + O(\Delta x, \Delta y)$	(3-49)
$\left.\frac{\partial^2 u}{\partial x\,\partial y}\right)_{i,j} =$	$\frac{1}{\Delta x}\left(\frac{u_{i+1,j+1} - u_{i+1,j-1}}{2\,\Delta y} - \frac{u_{i,j+1} - u_{i,j-1}}{2\,\Delta y}\right) + O[\Delta x, (\Delta y)^2]$	(3-50)
$\left.\frac{\partial^2 u}{\partial x\,\partial y}\right)_{i,j} =$	$\frac{1}{\Delta x}\left(\frac{u_{i,j+1} - u_{i,j-1}}{2\,\Delta y} - \frac{u_{i-1,j+1} - u_{i-1,j-1}}{2\,\Delta y}\right) + O[\Delta x, (\Delta y)^2]$	(3-51)
$\left.\frac{\partial^2 u}{\partial x\,\partial y}\right)_{i,j} =$	$\frac{1}{2\,\Delta x}\left(\frac{u_{i+1,j+1} - u_{i+1,j-1}}{2\,\Delta y} - \frac{u_{i-1,j+1} - u_{i-1,j-1}}{2\,\Delta y}\right) + O[(\Delta x)^2, (\Delta y)^2]$	(3-52)
$\left.\frac{\partial^2 u}{\partial x\,\partial y}\right)_{i,j} =$	$\frac{1}{2\,\Delta x}\left(\frac{u_{i+1,j+1} - u_{i+1,j}}{\Delta y} - \frac{u_{i-1,j+1} - u_{i-1,j}}{\Delta y}\right) + O[(\Delta x)^2, \Delta y]$	(3-53)
$\left.\frac{\partial^2 u}{\partial x\,\partial y}\right)_{i,j} =$	$\frac{1}{2\,\Delta x}\left(\frac{u_{i+1,j} - u_{i+1,j-1}}{\Delta y} - \frac{u_{i-1,j} - u_{i-1,j-1}}{\Delta y}\right) + O[(\Delta x)^2, \Delta y]$	(3-54)

$$u(x_0 + \Delta x, y_0 + \Delta y) = u(x_0, y_0) + \left(\Delta x \frac{\partial}{\partial x} + \Delta y \frac{\partial}{\partial y}\right) u(x_0, y_0)$$
$$+ \frac{1}{2!}\left(\Delta x \frac{\partial}{\partial x} + \Delta y \frac{\partial}{\partial y}\right)^2 u(x_0, y_0)$$
$$+ \cdots + \frac{1}{n!}\left(\Delta x \frac{\partial}{\partial x} + \Delta y \frac{\partial}{\partial y}\right)^n u(x_0 + \theta\,\Delta x, y_0 + \theta\,\Delta y)$$
$$0 \leqslant \theta \leqslant 1 \tag{3-37}$$

3-3 DIFFERENCE REPRESENTATION OF PARTIAL DIFFERENTIAL EQUATIONS

3-3.1 Truncation Error

As a starting point in our study of truncation error, let us consider the heat equation

$$\frac{\partial u}{\partial t} = \alpha \frac{\partial^2 u}{\partial x^2} \tag{3-55}$$

Using a forward-difference representation for the time derivative and a central-difference representation for the second derivative, we can approximate the heat equation by

$$\frac{u_j^{n+1} - u_j^n}{\Delta t} = \frac{\alpha}{(\Delta x)^2}(u_{j+1}^n - 2u_j^n + u_{j-1}^n) \tag{3-56a}$$

However, we noted in Section 3-2 that truncation errors were associated with the forward- and central-difference representations used in Eq. (3-56*a*). If we rearrange Eq. (3-55) to put zero on the right-hand side and include the truncation errors associated with the difference representation of the derivatives we obtain

$$\underbrace{\frac{\partial u}{\partial t} - \alpha \frac{\partial^2 u}{\partial x^2}}_{\text{PDE}} = \underbrace{\frac{u_j^{n+1} - u_j^n}{\Delta t} - \frac{\alpha}{(\Delta x)^2}(u_{j+1}^n - 2u_j^n + u_{j-1}^n)}_{\text{FDE}}$$
$$+ \underbrace{\left[-\left(\frac{\partial^2 u}{\partial t^2}\right)_{n,j} \frac{\Delta t}{2} + \alpha \left(\frac{\partial^4 u}{\partial x^4}\right)_{n,j} \frac{(\Delta x)^2}{12} + \cdots \right]}_{\text{T.E.}} \tag{3-56b}$$

The truncation errors associated with all derivatives in any one PDE should be obtained by expanding about the same point (n, j in the above discussion).

The difference representation given by Eq. (3-56*a*) will be referred to as the *simple explicit scheme* for the heat equation. An *explicit* scheme is one for which only one unknown appears in the difference equation in a manner which permits evaluation in terms of known quantities. Since the parabolic heat equation governs a marching problem for which an initial distribution of u must be specified, u's at the time level n can be considered as known. If the second derivative term in the heat equation was approximated by u's at the $n+1$ time level, three unknowns would appear in the difference equation and the procedure would be known as *implicit*, indicating that the algebraic formulation would require the simultaneous solution of several equations involving the unknowns. The differences between implicit and explicit schemes will be discussed further in Chapter 4.

The quantity in brackets (note that only the leading terms have been written out utilizing Taylor-series expansions) in Eq. (3-56*b*) is identified as the *truncation error* for this finite-difference representation of the heat equation and is defined as the difference between the partial differential equation and the difference approximation to it. That is, T.E. = PDE − FDE. The *order* of the truncation error in this case is $O(\Delta t) + O[(\Delta x)^2]$ which is frequently expressed in the form $O[\Delta t, (\Delta x)^2]$. Naturally, we solve only the finite-difference equations and hope that the truncation error is small. If we don't feel a little uneasy at this point, perhaps we should. How do we know that our difference representation is acceptable and that a marching solution technique will work in the sense of giving us an approximate solution to the PDE? In order to be acceptable, our difference representation for this marching problem needs to meet the conditions of *consistency* and *stability*.

3-3.2 Consistency

Consistency deals with the extent to which the finite-difference equations approximate the partial differential equations. The difference between the PDE and the finite-difference approximation has already been defined as the truncation error of the difference representation. A finite-difference representation of a PDE is said to be consistent if we can show that the difference between the PDE and its difference representation vanishes as the mesh is refined, i.e., $\lim_{\text{mesh}\to 0} (\text{PDE} - \text{FDE}) = \lim_{\text{mesh}\to 0} (\text{T.E.}) = 0$. This should always be the case if the order of the truncation error vanishes under grid refinement [i.e., $O(\Delta t)$, $O(\Delta x)$, etc.]. An example of a questionable scheme would be one for which the truncation error was $O(\Delta t/\Delta x)$ where the scheme would not formally be consistent unless the mesh were refined in a manner such that $\Delta t/\Delta x \to 0$. The DuFort-Frankel (DuFort and Frankel, 1953) differencing of the heat equation,

$$\frac{u_j^{n+1} - u_j^{n-1}}{2\,\Delta t} = \frac{\alpha}{(\Delta x)^2}(u_{j+1}^n - u_j^{n+1} - u_j^{n-1} + u_{j-1}^n) \tag{3-57}$$

for which the leading terms in the truncation error are

$$+\frac{\alpha}{12}\left.\frac{\partial^4 u}{\partial x^4}\right)_{n,j}(\Delta x)^2 - \alpha\left.\frac{\partial^2 u}{\partial t^2}\right)_{n,j}\left(\frac{\Delta t}{\Delta x}\right)^2 - \frac{1}{6}\left.\frac{\partial^3 u}{\partial t^3}\right)_{n,j}(\Delta t)^2$$

serves as an example. All is well if

$$\lim_{\Delta t, \Delta x \to 0}\left(\frac{\Delta t}{\Delta x}\right) = 0$$

but if Δt and Δx were to approach zero at the same rate such that $\Delta t/\Delta x = \beta$, then the DuFort-Frankel scheme is consistent with the hyperbolic equation,

$$\frac{\partial u}{\partial t} + \alpha\beta^2 \frac{\partial^2 u}{\partial t^2} = \alpha \frac{\partial^2 u}{\partial x^2}$$

3-3.3 Stability

Numerical stability is a concept applicable in the strict sense only to marching problems. A stable numerical scheme is one for which errors from any source (round-off, truncation, mistakes) are not permitted to grow in the sequence of numerical procedures as the calculation proceeds from one marching step to the next. Generally, concern over stability occupies much more of our time and energy than does concern over consistency. Consistency is relatively easy to check and most schemes which are conceived will be consistent just due to the methodology employed in their development. Stability is much more subtle and usually a bit of hard work is required in order to establish analytically that a scheme is stable. Much more will be presented in Section 3-6 on stability and some very workable methods will be developed for establishing the stability limits for linear partial differential equations. It will be possible to extend these guidelines to nonlinear equations in an approximate sense.

Using these guidelines, the DuFort-Frankel scheme, Eq. (3-57), for the heat equation would be found to be unconditionally stable whereas the simple explicit scheme would be stable only if $r = [\alpha \Delta t/(\Delta x)^2] \leqslant \frac{1}{2}$. This restriction would limit the size of the marching step permitted for any specified spatial mesh.

A scheme using a central time difference and having a more favorable truncation error $O[(\Delta t)^2, (\Delta x)^2]$,

$$\frac{u_j^{n+1} - u_j^{n-1}}{2\,\Delta t} = \frac{\alpha}{(\Delta x)^2}(u_{j+1}^n - 2u_j^n + u_{j-1}^n) \tag{3-58}$$

is unconditionally unstable and therefore cannot be used for real calculations despite the fact that it looks to be more accurate, in terms of truncation error, than the ones given previously which will work.

Sometimes instability can be identified with a physical implausibility. That is, conditions which would result in an unstable numerical procedure would also imply unacceptable modeling of physical processes. To illustrate this, we rearrange the simple explicit representation of the heat equation, Eq. (3-56*a*), so that the unknown appears on the left. Letting $r = \alpha\,\Delta t/(\Delta x)^2$, our difference equation becomes

$$u_j^{n+1} = r(u_{j+1}^n + u_{j-1}^n) + (1 - 2r)u_j^n \tag{3-59}$$

Suppose that at time t, $u_{j+1}^n = u_{j-1}^n = 100°\text{C}$ and $u_j^n = 0°\text{C}$. This arrangement is shown in Fig. 3-2. If $r > \frac{1}{2}$ we see that the temperature at point j at time level $n + 1$ will exceed the temperature at the two surrounding points at time level n. This seems unreasonable since we expect heat to flow from the warmer region to a colder region but not vice versa. The maximum temperature which we would expect to find at point j at time level $n + 1$ is 100°C. If $r = 1$, for example, u_j^{n+1} would equal 200°C by Eq. (3-59).

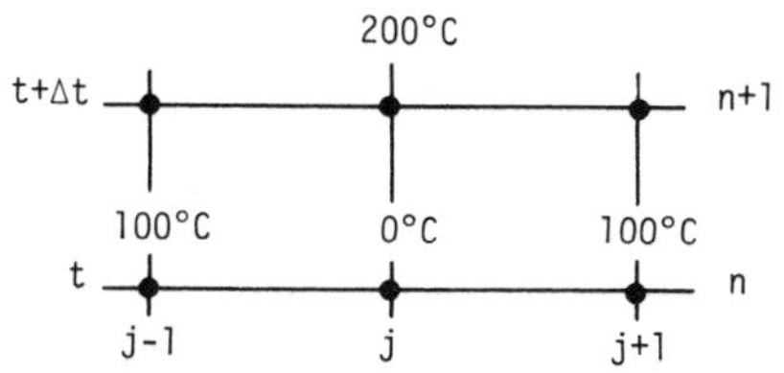

Figure 3-2 Physical implausibility resulting from $r = 1$.

3-3.4 Convergence for Marching Problems

Generally, we find that a consistent, stable scheme is convergent. Convergence here means that the solution to the finite-difference equation approaches the true solution to the PDE having the same initial and boundary conditions as the mesh is refined. A proof of this is available for initial value (marching) problems governed by linear PDE's. The theorem, due to Lax (see Richtmyer and Morton, 1967) is stated here without proof.

Lax's Equivalence Theorem. Given a properly posed initial value problem and a finite-difference approximation to it that satisfies the consistency condition, stability is the necessary and sufficient condition for convergence.

We might add that most computational work proceeds as though this theorem applies also to nonlinear PDE's although the theorem has never been proven for this more general category of equations.

3-3.5 Round-off and Discretization Errors

Any computed solution, including sometimes an "exact" analytic solution to a PDE, may be affected by rounding to a finite number of digits in the arithmetic operations. These errors are called *round-off errors* and we are especially aware of their existence in obtaining machine solutions to finite-difference equations because of the large number of dependent, repetitive operations which are usually involved. In some types of calculations, the magnitude of the round-off error is proportional to the number of grid points in the problem domain. In these cases, refining the grid may decrease the truncation error but increase the round-off error.

Discretization error is the error in the solution to the PDE caused by replacing the continuous problem by a discrete one and is defined as the difference between the exact solution of the PDE (round-off free) and the exact solution of the finite-difference equations (round-off free). In terms of the definitions developed thus far, the difference between the exact solution of the PDE and the computer solution to the finite-difference equations would be equal to the sum of the discretization error and the round-off error associated with the finite-difference calculation. We can also observe that the discretization error is the error in the solution which is caused by the truncation error in the difference representation of the PDE plus any errors introduced by the treatment of boundary conditions.

3-3.6 A Comment on Equilibrium Problems

Throughout our discussion of stability and convergence, the focus was on marching problems (parabolic and hyperbolic PDE's). Despite this emphasis on initial value problems, most of the material presented in this chapter also applies to equilibrium problems. The exception is the concept of stability. We should observe, however, that the important concept of consistency applies to difference representations of PDE's of all classes.

The "convergence" of the solution of the difference equation to the exact solution of the PDE might be aptly termed truncation or discretization convergence. The solution to equilibrium problems (elliptic equations) leads us to a system of simultaneous algebraic equations which needs to be solved only once, rather than in a marching manner. Thus, the concept of stability developed previously is not directly applicable as stated. To achieve "truncation convergence" for equilibrium problems it would seem that it is only necessary to devise a solution scheme in which the error in solving the simultaneous algebraic equations can be controlled as the mesh size is refined without limit. Many common schemes are iterative (Gauss-Seidel iteration is one example) in nature and for these we want to ensure that the iterative process converges. Here convergence means that the iterative process is repeated until the magnitude of the difference between the function at the $k + 1$ and the k iteration levels is as small as we wish for each grid point, i.e., $|u_{i,j}^{k+1} - u_{i,j}^{k}| < \epsilon$. This is known as *iteration convergence.* It would appear that (no proof can be cited) truncation convergence will be assumed for a consistent representation to an equilibrium problem if it can be shown that the iterative method of solution converges even for arbitrarily small choices of mesh sizes.

It is possible to use direct (noniterative) methods to solve the algebraic equations associated with equilibrium problems. For these methods we would want to be sure that the errors inherent in the method, especially round-off errors, do not get out of control as the mesh is refined and the number of points tends toward infinity.

In closing this section we should mention that there are aspects to the iterative solution of equilibrium problems which resemble the marching process in initial value problems and a sense in which stability concerns in the marching problems correspond to iterative convergence concerns in the solution to equilibrium problems.

3-3.7 Conservation Form and Conservative Property

Two different ideas will be discussed in this section. The first has to do with the PDE's themselves. The terms "conservation form," "conservation-law form," "conservative form," and "divergence form" are all equivalent and partial differential equations having this form have the property that the coefficients of the derivative terms are either constant or, if variable, their derivatives appear nowhere in the equation. Normally, for the PDE's which represent a physical conservation statement, this means that the divergence of a physical quantity can be identified in the equation. As an example, the conservative form of the equation for mass conservation (continuity equation) is

$$\frac{\partial \rho}{\partial t} + \frac{\partial \rho u}{\partial x} + \frac{\partial \rho v}{\partial y} + \frac{\partial \rho w}{\partial z} = 0 \tag{3-60}$$

which can be written in vector notation as

$$\frac{\partial \rho}{\partial t} + \nabla \cdot \rho \mathbf{V} = 0$$

A nonconservative or nondivergence form would be

$$\frac{\partial \rho}{\partial t} + u \frac{\partial \rho}{\partial x} + \rho \frac{\partial u}{\partial x} + v \frac{\partial \rho}{\partial y} + \rho \frac{\partial v}{\partial y} + w \frac{\partial \rho}{\partial z} + \rho \frac{\partial w}{\partial z} = 0 \tag{3-61}$$

As a second example, we consider the one-dimensional heat conduction equation for a substance whose density, ρ, specific heat, c, and thermal conductivity, k, all vary with position. The conservative form of this equation is

$$\rho c \frac{\partial T}{\partial t} = \frac{\partial}{\partial x}\left(k \frac{\partial T}{\partial x}\right) \tag{3-62}$$

whereas a nonconservative form would be

$$\rho c \frac{\partial T}{\partial t} = k \frac{\partial^2 T}{\partial x^2} + \frac{\partial k}{\partial x} \frac{\partial T}{\partial x} \tag{3-63}$$

In Eq. (3-62) the right-hand side can be identified as the negative of the divergence of the heat flux vector specialized for one-dimensional conduction. A difference formulation based on a PDE in nondivergence form may lead to numerical difficulties in situations where the coefficients may be discontinuous as in flows containing shock waves.

The second idea to be developed in this section is that of the *conservative property of a finite-difference representation.* The PDE's of interest in this book all have their basis in physical laws such as the conservation of mass, momentum, and energy. Such a PDE represents a conservation statement at a point. We strive to construct finite-difference representations which provide a good approximation to the PDE in a small, local neighborhood involving a few grid points. The same conservation principles which gave rise to the PDE's also apply to arbitrarily large regions (control volumes). In fact, in deriving the PDE's we usually start with the control-volume form of the conservation statement. If our finite-difference representation approximates the PDE closely in the neighborhood of each grid point, then we have reason to expect that the related conservation statement will be approximately enforced over a larger control volume containing a large number of grid points in the interior. Those finite-difference schemes which maintain the discretized version of the conservation statement exactly (except for round-off errors) for any mesh size over an arbitrary finite region containing any number of grid points is said to have the *conservative property.* For some problems this property is crucial.

The key word in the definition above is "exactly." All consistent schemes should approximately enforce the appropriate conservation statement over large regions, but schemes having the conservative property do so exactly (except for round-off errors), because of exact cancellation of terms. To illustrate this concept, we will consider a problem requiring the solution of the continuity equation for steady flow. The PDE can be written as

$$\nabla \cdot \rho \mathbf{V} = 0$$

We will assume that the PDE is approximated by a suitable finite-difference representation and solved throughout the flow. For an arbitrary control volume which could include the entire problem domain or any fraction of it, conservation of mass for steady flow requires that the net mass efflux be zero (mass flow rate in equals mass flow rate out). This is observed formally by applying the divergence theorem to the governing PDE,

$$\iiint_R \nabla \cdot \rho \mathbf{V} \, dR = \iint_S \rho \mathbf{V} \cdot \mathbf{n} \, dS = 0$$

To see if the finite-difference representation for the PDE has the conservative property, we must establish that the discretized version of the divergence theorem is satisfied. We normally check this for a control volume consisting of the entire problem domain. To do this the integral on the left is evaluated by summing the difference representation of the PDE at all grid points. If the difference scheme has the conservative property, all terms will cancel except those which represent fluxes at the boundaries. It should be possible to rearrange the remaining terms to obtain identically a finite-difference representation of the integral on the right. For this example the result will be a verification that the mass flux into the control volume equals the mass flux out. If the difference scheme used for the PDE is not conservative, the numerical solution may permit the existence of small mass sources or sinks.

Schemes having the conservative property occur in a natural way when differencing starts with the divergence form of the PDE. For some equations and problems, the divergence form is not an appropriate starting point. For these situations, use of a control volume method (Section 3-4.4) for obtaining the difference scheme is helpful. This difference representation will usually have the conservative property if care is taken to ensure that the expressions used to represent fluxes across the interface of two adjacent control volumes are the same in the difference form of the conservation statement for each of the two control volumes.

The conservative property issue has been actively discussed and debated over the short history of computational fluid mechanics and heat transfer. However, the conservative property is not the only important figure of merit for a difference representation. PDE's represent more than a conservation statement at a point. As shown by solution forms in Chapter 2, PDE's also contain information on characteristic directions and domains of dependence. Proper representation of this information is also important. Many useful finite-difference equations do not have the conservative property and in a few instances, prove to be more accurate in some sense than those that do. The importance of maintaining the conservation statement with high accuracy over a finite region is highly problem dependent. All consistent formulations, whether or not they have the conservative property, can provide an adequate representation for most problems if the grid is refined sufficiently.

3-4 FURTHER EXAMPLES OF METHODS FOR OBTAINING FINITE-DIFFERENCE EQUATIONS

As we start with a given PDE and a finite-difference mesh, several procedures are available to us for developing finite-difference equations. Among these are:

a. Taylor-series expansions
b. polynomial fitting
c. integral method (called the micro-integral method by some)
d. control-volume approach

It's sometimes possible to obtain exactly the same finite-difference representation by using all four methods. In our introduction to the subject, we'll lean most heavily on the use of Taylor-series expansions utilizing polynomial fitting on occasion in treating boundary conditions.

3-4.1 Use of Taylor Series

We now demonstrate how one might proceed on a slightly more formal basis with Taylor-series expansions to develop difference expressions satisfying specified constraints. Suppose we want to develop a difference approximation for $\partial u/\partial x)_{i,j}$ having a truncation error of $O(\Delta x)^2$ using at most values $u_{i-2,j}$, $u_{i-1,j}$, and $u_{i,j}$. With these constraints and objectives, it would appear logical to write Taylor-series expressions for $u_{i-2,j}$ and $u_{i-1,j}$ expanding about the point (i, j) and attempt to solve for $\partial u/\partial x)_{i,j}$ from the resulting equations in such a way as to obtain a truncation error of $O(\Delta x)^2$:

$$u_{i-2,j} = u_{i,j} + \left.\frac{\partial u}{\partial x}\right)_{i,j}(-2\,\Delta x) + \left.\frac{\partial^2 u}{\partial x^2}\right)_{i,j}\frac{(2\,\Delta x)^2}{2!} + \left.\frac{\partial^3 u}{\partial x^3}\right)_{i,j}\frac{(-2\,\Delta x)^3}{3!} + \cdots \tag{3-64}$$

$$u_{i-1,j} = u_{i,j} + \left.\frac{\partial u}{\partial x}\right)_{i,j}(-\Delta x) + \left.\frac{\partial^2 u}{\partial x^2}\right)_{i,j}\frac{(\Delta x)^2}{2!} + \left.\frac{\partial^3 u}{\partial x^3}\right)_{i,j}\frac{(-\Delta x)^3}{3!} + \cdots \tag{3-65}$$

It's often possible to determine the required form of the difference representation by inspection or simple substitution. To proceed by substitution, we will rearrange Eq. (3-64) to put $\partial u/\partial x)_{i,j}$ on the left-hand side such that

$$\left.\frac{\partial u}{\partial x}\right)_{i,j} = \frac{u_{i,j}}{2\,\Delta x} - \frac{u_{i-2,j}}{2\,\Delta x} + \frac{\partial^2 u}{\partial x^2}\Delta x + O(\Delta x)^2$$

As is, the representation is $O(\Delta x)$ because of the term $(\partial^2 u/\partial x^2)\Delta x$. We can substitute for $\partial^2 u/\partial x^2$ in the above equation using Eq. (3-65) to obtain the desired result. A more formal procedure to obtain the desired expression is sometimes useful. To proceed more formally we first multiply Eq. (3-64) by a and Eq. (3-65) by b and add the two equations. If $-2a - b = 1$, then the coefficient of $\partial u/\partial x)_{i,j}\,\Delta x$ will be 1 after the addition, and if $2a + b/2 = 0$, then the terms involving $\partial^2 u/\partial x^2)_{i,j}$ which would contribute a T.E. of $O(\Delta x)$ to the final result will be eliminated. A solution to the equations

$$-2a - b = 1 \qquad 2a + \frac{b}{2} = 0$$

is given by $a = \frac{1}{2}$, $b = -2$. Thus, if we multiply Eq. (3-64) by $\frac{1}{2}$, Eq. (3-65) by -2, add the results and solve for $\partial u/\partial x)_{i,j}$ we obtain

$$\left.\frac{\partial u}{\partial x}\right)_{i,j} = \frac{u_{i-2,j} - 4u_{i-1,j} + 3u_{i,j}}{2\,\Delta x} + O[(\Delta x)^2]$$

which can be recognized as Eq. (3-30). A careful check on the details of this example will reveal that it was really necessary to include terms involving $\partial^3 u/\partial x^3)_{i,j}$ in the Taylor-series expansions in order to determine whether or not these terms would cancel in the algebraic operations and reduce the truncation error even further to $O(\Delta x)^3$. Fortuitous cancellation of terms occurs frequently enough to warrant close attention to this point.

We should observe that it is sometimes necessary to carry out the inverse of the above process. That is, suppose we had obtained the approximation represented by Eq. (3-30) by some other means and we wanted to investigate the consistency and truncation error of such an expression. For this, the use of Taylor-series expansions would be invaluable and the recommended procedure would be to substitute the Taylor-series expressions from Eq. (3-64) and Eq. (3-65) above for $u_{i-2,j}$ and $u_{i-1,j}$ into the difference representation to obtain an expression of the form $\partial u/\partial x)_{i,j}$ + T.E. on the right-hand side. At this point the T.E. has been identified and if $\lim_{\Delta x \to 0}$ (T.E.) $= 0$ the difference representation is consistent.

As a slightly more complex example, we will develop a finite-difference approximation with truncation error $O(\Delta y)^2$ for $\partial u/\partial y$ at point (i, j) using at most $u_{i,j}$, $u_{i,j+1}$, $u_{i,j-1}$ when the grid spacing is not uniform. We will adopt the notation that $\Delta y_+ = y_{i,j+1} - y_{i,j}$ and $\Delta y_- = y_{i,j} - y_{i,j-1}$ as indicated in Fig. 3-3.

We recall that for equal spacing, the central-difference representation for a first derivative was equivalent to the arithmetic average of a forward and backward representation. That is, for $\Delta y_+ = \Delta y_- = \Delta y$

$$\left.\frac{\partial u}{\partial y}\right)_{i,j} = \frac{\bar{\delta}_y u_{i,j}}{2\,\Delta y} = \frac{\Delta_y u_{i,j} + \nabla_y u_{i,j}}{2\,\Delta y} + O[(\Delta y)^2]$$

We might wonder if, for unequal spacing, use of a geometrically weighted average will preserve the second-order accuracy:

$$\left.\frac{\partial u}{\partial y}\right)_{i,j} \stackrel{?}{=} \frac{\Delta_y u_{i,j}}{\Delta y_+}\left(\frac{\Delta y_-}{\Delta y_+ + \Delta y_-}\right) + \frac{\nabla_y u_{i,j}}{\Delta y_-}\left(\frac{\Delta y_+}{\Delta y_+ + \Delta y_-}\right) + O[(\Delta y)^2] \tag{3-66}$$

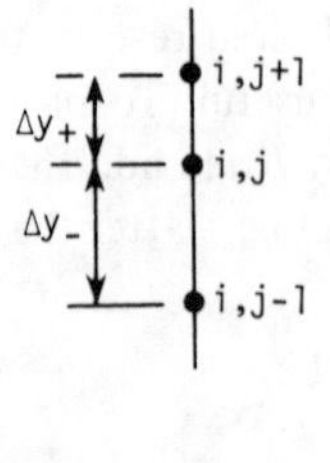

Figure 3-3 Notation for unequal y spacing.

The truth of the above statement may be evident to some, but it can be verified from basics by use of Taylor-series expansions about point (i, j). Letting $\Delta y_+/\Delta y_- = \alpha$, and adopting the more compact subscript notation to denote differentiation, $u_y = \partial u/\partial y)_{i,j}$, $u_{yy} = \partial^2 u/\partial y^2)_{i,j}$, etc., we obtain

$$u_{i,j+1} = u_{i,j} + u_y \alpha \Delta y_- + u_{yy} \frac{(\alpha \Delta y_-)^2}{2!} + u_{yyy} \frac{(\alpha \Delta y_-)^3}{3!} + u_{yyyy} \frac{(\alpha \Delta y_-)^4}{4!} + \cdots \tag{3-67}$$

$$u_{i,j-1} = u_{i,j} + u_y(-\Delta y_-) + u_{yy} \frac{(-\Delta y_-)^2}{2!} + u_{yyy} \frac{(-\Delta y_-)^3}{3!} + u_{yyyy} \frac{(-\Delta y_-)^4}{4!} + \cdots \tag{3-68}$$

As before, we will multiply Eq. (3-67) by a and Eq. (3-68) by b, add the results and solve for $\partial u/\partial y)_{i,j}$. Requiring that the coefficient of $\partial u/\partial y)_{i,j}\, \Delta y_-$ be equal to one after the addition, gives $a\alpha - b = 1$. For the final result to have a truncation error $O(\Delta y)^2$ or better, the coefficient of u_{yy} must be zero after the addition which requires that $\alpha^2 a + b = 0$. A solution to these two algebraic equations can be obtained readily as $a = 1/\alpha(\alpha + 1)$, $b = -\alpha/(\alpha + 1)$. Thus

$$\left.\frac{\partial u}{\partial y}\right)_{i,j} = \frac{a \times \text{Eq. (3-67)} + b \times \text{Eq. (3-68)}}{\Delta y_-} + O(\Delta y)^2$$

The final result can be written as

$$\left.\frac{\partial u}{\partial y}\right)_{i,j} = \frac{u_{i,j+1} + (\alpha^2 - 1)u_{i,j} - \alpha^2 u_{i,j-1}}{\alpha(\alpha + 1)\,\Delta y_-} \tag{3-69}$$

which can be rearranged further into the form given by Eq. (3-66).

Our Taylor series examples thus far have illustrated procedures for obtaining a finite-difference approximation to a single derivative. However, our main interest is in correctly approximating an entire PDE at an arbitrary point in the problem domain. For this reason, we must be careful to use the same expansion point in approximating all derivatives in the PDE by the Taylor-series method. If this is done, then the truncation error for the entire equation can be obtained by adding the truncation error for each derivative.

There is no requirement that the expansion point be (i, j) as indicated by the following examples where the order of the truncation error and the most convenient expansion points are indicated. The geometric arrangement of points used in the difference equation is indicated by the sketch of the difference "molecule."

Fully implicit form for the heat equation, Eq. (3-55):

$$\frac{u_j^{n+1} - u_j^n}{\Delta t} = \frac{\alpha}{(\Delta x)^2}(u_{j+1}^{n+1} - 2u_j^{n+1} + u_{j-1}^{n+1}) \qquad \text{T.E.} = O[\Delta t, (\Delta x)^2] \tag{3-70}$$

The difference molecule for this scheme is shown in Fig. 3-4, and point $(n + 1, j)$ is indicated as the most convenient expansion point.

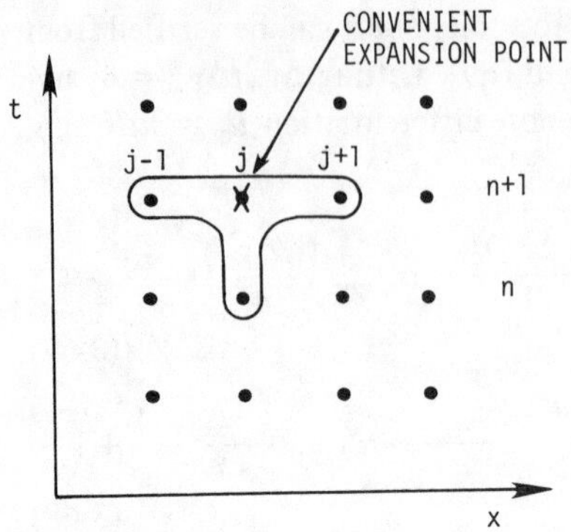

Figure 3-4 Difference molecule, fully implicit form for heat equation.

Crank-Nicolson form for the heat equation:

$$\frac{u_j^{n+1}-u_j^n}{\Delta t}=\frac{\alpha}{2(\Delta x)^2}\left[u_{j+1}^{n+1}+u_{j+1}^n-2(u_j^{n+1}+u_j^n)+u_{j-1}^{n+1}+u_{j-1}^n\right] \quad (3\text{-}71a)$$

$$\text{T.E.}=O[(\Delta t)^2,(\Delta x)^2]$$

The difference molecule for the Crank-Nicolson scheme is shown in Fig. 3-5, and point $(n+\frac{1}{2},j)$ is designated as the most convenient expansion point.

It is interesting to note that the *order* of the truncation error for difference representations of a complete PDE (not a single derivative term, however) is not dependent upon the choice of expansion point in the evaluation of this error by the Taylor-series method. We will demonstrate this point by considering the Crank-Nicolson scheme. The truncation error for the Crank-Nicolson scheme was most conveniently determined by expanding about the point $(n+\frac{1}{2}, j)$ to obtain the results stated above. Using this point resulted in the elimination of the maximum number of terms from the Taylor series by cancellation. Had we used point (n,j) or even $(n-1,j)$ as the expansion point, the conclusion on the order of the truncation error would have been the same. To reach this conclusion, however, we often must examine the truncation error very carefully. To illustrate, evaluating the truncation error of

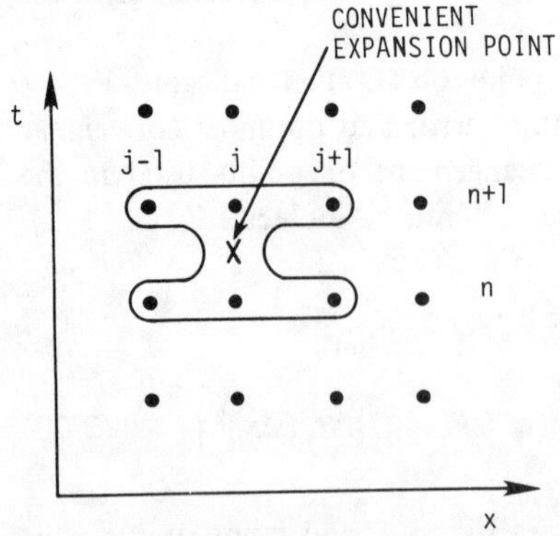

Figure 3-5 Difference molecule, Crank-Nicolson form for heat equation.

the Crank-Nicolson scheme by using expansions for u_{j-1}^n, u_{j+1}^n, u_{j-1}^{n+1}, u_{j+1}^{n+1}, u_j^{n+1} about point (n,j) in Eq. (3-71*a*) gives, after rearrangement,

$$u_t - \alpha u_{xx} = -u_{tt}\frac{\Delta t}{2} + \alpha u_{txx}\frac{\Delta t}{2} + O(\Delta x)^2 + O(\Delta t)^2 \tag{3-71b}$$

At first glance, we are tempted to conclude that the truncation error for the Crank-Nicolson scheme becomes $O(\Delta t) + O(\Delta x)^2$, when evaluated by expanding about point (n, j), because of the appearance of the terms $-u_{tt}\Delta t/2$ and $\alpha u_{txx}\Delta t/2$. However, we can recognize these two terms as $-\Delta t/2(\partial/\partial t)(u_t - \alpha u_{xx})$ where the quantity in the second set of parentheses is the left-hand side of Eq. (3-71*b*). Thus, we can differentiate Eq. (3-71*b*) with respect to t and multiply both sides by $-\Delta t/2$ to learn that $-\Delta t/2(\partial/\partial t)(u_t - \alpha u_{xx}) = O(\Delta t)^2 + O(\Delta x)^2$. From this we conclude that the truncation error for the Crank-Nicolson scheme is $O(\Delta t)^2 + O(\Delta x)^2$ when evaluated about either point (n, j) or point $(n + \frac{1}{2}, j)$. Use of other points will give the same results for the order of the truncation error. This example illustrates that the leading terms in the truncation error should be examined very carefully to see if they can be identified as a multiple of a derivative of the original PDE. If they can, they should be replaced by expressions of higher order.

3-4.2 Use of Polynomial Fitting

Many applications of polynomial fitting are observed in computational fluid mechanics and heat transfer. The technique can be used to develop the entire finite-difference representation for a PDE. However, the technique is perhaps most commonly employed in the treatment of boundary conditions or in gleaning information from the solution in the neighborhood of the boundary.

Let's consider some specific examples.

Example 3-1 In this example, the derivative approximations needed to represent a PDE will be obtained by assuming that the solution to the PDE can be approximated locally by a polynomial. The polynomial is then "fitted" to the points surrounding the general point (i, j) utilizing values of the function at the grid points. A sufficient number of points can be used to determine the coefficients in the polynomial exactly. The polynomial can then be differentiated to obtain the desired approximation to the derivatives. Let's consider Laplace's equation which governs the two-dimensional temperature distribution in a solid under steady-state conditions:

$$\frac{\partial^2 T}{\partial x^2} + \frac{\partial^2 T}{\partial y^2} = 0 \tag{3-72}$$

We suppose that both the x and y dependency of temperature can be expressed by a second-degree polynomial. For example, holding y fixed, we assume that temperatures at various x locations in the neighborhood of point (i, j) can be determined from

$$T(x, y_0) = a + bx + cx^2$$

For convenience we let $x = 0$ at point (i, j), and Δx = constant. Clearly,

$$\left.\frac{\partial T}{\partial x}\right)_{i,j} = b$$

$$\left.\frac{\partial^2 T}{\partial x^2}\right)_{i,j} = 2c$$

The coefficients a, b, and c can be evaluated in terms of temperatures at specific grid points and Δx. To do so, we must make some choices as to which neighboring grid points to use and this choice determines the geometrical arrangement of the difference molecule; that is, whether the resulting derivative approximations are central, forward, or backward differences. Here we'll choose points $(i-1, j)$, (i, j), and $(i+1, j)$ and obtain:

$$T(i, j) = a$$

$$T(i+1, j) = a + b\,\Delta x + c(\Delta x)^2$$

$$T(i-1, j) = a - b\,\Delta x + c(\Delta x)^2$$

from which we determine that

$$b = \left.\frac{\partial T}{\partial x}\right)_{i,j} = \frac{T_{i+1,j} - T_{i-1,j}}{2\,\Delta x}$$

$$c = \left.\frac{1}{2}\frac{\partial^2 T}{\partial x^2}\right)_{i,j} = \frac{T_{i+1,j} - 2T_{i,j} + T_{i-1,j}}{2(\Delta x)^2}$$

Thus,

$$\left.\frac{\partial^2 T}{\partial x^2}\right)_{i,j} = \frac{T_{i+1,j} - 2T_{i,j} + T_{i-1,j}}{(\Delta x)^2} \tag{3-73}$$

This represents an exact result if indeed a second-degree polynomial expresses the correct variation of temperature with x. In the general case, we only suppose that the second-degree polynomial is a good approximation to the solution. The truncation error of the expression, Eq. (3-73), can be determined by substituting Taylor-series expansions about point (i, j) for $T_{i+1,j}$ and $T_{i-1,j}$ into Eq. (3-73). The truncation error is found to be $O(\Delta x)^2$ and will involve only fourth-order and higher derivatives, which are equal to zero when the temperature variation is given by a second-degree polynomial.

A finite-difference approximation for $\partial^2 T/\partial y^2$ can be found in a like manner. We notice that arbitrary decisions need to be made in the process of polynomial fitting which will influence the form and truncation error of the result; particularly, these decisions influence which of the neighboring points will appear in the difference expression. We also observe that there is nothing unique about the procedure of polynomial fitting which guarantees that the difference approximation for the PDE is the best in any sense or that the numerical scheme is stable (when used for a marching problem).

Example 3-2 Suppose we have solved the finite-difference form of the energy equation for the temperature distribution near a solid boundary and we need to estimate the heat flux at the location. Our finite-difference solution gives us only the temperature at discrete grid points. From Fourier's law, the boundary heat flux is given by $q_w = -k\, \partial T/\partial y)_{y=0}$. Thus, we need to approximate $\partial T/\partial y)_{y=0}$ by a difference representation which uses the temperature obtained from the finite-difference solution to the energy equation. One way to proceed is to assume that the temperature distribution near the boundary is a polynomial and to "fit" such a polynomial, i.e., straight line, parabola, or third-degree polynomial, etc. to the finite-difference solution which has been determined at discrete points. By requiring that the polynomial match the finite-difference solution for T at certain discrete points, the unknown coefficients in the polynomial can be determined.

For example, if we assume that the temperature distribution near the boundary is again a second-degree polynomial of the form $T = a + by + cy^2$, then, referring to Fig. 3-6, we note that $\partial T/\partial y)_{y=0} = b$. Further, for equally spaced mesh points we can write

$$T_1 = a$$

$$T_2 = a + b\,\Delta y + c(\Delta y)^2$$

$$T_3 = a + b(2\,\Delta y) + c(2\,\Delta y)^2$$

from which we can determine that

$$a = T_1$$

$$b = \frac{-3T_1 + 4T_2 - T_3}{2\,\Delta y}$$

and

$$c = \frac{T_1 - 2T_2 + T_3}{2(\Delta y)^2}$$

Thus, we can evaluate the wall heat flux by the approximation

$$q_w = -k \left.\frac{\partial T}{\partial y}\right)_{y=0} \simeq -kb = \frac{k}{2\,\Delta y}(3T_1 - 4T_2 + T_3)$$

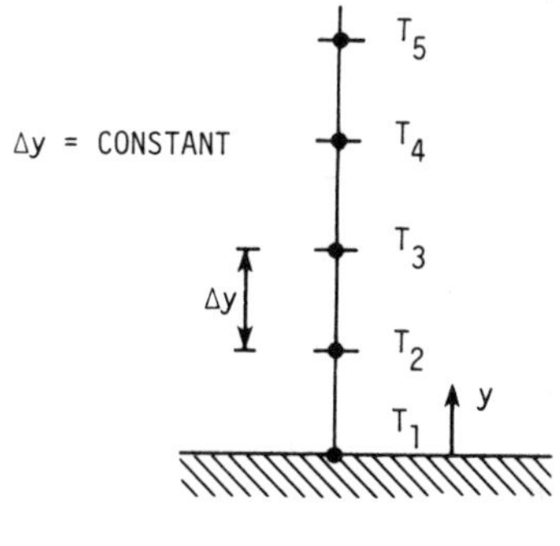

Figure 3-6 Finite-difference grid near wall.

It is natural to inquire about the truncation error of this approximation for $\partial T/\partial y)_{y=0}$. This may be established by expressing T_2 and T_3 in terms of Taylor-series expansions about the boundary point and substituting these evaluations into the difference expression for $\partial T/\partial y)_{y=0}$. Alternatively, we can identify the second-degree polynomial as a truncated Taylor-series expansion about $y = 0$:

Second-degree polynomial:

$$T = a + by + cy^2$$

Taylor series:

$$T = T(0) + \left.\frac{\partial T}{\partial y}\right)_0 y + \left.\frac{\partial^2 T}{\partial y^2}\right)_0 \frac{y^2}{2!} + \underbrace{\left.\frac{\partial^3 T}{\partial y^3}\right)_0 \frac{y^3}{3!} + \cdots}_{\text{T.E.}}$$

Thus, the approximation $T \simeq a + by + cy^2$ is equivalent to utilizing the first three terms of a Taylor-series expansion with the resulting T.E. in the expression for T being $O(\Delta y)^3$. Solving the Taylor series for an expression for $\partial T/\partial y)_{y=0}$ involves division by Δy which reduces the truncation error in the expression for $\partial T/\partial y)_{y=0}$ to $O(\Delta y)^2$.

Example 3-3 Suppose that the energy equation is being solved for the temperature distribution near the wall as in Example 3-2, but now the wall heat flux is specified as a boundary condition. We may then want to use polynomial fitting to obtain an expression for the boundary temperature which is called for in the difference equations for internal points. In other words, if $q_w = -k\,\partial T/\partial y)_{y=0}$ is given, how can we evaluate T at $y = 0$, i.e., (T_1) in terms of q_w/k and T_2, T_3, etc.? Here we might assume that $T = a + by + cy^2 + dy^3$ near the wall and that $\partial T/\partial y)_{y=0} = b = -q_w/k$ (given). Our objective is to evaluate T_1 which in this case equals a. Referring to Fig. 3-6, we can write

$$T_2 = a - \frac{q_w}{k}\,\Delta y + c(\Delta y)^2 + d(\Delta y)^3$$

$$T_3 = a - \frac{q_w}{k}\,(2\,\Delta y) + c(2\,\Delta y)^2 + d(2\,\Delta y)^3$$

$$T_4 = a - \frac{q_w}{k}\,(3\,\Delta y) + c(3\,\Delta y)^2 + d(3\,\Delta y)^3$$

These three equations can be solved for a, c, and d in terms of T_2, T_3, T_4, q_w/k, and Δy. The desired result, T_1 as a function of T_2, T_3, q_w/k, and Δy, follows directly from $T_1 = a$ and is given by

$$T_1 = \frac{1}{11}\left(18T_2 - 9T_3 + 2T_4 + \frac{6\,\Delta y q_w}{k}\right) + O[(\Delta y)^4] \tag{3-74}$$

The truncation error in the above expression can be established by substituting Taylor-series expansions about (i, j) for the temperatures on the right-hand side or by identifying the polynomial as a truncated series by inspection. We will close this discussion on polynomial fitting by listing some expressions for wall values

Table 3-3 Some useful results from polynomial fitting

Polynomial degree	Wall value of function or derivative	Equation
1	$\left.\frac{\partial T}{\partial y}\right)_{i,j} = \frac{T_{i,j+1} - T_{i,j}}{h} + O(h)$	(3-75)
1	$T_{i,j} = T_{i,j+1} - h\left.\frac{\partial T}{\partial y}\right)_{i,j} + O(h^2)$	(3-76)
2	$\left.\frac{\partial T}{\partial y}\right)_{i,j} = \frac{1}{2h}(-3T_{i,j} + 4T_{i,j+1} - T_{i,j+2}) + O(h^2)$	(3-77)
2	$T_{i,j} = \frac{1}{3}\left[4T_{i,j+1} - T_{i,j+2} - 2h\left.\frac{\partial T}{\partial y}\right)_{i,j}\right] + O(h^3)$	(3-78)
3	$\left.\frac{\partial T}{\partial y}\right)_{i,j} = \frac{1}{6h}(-11T_{i,j} + 18T_{i,j+1} - 9T_{i,j+2} + 2T_{i,j+3}) + O(h^3)$	(3-79)
3	$T_{i,j} = \frac{1}{11}\left[18T_{i,j+1} - 9T_{i,j+2} + 2T_{i,j+3} - 6h\left.\frac{\partial T}{\partial y}\right)_{i,j}\right] + O(h^4)$	(3-80)
4	$\left.\frac{\partial T}{\partial y}\right)_{i,j} = \frac{1}{12h}(-25T_{i,j} + 48T_{i,j+1} - 36T_{i,j+2} + T_{i,j+3} - 3T_{i,j+4}) + O(h^4)$	(3-81)
4	$T_{i,j} = \frac{1}{25}\left[48T_{i,j+1} - 36T_{i,j+2} + 16T_{i,j+3} - 3T_{i,j+4} - 12h\left.\frac{\partial T}{\partial y}\right)_{i,j}\right] + O(h^5)$	(3-82)

of a function and its first derivative in terms of values of the function. These expressions are useful, for example, in extracting a value of the function at the wall, if the wall value of the first derivative is specified. The results in Table 3-3 were obtained from polynomial fitting, assuming that $T(y)$ can be expressed as a polynomial of degree up to the fourth, and that $\Delta y = h =$ constant.

3-4.3 The Integral Method

The integral method provides yet another means for developing difference approximations to PDE's. We consider again the heat equation as the specimen equation

$$\frac{\partial u}{\partial t} = \alpha \frac{\partial^2 u}{\partial x^2} \tag{3-83}$$

The strategy is to develop an algebraic relationship among the u's at neighboring grid points by integrating the heat equation with respect to the independent variables t and x over the local neighborhood of point (n, j). The point (n, j) will also be identified as point (t_0, x_0). Grid points are spaced at intervals of Δx and Δt. We arbitrarily decide to integrate both sides of the equation over the interval t_0 to $t_0 + \Delta t$ and $x_0 - \Delta x/2$ to $x_0 + \Delta x/2$. Choosing $t_0 - \Delta t/2$ to $t_0 + \Delta t/2$ would lead to an inherently unstable difference equation. Unfortunately, at this point we have no way of knowing which choice for the integration interval would be the right or wrong one relative to stability of the solution method. This can only be determined by a trial calculation or application of the methods for stability analysis to be presented in

Section 3-6. The order of integration is chosen for each side in a manner to take advantage of exact differentials:

$$\int_{x_0-\Delta x/2}^{x_0+\Delta x/2}\left(\int_{t_0}^{t_0+\Delta t}\frac{\partial u}{\partial t}\,dt\right)dx=\alpha\int_{t_0}^{t_0+\Delta t}\left(\int_{x_0-\Delta x/2}^{x_0+\Delta x/2}\frac{\partial^2 u}{\partial x^2}\,dx\right)dt \tag{3-84}$$

The inner level of integration can be done exactly giving

$$\int_{x_0-\Delta x/2}^{x_0+\Delta x/2}[u(t_0+\Delta t,x)-u(t_0,x)]\,dx$$

$$=\alpha\int_{t_0}^{t_0+\Delta t}\left[\frac{\partial u}{\partial x}\left(t,x_0+\frac{\Delta x}{2}\right)-\frac{\partial u}{\partial x}\left(t,x_0-\frac{\Delta x}{2}\right)\right]dt \tag{3-85}$$

For the next level of integration, we take advantage of the mean-value theorem for integrals which assures us that for a continuous function $f(y)$

$$\int_{y_1}^{y_1+\Delta y}f(y)\,dy=f(\bar{y})\,\Delta y \tag{3-86}$$

where $\bar{y}$ is some value of y in the interval $y_1\leqslant\bar{y}\leqslant y_1+\Delta y$. Thus, any value of y on the interval will provide an approximation to the integral and we can write

$$\int_{y_1}^{y_1+\Delta y}f(y)\,dy\simeq f(\tilde{y})\,\Delta y \qquad \text{for } y_1\leqslant\tilde{y}\leqslant y_1+\Delta y$$

As we invoke the mean-value theorem to further simplify Eq. (3-85), we arbitrarily select x_0 on the left-hand side and $t_0+\Delta t$ on the right-hand side as the locations within the intervals of integration at which to evaluate the integrands:

$$[u(t_0+\Delta t,x_0)-u(t_0,x_0)]\,\Delta x=\alpha\left[\frac{\partial u}{\partial x}\left(t_0+\Delta t,x_0+\frac{\Delta x}{2}\right)\right.$$

$$\left.-\frac{\partial u}{\partial x}\left(t_0+\Delta t,x_0-\frac{\Delta x}{2}\right)\right]\Delta t \tag{3-87}$$

To express the result in purely algebraic terms requires that the first derivatives, $\partial u/\partial x$, on the right-hand side be approximated by finite differences. We could achieve this by falling back on our experience to date and simply utilize central differences. Alternatively, we can continue to pursue a purely integral approach and invoke the mean-value theorem for integrals again observing that

$$u(t_0+\Delta t,x_0+\Delta x)=u(t_0+\Delta t,x_0)+\int_{x_0}^{x_0+\Delta x}\frac{\partial u}{\partial x}(t_0+\Delta t,x)\,dx$$

$$\simeq u(t_0+\Delta t,x_0)+\frac{\partial u}{\partial x}\left(t_0+\Delta t,x_0+\frac{\Delta x}{2}\right)\Delta x \tag{3-88}$$

from which we can write

$$\frac{\partial u}{\partial x}\left(t_0 + \Delta t, x_0 + \frac{\Delta x}{2}\right) \simeq \frac{u(t_0 + \Delta t, x_0 + \Delta x) - u(t_0 + \Delta t, x_0)}{\Delta x} \tag{3-89}$$

In evaluating the integral in Eq. (3-88) through the mean-value theorem, we have arbitrarily evaluated the integrand at the midpoint of the interval. Hence, the final result is only an approximation. Treating the other first derivative in a similar manner permits the approximation to the heat equation to be written as

$$[u(t_0 + \Delta t, x_0) - u(t_0, x_0)]\ \Delta x = \frac{\alpha}{\Delta x}\,[u(t_0 + \Delta t, x_0 + \Delta x) - 2u(t_0 + \Delta t, x_0) + u(t_0 + \Delta t, x_0 - \Delta x)]\ \Delta t \tag{3-90}$$

Reverting back to the n, j notation whereby n denotes time (t) and j denotes space (x), we can rearrange the above in the form

$$\frac{u_j^{n+1} - u_j^n}{\Delta t} = \frac{\alpha}{(\Delta x)^2}\,(u_{j+1}^{n+1} - 2u_j^{n+1} + u_{j-1}^{n+1}) \tag{3-91}$$

which can be recognized as the fully implicit representation of the heat equation given in Section 3-4.1. The choice of $t_0 + \Delta t$ as the location to use in utilizing the mean-value theorem for the second integration on the right-hand side is responsible for the implicit form. If t_0 had been chosen instead, an explicit formulation would have resulted. We note that a statement of the truncation error does not evolve naturally as part of this method for developing difference equations but must be determined as a separate step.

3-4.4 Control-Volume Approach

A distinctly different point of view is taken in utilizing the control-volume approach than is adopted in working with any of the other methods considered thus far. In the Taylor-series and integral methods we accepted the PDE as the correct and appropriate form of the conservation principle (physical law) governing our problem and merely turned to mathematical tools to develop algebraic approximations to derivatives. We never considered again the physical law represented by the PDE. Physical reasoning had been used previously in deriving the PDE but then put aside. The Taylor-series and integral methods then proceed in a rather formal, mechanical way operating on the PDE.

In the control-volume method, we examine the PDE governing the problem at hand and recall the physical law or conservation statement which the PDE represents. In general, we translate the PDE to a statement in words as it might apply to a control volume in the neighborhood of a grid point in a finite-difference mesh. Here we are recognizing the discrete nature of the finite-difference model at the outset. We now proceed to work out a mathematical statement of the physical conservation principle following steps somewhat reminiscent of the procedures used by some to derive PDE's from physical laws except that we do not take the limit of shrinking the control volume to a point. When the governing PDE can be written in divergence form, we can be guided in this process by employing the Gauss divergence theorem to obtain the correct mathematical formulation for the physical law for a control volume. In practice, the control-volume method has a history of leading quickly to expressions that prove to be more accurate than other possibilities near boundaries, probably because the method keeps the discrete nature of the solution method in view at all times.

As an example, consider two-dimensional steady-state conduction in a solid where for constant thermal conductivity we find that the temperature distribution must satisfy Laplace's equation, Eq. (3-72).

A grid is first established on the problem domain. Points should be placed on the boundaries since temperatures at boundary points arise in the specification of boundary conditions. The problem domain is then equitably divided up into control volumes about each grid point. It will be most convenient to establish the boundaries of the control volumes halfway between adjacent grid points. Following this procedure, internal grid points will only be in the geometric center of the control volumes if the mesh spacing is constant, i.e., $\Delta x = c_1$, $\Delta y = c_2$.

We'll first consider the control volume labeled A in Fig. 3-7 which is representative of all internal (nonboundary) points. In steady state, the net rate at which heat flows into control volume A must be zero. This is the conservation statement which leads to the establishment of Laplace's equation as governing the temperature distribution at a point in the problem domain. We can also work backwards from the PDE to obtain the proper statement for a control volume by using the divergence theorem. First, we must recognize that the heat-flux vector is related to the temperature distribution through Fourier's law

$$\mathbf{q} = -k\,\nabla T$$

Thus, when k is constant, Eq. (3-72) can be interpreted as

$$-\nabla \cdot \mathbf{q} = \nabla \cdot (k\,\nabla T) = 0$$

Applying the divergence theorem gives

$$\iiint_R \nabla \cdot (k\,\nabla T)\,dR = \iint_S (k\,\nabla T) \cdot \mathbf{n}\,dS = 0$$

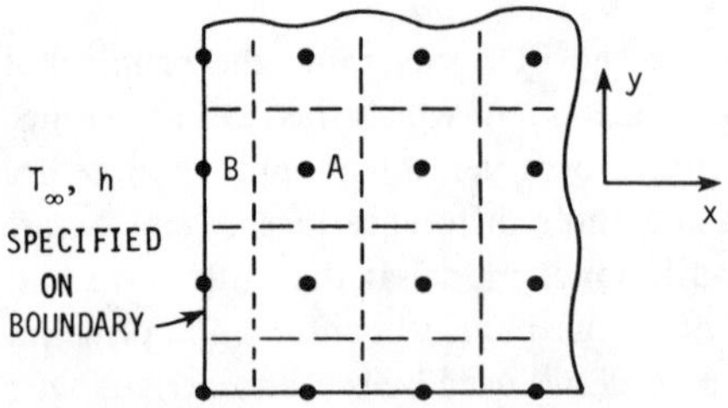

Figure 3-7 Finite-difference grid for control-volume method.

The integral on the right represents the net flow of heat into any arbitrary fixed region. Applying this to evaluate the inflow of heat through each of the four boundaries of the control volume about point (i, j) gives

$$-k\,\Delta y \left.\frac{\partial T}{\partial x}\right)_{i-1/2,j} + k\,\Delta y \left.\frac{\partial T}{\partial x}\right)_{i+1/2,j} - k\,\Delta x \left.\frac{\partial T}{\partial y}\right)_{i,j-1/2} + k\,\Delta x \left.\frac{\partial T}{\partial y}\right)_{i,j+1/2} = 0$$

The 1/2 in the subscripts refers to evaluation at the boundaries of the control volume which are halfway between mesh points. The energy balance is exact if the derivatives represent suitable average values for the boundaries concerned. Approximating the derivatives by central differences we find that

$$k\,\Delta y \frac{(T_{i-1,j} - T_{i,j})}{\Delta x} + k\,\Delta y \frac{(T_{i+1,j} - T_{i,j})}{\Delta x} + k\,\Delta x \frac{(T_{i,j-1} - T_{i,j})}{\Delta y}$$
$$+ k\,\Delta x \frac{(T_{i,j+1} - T_{i,j})}{\Delta y} = 0$$

Dividing through by $k\,\Delta x\,\Delta y$ we can write the result as

$$\frac{T_{i-1,j} - 2T_{i,j} + T_{i+1,j}}{(\Delta x)^2} + \frac{T_{i,j-1} - 2T_{i,j} + T_{i,j+1}}{(\Delta y)^2} = 0 \qquad (3\text{-}92)$$

Each of these two terms can be recognized as approximations to second derivatives, ($\partial^2 T/\partial x^2$ for the first and $\partial^2 T/\partial y^2$ for the second) which were obtained earlier by the Taylor-series expansion method.

Let's now consider the control volume on the boundary, labeled B in Fig. 3-7. In this example we will assume that the boundary conditions are convective. For the continuous (nondiscrete) problem this is formulated mathematically by $h(T_\infty - T_{i,j}) = -k\,\partial T/\partial x)_{i,j}$ where the point (i, j) is the point on the physical boundary associated with control volume B. If we were to proceed with the Taylor-series approach to this boundary condition, we would likely next seek a difference representation for $\partial T/\partial x)_{i,j}$. If a simple forward difference is used, the difference equation would be

$$h(T_\infty - T_{i,j}) = \frac{k}{\Delta x}(T_{i,j} - T_{i+1,j}) \qquad (3\text{-}93)$$

In the control-volume approach, however, we are forced to observe that there is some material associated with the boundary point and conduction may occur along the boundary. The energy balance on the control volume would account for possible transfer across all four boundaries

$$h\,\Delta y(T_\infty - T_{i,j}) + k\,\Delta y \left.\frac{\partial T}{\partial x}\right)_{i+1/2,j} + \frac{k\,\Delta x}{2} \left.\frac{\partial T}{\partial y}\right)_{i,j+1/2}$$
$$- \frac{k\,\Delta x}{2} \left.\frac{\partial T}{\partial y}\right)_{i,j-1/2} = 0 \qquad (3\text{-}94)$$

Using central differences to approximate the derivatives, we can write

$$h\,\Delta y(T_\infty - T_{i,j}) + k\,\Delta y\,\frac{(T_{i+1,j} - T_{i,j})}{\Delta x} + \frac{k\,\Delta x}{2}\,\frac{(T_{i,j+1} - T_{i,j})}{\Delta y}$$

$$+ \frac{k\,\Delta x}{2}\,\frac{(T_{i,j-1} - T_{i,j})}{\Delta y} = 0 \tag{3-95}$$

Dividing through by k, we can write the result as

$$\frac{h\,\Delta y}{k}\,T_\infty + \frac{\Delta y}{\Delta x}\,T_{i+1,j} + \frac{\Delta x}{2\,\Delta y}\,(T_{i,j+1} + T_{i,j-1}) - \left(\frac{h\,\Delta y}{k} + \frac{\Delta y}{\Delta x} + \frac{\Delta x}{\Delta y}\right)T_{i,j} = 0 \tag{3-96}$$

which is somewhat different than Eq. (3-93) which followed from the most obvious application of the Taylor-series method to approximate the derivative $\partial T/\partial x)_{i,j}$ which occurred in the formal mathematical statement of the boundary condition.

Looking back over the methodology of the control-volume and Taylor-series methods, we can note that the Taylor-series method readily provided difference approximations to derivatives and the representation for the complete PDE was made up from the addition of several such representations. In contrast, the control-volume method employs the conservation statement or physical law represented by the entire PDE and appears incapable of providing a finite-difference representation just to a derivative alone. That is, we would not know how to respond to a request to develop a difference representation to $\partial u/\partial x$ alone by the control-volume method. The distinctive characteristic of the control-volume approach is that a "balance" of some physical quantity is made on the region in the neighborhood of a grid point. The discrete nature of the problem domain is always taken into account in the control-volume approach which ensures that the physical law is satisfied over a finite region rather than only at a point as the mesh is shrunk to zero. It would appear that difference equations developed by the control-volume approach would almost certainly have the conservative property.

It is difficult to appreciate the subtle differences which may occur in the difference representations obtained for the same PDE by using the four different methods discussed in this section without working a large number of examples. In many cases, and especially for simple, linear equations, the resulting difference equations can be identical. That is, four different approaches can give the same result. There is no guarantee that difference equations developed by any of the methods will be numerically stable so that the same difference scheme developed by four methods could turn out to be worthless. The differences in the results obtained from using the different methods are more likely to become evident in coordinate systems other than rectangular and when the PDE's being approximated are in nonconservative form.

3-5 INTRODUCTION TO THE USE OF IRREGULAR MESHES

Naturally, it's most convenient to let the mesh increments such as Δx and Δy be constant throughout the computational domain. However, in many instances this is not possible due to boundaries which do not coincide with the regular mesh or the need to reduce the mesh spacing in certain regions in order to maintain the desired level of accuracy. These irregularities occur frequently enough in physical problems as to command a significant amount of attention from workers in computational fluid mechanics and heat transfer. Prior to resorting to the use of irregular meshes, it is well to look for possible analytical procedures such as a coordinate transformation to bring the boundaries and the finite-difference mesh into alignment. Coordinate transformations will be discussed in detail in Section 5-6.

3-5.1 Irregular Mesh Due to Shape of a Boundary

To illustrate this problem we will assume that the regular mesh used for a problem is square, i.e., $\Delta x = \Delta y =$ constant but that one of the boundaries is curved causing unequal spacing between the boundary and some internal grid points as illustrated in Fig. 3-8. We will assume that the governing PDE is Laplace's equation and that u is specified on the boundaries. Some of the procedures which can be used in dealing with irregular mesh points near boundaries are listed below:

1. Use an especially fine but regular mesh near the boundary and define the grid point closest to the actual boundary as the boundary point for computational purposes. Unless a coarser mesh is used away from the boundary resulting in irregular mesh problems where the transition in spacing is made, this method could require a very large number of grid points to achieve reasonable accuracy.
2. Use linear interpolation to assign values of u to any internal point which is less than a regular mesh increment from the boundary. The interpolation is between the specified boundary values of u and values of u determined at neighboring points by the finite-difference equations applicable to internal points in the regular mesh. Referring to Fig. 3-8 we might interpolate to obtain u_p by either

$$u_{p_1} = u_A + \frac{\Delta x}{\Delta x + \alpha \Delta x}(u_C - u_A)$$

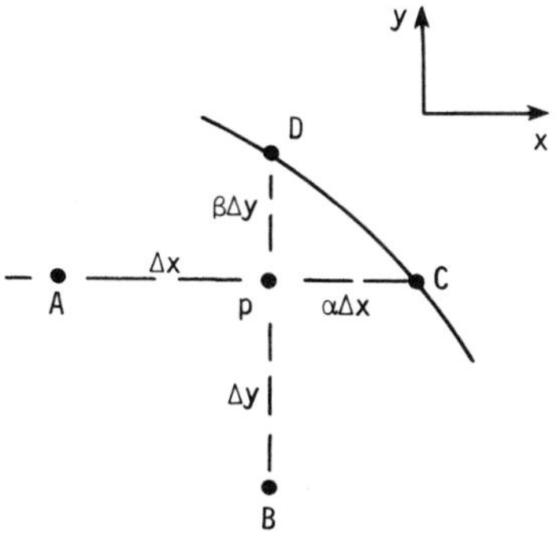

Figure 3-8 Irregular mesh caused by the shape of a boundary.

or

$$u_{p_2} = u_B + \frac{\Delta y}{\Delta y + \beta\, \Delta y}(u_D - u_B)$$

We could also use the average of the two interpolated values, $u_p = (u_{p_1} + u_{p_2})/2$.

3. Develop a finite-difference approximation to the governing PDE which is valid at internal points even when the mesh is irregular.

A difference expression valid on an arbitrary irregular mesh can be developed quite readily through the integral method by integrating about point x_0, y_0 and letting each integration interval extend halfway to a neighboring point. The mesh notation used is defined in Fig. 3-9. The starting point for the integral development of the difference expression is

$$\int_{y_0-\Delta y_-/2}^{y_0+\Delta y_+/2}\left(\int_{x_0-\Delta x_-/2}^{x_0+\Delta x_+/2}\frac{\partial^2 u}{\partial x^2}\,dx\right)dy + \int_{x_0-\Delta x_-/2}^{x_0+\Delta x_+/2}\left(\int_{y_0-\Delta y_-/2}^{y_0+\Delta y_+/2}\frac{\partial^2 u}{\partial y^2}\,dy\right)dx = 0 \tag{3-97}$$

Using the definition of an exact differential, this can be written as

$$\int_{y_0-\Delta y_-/2}^{y_0+\Delta y_+/2}\left[\frac{\partial u}{\partial x}\left(x_0+\frac{\Delta x_+}{2}, y\right) - \frac{\partial u}{\partial x}\left(x_0-\frac{\Delta x_-}{2}, y\right)\right]dy$$

$$+ \int_{x_0-\Delta x_-/2}^{x_0+\Delta x_+/2}\left[\frac{\partial u}{\partial y}\left(x, y_0+\frac{\Delta y_+}{2}\right) - \frac{\partial u}{\partial y}\left(x, y_0-\frac{\Delta y_-}{2}\right)\right]dx = 0$$

Employing the mean-value theorem for integrals and using the central point of the interval to evaluate the integrands gives

$$\left[\frac{\partial u}{\partial x}\left(x_0+\frac{\Delta x_+}{2}, y_0\right) - \frac{\partial u}{\partial x}\left(x_0-\frac{\Delta x_-}{2}, y_0\right)\right]\frac{\Delta y_+ + \Delta y_-}{2}$$

$$+ \left[\frac{\partial u}{\partial y}\left(x_0, y_0+\frac{\Delta y_+}{2}\right) - \frac{\partial u}{\partial y}\left(x_0, y_0-\frac{\Delta y_-}{2}\right)\right]\frac{\Delta x_+ + \Delta x_-}{2} = 0 \tag{3-98}$$

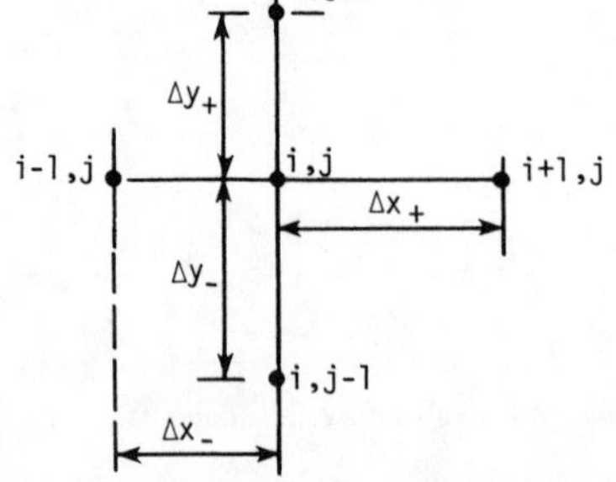

Figure 3-9 Notation for arbitrary irregular mesh.

Approximating these derivatives centrally as was done in Section 3-4.3 gives, after rearrangement, the following approximation for Laplace's equation in subscript notation:

$$\frac{2}{\Delta x_+ + \Delta x_-}\left(\frac{u_{i+1,j}-u_{i,j}}{\Delta x_+} - \frac{u_{i,j}-u_{i-1,j}}{\Delta x_-}\right) + \frac{2}{\Delta y_+ + \Delta y_-}\left(\frac{u_{i,j+1}-u_{i,j}}{\Delta y_+} - \frac{u_{i,j}-u_{i,j-1}}{\Delta y_-}\right) = 0 \tag{3-99}$$

When the above is specialized to the points near the irregular boundary depicted in Fig. 3-8, the derivative approximations appear as

$$\begin{aligned} \left.\frac{\partial^2 u}{\partial x^2}\right)_p &\simeq \frac{2}{\Delta x(1+\alpha)}\left(\frac{u_{\mathrm{C}}-u_p}{\alpha\,\Delta x} - \frac{u_p-u_{\mathrm{A}}}{\Delta x}\right) \\ \left.\frac{\partial^2 u}{\partial y^2}\right)_p &\simeq \frac{2}{\Delta y(1+\beta)}\left(\frac{u_D-u_p}{\beta\,\Delta y} - \frac{u_p-u_{\mathrm{B}}}{\Delta y}\right) \end{aligned} \tag{3-100}$$

Equation (3-99) can also be developed by utilizing Taylor-series expansions. The unequal spacing makes the Taylor-series method noticeably more laborious whereas the integral approach proceeds for unequal spacing with no increase in effort. However, Taylor-series expansions about (i, j) should be substituted into Eq. (3-99) to establish the consistency and truncation error of these approximations. This will be left as an exercise for the reader. As a note of warning, we might recall that our second-derivative approximations on a regular mesh acquired second-order truncation errors only through fortuitous cancellation of terms from the forward and backward Taylor-series expansions. This cancellation will not occur if the mesh increments are unequal.

When approximately the same number of grid points are being used, we might expect this third method of treating irregular points near boundaries to be the most accurate because the governing PDE is being approximated at each internal point [not the case for method (2)] and the location of the boundary is not being altered as was done in method (1).

Irregular boundaries with derivative (Neumann) boundary conditions can also be treated, but the algebraic relationships become more complex. Some elementary examples illustrating the application of derivative boundary conditions to irregular boundaries can be found in Forsythe and Wasow (1960), James et al. (1967), and Allen (1954).

3-5.2 Irregular Mesh Not Caused by Shape of a Boundary

In some flow problems it is necessary to use especially small grid spacing near solid boundaries or shock waves where gradients in the dependent variables are especially large in order to obtain the desired accuracy or "resolution." In the interest of computational economy we strive to use a coarser grid away from these critical regions so that we are forced to consider ways of treating an irregular or variable mesh in these problems. We can cite at least two ways to proceed:

1. We can employ a coordinate transformation so that unequal spacing in the original coordinate system becomes equal spacing in the new system but the PDE becomes altered somewhat in form. This will be described in detail later in our course of study.
2. The difference equation can be formulated in such a way that it remains valid when the spacing is irregular. Actually this is the same procedure as method (3) used above in connection with the irregular mesh caused by curved boundaries. Such a formulation for Laplace's equation is given as Eq. (3-99).

3-5.3 Concluding Remarks

The purpose of this section has been to introduce some of the problems and applicable solution procedures associated with irregular boundaries and unequal mesh spacing in general. Our coverage of the topic has been by no means complete. More advanced considerations on this topic tend to quickly become quite specialized and detailed. Good pedagogy suggests that we move on and see more of the forest before we spend any more time studying this tree. Some ideas on this topic will be developed further in connection with specific problems in fluid mechanics and heat transfer.

3-6 STABILITY CONSIDERATIONS

A finite-difference approximation to a partial differential equation may be consistent but the solution will not necessarily converge to the solution of the PDE. The Lax Equivalence theorem (see Section 3-3.4) states that a stable numerical method must also be used. We will address the question of stability in this section.

The problem of stability in numerical analysis is similar to the problem of stability encountered in a modern control system. The transfer function in a control system plays the role of the difference operator. Consider a marching problem in which initial values at time level n are known and values of the unknown at time level $n + 1$ are required. The difference operator may be viewed as a "black box" which has a certain transfer function. A schematic representation would appear as shown in Fig. 3-10. The stability of such a system depends upon the operations performed by the black box on the input data. A control systems engineer would require that the transfer function have no poles in the right-half plane. Without this requirement, input signals would be falsely amplified and the output would be useless; in fact, it would grow without bound. Similarly, the way in which the difference operator alters the input information to produce the solution at the next time level is the central concern of stability analysis.

Figure 3-10 Schematic diagram of stability.

As a starting point for stability analysis, consider the simple explicit approximation to the heat equation

$$\frac{u_j^{n+1} - u_j^n}{\Delta t} = \frac{\alpha}{(\Delta x)^2} (u_{j+1}^n - 2u_j^n + u_{j-1}^n)$$

This may be solved for u_j^{n+1} to yield:

$$u_j^{n+1} = u_j^n + \alpha \frac{\Delta t}{(\Delta x)^2} (u_{j+1}^n - 2u_j^n + u_{j-1}^n) \tag{3-101}$$

Let the exact solution of this equation be denoted by D. This is the solution that would be obtained using a computer with infinite accuracy. Similarly, denote the numerical solution of Eq. (3-101) computed using a real machine with finite accuracy by N. If the analytical solution of the partial differential equation is A, then we may write

$$\text{Discretization error} = A - D$$

$$\text{Round-off error} = N - D$$

The question of stability of a numerical method examines the error growth while computations are being performed. O'Brien et al. (1951) pose the question of stability in the following manner:

1. Does the overall error due to round-off

$$\begin{bmatrix} \text{Grow} \\ \text{Not grow} \end{bmatrix} \Rightarrow \text{strong} \begin{bmatrix} \text{instability} \\ \text{stability} \end{bmatrix}$$

2. Does a single general round-off error

$$\begin{bmatrix} \text{Grow} \\ \text{Not grow} \end{bmatrix} \Rightarrow \text{weak} \begin{bmatrix} \text{instability} \\ \text{stability} \end{bmatrix}$$

The second question is the one most frequently answered because it can be treated much more easily from a practical point of view. The question of weak stability is usually answered by using a Fourier analysis. This method is also referred to as a von Neumann analysis. It is assumed that proof of weak stability using this method implies strong stability.

3-6.1 Fourier or von Neumann Analysis

Consider the finite-difference Eq. (3-101). Let ϵ represent the error in the numerical solution due to round-off errors. The numerical solution actually computed may be written

$$N = D + \epsilon \tag{3-102}$$

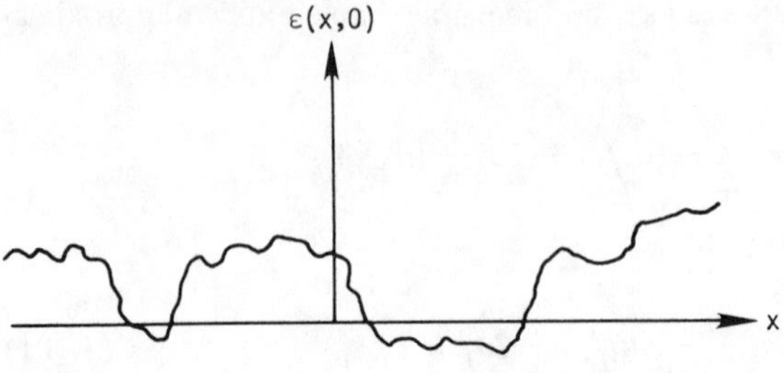

Figure 3-11 Initial error distribution.

This computed numerical solution must satisfy the difference equation. Substituting Eq. (3-102) into the difference Eq. (3-101) yields:

$$\frac{D_j^{n+1} + \epsilon_j^{n+1} - D_j^n - \epsilon_j^n}{\Delta t} = \alpha\left(\frac{D_{j+1}^n + \epsilon_{j+1}^n - 2D_j^n - 2\epsilon_j^n + D_{j-1}^n + \epsilon_{j-1}^n}{\Delta x^2}\right)$$

Since the exact solution D must satisfy the difference equation, the same is true of the error, i.e.,

$$\frac{\epsilon_j^{n+1} - \epsilon_j^n}{\Delta t} = \alpha\left(\frac{\epsilon_{j+1}^n - 2\epsilon_j^n + \epsilon_{j-1}^n}{\Delta x^2}\right) \tag{3-103}$$

In this case, the exact solution D and the error ϵ must both satisfy the same difference equation. This means that the numerical error and the exact numerical solution both possess the same growth property in time. Any perturbation of the input values at the nth time level will either be prevented from growing without bound for a stable system or will grow larger for an unstable system.

Consider a distribution of errors at any time in a mesh. We choose to view this distribution at time $t = 0$ for convenience. This error distribution is shown schematically in Fig. 3-11. We assume the error $\epsilon(x, t)$ can be written as a series of the form

$$\epsilon(x, t) = \sum_m b_m(t) e^{ik_m x} \tag{3-104}$$

where the period of the fundamental frequency ($m = 1$) is assumed to be $2L$. For the interval L units in length, the wave number may be written

$$k_m = \frac{m\pi}{L} \qquad m = 0, 1, 2, \ldots, M$$

where M is the number of intervals Δx units long contained in length L. For instance, if an interval of length $2L$ is subdivided using five points, the value of M is 2 and the corresponding frequencies are

$$f_m = \frac{k_m}{2\pi} = \frac{m}{2L}$$

$$f_0 = 0 \qquad m = 0$$

$$f_1 = \frac{1}{2L} \qquad m = 1$$

$$f_2 = \frac{1}{L} \qquad m = 2$$

The frequency measures the number of wave lengths in each $2L$ units of length. When $m = 0, f_0 = 0$ and this corresponds to a steady term in the assumed expansion.

Since the difference equation is linear, superposition may be used and we may examine the behavior of a single term of the series given in Eq. (3-104). Consider the term

$$\epsilon_m(x, t) = b_m(t)e^{ik_m x}$$

We seek solutions of the form

$$z^n e^{ik_m x}$$

which reduces to $e^{ik_m x}$ when $t = 0$ $(n = 0)$. Toward this end, let

$$z = e^{a\,\Delta t}$$

so that

$$z^n = e^{an\,\Delta t} = e^{at}$$

and
$$\epsilon_m(x, t) = e^{at}e^{ik_m x} \tag{3-105}$$

where k_m is real but a may be complex.

If Eq. (3-105) is substituted into Eq. (3-103) we obtain

$$e^{a(t+\Delta t)}e^{ik_m x} - e^{at}e^{ik_m x} = r(e^{at}e^{ik_m(x+\Delta x)} - 2e^{at}e^{ik_m x} + e^{at}e^{ik_m(x-\Delta x)})$$

where $r = \alpha\,\Delta t/(\Delta x)^2$. If we divide by $e^{at}e^{ik_m x}$ and utilize the relation

$$\cos\beta = \frac{e^{i\beta} + e^{-i\beta}}{2}$$

the above expression becomes

$$e^{a\,\Delta t} = 1 + 2r(\cos\beta - 1)$$

where $\beta = k_m\,\Delta x$. Employing the trigonometric identity

$$\sin^2\frac{\beta}{2} = \frac{1 - \cos\beta}{2}$$

the final expression is

$$e^{a\,\Delta t} = 1 - 4r\sin^2\frac{\beta}{2} \tag{3-106}$$

Furthermore, since $\epsilon_j^{n+1} = e^{a\,\Delta t}\epsilon_j^n$ for each frequency present in the solution for the error, it is clear that if $|e^{a\,\Delta t}|$ is less than or equal to one, a general component of the error will not grow from one time step to the next. This requires that

$$\left| 1 - 4r \sin^2 \frac{\beta}{2} \right| \leqslant 1 \tag{3-107}$$

The factor $1 - 4r \sin^2 \beta/2$ (representing $\epsilon_j^{n+1}/\epsilon_j^n$) is called the *amplification factor* and will be denoted by G. Clearly, the influence of boundary conditions is not included in this analysis. In general, the Fourier stability analysis assumes that we have imposed periodic boundary conditions.

In evaluating the inequality Eq. (3-107), two possible cases must be considered:

1. Suppose $\left(1 - 4r \sin^2 \frac{\beta}{2}\right) > 0$ then $4r \sin^2 \frac{\beta}{2} > 0$

2. Suppose $\left(1 - 4r \sin^2 \frac{\beta}{2}\right) < 0$ then $4r \sin^2 \frac{\beta}{2} - 1 \leqslant 1$

The first condition is always satisfied since r is positive. The second inequality is satisfied only if $r \leqslant \frac{1}{2}$ which is the stability requirement for this method. This numerically places a constraint on the size of the time step relative to the size of the mesh spacing. The reason for the physically implausible temperatures calculated in the example at the end of Section 3-3.3 is now very clear. The step size Δt selected was too large by a factor of 2 and the solution began to diverge immediately. The stability of the calculation with $\alpha(\Delta t/\Delta x^2) = \frac{1}{2}$ can easily be verified. It should be noted that the amplification factor given by Eq. (3-106) could have been deduced by substituting a general form given by Eq. (3-104) into the difference equation. The proof is left as an exercise for the reader.

The application of the von Neumann or Fourier stability method is equally straightforward for hyperbolic equations. As an example, the first-order wave equation in one dimension is:

$$\frac{\partial u}{\partial t} + c \frac{\partial u}{\partial x} = 0 \tag{3-108}$$

where c is the wave speed. This equation has one characteristic given by a solution of $x_t = c$. The solution of Eq. (3-108) is given by

$$u(x - ct) = \text{constant}$$

This solution requires the initial data prescribed at $t = 0$ to be propagated along the characteristics.

Lax (1954) proposed the following first-order method for solving equations of this form

$$u_j^{n+1} = \frac{u_{j+1}^n + u_{j-1}^n}{2} - c \frac{\Delta t}{\Delta x} \left(\frac{u_{j+1}^n - u_{j-1}^n}{2} \right) \tag{3-109}$$

The first term on the right-hand side represents an average value of the unknown at the previous time level while the second term is the difference form of the first derivative. If a term of the form

$$u_j^n = e^{at} e^{ik_m x}$$

is substituted into the difference equation, the amplification factor becomes

$$e^{a\Delta t} = \cos\beta - i\nu \sin\beta$$

The stability requirement is

$$|\cos\beta - i\nu\sin\beta| \leqslant 1$$

where $\nu = c\,\Delta t/\Delta x$ is called the *Courant number.* Since the square of the absolute value of a complex number is the sum of the squares of the real and imaginary parts, the method is stable if

$$|\nu| \leqslant 1 \tag{3-110}$$

Again a conditional stability requirement must be placed on the time step and the spatial mesh spacing. This is called the Courant-Friedrichs-Lewy (CFL) condition and was discussed at length relative to the concepts of convergence and stability in an historically important paper by Courant et al. (1928). Some authorities consider this paper to be the starting point for the development of modern numerical methods for partial differential equations.

The amplification factor or growth factor for a particular numerical method depends upon mesh size and wave number or frequency. The amplification factor for the Lax finite-difference method may be written

$$G = \cos\beta - i\nu\sin\beta = |G|e^{i\phi} = \sqrt{\cos^2\beta + \nu^2\sin^2\beta}\; e^{i\tan^{-1}(-\nu\tan\beta)} \tag{3-111}$$

where ϕ is the phase angle. Clearly the magnitude of G changes with Courant number ν and frequency parameter β. A good understanding of the amplification factor can be obtained from a polar plot. Figure 3-12 is a plot of Eq. (3-111) for several different

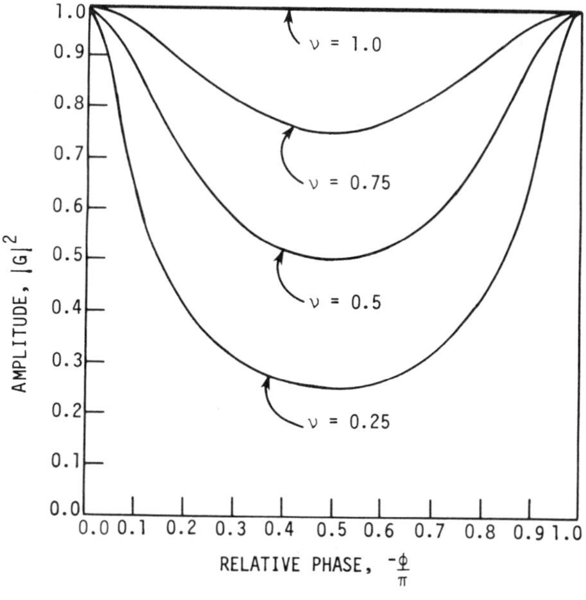

Figure 3-12 Amplitude-phase plot for the amplification factor of the Lax scheme.

Courant numbers. Several interesting results can be deduced by a careful examination of this plot. The phase angle for the Lax method varies from 0 for the low frequencies to $-\pi$ for the high frequencies. This may be seen by computing the phase for both cases. For a Courant number of one, all frequency components are propagated without attenuation in the mesh. For Courant numbers less than one, the low and high frequency components are only mildly altered while the mid-range frequency signal content is severely attenuated. The phase is also shown and we can determine the phase error for any frequency from these curves.

A physical interpretation of the results provided by Eq. (3-110) for hyperbolic equations is important. Consider the second-order wave equation

$$u_{tt} - c^2 u_{xx} = 0 \tag{3-112}$$

This equation has characteristics

$$x + ct = \text{constant} = c_1$$
$$x - ct = \text{constant} = c_2$$

A solution at a point (x, t) depends upon data contained between the characteristics which intersect that point as sketched in Fig. 3-13. The analytic solution at (x, t) is influenced only by information contained between c_1 and c_2.

The numerical stability requirement for many explicit numerical methods for solving hyperbolic PDE's is the CFL condition which is

$$\left| c \frac{\Delta t}{\Delta x} \right| \leqslant 1$$

This is the same as given in Eq. (3-110) and may be written as

$$\left(\frac{\Delta t}{\Delta x} \right)^2 \leqslant \frac{1}{c^2}$$

The characteristic slopes are given by $dt/dx = \pm 1/c$. The CFL condition requires that the analytic domain of influence lie within the numerical domain of influence. The numerical domain may include more than, but not less than, the analytical zone. Another interpretation is that the slope of the lines connecting $(j \pm 1, n)$ and $(j, n+1)$ must be smaller in absolute value (flatter) than the characteristics. The CFL require-

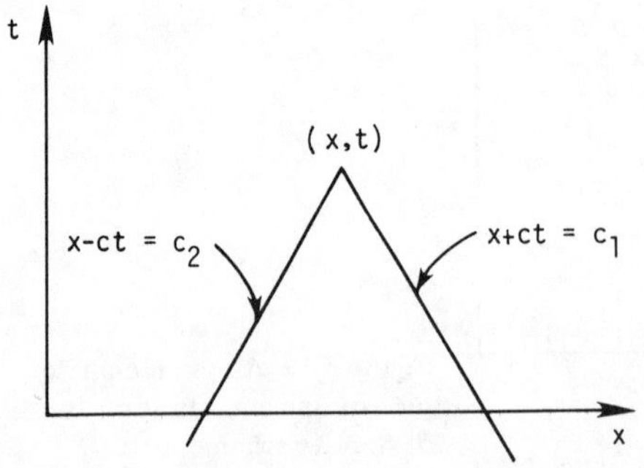

Figure 3-13 Characteristics of the second-order wave equation.

ment makes sense from a physical point of view. One would also expect the numerical solution to be degraded if too much unnecessary information is included by allowing $c(\Delta t/\Delta x)$ to become greatly different from unity. This is in fact what occurs numerically. The best results for hyperbolic systems using the most common explicit methods are obtained with Courant numbers near unity. This is consistent with our observations about attenuation associated with the Lax method as shown in Fig. 3-12.

Before we begin our study of stability for systems of equations, an example demonstrating the application of the von Neumann method to higher dimensional problems is in order.

Example 3-4 A solution of the two-dimensional heat equation

$$\frac{\partial u}{\partial t} = \alpha \frac{\partial^2 u}{\partial x^2} + \alpha \frac{\partial^2 u}{\partial y^2}$$

is desired using the simple explicit scheme. What is the stability requirement for the method?

The finite-difference equation for this problem is:

$$u_{j,k}^{n+1} = u_{j,k}^n + r_x(u_{j+1,k}^n - 2u_{j,k}^n + u_{j-1,k}^n) + r_y(u_{j,k+1}^n - 2u_{j,k}^n + u_{j,k-1}^n)$$

where $r_x = \alpha(\Delta t/\Delta x^2)$ and $r_y = \alpha(\Delta t/\Delta y^2)$. In this case a Fourier component of the form

$$u_{j,k}^n = e^{at} e^{ik_x x} e^{ik_y y}$$

is assumed. If $\beta_1 = k_x \Delta x$ and $\beta_2 = k_y \Delta y$ we obtain

$$e^{a\Delta t} = 1 + 2r_x(\cos\beta_1 - 1) + 2r_y(\cos\beta_2 - 1)$$

If the identity $\sin^2(\beta/2) = (1 - \cos\beta)/2$ is used, the amplification factor is

$$G = 1 - 4r_x \sin^2\frac{\beta_1}{2} - 4r_y \sin^2\frac{\beta_2}{2}$$

Thus for stability $|1 - 4r_x \sin^2(\beta_1/2) - 4r_y \sin^2(\beta_2/2)| \leqslant 1$, which is true only if $4r_x \sin^2\beta/2 + 4r_y \sin^2\beta/2 \leqslant 2$. The stability requirement is then $r_x + r_y \leqslant \frac{1}{2}$ or $\alpha\Delta t[1/(\Delta x)^2 + 1/(\Delta y)^2] \leqslant \frac{1}{2}$. This is similar to the analysis of the same method for the one-dimensional case but shows that the effective time step in two dimensions is reduced. This example was easily completed, but in general a stability analysis in more than a single space dimension and time is difficult. Frequently, the stability must be determined by computing the magnitude of the amplification factor for different values of r_x and r_y.

3-6.2 Stability Analysis for Systems of Equations

The previous discussion illustrates how the von Neumann analysis can be used to evaluate stability for a single equation. The basic idea used in this technique also provides a useful method of viewing stability for systems of equations. Systems of equations encountered in fluid mechanics and heat transfer can usually be written in the form

$$\frac{\partial \mathbf{E}}{\partial t} + \frac{\partial \mathbf{F}}{\partial x} = 0 \tag{3-113}$$

where **E** and **F** are vectors and $\mathbf{F} = \mathbf{F}(\mathbf{E})$. In general this system of equations is nonlinear. In order to perform a linear stability analysis, we rewrite the system as

$$\frac{\partial \mathbf{E}}{\partial t} + \left[\frac{\partial F}{\partial E}\right] \frac{\partial \mathbf{E}}{\partial x} = 0 \tag{3-114}$$

or

$$\frac{\partial \mathbf{E}}{\partial t} + [A] \frac{\partial \mathbf{E}}{\partial x} = 0$$

where $[A]$ is the Jacobian matrix $[\partial F/\partial E]$. We locally linearize the system by holding $[A]$ constant while the **E** vector is advanced through a single time step. A similar linearization is used for a single nonlinear equation permitting the application of the von Neumann method of the previous section.

For the sake of discussion, let us apply the Lax method to this system. The result is

$$\mathbf{E}_j^{n+1} = \frac{1}{2}\left([I] + \frac{\Delta t}{\Delta x}[A]^n\right)\mathbf{E}_{j-1}^n + \frac{1}{2}\left([I] - \frac{\Delta t}{\Delta x}[A]^n\right)\mathbf{E}_{j+1}^n \tag{3-115}$$

where the notation is as previously defined and $[I]$ is the identity matrix. The stability of the difference equation can again be evaluated by applying the Fourier or von Neumann method. If a typical term of a Fourier series is substituted into Eq. (3-115), an expression is obtained

$$\mathbf{e}^{n+1}(k) = [G(\Delta t, k)]\,\mathbf{e}^n(k) \tag{3-116}$$

where

$$[G] = [I]\cos\beta - i\frac{\Delta t}{\Delta x}[A]\sin\beta \tag{3-117}$$

and $\mathbf{e}^n$ represents the Fourier coefficients of the typical term. The $[G]$ matrix is called the amplification matrix. This matrix is now dependent upon step size and frequency or wave number, i.e., $[G] = [G(\Delta t, k)]$. For a stable finite-difference calculation, the largest eigenvalue of $[G]$, $\sigma_{\max}$, must obey

$$|\sigma_{\max}| \leqslant 1 \tag{3-118}$$

This leads to the requirement that

$$\left|\lambda_{\max}\frac{\Delta t}{\Delta x}\right| \leqslant 1 \tag{3-119}$$

where $\lambda_{\max}$ is the largest eigenvalue of the $[A]$ matrix, i.e., the Jacobian matrix of the system. A simple example to demonstrate this is of value.

Example 3-5 Determine the stability requirement necessary in solving the system of first-order equations

$$\frac{\partial u}{\partial t} + c \frac{\partial v}{\partial x} = 0$$

$$\frac{\partial v}{\partial t} + c \frac{\partial u}{\partial x} = 0$$

using the Lax method. In this problem

$$\mathbf{E} = \begin{bmatrix} u \\ v \end{bmatrix}$$

and

$$\frac{\partial \mathbf{E}}{\partial t} + [A] \frac{\partial \mathbf{E}}{\partial x} = 0$$

where

$$[A] = \begin{bmatrix} 0 & c \\ c & 0 \end{bmatrix}$$

Thus, the maximum eigenvalue is c and the stability requirement is the usual CFL condition

$$\left| c \frac{\Delta t}{\Delta x} \right| \leqslant 1$$

It should be noted that the stability analysis presented above does not include the effect of boundary conditions even though a matrix notation for the system is used. The influence of boundary conditions is easily included for systems of difference equations.

Equation (3-116) shows that the stability of a finite-difference operator is related to the amplification matrix. We may also write Eq. (3-116) as

$$\mathbf{e}^{n+1}(k) = [G(\Delta t, k)]^n [\mathbf{e}^1(k)] \tag{3-120}$$

The stability condition (Richtmyer and Morton, 1967) requires that for some positive τ, the matrices $[G(\Delta t, k)]^n$ be uniformly bounded for

$$0 < \Delta t < \tau$$

$$0 \leqslant n\,\Delta t \leqslant T$$

for all k where T is the maximum time. This leads to the *von Neumann necessary condition* for stability which is

$$|\sigma_i(\Delta t, k)| \leqslant 1 + O(\Delta t) \tag{3-121}$$

for
$$0 < \Delta t < \tau$$

for each eigenvalue and wave number where σ_i represents the eigenvalues of $[G(\Delta t, k)]$.

The stability requirement used in previous examples required that the maximum eigenvalue have a modulus less than or equal to one. Clearly that requirement is more stringent than Eq. (3-121). The von Neumann necessary condition provides that local growth $c\,\Delta t$ can be acceptable and, in fact, must be possible in many physical problems. The classical example illustrating this point is the heat equation with a source term.

Example 3-6 Suppose we wish to solve the heat equation with a source term

$$\frac{\partial u}{\partial t} = \alpha \frac{\partial^2 u}{\partial x^2} + cu$$

using the simple explicit finite-difference method. If a Fourier stability analysis is performed, the amplification factor is

$$G = 1 - 4r \sin^2 \frac{\beta}{2} + c\,\Delta t$$

This shows that the solution of the difference equation may grow with time and still satisfy the von Neumann necessary condition. Physical insight must be used when the stability of a finite-difference method is investigated. One must recognize that for hyperbolic systems the strict condition less than or equal to one should be used. Hyperbolic equations are wave-like and do not possess solutions which increase exponentially with time.

We have investigated stability of various finite-difference methods by using the von Neumann method. If the influence of boundary conditions on stability is desired, we must use the *matrix method.* This is most easily demonstrated by applying the Lax method to solve the one-dimensional linear wave equation

$$\frac{\partial u}{\partial t} + c \frac{\partial u}{\partial x} = 0$$

Assume that an array of m points is used to solve this problem and that the boundary conditions are periodic, i.e.,

$$u_{m+1}^n = u_1^n \tag{3-122}$$

If the Lax method is applied to this problem, a system of algebraic equations is generated which has the form

$$\mathbf{u}^{n+1} = [X]\mathbf{u}^n \tag{3-123}$$

where

$$\mathbf{u}^n = [u_1, u_2, \ldots, u_m]^T \tag{3-124}$$

and

$$[X] = \begin{bmatrix} 0 & \frac{1-\nu}{2} & 0 & \cdot & \cdot & \cdot & \frac{1+\nu}{2} \\ \frac{1+\nu}{2} & 0 & \frac{1-\nu}{2} & \cdot & \cdot & \cdot & 0 \\ 0 & \frac{1+\nu}{2} & \ddots & \ddots & & & \cdot \\ \cdot & 0 & \ddots & \ddots & \ddots & & \cdot \\ \cdot & \cdot & & \ddots & \ddots & \ddots & \cdot \\ \cdot & \cdot & & & \ddots & \ddots & \frac{1-\nu}{2} \\ \frac{1-\nu}{2} & 0 & \cdot & \cdot & \cdot & \frac{1+\nu}{2} & 0 \end{bmatrix} \tag{3-125}$$

The stability of the finite-difference calculation in Eq. (3-123) is governed by the eigenvalue structure of $[X]$. Since $[X]$ was formed assuming periodic boundary conditions, only the three diagonals noted in Eq. (3-125) and the two corner elements contribute to the calculation. This matix is called an aperiodic matrix. For matrices of the form

$$\begin{bmatrix} a_1 & a_2 & 0 & \cdot & \cdot & \cdot & a_0 \\ a_0 & a_1 & a_2 & 0 & \cdot & \cdot & 0 \\ 0 & a_0 & \ddots & \ddots & & & \cdot \\ \cdot & \cdot & & \ddots & \ddots & & \cdot \\ \cdot & \cdot & & & \ddots & \ddots & \cdot \\ \cdot & \cdot & & & & \ddots & a_2 \\ a_2 & 0 & \cdot & \cdot & \cdot & a_0 & a_1 \end{bmatrix} \tag{3-126}$$

the eigenvalues are given by

$$\lambda_j = a_1 + (a_0 + a_2)\cos\frac{2\pi}{m}(j-1) + i(a_0 - a_2)\sin\frac{2\pi}{m}(j-1)$$

$$j = 1, m$$

In this case a_0, a_1, and a_2 have the values

$$a_0 = \frac{1+\nu}{2} \qquad a_1 = 0 \qquad a_2 = \frac{1-\nu}{2}$$

and the eigenvalues are

$$\lambda_j = \cos \frac{2\pi}{m}(j-1) + i\nu \sin \frac{2\pi}{m}(j-1) \tag{3-127}$$

The numerical method is thus stable if $|\nu| \leqslant 1$, i.e., if the CFL condition is satisfied. This shows that an analysis based upon the matrix operator associated with the Lax method yields the same stability requirement as previously derived for the simple wave equation. For periodic boundary conditions, the Fourier and matrix method yield virtually identical results. Another example is needed in order to demonstrate the effect of boundary conditions and the discreteness of the mesh.

As in the previous example, assume that the Lax method is used to solve the first-order linear wave equation. If a four-point mesh is used, special treatment is needed to enforce the boundary conditions at the first and fourth points. For simplicity we set u at the first point equal to a constant value for all time so the equation for the first point reads

$$u_1^{n+1} = u_1^n$$

Since we are computing a solution to the wave equation, the value of u_4 cannot be arbitrarily chosen. It must be consistent with the way the solution is propagated. We elect to set

$$u_4^{n+1} = u_3^n$$

which determines the boundary value from the interior solution. With this boundary condition treatment the $[X]$ matrix becomes

$$\begin{bmatrix} 1 & 0 & 0 & 0 \\ \dfrac{1+\nu}{2} & 0 & \dfrac{1-\nu}{2} & 0 \\ 0 & \dfrac{1+\nu}{2} & 0 & \dfrac{1-\nu}{2} \\ 0 & 0 & 1 & 0 \end{bmatrix}$$

The eigenvalues are easily computed and are

$$\lambda_1 = 1$$

$$\lambda_2 = 0$$

$$\lambda_{3,4} = \pm\tfrac{1}{2}\sqrt{(1-\nu)(3+\nu)}$$

Using the requirement that $|\lambda| \leqslant 1$ for stability, the restriction on ν is not the usual CFL condition but is

$$(-\sqrt{8}-1) \leqslant \nu \leqslant (\sqrt{8}-1)$$

The CFL condition is altered by the boundary conditions in this example, as is normally the case.

It is clear that the boundary conditions on the mesh are included in the matrix method. This means that the influence of boundary conditions on stability is automatically included if the matrix analysis is used. Unfortunately, a closed-form solution for the eigenvalues is usually not available for arbitrary end boundary conditions.

The treatment of stability presented in this section has included the Fourier (von Neumann) method and the matrix method of analysis. These two techniques are probably the most widely used to determine the stability of finite-difference schemes. Other methods of analyzing stability have been devised and are frequently very convenient to use. The works of Hirt (1968) and Warming and Hyett (1974) are typical of these techniques. A more comprehensive mathematical analysis of stability including many theorems and proofs is contained in the book by Richtmyer and Morton (1967).

PROBLEMS

3-1 Verify that

$$\left(\frac{\partial^3 u}{\partial x^3}\right)_{i,j} = \frac{\Delta_x^3 u_{i,j}}{(\Delta x)^3} + O(\Delta x)$$

3-2 Consider the function $f(x) = e^x$. Using a mesh increment $\Delta x = 0.1$ determine $f'(x)$ at $x = 2$ with the forward difference formula, Eq. (3-26), the central-difference formula, Eq. (3-28), and the second-order three-point formula, Eq. (3-29). Compare the results with the exact value. Repeat the comparisons for $\Delta x = 0.2$. Have the order estimates for truncation errors been a reliable guide? Discuss this point.

3-3 Verify whether or not the following difference representation for the continuity equation for a two-dimensional steady, incompressible flow has the conservative property:

$$\frac{(u_{i+1,j} + u_{i+1,j-1} - u_{i,j} - u_{i,j-1})}{2\,\Delta x} + \frac{(v_{i+1,j} - v_{i+1,j-1})}{\Delta y} = 0$$

where u and v are the x and y components of velocity, respectively.

3-4 Work Prob. 3-3, doing the evaluation for the following difference representation for the continuity equation:

$$\frac{(u_{i+1,j} - u_{i-1,j})}{2\,\Delta x} + \frac{(v_{i,j+1} - v_{i,j-1})}{2\,\Delta y} = 0$$

3-5 Consider the nonlinear equation

$$u\,\frac{\partial u}{\partial x} = \mu\,\frac{\partial^2 u}{\partial y^2}$$

where μ is a constant.

(*a*) Is this equation in conservative form? If not, can you suggest a conservative form for the equation?

(*b*) Develop a finite-difference formulation for this equation using the integral approach.

3-6 Verify the approximation to $\partial^2 u/\partial x\,\partial y$ given by Eq. (3-50).

3-7 Verify the approximation to $\partial^2 u/\partial x^2$ given by Eq. (3-40).

3-8 Verify Eq. (3-79) in Table 3.3.

3-9 Verify Eq. (3-80) in Table 3.3.

3-10 Verify the following finite-difference approximation for use in two dimensions at the point (i, j). Assume $\Delta x = \Delta y = h$.

$$\frac{\partial^2 u}{\partial x^2} + \frac{\partial^2 u}{\partial y^2} = \frac{u_{i+1,j-1} + u_{i+1,j+1} + u_{i-1,j-1} + u_{i-1,j+1} - 4u_{i,j}}{2h^2} + O(h^2)$$

3-11 Develop a finite-difference approximation with truncation error $O(\Delta y)$ for $\partial^2 u/\partial y^2$ at point (i, j) using $u_{i,j}$, $u_{i,j+1}$, $u_{i,j-1}$ when the grid spacing is *not* uniform. Use the Taylor-series method. Can you devise a three-point scheme with second-order accuracy with unequal spacing? Before you draw your final conclusions, consider the use of compact implicit representations.

3-12 Establish the truncation error of the following finite-difference approximation to $\partial u/\partial y$ at the point (i, j) for a uniform mesh:

$$\frac{\partial u}{\partial y} \simeq \frac{-3u_{i,j} + 4u_{i,j+1} - u_{i,j+2}}{2\,\Delta y}$$

What is the order of the truncation error?

3-13 Investigate the truncation error of the following finite-difference approximation for a uniform mesh:

$$\left.\frac{\partial u}{\partial x}\right)_{i,j} \simeq \frac{1}{2h}\,\frac{\bar{\delta}_x u_{i,j}}{1 + \delta_x^2/6}$$

3-14 Utilize Taylor-series expansions about the point $(n + \frac{1}{2}, j)$ to determine the truncation error of the Crank-Nicolson representation of the heat equation, Eq. (3-71). Compare these results with the truncation error obtained from Taylor-series expansions about point n, j.

3-15 Develop a finite-difference approximation with truncation error $O(\Delta y)^2$ for $\partial T/\partial y$ at point (i, j) using $T_{i,j}$, $T_{i,j+1}$, and $T_{i,j+2}$ when the grid spacing is *not* uniform.

3-16 Determine the truncation error of the following finite-difference approximation for $\partial u/\partial x$ at point (i, j) when the grid spacing is *not* uniform.

$$\left.\frac{\partial u}{\partial x}\right)_{i,j} \simeq \frac{u_{i+1,j} - (\Delta x_+/\Delta x_-)^2 u_{i-1,j} - [1 - (\Delta x_+/\Delta x_-)^2]u_{i,j}}{\Delta x_-(\Delta x_+/\Delta x_-)^2 + \Delta x_+}$$

3-17 Suppose that a finite-difference solution has been obtained for the temperature T, near but not at an adiabatic boundary (i.e., $\partial T/\partial y = 0$ at the boundary) (Fig. P3.1). In most instances, it would be necessary or desirable to evaluate the temperature at the boundary point itself. For this case of an adiabatic boundary, develop expressions for the temperature at the boundary, T_1, in terms of temperatures at neighboring points T_2, T_3, etc. by assuming that the temperature distribution in the neighborhood of the boundary is

(*a*) a straight line
(*b*) a second degree polynomial
(*c*) a cubic polynomial (you only need to indicate how you would derive this one).

Indicate the order of the truncation error in each of the above approximations used to evaluate T_1.

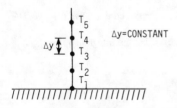

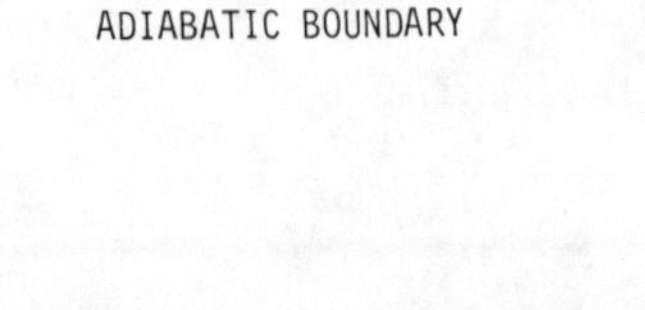

Figure P3.1

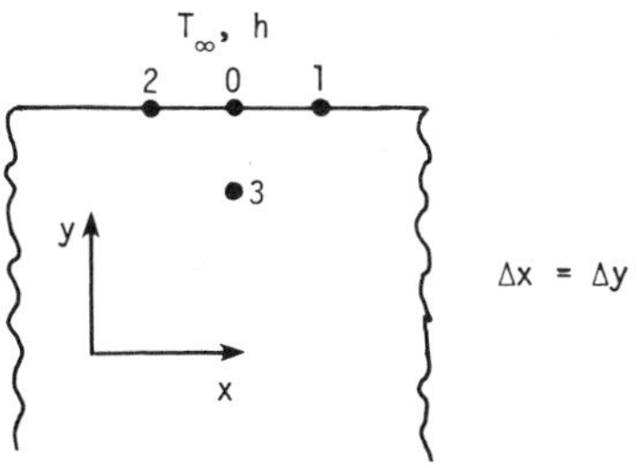

Figure P3.2

3-18 Let's consider a steady-state conduction problem governed by Laplace's equation with convective boundary conditions (see Fig. P3.2). The formal statement of the boundary condition is $-k\,\partial T/\partial y)_{\text{bdy}} = h(T_w - T_\infty)$ which can be readily cast into finite-difference form as $-k[(T_0 - T_\infty)/\Delta y] + O(\Delta y) = h(T_0 - T_\infty)$. Use the control-volume approach to develop an expression for the boundary condition at point 0. Evaluate the truncation error in this expression assuming that Laplace's equation applies at the boundary point.

3-19 Consider a heat conduction problem governed by $\partial T/\partial t = \alpha(\partial^2 T/\partial x^2)$. Develop a finite-difference representation for this equation by the control-volume approach. Do not assume that the grid is uniform.

3-20 For two-dimensional steady-state conduction in a solid, apply the control-volume method to derive an appropriate difference expression for the boundary temperature in control volume B in Fig. 3-7 for *adiabatic wall* boundary conditions.

3-21 Solve the one-dimensional heat equation using forward-time centered-space differences with $\alpha(\Delta t/\Delta x^2) = \frac{1}{2}$. Let the grid consist of five points including three interior and two boundary points. Assume a constant unity wall temperature and a zero initial temperature on the interior. Complete this calculation for ten integration steps. Compare your results with those obtained in the example of Section 3-3.3.

3-22 Show that the amplification factor derived for the finite-difference solution of the heat equation Eq. (3-101) could be obtained by direct substitution of a solution of the form

$$u_j^n = \sum_{-\infty}^{+\infty} C_m g_m^n e^{ik_x x}$$

In this form the C_m's are the Fourier coefficients of the initial error distribution and g_m is the amplification factor. Identify g_m with Eq. (3-106). Discuss the convergence of the solution and relate your conclusions to the Lax equivalence theorem.

3-23 Use a von Neumann stability analysis to show that a simple explicit Euler predictor using central differencing in space is unstable. The difference equation is:

$$u_j^{n+1} = u_j^n - c\,\frac{\Delta t}{\Delta x}\left(\frac{u_{j+1}^n - u_{j-1}^n}{2}\right)$$

Now show that the same difference method is stable when written as the implicit formula

$$u_j^{n+1} = u_j^n - c\,\frac{\Delta t}{\Delta x}\left(\frac{u_{j+1}^{n+1} - u_{j-1}^{n+1}}{2}\right)$$

3-24 The DuFort-Frankel method for solving the heat equation requires solution of the difference equation:

$$\frac{u_j^{n+1} - u_j^{n-1}}{2\,\Delta t} = \frac{\alpha}{(\Delta x)^2}\,(u_{j+1}^n - u_j^{n+1} - u_j^{n-1} + u_{j-1}^n)$$

Develop the stability requirements necessary for the solution of this equation.

3-25 Prove that the CFL condition is the stability requirement when the Lax-Wendroff method is applied to solve the simple one-dimensional wave equation. The difference equation is of the form:

$$u_j^{n+1} = u_j^n - \frac{c\,\Delta t}{2\,\Delta x}(u_{j+1}^n - u_{j-1}^n) + \frac{c^2(\Delta t)^2}{2(\Delta x)^2}(u_{j+1}^n - 2u_j^n + u_{j-1}^n)$$

3-26 Determine the stability requirement necessary to solve the one-dimensional heat equation with a source term

$$\frac{\partial u}{\partial t} = \alpha \frac{\partial^2 u}{\partial x^2} + ku$$

Use the central-space, forward-time difference method. Does the von Neumann necessary condition Eq. (3-121) make physical sense for this type of computational problem?

3-27 Use the matrix method to determine the stability of the Lax method used to solve the first-order wave equation on a mesh with two interior points and two boundary points. Assume the boundaries are held at constant values $u_{\text{left}} = 1, u_{\text{right}} = 0$.

3-28 Use the matrix method and evaluate the stability of the numerical method used in Prob. 3-21 for the heat equation using a five-point mesh. How many frequencies must one be concerned with in this case?

3-29 In attempting to solve a simple PDE, a system of finite-difference equations of the form $\mathbf{u}_j^{n+1} = [A]\mathbf{u}_j^n$ has evolved where

$$[A] = \begin{bmatrix} 1+\nu & \nu & 0 \\ 0 & 1+\nu & \nu \\ -\nu & 0 & 1+\nu \end{bmatrix}$$

Investigate the stability of this scheme.

CHAPTER

FOUR

APPLICATION OF FINITE-DIFFERENCE METHODS TO SELECTED MODEL EQUATIONS

In this chapter we examine in detail various finite-difference schemes which can be used to solve simple model equations. These model equations include the first-order wave equation, the heat equation, Laplace's equation and Burgers' equation. These equations are called model equations because they can be used to "model" the behavior of more complicated partial differential equations. For example, the heat equation can serve as a model equation for other parabolic partial differential equations such as the boundary-layer equations. All of the present model equations have exact solutions for certain boundary and initial conditions. We can use this knowledge to quickly evaluate and compare finite-difference methods which we might wish to apply to more complicated partial differential equations. The various methods discussed in this chapter were selected because they illustrate the basic properties of finite-difference algorithms. Each of the methods exhibit certain distinctive features that are characteristic of a class of methods. Some of these features may not be desirable, but the method is included anyway for pedagogical reasons. Other very useful methods have been omitted because they are similar to those that are included. Space does not permit a discussion of all possible finite-difference methods that could be used.

4-1 WAVE EQUATION

The one-dimensional wave equation is a second-order hyperbolic partial differential equation given by

$$\frac{\partial^2 u}{\partial t^2} = c^2 \frac{\partial^2 u}{\partial x^2} \tag{4-1}$$

This equation governs the propagation of sound waves traveling at a wave speed c in a uniform medium. A first-order equation which has properties similar to those of Eq. (4-1) is given by

$$\frac{\partial u}{\partial t} + c\,\frac{\partial u}{\partial x} = 0 \qquad c > 0 \tag{4-2}$$

Note that Eq. (4-1) can be obtained from Eq. (4-2). We will use Eq. (4-2) as our model equation and refer to it as the first-order, one-dimensional wave equation or simply the "wave equation." This linear hyperbolic equation describes a wave propagating in the x direction with a velocity c and it can be used to "model" in a rudimentary fashion the nonlinear equations governing inviscid flow. Although we will refer to Eq. (4-2) as the wave equation, the reader is cautioned to be aware of the fact that Eq. (4-1) is the classical wave equation. More appropriately, Eq. (4-2) is often called the one-dimensional linear convection equation.

The exact solution of the wave equation [Eq. (4-2)] for the pure initial value problem with initial data

$$u(x, 0) = F(x) \qquad (-\infty < x < \infty) \tag{4-3}$$

is given by

$$u(x, t) = F(x - ct) \tag{4-4}$$

Let us now examine some finite-difference schemes which could be used to solve the wave equation.

4-1.1 Euler Explicit Methods

The following simple explicit, one-step methods

$$\frac{u_j^{n+1} - u_j^n}{\Delta t} + c\,\frac{u_{j+1}^n - u_j^n}{\Delta x} = 0 \tag{4-5}$$

$$\frac{u_j^{n+1} - u_j^n}{\Delta t} + c\,\frac{u_{j+1}^n - u_{j-1}^n}{2\,\Delta x} = 0 \tag{4-6}$$

have respectively, truncation errors of $O[\Delta t, \Delta x]$ and $O[\Delta t, (\Delta x)^2]$. We refer to these schemes as being first-order accurate since the lowest-order term in the truncation error is first order, i.e., Δt and Δx for Eq. (4-5) and Δt for Eq. (4-6). These schemes are explicit since only one unknown u_j^{n+1} appears in each equation. Unfortunately, when the von Neumann stability analysis is applied to these schemes we find that they are unconditionally unstable. These simple schemes, therefore, prove to be worthless in solving the wave equation. Let us now proceed to look at methods which have more utility.

4-1.2 Upstream (Windward) Differencing Method

The simple Euler method, Eq. (4-5), can be made stable by replacing the forward space difference by a backward space difference provided that the wave speed c is positive. If the wave speed is negative, a forward difference must be used to assure stability. This point will be more carefully explored in Chapter 6 when the split coefficient matrix (SCM) methods are studied. If a backward difference is used, the following algorithm results:

$$\frac{u_j^{n+1} - u_j^n}{\Delta t} + c\,\frac{u_j^n - u_{j-1}^n}{\Delta x} = 0 \qquad c > 0 \tag{4-7}$$

This is a first-order accurate method with truncation error of $O[\Delta t, \Delta x]$. The von Neumann stability analysis shows that this method is stable provided that

$$0 \leqslant \nu \leqslant 1 \tag{4-8}$$

where $\nu = c\,\Delta t/\Delta x$.

Let us substitute Taylor-series expansions into Eq. (4-7) for u_j^{n+1} and u_{j-1}^n. The following equation results

$$\frac{1}{\Delta t}\left\{\left[u_j^n + \Delta t u_t + \frac{(\Delta t)^2}{2} u_{tt} + \frac{(\Delta t)^3}{6} u_{ttt} + \cdots\right] - u_j^n\right\}$$
$$+ \frac{c}{\Delta x}\left\{u_j^n - \left[u_j^n - \Delta x u_x + \frac{(\Delta x)^2}{2} u_{xx} - \frac{(\Delta x)^3}{6} u_{xxx} + \cdots\right]\right\} = 0 \tag{4-9}$$

Equation (4-9) simplifies to

$$u_t + cu_x = -\frac{\Delta t}{2} u_{tt} + \frac{c\,\Delta x}{2} u_{xx} - \frac{(\Delta t)^2}{6} u_{ttt} - c\,\frac{(\Delta x)^2}{6} u_{xxx} + \cdots \tag{4-10}$$

Note that the left-hand side of this equation corresponds to the wave equation and the right-hand side is the truncation error which is generally not zero. The significance of terms in the truncation error can be more easily interpreted if the time derivative terms are replaced by spatial derivatives. In order to replace u_{tt} by a spatial derivative term, we take the partial derivative of Eq. (4-10) with respect to time, to obtain

$$u_{tt} + cu_{xt} = -\frac{\Delta t}{2} u_{ttt} + \frac{c\,\Delta x}{2} u_{xxt} - \frac{(\Delta t)^2}{6} u_{tttt} - \frac{c(\Delta x)^2}{6} u_{xxxt} + \cdots \tag{4-11}$$

and take the partial derivative of Eq. (4-10) with respect to x and multiply by $-c$

$$-cu_{tx} - c^2 u_{xx} = \frac{c\,\Delta t}{2} u_{ttx} - \frac{c^2\,\Delta x}{2} u_{xxx} + \frac{c(\Delta t)^2}{6} u_{tttx} + \frac{c^2(\Delta x)^2}{6} u_{xxxx} + \cdots \tag{4-12}$$

Adding Eqs. (4-11) and (4-12) gives

$$u_{tt} = c^2 u_{xx} + \Delta t\left(\frac{-u_{ttt}}{2} + \frac{c}{2} u_{ttx} + O[\Delta t]\right) + \Delta x\left(\frac{c}{2} u_{xxt} - \frac{c^2}{2} u_{xxx} + O[\Delta x]\right) \tag{4-13}$$

In a similar manner we can obtain the following expressions for u_{ttt}, u_{ttx}, and u_{xxt}:

$$u_{ttt} = -c^3 u_{xxx} + O[\Delta t, \Delta x] \tag{4-14}$$

$$u_{ttx} = c^2 u_{xxx} + O[\Delta t, \Delta x] \tag{4-14}$$

$$u_{xxt} = -c u_{xxx} + O[\Delta t, \Delta x] \qquad (Cont.)$$

Combining Eqs. (4-10), (4-13), and (4-14) leaves

$$u_t + c u_x = \frac{c\,\Delta x}{2}(1 - \nu)u_{xx} - \frac{c(\Delta x)^2}{6}(2\nu^2 - 3\nu + 1)u_{xxx}$$

$$+ O[(\Delta x)^3, (\Delta x)^2\,\Delta t, \Delta x(\Delta t)^2, (\Delta t)^3] \tag{4-15}$$

An equation, such as Eq. (4-15), is called a *modified equation* (Warming and Hyett, 1974). It is the partial differential equation which is actually solved when a finite-difference method is applied to a PDE. It is important to emphasize that the equation obtained after substitution of the Taylor-series expansion, i.e., Eq. (4-10), must be used to eliminate the higher-order time derivatives rather than the original PDE, Eq. (4-2). This is due to the fact that a solution of the original PDE does not in general satisfy the difference equation, and since the modified equation represents the difference equation, it is obvious that the original PDE should not be used to eliminate the time derivatives.

The process of eliminating time derivatives can be greatly simplified if a table is constructed as shown in Table 4-1. The coefficients of each term in Eq. (4-10) are placed in the first row of the table. Note that all terms have been moved to the left-hand side of the equation. The u_{tt} term is then eliminated by multiplying Eq. (4-10) by the operator

$$-\frac{\Delta t}{2}\frac{\partial}{\partial t}$$

and adding the result to the first row, i.e., Eq. (4-10). This introduces the term $-(c\,\Delta t/2)u_{tx}$ which is eliminated by multiplying Eq. (4-10) by the operator

$$\frac{c\,\Delta t}{2}\frac{\partial}{\partial x}$$

and adding the result to the first two rows of the table. This procedure is continued until the desired number of time derivatives are eliminated. Each coefficient in the modified equation is then obtained by simply adding the coefficients in the corresponding column of the table. The algebra required to derive the modified equation can be programmed on a digital computer using a language code such as FORMAC (Fike, 1970).

The right-hand side of the modified equation [Eq. (4-15)] is the truncation error since it represents the difference between the original PDE and the finite-difference approximation to it. Consequently, the lowest-order term on the right-hand side of the modified equation gives the order of the method. In the present case, the method is first-order accurate since the lowest-order term is of $O[\Delta t, \Delta x]$. If $\nu = 1$, the right-hand side of the modified equation becomes zero and the wave equation is solved exactly. In this case, the upstream differencing scheme reduces to

$$u_j^{n+1} = u_{j-1}^n$$

Table 4-1 Procedure for determining modified equation

	u_t	u_x	u_{tt}	u_{tx}	u_{xx}	u_{ttt}	u_{ttx}	u_{txx}	u_{xxx}	u_{tttt}	u_{tttx}	u_{ttxx}	u_{txxx}	u_{xxxx}
Coefficients of Eq. (4-10)	1	c	$\frac{\Delta t}{2}$	0	$-\frac{c\,\Delta x}{2}$	$\frac{\Delta t^2}{6}$	0	0	$\frac{c\,\Delta x^2}{6}$	$\frac{\Delta t^3}{24}$	0	0	0	$-\frac{c\,\Delta x^3}{24}$
$-\frac{\Delta t}{2}\frac{\partial}{\partial t}$ Eq. (4-10)			$-\frac{\Delta t}{2}$	$-\frac{c\,\Delta t}{2}$	0	$-\frac{\Delta t^2}{4}$	0	$\frac{c\,\Delta t\,\Delta x}{4}$	0	$-\frac{\Delta t^3}{12}$	0	0	$-\frac{c\,\Delta t\,\Delta x^2}{12}$	0
$\frac{c}{2}\Delta t\frac{\partial}{\partial x}$ Eq. (4-10)				$\frac{c\,\Delta t}{2}$	$\frac{c^2}{2}\Delta t$	0	$\frac{c\,\Delta t^2}{4}$	0	$-\frac{c^2\,\Delta t\,\Delta x}{4}$	0	$\frac{c\,\Delta t^3}{12}$	0	0	$\frac{c^2\,\Delta t\,\Delta x^2}{12}$
$\frac{1}{12}\Delta t^2\frac{\partial^2}{\partial t^2}$ Eq. (4-10)						$\frac{1}{12}\Delta t^2$	$\frac{c\,\Delta t^2}{12}$	0	0	$\frac{1}{24}\Delta t^3$	0	$-\frac{c}{24}\Delta x\,\Delta t^2$	0	0
$-\frac{1}{3}c\,\Delta t^2\frac{\partial^2}{\partial t\,\partial x}$ Eq. (4-10)							$-\frac{1}{3}c\,\Delta t^2$	$-\frac{1}{3}c^2\,\Delta t^2$	0	0	$-\frac{1}{6}c\,\Delta t^3$	0	$+\frac{c^2}{6}\Delta x\,\Delta t^2$	0
$\left(\frac{1}{3}c^2\,\Delta t^2-\frac{c\,\Delta t\,\Delta x}{4}\right)\frac{\partial^2}{\partial x^2}$ Eq. (4-10)								$\frac{1}{3}c^2\,\Delta t^2$ $-\frac{c\,\Delta t\,\Delta x}{4}$	$\frac{1}{3}c^3\,\Delta t^2$ $-\frac{c^2\,\Delta t\,\Delta x}{4}$	0	0	$\frac{1}{6}c^2\,\Delta t^3$ $-\frac{c\,\Delta t^2\,\Delta x}{8}$	0	$-\frac{1}{6}c^3\,\Delta t^2\,\Delta x$ $+\frac{c^2}{8}\Delta t\,\Delta x^2$
$\frac{1}{12}c\,\Delta t^3\frac{\partial^3}{\partial t^2\,\partial x}$ Eq. (4-10)											$\frac{1}{12}c\,\Delta t^3$	$\frac{c^2}{12}\Delta t^3$	0	0
$\left(\frac{1}{6}c\,\Delta x\,\Delta t^2-\frac{1}{4}c^2\,\Delta t^3\right)\frac{\partial^3}{\partial t\,\partial x^2}$ Eq. (4-10)												$\frac{1}{6}c\,\Delta x\,\Delta t^2$ $-\frac{1}{4}c^2\,\Delta t^3$	$\frac{1}{6}c^2\,\Delta x\,\Delta t^2$ $-\frac{1}{4}c^3\,\Delta t^3$	0
$\left(\frac{c}{12}\Delta t\,\Delta x^2-\frac{1}{3}c^2\,\Delta x\,\Delta t^2+\frac{1}{4}c^3\,\Delta t^3\right)\frac{\partial^3}{\partial x^3}$ Eq. (4-10)													$\frac{c}{12}\Delta t\,\Delta x^2$ $-\frac{1}{3}c^2\,\Delta x\,\Delta t^2$ $+\frac{1}{4}c^3\,\Delta t^3$	$\frac{c^2}{12}\Delta t\,\Delta x^2$ $-\frac{1}{3}c^3\,\Delta x\,\Delta t^2$ $+\frac{1}{4}c^4\,\Delta t^3$
......														
Sum of coefficients	1	c	0	0	$\frac{c\,\Delta x}{2}(\nu-1)$	0	0	0	$\frac{c\,\Delta x^2}{6}(2\nu^2-3\nu+1)$	0	0	0	0	$\frac{c\,\Delta x^3}{24}(6\nu^3-12\nu^2+7\nu-1)$

which is equivalent to solving the wave equation exactly using the method of characteristics. Finite-difference algorithms which exhibit this behavior are said to satisfy the *"shift" condition* (Kutler and Lomax, 1971). Note that the difference between the exact solution of the modified equation and the exact solution of the wave equation represents the discretization error (for the case of periodic boundary conditions).

The lowest-order term of the truncation error in the present case contains the partial derivative u_{xx} which makes this term similar to the viscous term in one-dimensional fluid flow equations. For example, the viscous term in the one-dimensional Navier-Stokes equation (see Chapter 5) may be written as

$$\frac{\partial}{\partial x}(\tau_{xx}) = \frac{4}{3}\mu u_{xx} \tag{4-16}$$

if a constant coefficient of viscosity is assumed. Thus, when $\nu \neq 1$ the upstream differencing scheme introduces an *"artificial viscosity"* into the solution. This is often called implicit artificial viscosity as opposed to explicit artificial viscosity which is purposely added to a difference scheme. Artificial viscosity tends to reduce all gradients in the solution whether physically correct or numerically induced. This effect, which is the direct result of even derivative terms in the truncation error, is called *dissipation.*

Another quasi-physical effect of numerical schemes is called *dispersion.* This is the direct result of the odd derivative terms which appear in the truncation error. As a result of dispersion, phase relations between various waves are distorted. The combined effect of dissipation and dispersion is sometimes referred to as *diffusion.* Diffusion tends to spread out sharp dividing lines which may appear in the computational region. Figure 4-1 illustrates the effects of dissipation and dispersion on the computation of a discontinuity. In general, if the lowest-order term in the truncation error contains an even derivative, the resulting solution will predominately exhibit dissipative errors. On the other hand, if the leading term is an odd derivative, the resulting solution will predominately exhibit dispersive errors.

In Chapter 3 we discussed a technique for finding the relative errors in both amplitude (dissipation) and phase (dispersion) from the amplification factor. At this point it seems natural to ask if the amplification factor is related to the modified equation. The answer is definitely yes! Warming and Hyett (1974) have developed

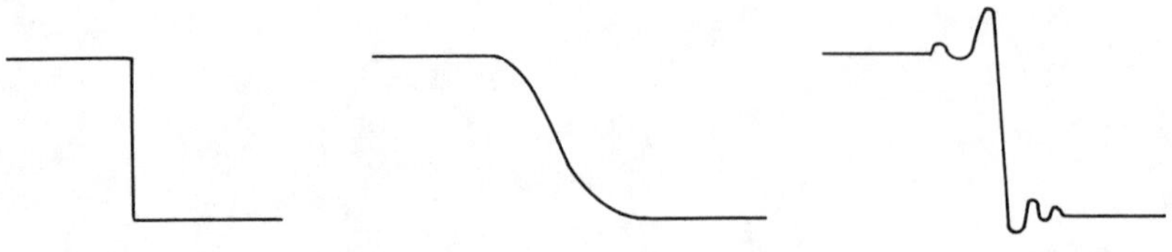

(a) (b) (c)

Figure 4-1 Effects of dissipation and dispersion. (a) Exact solution. (b) Numerical solution distorted primarily by dissipation errors (typical of first-order methods). (c) Numerical solution distorted primarily by dispersion errors (typical of second-order methods).

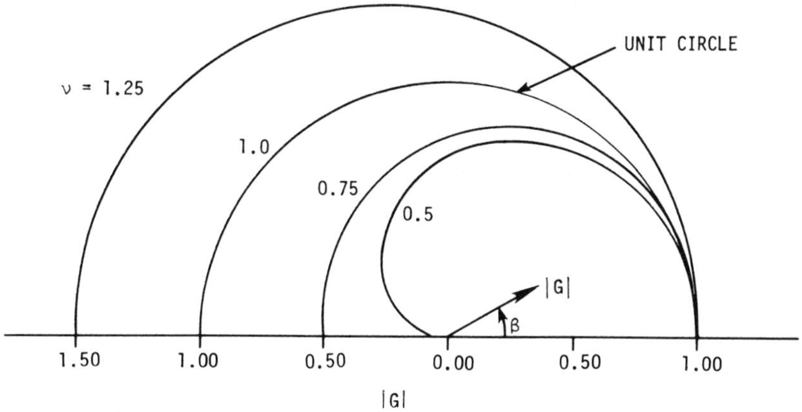

Figure 4-2 Amplification factor modulus for upstream differencing scheme.

a "heuristic" stability theory based on the even derivative terms in the modified equation and have determined the phase shift error by examining the odd derivative terms. Before showing the correspondence between the modified equation and the amplification factor, let us first examine the amplification factor of the present upstream differencing scheme

$$G = (1 - \nu + \nu \cos \beta) - i(\nu \sin \beta) \tag{4-17}$$

The modulus of this amplification factor

$$|G| = [(1 - \nu + \nu \cos \beta)^2 + (-\nu \sin \beta)^2]^{1/2}$$

is plotted in Fig. 4-2 for several values of ν. It is clear from this plot that ν must be less than or equal to 1 if the von Neumann stability condition $|G| \leqslant 1$ is to be met.

The amplification factor, Eq. (4-17), can also be expressed in the exponential form for a complex number

$$G = |G| e^{i\phi}$$

where ϕ is the phase angle given by

$$\phi = \tan^{-1} \frac{\text{Im}(G)}{\text{Re}(G)} = \tan^{-1} \left[\frac{-\nu \sin \beta}{1 - \nu + \nu \cos \beta} \right]$$

The phase angle for the exact solution of the wave equation (ϕ_e) is determined in a similar manner once the amplification factor of the exact solution is known. In order to find the exact amplification factor we substitute the elemental solution

$$u = e^{\alpha t} e^{i k_m x}$$

into the wave equation and find that $\alpha = -i k_m c$ which gives

$$u = e^{i k_m (x - ct)}$$

The exact amplification factor is then

$$G_e = \frac{u(t+\Delta t)}{u(t)} = \frac{e^{ik_m[x-c(t+\Delta t)]}}{e^{ik_m(x-ct)}}$$

which reduces to

$$G_e = e^{-ik_m c\,\Delta t} = e^{i\phi_e}$$

where

$$\phi_e = -k_m c\,\Delta t = -\beta\nu$$

and

$$|G_e| = 1$$

Thus, the total dissipation (amplitude) error which accrues from applying the upstream differencing method to the wave equation for n steps is given by

$$(1 - |G|^n)A_0$$

where A_0 is the initial amplitude of the wave. Likewise, the total dispersion (phase) error can be expressed as $n(\phi_e - \phi)$. The relative phase shift error after one time step is given by

$$\frac{\phi}{\phi_e} = \frac{\tan^{-1}[(-\nu\sin\beta)/(1-\nu+\nu\cos\beta)]}{-\beta\nu} \tag{4-18}$$

and is plotted in Fig. 4-3 for several values of ν. For small wave numbers (i.e., small β), the relative phase error reduces to

$$\frac{\phi}{\phi_e} \cong 1 - \frac{1}{6}(2\nu^2 - 3\nu + 1)\beta^2 \tag{4-19}$$

If the relative phase error exceeds 1 for a given value of β, the corresponding Fourier component of the numerical solution has a wave speed greater than the exact solution and this is a leading phase error. If the relative phase error is less than 1, the wave speed of the numerical solution is less than the exact wave speed and this is a lagging phase error. The upstream differencing scheme has a leading phase error for $0.5 < \nu < 1$ and a lagging phase error for $\nu < 0.5$.

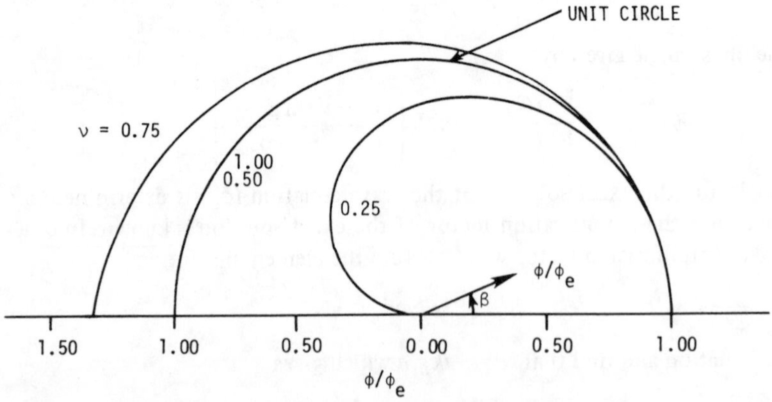

Figure 4-3 Relative phase error of upstream differencing scheme.

Example 4-1 Suppose the upstream differencing scheme is used to solve the wave equation ($c = 0.75$) with the initial condition

$$u(x, 0) = \sin(6\pi x) \qquad 0 \leqslant x \leqslant 1$$

and periodic boundary conditions. Determine the amplitude and phase errors after ten steps if $\Delta t = 0.02$ and $\Delta x = 0.02$.

Solution: In this problem, a unique value of β can be determined because the exact solution of the wave equation (for the present initial condition) is represented by a single term of a Fourier series. Since the amplification factor is also determined using a single term of a Fourier series which satisfies the wave equation, the frequency of the exact solution is identical to the frequency associated with the amplification factor, i.e., $f_m = k_m/2\pi$. Thus, the wave number for the present problem is given by

$$k_m = \frac{2m\pi}{2L} = \frac{6\pi}{1} = 6\pi$$

and β can be calculated as

$$\beta = k_m \, \Delta x = (6\pi)(0.02) = 0.12\pi$$

Using the Courant number,

$$\nu = \frac{c\,\Delta t}{\Delta x} = \frac{(0.75)(0.02)}{(0.02)} = 0.75$$

the modulus of the amplification factor becomes

$$|G| = [(1 - \nu + \nu \cos\beta)^2 + (-\nu \sin\beta)^2]^{1/2} = 0.986745$$

and the resulting amplitude error after ten steps is

$$(1 - |G|^n)A_0 = (1 - |G|^{10})(1) = 1 - 0.8751 = 0.1249$$

The phase angle (ϕ) after one step

$$\phi = \tan^{-1} \frac{-\nu \sin\beta}{1 - \nu + \nu\cos\beta} = -0.28359$$

can be compared with the exact phase angle (ϕ_e) after one step

$$\phi_e = -\beta\nu = -0.28274$$

to give the phase error after ten steps

$$10(\phi_e - \phi) = 0.0084465$$

Let us now compare the exact and numerical solutions after ten steps where the time is

$$t = 10\,\Delta t = 0.2$$

The exact solution is given by

$$u(x, 0.2) = \sin [6\pi(x - 0.15)]$$

and the numerical solution that results after applying the upstream differencing scheme for ten steps is

$$u(x, 0.2) = (0.8751) \sin [6\pi(x - 0.15) - 0.0084465]$$

In order to show the correspondence between the amplification factor and the modified equation, we write the modified equation [Eq. (4-15)] in the following form

$$u_t + cu_x = \sum_{n=1}^{\infty} \left(C_{2n} \frac{\partial^{2n} u}{\partial x^{2n}} + C_{2n+1} \frac{\partial^{2n+1} u}{\partial x^{2n+1}} \right) \tag{4-20}$$

where C_{2n} and C_{2n+1} represent, respectively, the coefficients of the even and odd spatial derivative terms. Warming and Hyett have shown that a necessary condition for stability is

$$(-1)^{l-1} C_{2l} > 0 \tag{4-21}$$

where C_{2l} represents the coefficient of the lowest-order even derivative term. This is analogous to the requirement that the coefficient of viscosity in viscous flow equations be greater than zero. In Eq. (4-15) the coefficient of the lowest-order even derivative term is

$$C_2 = \frac{c\,\Delta x}{2} (1 - \nu) \tag{4-22}$$

and therefore the stability condition becomes

$$\frac{c\,\Delta x}{2} (1 - \nu) > 0 \tag{4-23}$$

or $\nu < 1$ which was obtained earlier from the amplification factor. It should be remembered that the "heuristic" stability analysis, i.e. Eq. (4-21), can only provide a necessary condition for stability. Thus, for some finite-difference algorithms only partial information about the complete stability bound is obtained and for others (such as algorithms for the heat equation) a more complete theory must be employed.

Warming and Hyett have also shown that the relative phase error for difference schemes applied to the wave equation is given by

$$\frac{\phi}{\phi_e} = 1 - \frac{1}{c} \sum_{n=1}^{\infty} (-1)^n (k_m)^{2n} C_{2n+1} \tag{4-24}$$

where $k_m = \beta/\Delta x$ is the wave number. For small wave numbers, we need only retain the lowest-order term. For the upstream differencing scheme we find that

$$\frac{\phi}{\phi_e} \cong 1 - \frac{1}{c} (-1) \left(\frac{\beta}{\Delta x} \right)^2 C_3 = 1 - \frac{1}{6} (2\nu^2 - 3\nu + 1)\beta^2 \tag{4-25}$$

which is identical to Eq. (4-19). Thus, we have demonstrated that the amplification factor and the modified equation are directly related.

4-1.3 Lax Method

The Euler method, Eq. (4-6), can be made stable by replacing u_j^n with the averaged term $(u_{j+1}^n + u_{j-1}^n)/2$. The resulting algorithm is the well-known Lax method (Lax, 1954) which was presented earlier

$$\frac{u_j^{n+1} - (u_{j+1}^n + u_{j-1}^n)/2}{\Delta t} + c\,\frac{u_{j+1}^n - u_{j-1}^n}{2\,\Delta x} = 0 \tag{4-26}$$

This explicit, one-step scheme is first-order accurate with truncation error of $O[\Delta t, (\Delta x)^2/\Delta t]$ and is stable if $|\nu| \leqslant 1$. The modified equation is given by

$$u_t + cu_x = \frac{c\,\Delta x}{2}\left(\frac{1}{\nu} - \nu\right) u_{xx} + \frac{c(\Delta x)^2}{3}(1-\nu^2)u_{xxx} + \cdots \tag{4-27}$$

Note that this method is not uniformly consistent since $(\Delta x)^2/\Delta t$ may not approach zero in the limit as Δt and Δx go to zero. However, if ν is held constant as Δt and Δx approach zero, the method is consistent. The Lax method is known for its large dissipation error when $\nu \neq 1$. This large dissipation is readily apparent when we compare the coefficient of the u_{xx} term in Eq. (4-27) with the same coefficient in the modified equation of the upstream differencing scheme for various values of ν. The large dissipation can also be observed in the amplification factor

$$G = \cos\beta - i\nu\sin\beta \tag{4-28}$$

which was described in Section 3-6.1. The modulus of the amplification factor is plotted in Fig. 4-4a. The relative phase error is given by

$$\frac{\phi}{\phi_e} = \frac{\tan^{-1}(-\nu\tan\beta)}{-\beta\nu}$$

which produces a leading phase error as seen in Fig. 4-4b.

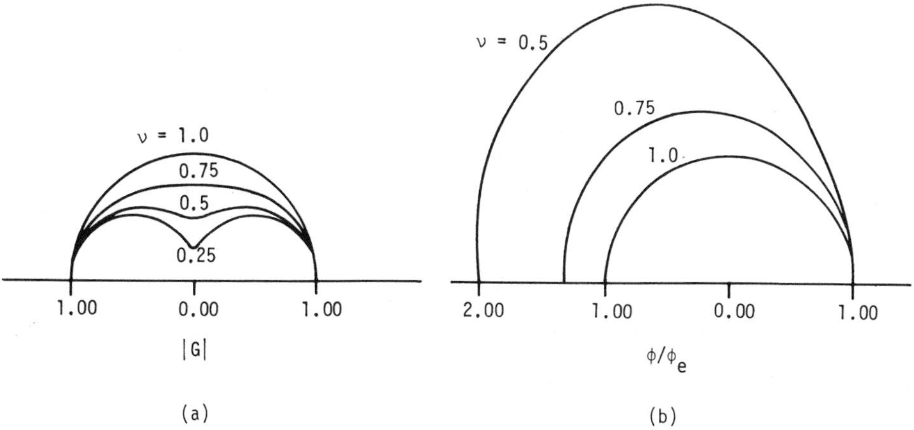

Figure 4-4 Lax method. (a) Amplification factor modulus. (b) Relative phase error.

4-1.4 Euler Implicit Method

The algorithms discussed previously have all been explicit. The following implicit scheme

$$\frac{u_j^{n+1} - u_j^n}{\Delta t} + \frac{c}{2\,\Delta x}\left(u_{j+1}^{n+1} - u_{j-1}^{n+1}\right) = 0 \tag{4-29}$$

is first-order accurate with truncation error of $O[\Delta t, (\Delta x)^2]$ and, according to a Fourier stability analysis, is unconditionally stable for all time steps. However, a system of algebraic equations must be solved at each new time level. To illustrate this, let us rewrite Eq. (4-29) so that the unknowns at time level $(n+1)$ appear on the left-hand side of the equation and the known quantity u_j^n appears on the right-hand side. This gives

$$\frac{\nu}{2}\,u_{j+1}^{n+1} + (1)u_j^{n+1} - \frac{\nu}{2}\,u_{j-1}^{n+1} = u_j^n \tag{4-30}$$

or

$$a u_{j+1}^{n+1} + d u_j^{n+1} + b u_{j-1}^{n+1} = C \tag{4-31}$$

where $a = \nu/2$, $d = 1$, $b = -\nu/2$, and $C = u_j^n$. Consider the computational mesh shown in Fig. 4-5, which contains $M + 2$ grid points in the x direction and known initial conditions at $n = 0$. Along the left boundary, u_0^{n+1} has a fixed value of u_0. Along the right boundary, u_{M+1}^{n+1} can be computed as part of the solution using characteristic theory. For example, if $\nu = 1$, then $u_{M+1}^{n+1} = u_M^n$. Applying Eq. (4-31) to the grid shown in Fig. 4-5, we find that the following system of M linear algebraic equations must be solved at each $(n + 1)$ time level:

$$\underbrace{\begin{bmatrix} d_1 & a_1 & 0 & \cdot & \cdot & \cdot & \cdot & 0 \\ b_2 & d_2 & a_2 & & & & & \cdot \\ 0 & b_3 & d_3 & a_3 & & & & \cdot \\ \cdot & & \ddots & \ddots & \ddots & & & \cdot \\ \cdot & & & \ddots & \ddots & \ddots & & 0 \\ \cdot & & & & & b_{M-1} & d_{M-1} & a_{M-1} \\ 0 & \cdot & \cdot & \cdot & \cdot & 0 & b_M & d_M \end{bmatrix}}_{[A]} \underbrace{\begin{bmatrix} u_1^{n+1} \\ u_2^{n+1} \\ \cdot \\ \cdot \\ \cdot \\ u_{M-1}^{n+1} \\ u_M^{n+1} \end{bmatrix}}_{[u]} = \underbrace{\begin{bmatrix} C_1 \\ C_2 \\ \cdot \\ \cdot \\ \cdot \\ C_{M-1} \\ C_M \end{bmatrix}}_{[C]} \tag{4-32}$$

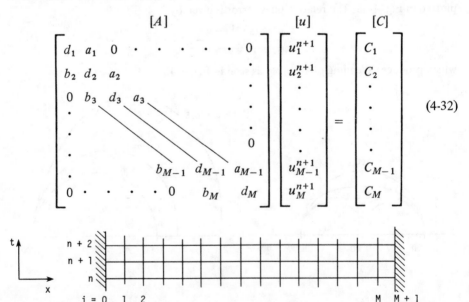

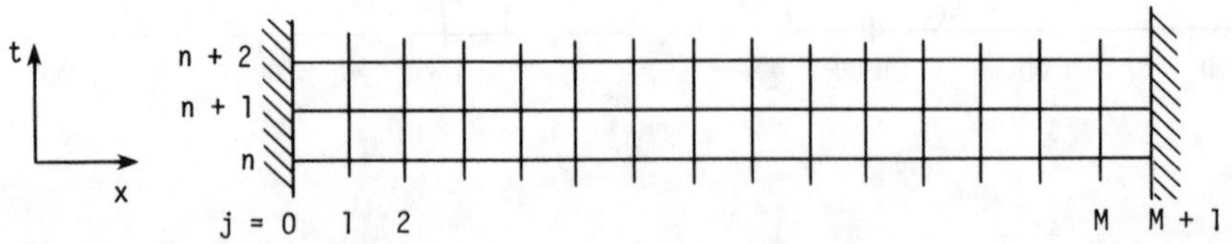

Figure 4-5 Computational mesh.

In Eq. (4-32), C_1 and C_M are given by

$$
\begin{aligned}
C_1 &= u_1^n - b u_0^{n+1} \\
C_M &= u_M^n - a u_{M+1}^{n+1}
\end{aligned}
\tag{4-33}
$$

where u_0^{n+1} and u_{M+1}^{n+1} are the known boundary conditions.

Matrix $[A]$ in Eq. (4-32) is a tridiagonal matrix. A technique for rapidly solving a tridiagonal system of linear algebraic equations is due to Thomas (1949) and is called the Thomas algorithm. In this algorithm, the system of equations is put into upper triangular form by replacing the diagonal elements d_i with

$$d_i - \frac{b_i}{d_{i-1}} a_{i-1} \qquad i = 2, 3, \ldots, M$$

and the C_i's with

$$C_i - \frac{b_i}{d_{i-1}} C_{i-1} \qquad i = 2, 3, \ldots, M$$

The unknowns are then computed using back substitution starting with

$$u_M^{n+1} = \frac{C_M}{d_M}$$

and continuing with

$$u_j^{n+1} = \frac{C_j - a_j u_{j+1}^{n+1}}{d_j} \qquad j = M-1, M-2, \ldots, 1$$

Further details of the Thomas algorithm are given in Section 4-3.3.

In general, implicit schemes require more computation time per time step, but of course permit a larger time step since they are usually unconditionally stable. However, the solution may become meaningless if too large a time step is taken. This is due to the fact that a large time step produces large truncation errors. The modified equation for the Euler implicit scheme is

$$u_t + c u_x = (\tfrac{1}{2} c^2 \Delta t) u_{xx} - [\tfrac{1}{6} c (\Delta x)^2 + \tfrac{1}{3} c^3 (\Delta t)^2] u_{xxx} + \cdots \tag{4-34}$$

which does not satisfy the shift condition. The amplification factor

$$G = \frac{1 - i\nu \sin \beta}{1 + \nu^2 \sin^2 \beta} \tag{4-35}$$

and the relative phase error

$$\frac{\phi}{\phi_e} = \frac{\tan^{-1}(-\nu \sin \beta)}{-\beta \nu} \tag{4-36}$$

are plotted in Fig. 4-6. The Euler implicit scheme is very dissipative for intermediate wave numbers and has a large lagging phase error for high wave numbers.

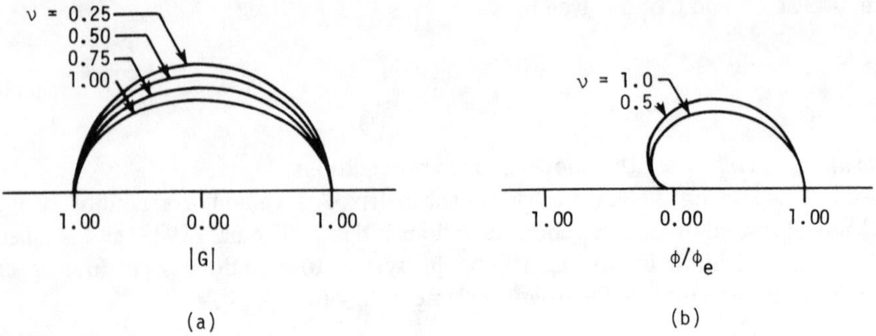

Figure 4-6 Euler implicit method. (a) Amplification factor modulus. (b) Relative phase error.

4-1.5 Leap Frog Method

The finite-difference schemes which have been presented in this chapter for solving the linear wave equation are all first-order accurate. In most cases, first-order schemes are not used to solve partial differential equations because of their inherent inaccuracy. The leap frog method is the simplest second-order accurate method. When applied to the first-order wave equation, this explicit, one-step, three-time level scheme becomes

$$\frac{u_j^{n+1} - u_j^{n-1}}{2\,\Delta t} + c\,\frac{u_{j+1}^n - u_{j-1}^n}{2\,\Delta x} = 0 \tag{4-37}$$

The leap frog method is referred to as a three-time level scheme since u must be known at time levels n and $n-1$ in order to find u at time level $n+1$. This method has a truncation error of $O[(\Delta t)^2, (\Delta x)^2]$ and is stable whenever $|\nu| \leqslant 1$. The modified equation is given by

$$u_t + cu_x = \frac{c(\Delta x)^2}{6}(\nu^2 - 1)u_{xxx} - \frac{c(\Delta x)^4}{120}(9\nu^4 - 10\nu^2 + 1)u_{xxxxx} + \cdots \tag{4-38}$$

The leading term in the truncation error contains the odd derivative u_{xxx} and hence the solution will predominantly exhibit dispersive errors. This is typical of second-order accurate methods. In this case, however, there are no even derivative terms in the modified equation so that the solution will not contain any dissipation error. As a consequence, the leap frog algorithm is neutrally stable and errors caused by improper boundary conditions or computer roundoff will not be damped (assuming periodic boundary conditions and $|\nu| \leqslant 1$). The amplification factor

$$G = \pm(1 - \nu^2 \sin^2 \beta)^{1/2} - i\nu \sin \beta \tag{4-39}$$

and the relative phase error

$$\frac{\phi}{\phi_e} = \frac{\tan^{-1}[-\nu \sin \beta/\pm(1 - \nu^2 \sin^2 \beta)^{1/2}]}{-\beta\nu} \tag{4-40}$$

are plotted in Fig. 4-7.

The leap frog method, while being second-order accurate with no dissipation error, does have its disadvantages. First of all, initial conditions must be specified at two-time levels. This difficulty can be circumvented by using a two-time level scheme for the first time step. A second disadvantage is due to the "leap frog" nature of the differencing (i.e., u_j^{n+1} does not depend on u_j^n) so that two independent solutions develop as the calculation proceeds. And finally, the leap frog method may require additional computer storage because it is a three-time level scheme. The required computer storage is reduced considerably if a simple overwriting procedure is employed whereby u_j^{n-1} is overwritten by u_j^{n+1}.

4-1.6 Lax-Wendroff Method

The Lax-Wendroff finite-difference scheme (Lax and Wendroff, 1960) can be derived from a Taylor-series expansion in the following manner:

$$u_j^{n+1} = u_j^n + \Delta t u_t + \tfrac{1}{2}(\Delta t)^2 u_{tt} + O[(\Delta t)^3] \tag{4-41}$$

Using the wave equations

$$\begin{aligned} u_t &= -cu_x \\ u_{tt} &= c^2 u_{xx} \end{aligned} \tag{4-42}$$

Equation (4-41) may be written as

$$u_j^{n+1} = u_j^n - c\,\Delta t u_x + \tfrac{1}{2}c^2(\Delta t)^2 u_{xx} + O[(\Delta t)^3] \tag{4-43}$$

And finally, if u_x and u_{xx} are replaced by second-order accurate central-difference expressions, the well-known Lax-Wendroff scheme is obtained:

$$u_j^{n+1} = u_j^n - \frac{c\,\Delta t}{2\,\Delta x}(u_{j+1}^n - u_{j-1}^n) + \frac{c^2(\Delta t)^2}{2(\Delta x)^2}(u_{j+1}^n - 2u_j^n + u_{j-1}^n) \tag{4-44}$$

This explicit, one-step scheme is second-order accurate with a truncation error of $O[(\Delta x)^2, (\Delta t)^2]$ and is stable whenever $|\nu| \leqslant 1$. The modified equation for this method is

$$u_t + cu_x = -c\frac{(\Delta x)^2}{6}(1-\nu^2)u_{xxx} - \frac{c(\Delta x)^3}{8}\nu(1-\nu^2)u_{xxxx} + \cdots \tag{4-45}$$

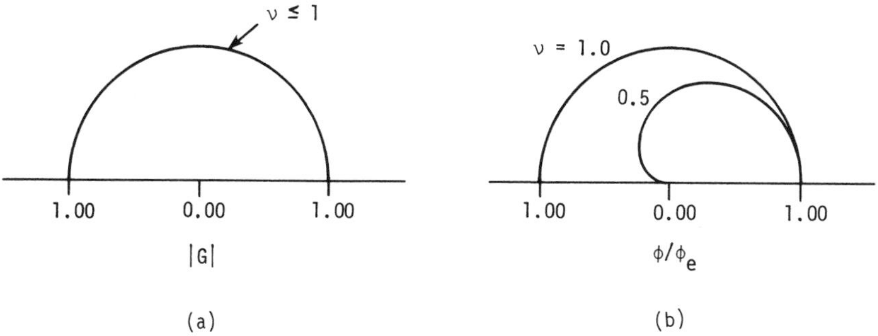

Figure 4-7 Leap frog method. (a) Amplification factor modulus. (b) Relative phase error.

The amplification factor

$$G = 1 - \nu^2(1 - \cos\beta) - i\nu\sin\beta \tag{4-46}$$

and the relative phase error

$$\frac{\phi}{\phi_e} = \frac{\tan^{-1}\{-\nu\sin\beta/[1-\nu^2(1-\cos\beta)]\}}{-\beta\nu} \tag{4-47}$$

are plotted in Fig. 4-8. The Lax-Wendroff scheme has a predominantly lagging phase error except for large wave numbers with $\sqrt{0.5} < \nu < 1$.

4-1.7 Two-Step Lax-Wendroff Method

For nonlinear equations such as the inviscid flow equations, a two-step variation of the original Lax-Wendroff method can be used. When applied to the wave equation, this explicit, two-step three-time level method becomes

Step 1:
$$\frac{u_{j+1/2}^{n+1/2} - (u_{j+1}^n + u_j^n)/2}{\Delta t/2} + c\,\frac{u_{j+1}^n - u_j^n}{\Delta x} = 0 \tag{4-48}$$

Step 2:
$$\frac{u_j^{n+1} - u_j^n}{\Delta t} + c\,\frac{u_{j+1/2}^{n+1/2} - u_{j-1/2}^{n+1/2}}{\Delta x} = 0 \tag{4-49}$$

This scheme is second-order accurate with a truncation error of $O[(\Delta x)^2, (\Delta t)^2]$ and is stable whenever $|\nu| \leqslant 1$. Step 1 is the Lax method applied at the midpoint $j + \frac{1}{2}$ for a half time step and Step 2 is the leap frog method for the remaining half time step. When applied to the linear wave equation, the two-step Lax-Wendroff scheme is equivalent to the original Lax-Wendroff scheme. This can be readily shown by substituting Eq. (4-48) into Eq. (4-49). Since the two schemes are equivalent, it follows that the modified equation and the amplification factor are the same for the two methods.

4-1.8 MacCormack Method

The MacCormack finite-difference method (MacCormack, 1969) is a widely used scheme for solving fluid flow equations. It is a variation of the two-step Lax-Wendroff

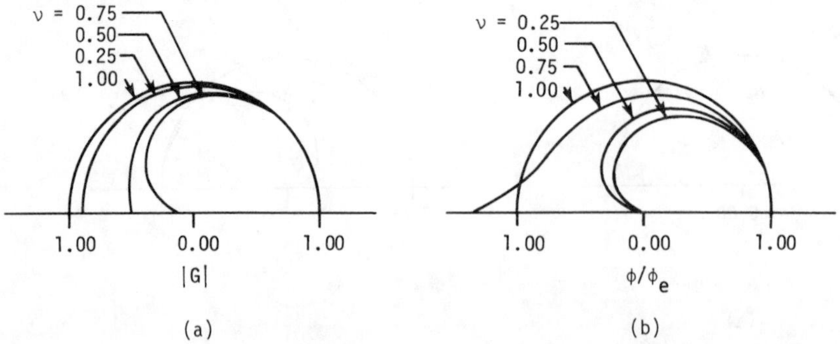

Figure 4-8 Lax-Wendroff method. (a) Amplification factor modulus. (b) Relative phase error.

scheme that removes the necessity of computing unknowns at the grid points $j+\frac{1}{2}$ and $j-\frac{1}{2}$. Because of this feature, the MacCormack method is particularly useful when solving nonlinear PDE's as will be shown in Section 4-4.3. When applied to the linear wave equation, this explicit, predictor-corrector method becomes

Predictor: $$\overline{u_j^{n+1}} = u_j^n - c\frac{\Delta t}{\Delta x}(u_{j+1}^n - u_j^n) \tag{4-50}$$

Corrector: $$u_j^{n+1} = \frac{1}{2}\left[u_j^n + \overline{u_j^{n+1}} - c\frac{\Delta t}{\Delta x}\left(\overline{u_j^{n+1}} - \overline{u_{j-1}^{n+1}}\right)\right] \tag{4-51}$$

The term $\overline{u_j^{n+1}}$ is a temporary "predicted" value of u at the time level $n+1$. The corrector equation provides the final value of u at the time level $n+1$. Note that in the predictor equation a forward difference is used for $\partial u/\partial x$, while in the corrector equation a backward difference is used. This differencing can be reversed and in some problems it is advantageous to do so. This is particularly true for problems involving moving discontinuities. For the present linear wave equation, the MacCormack scheme is equivalent to the original Lax-Wendroff scheme. Hence, the truncation error, stability limit, modified equation, and amplification factor are identical with those of the Lax-Wendroff scheme.

4-1.9 Upwind Method

The upwind method (Warming and Beam, 1975) is a variation of the MacCormack method which uses backward (upwind) differences in both the predictor and corrector steps for $c > 0$.

Predictor: $$\overline{u_j^{n+1}} = u_j^n - \frac{c\,\Delta t}{\Delta x}(u_j^n - u_{j-1}^n) \tag{4-52}$$

Corrector:

$$u_j^{n+1} = \frac{1}{2}\left[u_j^n + \overline{u_j^{n+1}} - \frac{c\,\Delta t}{\Delta x}\left(\overline{u_j^{n+1}} - \overline{u_{j-1}^{n+1}}\right) - \frac{c\,\Delta t}{\Delta x}(u_j^n - 2u_{j-1}^n + u_{j-2}^n)\right] \tag{4-53}$$

The addition of the second backward difference in Eq. (4-53) makes this scheme second-order accurate with truncation error of $O[(\Delta t)^2, (\Delta t)(\Delta x), (\Delta x)^2]$. If Eq. (4-52) is substituted into Eq. (4-53) the following one-step algorithm is obtained:

$$u_j^{n+1} = u_j^n - \nu(u_j^n - u_{j-1}^n) + \tfrac{1}{2}\nu(\nu-1)(u_j^n - 2u_{j-1}^n + u_{j-2}^n) \tag{4-54}$$

The modified equation for this scheme is

$$u_t + cu_x = \frac{c(\Delta x)^2}{6}(1-\nu)(2-\nu)u_{xxx} - \frac{(\Delta x)^4}{8\,\Delta t}\nu(1-\nu)^2(2-\nu)u_{xxxx} + \cdots \tag{4-55}$$

The upwind method satisfies the shift condition for both $\nu = 1$ and $\nu = 2$. The amplification factor is

$$G = 1 - 2\nu \left[\nu + 2(1-\nu)\sin^2\frac{\beta}{2} \right] \sin^2\frac{\beta}{2} - i\nu \sin\beta \left[1 + 2(1-\nu)\sin^2\frac{\beta}{2} \right] \tag{4-56}$$

and the resulting stability condition becomes $0 \leqslant \nu \leqslant 2$. The modulus of the amplification factor and the relative phase error are plotted in Fig. 4-9. The upwind method has a predominantly leading phase error for $0 < \nu < 1$ and a predominantly lagging phase error for $1 < \nu < 2$. We observe that the upwind method and the Lax-Wendroff method have opposite phase errors for $0 < \nu < 1$. This suggests that a considerable reduction in dispersive error would occur if a linear combination of the two methods were used. Fromm's method of zero-average phase error (Fromm, 1968) is based on this observation.

4-1.10 Time-Centered Implicit Method (Trapezoidal Differencing Method)

A second-order accurate implicit scheme can be obtained if the two Taylor-series expansions

$$\begin{aligned} u_j^{n+1} &= u_j^n + \Delta t (u_t)_j^n + \frac{(\Delta t)^2}{2}(u_{tt})_j^n + \frac{(\Delta t)^3}{6}(u_{ttt})_j^n + \cdots \\ u_j^n &= u_j^{n+1} - \Delta t (u_t)_j^{n+1} + \frac{(\Delta t)^2}{2}(u_{tt})_j^{n+1} - \frac{(\Delta t)^3}{6}(u_{ttt})_j^{n+1} + \cdots \end{aligned} \tag{4-57}$$

are subtracted and $(u_{tt})_j^{n+1}$ is replaced with

$$(u_{tt})_j^{n+1} = (u_{tt})_j^n + \Delta t (u_{ttt})_j^n + \cdots$$

The resulting expression becomes

$$u_j^{n+1} = u_j^n + \frac{\Delta t}{2}\left[(u_t)^n + (u_t)^{n+1}\right]_j + O[(\Delta t)^3] \tag{4-58}$$

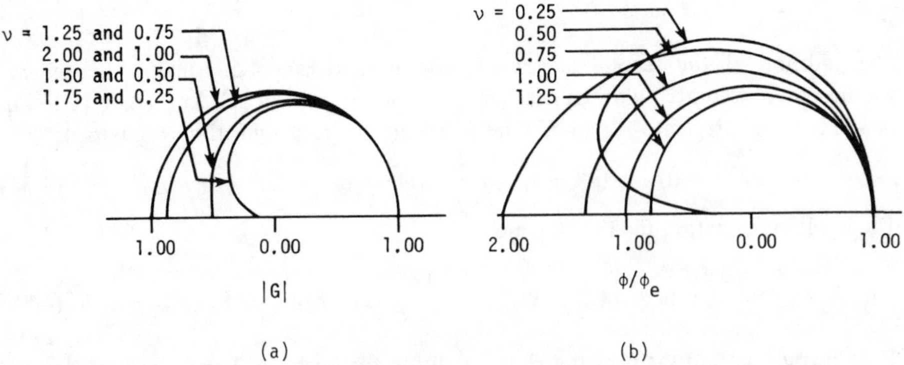

Figure 4-9 Upwind method. (a) Amplification factor modulus. (b) Relative phase error.

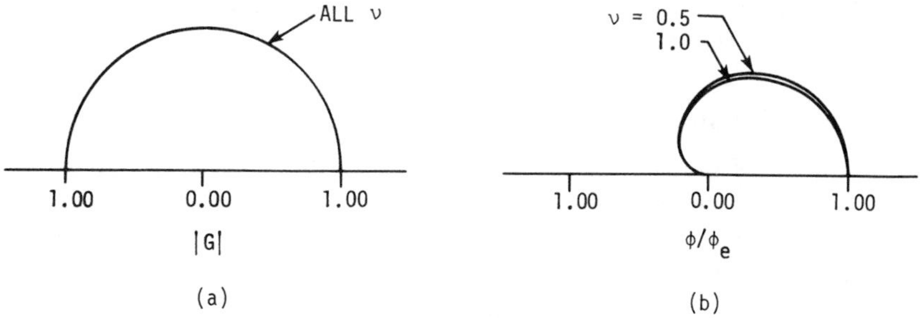

Figure 4-10 Time-centered implicit method. (a) Amplification factor modulus. (b) Relative phase error.

The time differencing in this equation is known as trapezoidal differencing or Crank-Nicolson differencing. Upon substituting the linear wave equation $u_t = -cu_x$ we obtain

$$u_j^{n+1} = u_j^n - \frac{c\,\Delta t}{2}\,[(u_x)^n + (u_x)^{n+1}]_j + O[(\Delta t)^3] \tag{4-59}$$

And finally, if the u_x terms are replaced by second-order central differences, the time-centered implicit method results

$$u_j^{n+1} = u_j^n - \frac{\nu}{4}\,(u_{j+1}^{n+1} + u_{j+1}^n - u_{j-1}^{n+1} - u_{j-1}^n) \tag{4-60}$$

This method has second-order accuracy with truncation error of $O[(\Delta x)^2, (\Delta t)^2]$ and is unconditionally stable for all time steps. However, a tridiagonal matrix must be solved at each new time level. The modified equation for this scheme is

$$u_t + cu_x = -\left[\frac{c^3(\Delta t)^2}{12} + \frac{c(\Delta x)^2}{6}\right]u_{xxx} - \left[\frac{c(\Delta x)^4}{120} + \frac{c^3(\Delta t)^2(\Delta x)^2}{24} + \frac{c^4(\Delta t)^4}{80}\right]u_{xxxxx} + \cdots \tag{4-61}$$

Note that the modified equation contains no even derivative terms so that the scheme has no implicit artificial viscosity. When this scheme is applied to the nonlinear fluid dynamic equations it often becomes necessary to add some explicit artificial viscosity to prevent the solution from "blowing up." The addition of explicit artificial viscosity (i.e., "smoothing" term) to this scheme will be discussed in Section 4-4.7. The modulus of the amplification factor

$$G = \frac{1 - (i\nu/2)\sin\beta}{1 + (i\nu/2)\sin\beta} \tag{4-62}$$

and the relative phase error are plotted in Fig. 4-10.

The time-centered implicit method can be made fourth-order accurate in space if the difference approximation given by Eq. (3-31) is used for u_x

$$(u_x)_j = \frac{1}{2\,\Delta x}\,\frac{\bar{\delta}_x}{1 + \delta_x^2/6}\,u_j + O[(\Delta x)^4] \tag{4-63}$$

The modified equation and phase error diagram for the resulting scheme can be found in Beam and Warming (1976).

4-1.11 Rusanov (Burstein-Mirin) Method

The methods presented thus far for solving the wave equation have either been first-order or second-order accurate. Only a small number of third-order methods have appeared in the literature. Rusanov (1970) and Burstein and Mirin (1970) simultaneously developed the following explicit, three-step method:

Step 1: $$u_{j+1/2}^{(1)} = \tfrac{1}{2}(u_{j+1}^n + u_j^n) - \tfrac{1}{3}\nu(u_{j+1}^n - u_j^n)$$

Step 2: $$u_j^{(2)} = u_j^n - \tfrac{2}{3}\nu(u_{j+1/2}^{(1)} - u_{j-1/2}^{(1)})$$

Step 3: $$u_j^{n+1} = u_j^n - \frac{1}{24}\nu(-2u_{j+2}^n + 7u_{j+1}^n - 7u_{j-1}^n + 2u_{j-2}^n) - \frac{3}{8}\nu(u_{j+1}^{(2)} - u_{j-1}^{(2)})$$

$$- \frac{\omega}{24}(u_{j+2}^n - 4u_{j+1}^n + 6u_j^n - 4u_{j-1}^n + u_{j-2}^n) \tag{4-64}$$

Step 3 contains the fourth-order difference term

$$\delta_x^4 u_j^n = u_{j+2}^n - 4u_{j+1}^n + 6u_j^n - 4u_{j-1}^n + u_{j-2}^n$$

which is multiplied by a free parameter ω. This term has been added to make the scheme stable. The need for this term is apparent when we examine the stability requirements for the scheme:

$$|\nu| \leqslant 1$$

and

$$4\nu^2 - \nu^4 \leqslant \omega \leqslant 3 \tag{4-65}$$

If the fourth-order difference term were not present (i.e., $\omega = 0$) we could not satisfy Eq. (4-65) for $0 < \nu \leqslant 1$. The modified equation for this method is

$$u_t + cu_x = -\frac{c(\Delta x)^3}{24}\left(\frac{\omega}{\nu} - 4\nu + \nu^3\right)u_{xxxx}$$

$$+ \frac{c(\Delta x)^4}{120}(-5\omega + 4 + 15\nu^2 - 4\nu^4)u_{xxxxx} + \cdots \tag{4-66}$$

In order to reduce the dissipation of this scheme, we can make the coefficient of the fourth derivative equal to zero by letting

$$\omega = 4\nu^2 - \nu^4 \tag{4-67}$$

In a like manner, we can reduce the dispersive error by setting the coefficient of the fifth derivative to zero which gives

$$\omega = \frac{(4\nu^2 + 1)(4 - \nu^2)}{5} \tag{4-68}$$

The amplification factor for this method is

$$G = 1 - \frac{\nu^2}{2}\sin^2\beta - \frac{2\omega}{3}\sin^4\frac{\beta}{2} - i\nu\sin\beta\left[1 + \frac{2}{3}(1 - \nu^2)\sin^2\frac{\beta}{2}\right] \tag{4-69}$$

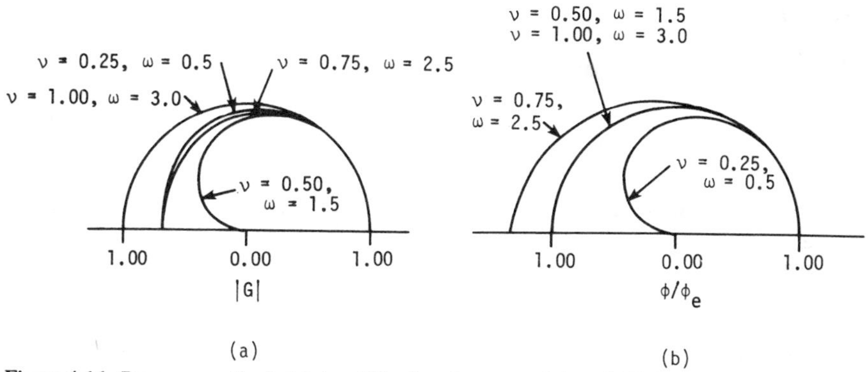

Figure 4-11 Rusanov method. (a) Amplification factor modulus. (b) Relative phase error.

The modulus of the amplification factor and the relative phase error are plotted in Fig. 4-11. This figure shows that the Rusanov method has a leading or a lagging phase error depending on the value of the free parameter ω.

4-1.12 Warming-Kutler-Lomax Method

Warming et al. (1973) developed a third-order method which uses MacCormack's method for the first two steps and has the same third step as the Rusanov method. This so-called WKL method is given by

Step 1: $$u_j^{(1)} = u_j^n - \tfrac{2}{3}\nu(u_{j+1}^n - u_j^n)$$

Step 2: $$u_j^{(2)} = \tfrac{1}{2}[u_j^n + u_j^{(1)} - \tfrac{2}{3}\nu(u_j^{(1)} - u_{j-1}^{(1)})]$$

Step 3: $$u_j^{n+1} = u_j^n - \frac{1}{24}\nu(-2u_{j+2}^n + 7u_{j+1}^n - 7u_{j-1}^n + 2u_{j-2}^n) - \frac{3}{8}\nu(u_{j+1}^{(2)} - u_{j-1}^{(2)})$$
$$- \frac{\omega}{24}(u_{j+2}^n - 4u_{j+1}^n + 6u_j^n - 4u_{j-1}^n + u_{j-2}^n) \qquad (4\text{-}70)$$

This method has the same stability bounds as the Rusanov method. In addition, the modified equation is identical to Eq. (4-66) for the present first-order wave equation. The WKL method has the same advantage over the Rusanov method that the MacCormack method has over the two-step Lax-Wendroff method.

The improved accuracy of third-order methods is at the expense of added computer time and additional complexity. These factors must be considered carefully when choosing a finite-difference scheme to solve a partial differential equation. In general, second-order accurate methods provide enough accuracy for most practical problems.

For the one-dimensional wave equation, the second-order accurate explicit schemes such as the Lax-Wendroff and upwind schemes give excellent results with a minimum of computational effort. An implicit scheme is probably not the optimum choice in this case because the solution is unsteady and intermediate results are typically desired at relatively small intervals.

4-2 HEAT EQUATION

The one-dimensional heat equation (diffusion equation)

$$\frac{\partial u}{\partial t} = \alpha \frac{\partial^2 u}{\partial x^2} \tag{4-71}$$

is a parabolic partial differential equation. In its present form, it is the governing equation for heat conduction or diffusion in a one-dimensional, isotropic medium. It can be used to "model" in a rudimentary fashion the parabolic boundary-layer equations. The exact solution of the heat equation for the initial condition

$$u(x, 0) = f(x)$$

and boundary conditions

$$u(0, t) = u(1, t) = 0$$

is

$$u(x, t) = \sum_{n=1}^{\infty} A_n e^{-\alpha k^2 t} \sin(kx) \tag{4-72}$$

where

$$A_n = 2 \int_0^1 f(x) \sin(kx)\, dx$$

and $k = n\pi$. Let us now examine some of the more important finite-difference algorithms which can be used to solve the heat equation.

4-2.1 Simple Explicit Method

The following explicit, one-step method

$$\frac{u_j^{n+1} - u_j^n}{\Delta t} = \alpha \frac{u_{j+1}^n - 2u_j^n + u_{j-1}^n}{(\Delta x)^2} \tag{4-73}$$

is first-order accurate with truncation error of $O[\Delta t, (\Delta x)^2]$. At steady state the accuracy is $O[(\Delta x)^2]$. As we have shown earlier, this scheme is stable whenever

$$0 \leqslant r \leqslant \tfrac{1}{2} \tag{4-74}$$

where

$$r = \frac{\alpha \Delta t}{(\Delta x)^2} \tag{4-75}$$

The modified equation is given by

$$u_t - \alpha u_{xx} = \left[-\frac{1}{2}\alpha^2 \Delta t + \frac{\alpha(\Delta x)^2}{12}\right] u_{xxxx} + \left[\frac{1}{3}\alpha^3 (\Delta t)^2 - \frac{1}{12}\alpha^2 \Delta t(\Delta x)^2 + \frac{1}{360}\alpha(\Delta x)^4\right] u_{xxxxxx} + \cdots \tag{4-76}$$

We note that if $r = \frac{1}{6}$ the truncation error becomes of $O[(\Delta t)^2, (\Delta x)^4]$. It is also interesting to note that no odd derivative terms appear in the truncation error. As a consequence, this scheme, as well as almost all other schemes for the heat equation, has no dispersive error. This fact can also be ascertained by examining the amplification factor for this scheme

$$G = 1 + 2r(\cos \beta - 1) \tag{4-77}$$

which has no imaginary part and hence no phase shift. The amplification factor is plotted in Fig. 4-12 for two values of r and is compared with the exact amplification factor of the solution. The exact amplification (decay) factor is obtained by substituting the elemental solution

$$u = e^{-\alpha k_m^2 t} e^{ik_m x}$$

into

$$G_e = \frac{u(t + \Delta t)}{u(t)}$$

which gives

$$G_e = e^{-\alpha k_m^2 \Delta t} \tag{4-78}$$

or

$$G_e = e^{-r\beta^2} \tag{4-79}$$

where $\beta = k_m \Delta x$. Hence the amplitude of the exact solution decreases by the factor $e^{-r\beta^2}$ during one time step assuming no boundary condition influence.

In Fig. 4-12, we observe that the simple explicit method is highly dissipative for large values of β when $r = \frac{1}{2}$. As expected, the amplification factor agrees closer with the exact decay factor when $r = \frac{1}{6}$.

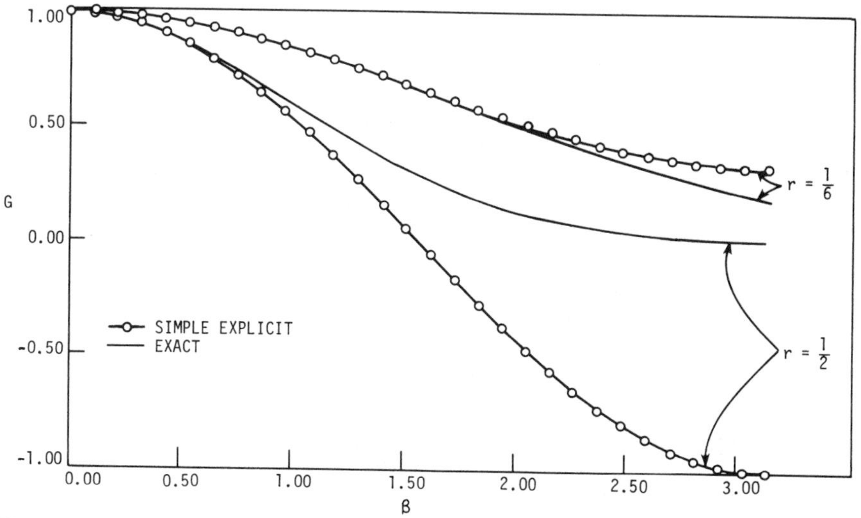

Figure 4-12 Amplification factor for simple explicit method.

The present simple explicit scheme marches the solution outward from the initial data line in much the same manner as the explicit schemes of the previous section. This is illustrated in Fig. 4-13. In this figure we see that the unknown u can be calculated at point P without any knowledge of the boundary conditions along AB and CD. We know, however, that point P should depend on the boundary conditions along AB and CD since the parabolic heat equation has the characteristic $t = \text{constant}$. From this we conclude that the present explicit scheme (with a finite Δt) does not properly model the physical behavior of a parabolic PDE. It would appear that an implicit method would be the more appropriate method for solving a parabolic PDE since an implicit method normally assimilates information from all grid points located on or below the characteristic $t = \text{constant}$. On the other hand, explicit schemes seem to provide a more natural finite-difference approximation for hyperbolic partial differential equations which possess limited zones of influence.

Example 4-2 Suppose the simple explicit method is used to solve the heat equation ($\alpha = 0.05$) with the initial condition

$$u(x, 0) = \sin(2\pi x) \qquad 0 \leqslant x \leqslant 1$$

and periodic boundary conditions. Determine the amplitude error after ten steps if $\Delta t = 0.1$ and $\Delta x = 0.1$.

Solution: A unique value of β can be determined in this problem for the same reason that was given in Example 4-1. Thus, the value of β becomes

$$\beta = k_m \, \Delta x = (2\pi)(0.1) = 0.2\pi$$

After computing r,

$$r = \frac{\alpha \, \Delta t}{(\Delta x)^2} = \frac{(0.05)(0.1)}{(0.1)^2} = 0.5$$

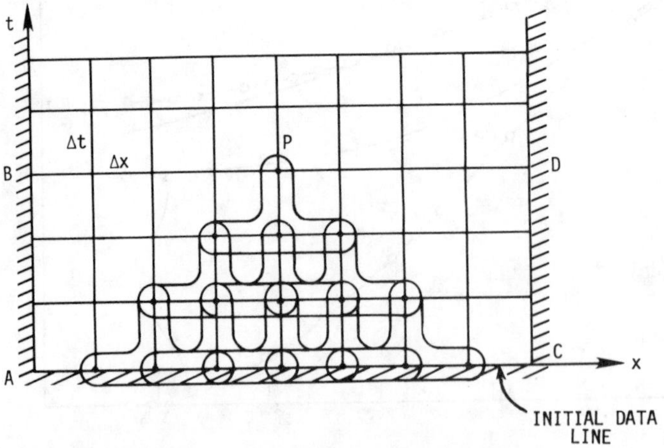

Figure 4-13 Zone of influence of simple explicit scheme.

the amplification factor for the simple explicit method is given by

$$G = 1 + 2r(\cos\beta - 1) = 0.809017$$

while the exact amplification factor becomes

$$G_e = e^{-r\beta^2} = 0.820869$$

As a result, the amplitude error is

$$A_0 |G_e^{10} - G^{10}| = (1)(0.1389 - 0.1201) = 0.0188$$

Using Eq. (4-72), the exact solution after ten steps ($t = 1.0$) is given by

$$u(x, 1) = e^{-\alpha 4\pi^2} \sin(2\pi x) = 0.1389 \sin(2\pi x)$$

which can be compared to the numerical solution

$$u(x, 1) = 0.1201 \sin(2\pi x)$$

4-2.2 Richardson's Method

Richardson (1910) proposed the following explicit, one-step, three-time level scheme for solving the heat equation:

$$\frac{u_j^{n+1} - u_j^{n-1}}{2\,\Delta t} = \alpha \frac{u_{j+1}^n - 2u_j^n + u_{j-1}^n}{(\Delta x)^2} \tag{4-80}$$

This scheme is second-order accurate with truncation error of $O[(\Delta t)^2, (\Delta x)^2]$. Unfortunately, this method proves to be unconditionally unstable and cannot be used to solve the heat equation. It is presented here for historic purposes only.

4-2.3 Simple Implicit (Laasonen) Method

A simple implicit scheme for the heat equation was proposed by Laasonen (1949). The algorithm for this scheme is

$$\frac{u_j^{n+1} - u_j^n}{\Delta t} = \alpha \frac{u_{j+1}^{n+1} - 2u_j^{n+1} + u_{j-1}^{n+1}}{(\Delta x)^2} \tag{4-81}$$

If we make use of the central difference operator

$$\delta_x^2 u_j^n = u_{j+1}^n - 2u_j^n + u_{j-1}^n$$

we can rewrite Eq. (4-81) in the simpler form

$$\frac{u_j^{n+1} - u_j^n}{\Delta t} = \alpha \frac{\delta_x^2 u_j^{n+1}}{(\Delta x)^2} \tag{4-82}$$

This scheme has first-order accuracy with a truncation error of $O[\Delta t, (\Delta x)^2]$ and is unconditionally stable. Upon examining Eq. (4-82), it is apparent that a tridiagonal system of linear algebraic equations must be solved at each time level $n + 1$.

The modified equation for this scheme is

$$u_t - \alpha u_{xx} = \left[\frac{1}{2}\alpha^2 \Delta t + \frac{\alpha(\Delta x)^2}{12}\right] u_{xxxx} + \left[\frac{1}{3}\alpha^3(\Delta t)^2 + \frac{1}{12}\alpha^2 \Delta t(\Delta x)^2 + \frac{1}{360}\alpha(\Delta x)^4\right] u_{xxxxxx} + \cdots \tag{4-83}$$

The amplification factor

$$G = [1 + 2r(1 - \cos\beta)]^{-1} \tag{4-84}$$

is plotted in Fig. 4-14 for $r = \frac{1}{2}$ and is compared with the exact decay factor.

4-2.4 Crank-Nicolson Method

Crank and Nicolson (1947) used the following implicit algorithm to solve the heat equation

$$\frac{u_j^{n+1} - u_j^n}{\Delta t} = \alpha \frac{\delta_x^2 u_j^n + \delta_x^2 u_j^{n+1}}{2(\Delta x)^2} \tag{4-85}$$

This unconditionally stable algorithm has become very well-known and is referred to as the Crank-Nicolson scheme. This scheme makes use of trapezoidal differencing to

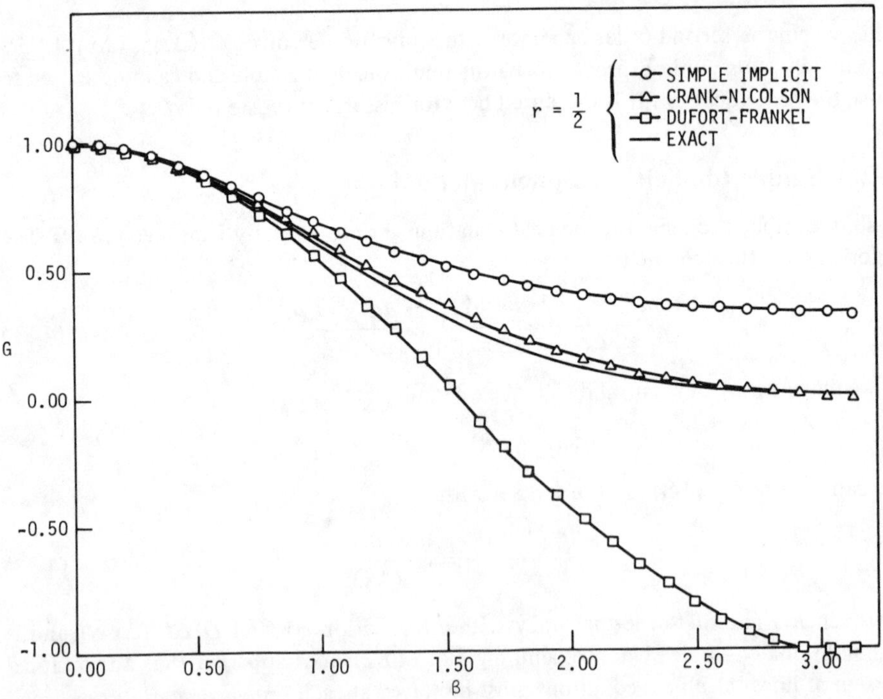

Figure 4-14 Amplification factors for several methods.

achieve second-order accuracy with a truncation error of $O[(\Delta t)^2, (\Delta x)^2]$. Once again, a tridiagonal system of linear algebraic equations must be solved at each time level $n+1$. The modified equation for the Crank-Nicolson method is

$$u_t - \alpha u_{xx} = \frac{\alpha(\Delta x)^2}{12} u_{xxxx} + \left[\frac{1}{12}\alpha^3(\Delta t)^2 + \frac{1}{360}\alpha(\Delta x)^4\right] u_{xxxxxx} + \cdots \tag{4-86}$$

The amplification factor

$$G = \frac{1 - r(1-\cos\beta)}{1 + r(1-\cos\beta)} \tag{4-87}$$

is plotted in Fig. 4-14 for $r = \frac{1}{2}$.

4-2.5 Combined Method A

The simple explicit, the simple implicit, and the Crank-Nicolson method are special cases of a general algorithm given by

$$\frac{u_j^{n+1} - u_j^n}{\Delta t} = \alpha \frac{\theta \delta_x^2 u_j^{n+1} + (1-\theta)\delta_x^2 u_j^n}{(\Delta x)^2} \tag{4-88}$$

where θ is a constant $(0 \leqslant \theta \leqslant 1)$. The simple explicit method corresponds to $\theta = 0$, the simple implicit method corresponds to $\theta = 1$, and the Crank-Nicolson method corresponds to $\theta = \frac{1}{2}$. This combined method has first-order accuracy with truncation error of $O[\Delta t, (\Delta x)^2]$ except for special cases such as:

(*a*) $\theta = \frac{1}{2}$ (Crank-Nicolson method) T.E. $= O[(\Delta t)^2, (\Delta x)^2]$

(*b*) $\theta = \frac{1}{2} - \frac{(\Delta x)^2}{12\alpha\,\Delta t}$ T.E. $= O[(\Delta t)^2, (\Delta x)^4]$

(*c*) $\theta = \frac{1}{2} - \frac{(\Delta x)^2}{12\alpha\,\Delta t}$ and $\frac{(\Delta x)^2}{\alpha\,\Delta t} = \sqrt{20}$ T.E. $= O[(\Delta t)^2, (\Delta x)^6]$

The truncation errors of these special cases can be obtained by examining the modified equation

$$u_t - \alpha u_{xx} = \left[\left(\theta - \frac{1}{2}\right)\alpha^2\,\Delta t + \frac{\alpha(\Delta x)^2}{12}\right] u_{xxxx} + \left[\left(\theta^2 - \theta + \frac{1}{3}\right)\alpha^3(\Delta t)^2 + \frac{1}{6}\left(\theta - \frac{1}{2}\right)\alpha^2\,\Delta t(\Delta x)^2 + \frac{1}{360}\alpha(\Delta x)^4\right] u_{xxxxxx} + \cdots \tag{4-89}$$

The present combined method is unconditionally stable if $\frac{1}{2} \leqslant \theta \leqslant 1$. However, when $0 \leqslant \theta < \frac{1}{2}$ the method is stable only if

$$0 \leqslant r \leqslant \frac{1}{2 - 4\theta} \tag{4-90}$$

4-2.6 Combined Method B

Richtmyer and Morton (1967) present the following general algorithm for a three-time level implicit scheme:

$$(1+\theta)\frac{u_j^{n+1}-u_j^n}{\Delta t}-\theta\,\frac{u_j^n-u_j^{n-1}}{\Delta t}=\alpha\,\frac{\delta_x^2 u_j^{n+1}}{(\Delta x)^2} \tag{4-91}$$

This general algorithm has first-order accuracy with truncation error of $O[\Delta t,(\Delta x)^2]$ except for special cases

(*a*) $\theta=\frac{1}{2}$ T.E. $=O[(\Delta t)^2,(\Delta x)^2]$

(*b*) $\theta=\dfrac{1}{2}+\dfrac{(\Delta x)^2}{12\alpha\,\Delta t}$ T.E. $=O[(\Delta t)^2,(\Delta x)^4]$

which can be verified by examining the modified equation

$$u_t-\alpha u_{xx}=[-(\theta-\tfrac{1}{2})\alpha^2\,\Delta t+\tfrac{1}{12}\alpha(\Delta x)^2]u_{xxxx}+\cdots \tag{4-92}$$

4-2.7 DuFort-Frankel Method

The unstable Richardson method [Eq. (4-80)] can be made stable by replacing u_j^n with the time-averaged expression $(u_j^{n+1}+u_j^{n-1})/2$. The resulting explicit, three-time level scheme

$$\frac{u_j^{n+1}-u_j^{n-1}}{2\,\Delta t}=\alpha\,\frac{u_{j+1}^n-u_j^{n+1}-u_j^{n-1}+u_{j-1}^n}{(\Delta x)^2} \tag{4-93}$$

was first proposed by DuFort and Frankel (1953). Note that Eq. (4-93) can be re-written as

$$u_j^{n+1}(1+2r)=u_j^{n-1}+2r(u_{j+1}^n-u_j^{n-1}+u_{j-1}^n) \tag{4-94}$$

so that only one unknown u_j^{n+1} appears in the scheme and therefore it is explicit. The truncation error for the DuFort-Frankel method is of $O[(\Delta t)^2,(\Delta x)^2,(\Delta t/\Delta x)^2]$. Consequently, if this method is to be consistent then $(\Delta t/\Delta x)^2$ must approach zero as Δt and Δx approach zero. As pointed out in Chapter 3, if $\Delta t/\Delta x$ approaches a constant value γ, instead of zero, the DuFort-Frankel scheme is consistent with the hyperbolic equation

$$\frac{\partial u}{\partial t}+\alpha\gamma^2\,\frac{\partial^2 u}{\partial t^2}=\alpha\,\frac{\partial^2 u}{\partial x^2}$$

If we let r remain constant as Δt and Δx approach zero, the term $(\Delta t/\Delta x)^2$ becomes formally a first-order term of $O(\Delta t)$. The modified equation is given by

$$u_t-\alpha u_{xx}=\left[\frac{1}{12}\,\alpha(\Delta x)^2-\alpha^3\,\frac{(\Delta t)^2}{(\Delta x)^2}\right]u_{xxxx}+\left[\frac{1}{360}\,\alpha(\Delta x)^4-\frac{1}{3}\,\alpha^3(\Delta t)^2+2\alpha^5\,\frac{(\Delta t)^4}{(\Delta x)^4}\right]u_{xxxxxx}+\cdots \tag{4-95}$$

The amplification factor

$$G = \frac{2r \cos\beta \pm \sqrt{1 - 4r^2 \sin^2\beta}}{1 + 2r} \tag{4-96}$$

is plotted in Fig. 4-14 for $r = \frac{1}{2}$. The explicit DuFort-Frankel scheme has the unusual property of being unconditionally stable for $r \geqslant 0$.

4-2.8 Methods for the Two-dimensional Heat Equation

The two-dimensional (2-D) heat equation is given by

$$\frac{\partial u}{\partial t} = \alpha \left(\frac{\partial^2 u}{\partial x^2} + \frac{\partial^2 u}{\partial y^2} \right) \tag{4-97}$$

Since this partial differential equation is different than the 1-D equation, caution must be exercised when attempting to apply the previous finite-difference methods to this equation. The following two examples illustrate some of the difficulties. If we apply the simple explicit method to the 2-D heat equation, the following algorithm results

$$\frac{u_{i,j}^{n+1} - u_{i,j}^n}{\Delta t} = \alpha \left[\frac{u_{i+1,j}^n - 2u_{i,j}^n + u_{i-1,j}^n}{(\Delta x)^2} + \frac{u_{i,j+1}^n - 2u_{i,j}^n + u_{i,j-1}^n}{(\Delta y)^2} \right] \tag{4-98}$$

where $x = i\,\Delta x$ and $y = j\,\Delta y$. As shown in Chapter 3, the stability condition is

$$\alpha\,\Delta t \left[\frac{1}{(\Delta x)^2} + \frac{1}{(\Delta y)^2} \right] \leqslant \frac{1}{2}$$

If $(\Delta x)^2 = (\Delta y)^2$, the stability condition reduces to $r \leqslant \frac{1}{4}$ which is twice as restrictive as the one-dimensional constraint $r \leqslant \frac{1}{2}$ and makes this method even more impractical.

When we apply the Crank-Nicolson scheme to the two-dimensional heat equation, we obtain

$$\frac{u_{i,j}^{n+1} - u_{i,j}^n}{\Delta t} = \frac{\alpha}{2} (\hat{\delta}_x^2 + \hat{\delta}_y^2)(u_{i,j}^{n+1} + u_{i,j}^n) \tag{4-99}$$

where the two-dimensional central-difference operators $\hat{\delta}_x^2$ and $\hat{\delta}_y^2$ are defined by

$$\begin{aligned} \hat{\delta}_x^2 u_{i,j}^n &= \frac{u_{i+1,j}^n - 2u_{i,j}^n + u_{i-1,j}^n}{(\Delta x)^2} = \frac{\delta_x^2 u_{i,j}^n}{(\Delta x)^2} \\ \hat{\delta}_y^2 u_{i,j}^n &= \frac{u_{i,j+1}^n - 2u_{i,j}^n + u_{i,j-1}^n}{(\Delta y)^2} = \frac{\delta_y^2 u_{i,j}^n}{(\Delta y)^2} \end{aligned} \tag{4-100}$$

As with the one-dimensional case, the Crank-Nicolson scheme is unconditionally stable when applied to the two-dimensional heat equation with periodic boundary conditions. Unfortunately, the resulting system of linear algebraic equations is no longer tridiagonal because of the five unknowns $u_{i,j}^{n+1}$, $u_{i+1,j}^{n+1}$, $u_{i-1,j}^{n+1}$, $u_{i,j+1}^{n+1}$, and $u_{i,j-1}^{n+1}$. The same is true for all the implicit schemes we have studied previously. In order to examine this further, let us rewrite Eq. (4-99) as

$$au_{i,j-1}^{n+1} + bu_{i-1,j}^{n+1} + cu_{i,j}^{n+1} + bu_{i+1,j}^{n+1} + au_{i,j+1}^{n+1} = d_{i,j}^n \tag{4-101}$$

where

$$a = -\frac{\alpha \Delta t}{2(\Delta y)^2} = -\frac{1}{2} r_y$$

$$b = -\frac{\alpha \Delta t}{2(\Delta x)^2} = -\frac{1}{2} r_x$$

$$c = 1 + r_x + r_y$$

$$d_{i,j}^n = u_{i,j}^n + \frac{\alpha \Delta t}{2} (\hat{\delta}_x^2 + \hat{\delta}_y^2) u_{i,j}^n$$

If we apply Eq. (4-101) to the two-dimensional (6 X 6) computational mesh shown in Fig. 4-15, the following system of 16 linear algebraic equations must be solved at each $(n + 1)$ time level.

$$\begin{bmatrix}
c & b & 0 & 0 & a & 0 & & & & & & & & & & 0 \\
b & c & b & & & a & & & & & & & & & & \\
0 & b & c & b & & & a & & & & & & & & & \\
0 & & b & c & 0 & & & a & & & & & & & & \\
a & & & 0 & c & b & & & a & & & & & & & \\
0 & a & & & b & c & b & & & a & & & & & & \\
 & & a & & & b & c & b & & & a & & & & & \\
 & & & a & & & b & c & 0 & & & a & & & & \\
 & & & & a & & & 0 & c & b & & & a & & & \\
 & & & & & a & & & b & c & b & & & a & & \\
 & & & & & & a & & & b & c & b & & & a & 0 \\
 & & & & & & & a & & & b & c & 0 & & & a \\
 & & & & & & & & a & & & 0 & c & b & & 0 \\
 & & & & & & & & & a & & & b & c & b & 0 \\
 & & & & & & & & & & a & & & b & c & b \\
0 & & & & & & & & & & 0 & a & 0 & 0 & b & c
\end{bmatrix}
\begin{bmatrix}
u_{2,2}^{n+1} \\ u_{3,2}^{n+1} \\ u_{4,2}^{n+1} \\ u_{5,2}^{n+1} \\ u_{2,3}^{n+1} \\ u_{3,3}^{n+1} \\ u_{4,3}^{n+1} \\ u_{5,3}^{n+1} \\ u_{2,4}^{n+1} \\ u_{3,4}^{n+1} \\ u_{4,4}^{n+1} \\ u_{5,4}^{n+1} \\ u_{2,5}^{n+1} \\ u_{3,5}^{n+1} \\ u_{4,5}^{n+1} \\ u_{5,5}^{n+1}
\end{bmatrix}
=
\begin{bmatrix}
d_{2,2}''' \\ d_{3,2}' \\ d_{4,2}' \\ d_{5,2}''' \\ d_{2,3}'' \\ d_{3,3} \\ d_{4,3} \\ d_{5,3}'' \\ d_{2,4}'' \\ d_{3,4} \\ d_{4,4} \\ d_{5,4}'' \\ d_{2,5}''' \\ d_{3,5}' \\ d_{4,5}' \\ d_{5,5}'''
\end{bmatrix}
\qquad (4\text{-}102)$$

where $d' = d - au_0$

$d'' = d - bu_0$

$d''' = d - (a + b)u_0$

A system of equations, like Eq. (4-102), requires substantially more computer time to solve than does a tridiagonal system. In fact, equations of this type are usually solved by iterative methods. These methods will be discussed in Section 4-3.

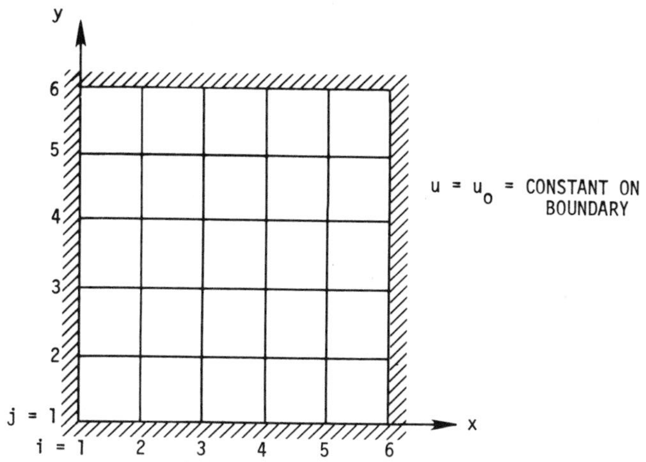

Figure 4-15 Two-dimensional computational mesh.

4-2.9 ADI Methods

The difficulties described above, which occur when attempting to solve the 2-D heat equation by conventional algorithms, led to the development of alternating-direction implicit (ADI) methods by Peaceman and Rachford (1955) and Douglas (1955). The usual ADI method is a two-step scheme given by

$$\text{Step 1:} \qquad \frac{u_{i,j}^{n+1/2} - u_{i,j}^{n}}{\Delta t/2} = \alpha(\hat{\delta}_x^2 u_{i,j}^{n+1/2} + \hat{\delta}_y^2 u_{i,j}^{n})$$

$$\text{Step 2:} \qquad \frac{u_{i,j}^{n+1} - u_{i,j}^{n+1/2}}{\Delta t/2} = \alpha(\hat{\delta}_x^2 u_{i,j}^{n+1/2} + \hat{\delta}_y^2 u_{i,j}^{n+1}) \qquad (4\text{-}103)$$

As a result of the "splitting" which is employed in this algorithm, only tridiagonal systems of linear algebraic equations must be solved. During Step 1, a tridiagonal matrix is solved for each j row of grid points and during Step 2, a tridiagonal matrix is solved for each i row of grid points. This procedure is illustrated in Fig. 4-16. The ADI method is second-order accurate with a truncation error of $O[(\Delta t)^2, (\Delta x)^2, (\Delta y)^2]$. Upon examining the amplification factor

$$G = \frac{[1 - r_x(1 - \cos\beta_x)]\,[1 - r_y(1 - \cos\beta_y)]}{[1 + r_x(1 - \cos\beta_x)]\,[1 + r_y(1 - \cos\beta_y)]}$$

we find this method to be unconditionally stable. The obvious extension of this method to three dimensions (making use of the time levels $n, n + \frac{1}{3}, n + \frac{2}{3}, n + 1$) leads to a conditionally stable method with truncation error of $O[(\Delta t, (\Delta x)^2, (\Delta y)^2, (\Delta z)^2]$. In order to circumvent this problem, Douglas and Gunn (1964) have developed a general method for deriving ADI schemes which are unconditionally stable and retain second-order accuracy. Using their approach, the Crank-Nicolson scheme

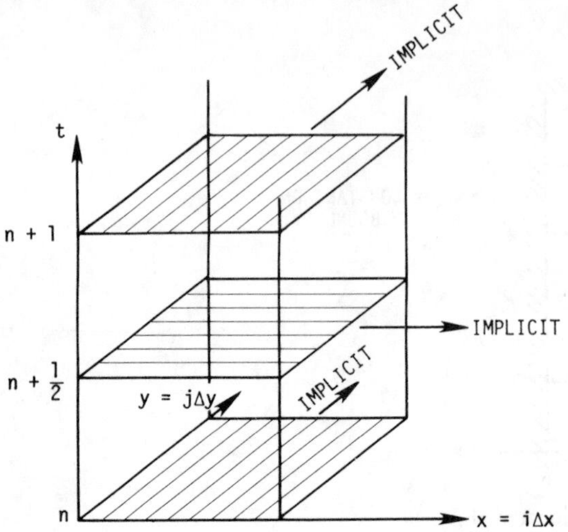

Figure 4-16 ADI calculation procedure.

can be extended to the three-dimensional heat equation and the following algorithm results

Step 1: $$u^* - u^n = \frac{r_x}{2}\,\delta_x^2(u^* + u^n) + r_y\,\delta_y^2 u^n + r_z\,\delta_z^2 u^n$$

Step 2: $$u^{**} - u^n = \frac{r_x}{2}\,\delta_x^2(u^* + u^n) + \frac{r_y}{2}\,\delta_y^2(u^{**} + u^n) + r_z\,\delta_z^2 u^n$$

Step 3: $$u^{n+1} - u^n = \frac{r_x}{2}\,\delta_x^2(u^* + u^n) + \frac{r_y}{2}\,\delta_y^2(u^{**} + u^n) + \frac{r_z}{2}\,\delta_z^2(u^{n+1} + u^n)$$

(4-104)

where the superscripts * and ** denote intermediate values and the subscripts i, j, k have been dropped from each term.

4-2.10 Splitting or Fractional-Step Methods

ADI methods are closely related and in some cases identical to the methods of fractional steps or methods of splitting, which were developed by Soviet mathematicians at about the same time as the ADI methods were developed in the United States. The basic idea of these methods is to split a finite-difference algorithm into a sequence of one-dimensional operations. For example, the simple implicit scheme applied to the 2-D heat equation could be split in the following manner

Step 1: $$\frac{u_{i,j}^{n+1/2} - u_{i,j}^n}{\Delta t} = \alpha\,\hat{\delta}_x^2 u_{i,j}^{n+1/2}$$

Step 2: $$\frac{u_{i,j}^{n+1} - u_{i,j}^{n+1/2}}{\Delta t} = \alpha\,\hat{\delta}_y^2 u_{i,j}^{n+1}$$

(4-105)

to give a first-order accurate method with a truncation error of $O[\Delta t, (\Delta x)^2, (\Delta y)^2]$. For further details on the method of fractional steps, the reader is urged to consult the book by Yanenko (1971).

4-2.11 ADE Methods

Another way of solving the two-dimensional heat equation is by means of an alternating-direction explicit (ADE) method. Unlike the ADI methods, the ADE methods do not require tridiagonal matrices to be "inverted." Since the ADE methods can also be used to solve the one-dimensional heat equation, we will apply the ADE algorithms to this equation, for simplicity.

The first ADE method was proposed by Saul'yev (1957). His two-step scheme is given by:

$$\text{Step 1:} \qquad \frac{u_j^{n+1} - u_j^n}{\Delta t} = \alpha \frac{u_{j-1}^{n+1} - u_j^{n+1} - u_j^n + u_{j+1}^n}{(\Delta x)^2}$$

$$\text{Step 2:} \qquad \frac{u_j^{n+2} - u_j^{n+1}}{\Delta t} = \alpha \frac{u_{j-1}^{n+1} - u_j^{n+1} - u_j^{n+2} + u_{j+1}^{n+2}}{(\Delta x)^2} \tag{4-106}$$

In the application of this method, Step 1 marches the solution from the left boundary to the right boundary. By marching in this direction, u_{j-1}^{n+1} is always known, and consequently, u_j^{n+1} can be determined "explicitly." In a like manner, Step 2 marches the solution from the right boundary to the left boundary, again resulting in an "explicit" formulation since u_{j+1}^{n+2} is always known. We assume that u is known on the boundaries. Although this scheme involves three time levels, only one storage array is required for u because of the unique way in which the calculation procedure sweeps through the mesh. This scheme is unconditionally stable and the truncation error is of $O[(\Delta t)^2, (\Delta x)^2, (\Delta t/\Delta x)^2]$. The scheme is formally first-order accurate due to the presence of the term $(\Delta t/\Delta x)^2$ in the truncation error.

Another ADE method was proposed by Barakat and Clark (1966). In this method the calculation procedure is simultaneously "marched" in both directions and the resulting solutions (p_j^{n+1} and q_j^{n+1}) are averaged to obtain the final value of u_j^{n+1}.

$$\frac{p_j^{n+1} - p_j^n}{\Delta t} = \alpha \frac{p_{j-1}^{n+1} - p_j^{n+1} - p_j^n + p_{j+1}^n}{(\Delta x)^2}$$

$$\frac{q_j^{n+1} - q_j^n}{\Delta t} = \alpha \frac{q_{j-1}^n - q_j^n - q_j^{n+1} + q_{j+1}^{n+1}}{(\Delta x)^2} \tag{4-107}$$

$$u_j^{n+1} = \tfrac{1}{2}(p_j^{n+1} + q_j^{n+1})$$

This method is unconditionally stable and the truncation error is approximately $O[(\Delta t)^2, (\Delta x)^2]$ because the simultaneous marching tends to cancel the $(\Delta t/\Delta x)^2$ terms. It has been observed that this method is about 18/16 times faster than the ADI method for the 2-D heat equation.

Larkin (1964) proposed a slightly different algorithm which replaces the p's and q's with u's whenever possible. His algorithm is

$$\frac{p_j^{n+1} - u_j^n}{\Delta t} = \alpha \frac{p_{j-1}^{n+1} - p_j^{n+1} - u_j^n + u_{j+1}^n}{(\Delta x)^2}$$

$$\frac{q_j^{n+1} - u_j^n}{\Delta t} = \alpha \frac{u_{j-1}^n - u_j^n - q_j^{n+1} + q_{j+1}^{n+1}}{(\Delta x)^2} \tag{4-108}$$

$$u_j^{n+1} = \tfrac{1}{2}(p_j^{n+1} + q_j^{n+1})$$

Numerical tests indicate that this method is usually less accurate than the Barakat and Clark scheme.

4-2.12 Keller Box and Modified Box Methods

The Keller box method (Keller, 1970) is a widely used method for solving parabolic PDE's such as the two-dimensional heat equation and the boundary-layer equations. Both the Keller box and the modified box methods are discussed in Section 7-3.5 where they are applied to the heat equation.

4-2.13 Hopscotch Method

As our final algorithm for solving the two-dimensional heat equation, let us examine the hopscotch method. This method is an explicit procedure which is unconditionally stable. The calculation procedure, illustrated in Fig. 4-17, involves two sweeps through the mesh. For the first sweep, $u_{i,j}^{n+1}$ is computed at each grid point (for which $i + j + n$ is even) by the simple explicit scheme

$$\frac{u_{i,j}^{n+1} - u_{i,j}^n}{\Delta t} = \alpha(\hat{\delta}_x^2 u_{i,j}^n + \hat{\delta}_y^2 u_{i,j}^n) \tag{4-109}$$

For the second sweep, $u_{i,j}^{n+1}$ is computed at each grid point (for which $i + j + n$ is odd) by the simple implicit scheme

$$\frac{u_{i,j}^{n+1} - u_{i,j}^n}{\Delta t} = \alpha(\hat{\delta}_x^2 u_{i,j}^{n+1} + \hat{\delta}_y^2 u_{i,j}^{n+1}) \tag{4-110}$$

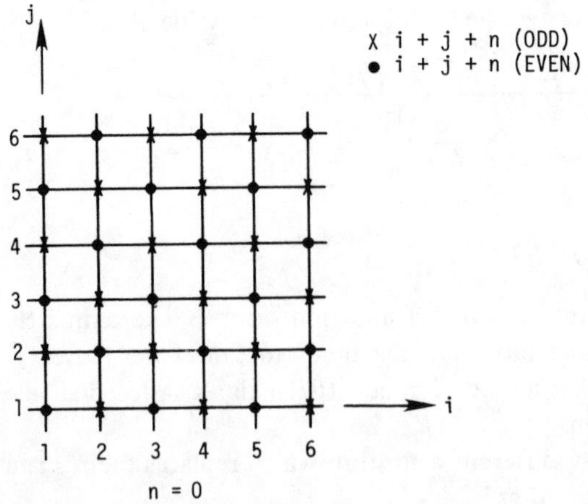

Figure 4-17 Hopscotch calculation procedure.

The second sweep appears to be implicit, but no simultaneous algebraic equations must be solved because $u_{i+1,j}^{n+1}$, $u_{i-1,j}^{n+1}$, $u_{i,j+1}^{n+1}$, and $u_{i,j-1}^{n+1}$ are known from the first sweep; hence, the algorithm is explicit. The truncation error for the hopscotch method is of $O[\Delta t, (\Delta x)^2, (\Delta y)^2]$.

4-2.14 Additional Comments

The selection of a best method for solving the heat equation is made difficult by the large variety of acceptable methods. In general, implicit methods are considered more suitable than explicit methods. For the one-dimensional heat equation, the Crank-Nicolson method is highly recommended because of its second-order temporal and spatial accuracy. For the two- and three-dimensional heat equations, both the ADI schemes of Douglas and Gunn and the Keller box and modified box methods give excellent results.

4-3 LAPLACE'S EQUATION

Laplace's equation is the model form for elliptic partial differential equations. For two-dimensional problems in Cartesian coordinates Laplace's equation is

$$\frac{\partial^2 u}{\partial x^2} + \frac{\partial^2 u}{\partial y^2} = 0 \tag{4-111}$$

Some of the important practical problems governed by a single elliptic equation include the steady-state temperature distribution in a solid and the incompressible irrotational ("potential") flow of a fluid.

The incompressible Navier-Stokes equations are an example of a more complicated system of equations which have an elliptic character. The steady Navier-Stokes equations are elliptic but in a coupled and complicated fashion since the pressure derivatives as well as velocity derivatives are sources of elliptic behavior. The time-dependent form of the incompressible Navier-Stokes equations is mixed parabolic-elliptic in type. The mixed parabolic-elliptic character of the time-dependent form of these equations becomes explicitly evident in the majority of the solution procedures observed to date as the equations are rearranged into a system of at least one parabolic equation and one elliptic Poisson equation of the form

$$\frac{\partial^2 u}{\partial x^2} + \frac{\partial^2 u}{\partial y^2} = f(x, y) \tag{4-112}$$

Thus, elliptic partial differential equations will be found very frequently to govern important problems in heat transfer and fluid mechanics. For this reason, we give serious attention to ways of solving our model elliptic equation.

4-3.1 Finite-Difference Representations for Laplace's Equation

The differences between "methods" for Laplace's equation and elliptic equations in general are not so much differences in the finite-difference representations (although these will vary) but, more often, differences in the techniques used for solving the resulting system of linear algebraic equations.

Five-point formula. By far the most common difference scheme for the two-dimensional Laplace equation is the five-point formula first used by Runge in 1908:

$$\frac{u_{i+1,j}-2u_{i,j}+u_{i-1,j}}{(\Delta x)^2}+\frac{u_{i,j+1}-2u_{i,j}+u_{i,j-1}}{(\Delta y)^2}=0 \tag{4-113}$$

which has a truncation error of $O[(\Delta x)^2, (\Delta y)^2]$.

The modified equation is

$$u_{xx}+u_{yy}=-\tfrac{1}{12}[u_{xxxx}(\Delta x)^2+u_{yyyy}(\Delta y)^2]+\cdots$$

Nine-point formula. This formula appears to be a logical choice when greater accuracy is desired for Laplace's equation. Here we let $\Delta x = h$, $\Delta y = k$:

$$u_{i+1,j+1}+u_{i-1,j+1}+u_{i+1,j-1}+u_{i-1,j-1}-2\frac{h^2-5k^2}{h^2+k^2}(u_{i+1,j}+u_{i-1,j})$$

$$+2\frac{5h^2-k^2}{h^2+k^2}(u_{i,j+1}+u_{i,j-1})-20u_{i,j}=0 \tag{4-114}$$

The truncation error for this scheme is of $O(h^4, k^4)$ but becomes $O(h^6)$ on a square mesh. Details of the truncation error and modified equation for this scheme are left as an exercise. Although the nine-point formula appears to be very attractive for Laplace's equation because of the favorable truncation error, this error may only be $O(h^2, k^2)$ when applied to a more general elliptic equation (including the Poisson equation) containing other terms.

Other representations for Laplace's equation. Since the Laplacian operator is invariant to a coordinate rotation, it is not surprising that for a square mesh ($\Delta x = \Delta y = h$), the four points obtained by rotating the grid points in Eq. (4-113) by 45 degrees about (i, j) and increasing the spacing to $\sqrt{2}h$ can be used to obtain a "diagonal" five-point representation for Laplace's equation:

$$u_{i+1,j+1}+u_{i+1,j-1}+u_{i-1,j-1}+u_{i-1,j+1}-4u_{i,j}=0 \tag{4-115}$$

A comparison of the geometrical arrangement of the two five-point schemes can be seen in Fig. 4-18. The diagonal five-point scheme has truncation error $O(h^2)$. (See Prob. 3-10.)

Other finite-difference schemes for Laplace's equation can be found in the literature (see, for example, Thom and Apelt, 1961) but none seems to offer significant advantages over the five- and nine-point schemes given above. To obtain smaller formal truncation error in these schemes, more grid points must be used in the difference molecules. High accuracy is difficult to maintain near boundaries with such schemes.

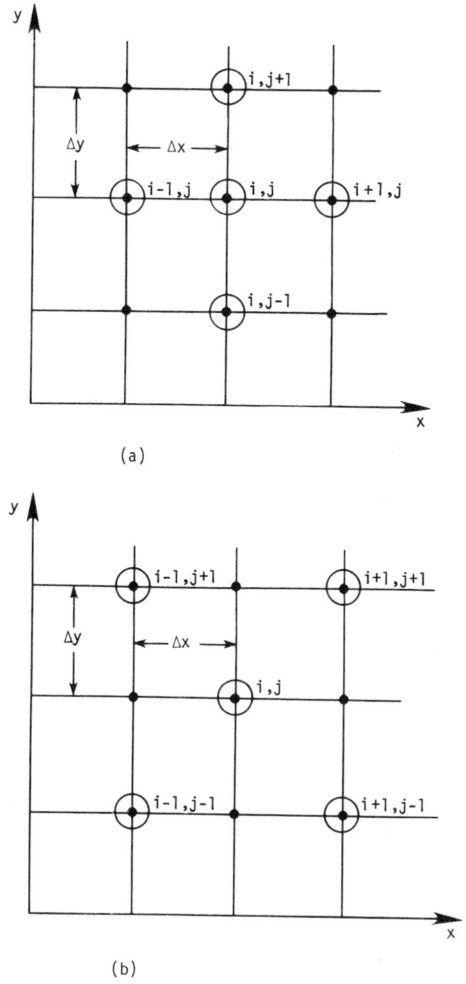

Figure 4-18 Geometrical arrangement of two five-point schemes for $\Delta x = \Delta y$. (a) Five-point scheme. (b) Diagonal five-point scheme.

4-3.2 Simple Example for Laplace's Equation

Let's consider how we might determine a function u satisfying $\partial^2 u/\partial x^2 + \partial^2 u/\partial y^2 = 0$ on the square domain $0 \leqslant x \leqslant 1, 0 \leqslant y \leqslant 1$ subject to Dirichlet boundary conditions.

Series solutions can be obtained for this problem (most readily by separation of variables) satisfying certain distributions of u at the boundaries. These are available in most textbooks that cover conduction heat transfer (Chapman, 1974) and can be used as test cases to verify the finite-difference formulation. Alternatively, we can choose a function such as $u = x^2 - y^2$ which satisfies Laplace's equation throughout the square and use it to establish boundary conditions for the finite-difference calculation. The resulting difference solution can then be compared with $u = x^2 - y^2$. In this example we will use the five-point scheme, Eq. (4-113) and let $\Delta x = \Delta y = 0.1$ result-

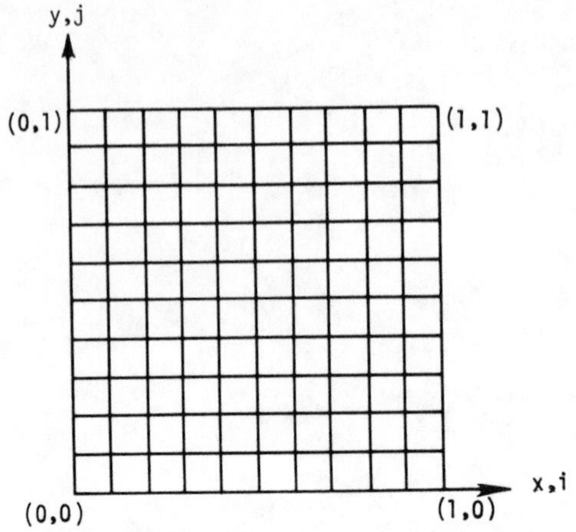

Figure 4-19 Finite-difference grid for Laplace's equation.

ing in a uniform 11 X 11 grid over the square problem domain. See Fig. 4-19. With $\Delta x = \Delta y$, the difference equation can be written as

$$u_{i+1,j} + u_{i-1,j} + u_{i,j+1} + u_{i,j-1} - 4u_{i,j} = 0 \tag{4-116}$$

for each point where u is unknown. In this example problem with Dirichlet boundary conditions we have 81 grid points where u is unknown. For each one of those points we can write the difference equation so that our problem is one of solving the system of 81 simultaneous linear algebraic equations for the 81 unknown $u_{i,j}$'s. Mathematically our problem can be expressed as

$$\begin{aligned} a_{11}u_1 + a_{12}u_2 + \cdots\cdots\cdots a_{1n}u_n &= c_1 \\ a_{21}u_1 + a_{22}u_2 + \cdots\cdots\cdots a_{2n}u_n &= c_2 \\ \vdots \qquad\qquad\qquad\qquad & \vdots \\ a_{n1}u_1 + \cdots\cdots\cdots\cdots\cdots a_{nn}u_n &= c_n \end{aligned} \tag{4-117}$$

or more compactly as $[A]\mathbf{u} = \mathbf{C}$ where $[A]$ is the matrix of known coefficients, $\mathbf{u}$ is the column vector of unknowns and $\mathbf{C}$ is a column vector of known quantities. It is worth noting that the matrix of coefficients will be very sparse since about 76 of the 81 a's in each row will be zero. To make our example algebraically as simple as possible we have let $\Delta x = \Delta y$. If $\Delta x \neq \Delta y$, the coefficients will be a little more involved but the algebraic equations will still be linear and can be represented by the general $[A]\mathbf{u} = \mathbf{C}$ system given above.

Methods for solving systems of linear algebraic equations can be readily classified as either direct or iterative. Direct methods are those which give the solution (exactly if round-off error does not exist) in a finite and predeterminable number of operations using an algorithm which is often quite complicated. Iterative methods consist of a repeated application of an algorithm which is usually quite simple. They yield the exact answer only as a limit of a sequence, but, if the iterative procedure converges, we can come within ϵ of the answer in a finite but usually not predeterminable number of operations. Some examples of both types of methods will be given.

4-3.3 Direct Methods for Solving Systems of Linear Algebraic Equations

Cramer's rule. This is one of the most elementary methods. All students have certainly heard of it and most are familiar with the workings of the procedure. Unfortunately the algorithm is immensely time consuming, the number of operations being approximately proportional to $(N + 1)!$, where N is the number of unknowns. A number of horror stories have been told about the large computation time required to solve systems of equations by Cramer's rule. A favorite one is recorded by Roache (1972) where it is pointed out that the multiplications required for the solution of 26 equations with 26 unknowns by Cramer's rule using the CDC 6600 would require 10^{16} years or 10^6 times the current estimate of the age of the universe! Apparently the choice of a method matters. Cramer's rule should never be used.

Gaussian elimination. Gaussian elimination is a very useful and efficient tool for solving many systems of algebraic equations, particularly for the special case of a tridiagonal system of equations. However, the method is not as fast as some others to be considered for more general systems of algebraic equations which arise in solving PDE's. Approximately N^3 multiplications are required in solving N equations. Also, round-off errors which can accumulate through the many algebraic operations sometimes cause deterioration of accuracy when N is large. Actually the accuracy of the method depends on the specific system of equations and the matter is too complex to resolve by a simple general statement. Rearranging the equations to the extent possible in order to put the coefficients which are largest in magnitude on the main diagonal (known as "pivoting") will tend to improve accuracy, however.

Since we will want to use an elimination scheme for tridiagonal systems of equations which arise in implicit difference schemes for marching problems, it would be well to gain some notion of how the basic Gaussian elimination procedure works. Consider the equations

$$
\begin{aligned}
a_{11}u_1 + a_{12}u_2 + \cdots\cdots\cdots &= c_1 \\
a_{21}u_1 + a_{22}u_2 + \cdots\cdots\cdots &= c_2 \\
\vdots \qquad\qquad\qquad &\ \ \vdots \\
a_{n1}u_n + \cdots\cdots\cdots\cdots\cdots &= c_n
\end{aligned}
\qquad (4\text{-}118)
$$

The objective is to transform the system into an upper triangular array by eliminating some of the unknowns from some of the equations by algebraic operations. To illustrate, we choose the first equation (row) as the "pivot" equation and use it to eliminate the u_1 term from each equation below it. This is done by multiplying the first equation by a_{21}/a_{11}† and subtracting it from the second equation to eliminate u_1 from the second equation. Multiplying the pivot equation by a_{31}/a_{11} and subtracting it from the third equation eliminates the first term from the third equation. This procedure can be continued to eliminate the u_1 from equations 2 through n. The system now appears as shown in Fig. 4-20.

†We must always interchange rows if necessary to avoid division by zero.

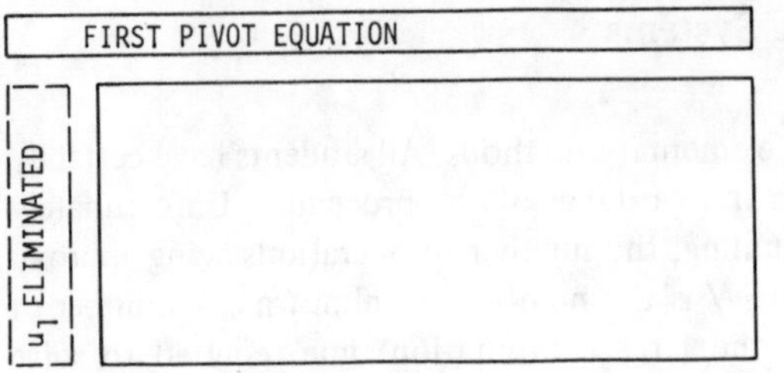

Figure 4-20 Gaussian elimination, u_1 eliminated below main diagonal.

Next, the second (as altered by the above procedure) is used as the pivot equation to eliminate u_2 from all equations below it leaving the system in the form shown in Fig. 4-21. The third equation in the altered system is then used as the next pivot equation and the process continues until only an upper triangular form remains:

$$\begin{aligned} a_{11}u_1 + a_{12}u_2 + \cdots\cdots\cdots &= c_1 \\ a'_{22}u_2 + a'_{23}u_3 + \cdots\cdots &= c'_2 \\ a'_{33}u_3 + \cdots\cdots &= c'_3 \\ &\vdots \\ a'_{nn}u_n &= c'_n \end{aligned} \tag{4-119}$$

At this point only one unknown appears in the last equation, two in the next to last equation, etc., so a solution can be obtained by back substitution.

Let's consider the following system of three equations as a specific numerical example:

$$\begin{aligned} U_1 + 4U_2 + U_3 &= 7 \\ U_1 + 6U_2 - U_3 &= 13 \\ 2U_1 - U_2 + 2U_3 &= 5 \end{aligned}$$

Using the top equation as a pivot we can eliminate U_1 from the lower two equations:

$$\begin{aligned} U_1 + 4U_2 + U_3 &= 7 \\ 2U_2 - 2U_3 &= 6 \\ -9U_2 + 0 &= -9 \end{aligned}$$

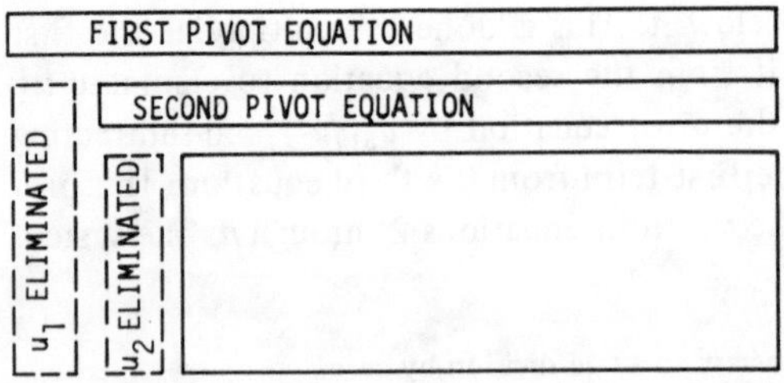

Figure 4-21 Gaussian elimination, u_1 and u_2 eliminated below main diagonal.

Now using the second equation as a pivot we obtain the upper triangular form:

$$U_1 + 4U_2 + U_3 = 7$$
$$2U_2 - 2U_3 = 6$$
$$-9U_3 = 18$$

Back substitution yields $U_3 = -2, U_2 = 1, U_1 = 5$.

Block-iterative methods for Laplace's equation (Section 4-3.4) lead to systems of simultaneous algebraic equations which have a tridiagonal matrix of coefficients. This was also observed earlier in Sections 4-1 and 4-2 for implicit formulations of PDE's for marching problems. To illustrate how Gaussian elimination can be efficiently modified to take advantage of the tridiagonal form of the coefficient matrix, we will consider the simple implicit scheme for the heat equation as an example:

$$\frac{\partial u}{\partial t} = \alpha \frac{\partial^2 u}{\partial x^2}$$

$$\frac{u_j^{n+1} - u_j^n}{\Delta t} = \frac{\alpha}{(\Delta x)^2}\left(u_{j+1}^{n+1} + u_{j-1}^{n+1} - 2u_j^{n+1}\right)$$

In terms of the format used before for algebraic equations, this can be rewritten as

$$bu_{j-1}^{n+1} + du_j^{n+1} + au_{j+1}^{n+1} = c$$

where

$$a = b = -\frac{\alpha\,\Delta t}{(\Delta x)^2} \qquad c = u_j^n \qquad d = 1 + \frac{2\alpha\,\Delta t}{(\Delta x)^2}$$

For Dirichlet boundary conditions, u_{j-1}^{n+1} is known at one boundary and u_{j+1}^{n+1} at the other. All known u's are collected into the c term so our system looks like

$$\begin{bmatrix} d_1 & a_1 & 0 & \cdot & \cdot & \cdot & \cdot & 0 \\ b_2 & d_2 & a_2 & 0 & \cdot & \cdot & \cdot & \cdot \\ 0 & b_3 & d_3 & a_3 & 0 & \cdot & \cdot & \cdot \\ 0 & 0 & b_4 & d_4 & a_4 & 0 & \cdot & \cdot \\ \cdot & \cdot & \cdot & \ddots & \ddots & \ddots & \cdot & \cdot \\ \cdot & \cdot & \cdot & \cdot & \ddots & \ddots & \ddots & 0 \\ \cdot & \cdot & \cdot & \cdot & \cdot & \ddots & \ddots & a_{NJ-1} \\ 0 & \cdot & \cdot & \cdot & \cdot & 0 & b_{NJ} & d_{NJ} \end{bmatrix} \begin{bmatrix} u_1^{n+1} \\ u_2^{n+1} \\ \cdot \\ \cdot \\ \cdot \\ \cdot \\ \cdot \\ u_{NJ}^{n+1} \end{bmatrix} = \begin{bmatrix} c_1 \\ c_2 \\ \cdot \\ \cdot \\ \cdot \\ \cdot \\ \cdot \\ c_{NJ} \end{bmatrix}$$

Even when other boundary conditions apply, the system can be cast into the above form although the first and last equations in the array may result from auxiliary relationships related to the boundary conditions and not the original difference equation which applies to nonboundary points.

For this tridiagonal system it is easy to modify the Gaussian elimination procedure to take advantage of the zeros in the matrix of coefficients. This modified procedure, suggested by Thomas (1949), was discussed briefly in Section 4-1.4.

Thomas algorithm. Referring to the tridiagonal matrix of coefficients above, the system is put into an upper triangular form by computing the new d_j by

$$d_j = d_j - \frac{b_j}{d_{j-1}} a_{j-1} \qquad j = 2, 3, \ldots, NJ$$

and the new c_j by

$$c_j = c_j - \frac{b_j}{d_{j-1}} c_{j-1} \qquad j = 2, 3, \ldots, NJ$$

then computing the unknowns from back substitution according to $u_{NJ} = c_{NJ}/d_{NJ}$ and then

$$u_k = \frac{c_k - a_k u_{k+1}}{d_k} \qquad k = NJ - 1, NJ - 2, \ldots, 1$$

In the above equations, the equals sign means "is replaced by" as in the FORTRAN programming language. A FORTRAN program for this procedure is given in Appendix A.

Some flexibility exists in the way in which boundary conditions are handled when the Thomas algorithm is used to solve for the unknowns. It is best that the reader develop an appreciation for these details through experience; however, a comment or two will be offered here by way of illustration. The main purpose of the elimination scheme is to determine values for the unknowns; therefore, for Dirichlet boundary conditions where boundary u's are known, the boundary u's in the problem need not be included in the list of unknowns. That is, u_1 in the elimination algorithm could correspond to the u at the first nonboundary point and u_{NJ} the u at the last nonboundary point. However, no harm is done, and programming may be made easier, by specializing a_1, d_1, b_{NJ}, and d_{NJ} to provide a redundant statement of the boundary conditions. That is, if we let $d_1 = 1$, $a_1 = 0$, $d_{NJ} = 1$, $b_{NJ} = 0$, $c_1 = u_1^{n+1}$ (given), and $c_{NJ} = d_{NJ}^{n+1}$ (given) the first and last algebraic equations become just a statement of the boundary conditions. As an example of how other boundary conditions fall easily into the tridiagonal format let's consider convective (mixed) boundary conditions for the heat equation:

$$h(u_\infty - u_{\text{bdy}}) = -k \left.\frac{\partial u}{\partial x}\right)_{\text{bdy}}$$

A control-volume analysis at the boundary where $j = 1$ leads to a difference equation which can be written as

$$d_1 u_1^{n+1} + a_1 u_2^{n+1} = c_1$$

where

$$d_1 = 1 + \frac{2\alpha\,\Delta t}{(\Delta x)^2}\left(1 + \frac{h\,\Delta x}{k}\right)$$

$$a_1 = \frac{-2\alpha\,\Delta t}{(\Delta x)^2} \qquad c_1 = \frac{2\alpha(\Delta t)h(\Delta x)}{(\Delta x)^2 k}u_\infty + u_1^n$$

which obviously fits the tridiagonal form for the first row.

Advanced-direct methods. Direct methods for solving systems of algebraic equations which are faster than Gaussian elimination certainly exist. Unfortunately, none of these methods are completely general. That is, they are applicable only to the algebraic equations arising from a special class of difference equations and associated boundary conditions. Many of these methods are "field size limited" (limited in applicability to relatively small systems of algebraic equations) due to the accumulation of round-off errors. As a class, the algorithms for fast direct procedures tend to be rather complicated and not easily adapted to irregular problem domains or complex boundary conditions. Somewhat more computer storage is usually required than for an iterative method suitable for a given problem. It seems that the simplest of these methods suffer from the field size limitations and are relatively restricted in their range of application and those that are not field size limited, have algorithms with very involved details which are beyond the scope of this text. For these reasons only a few of these methods will be mentioned here and none will be discussed in detail.

One of the simplest of the advanced-direct methods is the error vector propagation (EVP) method developed for the Poisson equation by Roache (1972). This method is field size limited; however, the concepts are straightforward and in some test cases where it was possible to control the growth of round-off errors, the method was found to be 10–100 times faster than the best iterative methods to be described next.

Two fast direct methods for the Poisson equation which are not limited due to accumulation of round-off errors are the "odd-even reduction" method of Buneman (1969) and the fast Fourier transform method of Hockney (1965, 1970). An interesting evaluation arising from the solution of the two-dimensional Navier-Stokes equations is given by Lugt and Ohring (1974), where they noted that the Buneman and Hockney procedures may be on the order of 10 to 20 times faster than the best iterative procedures. Helpful discussions of fast Fourier transforms as a tool in solving PDE's is given by LeBail (1972) and Buzbee et al. (1970). Applications of fast direct methods to problems in aerodynamics are described by Martin and Lomax (1975) and Schumann (1980).

Clearly, the fast direct methods should be considered for problems where the overriding consideration is expected to be computer execution time. Again we note that the algorithms are complex and the available forms of these procedures are likely to be set for specific and simple geometries and boundary conditions. Program development time may be large in adapting the procedures to a specific problem and this will have to be weighed against any possible savings in computer time. For many applications to elliptic problems, the iterative methods to be described below will be adequate.

4-3.4 Iterative Methods for Solving Systems of Linear Algebraic Equations

This class of methods is referred to by some as "relaxation methods" although to others, the word "relaxation" is reserved for use with reference to the particular scheme of residual relaxation suggested by Southwell many years ago. This class of methods can be further broken down into point- (or explicit-) iterative methods and block- (or implicit-) iterative methods. In brief, for point-iterative methods, the same simple algorithm is applied to each point where the unknown function is to be determined in successive iterative sweeps whereas in block iterative methods, subgroups of points are singled out for solution by elimination (direct) schemes in an overall iterative procedure.

Gauss-Seidel iteration. Although many different iterative methods have been suggested over the years, Gauss-Seidel iteration (often called Liebmann iteration when applied to the algebraic equation resulting from the differencing of an elliptic PDE) is one of the most efficient and useful point-iterative procedures for large systems of equations. The method is extremely simple but only converges under certain conditions related to "diagonal dominance" of the matrix of coefficients. Fortunately, the differencing of many steady-state conservation statements provides this diagonal dominance. The method makes explicit use of the sparseness of the matrix of coefficients.

The simplicity of the procedure will be demonstrated by an example prior to a concise statement regarding the sufficient condition for convergence. When the method can be used, the procedure for a general system of algebraic equations would be to (1) make initial guesses for all unknowns (a guessed value for one unknown will not be needed as seen in example below); (2) solve each equation for the unknown whose coefficient is largest in magnitude, using guessed values initially and the most recently computed values thereafter for the other unknowns in each equation; (3) repeat iteratively the solution of the equations in this manner until changes in the unknowns become "small," remembering to use the most recently computed value for each unknown when it appears on the right-hand side of an equation. As an example, let's consider the system

$$4x_1 - x_2 + x_3 = 4$$

$$x_1 + 6x_2 + x_3 = 9$$

$$-x_1 + 2x_2 + 5x_3 = 2$$

We would first rewrite the equations as

$$x_1 = \tfrac{1}{4}(4 + x_2 - x_3)$$

$$x_2 = \tfrac{1}{6}(9 - x_1 - x_3)$$

$$x_3 = \tfrac{1}{5}(2 + x_1 - 2x_2)$$

then make initial guesses for x_2 and x_3 (a guess for x_1 not needed) and compute x_1, x_2, x_3 iteratively as indicated above.

Sufficient condition for convergence of the Gauss-Seidel procedure. In order to provide a compact notation, we will order the equations, if possible, so that the coefficient largest in magnitude in each row is on the diagonal. Then if the system is irreducible (cannot be arranged so that some of the unknowns can be determined by solving less than n equations) and if

$$|a_{ii}| \geqslant \sum_{\substack{j=1 \\ j \neq i}}^{n} |a_{ij}| \tag{4-120}$$

for all i and if for at least one i,

$$|a_{ii}| > \sum_{\substack{j=1 \\ j \neq i}}^{n} |a_{ij}| \tag{4-121}$$

then the Gauss-Seidel iteration will converge. This is a *sufficient condition* which means that convergence may sometimes be observed when the above condition is not met. A necessary condition can be stated but it is impractical to evaluate. In words, the sufficient condition can be interpreted as requiring for each equation that the magnitude of the coefficient on the diagonal be greater than or equal to the sum of the magnitudes of the other coefficients in the equation with the greater than holding for at least one (usually corresponding to a point near a boundary for a physical problem) equation.

Perhaps we should now relate the above iterative convergence criteria to the system of equations which results from differencing Laplace's equation according to Eq. (4-113) above. First we observe that the coefficient largest in magnitude belongs to $u_{i,j}$. Since we apply Eq. (4-113) to each point where $u_{i,j}$ is unknown, we could clearly arrange all the equations in the system so that the coefficient largest in magnitude appeared on the diagonal. With the exercise of proper care in establishing difference representations, this type of diagonal dominance can normally be achieved for all elliptic equations. In terms of a linear difference equation for u, we would expect the Gauss-Seidel iterative procedure to converge if the finite-difference equation applicable to each point i, j where $u_{i,j}$ is unknown, is such that the magnitude of the coefficient of $u_{i,j}$ is greater than or equal to the sum of the magnitudes of the coefficients of the other unknowns in the equation. The greater than must hold for at least one equation.

We will not offer a proof for this sufficient condition for the convergence of the Gauss-Seidel iteration but hopefully, a simple example will suggest why it is true. If we look back to our simple three equation example for Gauss-Seidel iteration and consider that at any point our intermediate values of x's are the exact solution plus some ϵ, i.e., $x_1 = (x_1)_{\text{exact}} + \epsilon_1$ etc., then our condition of diagonal dominance is forcing the ϵ's to become smaller and smaller as the iteration is repeated cyclically. For one run through the iteration we could observe:

$$|\epsilon_1^2| \leq \tfrac{1}{4}|\epsilon_2^1| + \tfrac{1}{4}|\epsilon_3^1|$$

$$|\epsilon_2^2| \leq \tfrac{1}{6}|\epsilon_1^2| + \tfrac{1}{6}|\epsilon_3^1|$$

$$|\epsilon_3^2| \leq \tfrac{1}{5}|\epsilon_1^2| + \tfrac{2}{5}|\epsilon_2^2|$$

If ϵ_2^1 and ϵ_3^1 were initially each 10, $|\epsilon_1^2|$ would be ≤ 5 and $|\epsilon_1^3| \leq 1.446$. Here, superscripts denote iteration level.

Finally, we note for a general system of equations, the multiplications per iteration could be as great as n^2 but could be much less if the matrix was sparse.

Successive over-relaxation. Successive over-relaxation (SOR) is a technique which can be used in an attempt to accelerate any iterative procedure but we will propose it here primarily as a refinement to the Gauss-Seidel method. As to the origins of the method, one story (probably inaccurate) being passed around is that the method was suggested by a duck hunter who finally learned that if he pointed his gun ahead of the duck, he would score more hits than if he pointed the gun right at the duck. The duck is a moving target and if we anticipate its motion, we are more likely to hit it with the shot pattern. The duck hunter told his story to his neighbor who was a numerical analyst and SOR was born—or so the story goes.

As we apply Gauss-Seidel iteration to a system of simultaneous algebraic equations, we expect to make several recalculations or iterations before convergence to an acceptable level is achieved. Suppose that during this process we observe the change in the value of the unknown at a point between two successive iterations, note the direction of change and anticipate that the same trend will continue on to the next iteration. Why not go ahead and make a correction to the variable in the anticipated direction *before* the next iteration thereby, hopefully, accelerating the convergence? An arbitrary correction to the intermediate values of the unknowns from *any* iterative procedure (Gauss-Seidel iteration is of most interest to us at this point so we will use it as the representative iterative scheme) according to the form

$$u_{i,j}^{k+1'} = u_{i,j}^{k'} + \omega(u_{i,j}^{k+1} - u_{i,j}^{k'}) \tag{4-122}$$

is known as over-relaxation or successive over-relaxation (SOR). Here, k denotes iteration level and $u_{i,j}^{k+1}$ is the most recent value of $u_{i,j}$ calculated from the Gauss-Seidel procedure, $u_{i,j}^{k'}$ is the value from the previous iteration as adjusted by previous application of this formula if the over-relaxation is being applied successively (at each iteration) and $u_{i,j}^{k+1'}$ is the newly adjusted or "better guess" for $u_{i,j}$ at the $k+1$ iteration level. That is, we expect $u_{i,j}^{k+1'}$ to be closer to the final solution than the unaltered value $u_{i,j}^{k+1}$ from the Gauss-Seidel calculation. The formula is applied immediately at each point after $u_{i,j}^{k+1}$ has been obtained and $u_{i,j}^{k+1'}$ replaces $u_{i,j}^{k+1}$ in all subsequent calculations in the cycle. ω is the relaxation parameter and when $1 < \omega < 2$ *over-relaxation* is being employed. Over-relaxation can be likened to linear extrapolation based on values $u_{i,j}^{k'}$ and $u_{i,j}^{k+1}$. In some problems *under-relaxation* $0 < \omega < 1$ is employed. Under-relaxation appears to be most appropriate when the convergence at a point is taking on an oscillatory pattern and tending to "overshoot" the apparent final solution. For under-relaxation, the adjusted value, $u_{i,j}^{k+1'}$ is between $u_{i,j}^{k'}$ and $u_{i,j}^{k+1}$. Over-relaxation is usually appropriate for numerical solutions to Laplace's equation with Dirichlet boundary conditions. Under-relaxation is sometimes called for in elliptic problems, it seems, when the equations are nonlinear. Occasionally, for nonlinear problems, under-relaxation is even observed to be necessary for convergence.

We note that the relaxation parameter should be restricted to the range $0 < \omega < 2$. For convergence, we require that the magnitude of the changes in the u's from one iteration to the next become smaller. Use of $\omega \geqslant 2$ forces these changes to remain the same or increase, in contradiction to convergent behavior.

Two important remaining questions are, how can we properly determine a good or even the best value for ω and by how much does this procedure accelerate the convergence? No completely general answers to these questions are available but some guidelines can be drawn.

For Laplace's equation on a rectangular domain [represented by Eq. (4-113)] using a uniform mesh ($\Delta x = \Delta y = h$) and Dirichlet boundary conditions, an expression for the optimum ω (hereafter denoted by ω_{opt}) has been obtained by the theory developed by Young (1954) and Frankel (1950). This optimum ω is given by the smaller root of $t^2\omega^2 - 16\omega + 16 = 0$ where $t = \cos(\pi/p) + \cos(\pi/q)$ and p and q are the number of mesh increments on each side of the rectangular region. To cite one example, if $p = q = 45$, $\omega_{\text{opt}} = 1.87$. In general, however, for more complex elliptic problems it is not possible to determine ω_{opt} in advance. In these cases, some numerical experimentation should be helpful in identifying useful values for ω. Numerical examples and theory generally indicate that it is better to guess on the high side of ω_{opt} than on the low side. Forsythe and Wasow (1960) discuss considerations in the search for ω_{opt} in some detail, as do the books by Varga (1962) and Ames (1977).

Is the ω search worthwhile? The answer is emphatically *yes.* In some problems it is apparently possible to reduce the computation time by a factor of 30. This is significant! Occasionally, SOR may be found not to be of much help in accelerating convergence but it should always be considered and evaluated. The potential for savings in computation time is too great to ignore.

Since over-relaxation can be viewed as applying a correction to the values obtained from the Gauss-Seidel procedure based on extrapolation from previous iterates, it is natural to wonder if other, perhaps more accurate (in terms of truncation error of the extrapolation formula) extrapolation schemes can be used to accelerate the convergence of iterative procedures. In fact, other schemes such as Aitken and Richardson extrapolation have been used in this application. The details of these extrapolation schemes are covered in standard texts on numerical analysis, but as is perhaps expected, any advantage in accelerating the convergence of the iterative process by using more complex extrapolation schemes has to be weighed against any added computation costs due to requirements of additional storage or algebraic operations. SOR has simplicity in its favor and it can be programmed so that no additional arrays need to be stored.

Block-iterative methods. The Gauss-Seidel iteration method with SOR stands as the best all-around method for the finite-difference solution of elliptic equations discussed in detail thus far in this chapter. The number of iterations can usually be reduced even further by use of block-iterative concepts, but the number of algebraic operations required per iterative cycle generally increases, and whether the reduction in number of required iterative cycles compensates for the extra computation time per cycle is a matter that must be studied for each problem. However, several cases can be cited where the use of block-iterative methods has resulted in a net saving of computation time so that these procedures warrant serious attention. Ames (1977) and Forsythe and Wasow (1960) present useful discussions which compare the rates of convergence for several point- and block-iterative methods.

In block- (or group-) iterative methods, subgroups of the unknowns are singled out and their values modified simultaneously by obtaining a solution to the simultaneous algebraic equations by elimination methods. Thus, the block-iterative methods have an implicit nature and are sometimes known as implicit-iterative methods. In the most common block-iterative methods, the unknowns in the subgroups to be modified simultaneously are set up so that the matrix of coefficients will be tridiagonal in form permitting the Thomas algorithm to be used. The simplest block procedure is SOR by lines.

SOR by lines. Although this procedure is workable with almost any iterative algorithm, it makes the most sense to work within the framework of the Gauss-Seidel method with SOR. We can choose either rows or columns for grouping with equal ease. To illustrate the procedure, let's consider again the solution to Laplace's equation on a square domain with Dirichlet boundary conditions using the five-point scheme. This time, for more generality, let's consider the five-point scheme on a grid for which Δx is not necessarily equal to Δy. Letting $\beta = \Delta x/\Delta y$, the grid aspect ratio, the general equation for the Gauss-Seidel procedure can be written as

$$u_{i,j}^{k+1} = \frac{u_{i+1,j}^{k} + u_{i-1,j}^{k} + \beta^2 (u_{i,j+1}^{k} + u_{i,j-1}^{k+1})^{\dagger}}{2(1+\beta^2)} \tag{4-123}$$

where k denotes the iteration level, i denotes the column and j the row. If we agree to start at the bottom of the square and sweep up by rows, we could write for the general point:

$$u_{i,j}^{k+1} = \frac{u_{i+1,j}^{k+1} + u_{i-1,j}^{k+1} + \beta^2 (u_{i,j+1}^{k} + u_{i,j-1}^{k+1})}{2(1+\beta^2)} \tag{4-124}$$

If we study this equation carefully, we observe that only three unknowns are present since $u_{i,j-1}^{k+1}$ would be known from either the lower boundary conditions, if we were applying the equation to the first row of unknowns, or from the solution already obtained at the $k+1$ level from the row below. We have chosen to evaluate $u_{i,j+1}$ at the k iteration level rather than the $k+1$ level in order to obtain just three unknowns in the equation so that the efficient Thomas algorithm can be used. This configuration can be seen in Fig. 4-22.

The procedure is then to solve the system of $I-2$ simultaneous algebraic equa-

†In general, at least one unknown in each equation would already have been calculated at the $k+1$ level.

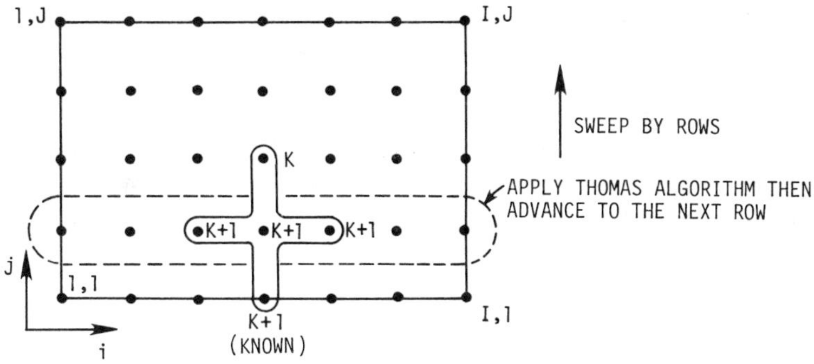

Figure 4-22 SOR by lines.

tions for the $I-2$ unknowns representing the values of $u_{i,j}$ at the $k+1$ iteration level. Successive over-relaxation can now be applied in the same manner as indicated previously before moving on to the next row. Some flexibility exists in the way successive over-relaxation is applied. After the Thomas algorithm is used to solve Eq. (4-124) for each row, the newly calculated values can be simply over-relaxed as indicated by Eq. (4-122) before the calculation is advanced to the next row.

Alternatively, the over-relaxation parameter, ω, can be introduced prior to solution of the simultaneous algebraic equations. This is accomplished by substituting the right-hand side of Eq. (4-124) into the right-hand side of Eq. (4-122) to replace $u_{i,j}^{k+1}$. The resulting equation

$$u_{i,j}^{k+1} = (1-\omega)u_{i,j}^{k} + \frac{\omega}{2(1+\beta^2)}\left[u_{i+1,j}^{k+1} + u_{i-1,j}^{k+1} + \beta^2\left(u_{i,j+1}^{k} + u_{i,j-1}^{k+1}\right)\right]$$

is then solved for each row by the Thomas algorithm. The over-relaxation has been accomplished as part of the row solution and not as a separate step. Since it is highly desirable to maintain diagonal dominance in the application of the Thomas algorithm, care should be taken when this latter procedure is used to ensure that $\omega \leqslant 1+\beta^2$.

In the SOR procedure by lines, one iterative cycle is completed when the tridiagonal inversion has been applied to all the rows. The process is then repeated until convergence has been achieved. In applying the method to a standard example problem with Dirichlet boundary conditions, Ames (1977) indicates that only $1/\sqrt{2}$ as many iterations would be required as for Gauss-Seidel iteration with SOR to reduce the initial errors by the same amount. On the other hand, use of the Thomas algorithm is expected to increase the computation time per iteration cycle somewhat.

The improved convergence rates observed for block-iterative methods compared with point-iterative methods might be thought of as being due to the greater influence exerted by the boundary values in each iterative pass. For example, in SOR by rows, the unknowns in each row are determined simultaneously so it is possible for the boundary values to influence all the unknowns in the row in one iteration. This is not the case for point-iterative methods such as the Gauss-Seidel procedure where in the first pass, at least one of the boundary points (details depend on sequence used in sweeping) only influences adjacent points.

ADI methods. The SOR method by lines proceeds by taking all lines in the same direction in a repetitive manner. The convergence rate can often be improved by following the sequence by rows, say, by a second sequence in the column direction. Thus, a complete iteration cycle would consist of a sweep over all rows followed by a sweep over the columns. Several closely related ADI forms are observed in practice. Perhaps the simplest procedure is to first employ Eq. (4-124) to sweep by rows. We will designate the values so determined by a $k+\frac{1}{2}$ superscript. This is followed by a sweep by columns using

$$u_{i,j}^{k+1} = \frac{u_{i+1,j}^{k+1/2} + u_{i-1,j}^{k+1} + \beta^2 (u_{i,j+1}^{k+1} + u_{i,j}^{k+1})}{2(1+\beta^2)}$$

This completes one iteration and over-relaxation can be achieved by applying Eq. (4-122) to all grid points as a second step before another iterative sweep is carried out. Alternatively, we can include the over-relaxation as part of the row and column sweeps using first for rows,

$$u_{i,j}^{k+1/2} = (1-\omega)u_{i,j}^{k} + \frac{\omega}{2(1+\beta^2)} [u_{i+1,j}^{k+1/2} + u_{i-1,j}^{k+1/2} + \beta^2 (u_{i,j+1}^{k} + u_{i,j-1}^{k+1/2})] \quad (4\text{-}125a)$$

and then, for columns,

$$u_{i,j}^{k+1} = (1-\omega)u_{i,j}^{k+1/2} + \frac{\omega}{2(1+\beta^2)} [u_{i+1,j}^{k+1/2} + u_{i-1,j}^{k+1} + \beta^2 (u_{i,j+1}^{k+1} + u_{i,j-1}^{k+1})] \quad (4\text{-}125b)$$

To preserve diagonal dominance in the Thomas algorithm requires $\omega \leqslant 1+\beta^2$ in the sweep by rows and $\omega \leqslant (1+\beta^2)/\beta^2$ in the sweep by columns.

Schemes patterned after the ADI procedures for the two-dimensional heat equation [Eq. (4-97)] are also very commonly used for obtaining solutions to Laplace's equations. If the boundary conditions for an unsteady problem governed by Eq. (4-97) are independent of time, the solution will asymptotically approach a steady-state distribution which satisfies Laplace's equation. Since we are only interested in the "steady-state" solution, the size of the time step can be selected with a view toward speeding convergence of the iterative process. Letting $\alpha\,\Delta t/2 = \rho_k$ in Eq. (4-103) we can write the Peaceman-Rachford ADI scheme for solving Laplace's equation as the two-step procedure,

Step 1: $$u_{i,j}^{k+1/2} = u_{i,j}^{k} + \rho_k(\hat{\delta}_x^2 u_{i,j}^{k+1/2} + \hat{\delta}_y^2 u_{i,j}^{k}) \quad (4\text{-}126a)$$

Step 2: $$u_{i,j}^{k+1} = u_{i,j}^{k+1/2} + \rho_k(\hat{\delta}_x^2 u_{i,j}^{k+1/2} + \hat{\delta}_y^2 u_{i,j}^{k+1}) \quad (4\text{-}126b)$$

where $\hat{\delta}_x^2$ and $\hat{\delta}_y^2$ are defined by Eq. (4-100).

Step 1 proceeds using the Thomas algorithm by rows and Step 2 completes the iteration cycle by applying the Thomas algorithm by columns. The ρ_k's are known as iteration parameters and Mitchell and Griffiths (1980) show that the Peaceman-Rachford iterative procedure for solving Laplace's equation in a square is convergent for any fixed value of ρ_k. On the other hand, for maximum computational efficiency, the iteration parameters should be varied with k, but the same ρ_k should be used in both steps of the iterative cycle. The key to using the ADI method for elliptic problems most efficiently lies in the proper choice of ρ_k's. Peaceman and Rachford (1955) suggested one procedure and another in common usage was suggested by Wachspress (1966). Although the evidence is not conclusive, some studies have suggested that the Wachspress parameters are superior to those suggested by Peaceman and Rachford. The reader is encouraged to study the literature regarding the selection of ρ_k's prior to using the Peaceman-Rachford ADI method.

It is difficult to compare the computational times required by point- and block-iterative methods with SOR because of the difficulty in establishing the optimum value of the over-relaxation factor. Conclusions are also very much dependent upon the specific problem considered, boundary conditions, and the number of grid points involved. The block-iterative methods as a class require fewer iterations than point-iterative methods but, as was mentioned earlier, more computational effort is required by each iteration. Experience suggests that SOR by lines will require very close to the same computation time as the Gauss-Seidel procedure with SOR for convergence to the same level for most problems. Use of an ADI procedure with SOR (fixed parameter) often provides a savings in computer time of 20–40% over that required by the Gauss-Seidel procedure with SOR. A greater savings can normally be observed if the iteration parameters are suitably varied in the ADI procedure.

Strongly-implicit methods. In recent years another type of block-iterative procedure has been gaining favor as an efficient method for solving the algebraic equations arising from the numerical solution of elliptic PDE's. To illustrate this approach, let us consider the system of algebraic equations arising from the use of the five-point difference scheme for Laplace's equation as

$$[A]\mathbf{u} = \mathbf{C}$$

where $[A]$ is the relatively sparse matrix of known coefficients, $\mathbf{u}$ is the column vector of unknowns, and $\mathbf{C}$ is a column vector of known quantities. Stone (1968) proposed the following "factorization" strategy which is known as the strongly-implicit procedure (SIP). The objective is to replace the sparse matrix $[A]$ by a modified form $[A+P]$ such that the modified matrix can be decomposed into upper and lower triangular sparse matrices denoted by $[U]$ and $[L]$, respectively. If the $[L]$ and $[U]$ matrices are not sparse, then very little will be gained in computational efficiency over the use of Gaussian elimination. Thus, the key to any computational advantage of the SIP procedure lies in the manner in which $[P]$ is selected. It is essential that the elements of $[P]$ be small in magnitude and permit the set of equations to remain more strongly implicit than for the ADI procedure. An iterative procedure is defined by writing $[A]\mathbf{u} = \mathbf{C}$ as

$$[A+P]\mathbf{u}^{n+1} = \mathbf{C} + [P]\mathbf{u}^n$$

Decomposing $[B] = [A+P]$ into the upper and lower triangular matrices $[U]$ and $[L]$ permits our system to be written as

$$[L][U]\mathbf{u}^{n+1} = \mathbf{C} + [P]\mathbf{u}^n$$

Defining an intermediate vector as $\mathbf{V}^{n+1} = [U]\mathbf{u}^{n+1}$, we form the following two-step algorithm

Step 1: $$[L]\mathbf{V}^{n+1} = \mathbf{C} + [P]\mathbf{u}^n \tag{4-127a}$$

Step 2: $$[U]\mathbf{u}^{n+1} = \mathbf{V}^{n+1} \tag{4-127b}$$

which is repeated iteratively. Step 1 consists simply of a forward substitution. This is followed by the backward substitution indicated by Step 2.

Stone (1968) selected $[P]$ so that $[L]$ and $[U]$ have only three nonzero diagonals with the principal diagonal of $[U]$ being the unity diagonal. Furthermore, the elements of $[L]$ and $[U]$ were determined such that the coefficients in the $[B]$ matrix in the locations of the nonzero entries of matrix $[A]$ are identical with those in $[A]$. Two additional nonzero diagonals appear in $[B]$. The elements of $[L]$, $[U]$, and $[P]$ can be determined from the defining equations established by forming the $[L][U]$ product. The details of this are given by Stone (1968). The procedure is implicit in both the x and y directions. Studies have indicated that, for a solution to Laplace's equations, the method requires only on the order of 50–60% of the computation time required for the ADI schemes.

Schneider and Zedan (1981) recently proposed an alternative procedure for establishing the $[L][U]$ matrices which is reported to reduce the computational cost for a converged solution to Laplace's equation by a factor of two to four over the procedures proposed by Stone (1968). They refer to their alternative procedure as the modified strongly-implicit (MSI) procedure. The basic two-step iterative sequence remains the same as given above in Eqs. (4-127*a*) and (4-127*b*). The improvement apparently results from extending the approach of Stone to a nine-point formulation. The MSI procedure then easily treats five-point difference representations as a special case and the great reduction in computational cost mentioned above (a factor of two to four) applies to use of the five-point representation. This new scheme appears to hold great promise as a very efficient and general procedure. Further details on the MSI procedure are given in Appendix C.

4-4 BURGERS' EQUATION (INVISCID)

We have discussed finite-difference methods and have applied them to simple linear problems. This has provided an understanding of the various techniques and acquainted us with the peculiarities of each approach. Unfortunately, the usual fluid mechanics problem is highly nonlinear. The governing partial differential equations form a nonlinear system which must be solved for the unknown pressures, densities, temperatures, and velocities.

A single equation that could serve as a nonlinear analog of the fluid mechanics equations would be very useful. This single equation must have terms which closely duplicate the physical properties of the fluid equations, i.e., the model equation should have a convective term, a diffusive or dissipative term, and a time-dependent term. Burgers (1948) introduced a simple nonlinear equation which meets these requirements

$$\underbrace{\frac{\partial u}{\partial t}}_{\substack{\text{Unsteady}\\ \text{term}}} + \underbrace{u\frac{\partial u}{\partial x}}_{\substack{\text{Convective}\\ \text{term}}} = \underbrace{\mu\frac{\partial^2 u}{\partial x^2}}_{\substack{\text{Viscous}\\ \text{term}}} \tag{4-128}$$

Equation (4-128) is parabolic when the viscous term is included. If the viscous term is neglected, the remaining equation is composed of the unsteady term and a nonlinear convection term. The resulting hyperbolic equation

$$\frac{\partial u}{\partial t} + u\frac{\partial u}{\partial x} = 0 \tag{4-129}$$

may be viewed as a simple analog of the Euler equations for the flow of an inviscid fluid. Equation (4-129) is a nonlinear convection equation and possesses properties which need to be examined in some detail. Methods for solving the inviscid Burgers equation will be presented in this section. Typical results for a number of commonly used finite-difference methods are included and the effects of the nonlinear terms are discussed. A discussion of the viscous Burgers equation follows in Section 4-5.

Equation (4-129) may be viewed as a nonlinear wave equation where each point on the wave front can propagate with a different speed. In contrast, the speed of propagation of all signals or waves was constant for the linear, one-dimensional convection equation, Eq. (4-2). A consequence of the changing wave speed is the coalescence of characteristics and the formation of discontinuous solutions similar to shock waves in fluid mechanics. This means the class of solutions which include discontinuities can be studied with this simple one-dimensional model.

Nonlinear hyperbolic partial differential equations exhibit two types of solutions according to Lax (1954). For simplicity, we consider a simple scalar equation

$$\frac{\partial u}{\partial t} + \frac{\partial F}{\partial x} = 0 \tag{4-130}$$

For the general case, both the unknown u and the variable $F(u)$ are vectors. We may write Eq. (4-130) as

$$\frac{\partial u}{\partial t} + A\frac{\partial u}{\partial x} = 0 \tag{4-131}$$

where $A = A(u)$ is the Jacobian matrix $\partial F_i/\partial u_j$ for the general case and is dF/du for our simple equation. Our equation or system of equations is hyperbolic which means that the eigenvalues of the matrix A are all real. A *genuine solution* of Eq. (4-131) is one in which u is continuous but bounded discontinuities in the derivatives of u may occur (Lipschitz continuous). A *weak solution* of Eq. (4-131) is a solution which is genuine except along a surface in (x, t) space across which the function u may be discontinuous. A constraint is placed upon the jump in u across the discontinuity in the domain of interest. If w is a test vector which is continuous and has continuous first derivatives but vanishes outside some bounded set, then u is termed a weak solution of Eq. (4-130) if

$$\iint_D (w_t u + w_x F)\, dx\, dt + \int w(x, 0)\phi(x)\, dx = 0 \tag{4-132}$$

where $\phi(x) = u(x, 0)$. A genuine solution is a weak solution and a weak solution which is continuous is a genuine solution. A complete discussion of the weak solution concept may be found in the excellent texts by Whitham (1974) and Jeffrey and Taniuti (1964). The mathematical theory of weak solutions for hyperbolic equations is a relatively recent development. Clearly the existence of shock waves in inviscid supersonic flow is an example of a weak solution. It is interesting to recognize that the shock solutions in inviscid supersonic flow were known fifty to one hundred years before the theory of weak solutions for hyperbolic systems was developed.

Let us return to the study of the inviscid Burgers equation and develop the requirements for a weak solution, i.e., the requirements necessary for the existence of a solution with a discontinuity such as that shown in Fig. 4-23.

Let $w(x, t)$ be an arbitrary test function which is continuous and has continuous first derivatives. Let $w(x, t)$ vanish on the boundary B of the domain D and everywhere outside D (complement of D). D is an arbitrary rectangular domain in the (x, t) plane. Clearly we may write

$$\iint_D \left(\frac{\partial u}{\partial t} + \frac{\partial F}{\partial x}\right) w(x, t)\, dx\, dt = 0 \tag{4-133}$$

or

$$\iint_D (u w_t + F w_x)\, dx\, dt = 0 \tag{4-134}$$

Equations (4-133) and (4-134) are equivalent when both u and F are continuous and have continuous first derivatives. The second integral of Eq. (4-132) does not appear since the function w vanishes on the boundary. Functions $u(x, t)$, which satisfy Eq. (4-134) for all test functions w, are called weak solutions of the inviscid Burgers equation. We do not require that u be differentiable in order to satisfy Eq. (4-134).

Suppose our domain D is now a rectangular region in the (x, t) plane which is separated by a curve $\tau(x, t) = 0$ across which u is discontinuous. We assume that u is

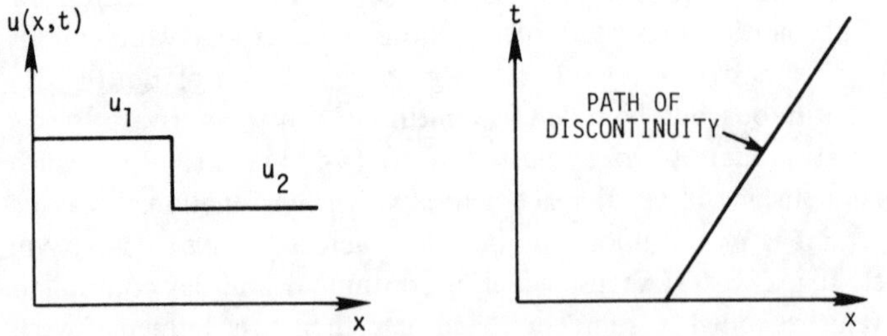

Figure 4-23 Typical traveling discontinuity problem for Burgers equation.

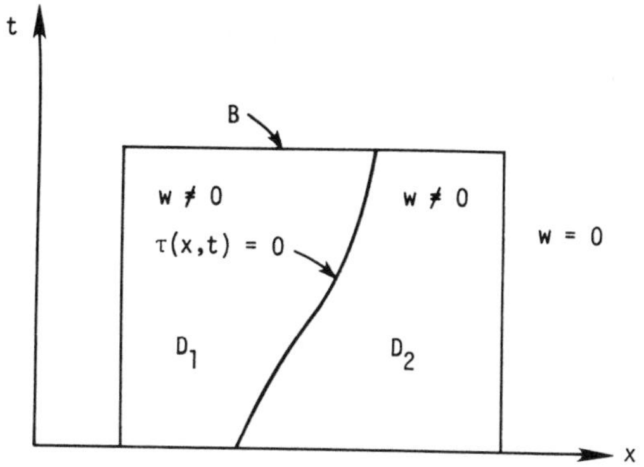

Figure 4-24 Schematic representation of an arbitrary domain with a discontinuity.

continuous and has continuous derivatives to the left of $\tau(D_1)$ and to the right of $\tau(D_2)$. Let the test function vanish on the boundary of D and outside of D. With these restrictions, Eq. (4-134) can be integrated by parts to yield

$$\iint_{D_1} \left(\frac{\partial u}{\partial t} + \frac{\partial F}{\partial x} \right) w \, dx \, dt + \iint_{D_2} \left(\frac{\partial u}{\partial t} + \frac{\partial F}{\partial x} \right) w \, dx \, dt$$

$$+ \int_{\tau} ([u] \cos \alpha_1 + [F] \cos \alpha_2) \, ds = 0 \tag{4-135}$$

The last integrand is evaluated along the curve $\tau(x, t) = 0$ separating the two regions D_1 and D_2. This integral occurs through the limits of the integration by parts on the discontinuity surface $\tau(x, t) = 0$. The bracket [ ] denotes the jump in the quantity across the discontinuity and $\cos \alpha_1$, $\cos \alpha_2$ are the cosines of the angles between the normal to $\tau(x, t) = 0$ and the t and x directions, respectively. The problem is illustrated in Fig. 4-24.

The integrals in Eq. (4-135) over D_1 and D_2 are zero by Eq. (4-133). We conclude that since the last integral vanishes for all test functions w with the required properties, we must have

$$[u] \cos \alpha_1 + [F] \cos \alpha_2 = 0 \tag{4-136}$$

This is the condition that u be a weak solution for Burgers' equation. Let us apply this condition to a moving discontinuity. Suppose initial data are prescribed for $u(x, 0)$ as shown in Fig. 4-23 where u_1 and u_2 denote the values to the left and to the right of the discontinuity. In one dimension we may write the equation of the surface $\tau(x, t) = 0$ as $t - t_1(x) = 0$. The direction cosines as required in Eq. (4-136) become

$$\cos \alpha_1 = \frac{1}{[1 + t_1'^2]^{1/2}} \qquad \cos \alpha_2 = -\frac{t_1'}{[1 + t_1'^2]^{1/2}}$$

where the prime denotes differentiation with respect to x. Thus,

$$\frac{[u]}{[1 + t'^2]^{1/2}} - \frac{[F]t'}{[1 + t'^2]^{1/2}} = 0$$

or

$$u_2 - u_1 = \frac{u_2^2 - u_1^2}{2} \frac{dt}{dx}$$

Therefore

$$\frac{dx}{dt} = \frac{u_1 + u_2}{2} \tag{4-137}$$

which shows that the discontinuity travels at the average value of the u function across the wave front. Since we now see that a discontinuity in u simply propagates at constant speed $(u_1 + u_2)/2$ with uniform states on each side, a numerical solution of a similar problem for a discontinuity can be compared with the exact solution. These comparisons are presented for a number of finite-difference methods in this section.

Rarefactions are as prevalent in high-speed flows as shock waves and the exact solution of Burgers' equation for a rarefaction is known. Consider initial data $u(x, 0)$ as shown in Fig. 4-25. The characteristic for Burgers' equation is given by

$$\frac{dt}{dx} = \frac{1}{u} \tag{4-138}$$

Figure 4-26 shows the characteristic diagram plotted in the (x, t) plane. In the left half plane, the characteristics are simply vertical lines while they are lines at an angle of $\pi/4$ radians to the right of the characteristic which bounds the expansion. This particular problem is similar to a centered expansion wave in compressible flow. Here the expansion is bounded by the $x = 0$ axis and the characteristic originating at the origin is denoted by the dashed line. The solution for this problem may be written

$$u = 0 \qquad x \leqslant 0$$

$$u = \frac{x}{t} \qquad 0 < x < t$$

$$u = 1 \qquad x \geqslant t$$

The initial distribution of u forms a centered expansion where the width of the expansion grows linearly with time.

We have examined two problems, shocks and rarefactions, which are frequently

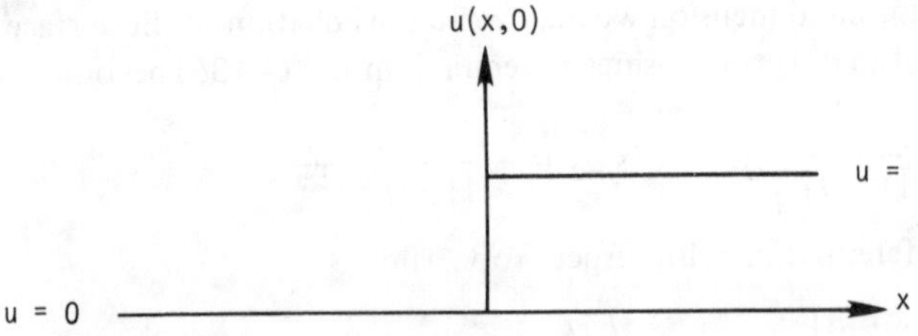

Figure 4-25 Initial data for rarefaction wave.

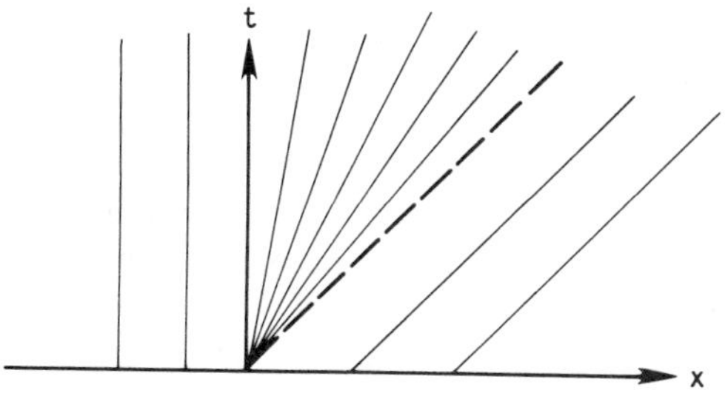

Figure 4-26 Characteristics for centered expansion.

encountered in high-speed flows by using the simple analog provided by Burgers equation. Clearly these types of solutions can occur in systems of nonlinear equations of the hyperbolic type. Armed with simple analytic solutions for these two important cases, let us examine the application of some finite-difference algorithms to the nonlinear, inviscid Burgers equation.

4-4.1 Lax Method

First-order methods for solving hyperbolic equations are infrequently used. The Lax method (1954) is presented as a typical first-order method to demonstrate the application to a nonlinear equation and the dissipative character of the result.

The conservation form of the basic partial differential equation

$$\frac{\partial u}{\partial t} + \frac{\partial F}{\partial x} = 0$$

is used for all examples which follow. For the Lax method, we expand in a Taylor series about the point (x, t) retaining only the first two terms

$$u(x, t + \Delta t) = u(x, t) + \Delta t \left(\frac{\partial u}{\partial t}\right)_{x,t} + \cdots$$

and substitute for the time derivative

$$u(x, t + \Delta t) = u(x, t) - \Delta t \left(\frac{\partial F}{\partial x}\right)_{x,t} + \cdots$$

Using centered differences and averaging the first term yields the Lax method (see Section 4-1.3)

$$u_j^{n+1} = \frac{u_{j+1}^n + u_{j-1}^n}{2} - \frac{\Delta t}{\Delta x}\,\frac{F_{j+1}^n - F_{j-1}^n}{2} \tag{4-139}$$

In Burgers equation $F = u^2/2$. The amplification factor in this case is

$$G = \cos\beta - i\,\frac{\Delta t}{\Delta x} A \sin\beta \tag{4-140}$$

where A is the Jacobian dF/du which is just the single element, u, for Burgers' equation. The stability requirement for this method is

$$\left| \frac{\Delta t}{\Delta x} u_{\max} \right| \leqslant 1 \tag{4-141}$$

because $u_{\max}$ is the maximum eigenvalue of the A matrix with the single element u.

The Lax method applied to a 1-0 right-moving discontinuity produces the solutions shown in Fig. 4-27. The location of the moving discontinuity is correctly predicted but the dissipative nature of the method is evident in the smearing of the discontinuity over several mesh intervals. As previously noted, this smearing becomes worse as the Courant number decreases. It is of interest to note that the application of the Lax method to Burgers' equation with a discontinuity produces the double point solutions as shown. A further comment on these results is in order. Notice that the computed solutions are monotone, i.e., the solution does not oscillate. Godunov (1959) has shown that monotone behavior of a solution cannot be assured for finite-difference methods with more than first-order accuracy. This monotone property is very desirable when discontinuities are computed as part of the solution. Unfortunately, the desirability of monotone behavior must be reconciled with the highly dissipative character of the results. The relative importance of these properties must be carefully evaluated for each case.

4-4.2 The Lax-Wendroff Method

The Lax-Wendroff method (Lax and Wendroff, 1960) was one of the first second-order, finite-difference methods for hyperbolic partial differential equations. The development of the Lax-Wendroff scheme for nonlinear equations again follows from a Taylor series:

$$u(x, t + \Delta t) = u(x, t) + \Delta t \left(\frac{\partial u}{\partial t}\right)_{x,t} + \frac{\Delta t^2}{2} \left(\frac{\partial^2 u}{\partial t^2}\right)_{x,t} + \cdots$$

The first time derivative can be directly replaced using the differential equation but we need to examine the second derivative term in more detail. We consider the original equation in the form

$$\frac{\partial u}{\partial t} = -\frac{\partial F}{\partial x}$$

Figure 4-27 Numerical solution of Burgers' equation using Lax method.

Taking the time derivative of this expression yields

$$\frac{\partial^2 u}{\partial t^2} = -\frac{\partial^2 F}{\partial t\,\partial x} = -\frac{\partial^2 F}{\partial x\,\partial t} = -\frac{\partial}{\partial x}\left(\frac{\partial F}{\partial t}\right)$$

where the order of differentiation on F has been interchanged. Now $F = F(u)$ which permits us to write

$$\frac{\partial u}{\partial t} = -\frac{\partial F}{\partial x} = -\frac{\partial F}{\partial u}\,\frac{\partial u}{\partial x} = -A\,\frac{\partial u}{\partial x}$$

and

$$\frac{\partial F}{\partial t} = \frac{\partial F}{\partial u}\,\frac{\partial u}{\partial t} = A\,\frac{\partial u}{\partial t}$$

Hence, we may replace $\partial F/\partial t$ with

$$\frac{\partial F}{\partial t} = -A\,\frac{\partial F}{\partial x}$$

so that

$$\frac{\partial^2 u}{\partial t^2} = \frac{\partial}{\partial x}\left(A\,\frac{\partial F}{\partial x}\right)$$

The Jacobian, A, contains a single element for Burgers' equation. It is clear that A is a matrix when u and F are vectors in treating a system of equations. Making the appropriate substitutions in the Taylor-series expansion for u, we obtain

$$u(x, t + \Delta t) = u(x, t) - \Delta t\,\frac{\partial F}{\partial x} + \frac{(\Delta t)^2}{2}\,\frac{\partial}{\partial x}\left(A\,\frac{\partial F}{\partial x}\right) + \cdots$$

After using central differencing, the Lax-Wendroff method is obtained

$$u_j^{n+1} = u_j^n - \frac{\Delta t}{\Delta x}\,\frac{F_{j+1}^n - F_{j-1}^n}{2} + \frac{1}{2}\left(\frac{\Delta t}{\Delta x}\right)^2 [A_{j+1/2}^n (F_{j+1}^n - F_j^n) - A_{j-1/2}^n (F_j^n - F_{j-1}^n)] \tag{4-142}$$

The Jacobian matrix is evaluated at the half interval, i.e.,

$$A_{j+1/2} = A\left(\frac{u_j + u_{j+1}}{2}\right)$$

In Burgers equation $F = u^2/2$ and $A = u$. In this case $A_{j+1/2} = (u_j + u_{j+1})/2$ and $A_{j-1/2} = (u_j + u_{j-1})/2$. The amplification factor for this method is

$$G = 1 - 2\left(\frac{\Delta t}{\Delta x}\,A\right)^2 (1 - \cos\beta) - 2i\,\frac{\Delta t}{\Delta x}\,A\,\sin\beta \tag{4-143}$$

and the stability requirement reduces to $|(\Delta t/\Delta x)u_{max}| \leqslant 1$.

The results obtained when the Lax-Wendroff method is applied to our example problem is shown in Fig. 4-28. The right-moving discontinuity is correctly positioned

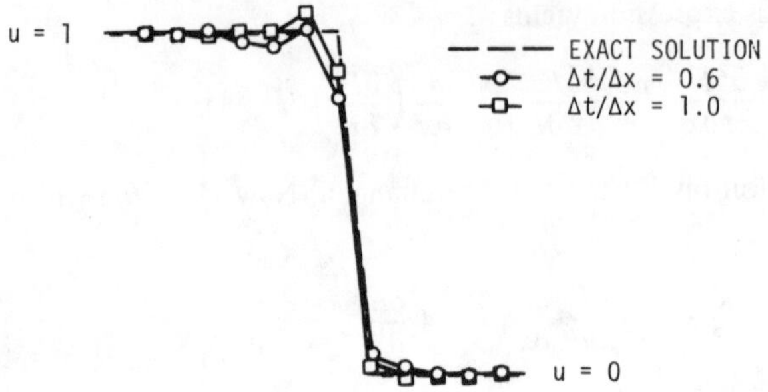

Figure 4-28 Application of the Lax-Wendroff method to the inviscid Burgers equation.

and is sharply defined. The dispersive nature of this method is evidenced through the presence of oscillations near the discontinuity. Even though the method uses central differences, some asymmetry will occur since the wave is moving. The solution shows more oscillations when a Courant number of 0.6 is used than for a Courant number of 1.0. In general, as the Courant number is reduced, the quality of the solution will be degraded (see Section 4-1.6).

4-4.3 The MacCormack Method

MacCormack's method is a predictor-corrector version of the Lax-Wendroff scheme as has been discussed in Section 4-1.8. This method is much easier to apply than the Lax-Wendroff scheme because the Jacobian does not appear. When applied to the inviscid Burgers equation, the MacCormack method becomes

$$u_j^{\overline{n+1}} = u_j^n - \frac{\Delta t}{\Delta x}(F_{j+1}^n - F_j^n)$$

$$u_j^{n+1} = \frac{1}{2}\left[u_j^n + u_j^{\overline{n+1}} - \frac{\Delta t}{\Delta x}(F_j^{\overline{n+1}} - F_{j-1}^{\overline{n+1}})\right] \tag{4-144}$$

The amplification factor and stability requirement are the same as presented for the Lax-Wendroff method. The results of applying this method are shown in Fig. 4-29. Again the right-moving wave is well defined. We note that the solutions obtained for the same problem at the same Courant number are different from those obtained using the Lax-Wendroff scheme. This is due both to the switched differencing in the

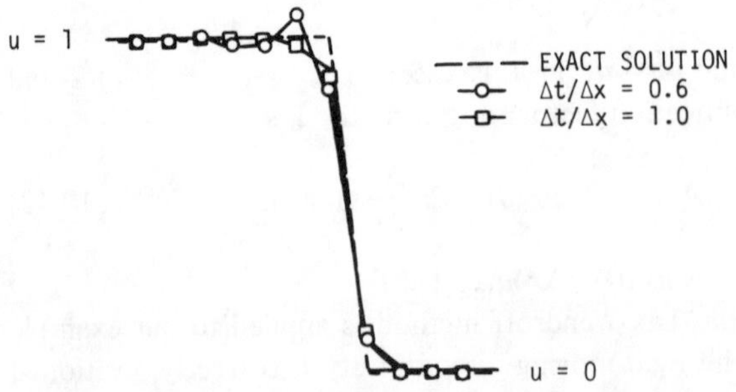

Figure 4-29 Solution of Burgers' equation using MacCormack's method.

predictor and the corrector and the nonlinear nature of the governing PDE. One should expect results which show some differences even though both methods are equivalent for linear problems.

In general, the MacCormack method provides very good resolution at discontinuities. It should be noted in passing that reversing the differencing in the predictor and corrector steps leads to quite different results. The best resolution of discontinuities occurs when the difference in the predictor is in the direction of propagation of the discontinuity. This will be apparent when problems at the end of the chapter are completed.

4-4.4 The Rusanov or Burstein-Mirin Method

The third-order Rusanov or Burstein-Mirin method was discussed in Section 4-1.11. This method uses central differencing and when applied to Eq. (4-130) becomes

$$u_{j+1/2}^{(1)} = \frac{1}{2}(u_{j+1}^n + u_j^n) - \frac{1}{3}\frac{\Delta t}{\Delta x}(F_{j+1}^n - F_j^n)$$

$$u_j^{(2)} = u_j^n - \frac{2}{3}\frac{\Delta t}{\Delta x}(F_{j+1/2}^{(1)} - F_{j-1/2}^{(1)})$$

$$u_j^{n+1} = u_j^n - \frac{1}{24}\frac{\Delta t}{\Delta x}(-2F_{j+2}^n + 7F_{j+1}^n - 7F_{j-1}^n + 2F_{j-2}^n) - \frac{3}{8}\frac{\Delta t}{\Delta x}(F_{j+1}^{(2)} - F_{j-1}^{(2)})$$

$$- \frac{\omega}{24}(u_{j+2}^n - 4u_{j+1}^n + 6u_j^n - 4u_{j-1}^n + u_{j-2}^n) \tag{4-145}$$

The last term in the third step represents a fourth derivative term

$$(\Delta x)^4 \frac{\partial^4 u}{\partial x^4}$$

and is added for stability. The third-order accuracy of the method is unaffected since this added term is of $O[(\Delta x)^4]$. A stability analysis of this method shows that the amplification factor is

$$G = 1 - \left(\frac{\Delta t}{\Delta x}u\right)^2 \frac{\sin^2\beta}{2} - \frac{\omega}{6}(1-\cos\beta) + i\frac{\Delta t}{\Delta x}u\sin\beta$$

$$\times \left\{1 + \frac{1}{3}(1-\cos\beta)\left[1 - \left(\frac{\Delta t}{\Delta x}u\right)^2\right]\right\} \tag{4-146}$$

It follows that stability is assured for Burgers' equation if

$$|\nu| \leqslant 1 \qquad \text{or} \qquad \left|\frac{\Delta t}{\Delta x}u_{\max}\right| \leqslant 1$$

and

$$4\nu^2 - \nu^4 \leqslant \omega \leqslant 3 \tag{4-147}$$

Application of this method to Burgers' equation for a right-moving shock produces the results shown in Fig. 4-30. The magnitude and position of the discontinuity are

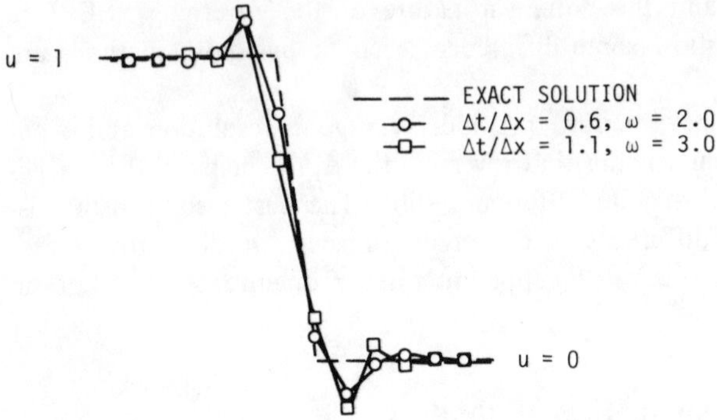

Figure 4-30 Rusanov method applied to Burgers' equation.

correctly produced but the results show an overshoot on both sides of the shock front. A schematic showing the numerical solution as it is computed from the base points is shown in Fig. 4-31.

4-4.5 The Warming-Kutler-Lomax Method

Warming et al. (1973) developed a third-order scheme using noncentered differences. This technique uses the MacCormack method for the first two levels evaluated at $\frac{2}{3}\,\Delta t$. The advantage of this method over the Rusanov technique is that only values at integral mesh points are required in the calculation.

The WKL method is

$$u_j^{(1)} = u_j^n - \frac{2}{3}\frac{\Delta t}{\Delta x}(F_{j+1}^n - F_j^n)$$

$$u_j^{(2)} = \frac{1}{2}\left[u_j^n + u_j^{(1)} - \frac{2}{3}\frac{\Delta t}{\Delta x}(F_j^{(1)} - F_{j-1}^{(1)})\right]$$

$$u_j^{n+1} = u_j^n - \frac{1}{24}\frac{\Delta t}{\Delta x}(-2F_{j+2}^n + 7F_{j+1}^n - 7F_{j-1}^n + 2F_{j-2}^n) - \frac{3}{8}\frac{\Delta t}{\Delta x}(F_{j+1}^{(2)} - F_{j-1}^{(2)})$$

$$- \frac{\omega}{24}(u_{j+2}^n - 4u_{j+1}^n + 6u_j^n - 4u_{j-1}^n + u_{j-2}^n) \tag{4-148}$$

The third level for the WKL scheme is exactly the same as that used in the Rusanov technique. We should note that different third-order schemes can be generated by

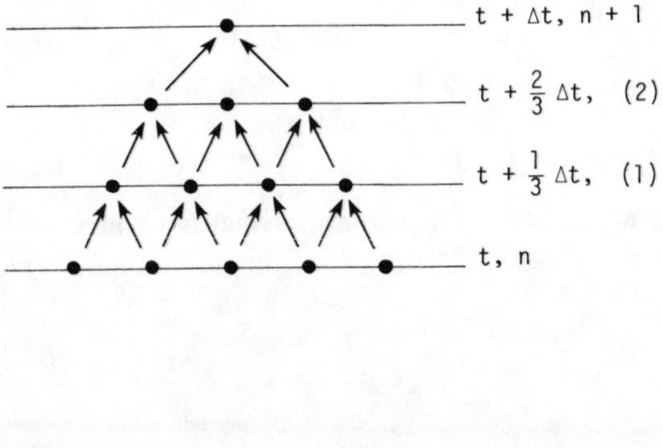

Figure 4-31 Point pyramid for the Rusanov method.

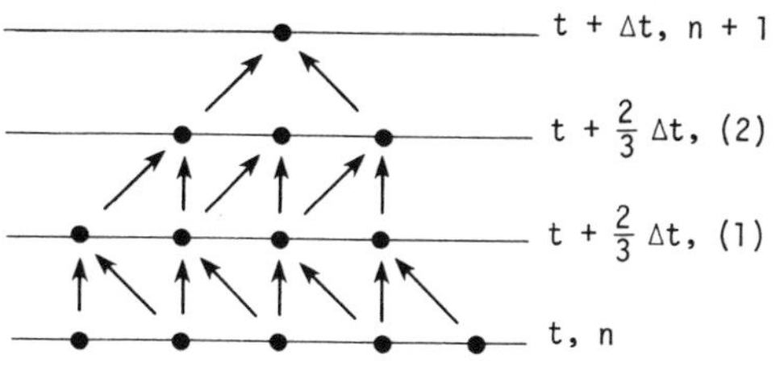

Figure 4-32 Point schematic for WKL method.

altering the first two steps. Burstein and Mirin have shown that any second-order method may be used to generate $u_j^{(2)}$. The linear stability bound for the WKL and the Rusanov methods is the same as given in Eq. (4-147). The schematic illustrating the grid points used in the WKL method is presented in Fig. 4-32. Notice that the preferential treatment of the first two levels is readily apparent in this diagram. The differencing in the first two levels can be reversed or even cycled from time step to time step.

The results of using the WKL method to solve Burgers' equation for a right-moving discontinuity are shown in Fig. 4-33. The solution is nearly the same as that obtained in the previous section. Based upon the calculated results, either of the third-order methods may be used with approximately equal accuracy.

4-4.6 Tuned Third-Order Methods

The parameter ω, which appears in the third level of the methods of the previous two sections, can be chosen arbitrarily as long as the stability bound is not violated. Once ω is selected at the beginning of a calculation, it retains the same value throughout the mesh. However, if the numerical damping term is written in conservation-law form for the third level, i.e.,

$$\frac{\partial}{\partial x}\left(\omega \frac{\partial^3 u}{\partial x^3}\right)$$

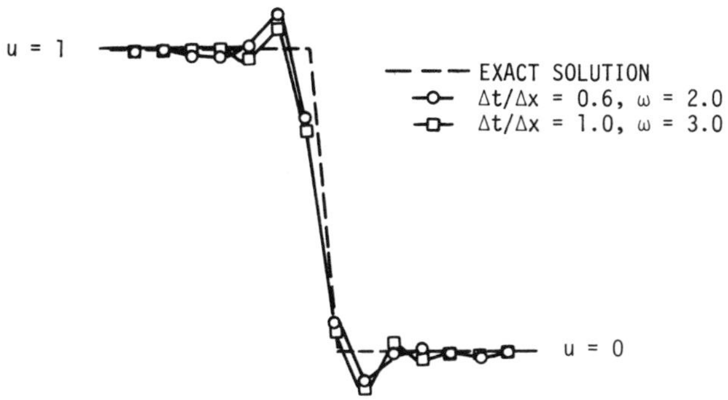

Figure 4-33 Burgers' equation solution using the WKL method.

then ω may be altered from point to point in the calculation and correct flux conservation in the mesh is assured. Using this approach, the ω term in the last level of either the Rusanov or WKL method can be written as

$$-\frac{\omega_{j+1/2}^n}{24}(u_{j+2}^n - 3u_{j+1}^n + 3u_j^n - u_{j-1}^n) + \frac{\omega_{j-1/2}^n}{24}(u_{j+1}^n - 3u_j^n + 3u_{j-1}^n - u_{j-2}^n) \tag{4-149}$$

The $\omega_{j\pm 1/2}^n$ values are now varied according to the effective Courant number in the mesh. Warming et al. (1973) suggest that these parameters be calculated at each point in the mesh to minimize either the dispersive error or the dissipative error.

A discussion of the modified equation for third-order methods was presented in Section 4-1.11. If the minimum dispersive error is desired, then according to Eq. (4-68) we should choose

$$\omega_{j\pm 1/2}^n = \frac{(4\nu_{j\pm 1/2}^2 + 1)(4 - \nu_{j\pm 1/2}^2)}{5} \tag{4-150}$$

It remains to arrive at a rational method to determine the effective Courant numbers, $\nu_{j\pm 1/2}$. Warming et al. (1973) suggest that the effective Courant numbers, used to determine the $\omega_{j\pm 1/2}$ parameters, be the average value at the mesh points used in the difference formula. Since the term containing $\omega_{j+1/2}$ involves points $j+2, j+1, j$, and $j-1$, we can write

$$\nu_{j+1/2} = \frac{1}{4}(\lambda_{j+2} + \lambda_{j+1} + \lambda_j + \lambda_{j-1})\frac{\Delta t}{\Delta x} \tag{4-151}$$

and similarly

$$\nu_{j-1/2} = \frac{1}{4}(\lambda_{j+1} + \lambda_j + \lambda_{j-1} + \lambda_{j-2})\frac{\Delta t}{\Delta x}$$

where λ is the local eigenvalue. For Burgers' equation, λ is just the unknown u. Results obtained using this variable ω or tuned approach are shown in Fig. 4-34. This shows that both third-order methods provide satisfactory solutions for the minimum disper-

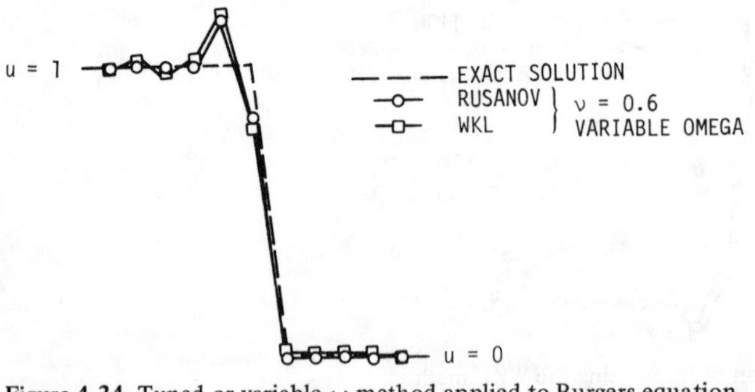

Figure 4-34 Tuned or variable ω method applied to Burgers equation.

sion case. A slightly larger overshoot occurs at the left of the discontinuity but a nearly exact solution is obtained on the right. The minimum dissipative method of computing $\omega_{j\pm 1/2}$ is not recommended. The ω parameter was added to provide stability and when the dissipation is minimized, stability problems can occur. Even for stable solutions, large oscillations may be present. It should be noted that the parameters $\omega_{j\pm 1/2}$ may be computed using any technique which does not violate the stability bound. Clearly, a different computed solution will be obtained for each way of computing these parameters.

4-4.7 Implicit Methods

The time-centered implicit method (trapezoidal method) was presented in Section 4-1.10. This scheme is based upon Eq. (4-57). If we substitute into Eq. (4-58) for the time derivatives using our model equation, we obtain

$$u_j^{n+1} = u_j^n - \frac{\Delta t}{2}\left[\left(\frac{\partial F}{\partial x}\right)^n + \left(\frac{\partial F}{\partial x}\right)^{n+1}\right] \tag{4-152}$$

It is immediately apparent that we now have a nonlinear problem and some sort of linearization or iteration technique must be used. Beam and Warming (1976) have suggested that we write

$$F^{n+1} \approx F^n + \left(\frac{\partial F}{\partial u}\right)^n (u^{n+1} - u^n) = F^n + A^n(u^{n+1} - u^n)$$

Thus

$$u_j^{n+1} = u_j^n - \frac{\Delta t}{2}\left\{2\left(\frac{\partial F}{\partial x}\right)^n + \frac{\partial}{\partial x}\,[A(u_j^{n+1} - u_j^n)]\right\}$$

If the x derivatives are replaced by second-order central differences, then

$$-\frac{\Delta t A_{j-1}^n}{4\,\Delta x}\,u_{j-1}^{n+1} + u_j^{n+1} + \frac{\Delta t A_{j+1}^n}{4\,\Delta x}\,u_{j+1}^{n+1} = -\frac{\Delta t}{\Delta x}\,\frac{F_{j+1}^n - F_{j-1}^n}{2}$$
$$-\frac{\Delta t A_{j-1}^n u_{j-1}^n}{4\,\Delta x} + u_j^n + \frac{\Delta t A_{j+1}^n}{4\,\Delta x}\,u_{j+1}^n \tag{4-153}$$

The Jacobian A has the single element u for Burgers' equation and a further simplification of the right side is possible. We see that the linearization applied by Beam and Warming leads to a linear system of algebraic equations at the next time level. This is a tridiagonal system and may be solved using the Thomas algorithm.

As pointed out in Section 4-1.10, this method is stable for any time step. It should be noted that the roots of the characteristic equation always lie on the unit circle. This is consistent with the fact that the modified equation contains no even derivative terms. Consequently, artificial smoothing is added to the scheme. The usual fourth difference

$$-\frac{\omega}{8}\,(u_{j+2}^n - 4u_{j+1}^n + 6u_j^n - 4u_{j-1}^n + u_{j-2}^n) \tag{4-154}$$

may be added to Eq. (4-153) and the formal accuracy of the method is unaltered. According to Beam and Warming, the implicit formula Eq. (4-153) with explicit damping added is stable if

$$0 < \omega \leqslant 1 \tag{4-155}$$

Figure 4-35 shows the results of applying the time-centered implicit formula to a right-moving discontinuity. The solution with no damping is clearly unacceptable. When explicit damping given by Eq. (4-154) is added, better results are obtained.

In addition to the trapezoidal formula just presented, Beam and Warming (1976) developed a three-point-backward, implicit and an Euler implicit method as part of a family of techniques. The Beam and Warming version of the Euler implicit scheme follows from the backward Euler formula

$$u^{n+1} = u^n + \Delta t \left(\frac{\partial u}{\partial t}\right)^{n+1}$$

which for our nonlinear equation becomes

$$u^{n+1} = u^n - \Delta t \left(\frac{\partial F}{\partial x}\right)^{n+1}$$

If the same linearization is applied we obtain

$$-\frac{\Delta t A_{j-1}^n}{2\,\Delta x} u_{j-1}^{n+1} + u_j^{n+1} + \frac{\Delta t A_{j+1}^n}{2\,\Delta x} u_{j+1}^{n+1} = -\frac{\Delta t}{\Delta x}\frac{F_{j+1}^n - F_{j-1}^n}{2} - \frac{\Delta t A_{j-1}^n}{2\,\Delta x} u_{j-1}^n + u_j^n + \frac{\Delta t A_{j+1}^n u_{j+1}^n}{2\,\Delta x} \tag{4-156}$$

This is again a tridiagonal system and is easily solved. We note that this scheme is unconditionally stable but damping must be added such as that given in Eq. (4-154), to insure a usable numerical result.

A simpler form of the implicit algorithms presented in this section can be obtained if they are written in "delta" form. This form uses the increments in the con-

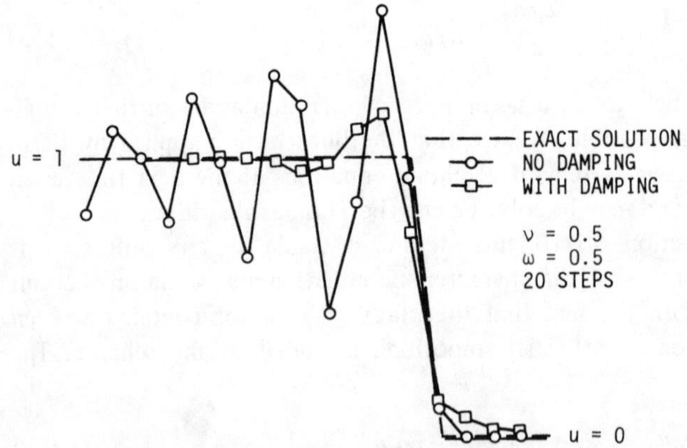

Figure 4-35 Solution of Burgers' equation using Beam-Warming (trapezoidal) method.

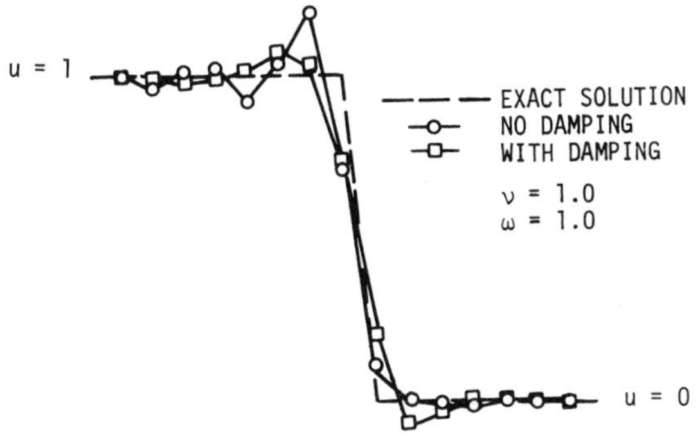

Figure 4-36 Solution for right-moving discontinuity time-centered implicit method, delta form.

served variables and fluxes. In multidimensional problems, it has the advantage of providing a steady-state solution which is independent of the time step in problems which possess a steady-state solution. Let us develop the time-centered implicit method using the delta form. Let $\Delta u_j = u_j^{n+1} - u_j^n$. The trapezoidal formula Eq. (4-152) may be written

$$\Delta u_j = -\frac{\Delta t}{2}\left[\left(\frac{\partial F}{\partial x}\right)^n + \left(\frac{\partial F}{\partial x}\right)^{n+1}\right]$$

Again a local linearization is used to obtain

$$F_j^{n+1} = F_j^n + A_j^n \, \Delta u_j$$

The final form of the difference equation becomes

$$-\frac{\Delta t A_{j-1}^n}{4\,\Delta x}\,\Delta u_{j-1} + \Delta u_j + \frac{\Delta t A_{j+1}^n}{4\,\Delta x}\,\Delta u_{j+1} = -\frac{\Delta t}{2\,\Delta x}\,(F_{j+1}^n - F_{j-1}^n) \tag{4-157}$$

This is much simpler than Eq. (4-153). The tridiagonal form is still retained but the right side does not require the multiplications of the original algorithm. This can be important for systems of equations where the operation count is large. The solution of Eq. (4-157) provides the incremental changes in the unknowns between two time levels. As noted previously, the stability of the delta form is unrestricted but the usual higher-order damping terms must be added. Results obtained using the delta form for our simple right-moving shock are shown in Fig. 4-36. The solutions with and without damping are essentially identical to those obtained using the expanded form, as should be expected. The delta form of the time-centered implicit scheme is recommended over the expanded version. In problems with time asymptotic solutions, the Δu terms approach zero and in all cases, matrix multiplications are reduced.

Solutions of the inviscid Burgers equation computed with an implicit scheme are generally inferior to those calculated with explicit techniques and more computational effort per integration step is required. In addition, transient results are usually desired and the larger step sizes permitted by implicit schemes are not of major significance. When discontinuities are present, results produced with explicit methods are superior to those produced with implicit techniques using central differences. For these reasons, explicit finite-difference methods such as MacCormack's scheme are recommended for solving the inviscid Burgers equation.

4-5 BURGERS' EQUATION (VISCOUS)

The complete nonlinear Burgers equation

$$\frac{\partial u}{\partial t} + u\,\frac{\partial u}{\partial x} = \mu\,\frac{\partial^2 u}{\partial x^2} \tag{4-158}$$

is a parabolic PDE which can serve as a model equation for the boundary-layer equations, the "parabolized" Navier-Stokes equations and the complete Navier-Stokes equations. In order to better model the steady boundary-layer and "parabolized" Navier-Stokes equations, the independent variables t and x can be replaced by x and y to give

$$\frac{\partial u}{\partial x} + u\,\frac{\partial u}{\partial y} = \mu\,\frac{\partial^2 u}{\partial y^2} \tag{4-159}$$

where x is the marching direction.

As with previous model equations, Burgers' equation has exact analytical solutions for certain boundary and initial conditions. These exact solutions are useful when comparing finite-difference methods. The exact steady-state solution [i.e., $\lim_{t\to\infty} u(x, t)$] of Eq. (4-158) for the boundary conditions

$$u(0, t) = u_0 \tag{4-160}$$

$$u(L, t) = 0 \tag{4-161}$$

is given by

$$u = u_0 \bar{u} \left\{ \frac{1 - \exp\,[\bar{u}\,\mathrm{Re}_L\,(x/L - 1)]}{1 + \exp\,[\bar{u}\,\mathrm{Re}_L\,(x/L - 1)]} \right\} \tag{4-162}$$

where

$$\mathrm{Re}_L = \frac{u_0 L}{\mu} \tag{4-163}$$

and $\bar{u}$ is a solution of the equation

$$\frac{\bar{u} - 1}{\bar{u} + 1} = \exp\,(-\bar{u}\,\mathrm{Re}_L) \tag{4-164}$$

For simplicity, the linearized Burgers equation

$$\frac{\partial u}{\partial t} + c\,\frac{\partial u}{\partial x} = \mu\,\frac{\partial^2 u}{\partial x^2} \tag{4-165}$$

is often used in place of Eq. (4-158). Note that if $\mu = 0$, the wave equation is obtained. If $c = 0$, the heat equation is obtained. The exact steady-state solution of Eq. (4-165) for the boundary conditions given by Eqs. (4-160) and (4-161) is

$$u = u_0 \left\{ \frac{1 - \exp\left[R_L(x/L - 1)\right]}{1 - \exp(-R_L)} \right\} \tag{4-166}$$

where

$$R_L = \frac{cL}{\mu}$$

The exact unsteady solution of Eq. (4-165) for the initial condition

$$u(x, 0) = \sin(kx)$$

and periodic boundary conditions is

$$u(x, t) = \exp(-k^2 \mu t) \sin k(x - ct) \tag{4-167}$$

This latter exact solution is useful in evaluating the temporal accuracy of a method.

Equations (4-158) and (4-165) can be combined into a generalized equation (Rakich, 1978)

$$u_t + (c + bu)u_x = \mu u_{xx} \tag{4-168}$$

where c and b are free parameters. If $b = 0$, the linearized Burgers equation is obtained and if $c = 0$ and $b = 1$, the nonlinear Burgers equation is obtained. If $c = \frac{1}{2}$ and $b = -1$, the generalized Burgers equation has the stationary solution

$$u = -\frac{c}{b}\left[1 + \tanh \frac{c(x - x_0)}{2\mu}\right] \tag{4-169}$$

which is shown in Fig. 4-37 for $\mu = \frac{1}{4}$. Hence, if the initial distribution of u is given by Eq. (4-169), the exact solution does not vary with time but remains fixed at the initial distribution. Additional exact solutions of Burgers equation can be found in the paper by Benton and Platzman (1972) which describes 35 different exact solutions.

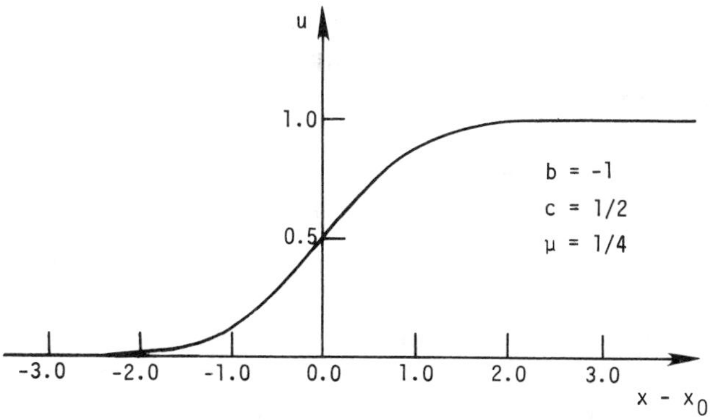

Figure 4-37 Exact solution of Eq. (4-168).

Equation (4-168) can be put into conservation form

$$u_t + \bar{F}_x = 0 \tag{4-170}$$

where $\bar{F}$ is defined by

$$\bar{F} = cu + \frac{bu^2}{2} - \mu u_x \tag{4-171}$$

Alternatively, Eq. (4-168) can be rewritten as

$$u_t + F_x = \mu u_{xx} \tag{4-172}$$

where F is defined by

$$F = cu + \frac{bu^2}{2} \tag{4-173}$$

For the linearized case ($b = 0$), F reduces to

$$F = cu$$

If we let A equal $\partial F/\partial u$, then Eq. (4-172) becomes

$$u_t + Au_x = \mu u_{xx} \tag{4-174}$$

where A equals u for the nonlinear Burgers equation ($c = 0, b = 1$) and A equals c for the linear Burgers equation ($b = 0$). We will use either Eq. (4-172) or Eq. (4-174) to represent Burgers' equation in the following discussion of applicable finite-difference schemes.

4-5.1 FTCS Method

Roache (1972) has given the name FTCS method to the scheme obtained by applying forward-time and centered-space difference to the linearized Burgers' equation [i.e., Eq. (4-174) with $A = c$]. The resulting algorithm is

$$\frac{u_j^{n+1} - u_j^n}{\Delta t} + c\,\frac{u_{j+1}^n - u_{j-1}^n}{2\,\Delta x} = \mu\,\frac{u_{j+1}^n - 2u_j^n + u_{j-1}^n}{(\Delta x)^2} \tag{4-175}$$

This is a first-order, explicit, one-step scheme with a truncation error of $O[\Delta t, (\Delta x)^2]$. The modified equation can be written as

$$u_t + cu_x = \left(\mu - \frac{c^2\,\Delta t}{2}\right) u_{xx} + \frac{c(\Delta x)^2}{3}\left(3r - \nu^2 - \frac{1}{2}\right) u_{xxx}$$
$$+ \frac{c(\Delta x)^3}{12}\left(\frac{r}{\nu} - \frac{3r^2}{\nu} - 2\nu + 10\nu r - 3\nu^3\right) u_{xxxx} + \dots \tag{4-176}$$

where r is defined as $\mu\,\Delta t/(\Delta x)^2$ for the viscous Burgers equation, and $\nu = c\,\Delta t/\Delta x$. Note that if $r = \frac{1}{2}$ and $\nu = 1$ the coefficients of the first two terms on the right-hand side of the modified equation become zero. Unfortunately, this eliminates the viscous term (μu_{xx}) that appears in the PDE we wish to solve. Thus, the FTCS method with

$r = \frac{1}{2}$ and $\nu = 1$, which incidentally reduces to $u_j^{n+1} = u_{j-1}^n$, is an unacceptable difference representation for Burgers' equation.

The "heuristic" stability analysis, described earlier, requires that the coefficient on u_{xx} be greater than zero. Hence,

$$\frac{c^2\,\Delta t}{2} \leqslant \mu$$

or

$$\frac{c^2(\Delta t)^2}{(\Delta x)^2} \leqslant 2\mu\,\frac{\Delta t}{(\Delta x)^2}$$

which can be rewritten as

$$\nu^2 \leqslant 2r \tag{4-177}$$

A very useful parameter which arises naturally when solving Burgers' equation is the *mesh Reynolds number* which is defined by

$$\mathrm{Re}_{\Delta x} = \frac{c\,\Delta x}{\mu} \tag{4-178}$$

This nondimensional parameter gives the ratio of convection to diffusion and plays an important role in determining the character of the solution for Burgers' equation. The mesh (cell) Reynolds number (also called Peclet number) can be expressed in terms of ν and r in the following manner

$$\mathrm{Re}_{\Delta x} = \frac{c\,\Delta x}{\mu} = \frac{c\,\Delta t}{\Delta x}\,\frac{(\Delta x)^2}{\mu\,\Delta t} = \frac{\nu}{r}$$

Thus, the stability condition given by Eq. (4-177) becomes

$$\mathrm{Re}_{\Delta x} \leqslant \frac{2}{\nu} \tag{4-179}$$

As pointed out earlier, the "heuristic" stability analysis does not always give the complete stability restrictions for a given numerical scheme and this happens in the present case. In order to obtain all of the stability conditions it is necessary to use the Fourier stability analysis. For the FTCS method, the amplification factor is

$$G = 1 + 2r(\cos\beta - 1) - i\nu(\sin\beta) \tag{4-180}$$

which is plotted in Fig. 4-38a for a given ν and r. The equation for G describes an ellipse which is centered on the positive real axis at $(1 - 2r)$ and has semi-major and semi-minor axes given by $2r$ and ν, respectively. In addition, the ellipse is tangent to the unit circle at the point where the positive real axis intersects the unit circle. For stability it is necessary that $|G| \leqslant 1$ which requires that the ellipse be entirely within the unit circle. This leads to the following necessary stability restrictions which are based on the lengths of the semi-major and semi-minor axes

$$\nu \leqslant 1 \qquad 2r \leqslant 1 \tag{4-181}$$

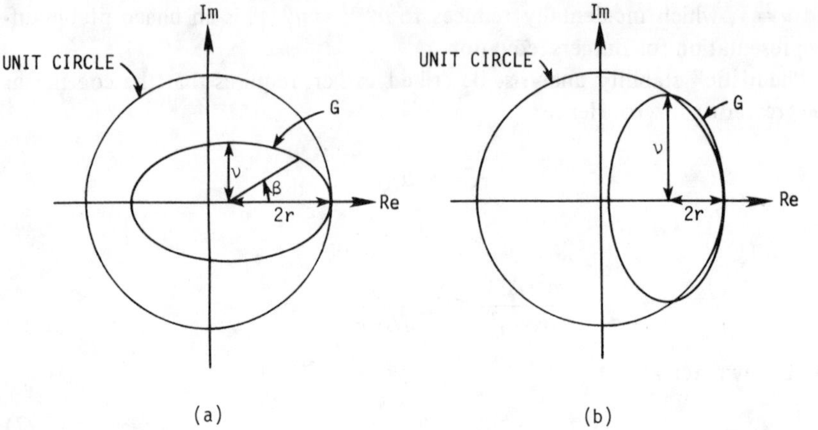

Figure 4-38 Stability of FTCS method. (a) $\nu < 1, r < \frac{1}{2}, \nu^2 < 2r$. (b) $\nu < 1, r < \frac{1}{2}, \nu^2 > 2r$.

It is possible, however, for these restrictions to be satisfied and the solution to still be unstable as can be seen in Fig. 4-38b. Of course, the complete stability limitations can be obtained by examining the modulus of the amplification factor in the usual manner. This analysis yields

$$\nu^2 \leqslant 2r \qquad r \leqslant \tfrac{1}{2} \tag{4-182}$$

Note that the first restriction was obtained previously by the "heuristic" stability analysis and that the two inequalities can be combined to yield a third inequality

$$\nu \leqslant 1$$

which was obtained by graphical considerations. In terms of the mesh Reynolds number, the stability restrictions become

$$2\nu \leqslant \mathrm{Re}_{\Delta x} \leqslant \frac{2}{\nu} \tag{4-183}$$

It should be mentioned that the right-hand inequality is incorrectly given as $\mathrm{Re}_{\Delta x} \leqslant 2$ in some references.

An important characteristic of finite-difference schemes which are used to solve Burgers' equation is whether they produce oscillations (wiggles) in the solution. Obviously, we do not want these oscillations to occur in our solutions of fluid flow problems. The FTCS method will produce oscillations in the solution of Burgers' equation for mesh Reynolds number in the range

$$2 \leqslant \mathrm{Re}_{\Delta x} \leqslant \frac{2}{\nu}$$

For mesh Reynolds numbers slightly above $2/\nu$ the oscillations will eventually cause the solution to "blow up" as expected from our previous stability analysis. In order to explain the origin of the wiggles, let us rewrite Eq. (4-175) in the following form

$$u_j^{n+1} = \left(r - \frac{\nu}{2}\right) u_{j+1}^n + (1 - 2r)u_j^n + \left(r + \frac{\nu}{2}\right) u_{j-1}^n \tag{4-184}$$

which is equivalent to

$$u_j^{n+1} = \frac{r}{2}(2 - \mathrm{Re}_{\Delta x})u_{j+1}^n + (1 - 2r)u_j^n + \frac{r}{2}(2 + \mathrm{Re}_{\Delta x})u_{j-1}^n \qquad (4\text{-}185)$$

Furthermore, let's assume we are trying to find the steady-state solution of Burgers' equation for the initial condition

$$u(x, 0) = 0 \qquad 0 \leqslant x < 1$$

and boundary conditions

$$u(0, t) = 0$$

$$u(1, t) = 1$$

using an 11-point mesh. For the first time step, the values of u at time level $n + 1$ are all zero except at $j = 10$ where

$$u_{10}^{n+1} = \frac{r}{2}(2 - \mathrm{Re}_{\Delta x})(1) + (1 - 2r)(0) + \frac{r}{2}(2 + \mathrm{Re}_{\Delta x})(0) = \frac{r}{2}(2 - \mathrm{Re}_{\Delta x})$$

and at the boundary ($j = 11$) where u_{11} is fixed at 1. If $\mathrm{Re}_{\Delta x}$ is greater than 2, the value of u_{10}^{n+1} will be negative which will initiate an oscillation as shown in Fig. 4-39a. This figure is drawn for the conditions

$$\nu = 0.4$$

$$r = 0.1$$

$$\mathrm{Re}_{\Delta x} = 4 < \frac{2}{\nu}$$

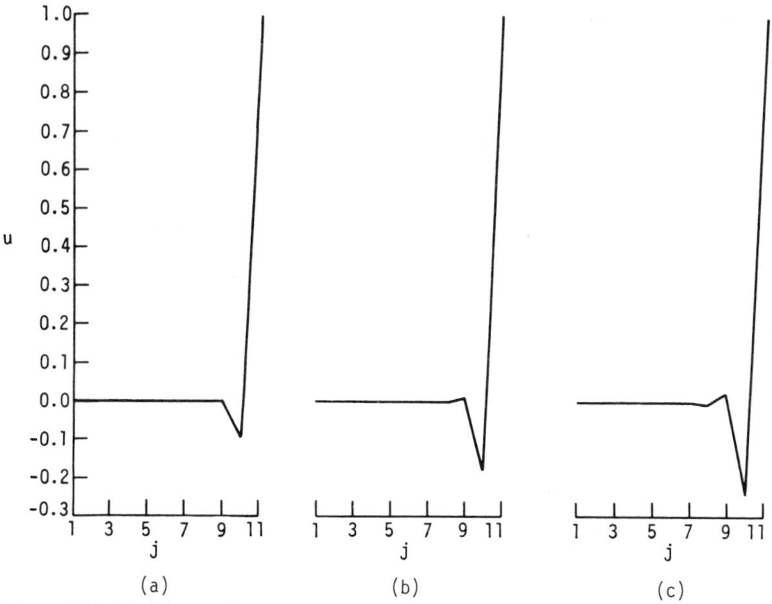

Figure 4-39 Oscillations in numerical solution of Burgers' equation. (a) $n + 1$ time level; (b) $n + 2$ time level; (c) $n + 3$ time level.

which make u_{10}^{n+1} equal to -0.1. During the next time step, the oscillation propagates one grid point further from the right-hand boundary. The values of u at $j = 9$ and $j = 10$ become

$$u_9^{n+2} = +0.01$$
$$u_{10}^{n+2} = -0.18$$

and the resulting solution is shown in Fig. 4-39b. The wiggles will eventually propagate to the other boundary but will remain bounded throughout the iteration to steady state. The oscillations that occur in this case are similar to the oscillations which appear when a second-order (or higher) scheme is used to solve the inviscid Burgers equation for a propagating discontinuity.

Additional insight into the origin of the wiggles can be obtained by examining the coefficients in Eq. (4-185) from a physical standpoint. We observe that when $\mathrm{Re}_{\Delta x}$ is greater than two the coefficient in front of u_{j+1}^n becomes negative. Hence, the larger the value for u_{j+1}^n the smaller the value for u_j^{n+1}. This represents a nonphysical behavior for a viscous problem since we would expect a greater "pull" on u_j^{n+1} (i.e., increased value) because of viscosity as u_{j+1}^n is increased. As a consequence of this nonphysical behavior, oscillations are produced in the solution.

The oscillations of the FTCS method can be eliminated if the second-order central difference used for the convective term cu_x is replaced by a first-order upwind difference. The resulting algorithm for $c > 0$ becomes

$$\frac{u_j^{n+1} - u_j^n}{\Delta t} + c\,\frac{u_j^n - u_{j-1}^n}{\Delta x} = \mu\,\frac{u_{j+1}^n - 2u_j^n + u_{j-1}^n}{(\Delta x)^2} \tag{4-186}$$

This first-order scheme eliminates the oscillations by adding additional dissipation to the solution. Unfortunately, the amount of dissipation causes the resulting solution to be sufficiently inaccurate to exclude Eq. (4-186) as a viable difference scheme for Burgers' equation. The large amount of dissipation is evident when we examine the modified equation for the scheme

$$u_t + cu_x = \left[\mu\left(1 + \frac{\mathrm{Re}_{\Delta x}}{2}\right) - \frac{c^2\,\Delta t}{2}\right] u_{xx} + \cdots \tag{4-187}$$

and compare it to the modified equation of the FTCS method. Equation (4-187) has the additional term $\mu\,\mathrm{Re}_{\Delta x}/2$ appearing in the coefficient of u_{xx}. Hence, if $\mathrm{Re}_{\Delta x}$ is greater than 2 this additional term produces more dissipation (diffusion) than is present in the original problem governed by Burgers' equation. In order to reduce dispersive errors without adding a large amount of artificial viscosity, Leonard (1979a, 1979b) has suggested using a third-order upstream (upwind) difference for the convective term. The resulting algorithm for $c > 0$ is

$$\frac{u_j^{n+1} - u_j^n}{\Delta t} + c\left(\frac{u_{j+1}^n - u_{j-1}^n}{2\,\Delta x} - \frac{u_{j+1}^n - 3u_j^n + 3u_{j-1}^n - u_{j-2}^n}{6\,\Delta x}\right)$$
$$= \mu\,\frac{u_{j+1}^n - 2u_j^n + u_{j-1}^n}{(\Delta x)^2} \tag{4-188}$$

and for $c < 0$ the algorithm becomes

$$\frac{u_j^{n+1} - u_j^n}{\Delta t} + c\left(\frac{u_{j+1}^n - u_{j-1}^n}{2\,\Delta x} - \frac{u_{j+2}^n - 3u_{j+1}^n + 3u_j^n - u_{j-1}^n}{6\,\Delta x}\right)$$
$$= \mu\,\frac{u_{j+1}^n - 2u_j^n + u_{j-1}^n}{(\Delta x)^2} \tag{4-189}$$

4-5.2 Leap Frog/DuFort-Frankel Method

We have noted earlier that the linearized Burgers equation is a combination of the first-order wave equation and the heat equation. This suggests that we might be able to combine some of the algorithms given previously for the wave equation and the heat equation. The leap frog/DuFort-Frankel method is one such example. When applied to Eq. (4-174), this method becomes

$$\frac{u_j^{n+1} - u_j^{n-1}}{2\,\Delta t} + A_j^n\,\frac{u_{j+1}^n - u_{j-1}^n}{2\,\Delta x} = \mu\,\frac{u_{j+1}^n - u_j^{n+1} - u_j^{n-1} + u_{j-1}^n}{(\Delta x)^2} \tag{4-190}$$

This explicit, one-step scheme is first-order accurate with a truncation error of $O[(\Delta t/\Delta x)^2, (\Delta t)^2, (\Delta x)^2]$. The modified equation for the linear case ($A = c$) can be written as

$$u_t + cu_x = \mu(1 - \nu^2)u_{xx} + \left[\frac{2\mu^2 c(\Delta t)^2}{(\Delta x)^2} - \frac{1}{6}\,c(\Delta x)^2\right.$$
$$\left. + \frac{1}{6}\,c^3(\Delta t)^2 - \frac{2\mu^2 c^3(\Delta t)^4}{(\Delta x)^4}\right] u_{xxx} + \cdots \tag{4-191}$$

Also for the linear case, a Fourier stability analysis can be performed which gives the stability condition

$$\nu \leqslant 1$$

Note that this stability condition is independent of the viscosity coefficient μ because of the DuFort-Frankel type of differencing used for the viscous term. However, because consistency requires that $(\Delta t/\Delta x)^2$ approach zero as Δt and Δx approach zero, a much smaller time step than allowed by $\nu \leqslant 1$ is implied. For this reason, the leap frog/DuFort-Frankel scheme seems better suited for the calculation of steady solutions (where time accuracy is unimportant) than for unsteady solutions according to Peyret and Viviand (1975). In the nonlinear case, this scheme is unstable if $\mu = 0$.

4-5.3 Brailovskaya Method

The following two-step explicit method for Eq. (4-172) was proposed by Brailovskaya (1965):

$$\text{Predictor:}\quad \overline{u_j^{n+1}} = u_j^n - \frac{\Delta t}{2\,\Delta x}(F_{j+1}^n - F_{j-1}^n) + r(u_{j+1}^n - 2u_j^n + u_{j-1}^n)$$

$$\text{Corrector:}\quad u_j^{n+1} = u_j^n - \frac{\Delta t}{2\,\Delta x}(\overline{F_{j+1}^{n+1}} - \overline{F_{j-1}^{n+1}}) + r(u_{j+1}^n - 2u_j^n + u_{j-1}^n) \tag{4-192}$$

This scheme is formally first-order accurate with a truncation error $O[\Delta t, (\Delta x)^2]$. If only a steady-state solution is desired, the first-order temporal accuracy is not important. For the linear Burgers equation, the von Neumann necessary condition for stability is

$$|G|^2 = 1 - \{\nu^2 \sin^2\beta(1 - \nu^2\sin^2\beta) + 4r(1 - \cos\beta) \times [1 - r(1 - \cos\beta)(1 + \nu^2 \sin^2\beta)]\} \leqslant 1 \tag{4-193}$$

If we ignore viscous effects (i.e., set $r = 0$), the stability condition becomes

$$\nu \leqslant 1$$

On the other hand, if we ignore the convection term, i.e., set $\nu = 0$, the stability condition becomes

$$r \leqslant \tfrac{1}{2}$$

Based on these observations, Carter (1971) has suggested the following stability criterion for the Brailovskaya scheme:

$$\Delta t \leqslant \min\left[\frac{(\Delta x)^2}{2\mu}, \frac{(\Delta x)}{|A|}\right] \tag{4-194}$$

An attractive feature of this scheme is that the viscous term remains the same in both predictor and corrector steps and needs to be computed only once.

4-5.4 Allen-Cheng Method

Allen and Cheng (1970) modified the Brailovskaya scheme to eliminate the stability restriction on r. Their scheme is given by

$$\text{Predictor:}\quad \overline{u_j^{n+1}} = u_j^n - \frac{\Delta t}{2\,\Delta x}(F_{j+1}^n - F_{j-1}^n) + r(u_{j+1}^n - 2\overline{u_j^{n+1}} + u_{j-1}^n)$$

$$\text{Corrector:}\quad u_j^{n+1} = u_j^n - \frac{\Delta t}{2\,\Delta x}(\overline{F_{j+1}^{n+1}} - \overline{F_{j-1}^{n+1}}) + r(\overline{u_{j+1}^{n+1}} - 2u_j^{n+1} + \overline{u_{j-1}^{n+1}}) \tag{4-195}$$

The unconventional differencing of the viscous term eliminates the stability restriction on r so that the stability condition becomes

$$\nu \leqslant 1$$

for the linear Burgers equation. As a result, when μ is large, this method permits a much larger time step to be taken than does the Brailovskaya scheme. The Allen-Cheng method is formally first-order accurate with a truncation error of $O[\Delta t, (\Delta x)^2]$.

4-5.5 Lax-Wendroff Method

We have previously applied the two-step Lax-Wendroff method to the wave equation. When applied to the complete Burgers equation, several different variations of the method are possible including the following:

Step 1:

$$u_j^{n+1/2} = \frac{1}{2}(u_{j+1/2}^n - u_{j-1/2}^n) - \frac{\Delta t}{\Delta x}(F_{j+1/2}^n - F_{j-1/2}^n) + r[(u_{j-3/2}^n - 2u_{j-1/2}^n + u_{j+1/2}^n) + (u_{j+3/2}^n - 2u_{j+1/2}^n + u_{j-1/2}^n)]$$

Step 2:

$$u_j^{n+1} = u_j^n - \frac{\Delta t}{\Delta x}(F_{j+1/2}^{n+1/2} - F_{j-1/2}^{n+1/2}) + r(u_{j+1}^n - 2u_j^n + u_{j-1}^n) \tag{4-196}$$

This version is based on the Lax-Wendroff scheme used by Thommen (1966) to solve the Navier-Stokes equations. An alternate version has been proposed by Palumbo and Rubin (1972) which computes provisional values at time level $n+1$ instead of $n+\frac{1}{2}$. The present version is formally first-order accurate with a truncation error of $O[\Delta t, (\Delta x)^2]$. The exact linear stability condition is

$$\frac{\Delta t}{(\Delta x)^2}(A^2\,\Delta t + 2\mu) \leqslant 1 \tag{4-197}$$

4-5.6 MacCormack Method

The MacCormack method for the complete Burgers equation (4-172) is

Predictor:

$$\overline{u_j^{n+1}} = u_j^n - \frac{\Delta t}{\Delta x}(F_{j+1}^n - F_j^n) + r(u_{j+1}^n - 2u_j^n + u_{j-1}^n)$$

Corrector:

$$u_j^{n+1} = \frac{1}{2}\left[u_j^n + \overline{u_j^{n+1}} - \frac{\Delta t}{\Delta x}(\overline{F_j^{n+1}} - \overline{F_{j-1}^{n+1}}) + r(\overline{u_{j+1}^{n+1}} - 2\overline{u_j^{n+1}} + \overline{u_{j-1}^{n+1}})\right] \tag{4-198}$$

which is second-order accurate in both time and space. In this version of the MacCormack scheme, a forward difference is employed in the predictor step for $\partial F/\partial x$ and a backward difference is used in the corrector step. The alternate version of the MacCormack scheme employs a backward difference in the predictor step and a forward difference in the corrector step. Both variants of the MacCormack scheme are second-order accurate. It is not possible to obtain a simple stability criterion for the MacCormack scheme applied to the Burgers equation. However, either the condition given by Eq. (4-194) or the empirical formula (Tannehill et al., 1975)

$$\Delta t \leqslant \frac{(\Delta x)^2}{|A|\,\Delta x + 2\mu} \tag{4-199}$$

can be used with an appropriate safety factor. The latter formula reduces to the usual viscous condition $r \leqslant \frac{1}{2}$ when $|A|$ is set equal to zero, and reduces to the usual inviscid condition $|A|\Delta t/\Delta x \leqslant 1$ when μ is set equal to zero. The MacCormack method has been widely used to solve not only the Euler equations but also the Navier-Stokes equations for laminar flow. For multidimensional problems, a time-split version of the MacCormack scheme has been developed and will be described in Section 4-5.8. For high Reynolds number problems, MacCormack has devised a rapid-solver method (MacCormack, 1976) as well as an implicit method (MacCormack, 1981) which will be discussed in Chapter 9.

An interesting variation of the regular MacCormack scheme is obtained when over-relaxation is applied to both predicted and corrected values (Désidéri and Tannehill, 1977a) in the following manner:

Predictor:

$$\begin{aligned} v_j^{\overline{n+1}} &= u_j^n - \frac{\Delta t}{\Delta x}(F_{j+1}^n - F_j^n) + r(u_{j+1}^n - 2u_j^n + u_{j-1}^n) \\ u_j^{\overline{n+1}} &= u_j^{\bar{n}} + \bar{\omega}(v_j^{\overline{n+1}} - u_j^{\bar{n}}) \end{aligned} \tag{4-200}$$

Corrector:

$$\begin{aligned} v_j^{n+1} &= u_j^{\overline{n+1}} - \frac{\Delta t}{\Delta x}(F_j^{\overline{n+1}} - F_{j-1}^{\overline{n+1}}) + r(u_{j+1}^{\overline{n+1}} - 2u_j^{\overline{n+1}} + u_{j-1}^{\overline{n+1}}) \\ u_j^{n+1} &= u_j^n + \omega(v_j^{n+1} - u_j^n) \end{aligned} \tag{4-201}$$

In these equations, the v's are intermediate quantities, the u's denote final predictions, $\bar{\omega}$ and ω are over-relaxation parameters and $u_j^{\bar{n}}$ represents the predicted value for u_j from the previous step. The regular MacCormack scheme is obtained by setting $\bar{\omega} = 1$ and $\omega = \frac{1}{2}$. In general, the over-relaxed MacCormack method is first-order accurate with a truncation error of $O[\Delta t, (\Delta x)^2]$. However, it can be shown (Désidéri and Tannehill, 1977b) that if

$$\omega\bar{\omega} = |\bar{\omega} - \omega| \tag{4-202}$$

the method is second-order accurate in time when applied to the linearized Burgers equation. The over-relaxation scheme accelerates the convergence over that of the regular MacCormack scheme by an approximate factor Ω given by

$$\Omega = \frac{2\bar{\omega}\omega}{1 - (\bar{\omega} - 1)(\omega - 1)} \tag{4-203}$$

A Fourier stability analysis applied to the linearized Burgers equation does not yield a necessary and sufficient stability condition in the form of an algebraic relation between the parameters ν, r, $\bar{\omega}$, and ω. However, a necessary condition of stability is

$$|(\bar{\omega} - 1)(\omega - 1)| \leqslant 1 \tag{4-204}$$

In general, the stability limitation must be computed numerically and it is usually more restrictive than the conditions $\bar{\omega} \leqslant 2$ and $\omega \leqslant 2$.

4-5.7 Briley-McDonald Method

The Briley-McDonald method (1973) is an implicit scheme which is often based on the following time differencing (Euler implicit) of Eq. (4-172):

$$\frac{u_j^{n+1} - u_j^n}{\Delta t} + \left(\frac{\partial F}{\partial x}\right)_j^{n+1} = \mu \left(\frac{\partial^2 u}{\partial x^2}\right)_j^{n+1} \tag{4-205}$$

The term $(\partial F/\partial x)_j^{n+1}$ is expanded as

$$\left(\frac{\partial F}{\partial x}\right)_j^{n+1} = \left(\frac{\partial F}{\partial x}\right)_j^n + \Delta t \left[\frac{\partial}{\partial t}\left(\frac{\partial F}{\partial x}\right)\right]_j^n + O[(\Delta t)^2] \tag{4-206}$$

thereby introducing $\partial/\partial t(\partial F/\partial x)$ which can be replaced by

$$\frac{\partial}{\partial t}\left(\frac{\partial F}{\partial x}\right) = \frac{\partial}{\partial x}\left(\frac{\partial F}{\partial t}\right) = \frac{\partial}{\partial x}\left(\frac{\partial F}{\partial u}\frac{\partial u}{\partial t}\right) = \frac{\partial}{\partial x}\left(A\frac{\partial u}{\partial t}\right) \tag{4-207}$$

Finally, if we combine Eqs. (4-205), (4-206), and (4-207) and employ forward-time differences and centered-spatial differences, the Briley-McDonald method is obtained

$$\frac{u_j^{n+1} - u_j^n}{\Delta t} + \frac{F_{j+1}^n - F_{j-1}^n}{2\,\Delta x} + \frac{A_{j+1}^n(u_{j+1}^{n+1} - u_{j+1}^n) - A_{j-1}^n(u_{j-1}^{n+1} - u_{j-1}^n)}{2\,\Delta x} = \mu\,\hat{\delta}_x^2 u_j^{n+1} \tag{4-208}$$

This scheme is formally first-order accurate with a truncation error of $O[\Delta t, (\Delta x)^2]$. However, at steady state the accuracy is of $O[(\Delta x)^2]$. The temporal accuracy can be increased by using trapezoidal differencing or by using additional time levels in the same manner as discussed earlier for the Beam-Warming scheme. For example, if we apply trapezoidal time differencing to Eq. (4-172) the following equation is obtained

$$\frac{u_j^{n+1} - u_j^n}{\Delta t} + \frac{1}{2}\left[\left(\frac{\partial F}{\partial x}\right)_j^n + \left(\frac{\partial F}{\partial x}\right)_j^{n+1}\right] = \frac{1}{2}\mu\left[\left(\frac{\partial^2 u}{\partial x^2}\right)_j^n + \left(\frac{\partial^2 u}{\partial x^2}\right)_j^{n+1}\right] \tag{4-209}$$

Proceeding as before, we find the resulting second-order accurate scheme to be

$$\frac{u_j^{n+1} - u_j^n}{\Delta t} + \frac{F_{j+1}^n - F_{j-1}^n}{2\,\Delta x} + \frac{A_{j+1}^n(u_{j+1}^{n+1} - u_{j+1}^n) - A_{j-1}^n(u_{j-1}^{n+1} - u_{j-1}^n)}{4\,\Delta x}$$

$$= \frac{\mu}{2(\Delta x)^2}\left[(\delta_x^2 u)_j^n + (\delta_x^2 u)_j^{n+1}\right] \tag{4-210}$$

Both of these schemes, Eq. (4-208) and Eq. (4-210), are unconditionally stable and produce tridiagonal systems of linear algebraic equations which can be solved using the Thomas algorithm.

The Briley-McDonald method is directly related to the method developed by Beam and Warming (1978) to solve the Navier-Stokes equations. In fact when the two methods are applied to Burgers' equation they can be reduced to the same form. In order to do this, the delta terms in the Beam-Warming method must be replaced by their equivalent expressions [i.e., Δu_j^n is replaced by $(u_j^{n+1} - u_j^n)$]. The Beam-Warming method for the Navier-Stokes equations is discussed in Chapter 9.

4-5.8 Time-Split MacCormack Method

In order to illustrate methods which are designed specifically for multidimensional problems, we introduce the 2-D Burgers' equation

$$\frac{\partial u}{\partial t}+\frac{\partial F}{\partial x}+\frac{\partial G}{\partial y}=\mu\left(\frac{\partial^2 u}{\partial x^2}+\frac{\partial^2 u}{\partial y^2}\right) \tag{4-211}$$

If we let A equal $\partial F/\partial u$ and B equal $\partial G/\partial u$, Eq. (4-211) can be rewritten as

$$u_t + Au_x + Bu_y = \mu(u_{xx} + u_{yy}) \tag{4-212}$$

The exact steady-state solution (derived by Rai, 1982) of the 2-D linearized Burgers equation

$$u_t + cu_x + du_y = \mu(u_{xx} + u_{yy}) \tag{4-213}$$

for the boundary conditions $(0 \leqslant t \leqslant \infty)$

$$\begin{aligned} u(x,0,t) &= \frac{1-\exp[(x-1)c/\mu]}{1-\exp(-c/\mu)} \qquad u(x,1,t)=0 \\ u(0,y,t) &= \frac{1-\exp[(y-1)d/\mu]}{1-\exp(-d/\mu)} \qquad u(1,y,t)=0 \end{aligned} \tag{4-214}$$

and the initial condition $(0 < x \leqslant 1, 0 < y \leqslant 1)$

$$u(x,y,0)=0$$

is given by

$$u(x,y)=\left\{\frac{1-\exp[(x-1)c/\mu]}{1-\exp(-c/\mu)}\right\}\left\{\frac{1-\exp[(y-1)d/\mu]}{1-\exp(-d/\mu)}\right\} \tag{4-215}$$

Note that the extension of this form of solution to the 3-D linearized Burgers equation is straightforward. All of the methods we have discussed for the 1-D Burgers equation can be readily extended to the 2-D Burgers equation. However, because of the more restrictive stability conditions in the explicit methods and the desire to maintain tridiagonal matrices in the implicit schemes, it is usually necessary to modify the previous algorithms for multidimensional problems. As an example, let us first consider the explicit time-split MacCormack method.

The time-split MacCormack method (MacCormack, 1971; MacCormack and Baldwin, 1975) "splits" the original MacCormack scheme into a sequence of one-dimensional operations thereby achieving a less restrictive stability condition. In other words, the splitting makes it possible to advance the solution in each direction with the maximum allowable time step. This is particularly advantageous if the allowable time steps $(\Delta t_x, \Delta t_y)$ are much different because of differences in the mesh spacings $(\Delta x, \Delta y)$. In order to explain this method, we will make use of the one-dimensional difference operators $L_x(\Delta t_x)$ and $L_y(\Delta t_y)$. The $L_x(\Delta t_x)$ operator applied to $u_{i,j}^n$

$$u_{i,j}^* = L_x(\Delta t_x)u_{i,j}^n \tag{4-216}$$

is by definition equivalent to the two-step formula

$$\begin{aligned} u_{i,j}^{\bar{*}} &= u_{i,j}^{n} - \frac{\Delta t_x}{\Delta x}(F_{i+1,j}^{n} - F_{i,j}^{n}) + \mu\,\Delta t_x\,\hat{\delta}_x^2 u_{i,j}^{n} \\ u_{i,j}^{*} &= \frac{1}{2}\left[u_{i,j}^{n} + u_{i,j}^{\bar{*}} - \frac{\Delta t_x}{\Delta x}(F_{i,j}^{\bar{*}} - F_{i-1,j}^{\bar{*}}) + \mu\,\Delta t_x\,\hat{\delta}_x^2 u_{i,j}^{\bar{*}}\right] \end{aligned} \qquad (4\text{-}217)$$

These expressions make use of a dummy time index which is denoted by *. The $L_y(\Delta t_y)$ operator is defined in a similar manner. That is

$$u_{i,j}^{*} = L_y(\Delta t_y)u_{i,j}^{n} \qquad (4\text{-}218)$$

is equivalent to

$$\begin{aligned} u_{i,j}^{\bar{*}} &= u_{i,j}^{n} - \frac{\Delta t_y}{\Delta y}(G_{i,j+1}^{n} - G_{i,j}^{n}) + \mu\,\Delta t_y\,\hat{\delta}_y^2 u_{i,j}^{n} \\ u_{i,j}^{*} &= \frac{1}{2}\left[u_{i,j}^{n} + u_{i,j}^{\bar{*}} - \frac{\Delta t_y}{\Delta y}(G_{i,j}^{\bar{*}} - G_{i,j-1}^{\bar{*}}) + \mu\,\Delta t_y\,\hat{\delta}_y^2 u_{i,j}^{\bar{*}}\right] \end{aligned} \qquad (4\text{-}219)$$

A second-order accurate scheme can be constructed by applying the L_x and L_y operators to $u_{i,j}^n$ in the following manner

$$u_{i,j}^{n+1} = L_y\left(\frac{\Delta t}{2}\right) L_x(\Delta t) L_y\left(\frac{\Delta t}{2}\right) u_{i,j}^{n} \qquad (4\text{-}220)$$

This scheme has a truncation error of $O[(\Delta t)^2, (\Delta x)^2, (\Delta y)^2]$. In general, a scheme formed by a sequence of these operators is: (1) stable, if the time step of each operator does not exceed the allowable step size for that operator; (2) consistent, if the sums of the time steps for each of the operators are equal; and (3) second-order accurate, if the sequence is symmetric. Other sequences which satisfy these criteria are given by

$$\begin{aligned} u_{i,j}^{n+1} &= L_y\left(\frac{\Delta t}{2}\right) L_x\left(\frac{\Delta t}{2}\right) L_x\left(\frac{\Delta t}{2}\right) L_y\left(\frac{\Delta t}{2}\right) u_{i,j}^{n} \\ u_{i,j}^{n+1} &= \left[L_y\left(\frac{\Delta t}{2m}\right)\right]^m L_x(\Delta t)\left[L_y\left(\frac{\Delta t}{2m}\right)\right]^m u_{i,j}^{n} \qquad m = \text{integer} \end{aligned} \qquad (4\text{-}221)$$

The last sequence is quite useful for the case where $\Delta y \ll \Delta x$.

4-5.9 ADI Methods

Polezhaev (1967) used an adaptation of the Peaceman-Rachford ADI scheme to solve the compressible Navier-Stokes equations. When applied to the 2-D Burgers equation, Eq. (4-212), this scheme becomes

$$\begin{aligned} \left[1 + \frac{\Delta t}{2}\left(A_{i,j}^{n}\frac{\bar{\delta}_x}{2\,\Delta x} - \mu\hat{\delta}_x^2\right)\right] u_{i,j}^{*} &= \left[1 - \frac{\Delta t}{2}\left(B_{i,j}^{n}\frac{\bar{\delta}_y}{2\,\Delta y} - \mu\hat{\delta}_y^2\right)\right] u_{i,j}^{n} \\ \left[1 + \frac{\Delta t}{2}\left(B_{i,j}^{*}\frac{\bar{\delta}_y}{2\,\Delta y} - \mu\hat{\delta}_y^2\right)\right] u_{i,j}^{n+1} &= \left[1 - \frac{\Delta t}{2}\left(A_{i,j}^{n}\frac{\bar{\delta}_x}{2\,\Delta x} - \mu\hat{\delta}_x^2\right)\right] u_{i,j}^{*} \end{aligned} \qquad (4\text{-}222)$$

This method is first-order accurate with a truncation error of $O[\Delta t, (\Delta x)^2, (\Delta y)^2]$ and is unconditionally stable for the linear case. Obviously, a tridiagonal system of algebraic equations must be solved during each step.

When the Briley-McDonald scheme, Eq. (4-208), is applied directly to the 2-D Burgers equation, a tridiagonal system of algebraic equations is no longer obtained. This difficulty can be avoided by applying the two-step ADI procedure of Douglas and Gunn (1964)

$$\left[1 + \Delta t \left(\frac{\bar{\delta}_x}{2\,\Delta x} A_{i,j}^n - \mu \hat{\delta}_x^2\right)\right] u_{i,j}^* = \left[1 - \Delta t \left(\frac{\bar{\delta}_y}{2\,\Delta y} B_{i,j}^n - \mu \hat{\delta}_y^2\right)\right] u_{i,j}^n + (\Delta t) S_{i,j}^n \tag{4-223}$$

$$\left[1 + \Delta t \left(\frac{\bar{\delta}_y}{2\,\Delta y} B_{i,j}^n - \mu \hat{\delta}_x^2\right)\right] u_{i,j}^{n+1} = u_{i,j}^n - \Delta t \left(\frac{\bar{\delta}_x}{2\,\Delta x} A_{i,j}^n - \mu \hat{\delta}_x^2\right) u_{i,j}^* + \Delta t S_{i,j}^n \tag{4-224}$$

where

$$S_{i,j}^n = -\frac{\bar{\delta}_x}{2\,\Delta x} F_{i,j}^n - \frac{\bar{\delta}_y}{2\,\Delta y} G_{i,j}^n + \frac{\bar{\delta}_x}{2\,\Delta x} (A_{i,j}^n u_{i,j}^n) + \frac{\bar{\delta}_y}{2\,\Delta y} (B_{i,j}^n u_{i,j}^n)$$

4-5.10 Predictor-Corrector, Multiple-Iteration Method

Rubin and Lin (1972) devised a predictor-corrector, multiple-iteration method to solve the 3-D "parabolized" Navier-Stokes equations. Their scheme eliminates cross-coupling of grid points in the normal (y) and lateral (z) directions and uses an iterative procedure to recover acceptable accuracy. In order to illustrate this method, let us use the following 3-D linear Burgers equation

$$u_x + cu_y + du_z = \mu(u_{yy} + u_{zz}) \tag{4-225}$$

as a model for the "parabolized" Navier-Stokes equations. The predictor-corrector, multiple-iteration method applied to this model equation is

$$\begin{aligned} u_{i+1,j,k}^{m+1} = u_{i,j,k} &- \frac{c\,\Delta x}{2\,\Delta y} (u_{i+1,\,j+1,k}^{m+1} - u_{i+1,\,j-1,k}^{m+1}) - \frac{d\,\Delta x}{2\,\Delta z} (u_{i+1,j,k+1}^m - u_{i+1,\,j,k-1}^m) \\ &+ \frac{\mu\,\Delta x}{(\Delta y)^2} (u_{i+1,\,j+1,k}^{m+1} - 2u_{i+1,\,j,k}^{m+1} + u_{i+1,\,j-1,k}^{m+1}) \\ &+ \frac{\mu\,\Delta x}{(\Delta z)^2} (u_{i+1,\,j,k+1}^m - 2u_{i+1,j,k}^{m+1} + u_{i+1,j,k-1}^m) \end{aligned} \tag{4-226}$$

where the superscript m indicates the iteration level and $x = i\,\Delta x$, $y = j\,\Delta y$, and $z = k\,\Delta z$. For the first iteration, m is set equal to zero and the corresponding terms are approximated by either linear replacement

$$u_{i+1,j,k}^0 = u_{i,j,k}$$

or by Taylor-series expansions such as

$$u^0_{i+1,j,k} = 2u_{i,j,k} - u_{i-1,j,k} + O[(\Delta x)^2]$$

As a result, Eq. (4-226) has three unknowns

$$m = 0 \quad \begin{cases} u^1_{i+1,j+1,k} \\ u^1_{i+1,j,k} \\ u^1_{i+1,j-1,k} \end{cases} \tag{4-227}$$

which produces a tridiagonal system of algebraic equations. The computation in the $i + 1$ plane proceeds outward from the known boundary conditions at $k = 1$ to the last k column of grid points. This completes the first iteration. For the next iteration ($m = 1$) the three unknowns in Eq. (4-226) are

$$m = 1 \quad \begin{cases} u^2_{i+1,j+1,k} \\ u^2_{i+1,j,k} \\ u^2_{i+1,j-1,k} \end{cases} \tag{4-228}$$

This iteration procedure is continued until the solution is converged in the $i + 1$ plane. Usually, only two iterations ($m = 0$, $m = 1$) are required to recover acceptable accuracy. The computation then advances to the $i + 2$ plane.

4-6 CONCLUDING REMARKS

In this chapter an attempt has been made to introduce basic finite-difference methods for solving simple model equations. It has not been the intent to include all numerical techniques which have been proposed for these equations. Some very useful methods have not been included. However, those which have been presented should provide a reasonable background for the more complex applications which follow in Chapters 6, 7, 8, and 9.

Based on the information presented on the various techniques, it is clear that many different numerical methods can be used to solve the same problem. The differences in the quality of the solutions produced using the applicable methods are frequently small, and the selection of an optimum technique becomes difficult. However, the selection process can be aided by the experience gained in programming the various methods to solve the model equations presented in this chapter.

PROBLEMS

4-1 Derive Eq. (4-19).

4-2 Derive the modified equation for the Lax method applied to the wave equation. Retain terms up to and including u_{xxxx}.

4-3 Repeat Prob. 4-2 for the Euler implicit scheme.

4-4 Derive the modified equation for the leap frog method. Retain terms up to and including u_{xxxxx}.

4-5 Repeat Prob. 4-4 for the Lax-Wendroff method.

4-6 Determine the errors in amplitude and phase for $\beta = 90°$ if the Lax method is applied to the wave equation for 10 time steps with $\nu = 0.5$.

4-7 Repeat Prob. 4-6 for the MacCormack scheme.

4-8 Suppose the Lax method is used to solve the wave equation ($c = \frac{1}{2}$) for the initial condition

$$u(x, 0) = \sin(2\pi x) \qquad 0 \leqslant x \leqslant 2$$

and periodic boundary conditions with $\Delta x = 0.02$ and $\Delta t = 0.02$.

(*a*) Use the amplification factor to find the amplitude error and the phase error after 20 steps.

(*b*) Use the modified equation to determine (approximately) the amplitude error after 20 steps.

Hint: The exact solution for the linearized Burgers equation

$$u_t + cu_x = \mu u_{xx}$$

for the initial condition

$$u(x, 0) = \sin(kx)$$

and periodic boundary conditions is

$$u(x, t) = \exp(-k^2 \mu t) \sin[k(x - ct)]$$

4-9 Derive the amplification factor for the leap frog method applied to the wave equation and determine the stability restriction for this scheme.

4-10 Repeat Prob. 4-9 for the upwind method.

4-11 Show that the Rusanov method applied to the wave equation is equivalent to the following one-step scheme:

$$u_j^{n+1} = u_j^n - \nu(\mu_x \delta_x)\left(1 - \frac{\delta_x^2}{6}\right)u_j^n + \nu^2 \delta_x^2 \left(\frac{1}{2} + \frac{\delta_x^2}{8}\right)u_j^n - \frac{\nu^3}{6}(\mu_x \delta_x^3)u_j^n - \frac{\omega}{24}\delta_x^4 u_j^n$$

4-12 Evaluate the stability of the Rusanov method applied to the wave equation using the Fourier stability analysis. Hint: See Prob. 4-11.

4-13 The following second-order accurate explicit scheme for the wave equation was proposed by Crowley (1967):

$$u_j^{n+1} = u_j^n - \nu(\mu_x \delta_x)u_j^n + \frac{\nu^2}{2}(\mu_x^2 \delta_x^2)u_j^n - \frac{1}{8}\nu^3(\mu_x \delta_x^3)u_j^n$$

(*a*) Derive the modified equation for this scheme. Retain terms up to and including u_{xxxxx}.

(*b*) Evaluate the necessary condition for stability.

(*c*) Determine the errors in amplitude and phase for $\beta = 90°$ if this scheme is applied to the wave equation for 10 time steps with $\nu = 1$.

4-14 Solve the wave equation $u_t + u_x = 0$ on a digital computer using

(*a*) Lax scheme

(*b*) Lax-Wendroff scheme

for the initial condition

$$u(x, 0) = \sin 2n\pi\left(\frac{x}{L}\right) \qquad 0 \leqslant x \leqslant L$$

and periodic boundary conditions. Choose a 41 grid point mesh with $\Delta x = 1$ and compute to $t = 18$. Solve this problem for $n = 1, 3$ and $\nu = 1.0, 0.6, 0.3$ and compare graphically with the exact solution. Determine β's for $n = 1$ and $n = 3$ and calculate the errors in amplitude and phase for each scheme with $\nu = 0.6$. Compare these errors with the errors appearing on the graphs.

4-15 Repeat Prob. 4-14 using the following schemes:

(*a*) Windward differencing scheme
(*b*) MacCormack scheme

4-16 Repeat Prob. 4-14 using the following schemes:

(*a*) MacCormack scheme
(*b*) Rusanov scheme ($\omega = 3$)

4-17 Solve the wave equation $u_t + u_x = 0$ on a digital computer using

(*a*) Windward differencing scheme
(*b*) MacCormack scheme

for the initial conditions

$$u(x, 0) = 1 \qquad x \leqslant 10$$

$$u(x, 0) = 0 \qquad x > 10$$

and Dirichlet boundary conditions. Choose a 41 grid point mesh with $\Delta x = 1$ and compute to $t = 18$. Solve this problem for $\nu = 1.0$, 0.6, and 0.3 and compare graphically with the exact solution.

4-18 Apply the windward differencing scheme to the two-dimensional wave equation

$$u_t + c(u_x + u_y) = 0$$

and determine the stability of the resulting scheme.

4-19 Derive the modified equation for the simple implicit method applied to the one-dimensional heat equation. Retain terms up to and including u_{xxxxxx}.

4-20 Evaluate the stability of the combined method B applied to the one-dimensional heat equation.

4-21 Determine the amplification factor of the ADE method of Saul'yev and examine the stability.

4-22 For the grid points $(i + j + n)$ even, show that the hopscotch method reduces to

$$u_{i,j}^{n+1} = 2u_{i,j}^{n+1} - u_{i,j}^{n}$$

4-23 Use the simple explicit method to solve the one-dimensional heat equation on the computational grid (Fig. P4-1) with boundary conditions

$$u_1^n = 2 = u_3^n$$

and initial conditions

$$u_1^1 = 2 = u_3^1 \qquad u_2^1 = 1$$

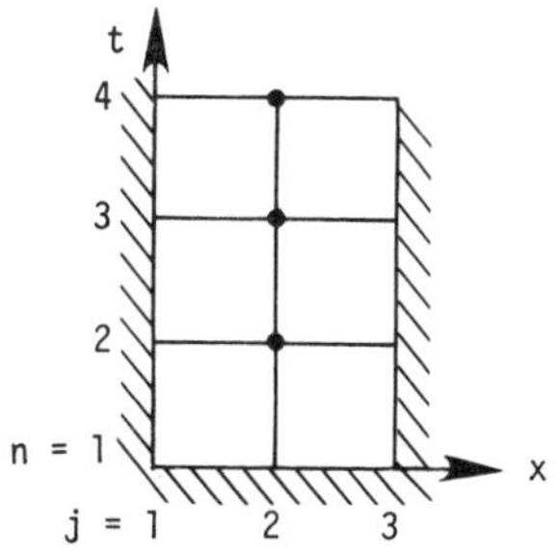

Figure P4-1

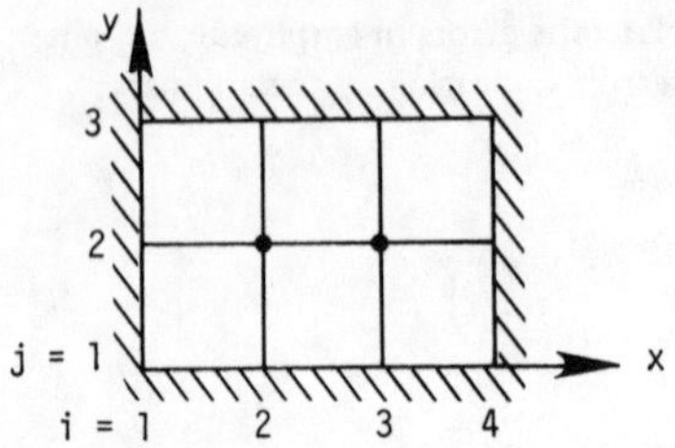

Figure P4-2

Show that if $r = \frac{1}{4}$, the steady-state value of u along $j = 2$ becomes

$$u_2^\infty = \lim_{n \to \infty} \sum_{k=1}^{n} \frac{1}{2^{k-1}}$$

Note that this infinite series is a geometric series which has a known sum.

4-24 Apply the ADI scheme to the two-dimensional heat equation and find u^{n+1} at the internal grid points in the mesh shown in Fig. P4-2 for $r_x = r_y = 2$. The initial conditions are

$$u^n = 1 - \frac{x}{3\,\Delta x} \qquad \text{along } y = 0$$

$$u^n = 1 - \frac{y}{2\,\Delta y} \qquad \text{along } x = 0$$

$$u^n = 0 \qquad \text{everywhere else}$$

and the boundary conditions remain fixed at their initial values.

4-25 Solve the heat equation $u_t = 0.2u_{xx}$ on a digital computer using

(*a*) Simple explicit method
(*b*) Barakat and Clark ADE method

for the initial condition

$$u(x, 0) = 100 \sin \frac{\pi x}{L} \qquad L = 1$$

and boundary conditions

$$u(0, t) = u(L, t) = 0$$

Compute to $t = 0.5$ using the parameters in Table P4-1 (if possible) and compare graphically with the exact solution.

4-26 Repeat Prob. 4-25 using the Crank-Nicolson scheme.

4-27 Repeat Prob. 4-25 using the DuFort-Frankel scheme.

Table P4-1

Case	Number of grid points	r
1	11	0.25
2	11	0.50
3	16	0.50
4	11	1.00
5	11	2.00

4-28 The heat equation

$$\frac{\partial T}{\partial t} = \alpha \frac{\partial^2 T}{\partial x^2}$$

governs the time-dependent temperature distribution in a homogeneous constant property solid under conditions where the temperature varies only in one space dimension. Physically this may be nearly realized in a long thin rod or very large (infinite) wall of finite thickness.

Consider a large wall of thickness L whose initial temperature is given by $T(t, x) = c \sin \pi x/L$. If the faces of the wall continue to be held at 0°, then a solution for the temperature at $t > 0$, $0 \leqslant x \leqslant L$ is

$$T(t, x) = c \exp\left(\frac{-\alpha\pi^2 t}{L^2}\right) \sin \frac{\pi x}{L}$$

For this problem let $c = 100°\mathrm{C}$, $L = 1$ m, $\alpha = 0.02\ \mathrm{m^2/hr}$. We will consider two explicit methods of solution. A. Simple explicit method, Eq. (4-73). Stability requires that $\alpha\,\Delta t/(\Delta x)^2 \leqslant \frac{1}{2}$ for this method. B. Alternating direction explicit (ADE) method, Eq. (4-107). This particular version of the ADE method was suggested by Barakat and Clark (1966). In this algorithm, the equation for p_j^{n+1} can be solved explicitly starting from the boundary at $x = 0$ whereas the equation for q_j^{n+1} should be solved starting at the boundary at $x = L$. There is no stability constraint on the size of the time step for this method. Develop computer programs to solve the problem described above by methods A and B. Also, you will want to provide a capability for evaluating the exact solution for purposes of comparison. Make at least the following comparisons:

1. For $\Delta x = 0.1$, $\Delta t = 0.1$ [resulting in $\alpha\,\Delta t/(\Delta x)^2 = 0.2$] compare the results from methods A and B and the exact solution for $t = 10$ h. A graphical comparison is suggested.
2. Repeat the above comparison after refining the space grid, i.e., let $\Delta x = 0.066667$ (15 increments). Is the reduction in error as suggested by $O(\Delta x)^2$?
3. For $\Delta x = 0.1$ choose Δt such that $\alpha\,\Delta t/(\Delta x)^2 = 0.5$ and compare the predictions of methods A and B and the exact solution for $t \approx 10$ h.
4. Demonstrate that method A does become unstable as $\alpha\,\Delta t/(\Delta x)^2$ exceeds 0.5. One suggestion is to plot the centerline temperature vs. time for $\alpha\,\Delta t/(\Delta x)^2 \simeq 0.6$ for 10-20 hours of problem time.
5. For $\Delta x = 0.1$, choose Δt such that $\alpha\,\Delta t/(\Delta x)^2 = 1.0$ and compare the results of method B and the exact solution for $t \approx 10$ h.
6. Increment $\alpha\,\Delta t/(\Delta x)^2$ to 2, then 3, etc., and repeat comparison 5 above until the agreement with the exact solution becomes noticeably poor.

4-29 Work Prob. 4-28 letting method B be the Crank-Nicolson scheme.

4-30 Work Prob. 4-28 letting method B be the simple implicit scheme.

4-31 Devise a way to solve the problem described in 4-28 utilizing the fourth-order accurate representation of the second derivative given by Eq. (3-35).

4-32 Use the difference scheme of Eq. (3-35) for second derivatives

$$\frac{\partial^2 u}{\partial x^2} \approx \frac{\delta_x^2 u_{i,j}}{h^2(1 + \delta_x^2/12)}$$

to develop a finite-difference representation for Laplace's equation where $\Delta x = \Delta y$. Write out the scheme explicitly in terms of u's on the finite-difference mesh. What is the truncation error of this representation?

4-33 Evaluate the truncation error of the difference scheme of Eq. (4-114) for Laplace's equation (*a*) when $\Delta x = \Delta y$, (*b*) when $\Delta x \neq \Delta y$.

4-34 What is the truncation error for the difference equation employing the nine-point scheme of Eq. (4-114) with $\Delta x = \Delta y$ for the Poisson equation $u_{xx} + u_{yy} = x + y$?

4-35 In the cross-section illustrated in Fig. P4-3, the surface 1-4-7 is insulated (adiabatic). The convective heat transfer coefficient at surface 1-2-3 is 28 $\mathrm{W/m^2\,°C}$. The thermal conductivity of

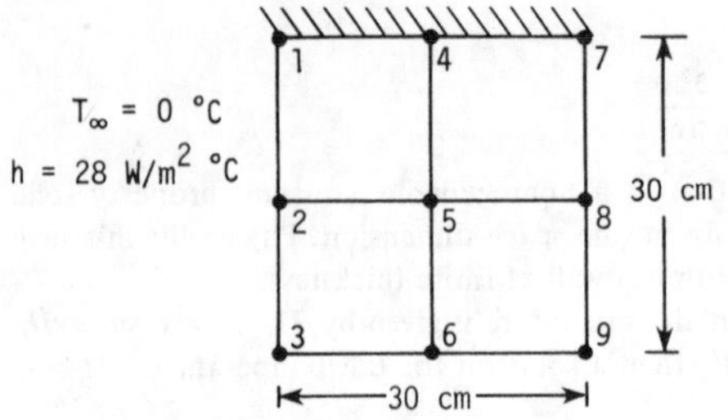

Figure P4-3

the solid material is 3.5 W/m °C. Using Gauss-Seidel iteration, compute the temperature at nodes 1, 2, 4, and 5.

4-36 A cylindrical pin fin (Fig. P4-4) is attached to a 200°C wall while its surface is exposed to a gas at 30°C. The convection heat transfer coefficient is 300 W/m² °C. The fin is made of stainless steel with a thermal conductivity of 18 W/m °C. Use 5 subdivisions and find the steady-state nodal temperatures by Gauss-Seidel iteration. Compute the total rate at which heat is transferred from the fin. You may neglect the heat loss from the outer end of the fin (i.e., assume end is adiabatic).

4-37 Solve the steady-state, two-dimensional heat conduction equation in the unit square, $0 < x < 1$, $0 < y < 1$. by finite-differences using mesh increments $\Delta x = \Delta y = 0.2$ and 0.1. Compare the center temperatures with the exact solution. Use boundary conditions:

$$T = 0 \qquad \text{at } x = 0, x = 1$$

$$\frac{\partial T}{\partial y} = 0 \qquad \text{at } y = 0$$

$$T = \sin(\pi x) \qquad \text{at } y = 1$$

4-38 Consider steady-state conduction governed by Laplace's equation in the two-dimensional domain shown in Fig. P4-5. The boundary conditions are shown in the figure. The mesh is square, i.e., $\Delta x = \Delta y = 0.02$ m.

(*a*) Develop an approximate difference equation for the boundary temperature at point G using the control-volume approach.

(*b*) After obtaining a suitable finite-difference representation for Laplace's equation, use Gauss-Seidel iteration to obtain the steady-state temperature distribution.

4-39 Solve Prob. 4-38 using the line iterative method.

4-40 It is required to estimate the temperature distribution in the two-dimensional wall of a combustion chamber at steady state. The geometry has been simplified for this preliminary analysis and is given in Fig. P4-6. Write a computer program using Gauss-Seidel iteration with SOR to solve this problem. Give careful attention to the equations at the boundaries. Use grid spacing of 2 cm ($\Delta x = \Delta y$) resulting in a 6 × 11 mesh, and use a thermal conductivity of 20 W/m² °C.

(*a*) Compute the steady-state temperature distribution.

(*b*) Compute the rate of heat transfer to the top and check to see how closely it matches the heat removed by the coolant.

(*c*) For the same convergence criteria, repeat the calculation for at least three values of the relaxation parameter, ω. If sufficient computer time is available, make a more detailed search for ω_{opt}.

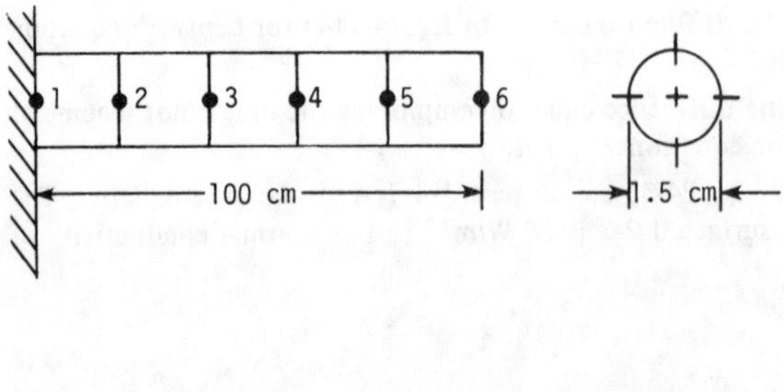

Figure P4-4

$$-k \left.\frac{\partial T}{\partial x}\right)_{x=0} = h\left(T_\infty - T_{bdy}\right)$$

$T_\infty = 300\,°C$

$h = 250 \text{ W/m}^2 \, °C$

$k = 5 \text{ W/m} \, °C$

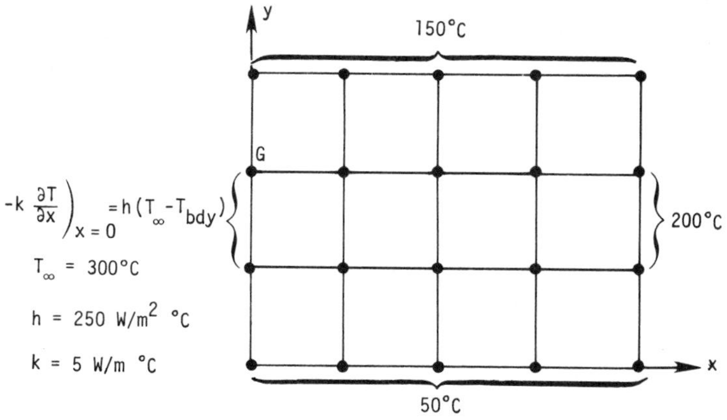

Figure P4-5

4-41 Solve Prob. 4-40 using the line iterative method.

4-42 Solve Prob. 4-40 using the ADI method.

4-43 Use the Lax method to solve the inviscid Burgers equation using a mesh with 51 points in the x direction. Solve this equation for a right propagating discontinuity with $u = 1$ on the first 11 mesh points and $u = 0$ at all other points. Repeat your calculations for Courant numbers of 1.0, 0.6, and 0.3 and compare your numerical solutions with the analytical solution at the same time.

4-44 Repeat Prob. 4-43 using MacCormack's method. Use both a forward-backward and a backward-forward predictor-corrector sequence.

4-45 Repeat Prob. 4-43 using the WKL method.

4-46 Repeat Prob. 4-43 using the Beam-Warming method.

4-47 Solve the inviscid Burgers equation for an expansion with initial data $u = 0$ for the first 21 mesh points and $u = 1$ elsewhere. Use MacCormack's method with both forward-backward and backward-forward predictor-corrector sequences. Compare your results at two different Courant numbers with the analytic solution.

4-48 Repeat Prob. 4-47 using the Beam-Warming method (trapezoidal) and the Euler implicit scheme.

$h_g = 1000 \text{ W/m}^2 \, °C$
$T_g = 2000 \, °C$

HOT GAS

ADIABATIC

10 cm

ADIABATIC

10 cm

20 cm

COOLING CHANNEL SURFACE
$h = 8000 \text{ W/m}^2 \, °C$
$T_b = 60 \, °C$

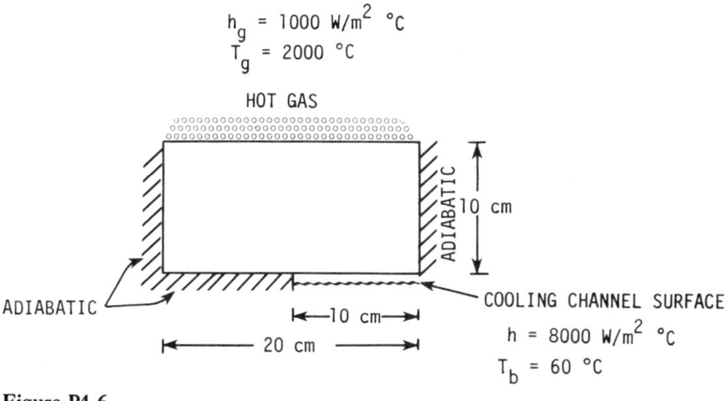

Figure P4-6

4-49 Solve the inviscid Burgers equation for a standing discontinuity. Initialize using $u = 1$ at the left end point and $u = -1$ at the right end point and zero everywhere else. Apply MacCormack's method to this problem.

4-50 Repeat Prob. 4-49 using the Beam-Warming scheme.

4-51 Show graphically the exact steady-state solution of Eq. (4-158) for the boundary conditions

$$u(0, t) = 1$$
$$u(1, t) = 0$$

and $\mu = 0.1$.

4-52 Verify that Eq. (4-169) is an exact stationary solution of Eq. (4-168).

4-53 Derive stability conditions for the FTCS method applied to the 1-D linearized Burgers equation.

4-54 Derive the stability conditions for the upwind difference scheme given by Eq. (4-186).

4-55 Use the FTCS method to solve the linearized Burgers equation for the initial condition

$$u(x, 0) = 0 \qquad 0 \leqslant x \leqslant 1$$

and the boundary conditions

$$u(0, t) = 100$$
$$u(1, t) = 0$$

on a 21 grid point mesh. Find the steady-state solution for the conditions

(*a*) $r = 0.50, \nu = 0.25$
(*b*) $r = 0.50, \nu = 1.00$
(*c*) $r = 0.10, \nu = 0.40$
(*d*) $r = 0.05, \nu = 0.50$

and compare the numerical solution with the exact solution.

4-56 Repeat Prob. 4-55 using the scheme proposed by Leonard.

4-57 Repeat Prob. 4-55 using the leap frog/DuFort-Frankel method.

4-58 Repeat Prob. 4-55 using the Allen-Cheng method.

4-59 Use the Fourier stability analysis to determine the stability limitations of the scheme proposed by Leonard, Eq. (4-188).

4-60 Determine the modified equation for the Allen-Cheng method. Retain terms up to and including u_{xxx}.

4-61 Apply the Brailovskaya scheme to the linearized Burgers equation on the computational grid shown in Fig. P4-7 and show that the steady-state value for u at $j = 2$ is

$$u_2^\infty = \lim_{n\to\infty} \sum_{i=1}^{n} \frac{1}{3^{n-i}} = \frac{3}{2}$$

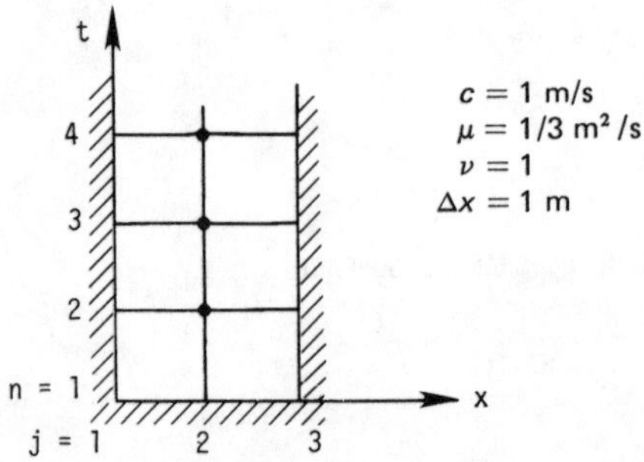

Figure P4-7

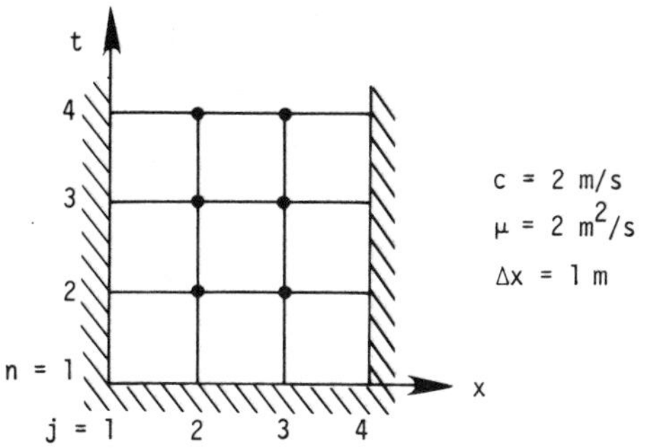

Figure P4-8

Boundary conditions are $u_1^n = \frac{3}{2} = u_3^n$ and the initial condition is $u_2^1 = 1$. Do not use a digital computer to solve this problem.

4-62 Apply the Beam-Warming scheme with Euler implicit time differencing to the linearized Burgers equation on the computational grid shown in Fig. P4-8 and determine the steady-state values for u at $j = 2$ and $j = 3$. The boundary conditions are $u_1^n = 1$, $u_4^n = 4$ and the initial conditions are $u_2^1 = 0 = u_3^1$. Do not use a digital computer to solve this problem.

4-63 Apply the two-step Lax-Wendroff method to the partial differential equation

$$u_t + F_x - uu_{xxx} = 0$$

where $F = F(u)$. Develop the final finite-difference equations.

4-64 Solve the linearized Burgers equation using

(*a*) FTCS method
(*b*) Upwind method, Eq. (4-186)
(*c*) Leonard method, Eq. (4-188)

for the initial condition

$$u(x, 0) = 0 \qquad 0 \leqslant x \leqslant 1$$

and the boundary conditions

$$u(0, t) = 100$$

$$u(1, t) = 0$$

on a 21 grid point mesh. Find the steady-state solution for $r = 0.10$ and $\nu = 0.40$ and compare the numerical solutions with the exact solution.

4-65 Repeat Prob. 4-64 using the following methods

(*a*) Leap frog/DuFort-Frankel method
(*b*) Allen-Cheng method
(*c*) MacCormack method, Eq. (4-198)

4-66 Repeat Prob. 4-64 using the Briley-McDonald method with Euler implicit time differencing.

PART TWO

APPLICATION OF FINITE-DIFFERENCE METHODS TO THE EQUATIONS OF FLUID MECHANICS AND HEAT TRANSFER

CHAPTER

FIVE

GOVERNING EQUATIONS OF FLUID MECHANICS AND HEAT TRANSFER

In this chapter, the governing equations of fluid mechanics and heat transfer (i.e., fluid dynamics) are described. Since the reader is assumed to have some background in this field, a complete derivation of the governing equations is not included. The equations are presented in order of decreasing complexity. For the most part, only the classical forms of the equations are given. Other forms of the governing equations which have been simplified primarily for computational purposes are presented in later chapters. Also included in this chapter is an introduction to turbulence modeling.

5-1 FUNDAMENTAL EQUATIONS

The fundamental equations of fluid dynamics are based on the following universal laws of conservation:

1. Conservation of Mass
2. Conservation of Momentum
3. Conservation of Energy

The equation that results from applying the Conservation of Mass law to a fluid flow is called the continuity equation. The Conservation of Momentum law is nothing more than Newton's Second Law. When this law is applied to a fluid flow it yields a vector equation known as the momentum equation. The Conservation of Energy law is identical to the First Law of Thermodynamics and the resulting fluid dynamic equation is named the energy equation. In addition to the equations developed from these universal laws, it is necessary to establish relationships between fluid properties in order to close the system of equations. An example of such a relationship is the equation of state which relates the thermodynamic variables pressure (p), density (ρ), and temperature (T).

Historically, there have been two different approaches taken to derive the equations of fluid dynamics. These are the phenomenological approach and the kinetic theory approach. In the phenomenological approach, certain relations between stress and rate of strain and heat flux and temperature gradient are postulated and the fluid dynamic equations are then developed from the conservation laws. The required constants of proportionality between stress and rate of strain and heat flux and temperature gradient (which are called transport coefficients) must be determined experimentally in this approach. In the kinetic theory approach (also called the mathematical theory of nonuniform gases), the fluid dynamic equations are obtained with the transport coefficients defined in terms of certain integral relations which involve the dynamics of colliding particles. The drawback to this approach is that the interparticle forces must be specified in order to evaluate the collision integrals. Thus, a mathematical uncertainty takes the place of the experimental uncertainty of the phenomenological approach. These two approaches will yield the same fluid dynamic equations if equivalent assumptions are made during their derivations.

The derivation of the fundamental equations of fluid dynamics will not be presented here. The derivation of the equations using the phenomenological approach is thoroughly treated by Schlichting (1968) and the kinetic theory approach is described in detail by Hirschfelder et al. (1954). The fundamental equations given in this chapter were derived for a uniform, homogeneous fluid without mass diffusion or finite-rate chemical reactions. In order to include these latter effects it is necessary to consider extra relations called the species continuity equations and to add terms to the energy equation to account for diffusion. Further information on reacting flows can be found in Dorrance (1962).

5-1.1 Continuity Equation

The Conservation of Mass law applied to a fluid passing through an infinitesimal, fixed control volume (see Fig. 5-1) yields the following equation of continuity

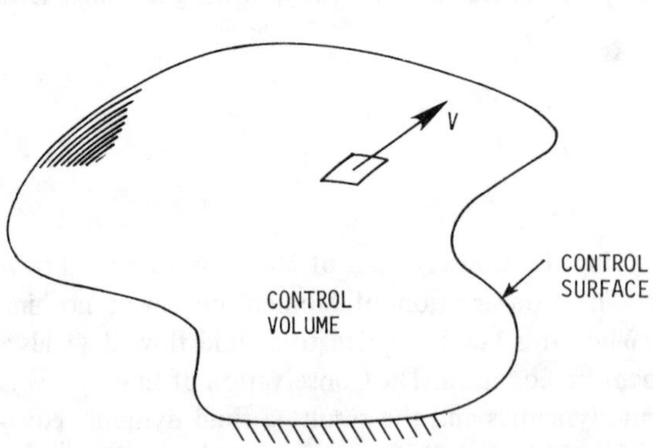

Figure 5-1. Control volume for Eulerian approach.

$$\frac{\partial \rho}{\partial t} + \nabla \cdot (\rho \mathbf{V}) = 0 \tag{5-1}$$

where ρ is the fluid density and $\mathbf{V}$ is the fluid velocity. The first term in this equation represents the rate of increase of the density in the control volume and the second term represents the rate of mass flux passing out of the control surface (which surrounds the control volume) per unit volume. It is convenient to use the substantial derivative

$$\frac{D(\ )}{Dt} \equiv \frac{\partial(\ )}{\partial t} + \mathbf{V} \cdot \nabla(\ ) \tag{5-2}$$

to change Eq. (5-1) into the form

$$\frac{D\rho}{Dt} + \rho(\nabla \cdot \mathbf{V}) = 0 \tag{5-3}$$

Equation (5-1) was derived using the *Eulerian approach.* In this approach, a fixed control volume is utilized and the changes to the fluid are recorded as the fluid passes through the control volume. In the alternative *Lagrangian approach,* the changes to the properties of a fluid element are recorded by an observer moving with the fluid element. In general, the Eulerian viewpoint is the appropriate choice for fluid mechanics.

For a Cartesian coordinate system, where u, v, w represent the x, y, z components of the velocity vector, Eq. (5-1) becomes

$$\frac{\partial \rho}{\partial t} + \frac{\partial}{\partial x}(\rho u) + \frac{\partial}{\partial y}(\rho v) + \frac{\partial}{\partial z}(\rho w) = 0 \tag{5-4}$$

Note that this equation is in conservation-law (divergence) form.

A flow in which the density of each fluid element remains constant is called *incompressible.* Mathematically, this implies that

$$\frac{D\rho}{Dt} = 0 \tag{5-5}$$

which reduces Eq. (5-3) to

$$\nabla \cdot \mathbf{V} = 0 \tag{5-6}$$

or

$$\frac{\partial u}{\partial x} + \frac{\partial v}{\partial y} + \frac{\partial w}{\partial z} = 0 \tag{5-7}$$

for the Cartesian coordinate system. For air flows with $V < 100$ m/s or $M < 0.3$ the assumption of incompressibility is a good approximation.

5-1.2 Momentum Equation

Newton's Second Law applied to a fluid passing through an infinitesimal, fixed control volume yields the following momentum equation:

$$\frac{\partial}{\partial t}(\rho \mathbf{V}) + \nabla \cdot \rho \mathbf{V}\mathbf{V} = \rho \mathbf{f} + \nabla \cdot \mathbf{\Pi}_{ij} \tag{5-8}$$

The first term in this equation represents the rate of increase of momentum per unit volume in the control volume. The second term represents the rate of momentum lost by convection (per unit volume) through the control surface. Note that $\rho\mathbf{VV}$ is a tensor so that $\nabla \cdot \rho\mathbf{VV}$ is not a simple divergence. This term can be expanded, however, as

$$\nabla \cdot \rho \mathbf{V}\mathbf{V} = \rho \mathbf{V} \cdot \nabla \mathbf{V} + \mathbf{V}(\nabla \cdot \rho \mathbf{V}) \tag{5-9}$$

When this expression for $\nabla \cdot \rho\mathbf{VV}$ is substituted into Eq. (5-8), and the resulting equation is simplified using the continuity equation, the momentum equation reduces to

$$\rho \frac{D\mathbf{V}}{Dt} = \rho \mathbf{f} + \nabla \cdot \mathbf{\Pi}_{ij} \tag{5-10}$$

The first term on the right-hand side of Eq. (5-10) is the body force per unit volume. Body forces act at a distance and apply to the entire mass of the fluid. The most common body force is the gravitational force. In this case, the force per unit mass (**f**) equals the acceleration of gravity vector **g**.

$$\rho \mathbf{f} = \rho \mathbf{g} \tag{5-11}$$

The second term on the right-hand side of Eq. (5-10) represents the surface forces per unit volume. These forces are applied by the external stresses on the fluid element. The stresses consist of normal stresses and shearing stresses and are represented by the components of the stress tensor $\mathbf{\Pi}_{ij}$.

The momentum equation given above is quite general and is applicable to both continuum and noncontinuum flows. It is only when approximate expressions are inserted for the shear stress tensor that Eq. (5-8) loses its generality. For all gases which can be treated as a continuum, and most liquids, it has been observed that the stress at a point is linearly dependent on the rates of strain (deformation) of the fluid. A fluid which behaves in this manner is called a *Newtonian fluid.* With this assumption, it is possible to derive (Schlichting, 1968) a general deformation law which relates the stress tensor to the pressure and velocity components. In compact tensor notation, this relationship becomes

$$\mathbf{\Pi}_{ij} = -p\delta_{ij} + \mu\left(\frac{\partial u_i}{\partial x_j} + \frac{\partial u_j}{\partial x_i}\right) + \delta_{ij}\mu' \frac{\partial u_k}{\partial x_k} \qquad (i,j,k = 1,2,3) \tag{5-12}$$

where δ_{ij} is the Kronecker delta function ($\delta_{ij} = 1$ if $i = j$ and $\delta_{ij} = 0$ if $i \neq j$); u_1, u_2, u_3 represent the three components of the velocity vector $\mathbf{V}$; x_1, x_2, x_3 represent the three components of the position vector; μ is the coefficient of viscosity (dynamic viscosity), and μ' is the second coefficient of viscosity. The two coefficients of viscosity are related to the coefficient of bulk viscosity (κ) by the expression

$$\kappa = \tfrac{2}{3}\mu + \mu' \tag{5-13}$$

In general, it is believed that the coefficient of bulk viscosity is negligible except in the study of the structure of shock waves and in the absorption and attenuation of acoustic waves. For this reason, we will ignore bulk viscosity for the remainder of the text. With $\kappa = 0$, the second coefficient of viscosity becomes

$$\mu' = -\tfrac{2}{3}\mu \tag{5-14}$$

and the stress tensor may be written as

$$\Pi_{ij} = -p\delta_{ij} + \mu\left[\left(\frac{\partial u_i}{\partial x_j} + \frac{\partial u_j}{\partial x_i}\right) - \frac{2}{3}\delta_{ij}\frac{\partial u_k}{\partial x_k}\right] \quad (i,j,k = 1,2,3) \tag{5-15}$$

The stress tensor is frequently separated in the following manner

$$\Pi_{ij} = -p\delta_{ij} + \tau_{ij} \tag{5-16}$$

where τ_{ij} represents the viscous stress tensor given by

$$\tau_{ij} = \mu\left[\left(\frac{\partial u_i}{\partial x_j} + \frac{\partial u_j}{\partial x_i}\right) - \frac{2}{3}\delta_{ij}\frac{\partial u_k}{\partial x_k}\right] \quad (i,j,k = 1,2,3) \tag{5-17}$$

Upon substituting Eq. (5-15) into Eq. (5-10), the famous *Navier-Stokes equation* is obtained:

$$\rho\frac{D\mathbf{V}}{Dt} = \rho\mathbf{f} - \nabla p + \frac{\partial}{\partial x_j}\left[\mu\left(\frac{\partial u_i}{\partial x_j} + \frac{\partial u_j}{\partial x_i}\right) - \frac{2}{3}\delta_{ij}\mu\frac{\partial u_k}{\partial x_k}\right] \tag{5-18}$$

For a Cartesian coordinate system, Eq. (5-18) can be separated into the following three scalar Navier-Stokes equations:

$$\begin{aligned}
\rho\frac{Du}{Dt} &= \rho f_x - \frac{\partial p}{\partial x} + \frac{\partial}{\partial x}\left[\frac{2}{3}\mu\left(2\frac{\partial u}{\partial x} - \frac{\partial v}{\partial y} - \frac{\partial w}{\partial z}\right)\right] + \frac{\partial}{\partial y}\left[\mu\left(\frac{\partial u}{\partial y} + \frac{\partial v}{\partial x}\right)\right] \\
&\quad + \frac{\partial}{\partial z}\left[\mu\left(\frac{\partial w}{\partial x} + \frac{\partial u}{\partial z}\right)\right] \\
\rho\frac{Dv}{Dt} &= \rho f_y - \frac{\partial p}{\partial y} + \frac{\partial}{\partial x}\left[\mu\left(\frac{\partial v}{\partial x} + \frac{\partial u}{\partial y}\right)\right] + \frac{\partial}{\partial y}\left[\frac{2}{3}\mu\left(2\frac{\partial v}{\partial y} - \frac{\partial u}{\partial x} - \frac{\partial w}{\partial z}\right)\right] \\
&\quad + \frac{\partial}{\partial z}\left[\mu\left(\frac{\partial v}{\partial z} + \frac{\partial w}{\partial y}\right)\right] \\
\rho\frac{Dw}{Dt} &= \rho f_z - \frac{\partial p}{\partial z} + \frac{\partial}{\partial x}\left[\mu\left(\frac{\partial w}{\partial x} + \frac{\partial u}{\partial z}\right)\right] + \frac{\partial}{\partial y}\left[\mu\left(\frac{\partial v}{\partial z} + \frac{\partial w}{\partial y}\right)\right] \\
&\quad + \frac{\partial}{\partial z}\left[\frac{2}{3}\mu\left(2\frac{\partial w}{\partial z} - \frac{\partial u}{\partial x} - \frac{\partial v}{\partial y}\right)\right]
\end{aligned} \tag{5-19}$$

Utilizing Eq. (5-8), these equations can be rewritten in conservation-law form as

$$\frac{\partial \rho u}{\partial t} + \frac{\partial}{\partial x}(\rho u^2 + p - \tau_{xx}) + \frac{\partial}{\partial y}(\rho uv - \tau_{xy}) + \frac{\partial}{\partial z}(\rho uw - \tau_{xz}) = \rho f_x \tag{5-20}$$

$$\frac{\partial \rho v}{\partial t}+\frac{\partial}{\partial x}(\rho uv-\tau_{xy})+\frac{\partial}{\partial y}(\rho v^2+p-\tau_{yy})+\frac{\partial}{\partial z}(\rho vw-\tau_{yz})=\rho f_y \qquad (5\text{-}20)$$

(Cont.)

$$\frac{\partial \rho w}{\partial t}+\frac{\partial}{\partial x}(\rho uw-\tau_{xz})+\frac{\partial}{\partial y}(\rho vw-\tau_{yz})+\frac{\partial}{\partial z}(\rho w^2+p-\tau_{zz})=\rho f_z$$

where the components of the viscous stress tensor τ_{ij} are given by

$$\tau_{xx}=\frac{2}{3}\mu\left(2\frac{\partial u}{\partial x}-\frac{\partial v}{\partial y}-\frac{\partial w}{\partial z}\right)$$

$$\tau_{yy}=\frac{2}{3}\mu\left(2\frac{\partial v}{\partial y}-\frac{\partial u}{\partial x}-\frac{\partial w}{\partial z}\right)$$

$$\tau_{zz}=\frac{2}{3}\mu\left(2\frac{\partial w}{\partial z}-\frac{\partial u}{\partial x}-\frac{\partial v}{\partial y}\right)$$

$$\tau_{xy}=\mu\left(\frac{\partial u}{\partial y}+\frac{\partial v}{\partial x}\right)=\tau_{yx}$$

$$\tau_{xz}=\mu\left(\frac{\partial w}{\partial x}+\frac{\partial u}{\partial z}\right)=\tau_{zx}$$

$$\tau_{yz}=\mu\left(\frac{\partial v}{\partial z}+\frac{\partial w}{\partial y}\right)=\tau_{zy}$$

The Navier-Stokes equations form the basis upon which the entire science of viscous flow theory has been developed. Strictly speaking, the term Navier-Stokes equations refers to the components of the viscous momentum equation [Eq. (5-18)]. However, it is common practice to include the continuity equation and the energy equation in the set of equations referred to as the Navier-Stokes equations.

If the flow is assumed incompressible and the coefficient of viscosity (μ) is assumed constant, Eq. (5-18) will reduce to the much simpler form

$$\rho\frac{D\mathbf{V}}{Dt}=\rho\mathbf{f}-\nabla p+\mu\nabla^2\mathbf{V} \qquad (5\text{-}21)$$

It should be remembered that Eq. (5-21) is derived by assuming a constant viscosity which may be a poor approximation for the nonisothermal flow of a liquid whose viscosity is highly temperature dependent. On the other hand, the viscosity of gases is only moderately temperature dependent and Eq. (5-21) is a good approximation for the incompressible flow of a gas.

5-1.3 Energy Equation

The First Law of Thermodynamics applied to a fluid passing through an infinitesimal, fixed control volume yields the following energy equation

$$\frac{\partial E_t}{\partial t}+\nabla\cdot E_t\mathbf{V}=\frac{\partial Q}{\partial t}-\nabla\cdot\mathbf{q}+\rho\mathbf{f}\cdot\mathbf{V}+\nabla\cdot(\Pi_{ij}\cdot\mathbf{V}) \qquad (5\text{-}22)$$

where E_t is the total energy per unit volume given by

$$E_t = \rho\left(e + \frac{V^2}{2} + \text{potential energy} + \cdots\right) \tag{5-23}$$

and e is the internal energy per unit mass. The first term on the left-hand side of Eq. (5-22) represents the rate of increase of total energy per unit volume in the control volume while the second term represents the rate of total energy lost by convection (per unit volume) through the control surface. The first term on the right-hand side of Eq. (5-22) is the rate of heat produced per unit volume by external agencies while the second term $(\nabla \cdot \mathbf{q})$ is the rate of heat lost by conduction (per unit volume) through the control surface. Fourier's law for heat transfer by conduction will be assumed so that the heat transfer $\mathbf{q}$ can be expressed as

$$\mathbf{q} = -k\,\nabla T \tag{5-24}$$

where k is the coefficient of thermal conductivity and T is the temperature. The third term on the right-hand side of Eq. (5-22) represents the work done on the control volume (per unit volume) by the body forces while the fourth term represents the work done on the control volume (per unit volume) by the surface forces. It should be obvious that Eq. (5-22) is simply the First Law of Thermodynamics applied to the control volume. That is, the increase of energy in the system is equal to heat added to the system plus the work done on the system.

For a Cartesian coordinate system, Eq. (5-22) becomes

$$\begin{aligned}\frac{\partial E_t}{\partial t} - \frac{\partial Q}{\partial t} - \rho(f_x u + f_y v + f_z w) + \frac{\partial}{\partial x}(E_t u + pu - u\tau_{xx} - v\tau_{xy} - w\tau_{xz} + q_x)\\ + \frac{\partial}{\partial y}(E_t v + pv - u\tau_{xy} - v\tau_{yy} - w\tau_{yz} + q_y)\\ + \frac{\partial}{\partial z}(E_t w + pw - u\tau_{xz} - v\tau_{yz} - w\tau_{zz} + q_z) = 0\end{aligned} \tag{5-25}$$

which is in conservation-law form. Using the continuity equation, the left-hand side of Eq. (5-22) can be replaced by the following expression

$$\rho\frac{D(E_t/\rho)}{Dt} = \frac{\partial E_t}{\partial t} + \nabla \cdot E_t\mathbf{V} \tag{5-26}$$

which is equivalent to

$$\rho\frac{D(E_t/\rho)}{Dt} = \rho\frac{De}{Dt} + \rho\frac{D(V^2/2)}{Dt} \tag{5-27}$$

if only internal energy and kinetic energy are considered significant in Eq. (5-23). Forming the scalar dot product of Eq. (5-10) with the velocity vector $\mathbf{V}$ allows one to obtain

$$\rho\frac{D\mathbf{V}}{Dt} \cdot \mathbf{V} = \rho\mathbf{f} \cdot \mathbf{V} - \nabla p \cdot \mathbf{V} + (\nabla \cdot \tau_{ij}) \cdot \mathbf{V} \tag{5-28}$$

Now if Eqs. (5-26), (5-27), and (5-28) are combined and substituted into Eq. (5-22), a useful variation of the original energy equation is obtained

$$\rho \frac{De}{Dt} + p(\nabla \cdot \mathbf{V}) = \frac{\partial Q}{\partial t} - \nabla \cdot \mathbf{q} + \nabla \cdot (\tau_{ij} \cdot \mathbf{V}) - (\nabla \cdot \tau_{ij}) \cdot \mathbf{V} \tag{5-29}$$

The last two terms in this equation can be combined into a single term since

$$\tau_{ij} \frac{\partial u_i}{\partial x_j} = \nabla \cdot (\tau_{ij} \cdot \mathbf{V}) - (\nabla \cdot \tau_{ij}) \cdot \mathbf{V} \tag{5-30}$$

This term is customarily called the *dissipation function* Φ and represents the heat equivalent of the rate at which mechanical energy is expended in the process of deformation of the fluid due to viscosity. After inserting the dissipation function, Eq. (5-29) becomes

$$\rho \frac{De}{Dt} + p(\nabla \cdot \mathbf{V}) = \frac{\partial Q}{\partial t} - \nabla \cdot \mathbf{q} + \Phi \tag{5-31}$$

Using the definition of enthalpy

$$h = e + \frac{p}{\rho} \tag{5-32}$$

and the continuity equation, Eq. (5-31) can be rewritten as

$$\rho \frac{Dh}{Dt} = \frac{Dp}{Dt} + \frac{\partial Q}{\partial t} - \nabla \cdot \mathbf{q} + \Phi \tag{5-33}$$

For a Cartesian coordinate system, the dissipation function, which is always positive if $\mu' = -(2/3)\mu$, becomes

$$\Phi = \mu \left[2\left(\frac{\partial u}{\partial x}\right)^2 + 2\left(\frac{\partial v}{\partial y}\right)^2 + 2\left(\frac{\partial w}{\partial z}\right)^2 + \left(\frac{\partial v}{\partial x} + \frac{\partial u}{\partial y}\right)^2 + \left(\frac{\partial w}{\partial y} + \frac{\partial v}{\partial z}\right)^2 \right.$$
$$\left. + \left(\frac{\partial u}{\partial z} + \frac{\partial w}{\partial x}\right)^2 - \frac{2}{3}\left(\frac{\partial u}{\partial x} + \frac{\partial v}{\partial y} + \frac{\partial w}{\partial z}\right)^2 \right] \tag{5-34}$$

If the flow is assumed incompressible, and if the coefficient of thermal conductivity is assumed constant, Eq. (5-31) reduces to

$$\rho \frac{De}{Dt} = \frac{\partial Q}{\partial t} + k \nabla^2 T + \Phi \tag{5-35}$$

5-1.4 Equation of State

In order to close the system of fluid dynamic equations it is necessary to establish relationships between the thermodynamic variables (p,ρ,T,e,h) as well as to relate the transport properties (μ,k) to the thermodynamic variables. For example, let us consider a compressible flow without external heat addition or body forces and use Eq.

(5-4) for the continuity equation, Eqs. (5-19) for the three momentum equations, and Eq. (5-25) for the energy equation. These five scalar equations contain seven unknowns ρ, p, e, T, u, v, w provided that the transport coefficients μ, k can be related to the thermodynamic properties in the list of unknowns. It is obvious that two additional equations are required to close the system. These two additional equations can be obtained by determining relationships that exist between the thermodynamic variables. Relations of this type are known as equations of state. According to the ***state principle*** of thermodynamics, the local thermodynamic state is fixed by any two independent thermodynamic variables provided that the chemical composition of the fluid is not changing due to diffusion or finite-rate chemical reactions. Thus for the present example, if we choose e and ρ as the two independent variables then equations of state of the form

$$p = p(e,\rho) \qquad T = T(e,\rho) \tag{5-36}$$

are required.

An example of an equation of state is the perfect gas equation of state

$$p = \rho R T \tag{5-37}$$

where R is the gas constant. Also for a perfect gas, the following relationships exist:

$$e = c_v T \qquad h = c_p T \qquad \gamma = \frac{c_p}{c_v} \qquad c_v = \frac{R}{\gamma - 1} \qquad c_p = \frac{\gamma R}{\gamma - 1} \tag{5-38}$$

where c_v is the specific heat at constant volume, c_p is the specific heat at constant pressure and γ is the ratio of specific heats. For air at standard conditions, $R = 287$ $\text{m}^2/\text{s}^2\ {}^\circ\text{K}$ and $\gamma = 1.4$. If we assume that the fluid in our example is a perfect gas, then Eqs. (5-36) become

$$p = (\gamma - 1)\rho e \qquad T = \frac{(\gamma - 1)e}{R} \tag{5-39}$$

For fluids which cannot be considered perfect, the required state relations can be found in the form of tables, charts, or curve fits.

The coefficients of viscosity and thermal conductivity have been related to the thermodynamic variables using kinetic theory. For example, Sutherland's formula for viscosity is given by

$$\mu = C_1 \frac{T^{3/2}}{T + C_2} \tag{5-40}$$

where C_1 and C_2 are constants for a given gas. For air at moderate temperatures, $C_1 = 1.458 \times 10^{-6}$ kg/(m s $\sqrt{{}^\circ\text{K}}$) and $C_2 = 110.4^\circ$K. The Prandtl number

$$\text{Pr} = \frac{c_p \mu}{k} \tag{5-41}$$

is often used to determine the coefficient of thermal conductivity k once μ is known. This is possible because the ratio (c_p/Pr) which appears in the expression

$$k = \frac{c_p}{\Pr}\mu \tag{5-42}$$

is approximately constant for most gases. For air at standard conditions $\Pr = 0.72$.

5-1.5 Vector Form of Equations

Before applying a finite-difference algorithm to the governing fluid dynamic equations, it is often convenient to combine the equations into a compact vector form. For example, the compressible Navier-Stokes equations in Cartesian coordinates without body forces or external heat addition can be written as

$$\frac{\partial \mathbf{U}}{\partial t} + \frac{\partial \mathbf{E}}{\partial x} + \frac{\partial \mathbf{F}}{\partial y} + \frac{\partial \mathbf{G}}{\partial z} = 0 \tag{5-43}$$

where **U**, **E**, **F**, and **G** are vectors given by

$$\mathbf{U} = \begin{bmatrix} \rho \\ \rho u \\ \rho v \\ \rho w \\ E_t \end{bmatrix}$$

$$\mathbf{E} = \begin{bmatrix} \rho u \\ \rho u^2 + p - \tau_{xx} \\ \rho u v - \tau_{xy} \\ \rho u w - \tau_{xz} \\ (E_t + p)u - u\tau_{xx} - v\tau_{xy} - w\tau_{xz} + q_x \end{bmatrix}$$

$$\mathbf{F} = \begin{bmatrix} \rho v \\ \rho u v - \tau_{xy} \\ \rho v^2 + p - \tau_{yy} \\ \rho v w - \tau_{yz} \\ (E_t + p)v - u\tau_{xy} - v\tau_{yy} - w\tau_{yz} + q_y \end{bmatrix} \tag{5-44}$$

$$\mathbf{G} = \begin{bmatrix} \rho w \\ \rho u w - \tau_{xz} \\ \rho v w - \tau_{yz} \\ \rho w^2 + p - \tau_{zz} \\ (E_t + p)w - u\tau_{xz} - v\tau_{yz} - w\tau_{zz} + q_z \end{bmatrix}$$

The first row of the vector Eq. (5-43) corresponds to the continuity equation as given by Eq. (5-4). Likewise, the second, third, and fourth rows are the momentum equations [Eqs. (5-20)] while the fifth row is the energy equation [Eq. (5-25)]. With the Navier-Stokes equations written in this form, it is often easier to code the desired finite-difference algorithm. Other fluid dynamic equations which are written in conservation-law form can also be placed in a similar vector form.

5-1.6 Nondimensional Form of Equations

The governing fluid dynamic equations are often put into nondimensional form. The advantage in doing this is that the characteristic parameters such as Mach number, Reynolds number, Prandtl number, etc., can be varied independently. Also, by nondimensionalizing the equations, the flow variables are "normalized" so that their values fall between certain prescribed limits such as 0 and 1. Many different nondimensionalizing procedures are possible. An example of one such procedure is

$$x^* = \frac{x}{L} \qquad y^* = \frac{y}{L} \qquad z^* = \frac{z}{L} \qquad t^* = \frac{t}{L/V_\infty}$$

$$u^* = \frac{u}{V_\infty} \qquad v^* = \frac{v}{V_\infty} \qquad w^* = \frac{w}{V_\infty} \qquad \mu^* = \frac{\mu}{\mu_\infty}$$

$$\rho^* = \frac{\rho}{\rho_\infty} \qquad p^* = \frac{p}{\rho_\infty V_\infty^2} \qquad T^* = \frac{T}{T_\infty} \qquad e^* = \frac{e}{V_\infty^2}$$

where the nondimensional variables are denoted by an asterisk, freestream conditions are denoted by ∞ and L is the reference length used in the Reynolds number

$$\mathrm{Re}_L = \frac{\rho_\infty V_\infty L}{\mu_\infty}$$

If this nondimensionalizing procedure is applied to the compressible Navier-Stokes equations given previously by Eqs. (5-43) and (5-44), the following nondimensional equations are obtained

$$\frac{\partial \mathbf{U}^*}{\partial t^*} + \frac{\partial \mathbf{E}^*}{\partial x^*} + \frac{\partial \mathbf{F}^*}{\partial y^*} + \frac{\partial \mathbf{G}^*}{\partial z^*} = 0 \tag{5-45}$$

where $\mathbf{U}^*$, $\mathbf{E}^*$, $\mathbf{F}^*$, and $\mathbf{G}^*$ are the vectors

$$\mathbf{U}^* = \begin{bmatrix} \rho^* \\ \rho^* u^* \\ \rho^* v^* \\ \rho^* w^* \\ E_t^* \end{bmatrix} \tag{5-46}$$

$$
\mathbf{E}^* = \begin{bmatrix} \rho^* u^* \\ \rho^* u^{*2} + p^* - \tau_{xx}^* \\ \rho^* u^* v^* - \tau_{xy}^* \\ \rho^* u^* w^* - \tau_{xz}^* \\ (E_t^* + p^*)u^* - u^*\tau_{xx}^* - v^*\tau_{xy}^* - w^*\tau_{xz}^* + q_x^* \end{bmatrix}
$$

$$
\mathbf{F}^* = \begin{bmatrix} \rho^* v^* \\ \rho^* u^* v^* - \tau_{xy}^* \\ \rho^* v^{*2} + p^* - \tau_{yy}^* \\ \rho^* v^* w^* - \tau_{yz}^* \\ (E_t^* + p^*)v^* - u^*\tau_{xy}^* - v^*\tau_{yy}^* - w^*\tau_{yz}^* + q_y^* \end{bmatrix} \qquad \text{(5-46) (Cont.)}
$$

$$
\mathbf{G}^* = \begin{bmatrix} \rho^* w^* \\ \rho^* u^* w^* - \tau_{xz}^* \\ \rho^* v^* w^* - \tau_{yz}^* \\ \rho^* w^{*2} + p^* - \tau_{zz}^* \\ (E_t^* + p^*)w^* - u^*\tau_{xz}^* - v^*\tau_{yz}^* - w^*\tau_{zz}^* + q_z^* \end{bmatrix}
$$

and

$$
E_t^* = \rho^* \left(e^* + \frac{u^{*2} + v^{*2} + w^{*2}}{2} \right)
$$

The components of the shear-stress tensor and the heat-flux vector in nondimensional form are given by

$$
\begin{aligned}
\tau_{xx}^* &= \frac{2\mu^*}{3\,\mathrm{Re}_L}\left(2\frac{\partial u^*}{\partial x^*} - \frac{\partial v^*}{\partial y^*} - \frac{\partial w^*}{\partial z^*}\right) \\
\tau_{yy}^* &= \frac{2\mu^*}{3\,\mathrm{Re}_L}\left(2\frac{\partial v^*}{\partial y^*} - \frac{\partial u^*}{\partial x^*} - \frac{\partial w^*}{\partial z^*}\right) \\
\tau_{zz}^* &= \frac{2\mu^*}{3\,\mathrm{Re}_L}\left(2\frac{\partial w^*}{\partial z^*} - \frac{\partial u^*}{\partial x^*} - \frac{\partial v^*}{\partial y^*}\right) \\
\tau_{xy}^* &= \frac{\mu^*}{\mathrm{Re}_L}\left(\frac{\partial u^*}{\partial y^*} + \frac{\partial v^*}{\partial x^*}\right) \\
\tau_{xz}^* &= \frac{\mu^*}{\mathrm{Re}_L}\left(\frac{\partial u^*}{\partial z^*} + \frac{\partial w^*}{\partial x^*}\right)
\end{aligned} \qquad \text{(5-47)}
$$

$$\tau_{yz}^* = \frac{\mu^*}{\text{Re}_L}\left(\frac{\partial v^*}{\partial z^*} + \frac{\partial w^*}{\partial y^*}\right)$$

$$q_x^* = -\frac{\mu^*}{(\gamma - 1)M_\infty^2 \text{ Re}_L \text{ Pr}} \frac{\partial T^*}{\partial x^*}$$

$$q_y^* = -\frac{\mu^*}{(\gamma - 1)M_\infty^2 \text{ Re}_L \text{ Pr}} \frac{\partial T^*}{\partial y^*} \qquad \text{(5-47) } (Cont.)$$

$$q_z^* = -\frac{\mu^*}{(\gamma - 1)M_\infty^2 \text{ Re}_L \text{ Pr}} \frac{\partial T^*}{\partial z^*}$$

where M_∞ is the freestream Mach number

$$M_\infty = \frac{V_\infty}{\sqrt{\gamma R T_\infty}}$$

and the perfect gas equations of state [Eqs. (5-39)] become

$$p^* = (\gamma - 1)\rho^* e^*$$

$$T^* = \frac{\gamma M_\infty^2 p^*}{\rho^*}$$

Note that the nondimensional form of the equations given by Eqs. (5-45) and (5-46) are identical (except for the asterisks) to the dimensional form given by Eqs. (5-43) and (5-44). For convenience, the asterisks can be dropped from the nondimensional equations and this is usually done.

5-1.7 Orthogonal Curvilinear Coordinates

The basic equations of fluid dynamics are valid for any coordinate system. We have previously expressed these equations in terms of a Cartesian coordinate system. For many applications it is more convenient to use a different orthogonal coordinate system. Let us define x_1, x_2, x_3 to be a set of generalized orthogonal curvilinear coordinates whose origin is at point P and let $\mathbf{i}_1, \mathbf{i}_2, \mathbf{i}_3$ be the corresponding unit vectors (see Fig. 5-2). The rectangular Cartesian coordinates are related to the generalized curvilinear coordinates by

$$x = x(x_1, x_2, x_3)$$

$$y = y(x_1, x_2, x_3) \qquad \text{(5-48)}$$

$$z = z(x_1, x_2, x_3)$$

so that if the Jacobian

$$\frac{\partial(x,y,z)}{\partial(x_1,x_2,x_3)}$$

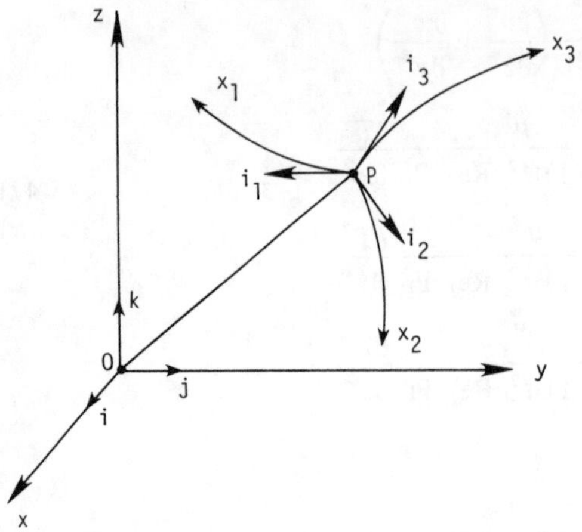

Figure 5-2 Orthogonal curvilinear coordinate system.

is nonzero, then

$$
\begin{aligned}
x_1 &= x_1(x,y,z) \\
x_2 &= x_2(x,y,z) \\
x_3 &= x_3(x,y,z)
\end{aligned}
\tag{5-49}
$$

The elemental arc length in Cartesian coordinates is

$$(ds)^2 = (dx)^2 + (dy)^2 + (dz)^2 \tag{5-50}$$

If Eq. (5-48) is differentiated and substituted into Eq. (5-50) the following result is obtained

$$(ds)^2 = (h_1\, dx_1)^2 + (h_2\, dx_2)^2 + (h_3\, dx_3)^2 \tag{5-51}$$

where

$$
\begin{aligned}
(h_1)^2 &= \left(\frac{\partial x}{\partial x_1}\right)^2 + \left(\frac{\partial y}{\partial x_1}\right)^2 + \left(\frac{\partial z}{\partial x_1}\right)^2 \\
(h_2)^2 &= \left(\frac{\partial x}{\partial x_2}\right)^2 + \left(\frac{\partial y}{\partial x_2}\right)^2 + \left(\frac{\partial z}{\partial x_2}\right)^2 \\
(h_3)^2 &= \left(\frac{\partial x}{\partial x_3}\right)^2 + \left(\frac{\partial y}{\partial x_3}\right)^2 + \left(\frac{\partial z}{\partial x_3}\right)^2
\end{aligned}
$$

If ϕ is an arbitrary scalar and **A** is an arbitrary vector, the expressions for the gradient, divergence, curl, and Laplacian operator in the generalized curvilinear coordinates become

$$\nabla\phi = \frac{1}{h_1}\frac{\partial \phi}{\partial x_1}\mathbf{i}_1 + \frac{1}{h_2}\frac{\partial \phi}{\partial x_2}\mathbf{i}_2 + \frac{1}{h_3}\frac{\partial \phi}{\partial x_3}\mathbf{i}_3 \tag{5-52}$$

$$\nabla\cdot\mathbf{A} = \frac{1}{h_1h_2h_3}\left[\frac{\partial}{\partial x_1}(h_2h_3A_1) + \frac{\partial}{\partial x_2}(h_3h_1A_2) + \frac{\partial}{\partial x_3}(h_1h_2A_3)\right] \tag{5-53}$$

$$\nabla\times\mathbf{A} = \frac{1}{h_1h_2h_3}\left\{h_1\left[\frac{\partial(h_3A_3)}{\partial x_2} - \frac{\partial(h_2A_2)}{\partial x_3}\right]\mathbf{i}_1 + h_2\left[\frac{\partial(h_1A_1)}{\partial x_3} - \frac{\partial(h_3A_3)}{\partial x_1}\right]\mathbf{i}_2 + h_3\left[\frac{\partial(h_2A_2)}{\partial x_1} - \frac{\partial(h_1A_1)}{\partial x_2}\right]\mathbf{i}_3\right\} \tag{5-54}$$

$$\nabla^2\phi = \frac{1}{h_1h_2h_3}\left[\frac{\partial}{\partial x_1}\left(\frac{h_2h_3}{h_1}\frac{\partial\phi}{\partial x_1}\right) + \frac{\partial}{\partial x_2}\left(\frac{h_3h_1}{h_2}\frac{\partial\phi}{\partial x_2}\right) + \frac{\partial}{\partial x_3}\left(\frac{h_1h_2}{h_3}\frac{\partial\phi}{\partial x_3}\right)\right] \tag{5-55}$$

The expression, $\mathbf{V}\cdot\nabla\mathbf{V}$, which is contained in the momentum equation term, $D\mathbf{V}/Dt$, can be evaluated as

$$\mathbf{V}\cdot\nabla\mathbf{V} = \left(\frac{u_1}{h_1}\frac{\partial}{\partial x_1} + \frac{u_2}{h_2}\frac{\partial}{\partial x_2} + \frac{u_3}{h_3}\frac{\partial}{\partial x_3}\right)(u_1\mathbf{i}_1 + u_2\mathbf{i}_2 + u_3\mathbf{i}_3)$$

where u_1, u_2, u_3 are the velocity components in the x_1, x_2, x_3 coordinate directions. After taking into account the fact that the unit vectors are functions of the coordinates, the final expanded form becomes

$$\begin{aligned}
\mathbf{V}\cdot\nabla\mathbf{V} = &\left(\frac{u_1}{h_1}\frac{\partial u_1}{\partial x_1} + \frac{u_2}{h_2}\frac{\partial u_1}{\partial x_2} + \frac{u_3}{h_3}\frac{\partial u_1}{\partial x_3} + \frac{u_1u_2}{h_1h_2}\frac{\partial h_1}{\partial x_2}\right.\\
&\left. + \frac{u_1u_3}{h_1h_3}\frac{\partial h_1}{\partial x_3} - \frac{u_2^2}{h_1h_2}\frac{\partial h_2}{\partial x_1} - \frac{u_3^2}{h_1h_3}\frac{\partial h_3}{\partial x_1}\right)\mathbf{i}_1\\
&+\left(\frac{u_1}{h_1}\frac{\partial u_2}{\partial x_1} + \frac{u_2}{h_2}\frac{\partial u_2}{\partial x_2} + \frac{u_3}{h_3}\frac{\partial u_2}{\partial x_3} - \frac{u_1^2}{h_1h_2}\frac{\partial h_1}{\partial x_2}\right.\\
&\left. + \frac{u_1u_2}{h_1h_2}\frac{\partial h_2}{\partial x_1} + \frac{u_2u_3}{h_2h_3}\frac{\partial h_2}{\partial x_3} - \frac{u_3^2}{h_2h_3}\frac{\partial h_3}{\partial x_2}\right)\mathbf{i}_2\\
&+\left(\frac{u_1}{h_1}\frac{\partial u_3}{\partial x_1} + \frac{u_2}{h_2}\frac{\partial u_3}{\partial x_2} + \frac{u_3}{h_3}\frac{\partial u_3}{\partial x_3} - \frac{u_1^2}{h_1h_3}\frac{\partial h_1}{\partial x_3}\right.\\
&\left. - \frac{u_2^2}{h_2h_3}\frac{\partial h_2}{\partial x_3} + \frac{u_1u_3}{h_1h_3}\frac{\partial h_3}{\partial x_1} + \frac{u_2u_3}{h_2h_3}\frac{\partial h_3}{\partial x_2}\right)\mathbf{i}_3
\end{aligned}$$

The components of the stress tensor given by Eq. (5-15) can be expressed in terms of the generalized curvilinear coordinate as

$$\begin{aligned}
\Pi_{x_1x_1} &= -p + \tfrac{2}{3}\mu(2e_{x_1x_1} - e_{x_2x_2} - e_{x_3x_3})\\
\Pi_{x_2x_2} &= -p + \tfrac{2}{3}\mu(2e_{x_2x_2} - e_{x_1x_1} - e_{x_3x_3})\\
\Pi_{x_3x_3} &= -p + \tfrac{2}{3}\mu(2e_{x_3x_3} - e_{x_1x_1} - e_{x_2x_2})\\
\Pi_{x_2x_3} &= \Pi_{x_3x_2} = \mu e_{x_2x_3}
\end{aligned} \tag{5-56}$$

$$\Pi_{x_1x_3} = \Pi_{x_3x_1} = \mu e_{x_1x_3} \qquad (5\text{-}56)$$
$$\Pi_{x_1x_2} = \Pi_{x_2x_1} = \mu e_{x_1x_2} \qquad (Cont.)$$

where the expressions for the strains are

$$\begin{aligned}
e_{x_1x_1} &= \frac{1}{h_1}\frac{\partial u_1}{\partial x_1} + \frac{u_2}{h_1h_2}\frac{\partial h_1}{\partial x_2} + \frac{u_3}{h_1h_3}\frac{\partial h_1}{\partial x_3} \\
e_{x_2x_2} &= \frac{1}{h_2}\frac{\partial u_2}{\partial x_2} + \frac{u_3}{h_2h_3}\frac{\partial h_2}{\partial x_3} + \frac{u_1}{h_1h_2}\frac{\partial h_2}{\partial x_1} \\
e_{x_3x_3} &= \frac{1}{h_3}\frac{\partial u_3}{\partial x_3} + \frac{u_1}{h_1h_3}\frac{\partial h_3}{\partial x_1} + \frac{u_2}{h_2h_3}\frac{\partial h_3}{\partial x_2} \qquad (5\text{-}57) \\
e_{x_3x_3} &= \frac{h_3}{h_2}\frac{\partial}{\partial x_2}\left(\frac{u_3}{h_3}\right) + \frac{h_2}{h_3}\frac{\partial}{\partial x_3}\left(\frac{u_2}{h_2}\right) \\
e_{x_1x_3} &= \frac{h_1}{h_3}\frac{\partial}{\partial x_3}\left(\frac{u_1}{h_1}\right) + \frac{h_3}{h_1}\frac{\partial}{\partial x_1}\left(\frac{u_3}{h_3}\right) \\
e_{x_1x_2} &= \frac{h_2}{h_1}\frac{\partial}{\partial x_1}\left(\frac{u_2}{h_2}\right) + \frac{h_1}{h_2}\frac{\partial}{\partial x_2}\left(\frac{u_1}{h_1}\right)
\end{aligned}$$

The components of $\nabla \cdot \boldsymbol{\Pi}_{ij}$ are

$$\begin{aligned}
x_1&: \frac{1}{h_1h_2h_3}\left[\frac{\partial}{\partial x_1}(h_2h_3\Pi_{x_1x_1}) + \frac{\partial}{\partial x_2}(h_1h_3\Pi_{x_1x_2}) + \frac{\partial}{\partial x_3}(h_1h_2\Pi_{x_1x_3})\right] \\
&+ \Pi_{x_1x_2}\frac{1}{h_1h_2}\frac{\partial h_1}{\partial x_2} + \Pi_{x_1x_3}\frac{1}{h_1h_3}\frac{\partial h_1}{\partial x_3} - \Pi_{x_2x_2}\frac{1}{h_1h_2}\frac{\partial h_2}{\partial x_1} - \Pi_{x_3x_3}\frac{1}{h_1h_3}\frac{\partial h_3}{\partial x_1} \\
x_2&: \frac{1}{h_1h_2h_3}\left[\frac{\partial}{\partial x_1}(h_2h_3\Pi_{x_1x_2}) + \frac{\partial}{\partial x_2}(h_1h_3\Pi_{x_2x_2}) + \frac{\partial}{\partial x_3}(h_1h_2\Pi_{x_2x_3})\right] \qquad (5\text{-}58) \\
&+ \Pi_{x_2x_3}\frac{1}{h_2h_3}\frac{\partial h_2}{\partial x_3} + \Pi_{x_1x_2}\frac{1}{h_1h_2}\frac{\partial h_2}{\partial x_1} - \Pi_{x_3x_3}\frac{1}{h_2h_3}\frac{\partial h_3}{\partial x_2} - \Pi_{x_1x_1}\frac{1}{h_1h_2}\frac{\partial h_1}{\partial x_2} \\
x_3&: \frac{1}{h_1h_2h_3}\left[\frac{\partial}{\partial x_1}(h_2h_3\Pi_{x_1x_3}) + \frac{\partial}{\partial x_2}(h_1h_3\Pi_{x_2x_3}) + \frac{\partial}{\partial x_3}(h_1h_2\Pi_{x_3x_3})\right] \\
&+ \Pi_{x_1x_3}\frac{1}{h_1h_3}\frac{\partial h_3}{\partial x_1} + \Pi_{x_2x_3}\frac{1}{h_2h_3}\frac{\partial h_3}{\partial x_2} - \Pi_{x_1x_1}\frac{1}{h_1h_3}\frac{\partial h_1}{\partial x_3} - \Pi_{x_2x_2}\frac{1}{h_2h_3}\frac{\partial h_2}{\partial x_3}
\end{aligned}$$

In generalized curvilinear coordinates, the dissipation function becomes

$$\begin{aligned}
\Phi = \mu[2(e^2_{x_1x_1} + e^2_{x_2x_2} + e^2_{x_3x_3}) + e^2_{x_2x_3} + e^2_{x_1x_3} + e^2_{x_1x_2} \\
-\tfrac{2}{3}(e_{x_1x_1} + e_{x_2x_2} + e_{x_3x_3})^2] \qquad (5\text{-}59)
\end{aligned}$$

The above formulas can now be used to derive the fluid dynamic equations in any orthogonal curvilinear coordinate system. Examples include:

1. Cartesian coordinates

$x_1 = x \qquad h_1 = 1 \qquad u_1 = u$

$x_2 = y \qquad h_2 = 1 \qquad u_2 = v$

$x_3 = z \qquad h_3 = 1 \qquad u_3 = w$

2. Cylindrical coordinates

$x_1 = r \qquad h_1 = 1 \qquad u_1 = u_r$

$x_2 = \theta \qquad h_2 = r \qquad u_2 = u_\theta$

$x_3 = z \qquad h_3 = 1 \qquad u_3 = u_z$

3. Spherical coordinates

$x_1 = r \qquad h_1 = 1 \qquad u_1 = u_r$

$x_2 = \theta \qquad h_2 = r \qquad u_2 = u_\theta$

$x_3 = \phi \qquad h_3 = r \sin\theta \qquad u_3 = u_\phi$

4. 2-D or axisymmetric body intrinsic coordinates

$x_1 = \xi \qquad h_1 = 1 + K(\xi)\eta \qquad u_1 = u$

$x_2 = \eta \qquad h_2 = 1 \qquad u_2 = v$

$x_3 = \phi \qquad h_3 = [r(\xi) + \eta \cos\alpha(\xi)]^m \qquad u_3 = w = 0$

where $K(\xi)$ is the local body curvature, $r(\xi)$ is the cylindrical radius, and

$$m = \begin{cases} 0 \text{ for 2-D flow} \\ 1 \text{ for axisymmetric flow} \end{cases}$$

These coordinate systems are illustrated in Fig. 5-3.

5-2 REYNOLDS EQUATIONS FOR TURBULENT FLOWS

5-2.1 Background

For more than 50 years it has been recognized that our understanding of turbulent flows is very incomplete. A quotation attributed to Sir Horace Lamb in 1932 might still be appropriate: "I am an old man now, and when I die and go to Heaven there are two matters on which I hope for enlightenment. One is quantum electrodynamics and the other is the turbulent motion of fluids. And about the former I am rather optimistic."

According to Hinze (1975), "Turbulent fluid motion is an irregular condition of flow in which the various quantities show a random variation with time and space coordinates so that statistically distinct average values can be discerned."

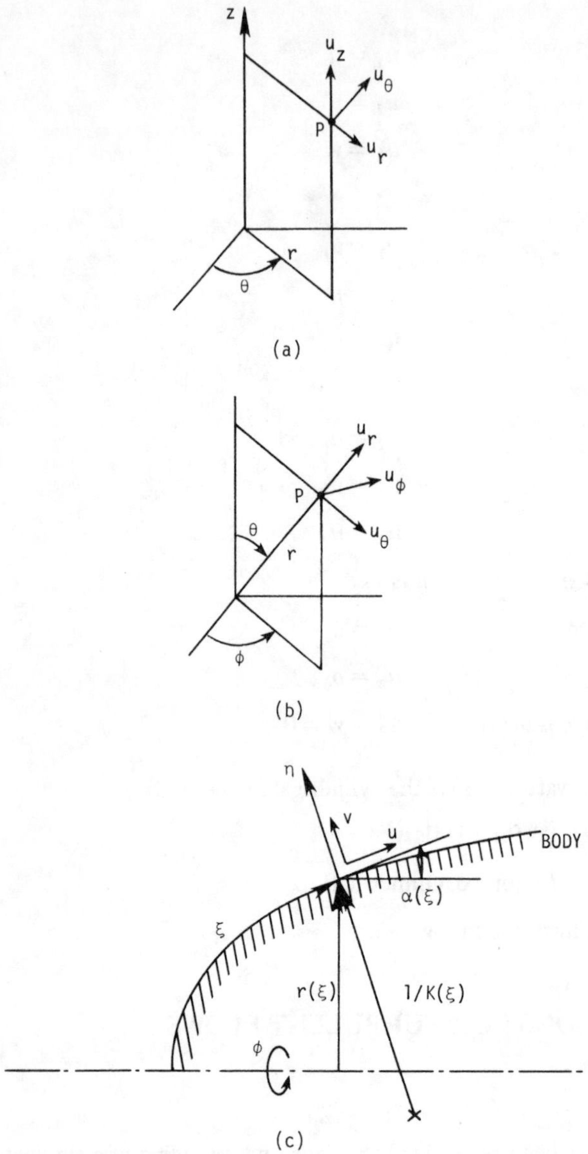

Figure 5-3 Curvilinear coordinate systems. (a) Cylindrical coordinates (r, θ, z); (b) spherical coordinates (r, θ, ϕ); (c) 2-D or axisymmetric body intrinsic coordinates (ξ, η, ϕ).

We are all familiar with some of the differences between laminar and turbulent flows. Usually, higher values of friction drag and pressure drop are associated with turbulent flows. The diffusion rate of a scalar quantity is usually greater in a turbulent flow than in a laminar flow (increased "mixing") and turbulent flows are usually noisier. A turbulent boundary layer can normally negotiate a more extensive region of unfavorable pressure gradient prior to separation than can a laminar boundary layer. Users of dimpled golf balls are well aware of this.

The unsteady Navier-Stokes equations are generally considered to govern turbulent flows in the continuum regime. If this is the case, then we might wonder why turbulent flows cannot be solved numerically as easily as laminar flows. Perhaps the wind tunnels can be dismantled once and for all. The main problem is that time and space scales of the turbulent motion are so small that the number of grid points required and the small size of the required time steps puts the practical computation of turbulent flows by this means outside the realm of possibility for present computers. Although estimates from various sources differ on the required mesh spacing, a common estimate is that at least 10 grid points would be required to adequately resolve the motion of a turbulent eddy. The scale of the smallest eddies are typically 10^{-3} times the size of the flow domain for flow along a solid surface. For a typical flow, 10^5 points may be required to resolve just 1 cm^3 of the flowfield.

Authorities disagree as to when computer technology will have advanced to the point where turbulent flow calculations can be made from first principles. Some claim that for practical problems it will never be possible to resolve all of the fine scale flow structure with numerical solutions of the unsteady Navier-Stokes equations. For the remainder of this century it is likely that the most advanced approach will involve solving the time-dependent Navier-Stokes equations for the evolution of the large eddies (which are responsible for most of the momentum transport) but will model the smallest (subgrid scale) eddies. This approach is generally referred to as "large eddy simulation." The paper by Chapman (1979) makes interesting reading on projections for future developments in computations for aerodynamic applications.

The main thrust of present day research in computational fluid mechanics and heat transfer in turbulent flows is through the time-averaged Navier-Stokes equations. These equations are also referred to as the Reynolds equations of motion or the Reynolds averaged equations. Time averaging the equations of motion gives rise to new terms which can be interpreted as "apparent" stress gradients and heat-flux quantities associated with the turbulent motion. These new quantities must be related to the mean flow variables through turbulence models. This process introduces further assumptions and approximations. Thus, this attack on the turbulent flow problem through solving the Reynolds equations of motion does not follow entirely from first principles since additional assumptions must be made to "close" the system of equations.

The Reynolds equations are derived by decomposing the dependent variables in the conservation equations into time mean (obtained over an appropriate time interval) and fluctuating components and then time averaging the entire equation. Two types of averaging are presently used, the classical Reynolds averaging and the mass-weighted averaging suggested by Favre (1965). For flows in which density fluctuations can be neglected, the two formulations become identical.

5-2.2 Averaging Procedures

In the conventional averaging procedure, following Reynolds, we define a time-averaged quantity $\bar{f}$ as

$$\bar{f} \equiv \frac{1}{\Delta t} \int_{t_0}^{t_0 + \Delta t} f\, dt \tag{5-60}$$

We require that Δt be large compared to the period of the random fluctuations associated with the turbulence, but small with respect to the time constant for any slow variations in the flowfield associated with ordinary unsteady flows. The Δt is sometimes indicated to approach infinity as a limit, but this should be interpreted as being relative to the characteristic fluctuation period of the turbulence. For practical measurements, Δt must be finite.

In the conventional Reynolds decomposition, the randomly changing flow variables are replaced by time averages plus fluctuations (see Fig. 5-4) about the average. For a Cartesian coordinate system we may write

$$\begin{array}{llll} u = \bar{u} + u' & v = \bar{v} + v' & w = \bar{w} + w' & \rho = \bar{\rho} + \rho' \\ p = \bar{p} + p' & h = \bar{h} + h' & T = \bar{T} + T' & H = \bar{H} + H' \end{array} \tag{5-61}$$

where total enthalpy H is defined by $H = h + u_i u_i/2$. Fluctuations in other fluid properties such as viscosity, thermal conductivity, and specific heat are usually small and will be neglected here.

By definition, the time average of a fluctuating quantity is zero:

$$\overline{f'} = \frac{1}{\Delta t} \int_{t_0}^{t_0 + \Delta t} f'\, dt \equiv 0 \tag{5-62}$$

It should be clear from these definitions that for symbolic flow variables f and g, the following relations hold

$$\overline{\bar{f}g'} = 0 \qquad \overline{\bar{f}g} = \bar{f}\bar{g} \qquad \overline{f + g} = \bar{f} + \bar{g} \tag{5-63}$$

It should also be clear that, whereas $\overline{f'} \equiv 0$, the time average of the product of two fluctuating quantities is, in general, not equal to zero, i.e., $\overline{f'f'} \neq 0$, In fact, the root mean square of the velocity fluctuations is known as the turbulence intensity.

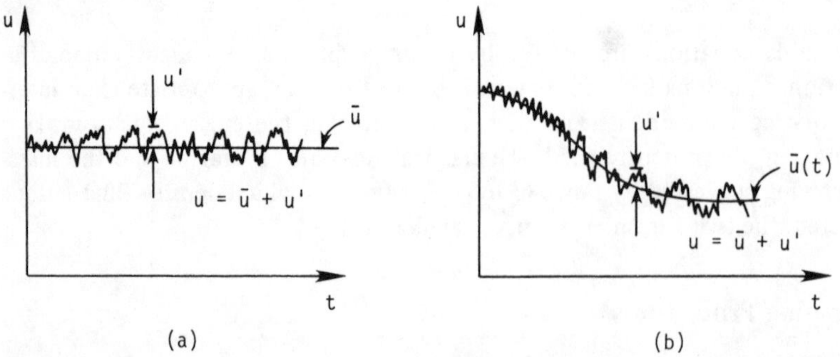

Figure 5-4 Relationship between u, $\bar{u}$, and u'. (a) Steady flow; (b) unsteady flow.

For treatment of compressible flows and mixtures of gases in particular, the mass-weighted averaging is convenient. In this approach we define mass-averaged variables according to $\tilde{f} = \overline{\rho f}/\bar{\rho}$. This gives

$$\tilde{u} = \frac{\overline{\rho u}}{\bar{\rho}} \qquad \tilde{v} = \frac{\overline{\rho v}}{\bar{\rho}} \qquad \tilde{w} = \frac{\overline{\rho w}}{\bar{\rho}} \qquad \tilde{h} = \frac{\overline{\rho h}}{\bar{\rho}} \qquad \tilde{T} = \frac{\overline{\rho T}}{\bar{\rho}} \qquad \tilde{H} = \frac{\overline{\rho H}}{\bar{\rho}} \tag{5-64}$$

We note that only the velocity components and thermal variables are mass averaged. Fluid properties such as density and pressure are treated as before.

To substitute into the conservation equations, we define new fluctuating quantities by

$$u = \tilde{u} + u'' \qquad v = \tilde{v} + v'' \qquad w = \tilde{w} + w'' \qquad h = \tilde{h} + h'' \qquad T = \tilde{T} + T''$$
$$H = \tilde{H} + H'' \tag{5-65}$$

It is very important to note that the time averages of the doubly primed fluctuating quantities ($\overline{u''}$, $\overline{v''}$, etc.) *are not* equal to zero, in general, unless $\rho' = 0$. In fact, it can be shown that $\overline{u''} = -\overline{\rho' u'}/\bar{\rho}$, $\overline{v''} = -\overline{\rho' v'}/\bar{\rho}$, etc. Instead, the time average of the doubly primed fluctuation multiplied by the density is equal to zero.

$$\overline{\rho f''} \equiv 0 \tag{5-66}$$

The above identity can be established by expanding $\overline{\rho f} = \overline{\rho(\tilde{f} + f'')}$ and using the definition of $\tilde{f}$.

5-2.3 Reynolds Form of the Continuity Equation

Starting with the continuity equation in the Cartesian coordinate system as given by Eq. (5-4), we first decompose the variables into the conventional time-averaged variables plus fluctuating components as given by Eqs. (5-61).

The entire equation is then time averaged yielding, in summation notation

$$\frac{\partial \bar{\rho}}{\partial t} + \cancelto{0}{\frac{\partial \bar{\rho}'}{\partial t}} + \frac{\partial}{\partial x_j}(\overline{\bar{\rho}\bar{u}_j}) + \frac{\partial}{\partial x_j}\cancelto{0}{(\overline{\rho' \bar{u}_j})} + \frac{\partial}{\partial x_j}\cancelto{0}{(\overline{\bar{\rho} u_j'})} + \frac{\partial}{\partial x_j}(\overline{\rho' u_j'}) = 0 \tag{5-67}$$

Three of the terms are identically zero as indicated because of the identity given by Eq. (5-62). Finally, the Reynolds form of the continuity equation in conventionally-averaged variables can be written

$$\frac{\partial \bar{\rho}}{\partial t} + \frac{\partial}{\partial x_j}(\bar{\rho}\bar{u}_j + \overline{\rho' u_j'}) = 0 \tag{5-68}$$

Substituting the mass-weighted averaged variables plus the doubly primed fluctuations given by Eqs. (5-65) into Eq. (5-4) and time averaging the entire equation gives

$$\frac{\partial \bar{\rho}}{\partial t} + \cancelto{0}{\frac{\partial \bar{\rho}'}{\partial t}} + \frac{\partial}{\partial x_j}(\overline{\bar{\rho}\tilde{u}_j}) + \frac{\partial}{\partial x_j}\cancelto{0}{(\overline{\rho' \tilde{u}_j})} + \frac{\partial}{\partial x_j}(\overline{\bar{\rho} u_j''}) + \frac{\partial}{\partial x_j}(\overline{\rho' u_j''}) = 0 \tag{5-69}$$

Two of the terms in Eq. (5-69) are obviously identically zero as indicated. In addition, the last two terms can be combined, i.e.,

$$\frac{\partial}{\partial x_j}(\overline{\bar{\rho} u_j''}) + \frac{\partial}{\partial x_j}(\overline{\rho' u_j''}) = \frac{\partial}{\partial x_j}\overline{\rho u_j''}$$

which is equal to zero by Eq. (5-66). This permits the continuity equation in mass-weighted variables to be written as

$$\frac{\partial \bar{\rho}}{\partial t} + \frac{\partial}{\partial x_j}(\bar{\rho}\tilde{u}_j) = 0 \tag{5-70}$$

We note that Eq. (5-70) is more compact in form than Eq. (5-68). For incompressible flows, $\rho' = 0$ and the differences between the conventional and mass-weighted variables vanish so that the continuity equation can be written as

$$\frac{\partial \bar{u}_j}{\partial x_j} = 0 \tag{5-71}$$

5-2.4 Reynolds Form of the Momentum Equations

The development of the Reynolds form of the momentum equations proceeds most easily when we start with the Navier-Stokes momentum equations in divergence or conservation-law form as in Eq. (5-20). Working first with the conventionally averaged variables, we replace the dependent variables in Eq. (5-20) with the time averages plus fluctuations according to Eq. (5-61). As an example, the resulting x component of Eq. (5-20) becomes after neglecting body forces

$$\frac{\partial}{\partial t}[(\bar{\rho}+\rho')(\bar{u}+u')] + \frac{\partial}{\partial x}[(\bar{\rho}+\rho')(\bar{u}+u')(\bar{u}+u') + (\bar{p}+p') - \tau_{xx}]$$

$$+ \frac{\partial}{\partial y}[(\bar{\rho}+\rho')(\bar{u}+u')(\bar{v}+v') - \tau_{yx}] + \frac{\partial}{\partial z}[(\bar{\rho}+\rho')(\bar{u}+u')(\bar{w}+w') - \tau_{zx}] = 0$$

Next, the entire equation is time averaged. Terms which are linear in fluctuating quantities become zero when time averaged as they did in the continuity equation. Several terms disappear in this manner while others can be grouped together and found equal to zero through use of the continuity equation. The resulting Reynolds x-momentum equation can be written as

$$\frac{\partial}{\partial t}(\bar{\rho}\bar{u} + \overline{\rho' u'}) + \frac{\partial}{\partial x}(\bar{\rho}\bar{u}\bar{u} + \bar{u}\overline{\rho' u'}) + \frac{\partial}{\partial y}(\bar{\rho}\bar{u}\bar{v} + \bar{u}\overline{\rho' v'}) + \frac{\partial}{\partial z}(\bar{\rho}\bar{u}\bar{w} + \bar{u}\overline{\rho' w'})$$

$$= -\frac{\partial \bar{p}}{\partial x} + \frac{\partial}{\partial x}\left[\mu\left(2\frac{\partial \bar{u}}{\partial x} - \frac{2}{3}\frac{\partial \bar{u}_k}{\partial x_k}\right) - \bar{u}\overline{\rho' u'} - \bar{\rho}\overline{u' u'} - \overline{\rho' u' u'}\right]$$

$$+ \frac{\partial}{\partial y}\left[\mu\left(\frac{\partial \bar{u}}{\partial y} + \frac{\partial \bar{v}}{\partial x}\right) - \bar{v}\overline{\rho' u'} - \bar{\rho}\overline{u' v'} - \overline{\rho' u' v'}\right]$$

$$+ \frac{\partial}{\partial z}\left[\mu\left(\frac{\partial \bar{u}}{\partial z} + \frac{\partial \bar{w}}{\partial x}\right) - \bar{w}\overline{\rho' u'} - \bar{\rho}\overline{u' w'} - \overline{\rho' u' w'}\right] \tag{5-72}$$

The complete Reynolds momentum equation (all three components) can be written

$$\frac{\partial}{\partial t}(\bar{\rho}\bar{u}_i + \overline{\rho' u_i'}) + \frac{\partial}{\partial x_j}(\bar{\rho}\bar{u}_i\bar{u}_j + \bar{u}_i\overline{\rho' u_j'}) = -\frac{\partial \bar{p}}{\partial x_i} + \frac{\partial}{\partial x_j}(\bar{\tau}_{ij} - \bar{u}_j\overline{\rho' u_i'} - \bar{\rho}\overline{u_i' u_j'} - \overline{\rho' u_i' u_j'}) \tag{5-73}$$

where

$$\bar{\tau}_{ij} = \mu\left[\left(\frac{\partial \bar{u}_i}{\partial x_j} + \frac{\partial \bar{u}_j}{\partial x_i}\right) - \frac{2}{3}\delta_{ij}\frac{\partial \bar{u}_k}{\partial x_k}\right] \tag{5-74}$$

To develop the Reynolds momentum equation in mass-weighted variables, we again start with Eq. (5-20) but use the decomposition indicated by Eq. (5-65) to represent the instantaneous variables. As an example, the resulting x component of Eq. (5-20) becomes,

$$\frac{\partial}{\partial t}[(\bar{\rho} + \rho')(\tilde{u} + u'')] + \frac{\partial}{\partial x}[(\bar{\rho} + \rho')(\tilde{u} + u'')(\tilde{u} + u'') + (\bar{p} + p') - \tau_{xx}]$$

$$+ \frac{\partial}{\partial y}[(\bar{\rho} + \rho')(\tilde{u} + u'')(\tilde{v} + v'') - \tau_{yx}] + \frac{\partial}{\partial z}[(\bar{\rho} + \rho')(\tilde{u} + u'')(\tilde{w} + w'') - \tau_{zx}] = 0 \tag{5-75}$$

Next, the entire equation is time averaged and the identity given by Eq. (5-66) is used to eliminate terms. The complete Reynolds momentum equation in mass-weighted variables becomes

$$\frac{\partial}{\partial t}(\bar{\rho}\tilde{u}_i) + \frac{\partial}{\partial x_j}(\bar{\rho}\tilde{u}_i\tilde{u}_j) = -\frac{\partial \bar{p}}{\partial x_i} + \frac{\partial}{\partial x_j}(\bar{\tau}_{ij} - \overline{\rho u_i'' u_j''}) \tag{5-76}$$

where, neglecting viscosity fluctuations, $\bar{\tau}_{ij}$ becomes

$$\bar{\tau}_{ij} = \mu\left[\left(\frac{\partial \tilde{u}_i}{\partial x_j} + \frac{\partial \tilde{u}_j}{\partial x_i}\right) - \frac{2}{3}\delta_{ij}\frac{\partial \tilde{u}_k}{\partial x_k}\right] + \mu\left[\left(\frac{\partial \overline{u_i''}}{\partial x_j} + \frac{\partial \overline{u_j''}}{\partial x_i}\right) - \frac{2}{3}\delta_{ij}\frac{\partial \overline{u_k''}}{\partial x_k}\right] \tag{5-77}$$

The momentum equation, Eq. (5-76) in mass-weighted variables is simpler in form than the corresponding equation using conventional variables. We note, however, that even when viscosity fluctuations are neglected, $\bar{\tau}_{ij}$ is more complex in Eq. (5-77) than the $\bar{\tau}_{ij}$ which appeared in the conventionally averaged equation [Eq. (5-74)]. In practice, the viscous terms involving the doubly primed fluctuations are expected to be small and are likely candidates for being neglected on the basis of order of magnitude arguments.

For incompressible flows the momentum equation can be written in the simpler form

$$\frac{\partial}{\partial t}(\rho\bar{u}_i) + \frac{\partial}{\partial x_j}(\rho\bar{u}_i\bar{u}_j) = -\frac{\partial \bar{p}}{\partial x_i} + \frac{\partial}{\partial x_j}(\bar{\tau}_{ij} - \rho\overline{u_i' u_j'}) \tag{5-78}$$

where $\bar{\tau}_{ij}$ takes on the reduced form

$$\bar{\tau}_{ij} = \mu\left(\frac{\partial \bar{u}_i}{\partial x_j} + \frac{\partial \bar{u}_j}{\partial x_i}\right) \tag{5-79}$$

As we noted in connection with the continuity equation, there is no difference between the mass-weighted and conventional variables for incompressible flow.

5-2.5 Reynolds Form of the Energy Equation

The thermal variables H, h, and T are all related and the energy equation takes on different forms depending upon which one is chosen to be the transported thermal variable. To develop one common form, we start with the energy equation as given by Eq. (5-22). The generation term, $\partial Q/\partial t$, will be neglected. Assuming that the total energy is comprised only of internal energy and kinetic energy, and replacing E_t by $\rho H - p$, we can write Eq. (5-22) in summation notation as

$$\frac{\partial}{\partial t}\rho H + \frac{\partial}{\partial x_j}(\rho u_j H + q_j - u_i \tau_{ij}) = \frac{\partial p}{\partial t} \tag{5-80}$$

To obtain the Reynolds energy equation in conventionally-averaged variables, we replace the dependent variables in Eq. (5-80) with the decomposition indicated by Eq. (5-61). After time averaging, the equation becomes

$$\begin{aligned}
&\frac{\partial}{\partial t}(\bar{\rho}\bar{H} + \overline{\rho' H'}) + \frac{\partial}{\partial x_j}\left(\bar{\rho}\bar{u}_j\bar{H} + \bar{\rho}\overline{u_j' H'} + \overline{\rho' u_j'}\bar{H} + \overline{\rho' u_j' H'} + \bar{u}_j\overline{\rho' H'} - k\frac{\partial \bar{T}}{\partial x_j}\right) \\
&= \frac{\partial \bar{p}}{\partial t} + \frac{\partial}{\partial x_j}\left[\bar{u}_i\left(-\frac{2}{3}\mu\,\delta_{ij}\frac{\partial \bar{u}_k}{\partial x_k}\right) + \mu\bar{u}_i\left(\frac{\partial \bar{u}_j}{\partial x_i} + \frac{\partial \bar{u}_i}{\partial x_j}\right) - \frac{2}{3}\mu\,\delta_{ij}\overline{u_i'\frac{\partial u_k'}{\partial x_k}}\right. \\
&\quad \left. + \mu\left(\overline{u_i'\frac{\partial u_j'}{\partial x_i}} + \overline{u_i'\frac{\partial u_i'}{\partial x_j}}\right)\right]
\end{aligned} \tag{5-81}$$

It is frequently desirable to utilize static temperature as a dependent variable in the energy equation. We will let $h = c_p T$ and write Eq. (5-33) in conservative form to provide a convenient starting point for the development of the Reynolds averaged form.

$$\frac{\partial}{\partial t}(\rho c_p T) + \frac{\partial}{\partial x_j}\left(\rho c_p u_j T - k\frac{\partial T}{\partial x_j}\right) = \frac{\partial p}{\partial t} + u_j\frac{\partial p}{\partial x_j} + \Phi \tag{5-82}$$

The dissipation function Φ [see Eq. (5-34)] can be written in terms of the velocity components using summation convention as

$$\Phi = \tau_{ij}\frac{\partial u_i}{\partial x_j} = \mu\left[-\frac{2}{3}\left(\frac{\partial u_k}{\partial x_k}\right)^2 + \frac{1}{2}\left(\frac{\partial u_j}{\partial x_i} + \frac{\partial u_i}{\partial x_j}\right)^2\right] \tag{5-83}$$

The variables in Eq. (5-83) are then replaced with the decomposition indicated by Eq. (5-61) and the resulting equation is time averaged. After eliminating terms known to be zero, the Reynolds energy equation in terms of temperature becomes

$$\frac{\partial}{\partial t}(c_p\bar{\rho}\bar{T} + c_p\overline{\rho' T'}) + \frac{\partial}{\partial x_j}(\bar{\rho}c_p\bar{T}\bar{u}_j) = \frac{\partial \bar{p}}{\partial t} + \bar{u}_j\frac{\partial \bar{p}}{\partial x_j} + \overline{u_j'\frac{\partial p'}{\partial x_j}}$$

$$+ \frac{\partial}{\partial x_j}\left(k\frac{\partial \bar{T}}{\partial x_j} - \bar{\rho}c_p\overline{T'u_j'} - c_p\overline{\rho' T' u_j'}\right) + \bar{\Phi} \tag{5-84}$$

where

$$\bar{\Phi} = \overline{\tau_{ij}\frac{\partial u_i}{\partial x_j}} = \bar{\tau}_{ij}\frac{\partial \bar{u}_i}{\partial x_j} + \overline{\tau_{ij}'\frac{\partial u_i'}{\partial x_j}} \tag{5-85}$$

The $\bar{\tau}_{ij}$ in Eq. (5-85) should be evaluated as indicated by Eq. (5-74).

To develop the Reynolds form of the energy equation in mass-weighted variables, we replace the dependent variables in Eq. (5-80) with the decomposition indicated by Eq. (5-65) and time average the entire equation. The result can be written

$$\frac{\partial}{\partial t}(\bar{\rho}\tilde{H}) + \frac{\partial}{\partial x_j}\left(\bar{\rho}\tilde{u}_j\tilde{H} + \overline{\rho u_j'' H''} - k\frac{\partial \bar{T}}{\partial x_j}\right) = \frac{\partial \bar{p}}{\partial t} + \frac{\partial}{\partial x_j}(\tilde{u}_i\bar{\tau}_{ij} + \overline{u_i''\tau_{ij}}) \tag{5-86}$$

where $\bar{\tau}_{ij}$ can be evaluated as given by Eq. (5-77) in terms of mass-weighted variables.

In terms of static temperature, the Reynolds energy equation in mass-weighted variables becomes

$$\frac{\partial}{\partial t}(\bar{\rho}c_p\tilde{T}) + \frac{\partial}{\partial x_j}(\bar{\rho}c_p\tilde{T}\tilde{u}_j) = \frac{\partial \bar{p}}{\partial t} + \tilde{u}_j\frac{\partial \bar{p}}{\partial x_j} + \overline{u_j''\frac{\partial p}{\partial x_j}}$$

$$+ \frac{\partial}{\partial x_j}\left(k\frac{\partial \tilde{T}}{\partial x_j} + k\frac{\partial \overline{T''}}{\partial x_j} - c_p\overline{\rho T'' u_j''}\right) + \bar{\Phi} \tag{5-87}$$

where

$$\bar{\Phi} = \overline{\tau_{ij}\frac{\partial u_i}{\partial x_j}} = \bar{\tau}_{ij}\frac{\partial \tilde{u}_i}{\partial x_j} + \overline{\tau_{ij}\frac{\partial u_i''}{\partial x_j}} \tag{5-88}$$

For incompressible flows the energy equation can be written in terms of total enthalpy as

$$\frac{\partial \rho\bar{H}}{\partial t} + \frac{\partial}{\partial x_j}\left(\rho u_j\bar{H} + \rho\overline{u_j' H'} - k\frac{\partial \bar{T}}{\partial x_j}\right) = \frac{\partial \bar{p}}{\partial t} + \frac{\partial}{\partial x_j}\left[\mu\bar{u}_i\left(\frac{\partial \bar{u}_j}{\partial x_i} + \frac{\partial \bar{u}_i}{\partial x_j}\right)\right.$$

$$\left. + \mu\left(\overline{u_i'\frac{\partial u_j'}{\partial x_i}} + \overline{u_i'\frac{\partial u_i'}{\partial x_j}}\right)\right] \tag{5-89}$$

and in terms of static temperature as

$$\frac{\partial}{\partial t}(\rho c_p\bar{T}) + \frac{\partial}{\partial x_j}(\rho c_p\bar{T}\bar{u}_j) = \frac{\partial \bar{p}}{\partial t} + \bar{u}_j\frac{\partial \bar{p}}{\partial x_j} + \overline{u_j'\frac{\partial p'}{\partial x_j}} + \frac{\partial}{\partial x_j}\left(k\frac{\partial \bar{T}}{\partial x_j} - \rho c_p\overline{T'u_j'}\right) + \bar{\Phi} \tag{5-90}$$

where $\bar{\Phi}$ is reduced slightly in complexity due to the vanishing of the volumetric dilatation term in $\bar{\tau}_{ij}$ for incompressible flow.

5-2.6 Comments on the Reynolds Equations

At first glance the Reynolds equations are likely to appear quite complex and we are tempted to question whether or not we have made any progress toward solving practical problems in turbulent flow. Certainly a major problem in fluid mechanics is that more equations can be written than can be solved. Fortunately, for many important flows, the Reynolds equations can be simplified. Before we turn to the task of simplifying the equations, let us examine the Reynolds equations further.

We will consider an incompressible turbulent flow first and interpret the Reynolds momentum equation in the form of Eq. (5-78). The equation governs the time-mean motion of the fluid and we recognize some familiar momentum flux and laminar-like stress terms plus some new terms involving fluctuations which must represent apparent turbulent stresses. These apparent turbulent stresses originated in the momentum flux terms of the Navier-Stokes equations. To put this another way, the equations of mean motion relate the particle acceleration to stress gradients and since we know how acceleration for the time-mean motion is expressed, anything new in these equations must be apparent stress gradients due to the turbulent motion. To illustrate, we will utilize the continuity equation to arrange Eq. (5-78) in a form in which the particle (substantial) derivative appears on the left-hand side,

$$\underbrace{\rho \frac{D\bar{u}_i}{Dt}}_{\substack{\text{Particle}\\ \text{acceleration}\\ \text{of mean motion}}} = \underbrace{-\frac{\partial \bar{p}}{\partial x_i}}_{\substack{\text{Mean pressure}\\ \text{gradient}}} + \underbrace{\frac{\partial (\bar{\tau}_{ij})_{\text{lam}}}{\partial x_j}}_{\substack{\text{Laminar-like}\\ \text{stress gradients}\\ \text{for the mean motion}}} + \underbrace{\frac{\partial (\bar{\tau}_{ij})_{\text{turb}}}{\partial x_j}}_{\substack{\text{Apparent stress}\\ \text{gradients due to}\\ \text{transport of}\\ \text{momentum by}\\ \text{turbulent fluctuations}}} \tag{5-91}$$

where $(\bar{\tau}_{ij})_{\text{lam}}$ is the same as Eq. (5-79) and has the same form in terms of the time-mean velocities as the stress tensor for a laminar incompressible flow. The apparent turbulent stresses can be written as

$$(\bar{\tau}_{ij})_{\text{turb}} = -\rho\overline{u_i'u_j'} \tag{5-92}$$

These apparent stresses are commonly called the Reynolds stresses.

For compressible turbulent flow, labeling the terms according to the acceleration of the mean motion and apparent stresses becomes more of a challenge. Using conventional averaging procedures, the presence of terms like $\overline{\rho' u_i'}$ can result in the flux of momentum across mean flow streamlines frustrating our attempts to categorize terms. The use of mass-weighted averaging eliminates the $\overline{\rho' u_i'}$ terms and provides a compact expression for the particle acceleration but complicates the separation of stresses into purely laminar-like and apparent turbulent categories. When conventionally averaged variables are used, the fluctuating components of $\bar{\tau}_{ij}$ vanish when the equations are time averaged. They do not vanish, however, when mass-weighted averaging is used. To illustrate, we will arrange Eq. (5-76) (using the continuity equation) in a form which utilizes the substantial derivative and label the terms as follows.

$$\underbrace{\bar{\rho}\frac{D\tilde{u}_i}{Dt}}_{\substack{\text{Particle}\\ \text{acceleration}\\ \text{of mean motion}}} = \underbrace{-\frac{\partial \bar{p}}{\partial x_i}}_{\substack{\text{Mean}\\ \text{pressure}\\ \text{gradient}}} + \underbrace{\frac{\partial(\bar{\tau}_{ij})_{\text{lam}}}{\partial x_j}}_{\substack{\text{Laminar-like}\\ \text{stress gradients}\\ \text{for the mean}\\ \text{motion}}} + \underbrace{\frac{\partial(\bar{\tau}_{ij})_{\text{turb}}}{\partial x_j}}_{\substack{\text{Apparent stress gradients}\\ \text{due to transport of}\\ \text{momentum by turbulent}\\ \text{fluctuations } and \text{ deformations}\\ \text{attributed to fluctuations}}} \tag{5-93a}$$

The form of Eq. (5-93*a*) is identical to that of Eq. (5-91) except that $\tilde{u}_i$ replaces the $\bar{u}_i$ used in Eq. (5-91). If we insist that $(\bar{\tau}_{ij})_{\text{lam}}$ have the same *form* as for a laminar flow, then the second half of the $\bar{\tau}_{ij}$ of Eq. (5-77) should be attributed to turbulent transport resulting in

$$(\bar{\tau}_{ij})_{\text{lam}} = \mu\left[\left(\frac{\partial \tilde{u}_i}{\partial x_j} + \frac{\partial \tilde{u}_j}{\partial x_i}\right) - \frac{2}{3}\,\delta_{ij}\,\frac{\partial \tilde{u}_k}{\partial x_k}\right] \tag{5-93b}$$

and

$$(\bar{\tau}_{ij})_{\text{turb}} = -\overline{\rho u_i'' u_j''} + \mu\left[\left(\frac{\partial \overline{u_i''}}{\partial x_j} + \frac{\partial \overline{u_j''}}{\partial x_i}\right) - \frac{2}{3}\,\delta_{ij}\,\frac{\partial \overline{u_k''}}{\partial x_k}\right] \tag{5-93c}$$

As before, viscosity fluctuations have been neglected in obtaining Eq. (5-93*a*). The second term in the expression for $(\bar{\tau}_{ij})_{\text{turb}}$ involving the molecular viscosity is expected to be much smaller than the $-\overline{\rho u_i'' u_j''}$ component.

We can perform a similar analysis on the Reynolds form of the energy equation and identify certain terms involving temperature or enthalpy fluctuations as apparent heat-flux quantities. For example, in Eq. (5-84) the molecular "laminar-like" heat-flux term is

$$-(\nabla\cdot\mathbf{q})_{\text{lam}} = \frac{\partial}{\partial x_j}\left(k\,\frac{\partial \bar{T}}{\partial x_j}\right) \tag{5-94a}$$

and the apparent turbulent (Reynolds) heat-flux component is

$$-(\nabla\cdot\mathbf{q})_{\text{turb}} = \frac{\partial}{\partial x_j}\left(-\bar{\rho}c_p\overline{T'u_j'} - c_p\overline{\rho' T' u_j'}\right) \tag{5-94b}$$

Further examples illustrating the form of the Reynolds stress and heat-flux terms will be given in sections which consider reduced forms of the Reynolds equations.

The Reynolds equations cannot be solved in the form given because the new apparent turbulent stresses and heat-flux quantities must be viewed as new unknowns. To proceed further we need to find additional equations involving the new unknowns or make assumptions regarding the relationship between the new apparent turbulent quantities and the time-mean flow variables. This is known as the closure problem which is most commonly handled through *turbulence modeling* which will be discussed in Section 5-4.

5-3 THE BOUNDARY-LAYER EQUATIONS

5-3.1 Background

The concept of a boundary layer originated with Ludwig Prandtl in 1904 (Prandtl, 1926). Prandtl reasoned from experimental evidence that for sufficiently large Reynolds numbers a thin region existed near a solid boundary where viscous effects were at least as important as inertia effects no matter how small the viscosity of the fluid might be. Prandtl deduced that a much reduced form of the governing equations could be used by systematically employing two constraints. These were that the viscous layer must be thin relative to the characteristic streamwise dimension of the object immersed in the flow, $\delta/L \ll 1$, and that the largest viscous term must be of the same approximate magnitude as any inertia (particle acceleration) term. Prandtl used what we now call an order of magnitude analysis to reduce the governing equations. Essentially, his conclusions were that second derivatives of the velocity components in the streamwise direction were negligible compared to corresponding derivatives transverse to the main flow direction and that the entire momentum equation for the transverse direction could be neglected.

In the years since 1904 we have found that a similar reduction can often be made in the governing equations for other flows for which a primary flow direction can be identified. These flows include jets, wakes, mixing layers, and the developing flow in pipes and other internal passages. Thus, the terminology boundary-layer flow or boundary-layer approximation has taken on a more general meaning which refers to circumstances which permit the neglect of the transverse momentum equation and the streamwise second derivative term in the remaining momentum equation (or equations in the case of three-dimensional flow). It is increasingly common to refer to these reduced equations as the "thin-shear-layer" equations. This terminology seems especially appropriate in light of the applicability of the equations to free-shear flows such as jets and wakes as well as flows along a solid boundary. We will use both designations, boundary layer and thin-shear layer, interchangeably in this book.

5-3.2 The Boundary-Layer Approximation for Steady Incompressible Flow

It is useful to review the methodology used to obtain the boundary-layer approximations to the Navier-Stokes and Reynolds equations for steady, two-dimensional, incompressible, constant property flow along an isothermal surface at temperature T_w. First we define the nondimensional variables (very much as was done in Section 5-1.6)

$$u^* = \frac{u}{u_\infty} \qquad v^* = \frac{v}{u_\infty} \qquad x^* = \frac{x}{L} \qquad y^* = \frac{y}{L} \qquad p^* = \frac{p}{\rho u_\infty^2} \qquad \theta = \frac{T - T_\infty}{T_w - T_\infty} \tag{5-95}$$

and introduce them into the Navier-Stokes equations by substitution. After rearrangement, the results can be written as

continuity:

$$\frac{\partial u^*}{\partial x^*} + \frac{\partial v^*}{\partial y^*} = 0 \tag{5-96}$$

x momentum:

$$u^* \frac{\partial u^*}{\partial x^*} + v^* \frac{\partial u^*}{\partial y^*} = -\frac{\partial p^*}{\partial x^*} + \frac{1}{\mathrm{Re}_L}\left(\frac{\partial^2 u^*}{\partial x^{*2}} + \frac{\partial^2 u^*}{\partial y^{*2}}\right) \tag{5-97}$$

y momentum:

$$u^* \frac{\partial v^*}{\partial x^*} + v^* \frac{\partial v^*}{\partial y^*} = -\frac{\partial p^*}{\partial y^*} + \frac{1}{\mathrm{Re}_L}\left(\frac{\partial^2 v^*}{\partial x^{*2}} + \frac{\partial^2 v^*}{\partial y^{*2}}\right) \tag{5-98}$$

energy:

$$u^* \frac{\partial \theta}{\partial x^*} + v^* \frac{\partial \theta}{\partial y^*} = \frac{1}{\mathrm{Re}_L \, \mathrm{Pr}}\left(\frac{\partial^2 \theta}{\partial x^{*2}} + \frac{\partial^2 \theta}{\partial y^{*2}}\right) + \mathrm{Ec}\left(u^* \frac{\partial p^*}{\partial x^*} + v^* \frac{\partial p^*}{\partial y^*}\right)$$

$$+ \frac{\mathrm{Ec}}{\mathrm{Re}_L}\left[2\left(\frac{\partial u^*}{\partial x^*}\right)^2 + 2\left(\frac{\partial v^*}{\partial y^*}\right)^2 + \left(\frac{\partial v^*}{\partial x^*} + \frac{\partial u^*}{\partial y^*}\right)^2\right] \tag{5-99}$$

In the above,

$$\mathrm{Re}_L = \text{Reynolds number} = \frac{\rho u_\infty L}{\mu}$$

$$\mathrm{Pr} = \text{Prandtl number} = \frac{c_p \mu}{k}$$

$$\mathrm{Ec} = \text{Eckert number} = 2\,\frac{T_0 - T_\infty}{T_w - T_\infty}$$

and u_∞, T_∞ are the freestream velocity and temperature, respectively, and T_0 is the stagnation temperature. The product Re Pr is also known as the Peclet number.

Following Prandtl we assume that the thickness of the viscous and thermal boundary layers are small relative to a characteristic length in the primary flow direction. That is, $\delta/L \ll 1$ and $\delta_t/L \ll 1$ (see Fig. 5-5). For convenience we let $\epsilon = \delta/L$ and $\epsilon_t = \delta_t/L$. Since ϵ and ϵ_t are both assumed to be small, we will take them to be of the same order of magnitude. We are assured that ϵ and ϵ_t are small over L if $\partial\delta/\partial x$ and

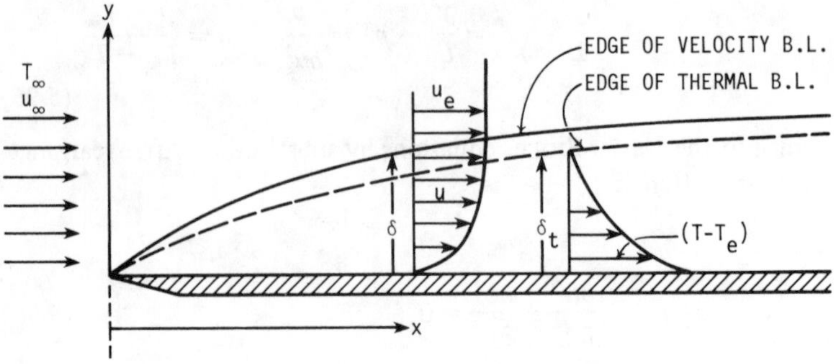

Figure 5-5 Notation and coordinate system for a boundary layer on a flat plate.

$\partial\, \delta_t/\partial x$ are everywhere small. At a distance L from the origin of the boundary layer we now estimate typical or expected sizes of terms in the equation.

As a general rule, we estimate sizes of derivatives by using the "mean value" provided by replacing the derivative by a finite difference over the expected range of the variables in the boundary-layer flow. For example, we estimate the size of $\partial u^*/\partial x^*$ by noting that for flow over a flat plate in a uniform stream, u^* ranges between one and zero as x^* ranges between zero and one; thus we say that we expect $\partial u^*/\partial x^*$ to be of the order of magnitude of 1. That is

$$\left|\frac{\partial u^*}{\partial x^*}\right| \approx \left|\frac{0-1}{1-0}\right| = 1$$

A factor of two or so does not matter in our estimates but a factor of 10–100 does and represents an order of magnitude. It should be noted that the velocity at the outer edge of the boundary layer may deviate somewhat from u_∞ (as would be the case for flows with a pressure gradient) without changing the order of magnitude of $\partial u^*/\partial x^*$. Having established $(\partial u^*/\partial x^*) \cong 1$, we now consider the $\partial v^*/\partial y^*$ term in the continuity equation. We require that this term be of the same order of magnitude as $\partial u^*/\partial x^*$ so that mass can be conserved. Since y^* ranges between 0 and ϵ in the boundary layer, we expect from the continuity equation that v^* will also range between 0 and ϵ. Thus, $v^* \cong \epsilon$. If $\partial\delta/\partial x$ should locally become large due to some perturbation, then the continuity equation suggests that v^* could also become large locally. The nondimensional thermal variable, θ, clearly ranges between zero and one for incompressible, constant property flow.

We are now in a position to establish the order of magnitudes for the terms in the Navier-Stokes equations. The estimates are noted below the terms in Eqs. (5-100)–(5-103).

continuity:

$$\frac{\partial u^*}{\partial x^*} + \frac{\partial v^*}{\partial y^*} = 0 \tag{5-100}$$

$$\quad 1 \qquad\quad 1$$

x momentum:

$$u^* \frac{\partial u^*}{\partial x^*} + v^* \frac{\partial u^*}{\partial y^*} = -\frac{\partial p^*}{\partial x^*} + \frac{1}{\mathrm{Re}_L}\left(\frac{\partial^2 u^*}{\partial x^{*2}} + \frac{\partial^2 u^*}{\partial y^{*2}}\right) \tag{5-101}$$

$$1 \quad 1 \qquad \epsilon \quad \frac{1}{\epsilon} \qquad 1 \qquad \epsilon^2 \qquad 1 \qquad \frac{1}{\epsilon^2}$$

y momentum:

$$u^* \frac{\partial v^*}{\partial x^*} + v^* \frac{\partial v^*}{\partial y^*} = -\frac{\partial p^*}{\partial y^*} + \frac{1}{\mathrm{Re}_L}\left(\frac{\partial^2 v^*}{\partial x^{*2}} + \frac{\partial^2 v^*}{\partial y^{*2}}\right) \tag{5-102}$$

$$1 \quad \epsilon \qquad \epsilon \quad 1 \qquad \epsilon \qquad \epsilon^2 \qquad \epsilon \qquad \frac{1}{\epsilon}$$

energy:

$$u^* \frac{\partial \theta}{\partial x^*} + v^* \frac{\partial \theta}{\partial y^*} = \frac{1}{\mathrm{Re}_L \, \mathrm{Pr}}\left(\frac{\partial^2 \theta}{\partial x^{*2}} + \frac{\partial^2 \theta}{\partial y^{*2}}\right) + \mathrm{Ec}\left(u^* \frac{\partial p^*}{\partial x^*} + v^* \frac{\partial p^*}{\partial y^*}\right)$$

$$1 \quad 1 \qquad \epsilon \quad \frac{1}{\epsilon} \qquad \epsilon^2 \qquad 1 \qquad \frac{1}{\epsilon^2} \qquad 1 \quad 1 \quad 1 \qquad \epsilon \quad \epsilon$$

$$+ \frac{\mathrm{Ec}}{\mathrm{Re}_L}\left[2\left(\frac{\partial u^*}{\partial x^*}\right)^2 + 2\left(\frac{\partial v^*}{\partial y^*}\right)^2 + \left(\frac{\partial v^*}{\partial x^*} + \frac{\partial u^*}{\partial y^*}\right)^2\right] \tag{5-103}$$

$$\epsilon^2 \qquad 1 \qquad 1 \qquad \epsilon^2 \quad 1 \quad \frac{1}{\epsilon^2}$$

Some comments are in order. In Eq. (5-101), the order of magnitude of the pressure gradient was established by the observation that the Navier-Stokes equations reduce to the Euler equations (see Section 5-5) at the outer edge of the viscous region. The pressure gradient must be capable of balancing the inertia terms. Hence, the pressure gradient and the inertia terms must be of the same order of magnitude. We are also requiring that the largest viscous term be of the same order of magnitude as the inertia terms. For this to be true, Re_L must be of the order of magnitude of $1/\epsilon^2$ as can be seen from Eq. (5-101).

The order of magnitude of all terms in Eq. (5-102) can be established in a straightforward manner except for the pressure gradient. Since the pressure gradient must be balanced by other terms in the equation, its order of magnitude cannot be greater than any of the others in Eq. (5-102). Accordingly, its maximum order of magnitude must be ϵ as recorded in Eq. (5-102).

In the energy equation, we have assumed somewhat arbitrarily that the Eckert number was of the order of magnitude of one. This should be considered a typical value. Ec can become an order of magnitude larger or smaller in certain applications. The order of magnitude of the Peclet number, Re Pr, was set at $1/\epsilon^2$. Since we have already assumed $\mathrm{Re}_L \cong (1/\epsilon^2)$ in dealing with the momentum equations, this suggests that $\mathrm{Pr} \cong 1$. This is consistent with our original hypothesis that ϵ and ϵ_t were both small, i.e., $\delta \cong \delta_t$. In other words, we are assuming that the Peclet number is of the same order of magnitude as the Reynolds number. We expect that the present results will be applicable to flows in which the Prandtl number does not vary from one by more than an order of magnitude. The exact limitation of the analysis must be determined by comparisons with experimental data. Three orders are specified for the last term in parentheses in Eq. (5-103) to account for the cross-product term which results from squaring the quantity in parentheses.

Carrying out the multiplications needed to establish the order of each term in Eqs. (5-101)–(5-103), we observe that all terms in the x-momentum equation are of order one in magnitude except for the streamwise second derivative (diffusion) term which is of order ϵ^2. No term in the y-momentum equation is larger than ϵ in estimated magnitude. Several terms in the energy equation are of order one in magnitude although several in the compression work and viscous dissipation terms are smaller. Keeping terms whose order of magnitude estimates are equal to one gives the boundary-layer equations. These are recorded below in terms of dimensional variables:

continuity:

$$\frac{\partial u}{\partial x} + \frac{\partial v}{\partial y} = 0 \tag{5-104}$$

momentum:

$$u\frac{\partial u}{\partial x} + v\frac{\partial u}{\partial y} = -\frac{1}{\rho}\frac{dp}{dx} + \nu\frac{\partial^2 u}{\partial y^2} \tag{5-105}$$

energy:

$$u\frac{\partial T}{\partial x} + v\frac{\partial T}{\partial y} = \alpha\frac{\partial^2 T}{\partial y^2} + \frac{\beta T u}{\rho c_p}\frac{dp}{dx} + \frac{\mu}{\rho c_p}\left(\frac{\partial u}{\partial y}\right)^2 \tag{5-106}$$

where ν is the kinematic viscosity, μ/ρ, and α is the thermal diffusivity, $k/\rho c_p$.

The form of the energy equation has been generalized to include nonideal gas behavior through the introduction of β, the volumetric expansion coefficient

$$\beta = -\frac{1}{\rho}\left.\frac{\partial \rho}{\partial T}\right)_p$$

For an ideal gas $\beta = 1/T$ where T is absolute temperature. It should be pointed out that the last two terms in Eq. (5-106) were retained from the order of magnitude analysis on the basis that Ec ~ 1. Should Ec become of the order of ϵ or smaller for a particular flow, neglecting these terms should be permissible.

To complete the mathematical formulation, initial and boundary conditions must be specified. The steady boundary-layer momentum and energy equations are parabolic with the streamwise direction being the marching direction. Initial distributions of u and T must be provided. The usual boundary conditions are

$$u(x,0) = v(x,0) = 0 \tag{5-107}$$

$$T(x,0) = T_w(x) \quad \text{or} \quad \left.\frac{\partial T}{\partial y}\right)_{y=0} = \frac{q(x)}{k} \tag{5-107 (Cont.)}$$

$$\lim_{y\to\infty} u(x,y) = u_e(x) \qquad \lim_{y\to\infty} T(x,y) = T_e(x)$$

where the subscript e refers to conditions at the edge of the boundary layer. The pressure gradient term in Eqs. (5-105) and (5-106) is to be evaluated from the given boundary information. With $u_e(x)$, specified, dp/dx can be evaluated from an application of the equations which govern the inviscid outer flow (Euler's equations) giving $dp/dx = -\rho u_e \, du_e/dx$.

It is not difficult to extend the boundary-layer equations to variable property and/or compressible flows. The constant property restriction was made only as a convenience as we set about the task of illustrating principles which can frequently be used to determine a reduced but approximate set of governing equations for a flow of interest. The compressible form of the boundary-layer equations, which will also account for property variations, will be presented in Section 5-3.3.

Before moving on from our order of magnitude deliberations for laminar flows, we should raise the question as to which terms neglected in the boundary-layer approximation should first become important as δ/L becomes larger and larger. Terms of order ϵ will next become important and then eventually, terms of order ϵ^2. We note that the second derivative term neglected in the streamwise momentum equation is of order ϵ^2 whereas most terms in the y-momentum equation are of order ϵ. This means that contributions through the transverse momentum equation are expected to become important before additional terms need to be considered in the x-momentum equation. The set of equations which results from retaining terms of both orders 1 and ϵ in the order of magnitude analysis while neglecting terms of order ϵ^2 and higher, has proven to be useful in computational fluid dynamics. Such steady flow equations, which neglect all streamwise second derivative terms, are known as the "parabolized" Navier-Stokes equations for supersonic applications and are known as the "partially parabolized" Navier-Stokes equations for subsonic applications. These are but two examples from a category of equations frequently called "parabolized" Navier-Stokes equations. These equations are intermediate in complexity between the Navier-Stokes and boundary-layer equations and will be discussed in Chapter 8.

We next consider extending the boundary-layer approximation to an incompressible, constant property two-dimensional turbulent flow. Under our incompressible assumption, $\rho' = 0$ and the Reynolds equations simplify considerably. We will nondimensionalize the incompressible Reynolds equations very much in the same manner as for the Navier-Stokes equations, letting

$$u^* = \frac{\bar{u}}{u_\infty} \quad v^* = \frac{\bar{v}}{u_\infty} \quad x^* = \frac{x}{L} \quad y^* = \frac{y}{L} \quad p^* = \frac{\bar{p}}{\rho u_\infty^2} \quad (u')^* = \frac{u'}{u_\infty} \quad (v')^* = \frac{v'}{u_\infty}$$

$$\theta = \frac{\bar{T} - T_\infty}{T_w - T_\infty} \quad H^* = \frac{\bar{H} - H_\infty}{H_w - H_\infty} \quad \theta' = \frac{T'}{T_w - T_\infty} \tag{5-108}$$

Asterisks appended to parentheses, ( )*, will indicate that all quantities within the parentheses are dimensionless; that is, instead of $\overline{u'^*v'^*}$, we will use the more convenient notation $(\overline{u'v'})^*$.

As before, we assume that $\delta/L \ll 1$, $\delta_t/L \ll 1$, and let $\epsilon = \delta/L \approx \delta_t/L$. We rely on experimental evidence to guide us in establishing the magnitude estimates for the Reynolds stress and heat-flux terms. Experiments indicate that the Reynolds stresses can be at least as large as the laminar counterparts. This requires that $(\overline{u'v'})^* \sim \epsilon$. Measurements suggest that $(\overline{u'^2})^*$, $(\overline{u'v'})^*$, $(\overline{v'^2})^*$, while differing in magnitudes and distribution somewhat, are nevertheless of the same order of magnitude in the boundary layer. That is, we cannot stipulate that the magnitudes of any of these terms are different by a factor of ten or more. A similar observation can be made for the energy equation leading to the conclusion that $(\overline{\theta'v'})^*$ and $(\overline{\theta'u'})^*$ are of the order of magnitude of ϵ. Triple correlations such as $(\overline{u'u'u'})^*$ are clearly expected to be smaller than double correlations and they will be taken to be of order ϵ^2 (Schubauer and Tchen, 1959). It will be expedient to invoke the boundary-layer approximation to the form of the energy equation which employs the total enthalpy as the transported thermal variable, Eq. (5-89). We will, however, substitute for H' according to

$$H' = c_p T' + u_i'\bar{u}_i + \frac{u_i'u_i'}{2}$$

The incompressible nondimensional Reynolds equations are given below along with the order of magnitude estimates for the individual terms.

continuity:

$$\frac{\partial u^*}{\partial x^*} + \frac{\partial v^*}{\partial y^*} = 0 \tag{5-109}$$

$$1 \qquad\qquad 1$$

x momentum:

$$u^*\frac{\partial u^*}{\partial x^*} + v^*\frac{\partial u^*}{\partial y^*} = -\frac{\partial p^*}{\partial x^*} + \frac{1}{\mathrm{Re}_L}\left(\frac{\partial^2 u^*}{\partial x^{*2}} + \frac{\partial^2 u^*}{\partial y^{*2}}\right) - \frac{\partial}{\partial y^*}(\overline{u'v'})^* - \frac{\partial}{\partial x^*}(\overline{u'^2})^*$$

$$1 \quad 1 \qquad \epsilon \quad \frac{1}{\epsilon} \qquad\quad 1 \qquad \epsilon^2 \qquad 1 \qquad \frac{1}{\epsilon^2} \qquad \frac{\epsilon}{\epsilon} \qquad\qquad \epsilon$$

(5-110)

y momentum:

$$u^*\frac{\partial v^*}{\partial x^*} + v^*\frac{\partial v^*}{\partial y^*} = -\frac{\partial p^*}{\partial y^*} + \frac{1}{\mathrm{Re}_L}\left(\frac{\partial^2 v^*}{\partial x^{*2}} + \frac{\partial^2 v^*}{\partial y^{*2}}\right) - \frac{\partial(\overline{v'u'})^*}{\partial x^*} - \frac{\partial(\overline{v'^2})^*}{\partial y^*}$$

$$1 \quad \epsilon \qquad \epsilon \quad 1 \qquad\quad 1 \qquad \epsilon^2 \qquad \epsilon \qquad \frac{1}{\epsilon} \qquad\qquad \epsilon \qquad\qquad 1$$

(5-111)

energy:

$$u^* \frac{\partial H^*}{\partial x^*} + v^* \frac{\partial H^*}{\partial y^*} = \frac{T_w - T_\infty}{T_w - T_0} \left[-\frac{\partial}{\partial x^*} (\overline{u'\theta'})^* - \frac{\partial}{\partial y^*} (\overline{v'\theta'})^* \right]$$

$$1 \quad 1 \quad \epsilon \quad \frac{1}{\epsilon} \qquad\qquad 1 \qquad\qquad \epsilon \qquad\qquad \frac{\epsilon}{\epsilon}$$

$$+ \frac{1}{\mathrm{Re}_L \, \mathrm{Pr}} \left(\frac{\partial^2 \theta}{\partial x^{*2}} + \frac{\partial^2 \theta}{\partial y^{*2}} \right) + \mathrm{Ec} \left[-\frac{\partial}{\partial x^*} (u\overline{u'u'})^* - \frac{\partial}{\partial x^*} (v\overline{v'u'})^* \right.$$

$$\epsilon^2 \qquad 1 \qquad \frac{1}{\epsilon^2} \qquad 1 \qquad\qquad \epsilon \qquad\qquad \epsilon^2$$

$$- \frac{\partial}{\partial y^*} (u\overline{u'v'})^* - \frac{\partial}{\partial y^*} (v\overline{v'v'})^* - \frac{1}{2} \frac{\partial}{\partial x^*} (\overline{u'u'u'})^* - \frac{1}{2} \frac{\partial}{\partial x^*} (\overline{v'v'u'})^*$$

$$\frac{\epsilon}{\epsilon} \qquad\qquad \frac{\epsilon^2}{\epsilon} \qquad\qquad \epsilon^2 \qquad\qquad \epsilon^2$$

$$\left. - \frac{1}{2} \frac{\partial}{\partial y^*} (\overline{u'u'v'})^* - \frac{1}{2} \frac{\partial}{\partial y^*} (\overline{v'v'v'})^* \right] + \frac{\mathrm{Ec}}{\mathrm{Re}_L} \left[2 \frac{\partial}{\partial x^*} \left(u^* \frac{\partial u^*}{\partial x^*} \right) \right.$$

$$\frac{\epsilon^2}{\epsilon} \qquad\qquad \frac{\epsilon^2}{\epsilon} \qquad\qquad \epsilon^2 \qquad\qquad 1$$

$$+ \frac{\partial}{\partial x^*} \left(v^* \frac{\partial u^*}{\partial y^*} \right) + \frac{\partial}{\partial x^*} \left(v^* \frac{\partial v^*}{\partial x^*} \right) + \frac{\partial}{\partial y^*} \left(u^* \frac{\partial v^*}{\partial y^*} \right) + \frac{\partial}{\partial y^*} \left(u^* \frac{\partial u^*}{\partial y^*} \right)$$

$$\frac{\epsilon}{\epsilon} \qquad\qquad \epsilon^2 \qquad\qquad \frac{\epsilon}{\epsilon^2} \qquad\qquad \frac{1}{\epsilon^2}$$

$$+ 2 \frac{\partial}{\partial y^*} \left(v^* \frac{\partial v^*}{\partial y^*} \right) + 2 \frac{\partial}{\partial x^*} \left(\overline{u' \frac{\partial u'}{\partial x}} \right)^* + \frac{\partial}{\partial x^*} \left(\overline{v' \frac{\partial u'}{\partial y}} \right)^* + \frac{\partial}{\partial x} \left(\overline{v' \frac{\partial v'}{\partial x}} \right)^*$$

$$\frac{\epsilon^2}{\epsilon^2} \qquad\qquad \epsilon \qquad\qquad \frac{\epsilon}{\epsilon} \qquad\qquad \epsilon$$

$$\left. + \frac{\partial}{\partial y^*} \left(\overline{u' \frac{\partial v'}{\partial x}} \right)^* + \frac{\partial}{\partial y^*} \left(\overline{u' \frac{\partial u'}{\partial y}} \right)^* + 2 \frac{\partial}{\partial y^*} \left(\overline{v' \frac{\partial v'}{\partial y}} \right)^* \right] \tag{5-112}$$

$$\frac{\epsilon}{\epsilon} \qquad\qquad \frac{\epsilon}{\epsilon^2} \qquad\qquad \frac{\epsilon}{\epsilon^2}$$

Again we assume that Pr and Ec are near one in order of magnitude. The two-dimensional boundary-layer equations are obtained by retaining only terms of order one. They can be written in dimensional variables as

continuity:

$$\frac{\partial \bar{u}}{\partial x} + \frac{\partial \bar{v}}{\partial y} = 0$$

momentum:

$$\rho \bar{u} \frac{\partial \bar{u}}{\partial x} + \rho \bar{v} \frac{\partial \bar{u}}{\partial y} = -\frac{d\bar{p}}{dx} + \mu \frac{\partial^2 \bar{u}}{\partial y^2} - \rho \frac{\partial}{\partial y}(\overline{u'v'}) \tag{5-113}$$

energy:

$$\rho \bar{u} \frac{\partial \bar{H}}{\partial x} + \rho \bar{v} \frac{\partial \bar{H}}{\partial y} = k \frac{\partial^2 \bar{T}}{\partial y^2} - \rho c_p \frac{\partial}{\partial y}(\overline{v'T'}) - \rho \frac{\partial}{\partial y}(\bar{u}\,\overline{v'u'}) + \mu \frac{\partial}{\partial y}\left(\bar{u} \frac{\partial \bar{u}}{\partial y}\right) \tag{5-114}$$

It should be noted that terms of order one do remain in the y-momentum equation for turbulent flow, namely,

$$\frac{1}{\rho} \frac{\partial \bar{p}}{\partial y} = -\frac{\partial}{\partial y}(\overline{v'^2})$$

These terms have not been listed above with the boundary-layer equations because they contribute no information about the mean velocities. The pressure variation across the boundary layer is of order ϵ (negligible in comparison with the streamwise variation). The boundary-layer energy equation can be easily written in terms of static temperature by substituting

$$c_p \bar{T} + \frac{\bar{u}^2}{2}$$

for $\bar{H}$ in Eq. (5-114). In doing this, we are neglecting $\bar{v}^2$ compared to $\bar{u}^2$ in the kinetic energy of the mean motion. Examination of the way in which $\bar{H}$ appears in Eq. (5-114) reveals that this is permissible in the boundary-layer approximation. Utilizing Eq. (5-113) to eliminate the kinetic energy terms permits the boundary-layer form of the energy equation to be written as

$$\rho \bar{u} c_p \frac{\partial \bar{T}}{\partial x} + \rho \bar{v} c_p \frac{\partial \bar{T}}{\partial y} = k \frac{\partial^2 \bar{T}}{\partial y^2} - \rho c_p \frac{\partial}{\partial y}(\overline{v'T'}) + \bar{u} \frac{d\bar{p}}{dx} + \left(\mu \frac{\partial \bar{u}}{\partial y} - \rho \overline{v'u'}\right) \frac{\partial \bar{u}}{\partial y} \tag{5-115}$$

The last two terms on the right-hand side of Eq. (5-115) can be neglected in some applications. However, it is not correct to categorically neglect these terms for incompressible flows. The last term on the right-hand side, for example, represents the viscous dissipation of energy which obviously is important in incompressible lubrication applications where the major heat transfer concern is to remove the heat generated by the viscous dissipation. In some instances it is also possible to neglect one or both of the last two terms on the right-hand side of Eq. (5-114). Both Eq. (5-114) and (5-115) can be easily treated with finite-difference methods in their entirety, so the temptation to impose further reductions should generally be resisted unless it is absolutely clear that the terms neglected will indeed be negligible. The boundary conditions remain unchanged for turbulent flow.

In closing this section on the development of the thin-shear-layer approximation, it is worthwhile noting that for turbulent flow, the largest term neglected in the streamwise momentum equation, the Reynolds normal stress term, was estimated to be an order of magnitude larger, ϵ, than the largest term neglected in the laminar flow analysis. We note also that only one Reynolds stress term and one Reynolds heat-flux term remain in the governing equations after the boundary-layer approximation is invoked.

For any steady internal flow application of the thin-shear-layer equations, it is possible to develop a global channel mass flow constraint. This permits the pressure gradient to be computed rather than requiring that it be given as in the case for external flow. This point will be developed further in Chapter 7.

5-3.3 The Boundary-Layer Equations for Compressible Flow

The order of magnitude reduction of the Reynolds equations to boundary-layer form is a lengthier process for compressible flow. Only the results will be presented here. Details of the arguments for elimination of terms are given by Schubauer and Tchen (1959), van Driest (1951), and Cebeci and Smith (1974). As was the case for incompressible flow, guidance must be obtained from experimental observations in assessing the magnitudes of turbulence quantities. An estimate must be made for $\rho'/\bar{\rho}$ for compressible flows.

Measurements in gases for Mach numbers less than about 5 indicate that temperature fluctuations are nearly isobaric for adiabatic flows. This suggests that $T'/\bar{T} \approx -\rho'/\bar{\rho}$. However, there is evidence that appreciable pressure fluctuations exist (8–10% of the mean wall static pressure) at $M_e = 5$ and it is speculated that $p'/\bar{p}$ increases with increasing Mach number. In the absence of specific experimental evidence to the contrary, it is common to base the order of magnitude estimates of fluctuating terms on the assumption that the pressure fluctuations are small. This appears to be a safe assumption for $M_e \leqslant 5$ and good predictions based on this assumption have been noted for Mach numbers as high as 7.5. We will adopt the isobaric assumption here. It is primarily the correlation terms involving the density fluctuations which may increase in magnitude with increasing Mach number above $M_e \approx 5$.

We find that the difference between $\tilde{u}$ and $\bar{u}$ vanishes under the boundary-layer approximation. This follows because $\overline{\rho' u'}$ is expected to be small compared to $\bar{\rho}\bar{u}$ and can be neglected in the momentum equation. We also find $\bar{T} = \tilde{T}$ and $\bar{H} = \tilde{H}$ to be consistent with the boundary-layer approximation. On the other hand, $\overline{\rho' v'}$ and $\bar{\rho}\bar{v}$ are both of about the same order of magnitude in a thin shear layer. Thus, $\bar{v} \neq \tilde{v}$. Below, the unsteady boundary-layer equations for a compressible fluid are written in a form applicable to both two-dimensional and axisymmetric turbulent flow. For convenience we will drop the use of bars over time-mean quantities and make use of $\tilde{v} = (\bar{\rho}\bar{v} + \overline{\rho' v'})/\bar{\rho}$. The equations are also valid for laminar flow when the terms involving fluctuating quantities are set equal to zero. The coordinate system is indicated in Fig. 5-6:

continuity:

$$\frac{\partial \rho}{\partial t} + \frac{\partial}{\partial x}(r^m \rho u) + \frac{\partial}{\partial y}(r^m \rho \tilde{v}) = 0 \tag{5-116}$$

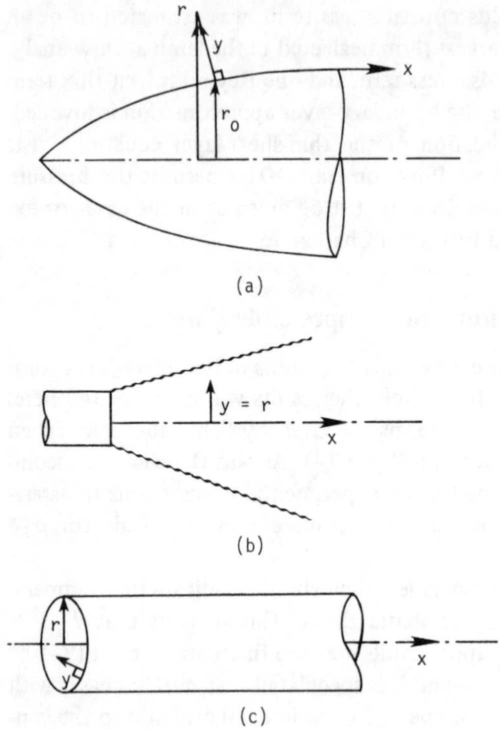

Figure 5-6 Coordinate system for axisymmetric thin-shear-layer equations. (a) External boundary layer; (b) axisymmetric free shear flow; (c) confined axisymmetric flow.

momentum:

$$\rho\frac{\partial u}{\partial t}+\rho u\frac{\partial u}{\partial x}+\rho\tilde{v}\frac{\partial u}{\partial y}=-\frac{dp}{dx}+\frac{1}{r^m}\frac{\partial}{\partial y}\left[r^m\left(\mu\frac{\partial u}{\partial y}-\rho\overline{u'v'}\right)\right] \tag{5-117}$$

energy:

$$\rho\frac{\partial H}{\partial t}+\rho u\frac{\partial H}{\partial x}+\rho\tilde{v}\frac{\partial H}{\partial y}=\frac{1}{r^m}\frac{\partial}{\partial y}\left(r^m\left\{\frac{\mu}{\mathrm{Pr}}\frac{\partial H}{\partial y}-\rho c_p\overline{v'T'}+u\left[\left(1-\frac{1}{\mathrm{Pr}}\right)\mu\frac{\partial u}{\partial y}-\rho\overline{v'u'}\right]\right\}\right)+\frac{\partial p}{\partial t} \tag{5-118}$$

state:

$$\rho=\rho(p,T) \tag{5-119}$$

In the above, m is a flow index equal to unity for axisymmetric flow ($r^m = r$) and equal to zero for two-dimensional flow ($r^m = 1$). Other forms of the energy equation will be noted in subsequent sections.

We note that the boundary-layer equations for compressible flow are not significantly more complex than for incompressible flow. Only one Reynolds stress and one heat-flux term appear regardless of whether the flow is compressible or incompressible. As for purely laminar flows, the main difference is in the property variations of μ, k, and ρ for the compressible case which nearly always requires that a solution be obtained for some form of the energy equation. When properties can be assumed constant (as for many incompressible flows), the momentum equation is independent of the energy equation and, as a result, the energy equation need not be solved for many problems of interest.

The boundary-layer approximation remains valid for a flow in which the turning of the mainstream results in a three-dimensional flow as long as velocity derivatives with respect to only one coordinate direction are large. That is, the three-dimensional boundary layer is a flow which remains "thin" with respect to only one coordinate direction. The three-dimensional unsteady boundary-layer equations in Cartesian coordinates, applicable to a compressible turbulent flow, are given below. The y direction is normal to the wall.

continuity:

$$\frac{\partial \rho}{\partial t} + \frac{\partial \rho u}{\partial x} + \frac{\partial \rho \tilde{v}}{\partial y} + \frac{\partial \rho w}{\partial z} = 0 \tag{5-120}$$

x momentum:

$$\frac{\partial u}{\partial t} + \rho u \frac{\partial u}{\partial x} + \rho \tilde{v} \frac{\partial u}{\partial y} + \rho w \frac{\partial u}{\partial z} = -\frac{\partial p}{\partial x} + \frac{\partial}{\partial y}\left(\mu \frac{\partial u}{\partial y} - \rho \overline{u'v'}\right) \tag{5-121}$$

z momentum:

$$\frac{\partial w}{\partial t} + \rho u \frac{\partial w}{\partial x} + \rho \tilde{v} \frac{\partial w}{\partial y} + \rho w \frac{\partial w}{\partial z} = -\frac{\partial p}{\partial z} + \frac{\partial}{\partial y}\left(\mu \frac{\partial w}{\partial y} - \rho \overline{w'v'}\right) \tag{5-122}$$

energy:

$$\frac{\partial H}{\partial t} + \rho u \frac{\partial H}{\partial x} + \rho \tilde{v} \frac{\partial H}{\partial y} + \rho w \frac{\partial H}{\partial z}$$

$$= \frac{\partial}{\partial y}\left[\frac{\mu}{\Pr}\frac{\partial H}{\partial y} - \rho c_p \overline{v'T'} + \mu\left(1 - \frac{1}{\Pr}\right)\left(u \frac{\partial u}{\partial y} + w \frac{\partial w}{\partial y}\right) - \rho u \overline{v'u'} - \rho w \overline{v'w'}\right] \tag{5-123}$$

For a three-dimensional flow, the boundary-layer approximation permits H to be written as

$$H = c_p T + \frac{u^2}{2} + \frac{w^2}{2}$$

The three-dimensional boundary-layer equations are used primarily for external flows. This usually permits the pressure gradient terms to be evaluated from the solution to the inviscid flow (Euler) equations. Three-dimensional internal flows are normally computed from slightly different equations to be discussed in Chapter 8.

It is common to employ body intrinsic curvilinear coordinates to compute the three-dimensional boundary layers occurring on wings and other shapes of practical interest. Often, this curvilinear coordinate system is nonorthogonal. An example of this can be found in Cebeci et al. (1977). The orthogonal system is somewhat more common (see, for example, Blottner and Ellis, 1973). One coordinate, x_2, is almost always taken to be orthogonal to the body surface. This convention will be followed here. Below we record the three-dimensional boundary-layer equations in the orthogonal curvilinear coordinate system described in Section 5-1.7. Typically, x_1 will be directed roughly in the primary flow direction and x_3 will be in the crossflow direction. The metric coefficients (h_1, h_2, h_3) are as defined in Section 5-1.7; however, h_2 will be taken as unity as a result of the boundary-layer approximation. In addition, we will make use of the geodesic curvatures of the surface coordinate lines,

$$K_1 = \frac{1}{h_1 h_3}\frac{\partial h_1}{\partial x_3} \quad \text{and} \quad K_3 = \frac{1}{h_1 h_3}\frac{\partial h_3}{\partial x_1} \tag{5-124}$$

With this notation, the boundary-layer form of the conservation equations for a compressible, turbulent flow can be written,

continuity:

$$\frac{\partial}{\partial x_1}(\rho h_3 u_1) + \frac{\partial}{\partial x_2}(h_1 h_3 \rho \tilde{u}_2) + \frac{\partial}{\partial x_3}(\rho h_1 u_3) = 0 \tag{5-125}$$

x_1 momentum:

$$\frac{\rho u_1}{h_1}\frac{\partial u_1}{\partial x_1} + \rho\tilde{u}_2\frac{\partial u_1}{\partial x_2} + \frac{\rho u_3}{h_3}\frac{\partial u_1}{\partial x_3} + \rho u_1 u_3 K_1 - \rho u_3^2 K_3 = -\frac{1}{h_1}\frac{\partial p}{\partial x_1} + \frac{\partial}{\partial x_2}\left(\mu\frac{\partial u_1}{\partial x_2} - \rho\overline{u_1' u_2'}\right) \tag{5-126}$$

x_3 momentum:

$$\frac{\rho u_1}{h_1}\frac{\partial u_3}{\partial x_1} + \rho\tilde{u}_2\frac{\partial u_3}{\partial x_2} + \frac{\rho u_3}{h_3}\frac{\partial u_3}{\partial x_3} + \rho u_1 u_3 K_3 - \rho u_1^2 K_1 = -\frac{1}{h_3}\frac{\partial p}{\partial x_3} + \frac{\partial}{\partial x_2}\left(\mu\frac{\partial u_3}{\partial x_2} - \rho\overline{u_3' u_2'}\right) \tag{5-127}$$

energy:

$$\frac{\rho u_1}{h_1}\frac{\partial H}{\partial x_1} + \rho\tilde{u}_2\frac{\partial H}{\partial x_2} + \frac{\rho u_3}{h_3}\frac{\partial H}{\partial x_3} = \frac{\partial}{\partial x_2}\left[\frac{\mu}{\mathrm{Pr}}\frac{\partial H}{\partial x_2} - \rho c_p\overline{u_2' T'} + \mu\left(1 - \frac{1}{\mathrm{Pr}}\right)\left(u_1\frac{\partial u_1}{\partial x_2} + u_3\frac{\partial u_3}{\partial x_2}\right) - \rho u_1\overline{u_2' u_1'} - \rho u_3\overline{u_2' u_3'}\right] \tag{5-128}$$

As always, an equation of state, $\rho = \rho(p,T)$ is needed to close the system of equations for a compressible flow. The above equations remain valid for a laminar flow when the fluctuating quantities are set equal to zero.

5-4 INTRODUCTION TO TURBULENCE MODELING

5-4.1 Background

The need for turbulence modeling was pointed out in Section 5-2. In order to predict turbulent flows by finite-difference solutions to the Reynolds equations, it becomes necessary to make closing assumptions about the apparent turbulent stress and heat-flux quantities. All presently known turbulence models have limitations; the ultimate turbulence model has yet to be developed. Some argue philosophically that we have a system of equations for turbulent flows which is both accurate and general in the Navier-Stokes set, and therefore, to hope to develop an alternative system having the same accuracy and generality (but being simpler to solve) through turbulence modeling is being overly optimistic. If this premise is accepted, then our expectations in turbulence modeling are reduced from seeking the ultimate to seeking models which have reasonable accuracy over a limited range of flow conditions.

It is important to remember that turbulence models must be verified by comparing predictions with experimental measurements. Care must be taken in interpreting predictions of models outside the range of conditions over which they have been verified by comparisons with experimental data.

The purpose of this section is to introduce the methodology commonly used in turbulence modeling. The intent is not to present all models in sufficient detail that they can be used without consulting the original references, but rather to outline the rationale for the evolution of modeling strategy. Simpler models will be described in sufficient detail to enable the reader to formulate a "base-line" model applicable to simple thin shear layers.

5-4.2 Modeling Terminology

Boussinesq (1877) suggested, more than one hundred years ago, that the apparent turbulent shearing stresses might be related to the rate of mean strain through an apparent scalar turbulent or "eddy" viscosity. For the general Reynolds stress tensor, the Boussinesq assumption gives

$$-\overline{\rho u_i' u_j'} = \mu_T \left(\frac{\partial u_i}{\partial x_j} + \frac{\partial u_j}{\partial x_i} \right) - \frac{2}{3} \delta_{ij} \left(\mu_T \frac{\partial u_k}{\partial x_k} + \rho \bar{k} \right) \tag{5-129}$$

where μ_T is the turbulent viscosity and $\bar{k}$ is the kinetic energy of turbulence, $\bar{k} = \overline{u_i' u_i'}/2$. Following the convention introduced in Section 5-3.2, we are omitting bars over the time-mean variables.

By analogy with kinetic theory by which the molecular viscosity for gases can be evaluated with reasonable accuracy, we might expect that the turbulent viscosity can be modeled as

$$\mu_T = \rho v_T l \tag{5-130}$$

where v_T and l are characteristic velocity and length scales of the turbulence, respectively. The problem, of course, is to find suitable means for evaluating v_T and l.

Turbulence models to close the Reynolds equations can be divided into two categories according to whether or not the Boussinesq assumption is used. Models using the Boussinesq assumption will be referred to as Category I or turbulent viscosity models. Most models currently employed in engineering calculations are of this type. Experimental evidence indicates that the turbulent viscosity hypothesis is a valid one in many flow circumstances. There are exceptions, however, and there is no physical requirement that it hold. Models which affect closure to the Reynolds equations without this assumption will be referred to as Category II models and include those known as Reynolds stress or stress-equation models.

The other common classification of models is according to the number of supplementary partial differential equations which must be solved in order to supply the modeling parameters. This number ranges from zero for the simplest algebraic models to twelve for the most complex of the Reynolds stress models (Donaldson and Rosenbaum, 1968). Reference is also sometimes made to the "order" of the closure. According to this terminology, a first-order closure evaluates the Reynolds stresses through functions of the mean velocity and geometry alone. A second-order closure employs a solution to a modeled form of a transport partial differential equation for one or more of the characteristics of turbulence.

Category III models will be defined as those that are not based entirely on the Reynolds equations. A promising computational approach known as "large eddy simulation" falls into this category. In this approach (Deardorff, 1970), an attempt is made to resolve the large scale turbulent motion from first principles by numerically solving a "filtered" set of equations governing this large scale, three-dimensional, time-dependent motion. Turbulence modeling is employed to approximate the effects of the "subgrid" scale turbulence. Such calculations have shown much promise, but the technique is much too costly at present to be considered as an engineering tool.

As we turn to examples of specific turbulence models, it will be helpful to keep in mind an example set of conservation equations for which turbulence modeling is needed. The thin-shear-layer equations, Eqs. (5-116)–(5-119), will serve this purpose reasonably well. In the two-dimensional or axisymmetric thin-shear-layer equations, the modeling task reduces to finding expressions for $-\rho\overline{v'u'}$ and $\rho c_p\overline{v'T'}$. We will give the highest priority to ways of treating these two terms and indicate extensions to more complex equations where possible.

5-4.3 Simple Algebraic or Zero-Equation Models

Algebraic turbulence models invariably utilize the Boussinesq assumption. One of the most successful of this type of model was suggested by Prandtl in the 1920's,

$$\mu_T = \rho l^2 \left|\frac{\partial u}{\partial y}\right| \tag{5-131}$$

where l, a "mixing length," can be thought of as a transverse distance over which particles maintain their original momentum, somewhat on the order of a mean free path for the collision or mixing of globules of fluid. The product $l|\partial u/\partial y|$ can be interpreted as the characteristic velocity of turbulence, v_T. In Eq. (5-131) u is the component of velocity in the primary flow direction and y is the coordinate transverse to the primary flow direction.

For three-dimensional thin shear layers, Prandtl's formula is usually interpreted as

$$\mu_T = \rho l^2 \left[\left(\frac{\partial u}{\partial y}\right)^2 + \left(\frac{\partial w}{\partial y}\right)^2\right]^{1/2} \tag{5-131a}$$

This formula treats the turbulent viscosity as a scalar and gives qualitatively correct trends, especially near the wall. There is increasing experimental evidence, however, that in the outer layer, the turbulent viscosity should be treated as a tensor (i.e., dependent upon the direction of strain) in order to provide the best agreement with measurements. For flows in corners or in other geometries where a single "transverse" direction is not clearly defined, Prandtl's formula must be modified further (see, for example, Patankar et al., 1979).

The evaluation of l in the mixing-length model varies with the type of flow being considered, wall boundary layer, jet, wake, etc. For flow along a solid surface (internal or external flow), good results are observed by evaluating l according to

$$l_i = \kappa y(1 - e^{-y^+/A^+}) \tag{5-132}$$

in the inner region closest to the solid boundaries and switching to

$$l_0 = C_1 \delta \tag{5-133}$$

when l_i predicted by Eq. (5-132) first exceeds l_0. The constant C_1 in Eq. (5-133) is usually assigned a value close to 0.089, and δ is the velocity boundary-layer thickness.

In Eq. (5-132) κ is the von Kármán constant usually taken as 0.41 and A^+ is the damping constant most commonly evaluated as 26. The quantity in parentheses is the van Driest damping function (van Driest, 1956) and is the most common expression used to bridge the gap between the fully turbulent region where $l = \kappa y$ and the viscous sublayer where $l \to 0$. The parameter y^+ is defined as

$$y^+ = \frac{y(|\tau_w|/\rho_w)^{1/2}}{\nu_w}$$

Numerous variations on the exponential function have been utilized in order to account for effects of property variations, pressure gradients, blowing, and surface roughness. A discussion of modifications to account for several of these effects can be found in Cebeci and Smith (1974). It appears reasonably clear from comparisons in the literature, however, that the inner layer model as stated [Eq. (5-132)] requires no modification to accurately predict the variable property flow of gases with moderate pressure gradients on smooth surfaces.

The expression for l_i, Eq. (5-132), is responsible for producing the inner, "law-of-the-wall" region of the turbulent flow and l_0 [Eq. (5-133)] produces the outer "wake-like" region. These two zones are indicated in Fig. 5-7 which depicts a typical velocity distribution for an incompressible turbulent boundary layer on a smooth impermeable plate using "law-of-the-wall" coordinates. Re_θ is the Reynolds number based on

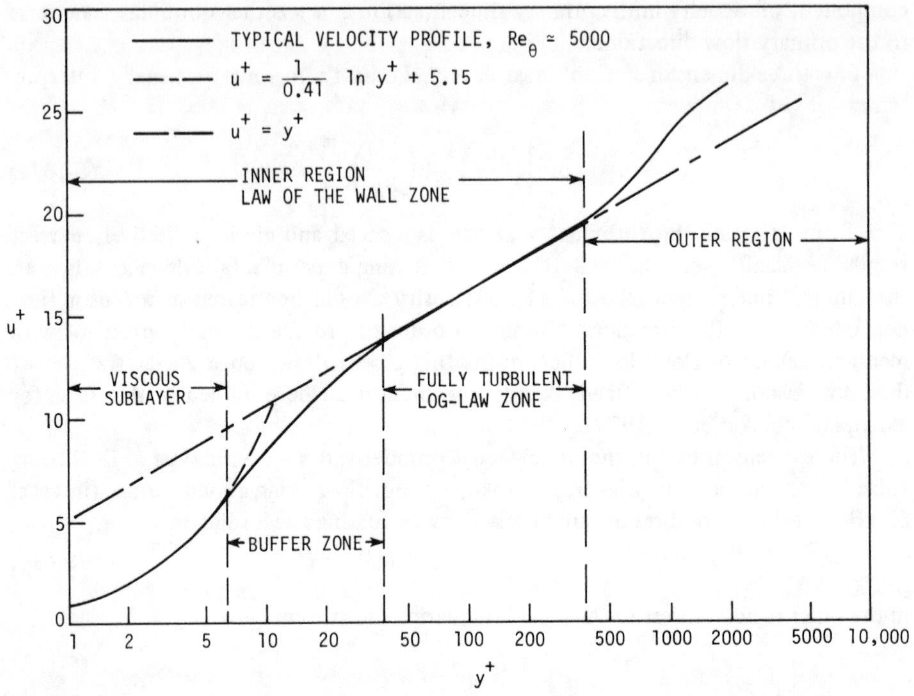

Figure 5-7 Zones in the turbulent boundary layer for a typical incompressible flow over a smooth flat plate.

momentum thickness, $\rho_e u_e \theta/\mu_e$, where for two-dimensional flow the momentum thickness is defined as

$$\theta = \int_0^\infty \frac{\rho u}{\rho_e u_e}\left(1 - \frac{u}{u_e}\right) dy$$

The nondimensional velocity u^+ is defined as $u^+ = u/(|\tau_w|/\rho_w)^{1/2}$. The inner and outer regions are indicated on the figure. Under normal conditions, the inner law-of-the-wall zone only includes about 20% of the boundary layer. The log-linear zone is the characteristic "signature" of a turbulent wall boundary layer although the law-of-the-wall plot changes somewhat in general appearance as the Reynolds and Mach numbers are varied.

It is worth noting that at low momentum thickness Reynolds numbers, i.e., relatively near the origin of the turbulent boundary layer, both inner and outer regions are tending toward zero and problems might be expected with the two region turbulence model employing Eqs. (5-132) and (5-133). The difficulty occurs because the smaller δ's occurring near the origin of the turbulent boundary layer are causing the switch to the outer model to occur before the wall damping effect has permitted the fully turbulent law-of-the-wall zone to develop. This causes the finite-difference scheme using such a model to underpredict the wall shear stress. The discrepancy is nearly negligible for incompressible flow but the effect is more serious for compressible flows, persisting at higher and higher Reynolds numbers as the Mach number increases due to the relative thickening of the viscous sublayer from thermal effects (Pletcher, 1976). Naturally, details of the effect are influenced by the level of wall cooling present in the compressible flow.

Predictions can be brought into good agreement with measurements at low Reynolds numbers by simply delaying the switch from the inner model, Eq. (5-132), to the outer, Eq. (5-133), until $y^+ \geqslant 50$. If, at $y^+ = 50$ in the flow, $l/\delta \leqslant 0.089$ then no adjustment is necessary. On the other hand if Eq. (5-132) predicts $l/\delta > 0.089$, then the mixing length becomes constant in the outer region at the value computed at $y^+ = 50$ by Eq. (5-132). This simple adjustment ensures the existence of the log-linear region in the flow which is in agreement with the preponderance of measurements.

Other modeling procedures have been used successfully for the inner and outer regions. Some workers advocate the use of wall functions based on a Couette flow assumption (Patankar and Spalding, 1970) in the near-wall region. This approach probably has not been quite as well-refined to account for variable properties, transpiration and other near-wall effects as the van Driest function.

An alternative treatment to Eq. (5-133) is often used to evaluate the turbulent viscosity in the outer region (Cebeci and Smith, 1974). This follows the Clauser formulation,

$$\mu_{T(\text{outer})} = \alpha \rho u_e |\delta_k^*| \tag{5-134}$$

where α accounts for low Reynolds number effects. Cebeci and Smith (1974) recommend

$$\alpha = 0.0168 \frac{(1.55)}{1 + \pi} \tag{5-135}$$

where $\pi = 0.55[1 - \exp(-0.243z^{1/2} - 0.298z)]$ and $z = \text{Re}_\theta/425 - 1$. For Re_θ greater than 5000, $\alpha \cong 0.0168$. The parameter δ_k^* is the kinematic displacement thickness defined as

$$\delta_k^* = \int_0^\infty \left(1 - \frac{u}{u_e}\right) dy \tag{5-136}$$

Closure for the Reynolds heat-flux term, $\rho c_p \overline{v'T'}$, is usually handled in algebraic models by a form of the Reynolds analogy which is based on the similarity between the transport of heat and momentum. The Reynolds analogy is applied to the apparent turbulent conductivity in the assumed Boussinesq form

$$\rho c_p \overline{v'T'} = -k_T \frac{\partial T}{\partial y}$$

In turbulent flow, this additional transport of heat is caused by the turbulent motion. Experiments confirm that the ratio of the diffusivities for the turbulent transport of heat and momentum, called the turbulent Prandtl number, $\text{Pr}_T = \mu_T c_p / k_T$, is a well-behaved function across the flow. Most algebraic turbulence models do well by letting the turbulent Prandtl number be a constant near one; most commonly, $\mathrm{Pr}_T = 0.9$. Experiments indicate that for wall shear flows Pr_T varies somewhat between about 0.6 ~ 0.7 at the outer edge of the boundary layer to about 1.5 near the wall, although the evidence is not conclusive. Several semi-empirical distributions for Pr_T have been proposed; a sampling is found in Cebeci and Smith (1974), Kays (1972), and Reynolds (1975). Using the turbulent Prandtl number, the apparent turbulent heat flux is related to the turbulent viscosity and mean flow variables as

$$-\rho c_p \overline{v'T'} = \frac{c_p \mu_T}{\mathrm{Pr}_T} \frac{\partial T}{\partial y} \tag{5-137}$$

and closure has been completed.

For other than thin shear flows, it may be necessary to model other Reynolds heat-flux terms. To do so, the turbulent conductivity, $k_T = c_p \mu_T / \mathrm{Pr}_T$, is normally considered as a scalar and the Boussinesq-type approximation is extended to other components of the temperature gradient. As an example, we would evaluate $-\rho c_p \overline{u'T'}$ as

$$-\rho c_p \overline{u'T'} = \frac{c_p \mu_T}{\mathrm{Pr}_T} \frac{\partial T}{\partial x}$$

To summarize, a recommended base-line algebraic model for wall boundary layers consists of evaluating the turbulent viscosity by Prandtl's mixing-length formula, Eq. (5-131), where l is given by Eq. (5-132) for the inner region and then using Eq. (5-133) with Eq. (5-131) for the outer region. Alternatively, the Clauser formulation, Eq. (5-134) could be used in the outer region. The apparent turbulent heat flux can be evaluated through Eq. (5-137) using a turbulent Prandtl number of 0.9. This simplest form of modeling has employed four empirical, adjustable constants as given in Table 5-1.

Algebraic models have accumulated an impressive record of good performance for relatively simple viscous flows but need to be modified in order to accurately predict flows with "complicating" features. It should be noted that compressible flows do not represent a "complication" in general. The turbulence structure of the flow

Table 5-1 Empirical constants employed in algebraic turbulence models for wall boundary layers

Symbol	Description
κ	von Kármán constant, used for inner layer ≈ 0.41
A^+	van Driest constant for damping function ≈ 26 but frequently modified to account for complicating effects
C_1 or α	Constant for outer region model $C_1 \approx 0.089$, $\alpha \approx 0.0168$ but usually includes $f(\mathrm{Re}_\theta)$ in α
Pr_T	Turbulent Prandtl number, $\mathrm{Pr}_T \approx 0.9$ is most common

Table 5-2 Effects requiring alterations or additions to simplest form of algebraic turbulence models

Effect	References
Low Reynolds number	Cebeci and Smith (1974), Pletcher (1976), Bushnell et al. (1975), Herring and Mellor (1968), Bushnell et al. (1976), McDonald (1970)
Roughness	Cebeci and Smith (1974), Bushnell et al. (1976), McDonald and Fish (1973), Healzer et al. (1974), Adams and Hodge (1977)
Transpiration	Cebeci and Smith (1974), Bushnell et al. (1976), Pletcher (1974), Baker and Launder (1974), Kays and Moffat (1975)
Strong pressure gradients	Cebeci and Smith (1974), Bushnell et al. (1976), Adams and Hodge (1977), Pletcher (1974), Baker and Launder (1974), Kays and Moffat (1975), Jones and Launder (1972), Kreskovsky et al. (1974), Horstman (1977)
Merging shear layers	Bradshaw et al. (1973), Stephenson (1976), Emery and Gessner (1976), Cebeci and Chang (1978), Malik and Pletcher (1978)

appears to remain essentially unchanged for Mach numbers up through at least 5. Naturally, the variation of density and other properties must be accounted for in the form of the conservation equations used with the turbulence model. Table 5-2 lists several flow conditions requiring alterations or extensions to the simplest form of algebraic models cited above. Some key references are also tabulated where such model modifications are discussed.

The above discussion of algebraic models for wall boundary layers is by no means complete. Over the years dozens of slightly different algebraic models have been suggested. Eleven algebraic models were compared in a study by McEligot et al. (1970) for turbulent pipe flow with heat transfer. None were found superior to the van Driest-damped mixing-length model presented here.

Somewhat less information is available in the literature on algebraic turbulence models for free shear flows. This category of flows has historically been more difficult to model than wall boundary layers, especially if model generality is included as a measure of merit. Some discussion of the status of the simple models for round jets can be found in the works of Madni and Pletcher (1975*b*, 1977*a*). The initial mixing region of round jets can be predicted fairly well using Prandtl's mixing-length formulation, Eq. (5-131), with

$$l = 0.0762\delta_m \tag{5-138}$$

where δ_m is the width of the mixing zone. This model does not perform well after the shear layers have merged and a switch at that point to models of the form (Hwang and Pletcher, 1978)

$$\nu_T = \gamma F y_{1/2}(u_{\max} - u_{\min}) \tag{5-139}$$

or (Madni and Pletcher, 1975*b*)

$$\upsilon_T = \frac{2F}{a} \int_y^\infty |u_e - u| y\, dy \tag{5-140}$$

has provided good agreement with measurements for round co-flowing jets. Equation (5-139) is a modification of the model suggested for jets by Prandtl (1926). In the above, a is the jet discharge radius and γ is an intermittency function

$$\gamma = 1 \qquad 0 \leqslant \frac{y}{y_{1/2}} \leqslant 0.8$$

$$\gamma = (0.5)^z \qquad \frac{y}{y_{1/2}} > 0.8 \qquad \text{where} \qquad z = \left(\frac{y}{y_{1/2}} - 0.8\right)^{2.5} \tag{5-141}$$

F is a function of the ratio (R) of the stream velocity to the jet discharge velocity given by $F = 0.015(1 + 2.13R^2)$. The distance y is measured from the jet centerline and $y_{1/2}$ is the "velocity half width," the distance from the centerline to the point at which the velocity has decreased to the average of the centerline and external stream velocities.

Philosophically, the strongest motivation for turning to more complex models is the observation that the algebraic model evaluates the turbulent viscosity only in terms of *local* flow parameters, yet we feel that a turbulence model ought to provide a mechanism by which effects upstream can influence the turbulence structure (and viscosity) downstream. Further, with the simplest models, ad hoc additions and corrections are frequently required to handle specific effects, and constants need to be changed to handle different classes of shear flows. To many investigators, it is appealing to develop a model general enough that specific modifications to the constants are not required to treat different classes of flows.

If we accept the general form for the turbulent viscosity, $\mu_T = \rho \upsilon_T l$, then a logical way to extend the generality of turbulent viscosity models is to permit υ_T and perhaps l to be more complex (and thus more general) functions of the flow capable of being influenced by upstream (historic) effects. This rationale serves to motivate several of the more complex turbulence models.

5-4.4 One-Half Equation Models

A *one-half equation model* will be defined as one in which the value of one model parameter (υ_T, l, or μ_T itself) is permitted to vary with the primary flow direction in a manner determined by the solution to an *ordinary* differential equation (ODE). The ordinary differential equation usually results from either neglecting or assuming the variation of the model parameter with one coordinate direction. Extended mixing-length models and relaxation models fall into this category. A *one-equation model* is one in which an additional *partial* differential equation is solved for a model parameter. The main features of several one-half equation models are tabulated in Table 5-3.

The first three models in Table 5-3 differ in detail although all three utilize an integral form of a transport equation for turbulence kinetic energy as a basis for letting

Table 5-3 Some one-half equation models

Model	Transport equation used as basis for ODE	Model parameter determined by ODE solution	References
5A	Turbulence kinetic energy	l_∞	McDonald and Camerata (1968), Kreskovsky et al. (1974), McDonald and Kreskovsky (1974)
5B	Turbulence kinetic energy	l_∞	Chan (1972)
5C	Turbulence kinetic energy	l_∞	Adams and Hodge (1977)
5D	Empirical ODE for $\mu_{T(\text{outer})}$	$\mu_{T(\text{outer})}$	Shang and Hankey (1975)
5E	Empirical ODE for $\mu_{T(\text{outer})}$	$\mu_{T(\text{outer})}$	Reyhner (1968)
5F	Empirical ODE for l_∞	l_∞	Malik and Pletcher (1978), Pletcher (1978)

flow history influence the turbulent viscosity. Models of this type have been refined to allow prediction of transition, roughness effects, transpiration, pressure gradients, and qualitative features of relaminarization. Most of the test cases reported for the models have involved external rather than channel flows.

Although models 5D, 5E, and 5F appear to be purely empirical relaxation or lag models, Birch (1976) shows that models of this type are actually equivalent to one-dimensional versions of transport partial differential equations for the quantities concerned except that these transport equations are not generally derivable from the Navier-Stokes equations. This is no serious drawback since transport equations cannot be solved without considerable empirical simplification and modeling of terms. In the end, these transport equations tend to have similar forms characterized by generation, dissipation, diffusion, and convection terms regardless of the origin of the equation.

To illustrate, for wall boundary layers, model 5F utilizes the expression for mixing length given by Eq. (5-132) for the inner region. In the outer part of the flow, the mixing length is calculated according to

$$l_0 = 0.12L \tag{5-142}$$

where L is determined by the solution to an ordinary differential equation. For a shear layer of constant width, the natural value of L is presumed to be the thickness, δ, of the shear layer. When δ is changing with the streamwise direction, x, L will lag δ in a manner controlled by the relaxation time for the large eddy structure, which is assumed to be equal to $\delta/\bar{u}_\tau$ where $\bar{u}_\tau$ is a characteristic turbulence velocity. If it is further assumed that the fluid in the outer part of the shear layer travels at velocity u_e, then the streamwise distance traversed by the flow during the relaxation time is $L^* = C_2 u_e \delta/\bar{u}_\tau$. A rate equation can be developed by assuming that L will tend toward δ according to

$$\frac{dL}{dx} = \frac{\delta - L}{L^*} \tag{5-143}$$

This model has been extended to free shear flows (Minaie and Pletcher, 1982) by interpreting δ as the distance between the location of the maximum shear stress and the outer edge of the shear flow and replacing u_e by the average streamwise velocity over the shear layer. The optimum evaluation for $\bar{u}_\tau$ appears unsettled. The expression $\bar{u}_\tau = (L/\delta)(|\tau_w|/\rho_w)^{1/2}$ has been employed successfully for flows along solid surfaces, whereas $\bar{u}_\tau = (\tau_{\max}/\rho_w)^{1/2}$ has proven satisfactory for free shear flows predicted to date. It might be speculated that this latter evaluation would work reasonably well for wall boundary layers also. The final form of the transport ODE for L used for separating wall boundary layers (Pletcher, 1978) and merging shear layers in annular passages (Malik and Pletcher, 1978) can be written as

$$u_e \frac{dL}{dx} = 1.25 \left| \frac{\tau_w}{\rho_w} \right|^{1/2} \left[\frac{L}{\delta} - \left(\frac{L}{\delta} \right)^2 \right] \tag{5-144}$$

5-4.5 One-Equation Models

One obvious shortcoming of algebraic viscosity models which normally evaluate v_T in the expression $\mu_T = \rho v_T l$ by $v_T = l|\partial u/\partial y|$ is that $\mu_T = k_T = 0$ whenever $\partial u/\partial y = 0$. This would suggest that μ_T and k_T would be zero at the centerline of a pipe, in regions near the mixing of a wall jet with a mainstream and in flow through an annulus or between parallel plates where one wall is heated and the other cooled. Measurements (and common sense) indicate that μ_T and k_T are *not* zero under all conditions whenever $\partial u/\partial y = 0$. The mixing-length models can be "fixed-up" to overcome this deficiency but this conceptual shortcoming provides motivation for considering other interpretations for μ_T and k_T. In fairness to the algebraic models we should mention that this defect is not always crucial because Reynolds stresses and heat fluxes are frequently small when $\partial u/\partial y = 0$. For some examples illustrating this point, see Malik and Pletcher (1981).

It was the suggestion of Prandtl and Kolmogorov in the 1940's to let v_T in $\mu_T = \rho v_T l$ be proportional to the square root of the kinetic energy of turbulence, $\bar{k} = \frac{1}{2}\overline{u_i' u_i'}$. Thus the turbulent viscosity can be evaluated as

$$\mu_T = C_k \rho l (\bar{k})^{1/2} \tag{5-145}$$

and μ_T no longer becomes equal to zero when $\partial u/\partial y = 0$. The kinetic energy of turbulence is a measurable quantity and is easily interpreted physically. We naturally inquire how we might predict $\bar{k}$.

A transport partial differential equation can be developed (Problem 5-18) for $\bar{k}$ from the Navier-Stokes equations. For incompressible two-dimensional thin-shear-layer flows, the equation takes the form

$$\rho \frac{D\bar{k}}{Dt} = \mu \frac{\partial^2 \bar{k}}{\partial y^2} - \frac{\partial}{\partial y} (\rho \overline{v'k'} + \overline{v'p'}) - \rho \overline{v'u'} \frac{\partial u}{\partial y}$$
$$- \mu \left[\overline{\left(\frac{\partial u'}{\partial y} \right)^2} + \overline{\left(\frac{\partial v'}{\partial y} \right)^2} + \overline{\left(\frac{\partial w'}{\partial y} \right)^2} \right] \tag{5-146}$$

which is commonly modeled as

$$\underbrace{\rho \frac{D\bar{k}}{Dt}}_{\substack{\text{Particle rate}\\ \text{of increase}\\ \text{of } \bar{k}}} = \underbrace{\frac{\partial}{\partial y}\left[\left(\mu + \frac{\mu_T}{\text{Pr}_k}\right)\frac{\partial \bar{k}}{\partial y}\right]}_{\substack{\text{Diffusion rate}\\ \text{for } \bar{k}}} + \underbrace{\mu_T \left(\frac{\partial u}{\partial y}\right)^2}_{\substack{\text{Generation}\\ \text{rate for } \bar{k}}} - \underbrace{\frac{C_D \rho (\bar{k})^{3/2}}{l}}_{\substack{\text{Dissipation}\\ \text{rate for } \bar{k}}} \tag{5-147}$$

The physical interpretation of the various terms is indicated above for Eq. (5-147). This modeled transport equation is then added to the system of PDE's to be solved for the problem at hand. Note that a length parameter, l, needs to be specified algebraically. In the above, Pr_k is a Prandtl number for turbulence kinetic energy (≈ 1.0) and $C_D \simeq 0.164$ if l is taken as the ordinary mixing length.

The above modeling for the $\bar{k}$ transport equation is only valid in the fully turbulent regime, i.e., away from any wall damping effects. For typical wall flows this means for y^+ greater than about 30. Inner boundary conditions for the $\bar{k}$ equation are often supplied through the use of wall functions (Launder and Spalding, 1974). Another way of treating the inner boundary condition for $\bar{k}$ is to make use of the experimental observation that very near the wall convection and diffusion of $\bar{k}$ are usually negligible. Thus, generation and dissipation of $\bar{k}$ are in balance and it can be shown (Problem 5-19) that the turbulence kinetic energy model reduces to Prandtl's mixing-length formulation, Eq. (5-131) under these conditions. At the location where the diffusion and convection are first neglected, we can establish (Problem 5-22) an inner boundary condition for $\bar{k}$ as

$$\bar{k}(x, y_c) = \frac{\tau(y_c)}{\rho C_D^{2/3}} \tag{5-148}$$

where y_c is a point within the region where the logarithmic law of the wall is expected to be valid. For $y < y_c$ the Prandtl-type algebraic inner region model [Eqs. (5-131) and (5-132)] can be used. Further details on one-equation models for incompressible flow can be found in Launder and Spalding (1972).

Recently, the one-equation model has been extended to compressible flows (Rubesin, 1976) and the results to date appear encouraging. Apparently for flows containing shock wave interactions which greatly affect the stream turbulence level, the predictions of Rubesin's one-equation model provide a definite improvement over those from algebraic models. On the whole, however, the performance of most one-equation models (for both incompressible and compressible flows) has been disappointing in that relatively few cases have been observed in which these models offer an improvement over the predictions of the algebraic models. In fact, several flows can be predicted more accurately by the one-half equation models than by the representative one-equation model of the Prandtl-Kolmogorov type which merely alters the velocity of turbulence used in the viscosity expression. The reason for this may be that in most flows, an improvement in the specification of a characteristic length scale, l, will have more effect than a change in the velocity of turbulence, v_T, and many of the one-half equation models listed in Table 5-3 offer an improvement in this length scale.

Other one-equation models have been suggested which deviate somewhat from the Prandtl-Kolmogorov pattern. The most notable of these is by Bradshaw et al. (1967). The turbulence energy equation is used in the Bradshaw model but the modeling is different in both the momentum equation, where the turbulent shearing stress is assumed proportional to $\bar{k}$, and in the turbulence energy equation. The details will not be given here, but an interesting feature of the Bradshaw method is that as a consequence of the form of modeling used for the turbulent transport terms, the system of equations becomes hyperbolic and can be solved by a procedure similar to the method of characteristics. The Bradshaw method has enjoyed good success in the prediction of the wall boundary layers. Even so, the predictions have not been notably superior to those of the algebraic models, one-half equation models, or other one-equation models.

5-4.6 One and One-Half and Two-Equation Models

One conceptual advance made by moving from a purely algebraic mixing-length model to a one-equation model was that the latter permitted one model parameter to vary throughout the flow, being governed by a PDE of its own. In the one-equation models, a length parameter still appears which is generally evaluated by an algebraic expression dependent upon only *local* flow parameters. Researchers in turbulent flow have long felt that the length scale in turbulence models should also depend upon the upstream "history" of the flow and not just local flow conditions. An obvious way to provide more complex dependence of l on the flow is to derive a transport equation for the variation of l. If the equation for l added to the system is an ordinary differential equation, such as given by Eq. (5-144) for model 5F, the resulting model might well be termed a *one and one-half equation model.* Such a model has been employed to predict separating external turbulent boundary layers (Pletcher, 1978), flow in annular passages with heat transfer (Malik and Pletcher, 1981), and plane and round jets (Minaie and Pletcher, 1982).

Frequently, the equation from which the length scale is obtained is a partial differential equation and the model is then referred to as a *two-equation turbulence model.*

Although a transport PDE can be developed for a length scale, the terms in this equation are not easily modeled, and some workers have experienced better success by solving a transport equation for a length scale related parameter rather than the length scale itself. This point is discussed by Launder and Spalding (1974).

One of the most frequently used two-equation models is the $\bar{k}$-ϵ model first proposed by Harlow and Nakayama (1968). The description here follows the papers of Jones and Launder (1972) and Launder and Spalding (1974). The parameter ϵ is a turbulence dissipation rate and is assumed to be related to other model parameters through $\epsilon = C(\bar{k})^{3/2}/l_\epsilon$ where l_ϵ is referred to as the dissipation length and C is a constant. The turbulent viscosity is related to ϵ through

$$\mu_T = \frac{C_\mu \rho(\bar{k})^2}{\epsilon} = c'\rho(\bar{k})^{1/2} l_\epsilon \tag{5-149}$$

Considerable experience has been accumulated by several investigators with this model but primarily for flows in which property variations were small.

Potential users of the model should study the literature carefully before doing so, but a representative form of the model is outlined here to demonstrate the spirit in which the two-equation models proceed. The turbulent viscosity is evaluated as given by Eq. (5-149). The turbulence kinetic energy is obtained by solving a transport equation of the form given by Eq. (5-147) where the last term is recognized as the density times the dissipation rate, $\rho\epsilon$. A parabolic transport equation for ϵ (the form given here has been used for incompressible flow) is added to close the system:

$$\rho \frac{D\epsilon}{Dt} = \frac{\partial}{\partial y}\left(\frac{\mu_T}{\mathrm{Pr}_\epsilon}\frac{\partial \epsilon}{\partial y}\right) + \frac{C_2 \mu_T \epsilon}{\bar{k}}\left(\frac{\partial u}{\partial y}\right)^2 - \frac{C_3 \rho \epsilon^2}{\bar{k}} \tag{5-150}$$

The terms on the right-hand side of Eq. (5-150) from left to right can be interpreted as the diffusion, generation and dissipation rates of ϵ. Typical values of the model constants are tabulated below in Table 5-4.

Numerous other two-equation models have been suggested, the most frequently used being the Ng-Spalding (1972) model and the Wilcox-Traci model (1976), the latter being a modification to the earlier Saffman-Wilcox model (1974). Rubesin (1977) summarizes the main differences between the models while Chambers and Wilcox (1976) explore the similarities and differences in more detail. All of these models employ a modeled form of the turbulence kinetic energy equation but the modeling for the gradient diffusion term is different. The most striking difference, however, is in the choice of dependent variable for the second model transport equation from which the length scale is determined.

Rubesin (1977) shows several comparisons between these models for incompressible flow, and overall they perform quite well; it is difficult to identify the best from the comparisons he has shown.

The above form of the transport equation for ϵ is not appropriate for the near-wall region, i.e., the viscous sublayer. This is just as noted for the turbulence kinetic energy equation [Eq. (5-147)] presented earlier. Inner boundary conditions for ϵ can be provided at the same point, y_c, used for imposing boundary conditions on $\bar{k}$ [see Eq. (5-148)]. At the point y_c, Prandtl's mixing-length formulation is assumed to be valid and

$$\epsilon = \frac{C_D \bar{k}^{3/2}}{l} = \frac{C_D[\bar{k}(y_c)]^{3/2}}{\kappa y}$$

The quantity $\bar{k}(y_c)$ can be evaluated as indicated in Eq. (5-148). Most applications of the $\bar{k}$-ϵ model have made use of wall functions (see Launder and Spalding, 1974) to treat the near-wall region. Alternatively, additional terms have been added to the $\bar{k}$ and

Table 5-4 Model constants for $\bar{k}$-ϵ two-equation model

C_μ	C_2	C_3	$\mathrm{Pr}_{,k}$	$\mathrm{Pr}_{,\epsilon}$	$\mathrm{Pr}_{,T}$
0.09	1.44	1.92	1.0	1.3	0.9

ϵ equations to extend their applicability to the viscous sublayer by Jones and Launder (1972), Wolfstein (1969), and others. In this connection, the viscous sublayer is often referred to as the region of low turbulence Reynolds number $[(\bar{k})^{1/2} l_\epsilon/\nu]$. This inner modeling is crucial for complex turbulent flows as for example those containing separated regions or severe property variations. The uncertainty of such inner region modeling for complex flows appears to limit the range of applicability of the $\bar{k}$-ϵ model (and nearly all other models) at the present time.

Additional modifications to the $\bar{k}$-ϵ model have been suggested to account for the effects of buoyancy and streamline curvature on the turbulence structure. The most common $\bar{k}$-ϵ closure for the Reynolds heat-flux terms utilizes the same turbulent Prandtl number formulation as used with algebraic models, Eq. (5-137).

Despite the enthusiasm which is noted from time to time over two-equation models, it is perhaps appropriate to point out again the two major restrictions on this type of model. First, two-equation models of the type discussed herein are merely turbulent *viscosity* models which assume that the Boussinesq approximation [Eq. (5-129)] holds. In algebraic models, μ_T is a local function whereas in two-equation models μ_T is a more general and complex function governed by two additional PDE's. If the Boussinesq approximation fails, then even two-equation models fail. Obviously, in many flows the Boussinesq approximation models reality closely enough for engineering purposes.

The second shortcoming of two-equation models is the need to make assumptions in evaluating the various terms in the model transport equations, especially in evaluating the third-order turbulent correlations. This same shortcoming, however, plagues all higher-order closure attempts. These model equations contain no magic; they only reflect the best understanding and intuition of the originators. We can be optimistic, however, that the models can be improved by improved modeling of these terms.

5-4.7 Reynolds Stress Models

By Reynolds stress models (sometimes called stress-equation models) we are referring to those Category II models which *do not* assume that the turbulent shearing stress is proportional to the rate of mean strain. That is, for a two-dimensional incompressible flow

$$-\rho\overline{u'v'} \neq \mu_T\left(\frac{\partial u}{\partial y} + \frac{\partial v}{\partial x}\right)$$

These models have been used to date largely as tools or subjects in turbulence research rather than to solve engineering problems. Exact transport equations can be derived (Problem 5-17) for the Reynolds stresses. Naturally, these equations contain terms which must be modeled in keeping with the postulate that in turbulent flows, an equation can be derived for about anything, but none of them can be solved exactly. Such modeling, which generally follows the pioneering work of Rotta (1951), requires the solution of at least three additional transport PDE's. For a flow in which normal stresses are important, five additional equations are usually required. An example is the Reynolds stress model of Daly and Harlow (1970), which makes use of the dissipation rate, ϵ, requiring solutions to

1. $\rho \dfrac{D\overline{v'u'}}{Dt} = ---$

2. $\rho \dfrac{D\overline{u'^2}}{Dt} = ---$

3. $\rho \dfrac{D\overline{v'^2}}{Dt} = ---$

4. $\rho \dfrac{D\overline{w'^2}}{Dt} = ---$

5. $\rho \dfrac{D\epsilon}{Dt} = ---$

in addition to solutions of the usual conservation equations of mass, momentum, and energy. Evaluation of the right-hand sides of the above equations require numerous modeling assumptions.

The most widely known Reynolds stress models at the present time are probably those of Hanjalić and Launder (1972), Daly and Harlow (1970), and Donaldson (1972). The status and prospects of Reynolds stress closures as of 1979 have been described by Launder (1979).

Reynolds stress models are not restricted by the Boussinesq approximation relating turbulent stresses to rates of mean strain and contain the greatest number of model PDE's and constants of the models discussed. Thus, it would seem that these models ought to have the best chance of emerging as "ultimate" turbulence models if success is to be achieved at all through the time-averaged Navier-Stokes equations. Nevertheless, these models still must utilize approximations and assumptions in modeling terms which presently cannot be measured. These Reynolds stress models are perhaps still in their infancy and it may be some time yet before they have been tested and refined to the point that they become commonplace in engineering calculations. Since simpler models perform adequately for many flows, the expectation is that the Reynolds stress models may only be used in engineering predictions where the flow complexity demands it. At the present time, the Reynolds stress models have not even been tested for many types of complex flows.

5-5 EULER EQUATIONS

Prandtl discovered in 1904 (see Section 5-3.1) that for sufficiently large Reynolds numbers the important viscous effects are confined to a thin boundary layer near the surface of a solid boundary. As a consequence of this discovery, the inviscid (nonviscous, nonconducting) portion of the flowfield can be solved independently of the boundary layer. Of course this is only true if the boundary layer is very thin compared to the characteristic length of the flowfield so that the interaction between the boundary layer and the inviscid portion of the flowfield is negligible. For flows in which the interaction is not negligible, it is still possible to use separate sets of equations for the two regions but the equations must be solved in an iterative fashion. This iterative procedure can be computationally inefficient and as a result it is sometimes desirable to use a single set of equations which remain valid throughout the flowfield. Equations of this latter type will be discussed in Chapter 8.

In the present section, a reduced set of equations will be discussed which are valid only in the inviscid portion of the flowfield. These equations are obtained by dropping both the viscous terms and the heat transfer terms from the complete Navier-Stokes equations. The resulting equations can be numerically solved (see Chapter 6) using much less computer time than is required for the complete Navier-Stokes equations. We will refer to these simplified equations as the *Euler equations* although strictly speaking Euler's name should be attached only to the inviscid momentum equation. In addition to the assumption of inviscid flow, it will also be assumed that there is no external heat transfer so that the heat generation term $\partial Q/\partial t$ in the energy equation can be dropped.

5-5.1 Continuity Equation

The continuity equation does not contain viscous terms or heat transfer terms so that the various forms of the continuity equation given in Section 5-1.1 cannot be simplified for an inviscid flow. However, if the steady form of the continuity equation reduces to two terms for a given coordinate system it becomes possible to discard the continuity equation by introducing the so-called *stream function* ψ. This holds true whether the flow is viscous or nonviscous. For example, the continuity equation for a two-dimensional, steady, compressible flow in Cartesian coordinates is

$$\frac{\partial}{\partial x}(\rho u) + \frac{\partial}{\partial y}(\rho v) = 0 \tag{5-151}$$

If the stream function ψ is defined such that

$$\begin{aligned} \rho u &= \frac{\partial \psi}{\partial y} \\ \rho v &= -\frac{\partial \psi}{\partial x} \end{aligned} \tag{5-152}$$

it can be seen by substitution that Eq. (5-151) is satisfied. Hence, the continuity equation does not need to be solved and the number of dependent variables is reduced by one. The disadvantage is that the velocity derivatives in the remaining equations are replaced using Eqs. (5-152) so that these remaining equations will now contain derivatives which are one order higher. The physical significance of the stream function is obvious when we examine:

$$\begin{aligned} d\psi &= \frac{\partial \psi}{\partial x}\,dx + \frac{\partial \psi}{\partial y}\,dy = -\rho v\,dx + \rho u\,dy \\ &= \rho \mathbf{V}\cdot d\mathbf{A} = d\dot{m} \end{aligned} \tag{5-153}$$

We see that lines of constants ψ ($d\psi = 0$) are lines across which there is no mass flow ($d\dot{m} = 0$). A *streamline* is defined as a line in the flowfield whose tangent at any point is in the same direction as the flow at that point. Hence, lines of constant ψ are streamlines and the difference between the values of ψ for any two streamlines represents the mass flow rate per unit width between those streamlines.

For an incompressible, two-dimensional flow the continuity equation in Cartesian coordinates is

$$\frac{\partial u}{\partial x} + \frac{\partial v}{\partial y} = 0 \tag{5-154}$$

and the stream function is defined by

$$u = \frac{\partial \psi}{\partial y} \qquad v = -\frac{\partial \psi}{\partial x} \tag{5-155}$$

For a steady, axially symmetric, compressible flow in cylindrical coordinates (see Section 5-1.7), the continuity equation is given by

$$\frac{1}{r}\frac{\partial}{\partial r}(r\rho u_r) + \frac{\partial}{\partial z}(\rho u_z) = 0 \tag{5-156}$$

and the stream function is defined by

$$\rho u_r = \frac{1}{r}\frac{\partial \psi}{\partial z} \qquad \rho u_z = -\frac{1}{r}\frac{\partial \psi}{\partial r} \tag{5-157}$$

For the case of three-dimensional flows, it is possible to use two stream functions to replace the continuity equation. However, the complexity of this approach usually makes it less attractive than using the continuity equation in its original form.

5-5.2 Inviscid Momentum Equations

When the viscous terms are dropped from the Navier-Stokes equations [Eq. (5-18)], the following equation results

$$\rho \frac{D\mathbf{V}}{Dt} = \rho\mathbf{f} - \nabla p \tag{5-158}$$

This equation was first derived by Euler in 1755 and has been named Euler's equation. If we neglect body forces and assume steady flow, Euler's equation reduces to

$$\mathbf{V}\cdot\nabla\mathbf{V} = -\frac{1}{\rho}\nabla p \tag{5-159}$$

Integrating this equation along a line in the flowfield gives

$$\int (\mathbf{V}\cdot\nabla\mathbf{V})\cdot d\mathbf{r} = -\int \frac{1}{\rho}\,\nabla p\cdot d\mathbf{r} \tag{5-160}$$

where $d\mathbf{r}$ is the differential length of the line. For a Cartesian coordinate system, $d\mathbf{r}$ is defined by

$$d\mathbf{r} = dx\,\mathbf{i} + dy\,\mathbf{j} + dz\,\mathbf{k} \tag{5-161}$$

Let us assume that the line is a streamline. Hence, $\mathbf{V}$ has the same direction as $d\mathbf{r}$ and we can simplify the integrand on the left side of Eq. (5-160) in the following manner

$$(\mathbf{V}\cdot\nabla\mathbf{V})\cdot d\mathbf{r} = V\,\frac{\partial \mathbf{V}}{\partial r}\cdot d\mathbf{r} = V\,\frac{\partial V}{\partial r}\,dr = V\,dV = d\left(\frac{V^2}{2}\right)$$

Likewise, the integrand on the right-hand side becomes

$$\frac{1}{\rho}\,\nabla p\cdot d\mathbf{r} = \frac{dp}{\rho}$$

and Eq. (5-160) reduces to

$$\frac{V^2}{2} + \int \frac{dp}{\rho} = \text{constant} \tag{5-162}$$

The integral in this equation can be evaluated if the flow is assumed *barotropic.* A barotropic fluid is one in which ρ is a function only of p (or a constant), i.e., $\rho = \rho(p)$. Examples of barotropic flows are

1. steady incompressible flow

$$\rho = \text{constant} \tag{5-163}$$

2. isentropic (constant entropy) flow (see Section 5-5.4)

$$\rho = (\text{constant})\,p^{1/\gamma} \tag{5-164}$$

Thus for an incompressible flow, the integrated Euler's equation [Eq. (5-162)] becomes

$$p + \tfrac{1}{2}\rho V^2 = \text{constant} \tag{5-165}$$

which is called *Bernoulli's equation.* For an isentropic, compressible flow, Eq. (5-162) can be expressed as

$$\frac{V^2}{2} + \frac{\gamma}{\gamma - 1}\,\frac{p}{\rho} = \text{constant} \tag{5-166}$$

which is sometimes referred to as the *compressible Bernoulli equation.* It should be remembered that Eqs. (5-165) and (5-166) are valid only along a given streamline since the constants appearing in these equations can vary between streamlines.

We will now show that Eqs. (5-165) and (5-166) can be made valid everywhere in the flowfield if the flow is assumed *irrotational.* An irrotational flow is one in which the fluid particles do not rotate about their axis. From the study of kinematics (see, for example, Owczarek, 1964), the vorticity ζ which is defined by

$$\zeta = \nabla \times \mathbf{V} \tag{5-167}$$

is equivalent to twice the angular velocity of a fluid particle. Thus, for an irrotational flow

$$\zeta = \nabla \times \mathbf{V} = 0 \tag{5-168}$$

and as a result we can express $\mathbf{V}$ as the gradient of a single-valued point function ϕ since

$$\nabla \times \mathbf{V} = \nabla \times (\nabla \phi) = 0 \tag{5-169}$$

The scalar ϕ is called the *velocity potential.* Also, from kinematics, the acceleration of a fluid particle, $D\mathbf{V}/Dt$, is given by

$$\frac{D\mathbf{V}}{Dt} = \frac{\partial \mathbf{V}}{\partial t} + \nabla\left(\frac{V^2}{2}\right) - \mathbf{V} \times \zeta \tag{5-170}$$

which is called *Lagrange's acceleration formula.* For an irrotational flow, this equation reduces to

$$\frac{D\mathbf{V}}{Dt} = \frac{\partial \mathbf{V}}{\partial t} + \nabla\left(\frac{V^2}{2}\right)$$

which can be substituted into Euler's equation to give

$$\frac{\partial \mathbf{V}}{\partial t} + \nabla\left(\frac{V^2}{2}\right) = \mathbf{f} - \frac{1}{\rho}\nabla p \tag{5-171}$$

If we again neglect body forces and assume steady flow, Eq. (5-171) can be rewritten as

$$\nabla\left(\frac{V^2}{2} + \int \frac{dp}{\rho}\right) = 0 \tag{5-172}$$

since

$$\nabla \int \frac{dp}{\rho} = \frac{\nabla p}{\rho}$$

Integrating Eq. (5-172) along any arbitrary line in the flowfield yields

$$\frac{V^2}{2} + \int \frac{dp}{\rho} = \text{constant} \tag{5-173}$$

The constant in this equation now has the same value everywhere in the flowfield since Eq. (5-173) was integrated along any arbitrary line. The incompressible Bernoulli equation [Eq. (5-165)] and the compressible Bernoulli equation [Eq. (5-166)] follow directly from Eq. (5-173) in the same manner as before. The only difference is that the resulting equations are now valid everywhere in the inviscid flowfield because of our additional assumption of irrotationality.

For the special case of an inviscid, incompressible, irrotational flow, the continuity equation

$$\nabla \cdot \mathbf{V} = 0 \tag{5-174}$$

can be combined with

$$\mathbf{V} = \nabla \phi \tag{5-175}$$

to give Laplace's equation

$$\nabla^2 \phi = 0 \tag{5-176}$$

5-5.3 Inviscid Energy Equations

The inviscid form of the energy equation given by Eq. (5-22) becomes

$$\frac{\partial E_t}{\partial t} + \nabla \cdot E_t \mathbf{V} = \rho \mathbf{f} \cdot \mathbf{V} - \nabla \cdot (p\mathbf{V}) \tag{5-177}$$

which is equivalent to

$$\frac{\partial}{\partial t}(\rho H) + \nabla \cdot (\rho H \mathbf{V}) = \rho \mathbf{f} \cdot \mathbf{V} + \frac{\partial p}{\partial t} \tag{5-178}$$

Additional forms of the inviscid energy equation can be obtained from Eq. (5-29)

$$\rho \frac{De}{Dt} + p(\nabla \cdot \mathbf{V}) = 0 \tag{5-179}$$

and from Eq. (5-33)

$$\rho \frac{Dh}{Dt} = \frac{Dp}{Dt} \tag{5-180}$$

If we use the continuity equation and ignore the body force term, Eq. (5-178) can be written as

$$\frac{DH}{Dt} = \frac{1}{\rho} \frac{\partial p}{\partial t} \tag{5-181}$$

which for a steady flow becomes

$$\mathbf{V} \cdot \nabla H = 0 \tag{5-182}$$

This equation can be integrated along a streamline to give

$$H = h + \frac{V^2}{2} = \text{constant} \tag{5-183}$$

The constant will remain the same throughout the inviscid flowfield for the special case of an isoenergetic (homenergic) flow.

For an incompressible flow, Eq. (5-179) reduces to

$$\frac{De}{Dt} = 0 \tag{5-184}$$

which for a steady flow implies that the internal energy is constant along a streamline.

5-5.4 Additional Equations

The conservation equations for an inviscid flow have been presented in this section. It is possible to derive additional relationships which prove to be quite useful in particular applications. In some cases, these auxiliary equations can be used to replace one or more of the conservation equations. Several of the auxiliary equations are based on the First and Second Laws of Thermodynamics which provide the relationship

$$T\,ds = de + p\,d\left(\frac{1}{\rho}\right) \tag{5-185}$$

where s is the entropy. Using the definition of enthalpy

$$h = e + \frac{p}{\rho}$$

it is possible to rewrite Eq. (5-185) as

$$T\,ds = dh - \frac{dp}{\rho} \tag{5-186}$$

This latter equation can also be written as

$$T\nabla s = \nabla h - \frac{\nabla p}{\rho}$$

since at any given instant a fluid particle can change its state to that of a neighboring particle. Upon combining this equation with Eqs. (5-170) and (5-158) and ignoring body forces, we obtain

$$\frac{\partial \mathbf{V}}{\partial t} - \mathbf{V} \times \boldsymbol{\zeta} = T\nabla s - \nabla h - \nabla\left(\frac{V^2}{2}\right)$$

or

$$\frac{\partial \mathbf{V}}{\partial t} - \mathbf{V} \times \boldsymbol{\zeta} = T\nabla s - \nabla H \tag{5-187}$$

which is called *Crocco's equation.* This equation provides a relationship between vorticity and entropy. For a steady flow it becomes

$$\mathbf{V} \times \boldsymbol{\zeta} = \nabla H - T\nabla s \tag{5-188}$$

We have shown earlier that for a steady, inviscid, adiabatic flow

$$\mathbf{V} \cdot \nabla H = 0$$

which if combined with Eq. (5-188) gives

$$\mathbf{V} \cdot \nabla s = 0$$

since $\mathbf{V} \times \boldsymbol{\zeta}$ is normal to $\mathbf{V}$. Thus, we have proved that entropy remains constant along a streamline for a steady, nonviscous, nonconducting, adiabatic flow. This is called an *isentropic* flow. If we also assume that the flow is irrotational and isoenergetic, then Crocco's equation tells us that the entropy remains constant everywhere (i.e., homentropic flow).

The thermodynamic relationship given by Eq. (5-185) involves only changes in properties since it does not contain path-dependent functions. For the isentropic flow of a perfect gas it can be written as

$$T\,ds = 0 = c_p\,dT - RT\frac{dp}{p}$$

or

$$\frac{dp}{p} = \frac{\gamma}{\gamma - 1}\frac{dT}{T}$$

The latter equation can be integrated to yield

$$\frac{p}{T^{\gamma/(\gamma-1)}} = \text{constant}$$

which becomes

$$\frac{p}{\rho^{\gamma}} = \text{constant} \tag{5-189}$$

after substituting the perfect gas equation of state. The latter isentropic relation was used earlier to derive the compressible Bernoulli equation [Eq. (5-166)]. It is interesting to note that the integrated energy equation, given by Eq. (5-183), can be made identical to Eq. (5-166) if the flow is assumed to be isentropic.

The speed of sound is given by

$$a = \sqrt{\left(\frac{\partial p}{\partial \rho}\right)_s} \tag{5-190}$$

where the subscript s indicates a constant entropy process. At a point in the flow of a perfect gas, Eqs. (5-189) and (5-190) can be combined to give

$$a = \sqrt{\frac{dp}{d\rho}} = \sqrt{\frac{\gamma p}{\rho}} = \sqrt{\gamma RT} \tag{5-191}$$

5-5.5 Vector Form of Euler Equations

The compressible Euler equations in Cartesian coordinates without body forces or external heat addition can be written in vector form as

$$\frac{\partial \mathbf{U}}{\partial t} + \frac{\partial \mathbf{E}}{\partial x} + \frac{\partial \mathbf{F}}{\partial y} + \frac{\partial \mathbf{G}}{\partial z} = 0 \tag{5-192}$$

where **U**, **E**, **F**, and **G** are vectors given by

$$\mathbf{U} = \begin{bmatrix} \rho \\ \rho u \\ \rho v \\ \rho w \\ E_t \end{bmatrix} \qquad \mathbf{E} = \begin{bmatrix} \rho u \\ \rho u^2 + p \\ \rho u v \\ \rho u w \\ (E_t + p)u \end{bmatrix}$$

$$\mathbf{F} = \begin{bmatrix} \rho v \\ \rho u v \\ \rho v^2 + p \\ \rho v w \\ (E_t + p)v \end{bmatrix} \qquad \mathbf{G} = \begin{bmatrix} \rho w \\ \rho u w \\ \rho v w \\ \rho w^2 + p \\ (E_t + p)w \end{bmatrix}$$

For a steady, isoenergetic flow of a perfect gas, it becomes possible to remove the energy equation from the vector set and use, instead, the algebraic form of the equation given by Eq. (5-166). This reduces the overall computation time since one less partial differential equation needs to be solved.

5-5.6 Simplified Forms of Euler Equations

The Euler equations can be simplified by making additional assumptions. If the flow is assumed steady, irrotational, and isentropic, the Euler equations can be combined into a single equation called the *velocity potential equation.* The velocity potential equation is derived in the following manner. In a Cartesian coordinate system, the continuity equation may be written as

$$\frac{\partial}{\partial x}(\rho \phi_x) + \frac{\partial}{\partial y}(\rho \phi_y) + \frac{\partial}{\partial z}(\rho \phi_z) = 0 \tag{5-193}$$

where the velocity components have been replaced by

$$u = \frac{\partial \phi}{\partial x} \qquad v = \frac{\partial \phi}{\partial y} \qquad w = \frac{\partial \phi}{\partial z} \tag{5-194}$$

The momentum (and energy) equations reduce to Eq. (5-162) with the assumptions of steady, irrotational, and isentropic flow. In differential form this equation becomes

$$dp = -\rho\, d\left(\frac{V^2}{2}\right) = -\rho\, d\left(\frac{\phi_x^2 + \phi_y^2 + \phi_z^2}{2}\right) \tag{5-195}$$

Combining Eqs. (5-190) and (5-195) yields the equation

$$dp = -\frac{\rho}{a^2}\, d\left(\frac{\phi_x^2 + \phi_y^2 + \phi_z^2}{2}\right) \tag{5-196}$$

which may be used to find the derivatives of ρ in each direction. After substituting these expressions for ρ_x, ρ_y, and ρ_z into Eq. (5-193) and simplifying, the velocity potential equation is obtained:

$$\left(1 - \frac{\phi_x^2}{a^2}\right)\phi_{xx} + \left(1 - \frac{\phi_y^2}{a^2}\right)\phi_{yy} + \left(1 - \frac{\phi_z^2}{a^2}\right)\phi_{zz} - \frac{2\phi_x\phi_y}{a^2}\phi_{xy}$$
$$- \frac{2\phi_x\phi_z}{a^2}\phi_{xz} - \frac{2\phi_y\phi_z}{a^2}\phi_{yz} = 0 \tag{5-197}$$

Note that for an incompressible flow ($a \to \infty$), the velocity potential equation reduces to Laplace's equation.

The Euler equations can be further simplified if we consider the flow over a slender body where the freestream is only slightly disturbed (perturbed). An example is the flow over a thin airfoil. The analysis of this type of flowfield is referred to as small-perturbation theory. In order to demonstrate how the velocity potential equation can be simplified for flows of this type, we assume that a slender body is placed in a two-dimensional flow. The body causes a disturbance of the uniform flow and the velocity components are written as

$$\begin{aligned} u &= U_\infty + u' \\ v &= v' \end{aligned} \tag{5-198}$$

where the prime denotes perturbation velocity. If we let ϕ' be the perturbation velocity potential, then

$$\begin{aligned} u &= \frac{\partial \phi}{\partial x} = U_\infty + \frac{\partial \phi'}{\partial x} \\ v &= \frac{\partial \phi}{\partial y} = \frac{\partial \phi'}{\partial y} \end{aligned} \tag{5-199}$$

Substituting these expressions along with Eq. (5-191) into Eq. (5-166) gives

$$a^2 = a_\infty^2 - \frac{\gamma - 1}{2}\left[2u'U_\infty + (u')^2 + (v')^2\right] \tag{5-200}$$

which can then be combined with the velocity potential equation to yield

$$(1 - M_\infty^2)\frac{\partial u'}{\partial x} + \frac{\partial v'}{\partial y} = M_\infty^2\left[(\gamma + 1)\frac{u'}{U_\infty} + \left(\frac{\gamma+1}{2}\right)\frac{(u')^2}{U_\infty^2} + \left(\frac{\gamma-1}{2}\right)\frac{(v')^2}{U_\infty^2}\right]\frac{\partial u'}{\partial x}$$
$$+ M_\infty^2\left[(\gamma - 1)\frac{u'}{U_\infty} + \left(\frac{\gamma+1}{2}\right)\frac{(v')^2}{U_\infty^2} + \left(\frac{\gamma-1}{2}\right)\frac{(u')^2}{U_\infty^2}\right]\frac{\partial v'}{\partial y}$$
$$+ M_\infty^2\frac{v'}{U_\infty}\left(1 + \frac{u'}{U_\infty}\right)\left(\frac{\partial u'}{\partial y} + \frac{\partial v'}{\partial x}\right) \tag{5-201}$$

Since the flow is only slightly disturbed from the freestream, we assume

$$\frac{u'}{U_\infty}, \frac{v'}{U_\infty} \ll 1$$

As a result, Eq. (5-200) simplifies to

$$a^2 = a_\infty^2 - (\gamma - 1)u'U_\infty \tag{5-202}$$

and Eq. (5-201) becomes

$$\left[\frac{1 - M_\infty^2}{M_\infty^2} - (\gamma + 1)\frac{u'}{U_\infty}\right] M_\infty^2 \phi'_{xx} + \phi'_{yy} = 0 \tag{5-203}$$

The latter equation is called the *transonic small disturbance equation.* This nonlinear equation is either elliptic or hyperbolic depending on whether the flow is subsonic or supersonic.

For flows in the subsonic or supersonic regimes, the magnitude of the term $M_\infty^2(\gamma + 1)(u'/U_\infty)\phi'_{xx}$ is small in comparison with $(1 - M_\infty^2)\phi'_{xx}$ and Eq. (5-203) reduces to the linear *Prandtl-Glauert equation*:

$$(1 - M_\infty^2)\phi'_{xx} + \phi'_{yy} = 0 \tag{5-204}$$

Once the perturbation velocity potential is known, the pressure coefficient can be determined from

$$C_p = \frac{p - p_\infty}{\frac{1}{2}\rho U_\infty^2} = \frac{2}{\gamma M_\infty^2}\left(\frac{p}{p_\infty} - 1\right) = -\frac{2u'}{U_\infty} \tag{5-205}$$

which is derived using Eqs. (5-166), (5-189), (5-198), and the binomial expansion theorem.

5-5.7 Shock Equations

A shock wave is a very thin region in a supersonic flow across which there is a large variation in the flow properties. Because these variations occur in such a short distance, viscosity and heat conductivity play a dominant role in the structure of the shock wave. However, unless one is interested in studying the structure of the shock wave, it is usually possible to consider the shock wave to be infinitesimally thin (i.e., a mathematical discontinuity) and use the Euler equations to determine the changes in flow properties across the shock wave. For example, let us consider the case of a stationary straight shock wave oriented perpendicular to the flow direction (i.e., a normal shock). The two-dimensional flow is in the positive x direction and the conditions upstream of the shock wave are designated with a subscript 1 while the conditions downstream are designated with a subscript 2. Since a shock wave is a weak solution to the hyperbolic Euler equations, we can apply the theory of weak solutions, described in Section 4-4, to Eq. (5-192). For the present discontinuity, this gives

$$[\mathbf{E}] = 0$$

or

$$\mathbf{E}_1 = \mathbf{E}_2$$

Thus,

$$\begin{aligned}
\rho_1 u_1 &= \rho_2 u_2 \\
p_1 + \rho_1 u_1^2 &= p_2 + \rho_2 u_2^2 \\
\rho_1 u_1 v_1 &= \rho_2 u_2 v_2 \\
(E_{t_1} + p_1)u_1 &= (E_{t_2} + p_2)u_2
\end{aligned}$$

Upon simplifying the above shock relations, we find that

$$\begin{aligned}
\rho_1 u_1 &= \rho_2 u_2 \\
p_1 + \rho_1 u_1^2 &= p_2 + \rho_2 u_2^2 \\
v_1 &= v_2 \\
h_1 + \frac{u_1^2}{2} &= h_2 + \frac{u_2^2}{2}
\end{aligned} \tag{5-206}$$

Solving these equations for the pressure ratio across the shock, we obtain

$$\frac{p_2}{p_1} = \frac{(\gamma + 1)\rho_2 - (\gamma - 1)\rho_1}{(\gamma + 1)\rho_1 - (\gamma - 1)\rho_2} \tag{5-207}$$

This latter equation relates thermodynamic properties across the shock wave and is called the *Rankine-Hugoniot equation.* The label, Rankine-Hugoniot relations, is frequently applied to all equations which relate changes across shock waves.

For shock waves inclined to the freestream (i.e. oblique shocks) the shock relations become

$$\begin{aligned}
\rho_1 V_{n_1} &= \rho_2 V_{n_2} \\
p_1 + \rho_1 V_{n_1}^2 &= p_2 + \rho_2 V_{n_2}^2 \\
V_{t_1} &= V_{t_2} \\
h_1 + \frac{V_1^2}{2} &= h_2 + \frac{V_2^2}{2}
\end{aligned} \tag{5-208}$$

where V_n and V_t are the normal and tangential components of the velocity vector, respectively. These equations also apply to moving shock waves if the velocity components are measured with respect to the moving shock wave. In this case, the normal component of the flow velocity ahead of the shock (measured with respect to the shock) can be related to the pressure behind the shock by manipulating the above equations to form

$$V_{n_1}^2 = \frac{\gamma + 1}{2} \frac{p_1}{\rho_1} \left(\frac{p_2}{p_1} + \frac{\gamma - 1}{\gamma + 1} \right) \tag{5-209}$$

This latter equation is useful when attempting to numerically treat moving shock waves as discontinuities as will be seen in Chapter 6. A comprehensive listing of shock relations is available in NACA Report 1135 (Ames Research Staff, 1953).

5-6 TRANSFORMATION OF GOVERNING EQUATIONS

The classical governing equations of fluid dynamics have been presented in this chapter. These equations have been written in either vector or tensor form. In Section 5-1.7, it was shown how these equations can be expressed in terms of any generalized orthogonal curvilinear coordinate system. For many applications, however, a nonorthogonal coordinate system is desirable. In this section we will show how the governing equations can be transformed from a Cartesian coordinate system to any general nonorthogonal (or orthogonal) coordinate system. In the process, we will demonstrate how simple transformations can be used to cluster grid points in regions of large gradients such as boundary layers and how to transform a nonrectangular computational region in the physical plane into a rectangular uniformally-spaced grid in the computational plane. These latter transformations are simple examples from a very important topic of computational fluid dynamics called grid generation. A complete discussion of grid generation will be presented in Chapter 10.

5-6.1 Simple Transformations

In this section, simple independent variable transformations are used to illustrate how the governing fluid dynamic equations are transformed. As a first example, we will consider the problem of clustering grid points near a wall. Refinement of the mesh near a wall is mandatory, in most cases, if the details of the boundary layer are to be properly resolved. Figure 5-8*a* shows a mesh above a flat plate in which grid points are clustered near the plate in the normal direction (y) while the spacing in the x direction is uniform. Because the spacing is not uniform in the y direction it is convenient to apply a transformation to the y coordinate so that the governing equations can be solved on a uniformly spaced grid in the computational plane $(\bar{x}, \bar{y})$ as seen in Fig. 5-8*b*. A suitable transformation for a two-dimensional boundary-layer type of problem is given by

Transformation 1:

$$\bar{x} = x$$

$$\bar{y} = 1 - \frac{\ln \{[\beta + 1 - (y/h)]/[\beta - 1 + (y/h)]\}}{\ln [(\beta + 1)/(\beta - 1)]} \qquad 1 < \beta < \infty \tag{5-210}$$

This stretching transformation clusters more points near $y = 0$ as the stretching parameter β approaches 1.

In order to apply this transformation to the governing fluid dynamic equations, the following partial derivatives are formed

$$\frac{\partial}{\partial x} = \frac{\partial \bar{x}}{\partial x}\frac{\partial}{\partial \bar{x}} + \frac{\partial \bar{y}}{\partial x}\frac{\partial}{\partial \bar{y}}$$
$$\frac{\partial}{\partial y} = \frac{\partial \bar{x}}{\partial y}\frac{\partial}{\partial \bar{x}} + \frac{\partial \bar{y}}{\partial y}\frac{\partial}{\partial \bar{y}} \tag{5-211}$$

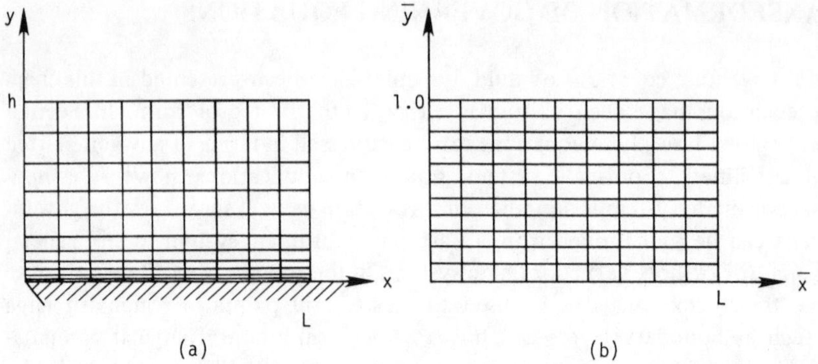

Figure 5-8 Grid clustering near a wall. (a) Physical plane (x, y); (b) computational plane $(\bar{x}, \bar{y})$.

where

$$\frac{\partial \bar{x}}{\partial x} = 1 \qquad \frac{\partial \bar{y}}{\partial x} = 0$$

$$\frac{\partial \bar{x}}{\partial y} = 0 \qquad \frac{\partial \bar{y}}{\partial y} = \frac{2\beta}{h\{\beta^2 - [1 - (y/h)]^2\} \ln [(\beta + 1)/(\beta - 1)]}$$

As a result, the partial derivatives simplify to

$$\frac{\partial}{\partial x} = \frac{\partial}{\partial \bar{x}}$$
$$\frac{\partial}{\partial y} = \left(\frac{\partial \bar{y}}{\partial y}\right) \frac{\partial}{\partial \bar{y}} \tag{5-212}$$

If we now apply this transformation to the steady, two-dimensional, incompressible continuity equation written in Cartesian coordinates

$$\frac{\partial u}{\partial x} + \frac{\partial v}{\partial y} = 0 \tag{5-213}$$

the following transformed equation is obtained

$$\frac{\partial u}{\partial \bar{x}} + \left(\frac{\partial \bar{y}}{\partial y}\right) \frac{\partial v}{\partial \bar{y}} = 0 \tag{5-214}$$

This transformed equation can now be differenced on the uniformly spaced grid in the computational plane. The grid spacing can be computed from

$$\Delta \bar{x} = \frac{L}{NI - 1}$$
$$\Delta \bar{y} = \frac{1}{NJ - 1} \tag{5-215}$$

where NI and NJ are the number of grid points in the x and y directions, respectively. We note that the expression for the metric $\partial \bar{y}/\partial y$ contains y so that we must be able to express y as a function of $\bar{y}$. This is referred to as the inverse of the transformation. For the present transformation, given by Eqs. (5-210), the inverse can be readily found as

$$x = \bar{x}$$
$$y = h\frac{(\beta+1)-(\beta-1)\{[(\beta+1)/(\beta-1)]^{1-\bar{y}}\}}{[(\beta+1)/(\beta-1)]^{1-\bar{y}}+1} \tag{5-216}$$

The stretching transformation discussed here is from the family of general stretching transformations proposed by Roberts (1971). Another transformation from this family refines the mesh near walls of a duct as seen in Fig. 5-9. This transformation is given by

Transformation 2:

$$\bar{x} = x$$
$$\bar{y} = \alpha + (1-\alpha)\frac{\ln(\{\beta + [y(2\alpha+1)/h] - 2\alpha\}/\{\beta - [y(2\alpha+1)/h] + 2\alpha\})}{\ln[(\beta+1)/(\beta-1)]} \tag{5-217}$$

For this transformation, if $\alpha = 0$ the mesh will be refined near $y = h$ only, whereas, if $\alpha = \frac{1}{2}$ the mesh will be refined equally near $y = 0$ and $y = h$. Roberts has shown that the stretching parameter β is related (approximately) to the nondimensional boundary-layer thickness (δ/h) by

$$\beta = \left(1 - \frac{\delta}{h}\right)^{-1/2} \qquad 0 < \frac{\delta}{h} < 1 \tag{5-218}$$

where h is the height of the mesh. The amount of stretching for various values of δ/h is illustrated in Fig. 5-10 for the case where $\alpha = 0$. For the transformation given by Eqs. (5-217), the metric $\partial\bar{y}/\partial y$ is

$$\frac{\partial \bar{y}}{\partial y} = \frac{2\beta(1-\alpha)(2\alpha+1)}{h\{\beta^2 - [y(2\alpha+1)/h - 2\alpha]^2\}\ln[(\beta+1)/(\beta-1)]} \tag{5-219}$$

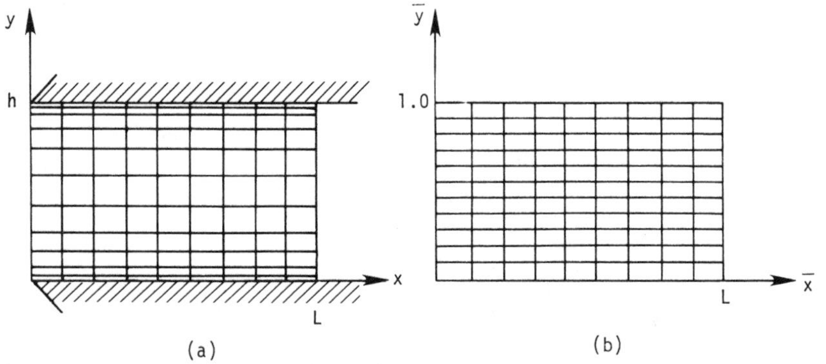

Figure 5-9 Grid clustering in a duct. (a) Physical plane (x, y); (b) computational plane $(\bar{x}, \bar{y})$.

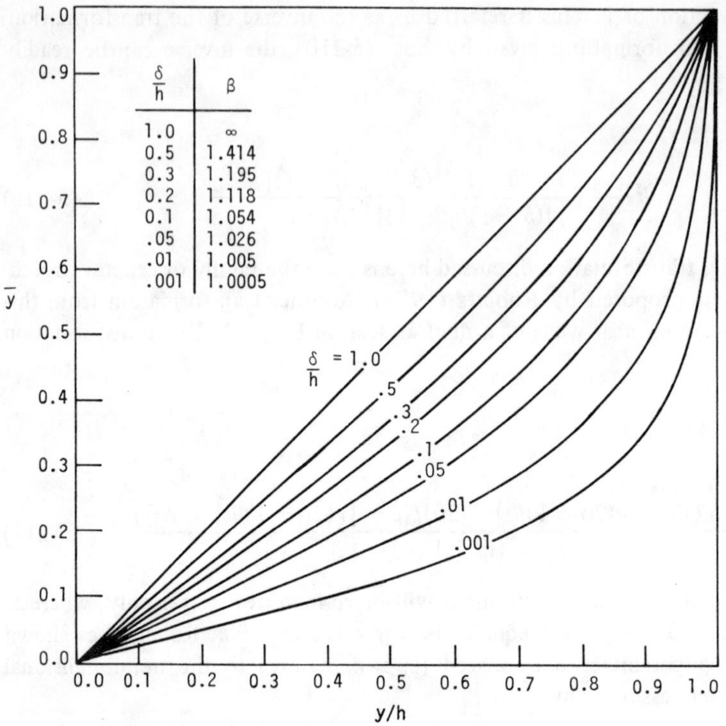

Figure 5-10 Roberts' stretching transformation ($\alpha = 0$).

and the inverse transformation becomes

$$x = \bar{x}$$

$$y = h\,\frac{(\beta + 2\alpha)[(\beta + 1)/(\beta - 1)]^{(\bar{y}-\alpha)/(1-\alpha)} - \beta + 2\alpha}{(2\alpha + 1)\{1 + [(\beta + 1)/(\beta - 1)]^{(\bar{y}-\alpha)/(1-\alpha)}\}} \tag{5-220}$$

A useful transformation for refining the mesh about some interior point y_c (see Fig. 5-11) is given by

Transformation 3:

$$\bar{x} = x$$

$$\bar{y} = B + \frac{1}{\tau}\sinh^{-1}\left[\left(\frac{y}{y_c} - 1\right)\sinh(\tau B)\right] \tag{5-221}$$

where

$$B = \frac{1}{2\tau}\ln\left[\frac{1 + (e^{\tau} - 1)(y_c/h)}{1 + (e^{-\tau} - 1)(y_c/h)}\right] \qquad 0 < \tau < \infty$$

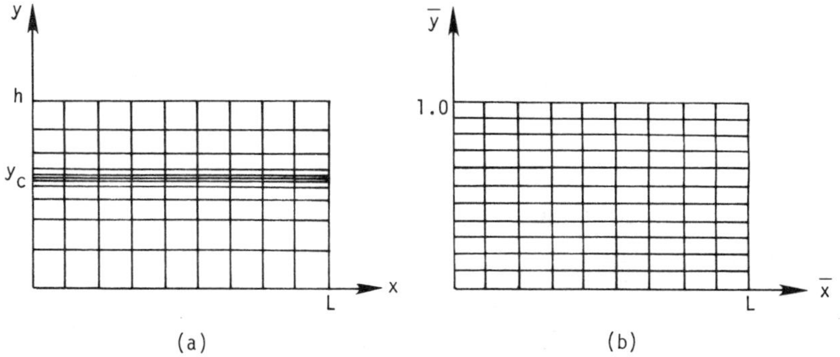

Figure 5-11 Grid clustering near an interior point. (a) Physical plane (x, y); (b) computational plane $(\bar{x}, \bar{y})$.

In this transformation, τ is the stretching parameter which varies from zero (no stretching) to large values which produce the most refinement near $y = y_c$. The metric $\partial \bar{y}/\partial y$ and y become

$$\frac{\partial \bar{y}}{\partial y} = \frac{\sinh(\tau B)}{\tau y_c \sqrt{1 + [(y/y_c) - 1]^2 \sinh^2(\tau B)}} \tag{5-222}$$

$$y = y_c \left\{ 1 + \frac{\sinh[\tau(\bar{y} - B)]}{\sinh(\tau B)} \right\} \tag{5-223}$$

For our final transformation, we will examine a simple transformation which can be used to transform a nonrectangular region in the physical plane into a rectangular region in the computational plane, as seen in Fig. 5-12. The required transformation is

Transformation 4:

$$\bar{x} = x$$

$$\bar{y} = \frac{y}{h(x)} \tag{5-224}$$

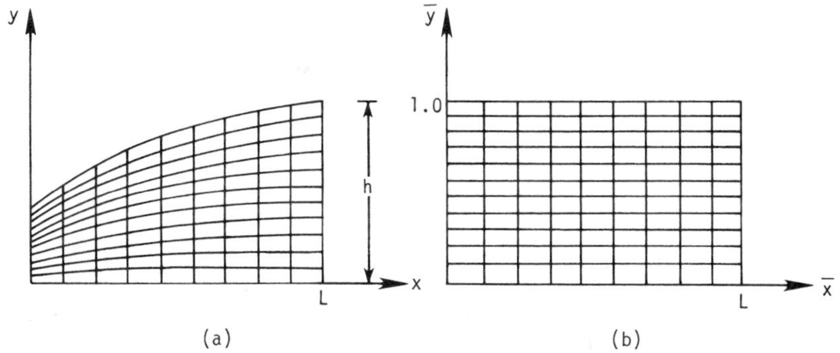

Figure 5-12 Rectangularization of computational grid. (a) Physical plane (x, y); (b) computational plane $(\bar{x}, \bar{y})$.

The known distance between the lower boundary and the upper boundary (measured along a x = constant line) is designated by $h(x)$. The required partial derivatives are

$$\frac{\partial}{\partial x} = \frac{\partial}{\partial \bar{x}} - \bar{y}\,\frac{h'(x)}{h(x)}\,\frac{\partial}{\partial \bar{y}}$$
$$\frac{\partial}{\partial y} = \frac{1}{h(x)}\,\frac{\partial}{\partial \bar{y}} \tag{5-225}$$

where $h'(x) = dh(x)/dx$. Hence, the steady, two-dimensional, incompressible continuity equation in Cartesian coordinates is transformed to

$$\frac{\partial u}{\partial \bar{x}} - \bar{y}\,\frac{h'(\bar{x})}{h(\bar{x})}\,\frac{\partial u}{\partial \bar{y}} + \frac{1}{h(\bar{x})}\,\frac{\partial v}{\partial \bar{y}} = 0 \tag{5-226}$$

5-6.2 Generalized Transformation

In the preceding section we examined simple independent variable transformations which make it possible to solve the governing equations on a uniformly spaced computational grid. Let us now consider a completely general transformation of the form

$$\begin{aligned} \xi &= \xi(x, y, z) \\ \eta &= \eta(x, y, z) \\ \zeta &= \zeta(x, y, z) \end{aligned} \tag{5-227}$$

which can be used to transform the governing equations from the physical domain (x, y, z) to the computational domain (ξ, η, ζ). Using the chain rule of partial differentiation, the partial derivatives become

$$\begin{aligned} \frac{\partial}{\partial x} &= \xi_x\,\frac{\partial}{\partial \xi} + \eta_x\,\frac{\partial}{\partial \eta} + \zeta_x\,\frac{\partial}{\partial \zeta} \\ \frac{\partial}{\partial y} &= \xi_y\,\frac{\partial}{\partial \xi} + \eta_y\,\frac{\partial}{\partial \eta} + \zeta_y\,\frac{\partial}{\partial \zeta} \\ \frac{\partial}{\partial z} &= \xi_z\,\frac{\partial}{\partial \xi} + \eta_z\,\frac{\partial}{\partial \eta} + \zeta_z\,\frac{\partial}{\partial \zeta} \end{aligned} \tag{5-228}$$

The metrics $(\xi_x, \eta_x, \zeta_x, \xi_y, \eta_y, \zeta_y, \xi_z, \eta_z, \zeta_z)$ appearing in these equations can be determined in the following manner. We first write the differential expressions

$$\begin{aligned} d\xi &= \xi_x\,dx + \xi_y\,dy + \xi_z\,dz \\ d\eta &= \eta_x\,dx + \eta_y\,dy + \eta_z\,dz \\ d\zeta &= \zeta_x\,dx + \zeta_y\,dy + \zeta_z\,dz \end{aligned} \tag{5-229}$$

which in matrix form become

$$\begin{bmatrix} d\xi \\ d\eta \\ d\zeta \end{bmatrix} = \begin{bmatrix} \xi_x & \xi_y & \xi_z \\ \eta_x & \eta_y & \eta_z \\ \zeta_x & \zeta_y & \zeta_z \end{bmatrix} \begin{bmatrix} dx \\ dy \\ dz \end{bmatrix} \tag{5-230}$$

In a like manner we can write

$$\begin{bmatrix} dx \\ dy \\ dz \end{bmatrix} = \begin{bmatrix} x_\xi & x_\eta & x_\zeta \\ y_\xi & y_\eta & y_\zeta \\ z_\xi & z_\eta & z_\zeta \end{bmatrix} \begin{bmatrix} d\xi \\ d\eta \\ d\zeta \end{bmatrix} \tag{5-231}$$

Therefore,

$$\begin{aligned} \begin{bmatrix} \xi_x & \xi_y & \xi_z \\ \eta_x & \eta_y & \eta_z \\ \zeta_x & \zeta_y & \zeta_z \end{bmatrix} &= \begin{bmatrix} x_\xi & x_\eta & x_\zeta \\ y_\xi & y_\eta & y_\zeta \\ z_\xi & z_\eta & z_\zeta \end{bmatrix}^{-1} \\ &= J \begin{bmatrix} y_\eta z_\zeta - y_\zeta z_\eta & -(x_\eta z_\zeta - x_\zeta z_\eta) & x_\eta y_\zeta - x_\zeta y_\eta \\ -(y_\xi z_\zeta - y_\zeta z_\xi) & x_\xi z_\zeta - x_\zeta z_\xi & -(x_\xi y_\zeta - x_\zeta y_\xi) \\ y_\xi z_\eta - y_\eta z_\xi & -(x_\xi z_\eta - x_\eta z_\xi) & x_\xi y_\eta - x_\eta y_\xi \end{bmatrix} \end{aligned} \tag{5-232}$$

Thus, the metrics are:

$$\begin{aligned} \xi_x &= J(y_\eta z_\zeta - y_\zeta z_\eta) \\ \xi_y &= -J(x_\eta z_\zeta - x_\zeta z_\eta) \\ \xi_z &= J(x_\eta y_\zeta - x_\zeta y_\eta) \\ \eta_x &= -J(y_\xi z_\zeta - y_\zeta z_\xi) \\ \eta_y &= J(x_\xi z_\zeta - x_\zeta z_\xi) \\ \eta_z &= -J(x_\xi y_\zeta - x_\zeta y_\xi) \\ \zeta_x &= J(y_\xi z_\eta - y_\eta z_\xi) \\ \zeta_y &= -J(x_\xi z_\eta - x_\eta z_\xi) \\ \zeta_z &= J(x_\xi y_\eta - x_\eta y_\xi) \end{aligned} \tag{5-233}$$

where J is the Jacobian of the transformation

$$J = \frac{\partial(\xi, \eta, \zeta)}{\partial(x, y, z)} = \begin{vmatrix} \xi_x & \xi_y & \xi_z \\ \eta_x & \eta_y & \eta_z \\ \zeta_x & \zeta_y & \zeta_z \end{vmatrix} \tag{5-234}$$

which can be evaluated in the following manner

$$\begin{aligned} J = 1/J^{-1} &= 1 \Big/ \frac{\partial(x, y, z)}{\partial(\xi, \eta, \zeta)} = 1 \Big/ \begin{vmatrix} x_\xi & x_\eta & x_\zeta \\ y_\xi & y_\eta & y_\zeta \\ z_\xi & z_\eta & z_\zeta \end{vmatrix} \\ &= 1/[x_\xi(y_\eta z_\zeta - y_\zeta z_\eta) - x_\eta(y_\xi z_\zeta - y_\zeta z_\xi) + x_\zeta(y_\xi z_\eta - y_\eta z_\xi)] \end{aligned} \tag{5-235}$$

The metrics can be readily determined if analytical expressions are available for the inverse of the transformation:

$$\begin{aligned} x &= x(\xi, \eta, \zeta) \\ y &= y(\xi, \eta, \zeta) \\ z &= z(\xi, \eta, \zeta) \end{aligned} \tag{5-236}$$

For cases where the transformation is the direct result of a grid generation scheme, the metrics can be computed numerically using central differences in the computational plane. A complete discussion on the proper way to compute metrics will be presented in Chapter 10.

If we apply the generalized transformation to the compressible Navier-Stokes equations written in vector form [Eqs. (5-44)] the following transformed equation is obtained

$$\mathbf{U}_t + \xi_x \mathbf{E}_\xi + \eta_x \mathbf{E}_\eta + \zeta_x \mathbf{E}_\zeta + \xi_y \mathbf{F}_\xi + \eta_y \mathbf{F}_\eta + \zeta_y \mathbf{F}_\zeta + \xi_z \mathbf{G}_\xi + \eta_z \mathbf{G}_\eta + \zeta_z \mathbf{G}_\zeta = 0 \tag{5-237}$$

Viviand (1974) and Vinokur (1974) have shown that the gas dynamic equations can be put back into strong conservation-law form after a transformation has been applied. In order to do this, the transformed equation is first divided by the Jacobian and is then rearranged into conservation-law form by adding and subtracting like terms. When this procedure is applied to Eq. (5-237), the following equation results

$$\begin{aligned} &\left(\frac{\mathbf{U}}{J}\right)_t + \left(\frac{\mathbf{E}\xi_x + \mathbf{F}\xi_y + \mathbf{G}\xi_z}{J}\right)_\xi + \left(\frac{\mathbf{E}\eta_x + \mathbf{F}\eta_y + \mathbf{G}\eta_z}{J}\right)_\eta \\ &\quad + \left(\frac{\mathbf{E}\zeta_x + \mathbf{F}\zeta_y + \mathbf{G}\zeta_z}{J}\right)_\zeta - \mathbf{E}\left[\left(\frac{\xi_x}{J}\right)_\xi + \left(\frac{\eta_x}{J}\right)_\eta + \left(\frac{\zeta_x}{J}\right)_\zeta\right] \\ &\quad - \mathbf{F}\left[\left(\frac{\xi_y}{J}\right)_\xi + \left(\frac{\eta_y}{J}\right)_\eta + \left(\frac{\zeta_y}{J}\right)_\zeta\right] - \mathbf{G}\left[\left(\frac{\xi_z}{J}\right)_\xi + \left(\frac{\eta_z}{J}\right)_\eta + \left(\frac{\zeta_z}{J}\right)_\zeta\right] = 0 \end{aligned} \tag{5-238}$$

The last three terms in brackets are all equal to zero and can be dropped. This can be verified by substituting the metrics given by Eqs. (5-233) into these terms. If we now define the quantities

$$\begin{aligned} \mathbf{U}_1 &= \frac{\mathbf{U}}{J} \\ \mathbf{E}_1 &= \frac{1}{J}(\mathbf{E}\xi_x + \mathbf{F}\xi_y + \mathbf{G}\xi_z) \\ \mathbf{F}_1 &= \frac{1}{J}(\mathbf{E}\eta_x + \mathbf{F}\eta_y + \mathbf{G}\eta_z) \\ \mathbf{G}_1 &= \frac{1}{J}(\mathbf{E}\zeta_x + \mathbf{F}\zeta_y + \mathbf{G}\zeta_z) \end{aligned} \tag{5-239}$$

and substitute them into Eq. (5-238), the final equation is in strong conservation-law form:

$$\frac{\partial \mathbf{U}_1}{\partial t} + \frac{\partial \mathbf{E}_1}{\partial \xi} + \frac{\partial \mathbf{F}_1}{\partial \eta} + \frac{\partial \mathbf{G}_1}{\partial \zeta} = 0 \tag{5-240}$$

It should be remembered that the vectors $\mathbf{E}_1$, $\mathbf{F}_1$, and $\mathbf{G}_1$ contain partial derivatives in the viscous and heat transfer terms. These partial derivative terms are to be transformed using Eqs. (5-228). For example, the shearing stress term, τ_{xy}, would be transformed to

$$\tau_{xy} = \mu \left(\xi_y \frac{\partial u}{\partial \xi} + \eta_y \frac{\partial u}{\partial \eta} + \zeta_y \frac{\partial u}{\partial \zeta} + \xi_x \frac{\partial v}{\partial \xi} + \eta_x \frac{\partial v}{\partial \eta} + \zeta_x \frac{\partial v}{\partial \zeta} \right) \tag{5-241}$$

The strong conservation-law form of the governing equations is a convenient form for applying finite-difference schemes. However, when using this form of the equations, caution must be exercised if the grid is changing. In this case, a constraint on the way the metrics are differenced, called the *geometric conservation law* (Thomas and Lombard, 1978), must be satisfied in order to prevent additional errors from being introduced into the solution. This will be discussed further in Chapter 10.

PROBLEMS

5-1 Verify Eq. (5-9).

5-2 Show that for an incompressible, constant property flow Eq. (5-18) reduces to Eq. (5-21).

5-3 Verify Eq. (5-30).

5-4 Using the nondimensionalization procedure described in Section 5-1.6, derive Eqs. (5-47).

5-5 Write the energy equation [Eq. (5-33)] in terms of axisymmetric body intrinsic coordinates.

5-6 Write the incompressible Navier-Stokes equation [Eq. (5-21)] in a spherical coordinate system.

5-7 Show that $\overline{\rho' u''} = \overline{\rho' u'}$.

5-8 Show that $\tilde{u} - \bar{u} = \overline{\rho' u'}/\bar{\rho}$.

5-9 Verify that $\overline{u''} = -\overline{\rho' u''}/\bar{\rho}$.

5-10 Starting with Eq. (5-80) show the steps in the development of Eq. (5-81).

5-11 Develop Eq. (5-84) by substitution (i.e., using $c_p \tilde{T} = \tilde{H} - \tilde{u}_i \tilde{u}_i/2$) starting with Eq. (5-81).

5-12 Show the steps in the derivation of Eq. (5-76) starting with the Navier-Stokes equations.

5-13 Apply an order of magnitude analysis to the incompressible two-dimensional Navier-Stokes equations for the case of a planar two-dimensional laminar jet. Indicate which terms in the Navier-Stokes equations can be neglected in this flow.

5-14 Explain why the boundary-layer equations may be applicable to the developing flow in a tube.

5-15 Determine the proper boundary conditions to apply to the thin-shear-layer equations for the two-dimensional shear layer formed by the merging of two infinite streams at uniform velocities U_a and U_b.

5-16 The boundary-layer equations, Eqs. (5-104)–(5-106), were developed for Prandtl numbers of the order of magnitude of one. For a laminar flow over a heated flat plate, indicate what alterations should be made in these equations to properly treat flows in which the Prandtl number becomes of the order of magnitude of (*a*) ϵ, (*b*) ϵ^2, (*c*) $1/\epsilon$, (*d*) $1/\epsilon^2$.

5-17 Using the Navier-Stokes equations, develop an exact Reynolds stress transport equation applicable to an incompressible turbulent boundary layer, i.e., obtain an expression for $\rho D\overline{u_i'u_j'}/Dt$. Show the steps in your development.

5-18 Using the expression for the transport of Reynolds stresses from Prob. 5-17, let $i = j$ to obtain an expression for the transport of turbulence kinetic energy.

5-19 Using the modeled form of the turbulence kinetic energy equation, Eq. (5-147), show that when convection and diffusion of turbulence kinetic energy are negligible, the kinetic energy turbulence model reduces to the Prandtl mixing-length formula.

5-20 Assuming that convection and diffusion of turbulence kinetic energy are negligible within the log-law region of a turbulent wall boundary layer, find an expression for the turbulence kinetic energy at the outer edge of the log-law region in terms of the wall shear stress. Compare this estimate with experimental measurements of $\bar{k}$ such as those of Klebanoff (see Hinze, 1975).

5-21 Assuming the validity of the Prandtl mixing-length formula for a turbulent wall boundary layer, obtain an expression for the ratio of the apparent turbulent viscosity to the molecular viscosity in the log-law region.

5-22 Verify the inner boundary condition for $\bar{k}$ stated in Eq. (5-148).

5-23 In a two-dimensional body intrinsic coordinate system, define the stream function for a steady compressible flow.

5-24 Obtain Eq. (5-220).

5-25 Verify Eqs. (5-222) and (5-223).

5-26 Transform the two-dimensional incompressible Navier-Stokes equation [Eq. (5-21)] using the transformation defined by Eqs. (5-217).

5-27 Show that the transformation defined by

$$x = r \cos \theta$$

$$y = r \sin \theta$$

$$z = z$$

will transform the three-dimensional compressible continuity equation expressed in cylindrical coordinates into the compressible continuity equation in Cartesian coordinates.

5-28 Apply in a successive manner the transformations given by Eqs. (5-224) and Eqs. (5-210) to the inviscid energy equation [Eq. (5-179)] written for a two-dimensional steady flow.

5-29 Transform the two-dimensional continuity equation

$$\frac{\partial \rho}{\partial t} + \frac{\partial \rho u}{\partial x} + \frac{\partial \rho v}{\partial y} = 0$$

to the (τ, ξ, η) computational domain using the transformation

$$\tau = t$$

$$\xi = \xi(t, x, y)$$

$$\eta = \eta(t, x, y)$$

Use the technique of Viviand to write the transformed equation in conservation-law form. Show all intermediate steps.

5-30 Transform the steady form of Euler's equations [Eqs. (5-192)] to the (ξ, η, ζ) computational domain using the transformation

$$\xi = x$$

$$\eta = \eta(x, y, z)$$

$$\zeta = \zeta(x, y, z)$$

Using the technique of Viviand, write the transformed equations in conservation-law form.

5-31 Consider the generalized transformation

$$\tau = t$$

$$\xi = \xi(t, x, y, z)$$

$$\eta = \eta(t, x, y, z)$$

$$\zeta = \zeta(t, x, y, z)$$

(*a*) Determine suitable expressions for the Jacobian of the transformation as well as the metrics.

(*b*) Apply this transformation to the compressible Navier-Stokes equations written in vector form [Eqs. (5-44)].

CHAPTER
SIX

NUMERICAL METHODS FOR INVISCID FLOW EQUATIONS

6-1 INTRODUCTION

The Navier-Stokes equations govern the flows commonly encountered in both internal and external applications. Computing a solution of the Navier-Stokes equations is frequently impossible or at least impractical and in many of these applications, unnecessary. Results obtained from a solution of the Euler equations are particularly useful in preliminary design work where information on pressure alone is desired. In problems where heat transfer and skin friction are required, a solution of the boundary-layer equations usually provides an adequate approximation. However, the outer edge conditions, including the pressure, must be specified from the inviscid solution as the first step in such an analysis.

The Euler equations are also of interest because many of the major elements of fluid dynamics are incorporated in them. For example, fluid flows frequently have internal discontinuities such as shock waves or contact surfaces. Solutions relating the end states across a shock are given by the Rankine-Hugoniot relations. These relations are contained in solutions of the Euler equations.

The Euler equations govern the motion of an inviscid nonheat-conducting gas and have a different character in different flow regimes. If the time-dependent terms are retained, the resulting unsteady equations are hyperbolic for all Mach numbers and solutions can be obtained using marching procedures. The situation is very different when a steady flow is assumed. In this case, the Euler equations are elliptic when the flow is subsonic and hyperbolic when the flow is supersonic. This change in character of the governing equations is the reason that development of methods for solving steady transonic flows has required many years. Many simplified versions of the Euler equations are used for inviscid fluid flows. When studying incompressible flows, it is often to our advantage to assume irrotationality. Under these conditions, a solution of Laplace's equation for the velocity potential provides the required data. Associated with the Euler equations are the companion set of small perturbation equations. In subsonic and supersonic flows, we observe that the Prandtl-Glauert equation provides the first-order theory for the potential function. In transonic flow the equation obtained for small perturbations is still a nonlinear equation. The classification of the various forms of the inviscid equations of motion is given in Table 6-1.

Many different methods are used to obtain solutions to the Euler equations or any of the various reduced forms of the Euler equations. The main goal of this chapter is to present the most commonly used methods for solving inviscid flow problems. Since we are primarily interested in finite-difference methods, many techniques which are extensively used to solve fluid flow problems will not be included. Most notable among these is the finite element method. This technique has received considerable use in computing incompressible flows about various configurations.

6-2 THE METHOD OF CHARACTERISTICS

Closed-form solutions of nonlinear hyperbolic partial differential equations do not exist for general cases. In order to obtain solutions to such equations we are required to use numerical methods. The method of characteristics is the oldest and most nearly exact method which is still used to solve hyperbolic PDE's. Even though this technique is being replaced by newer, more easily implemented finite-difference methods, a background in characteristic theory and its application is essential.

In our discussion in Chapter 2, we observed that certain directions or surfaces which bound the zones of influence are associated with hyperbolic equations. Signals are propagated along these particular surfaces influencing the solution at other points within the zone of influence. The method of characteristics is a technique which utilizes the known physical behavior of the solution at each point in the flow. A clear understanding of the essential elements of the method of characteristics can be obtained by studying a second-order linear partial differential equation.

6-2.1 Linear Systems

Consider the steady supersonic flow of an inviscid, nonheat-conducting perfect gas. Suppose the freestream flow is only slightly disturbed by a thin body so the fluid motion satisfies the small perturbation assumptions given by (see Section 5-5.6)

$$\frac{u}{U_\infty} \ll 1 \qquad \frac{v}{U_\infty} \ll 1$$

Table 6-1 Classification of the Euler equations

	Subsonic, $M < 1$	Sonic, $M = 1$	Supersonic, $M > 1$
Steady	Elliptic	Parabolic	Hyperbolic
Unsteady	Hyperbolic	Hyperbolic	Hyperbolic

where u and v are perturbation velocity components. If transonic and hypersonic flows are not considered, the governing partial differential equations reduce to the Prandtl-Glauert equation for supersonic flow. If the x axis is aligned with the freestream, this equation may be written

$$(1 - M_\infty^2)\phi_{xx} + \phi_{yy} = 0 \tag{6-1}$$

The freestream Mach number is denoted by M_∞ and the perturbation potential is denoted by ϕ. Initial data are specified along a smooth curve, C, which we choose to be x = constant in this case. Boundary conditions are prescribed at $y = 0$.

$$\begin{aligned} \frac{\partial \phi}{\partial y}(x, 0) &= U_\infty \left(\frac{dy}{dx}\right)_{\text{wall}} \\ \phi(0, y) &= 0 \end{aligned} \tag{6-2}$$

In order to present the formulation for a system of equations, it is advantageous to consider the nearly equivalent formulation introduced in Chapter 2. Using the perturbation velocity components

$$u = \frac{\partial \phi}{\partial x} \qquad v = \frac{\partial \phi}{\partial y}$$

and denoting $M_\infty^2 - 1$ by β^2, Eq. (6-1) may be written as the system

$$\begin{aligned} \beta^2 \frac{\partial u}{\partial x} - \frac{\partial v}{\partial y} &= 0 \\ \frac{\partial v}{\partial x} - \frac{\partial u}{\partial y} &= 0 \end{aligned} \tag{6-3}$$

with associated initial data and boundary conditions

$$\begin{aligned} &\left.\begin{aligned} u(0, y) &= 0 \\ v(0, y) &= 0 \end{aligned}\right\} \quad y > 0 \\ &v(x, 0) = v_{\text{wall}} \qquad y = 0 \end{aligned} \tag{6-4}$$

In order to use the method of characteristics, the system given by Eq. (6-3) is written along the characteristics. The differential equations of the characteristics are developed as the first step in this procedure.

Suppose the initial data for this problem are prescribed along an arbitrary smooth curve, C, and we consider methods for constructing a solution of Eq. (6-3) in the neighborhood of this curve. If the solution is sufficiently smooth, the first method that might be considered is to write a Taylor series about a point on C. Assume that our interest is in a small neighborhood and only terms through the first derivatives need to be retained. The solution for either u or v may then be written in the form

$$u(x + \Delta x, y + \Delta y) = u(x, y) + \Delta x \frac{\partial u}{\partial x}(x, y) + \Delta y \frac{\partial u}{\partial y}(x, y) + \cdots \tag{6-5}$$

In this expression, the coordinates (x, y) are on the initial data curve where u and v are known. However, we need to compute the first derivatives in the Taylor series. If s represents arc length along the curve C, we may write

$$\begin{aligned} \frac{du}{ds} &= \frac{\partial u}{\partial x}\frac{dx}{ds} + \frac{\partial u}{\partial y}\frac{dy}{ds} \\ \frac{dv}{ds} &= \frac{\partial v}{\partial x}\frac{dx}{ds} + \frac{\partial v}{\partial y}\frac{dy}{ds} \end{aligned} \tag{6-6}$$

The system of four equations in the unknown derivatives given by Eqs. (6-3) and (6-6) may be solved by any standard method such as Cramer's rule. It is clear that the determinant of the coefficients of the system must not vanish. (If the determinant of the coefficients vanishes, the direction of the curve C is along the characteristics of the system and, consistent with our discussion in Chapter 2, the derivatives may not be uniquely determined.) The differential equations of the characteristics are obtained by setting the determinant of this system equal to zero.

$$\begin{vmatrix} \beta^2 & 0 & 0 & -1 \\ 0 & -1 & 1 & 0 \\ \frac{dx}{ds} & \frac{dy}{ds} & 0 & 0 \\ 0 & 0 & \frac{dx}{ds} & \frac{dy}{ds} \end{vmatrix} = 0 \tag{6-7}$$

Expanding this determinant and solving the characteristic equation yields the expressions

$$\frac{dy}{dx} = \pm \frac{1}{\beta} \tag{6-8}$$

which are differential equations of the characteristics as illustrated in Fig. 6-1. Since β is constant, the characteristics can be obtained by integration and are given by

$$\begin{aligned} \xi &= x - \beta y \\ \eta &= x + \beta y \end{aligned} \tag{6-9}$$

The original differential equations written along the characteristics are called the compatibility equations. These compatibility equations may be derived by continuing

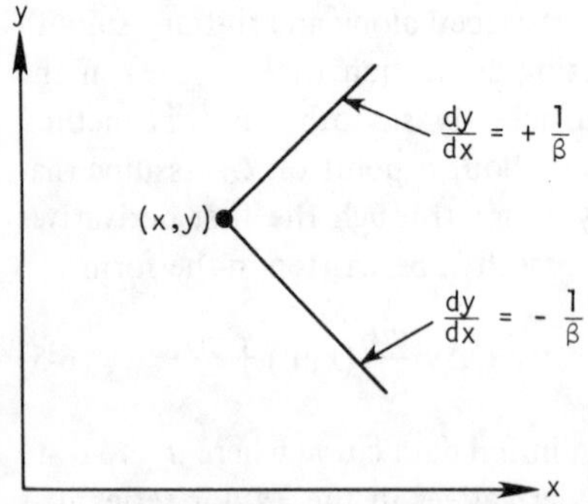

Figure 6-1 Characteristics of the Prandtl-Glauert equation.

to solve the original system of equations for the first derivatives. Along the characteristic directions, the determinant of the coefficients vanishes. If we solve for any of the first derivatives, for instance, $\partial u/\partial x$, and require that they are at least bounded, the determinant forming the numerator must also vanish. This may be written

$$\begin{vmatrix} 0 & 0 & 0 & -1 \\ 0 & -1 & 1 & 0 \\ \dfrac{du}{ds} & \dfrac{dy}{ds} & 0 & 0 \\ \dfrac{dv}{ds} & 0 & \dfrac{dx}{ds} & \dfrac{dy}{ds} \end{vmatrix} = 0 \tag{6-10}$$

If this determinant is expanded, the compatibility equations are given by

$$\frac{du}{ds} = \left(\frac{dy}{dx}\right)\frac{dv}{ds}$$

or

$$\frac{d}{ds}(\beta u + v) = 0 \tag{6-11}$$

along a right running characteristic where

$$\frac{dy}{dx} = -\frac{1}{\beta}$$

and

$$\frac{d}{ds}(\beta u - v) = 0 \tag{6-12}$$

along the left running characteristic

$$\frac{dy}{dx} = \frac{1}{\beta}$$

A more general procedure for deriving the characteristics is given by Whitham (1974). We will repeat the details of the procedure here and omit the derivation of the technique. In order to find the characteristics of the system [Eq. (6-3)], we write these equations in the vector form

$$\frac{\partial \mathbf{w}}{\partial x} + [A]\frac{\partial \mathbf{w}}{\partial y} = 0 \tag{6-13}$$

where

$$\mathbf{w} = \begin{bmatrix} u \\ v \end{bmatrix}$$

and

$$[A] = \begin{bmatrix} 0 & -\dfrac{1}{\beta^2} \\ -1 & 0 \end{bmatrix} \tag{6-14}$$

The eigenvalues of this system are the eigenvalues of $[A]$. These are obtained by extracting the roots of the characteristic equation of $[A]$. Thus we write

$$|[A] - \lambda[I]| = 0$$

or

$$\begin{vmatrix} -\lambda & -\dfrac{1}{\beta^2} \\ -1 & -\lambda \end{vmatrix} = 0$$

This produces the quadratic equation

$$\lambda^2 - \frac{1}{\beta^2} = 0$$

The roots of this equation are

$$\lambda_1 = \frac{1}{\beta}$$

$$\lambda_2 = -\frac{1}{\beta}$$

This pair of roots form the differential equations of the characteristics we have already derived in Eq. (6-8). Since our original Prandtl-Glauert equation for supersonic flow is just a wave equation in ϕ, we could have written the characteristic differential equations using the results from our discussion of the second-order PDE [Eq. (2-15)]. The next step is to determine the compatibility equations. Following Witham, these equations are obtained by premultiplying the system given by Eq. (6-13) by the left eigenvector of $[A]$. This effectively provides a method of writing the equations along the characteristics.

Let $\mathbf{L}^1$ represent the left eigenvector of $[A]$ corresponding to λ_1 and $\mathbf{L}^2$ represent the left eigenvector corresponding to λ_2. We derive the eigenvectors of $[A]$ by writing

$$[L^i]^T[A - \lambda_i I] = 0 \tag{6-15}$$

If we let

$$\mathbf{L}^1 = \begin{bmatrix} l_1 \\ l_2 \end{bmatrix}$$

then

$$[l_1^1, l_2^1]\begin{bmatrix} -\dfrac{1}{\beta} & -\dfrac{1}{\beta^2} \\ -1 & -\dfrac{1}{\beta} \end{bmatrix} = 0$$

This provides the equations

$$\frac{l_1^1}{\beta} + l_2^1 = 0 \qquad \frac{l_1^1}{\beta^2} + \frac{l_2^1}{\beta} = 0$$

which are equivalent as expected. Since we are only able to obtain the normalized components of $\mathbf{L}^1$, assume $l_1^1 = -\beta$. Then the solution for l_2^1 is

$$l_2^1 = 1$$

and

$$\mathbf{L}^1 = \begin{bmatrix} -\beta \\ 1 \end{bmatrix}$$

In a similar manner, the solution for $\mathbf{L}^2$ is

$$\mathbf{L}^2 = \begin{bmatrix} \beta \\ 1 \end{bmatrix}$$

The compatibility equations are now obtained by writing our system [Eq. (6-13)] along the characteristics. To do this we multiply Eq. (6-13) by the transpose of the left eigenvector.

$$[\mathbf{L}^i]^T [\mathbf{w}_x + [A]\mathbf{w}_y] = 0 \tag{6-16}$$

The term $[\mathbf{L}^i]^T[A]$ may be replaced by $[\mathbf{L}^i]^T\lambda_i[I]$ by substituting from Eq. (6-15). Thus, we may write Eq. (6-16) as

$$[\mathbf{L}^i]^T [\mathbf{w}_x + \lambda_i \mathbf{w}_y] = 0$$

The compatibility equation along λ_1 is obtained from

$$[-\beta, 1]\begin{bmatrix} u_x + \dfrac{1}{\beta} u_y \\ v_x + \dfrac{1}{\beta} v_y \end{bmatrix} = 0$$

Thus

$$\frac{\partial}{\partial x}(\beta u - v) + \frac{1}{\beta}\frac{\partial}{\partial y}(\beta u - v) = 0 \tag{6-17a}$$

In a similar manner, the compatibility equation along the right running characteristic in partial derivative form is

$$\frac{\partial}{\partial x}(\beta u + v) - \frac{1}{\beta}\frac{\partial}{\partial y}(\beta u + v) = 0 \tag{6-17b}$$

Equation (6-17*a*) is valid along the positive or left running characteristic. It expresses the fact that the quantity $(\beta u - v)$ is constant along λ_1. This can be demonstrated by letting s represent distance along the characteristic and writing

$$\frac{d}{ds}(\beta u - v) = \frac{\partial}{\partial x}(\beta u - v)\frac{dx}{ds} + \frac{\partial}{\partial y}(\beta u - v)\frac{dy}{ds}$$

However, if $(\beta u - v)$ is constant along the characteristic, we may write

$$\frac{d}{ds}(\beta u - v) = 0$$

or

$$\frac{\partial}{\partial x}(\beta u - v) + \left(\frac{dy}{dx}\right)\frac{\partial}{\partial y}(\beta u - v) = 0$$

which is the same as Eq. (6-17*a*). Therefore, we conclude that $(\beta u - v)$ is constant along λ_1 and $(\beta u + v)$ is constant along λ_2. The quantities $(\beta u - v)$ and $(\beta u + v)$ are called *Riemann invariants* (Garabedian, 1964). Since these two quantities are constant along opposite pairs of characteristics, it is easy to determine u and v at a given point. If at a point (x, y) we know $(\beta u + v)$ and $(\beta u - v)$, we can immediately compute both u and v. An example illustrating this is in order.

Example 6-1 A uniform inviscid supersonic flow ($M_\infty = \sqrt{2}$) encounters a one-period sine wave wrinkle in the metal skin of a wind tunnel. The geometry of this configuration is shown in Fig. 6-2. The maximum amplitude of the sine wave is ϵ/L and $\epsilon/L \ll 1$. Determine the solution for the perturbation velocities, u and v, using the method of characteristics.

Solution: Since the flow is assumed to satisfy the small perturbation assumption, the Prandtl-Glauert equation can be used. We choose to solve the system of

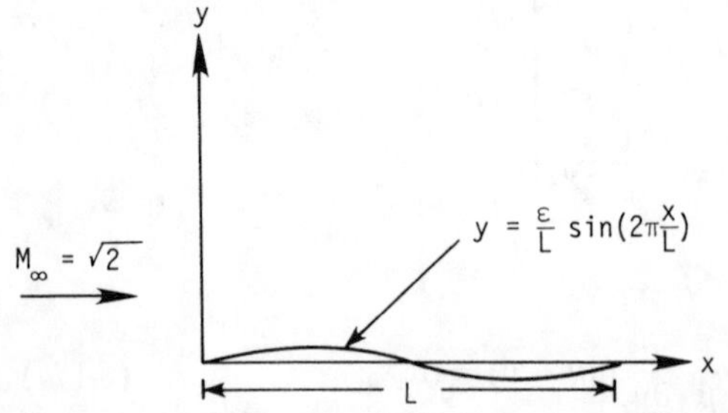

Figure 6-2 Wavy wall geometry.

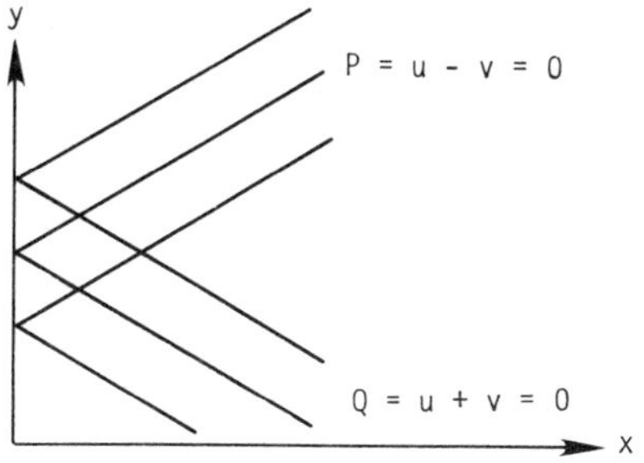

Figure 6-3 Initial data line.

equations [Eq. (6-3)] for the perturbation components u and v. In this case, $\beta^2 = 1$ and we elect to solve the system of partial differential equations

$$\frac{\partial u}{\partial x} - \frac{\partial v}{\partial y} = 0$$

$$\frac{\partial v}{\partial x} - \frac{\partial u}{\partial y} = 0$$

with initial data specified along $x = 0, y > 0$

$$u = 0$$

$$v = 0$$

subject to the surface boundary condition (see Section 6-4),

$$v = 2\pi U_\infty \frac{\epsilon}{L^2} \cos\left(2\pi \frac{x}{L}\right) \qquad 0 \leqslant x \leqslant L$$

Since the problem is two dimensional and obeys the small perturbation assumptions, we may apply the boundary conditions in the $y = 0$ plane. This makes the problem much easier.

We begin our characteristic solution by sketching the characteristics which originate at the initial data surface $x = 0$. Along the left running characteristics we know that

$$\frac{dy}{dx} = 1 \qquad u - v = P = \text{constant}$$

while along the other characteristic

$$\frac{dy}{dx} = -1 \qquad u + v = Q = \text{constant}$$

Therefore, we determine u and v at any point as

$$u = \frac{P + Q}{2} \qquad v = \frac{Q - P}{2}$$

Since the right running characteristics that strike the surface originate in the freestream, the Q variable is initially zero. It is also true that $P = 0$ for those characteristics which originate in the freestream (see Fig. 6-3).

Consider the characteristic that strikes the wavy wall. An up or left running characteristic is introduced at that point in such a way that the surface boundary condition is satisfied. Thus at any station, x_1, we have

$$Q = u + v = 0$$

$$v = \frac{2\pi\epsilon}{L^2} U_\infty \cos\left(2\pi \frac{x_1}{L}\right)$$

Therefore

$$u = -\frac{2\pi\epsilon}{L^2} U_\infty \cos\left(2\pi \frac{x_1}{L}\right)$$

and

$$P = u - v = -\frac{4\pi\epsilon}{L^2} U_\infty \cos\left(2\pi \frac{x_1}{L}\right)$$

The solution for u and v is constructed by marching outward from the initial data surface in the x direction. A grid with indices and the corresponding characteristics is shown in Fig. 6-4. The solution can now be obtained at the intersections of the characteristics. At point (1, 3)

$$P = 0$$

$$Q = 0$$

$$u = 0$$

$$v = 0$$

At (3, 2)

$$P = -\frac{4\pi\epsilon}{L^2} U_\infty \cos\left(2\pi \frac{x_1}{L}\right)$$

$$Q = 0$$

$$u = -\frac{2\pi\epsilon}{L^2} U_\infty \cos\left(2\pi \frac{x_1}{L}\right)$$

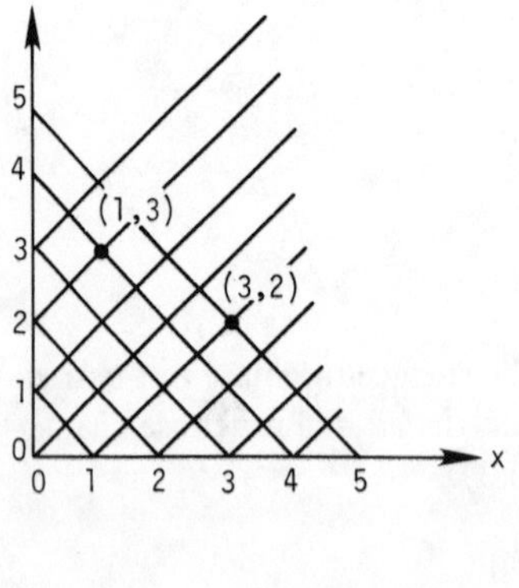

Figure 6-4 Characteristic net.

$$v = \frac{2\pi\epsilon}{L^2} \cos\left(2\pi \frac{x_1}{L}\right)$$

The solution is known everywhere in the domain of interest. The results of this example may be verified by solving the Prandtl-Glauert equation directly for the velocity potential and then computing the solution for u and v.

6-2.2 Nonlinear Systems

The development presented thus far is for a system of two linear equations and was chosen for its simplicity. In more complex nonlinear problems, the results are not as easily obtained. In the general case, the characteristic slopes are not constant but must vary as the fluid properties change. The governing partial differential equations may be nonhomogeneous. Clearly, the compatibility equations cannot be directly integrated in closed form along the characteristics in that case. For the general nonlinear problem, both the compatibility equations and the characteristic equations must be integrated numerically to obtain a complete flowfield solution. Not only are the flow variables unknown but the location in the field along the characteristics must be computed.

In order to illustrate the difference in applying the method of characteristics to a linear and a nonlinear problem, we consider the two-dimensional supersonic flow of a perfect gas over a flat surface. For simplicity we choose a rectangular coordinate system and write the Euler equations (see Chapter 5) governing this inviscid flow as the matrix system

$$\frac{\partial \mathbf{w}}{\partial x} + [A] \frac{\partial \mathbf{w}}{\partial y} = 0 \tag{6-18}$$

where

$$\mathbf{w} = \begin{bmatrix} u \\ v \\ p \\ \rho \end{bmatrix}$$

and

$$[A] = \frac{1}{u^2 - a^2} \begin{bmatrix} uv & -a^2 & -\dfrac{v}{\rho} & 0 \\ 0 & \dfrac{v}{u}(u^2 - a^2) & \dfrac{u^2 - a^2}{\rho u} & 0 \\ -\rho v a^2 & \rho u a^2 & uv & 0 \\ -\rho v & \rho u & \dfrac{v}{u} & \dfrac{v}{u}(u^2 - a^2) \end{bmatrix}$$

The initial data, **I**, are prescribed and may be written as

$$\mathbf{w}(0,y) = \mathbf{I}(y) \qquad 0 \leqslant y \leqslant h$$

and the boundary conditions are

$$v(x,0) = 0$$
$$u(x,h) = u_\infty$$
$$v(x,h) = v_\infty$$
$$p(x,h) = p_\infty$$
$$\rho(x,h) = \rho_\infty$$

The eigenvalues of $[A]$ determine the characteristic directions and must be found as the first step. These eigenvalues are

$$\lambda_1 = \frac{v}{u} \qquad \lambda_2 = \frac{v}{u}$$
$$\lambda_3 = \frac{uv + a\sqrt{u^2 + v^2 - a^2}}{u^2 - a^2} \qquad \lambda_4 = \frac{uv - a\sqrt{u^2 + v^2 - a^2}}{u^2 - a^2} \tag{6-19a}$$

The matrix of left eigenvectors associated with these values of λ may be written

$$[T]^{-1} = \begin{bmatrix} \dfrac{\rho u}{a^2} & \dfrac{\rho v}{a^2} & 0 & 1 \\ \rho u & \rho v & 1 & 0 \\ -\dfrac{1}{\sqrt{u^2+v^2-a^2}} & +\dfrac{u}{v}\dfrac{1}{\sqrt{u^2+v^2-a^2}} & \dfrac{1}{\rho v a} & 0 \\ \dfrac{1}{\sqrt{u^2+v^2-a^2}} & -\dfrac{u}{v}\dfrac{1}{\sqrt{u^2+v^2-a^2}} & \dfrac{1}{\rho v a} & 0 \end{bmatrix} \tag{6-19b}$$

We obtain the compatibility relations by premultiplying the original system by $[T]^{-1}$. These relations along the wave fronts are given by

$$-v\frac{du}{ds_3} + u\frac{dv}{ds_3} + \frac{\beta}{\rho}\frac{dp}{ds_3} = 0 \tag{6-20}$$

along

$$\frac{dy}{dx} = \lambda_3$$

and

$$v\frac{du}{ds_4} - u\frac{dv}{ds_4} + \frac{\beta}{\rho}\frac{dp}{ds_4} = 0 \tag{6-21}$$

along

$$\frac{dy}{dx} = \lambda_4$$

In these expressions

$$\beta = \sqrt{M^2 - 1} \qquad M^2 = \frac{u^2 + v^2}{a^2}$$

Equation (6-20) is an ordinary differential equation which holds along the characteristic with slope λ_3. Arc length along this characteristic is denoted by s_3. A similar result is expressed in Eq. (6-21). In contrast to our linear example using the Prandtl-Glauert equation, the analytic solution for the characteristics is not known for the general nonlinear problem. It is clear that we must numerically integrate to determine the shape of the characteristics in a step by step manner. Consider the characteristic defined by λ_3

$$\frac{dy}{dx} = \frac{uv + a\sqrt{u^2 + v^2 - a^2}}{u^2 - a^2}$$

Starting at an initial data surface, this expression can be integrated to obtain the coordinates of the next point on the curve. At the same time, the differential equation defining the other wave front characteristic can be integrated. For a simple first-order integration, this provides us with two equations for the wave front characteristics. From these expressions, we determine the coordinates of their intersection (point A in Fig. 6-5). Once the point A is known, the compatibility relations, Eqs. (6-20) and (6-21), are integrated along the characteristics to this point. This provides a system of equations for the unknowns at point A. Of course auxiliary relationships are required to complete the problem and these are provided by integrating the streamline compatibility equations or by using other valid equations relating the unknowns at A.

By using this procedure, a first-order estimate of both the location of point A and the associated flow variables can be obtained. These first-order estimates are usually used as a first step in a predictor-corrector scheme in calculating the solution to a system of hyperbolic PDE's using the method of characteristics. In the corrector step, a new intersection point B can be computed which now includes the nonlinear nature of the characteristic curves. In a similar manner, the dependent variables at B are computed.

The calculation of the solution at point B presents an interesting problem. Because the problem is nonlinear, the final intersection point B does not necessarily

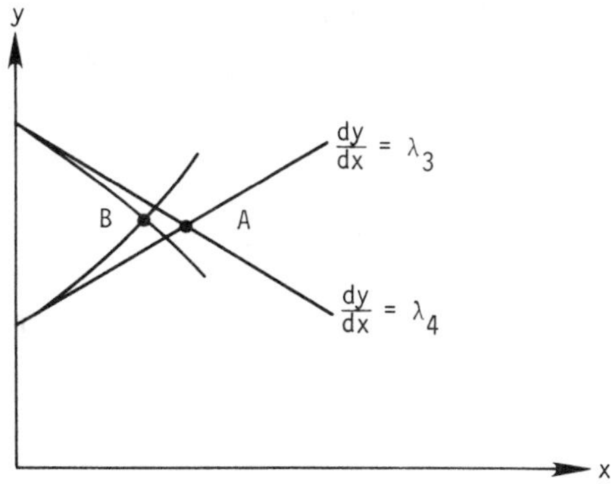

Figure 6-5 Characteristic solution point.

appear at the same value of x for all solution points. Consequently, the solution is usually interpolated onto an x = constant surface before the next integration step is started. This requires additional logic and adds considerably to the difficulty in structuring an accurate code.

The problem of integrating the compatibility equations and satisfying the boundary conditions at both permeable and impermeable boundaries is discussed in the next section. It should be clear that the wall boundary condition is iterative in the sense that we attempt to satisfy a particular boundary condition at a point on a surface with an initially unknown x coordinate.

The two-dimensional flow problem used in this section actually can be treated using characteristics in a much simpler setting (see Section 6-4). The main reason for this discussion is to present the ideas behind the numerical integration of the equations of motion using characteristic methods and to introduce some of the inherent difficulties in the general method. More complete descriptions are given by numerous authors including Owczarek (1964), Shapiro (1953), and Courant and Friedrichs (1948).

6-3 SHOCK-CAPTURING METHODS

Shock-capturing schemes are the most widely used techniques for computing inviscid flows with shocks. In this approach, the Euler equations are cast in conservation-law form and any shock waves or other discontinuities are predicted as part of the solution with no special treatment required. The shock waves predicted by these methods are indeed smeared over several mesh intervals but the simplicity of the approach may outweigh the slight compromise in results compared to shock fitting schemes.

Most flowfield solutions computed using shock-capturing methods employ shock fitting of the boundary. Since boundary shocks can be fit with any of the standard schemes discussed in this section and Section 6-4, the real advantage accrues when a complicated internal shock structure exists. In this case, the internal structure is captured and the special treatment of each shock wave is eliminated.

Lax (1954) has shown that shock wave speed and strength are correctly predicted when the conservative form of the Euler equations is used. This means that the physically correct weak solution corresponding to the Rankine-Hugoniot equations for shocks is obtained if the conservation-law form is used, and the equations are discretized in a conservative manner. In Section 6-4 we will obtain a weak solution of the Euler equations using nonconservative form and demonstrate that the discontinuity in this solution does not physically represent a shock wave. The form of the solution predicted depends upon the form of the equations used.

As an example of conservation form, consider the supersonic flow of a perfect gas over a two-dimensional surface. If we assume the x axis forms the body surface and is also the marching direction, the equations are given by the steady two-dimensional version of Eq. (5-192) and may be written

$$\frac{\partial \rho u}{\partial x} + \frac{\partial \rho v}{\partial y} = 0 \tag{6-22}$$

$$\frac{\partial}{\partial x}(p + \rho u^2) + \frac{\partial}{\partial y}(\rho uv) = 0 \tag{6-23}$$

$$\frac{\partial}{\partial x}(\rho uv) + \frac{\partial}{\partial y}(p + \rho v^2) = 0 \tag{6-24}$$

For a steady isoenergetic flow, the total enthalpy is constant. In this case, the differential energy equation is not numerically integrated in favor of using the analytically integrated form

$$H = \frac{\gamma}{\gamma - 1}\frac{p}{\rho} + \frac{u^2 + v^2}{2} = \text{constant} \tag{6-25}$$

The system formed by Eqs. (6-22), (6-23), and (6-24) in conjunction with the constant total enthalpy equation is hyperbolic for supersonic flow and a solution can be obtained by marching or integrating the equations in the x direction starting from an initial data surface. The geometry for such a marching problem is shown in Fig. 6-6. Initial data are prescribed along the line $x = 0$ and the solution is advanced in the x direction subject to wall boundary conditions and an appropriate condition at $y_{\max}$.

Equations (6-22)-(6-24) are of the form

$$\frac{\partial \mathbf{E}}{\partial x} + \frac{\partial \mathbf{F}}{\partial y} = 0 \tag{6-26}$$

where

$$\mathbf{E} = \begin{bmatrix} \rho u \\ p + \rho u^2 \\ \rho uv \end{bmatrix} \qquad \mathbf{F} = \begin{bmatrix} \rho v \\ \rho uv \\ p + \rho v^2 \end{bmatrix}$$

Equation (6-26) may be integrated with any of the methods presented in Chapter 4 for hyperbolic PDE's. Typically, MacCormack's scheme would be a good choice. The

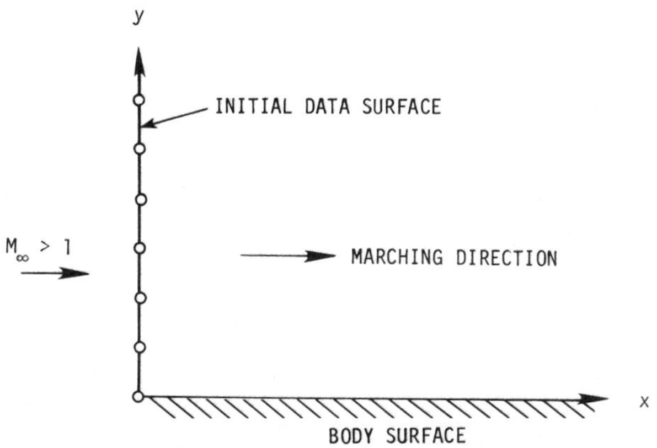

Figure 6-6 Coordinate system for marching problem.

forward predictor, backward corrector version of MacCormack's method applied to Eq. (6-26) may be written

$$\overline{\mathbf{E}_j^{n+1}} = \mathbf{E}_j^n - \frac{\Delta x}{\Delta y}(\mathbf{F}_{j+1}^n - \mathbf{F}_j^n)$$

$$\mathbf{E}_j^{n+1} = \frac{1}{2}\left[\mathbf{E}_j^n + \overline{\mathbf{E}_j^{n+1}} - \frac{\Delta x}{\Delta y}(\overline{\mathbf{F}_j^{n+1}} - \overline{\mathbf{F}_{j-1}^{n+1}})\right] \tag{6-27}$$

At the end of the predictor and corrector steps, **E** must be decoded to obtain the primitive variables. In this way, the new flux vector can be formed for the next integration step. After advancing the solution, the y component of velocity is immediately known as

$$v = \frac{E_3}{E_1}$$

where the subscripts denote elements of **E**. A quadratic equation must be solved for the x component of the velocity. If we combine E_2 with the energy equation to eliminate p we have

$$\rho = \frac{E_2}{u^2 + [(\gamma - 1)/2\gamma](2H - u^2 - v^2)}$$

We now eliminate ρ in favor of u using E_1, as

$$\rho = \frac{E_1}{u} \tag{6-28}$$

This yields a quadratic equation for u which has roots

$$u = \frac{\gamma}{\gamma + 1}\frac{E_2}{E_1} \pm \sqrt{\left(\frac{\gamma}{\gamma + 1}\frac{E_2}{E_1}\right)^2 - \frac{\gamma - 1}{\gamma + 1}(2H - v^2)} \tag{6-29}$$

The correct sign on the radical is typically positive. The density can now be computed from E_1 and the pressure from E_2 as

$$p = E_2 - \rho u^2 \tag{6-30}$$

Having completed this process, **F** can be recalculated and the next step in the integration can be implemented.

Example 6-2 Compute the flowfield produced by a two-dimensional wedge moving at a Mach number of 2.0 if the wedge half angle is 15°. Assume inviscid flow of a perfect gas.

Solution: The problem requires that we determine the shock wave location and strength as well as internal flow detail. The wedge and associated flow are shown in Fig. 6-7. In a two-dimensional wedge flow with an attached shock wave, the flow is conical. This means that flow properties along rays from the vertex of

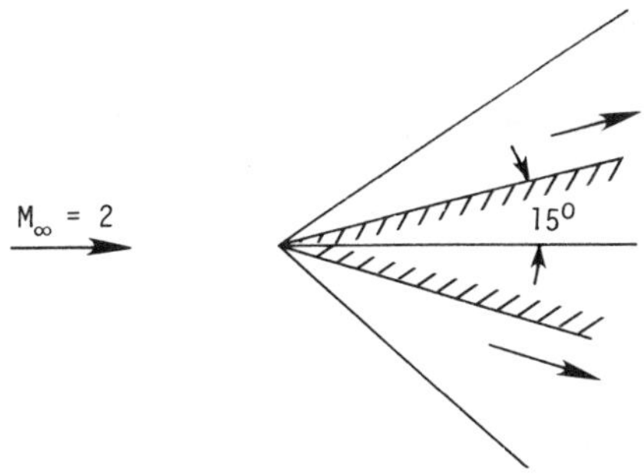

Figure 6-7 Wedge flow with attached shock.

the wedge are constant (Anderson, 1982). This results in a simplification of the problem.

For this problem, the governing PDE's are given by Eqs. (6-22)-(6-24) and the energy equation [Eq. (6-25)]. The boundary conditions are the surface tangency requirement at the wedge surface and freestream conditions outside the shock wave. We recognize that we can select the x axis along the wedge surface and march the equations in this direction so long as the shock layer Mach number is greater than one. Unfortunately, the shock layer expands as we move downstream, and this eventually causes our outer boundary point (at $y = y_{max}$) to interfere with the shock wave.

The problem can easily be solved utilizing the fact that the shock wave is straight and the thickness of the shock layer grows linearly with x. We introduce the independent variable transformation given by

$$\xi = x \qquad \eta = \frac{y}{x} \tag{6-31}$$

This provides the grid shown in Fig. 6-8. We can solve the wedge-flow problem with no difficulty now because the constant η lines grow linearly with x. Since the governing equations are hyperbolic in the ξ direction, initial data must be prescribed along some noncharacteristic surface. The line $\xi = 1$ is an easy choice. The PDE's are integrated in the ξ direction using arbitrarily assigned initial data.

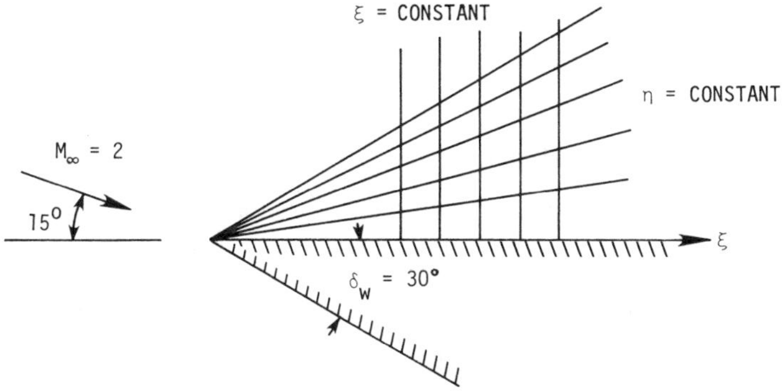

Figure 6-8 Wedge with transformed shock layer.

Since the solution to two-dimensional wedge flow is conical, the conical solution will be obtained for large ξ (asymptotically).

If the governing PDE's are transformed from (x, y) into (ξ, η) coordinates, they become

$$\frac{\partial \bar{\mathbf{E}}}{\partial \xi} + \frac{\partial \bar{\mathbf{F}}}{\partial \eta} = 0 \tag{6-32}$$

where $$\bar{\mathbf{E}} = \xi \mathbf{E}$$

and $$\bar{\mathbf{F}} = \mathbf{F} - \eta \mathbf{E}$$

An additional problem can be avoided by utilizing the conical flow property in this problem. The stability of the integration scheme used in solving Eq. (6-32) depends upon the eigenvalue structure of the $[A]$ matrix of the expanded system written in (ξ, η) coordinates.

$$\frac{\partial \mathbf{w}}{\partial \xi} + [A]\frac{\partial \mathbf{w}}{\partial \eta} + \mathbf{H} = 0 \tag{6-33}$$

In this expression $\mathbf{w}$ is the vector of primitive variables and $\mathbf{H}$ is a source term which occurs in this expanded form. If the eigenvalues of $[A]$ are evaluated, they are found to depend explicitly on the ξ coordinate. That is, ξ appears in the expressions for the eigenvalues. As the solution is marched downstream in ξ, the allowable step size must change as ξ increases if an explicit method such as MacCormack's is used. If the step size did not change as ξ increased, a stability problem would occur. This problem can be avoided if we elect to integrate the equations from $\xi = 1$ to $\xi = 1 + \Delta\xi$ in an iterative manner until a converged solution is obtained.

The application of boundary conditions requires careful consideration. We must include enough points in the η direction so that the shock wave can form naturally and not be interfered with by the fixed freestream conditions which are maintained at $\eta = \eta_{\max}$. For example, if our shock wave angle (measured from the wedge surface) is 20°, and we elect to use 10 points in the shock layer

$$\eta_{\text{shock}} = \tan(20^\circ) = 0.3640$$

$$\Delta\eta = \frac{0.3640}{10 - 1} = 0.0404$$

Suppose we add an additional 5 points using this computed $\Delta\eta$, then

$$\eta_{\max} = 0.0404(15 - 1) = 0.5662$$

and the last mesh point is at an angle of 29.52°. This should provide sufficient freedom for the shock wave to form without interference from the fixed boundary condition at $\eta_{\max}$.

If MacCormack's method is used, the predictor can be directly applied at the wall since forward differences are used. However, the corrector step must be modified. One way to assure satisfaction of surface tangency is to also use a forward corrector and overwrite the decoded value of v at the wall with the boundary condition $v = 0$. While the use of forward differences in both the predictor and corrector is generally unstable, the wall boundary condition alters the stability in such a way as to provide a stable solution.

Typical shock-capturing pressure results for wedge flow are presented in Fig. 6-9. These results show an excellent solution, at $\nu = 1.0$ with a sharp shock wave, and very little smearing with few oscillations. However, the same calculation at a Courant number of 0.7 demonstrates the dispersive behavior of second-order methods previously discussed in Chapter 4.

Before we leave this wedge-flow example it is worthwhile to note that a solution could also have been obtained using a time-dependent formulation. If the governing PDE's are written in polar coordinates including the time terms, they are of the form

$$\frac{\partial \mathbf{E}}{\partial t} + \frac{\partial \mathbf{F}}{\partial \theta} + \frac{\partial \mathbf{G}}{\partial R} + \mathbf{H} = 0 \tag{6-34}$$

where the origin is at the vertex of the wedge and the vectors are the appropriate polar forms. If we assume a priori that the flow is conical, a solution can be computed in a R = constant plane if the radial derivatives are discarded. This requires a solution of the system

$$\frac{\partial \mathbf{E}}{\partial t} + \frac{\partial \mathbf{F}}{\partial \theta} + \mathbf{H} = 0 \tag{6-35}$$

This system is hyperbolic in time and can be integrated to attain a steady wedge-flow solution. In some ways, the time-dependent set is easier to use. For example, the decoding procedure is much simpler.

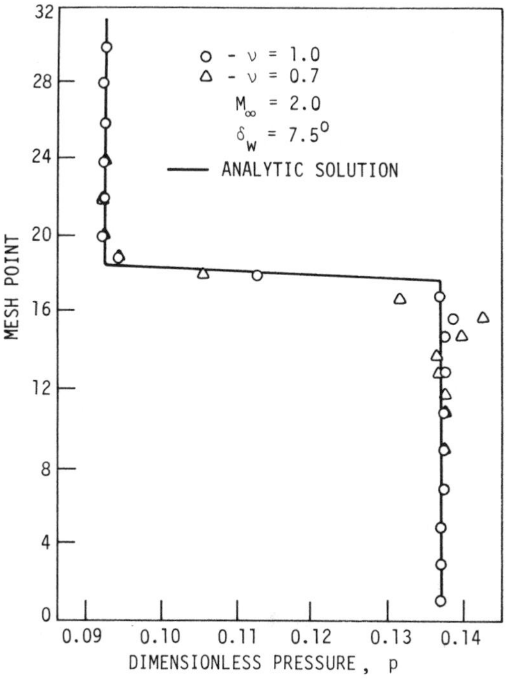

Figure 6-9 Shock-capturing pressure results for wedge flow.

As in Example 6-2, the equations of motion are usually transformed into a computational domain. One of the more frequently used transformations is that of Viviand (1974) and Vinokur (1974). This transformation (see Chapter 5) assures us that a system of equations in a strong conservation-law form can be written in the same form after changing the independent variables. There may be disadvantages to Viviand's transformed equation form because the Jacobian of the mapping always appears in the denominator of the conservative variable terms. In order to avoid the introduction of errors through the geometry, special care must be taken in forming the metrics. This point is discussed in detail in Chapter 10.

The difficulty encountered in using a simple rectangular mesh for Example 6-2 could have been eliminated if the shock wave was treated as a discontinuity. In fact, most shock-capturing codes fit boundary shock waves as discontinuities and capture interior shock waves as they develop. While the same philosophy of shock fitting holds for the steady flow marching problem as for time-dependent flows, a slightly different scheme is sometimes used to predict the interior or post shock pressure when the conservative form of the original equations is used. Consider a system of PDE's of the form given in Eq. (6-26). Suppose we make use of a normalizing transformation

$$(x, y) \rightarrow (\xi, \eta)$$
$$\xi = x \qquad \eta = \frac{y}{y_s(x)} \tag{6-36}$$

where $y - y_s(x) = 0$ is the equation of the shock wave. As shown in Fig. 6-10, the physical domain is now transformed into a computational domain with the shock wave at $\eta = 1.0$. The conservation form for the governing equations using such a transformation may be similar to Viviand's or any other form that conserves the appropriate flux terms. We again assume that the solution for the interior of the shock layer is advanced. At the shock wave, one-sided integration must be used to obtain an estimate for one of the variables. We assume initially that we know everything along an initial data surface including the shock slope. We advance the solution on the interior including the shock point. In addition, the shock slope equation, (dy_s/dx), is integrated providing an updated estimate of the new shock position. We now calculate the shock slope at the new location and the dependent variables other than pressure can then be obtained.

If the pressure on the downstream side of the shock is known, we clearly can

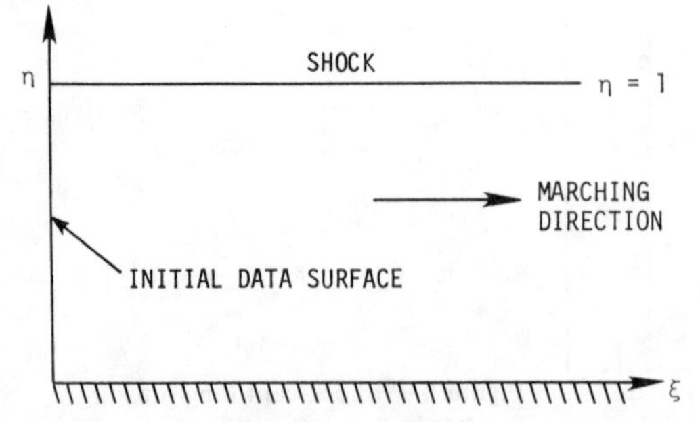

Figure 6-10 Normalizing transformation.

determine the density and both velocity components from the Rankine-Hugoniot equations. Our requirement is to develop the expression for shock slope. We write the surface equation of the shock wave as

$$y - y_s(x) = 0 \tag{6-37}$$

The shock normal is then written

$$\mathbf{n}_s = \frac{1}{[1 + (dy_s/dx)^2]^{1/2}} \left(-\mathbf{i}\frac{dy_s}{dx} + \mathbf{j} \right) \tag{6-38}$$

The normal component of velocity on the freestream side of the shock wave is given by

$$u_{\infty n} = \mathbf{n}_s \cdot \mathbf{V}_\infty = \frac{1}{[1 + (dy_s/dx)^2]^{1/2}} \left(-u_\infty \frac{dy_s}{dx} + v_\infty \right) \tag{6-39}$$

If this equation is solved for the shock slope we obtain

$$(u_{\infty n}^2 - u_\infty^2)\frac{dy_s}{dx} = -u_\infty v_\infty \pm \sqrt{u_\infty^2 v_\infty^2 - (u_{\infty n}^2 - u_\infty^2)(u_{\infty n}^2 - v_\infty^2)} \tag{6-40}$$

The term $u_{\infty n}^2$ required in Eq. (6-40) is known from the pressure ratio across the shock as given in Eq. (5-209) and is

$$u_{\infty n}^2 = \frac{\gamma - 1}{2} \frac{p_\infty}{\rho_\infty} \left(1 + \frac{\gamma + 1}{\gamma - 1} \frac{p_2}{p_\infty} \right) \tag{6-41}$$

After the shock slope is computed, all quantities are known at the new location. The same procedure is repeated for both the predictor and corrector steps. We have again performed the shock fitting assuming the post shock pressure (or other quantity) was known. This follows the approach suggested by Thomas et al. (1972).

Since we are examining methods for either time-dependent or steady supersonic inviscid flows, the governing equations are hyperbolic. Usually hyperbolic systems are solved using explicit methods. However, the step size for most explicit schemes is limited by the CFL condition. This can lead to unreasonably long computation times for some problems. The development and use of fully implicit methods for hyperbolic PDE's is a relatively recent phenomenon. Early efforts were usually partially explicit or iterative in nature. Recently noniterative algorithms have been developed by Lindemuth and Killeen (1973), Briley and McDonald (1973), and Beam and Warming (1976). The advantage of implicit methods lies in the unrestricted stability limit. Although more computational effort is required per time step compared to an explicit method, the overall time required to obtain a solution may be less. We will review the development of the basic scheme presented by Beam and Warming (1976) for the conservation form of the governing equations and then consider the split flux algorithm developed by Steger and Warming (1979).

The basic system under consideration is of the form given in Eq. (5-192) and is repeated here for convenience

$$\frac{\partial \mathbf{U}}{\partial t} + \frac{\partial \mathbf{E}}{\partial x} + \frac{\partial \mathbf{F}}{\partial y} = 0 \tag{6-42}$$

where **U** is the vector of conservative variables and **E** and **F** are vector functions of **U**. If the trapezoidal rule given by Eq. (4-58) is used as the basic integration scheme, the value of **U** at the advanced time level is given by

$$\mathbf{U}^{n+1} = \mathbf{U}^n + \frac{\Delta t}{2}\left[\left(\frac{\partial \mathbf{U}}{\partial t}\right)^n + \left(\frac{\partial \mathbf{U}}{\partial t}\right)^{n+1}\right]$$

or

$$\mathbf{U}^{n+1} = \mathbf{U}^n - \frac{\Delta t}{2}\left[\left(\frac{\partial \mathbf{E}}{\partial x} + \frac{\partial \mathbf{F}}{\partial y}\right)^n + \left(\frac{\partial \mathbf{E}}{\partial x} + \frac{\partial \mathbf{F}}{\partial y}\right)^{n+1}\right] \tag{6-43}$$

This expression provides a second-order integration algorithm for the unknown vector, $\mathbf{U}^{n+1}$, at the next time level. It is implicit because the derivatives of **U** as well as **U** appear at the advanced level thus coupling the unknowns at neighboring grid points. A local Taylor-series expansion of the derivatives of **E** and **F** is used to obtain a linear equation which can be solved for $\mathbf{U}^{n+1}$. Let

$$\begin{aligned} \mathbf{E}^{n+1} &= \mathbf{E}^n + [A](\mathbf{U}^{n+1} - \mathbf{U}^n) \\ \mathbf{F}^{n+1} &= \mathbf{F}^n + [B](\mathbf{U}^{n+1} - \mathbf{U}^n) \end{aligned} \tag{6-44}$$

where $[A]$ and $[B]$ are defined

$$[A] = \frac{\partial \mathbf{E}}{\partial \mathbf{U}} \qquad [B] = \frac{\partial \mathbf{F}}{\partial \mathbf{U}}$$

When the linearization given by Eq. (6-44) is substituted into Eq. (6-43), a linear system for $\mathbf{U}^{n+1}$ results and may be written as

$$\left\{[I] + \frac{\Delta t}{2}\left(\frac{\partial}{\partial x}[A]^n + \frac{\partial}{\partial y}[B]^n\right)\right\}\mathbf{U}^{n+1} = \left\{[I] + \frac{\Delta t}{2}\left(\frac{\partial}{\partial x}[A]^n + \frac{\partial}{\partial y}[B]^n\right)\right\}\mathbf{U}^n - \Delta t\left(\frac{\partial \mathbf{E}}{\partial x} + \frac{\partial \mathbf{F}}{\partial y}\right)^n \tag{6-45}$$

This is a linear system for the unknown $\mathbf{U}^{n+1}$. Direct solution of Eq. (6-45) is usually avoided due to the large operation count in treating multidimensional systems. The path usually chosen is to reduce the multidimensional problem into a sequence of one-dimensional inversions. This is done using the method of fractional steps (Yanenko, 1971) or the method of approximate factorization (Peaceman and Rachford, 1955; Douglas, 1955).

Equation (6-45) may be approximately factored into the equation

$$\left([I] + \frac{\Delta t}{2}\frac{\partial}{\partial x}[A]^n\right)\left([I] + \frac{\Delta t}{2}\frac{\partial}{\partial y}[B]^n\right)\mathbf{U}^{n+1} = \left([I] + \frac{\Delta t}{2}\frac{\partial}{\partial x}[A]^n\right) \times \left([I] + \frac{\Delta t}{2}\frac{\partial}{\partial y}[B]^n\right)\mathbf{U}^n - \Delta t\left(\frac{\partial \mathbf{E}}{\partial x} + \frac{\partial \mathbf{F}}{\partial y}\right)^n \tag{6-46}$$

This expression differs from the original Eq. (6-45) by a term which is of order $(\Delta t)^2$ and the formal accuracy of our implicit algorithm is maintained as second order. This factored scheme may be written as the alternating direction sequence

$$\left([I] + \frac{\Delta t}{2}\frac{\partial}{\partial x}[A]^n\right)\mathbf{U}' = \text{RHS [Eq. (6-46)]}$$

$$\left([I] + \frac{\Delta t}{2}\frac{\partial}{\partial y}[B]^n\right)\mathbf{U}^{n+1} = \mathbf{U}' \tag{6-47}$$

A simpler algorithm results if the delta form introduced in Chapter 4 is used. Since the operators on both sides of Eq. (6-46) are the same, define

$$\Delta\mathbf{U}^n = \mathbf{U}^{n+1} - \mathbf{U}^n$$

so that

$$\left([I] + \frac{\Delta t}{2}\frac{\partial}{\partial x}[A]^n\right)\left([I] + \frac{\Delta t}{2}\frac{\partial}{\partial y}[B]^n\right)\Delta\mathbf{U}^n = -\Delta t\left(\frac{\partial \mathbf{E}}{\partial x} + \frac{\partial \mathbf{F}}{\partial y}\right)^n \tag{6-48}$$

Again this may be replaced by the alternating direction sequence

$$\left([I] + \frac{\Delta t}{2}\frac{\partial}{\partial x}[A]^n\right)\Delta\mathbf{U}' = -\Delta t\left(\frac{\partial \mathbf{E}}{\partial x} + \frac{\partial \mathbf{F}}{\partial y}\right)^n$$

$$\left([I] + \frac{\Delta t}{2}\frac{\partial}{\partial y}[B]^n\right)\Delta\mathbf{U}^n = \Delta\mathbf{U}' \tag{6-49}$$

The solution of this system is not trivial. The x and y sweeps each require the solution of a block tridiagonal system of equations assuming the spatial derivatives are approximated by central differences. Each block is $m \times m$ if there are m elements in the unknown **U** vector (see Appendix B).

The implicit algorithm developed here used the trapezoidal rule. Generalized time differencing presented by Warming and Beam (1977) can be used to generate a number of implicit algorithms with varying accuracy. This point is discussed in Chapter 8. Additional consideration is presented on the required addition of artificial damping in conjunction with nondissipative schemes.

Steger and Warming (1979) developed an implicit algorithm using a splitting of **E** and **F** in the governing equations. While the precise splitting can be accomplished in a number of ways, the usual way is according to the signs of the eigenvalues of the system as is the case in the SCM method (Section 6-4). To illustrate the flux splitting concept, consider the one-dimensional system of hyperbolic PDE's

$$\frac{\partial \mathbf{U}}{\partial t} + \frac{\partial \mathbf{E}}{\partial x} = 0$$

This system can also be written in the form

$$\frac{\partial \mathbf{U}}{\partial t} + [A]\frac{\partial \mathbf{U}}{\partial x} = 0 \tag{6-50}$$

where $[A]$ is the Jacobian $\partial\mathbf{E}/\partial\mathbf{U}$. This system is hyperbolic if a similarity transformation exists so that

$$[T]^{-1}[A][T] = [\lambda] \tag{6-51}$$

where $[\lambda]$ is a diagonal matrix of eigenvalues of $[A]$, and $[T]^{-1}$ is the matrix whose rows are the left eigenvectors of $[A]$ taken in order. Now the flux vectors [**E** and **F** in Eq. (6-42)] of the Euler equations have the interesting property that

$$\mathbf{E} = [A]\mathbf{U} \tag{6-52}$$

This may be verified by simply multiplying the indicated matrices. According to Steger and Warming, if the equation of state is of the form

$$p = \rho f(e) \tag{6-53}$$

where e is the internal energy, then the flux vector $\mathbf{E}(\mathbf{U})$ is a homogeneous function of degree one in **U** which means that

$$\mathbf{E}(\alpha\mathbf{U}) = \alpha\mathbf{E}(\mathbf{U}) \tag{6-54}$$

for any α. We can use this property and the fact that the system is hyperbolic to achieve the desired split flux form.

Combining Eqs. (6-51) and (6-52), **E** may be written

$$\mathbf{E} = [A]\mathbf{U} = [T][\lambda][T]^{-1}\mathbf{U} \tag{6-55}$$

The matrix of eigenvalues is divided into two matrices, one with only positive elements and the other with negative elements. We write the $[A]$ matrix

$$[A] = [A]^+ + [A]^- = [T][\lambda^+][T]^{-1} + [T][\lambda^-][T]^{-1} \tag{6-56}$$

and define

$$\mathbf{E} = \mathbf{E}^+ + \mathbf{E}^- \tag{6-57}$$

so that

$$\mathbf{E}^+ = [A]^+\mathbf{U} \qquad \mathbf{E}^- = [A]^-\mathbf{U} \tag{6-58}$$

The original conservation-law form written using the split flux notation becomes

$$\frac{\partial \mathbf{U}}{\partial t} + \frac{\partial \mathbf{E}^+}{\partial x} + \frac{\partial \mathbf{E}^-}{\partial x} = 0 \tag{6-59}$$

where the plus indicates that a backward difference should be used, and a minus indicates that a forward difference is required.

The split flux idea can be used either for explicit or implicit algorithms. For example, a second-order upwind scheme may be written (Warming and Beam, 1975)

$$\overline{\mathbf{U}_j^{n+1}} = \mathbf{U}_j^n - \frac{\Delta t}{\Delta x}(\nabla \mathbf{E}_j^+ + \Delta \mathbf{E}_j^-)$$

$$\mathbf{U}_j^{n+1} = \frac{1}{2}\left[\mathbf{U}_j^n + \overline{\mathbf{U}_j^{n+1}} - \frac{\Delta t}{\Delta x}(\nabla^2 \mathbf{E}_j^{+n} + \nabla \overline{\mathbf{E}_j^{+n+1}}) + \frac{\Delta t}{\Delta x}(\Delta^2 \mathbf{E}_j^{-n} - \Delta \overline{\mathbf{E}_j^{-n+1}})\right] \tag{6-60}$$

An implicit algorithm using the trapezoidal rule is easily derived using the split flux idea and takes the form

$$\left\{[I] + \frac{\Delta t}{2\,\Delta x}\left(\nabla[A_j]^+ + \Delta[A_j]^-\right)\right\}\Delta \mathbf{U}_j^n = -\frac{\Delta t}{\Delta x}\left[\nabla \mathbf{E}^+ + \Delta \mathbf{E}^-\right] \tag{6-61}$$

This algorithm is first-order accurate in space even though it is second-order accurate in time. The spatial accuracy can be improved by simply increasing the order of the spatial difference operators. Frequently, interest is in the steady-state solution. If this is the case, the right-hand side can be modified to obtain second-order accuracy in space for the steady-state result without altering the block tridiagonal structure of the left-hand side.

It is interesting to note that an approximate factorization of the left-hand side of Eq. (6-61) is possible resulting in the product of two operators

$$\left([I] + \frac{\Delta t}{2\,\Delta x}\nabla[A_j]^+\right)\left([I] + \frac{\Delta t}{2\,\Delta x}\Delta[A_j]^-\right)\Delta U_j^n = \text{RHS [Eq. (6-61)]} \tag{6-62}$$

This permits the algorithm to be implemented in the sequence

$$\begin{aligned}\left([I] + \frac{\Delta t}{2\,\Delta x}\nabla[A_j]^+\right)\Delta \mathbf{U}_j' &= \text{RHS [Eq. (6-61)]}\\ \left([I] + \frac{\Delta t}{2\,\Delta x}\Delta[A_j]^-\right)\Delta \mathbf{U}_j^n &= \Delta \mathbf{U}_j'\end{aligned} \tag{6-63}$$

If Eqs. (6-63) are used, each one-dimensional sweep requires the solution of two block bidiagonal systems. The original system [Eq. (6-61)] requires the solution of a single block tridiagonal system for each time step. It is important to note that savings expected in using Eq. (6-63) may not be realized for all problems. Usually, the major advantage of using the split form with bidiagonal systems occurs in multidimensional cases.

The use of split flux techniques for shock-capturing applications produces somewhat better results than the standard central-difference schemes but problems remain even in this formulation. Using this method, shock waves are well-represented and are consistent with the better shock-capturing methods. Results produced using split flux may not be as good when a sonic line is encountered. Small oscillations occur when a sonic line is crossed because the flux splitting depends on the eigenvalues. These flux terms have a discontinuous first derivative when the eigenvalues change sign. Steger (1981) reports excellent success if the eigenvalues are redefined as

$$\lambda^{\pm} = \frac{\lambda \pm \sqrt{\lambda^2 + \epsilon^2}}{2} \tag{6-64}$$

where ϵ is a blending coefficient designed to insure a smooth transition when the λ's change sign.

Our discussion of shock-capturing methods has centered on marching problems both in space and in time. The time marching problems may either imply a study of some transient process or the calculation of a steady flowfield as a time asymptotic limit of a time-dependent problem. In the latter case, the time-dependent limit can be viewed as a relaxation process applied to the Euler equations with time included to provide a physical relaxation direction. True relaxation procedures are not time asymptotic but instead rely upon relaxing the steady governing equations to obtain the solution for a given flowfield. Papers treating the relaxation solution of the Euler equations include work by Steger (1981) and Johnson (1980). Steger used the split flux form of the steady Euler equations and a standard relaxation scheme to obtain the final solution. Johnson has used a novel approach which he refers to as the surrogate equation technique. In this scheme the Euler equations in conservation-law form are embedded in a higher-order system. The solution provides the solution of the Euler equations as one of a restricted set. The disadvantage is that a system which is higher order than the original must be solved.

We now address the problem of applying surface boundary conditions when shock-capturing methods are used. As Moretti (1969) has pointed out, the correct application of boundary conditions is a difficult task. Incorrect wall conditions can provide locally polluted results and in many cases can destroy a solution. Hyperbolic equations are particularly sensitive. Due to their wave-like nature, boundary errors are propagated throughout the mesh reflecting until actual instability can result.

In previous discussions, we have explained the simple overwriting concept of applying boundary conditions. We will discuss three additional boundary condition procedures including

1. Reflection
2. Abbett's method (steady, supersonic flow)
3. Kentzer's method

The idea of reflection is probably the oldest inviscid surface boundary condition scheme. The reflection concept has no basis other than to some order of approximation, it prohibits any normal flux of mass through a solid boundary.

In implementing reflection, the body surface lies on the next to last layer or row of mesh points and the last row is interior to the body. In order to understand the basic idea, suppose we are solving a steady two-dimensional supersonic flow problem by integrating the equations of motion. As shown in Fig. 6-11, the body surface ($j = 2$) is assumed to lie on the x axis, and the first row of points ($j = 1$) is interior to the body. If integration of the Euler equations is accomplished by a second-order method such as MacCormack's, the body surface values of the flow variables can directly be obtained by integration. Values of the primitive variables at the sublayer

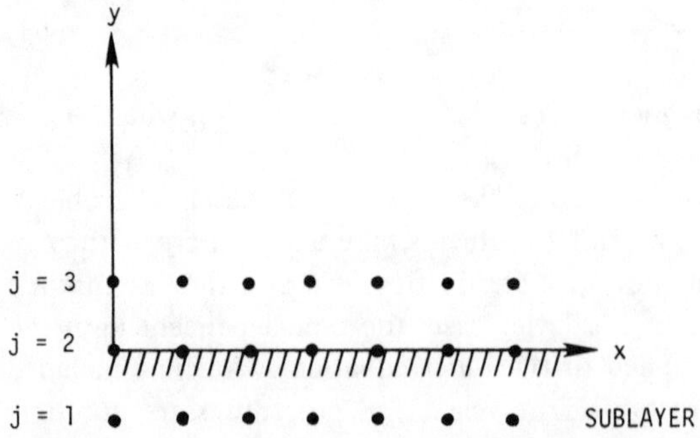

Figure 6-11 Sublayer required for reflection boundary conditions.

points are assigned using the reflection process. We assume that the pressure, density, and tangential velocity are even functions of the normal distance above the body while the normal velocity is assumed to be an odd function of the normal distance. The pressure, density, and tangential velocity at the sublayer point are set equal to their respective values at the first point above the body while the normal velocity is set equal to the negative of its value at $j = 3$.

For a steady two-dimensional marching problem, the body surface equation may be written

$$F(x,y) = y - f(x) = 0 \tag{6-65}$$

The inviscid surface tangency condition then becomes

$$v = u \frac{\partial f}{\partial x} \tag{6-66}$$

In general, reflection is implemented by developing expressions for the surface normal and the normal and tangential velocity components. The surface unit normal is given by

$$\mathbf{n} = \frac{\nabla F}{|\nabla F|}$$

and the velocity tangent to the surface is given by

$$\mathbf{u}_{\text{tan}} = \mathbf{n} \times \mathbf{V} \times \mathbf{n} \tag{6-67}$$

while the velocity normal to the surface is given by

$$\mathbf{u}_{\text{nor}} = (\mathbf{V} \cdot \mathbf{n})\mathbf{n} \tag{6-68}$$

The scalar magnitude of the velocities given by Eqs. (6-67) and (6-68) are reflected to provide the sublayer values. Practically, only one component of the tangential velocity is used as opposed to the scalar magnitude. The magnitude of the tangential velocity can become unwieldy and lead to an impractically long expression if the total tangential velocity is used. In general we develop a system of equations which must be simultaneously solved for sublayer values when this scheme is used (see Prob. 6-11). Although reflection is relatively easy to implement, it is not used with great regularity. Reflection is very inaccurate on bodies in regions with high curvature.

Abbett (1973) devised a boundary condition application procedure which uses as much physical reasoning as possible in enforcing the surface tangency condition. The basic idea used in Abbett's wave corrector method is that the velocity vector computed in the final step of the integration routine is not parallel to the body surface. A simple wave is introduced into the flow at the body to either compress or expand the gas and turn the flow so that the velocity is parallel to the surface of the body.

Consider the supersonic steady flow of a perfect gas. Suppose we are solving a steady marching problem and assume that an orthogonal coordinate system (x_1, x_2, x_3) is used with velocity vector

$$\mathbf{V} = \mathbf{i}_1 u + \mathbf{i}_2 v + \mathbf{i}_3 w \tag{6-69}$$

The velocity components in the respective directions are (u, v, w). Let the unit vector normal to the surface be

$$\mathbf{n} = \frac{\nabla F}{|\nabla F|}$$

where the body surface equation is given by

$$F(x_1, x_2, x_3) = x_1 - f(x_2, x_3) = 0 \tag{6-70}$$

Therefore the body surface normal is

$$\mathbf{n} = \frac{\mathbf{i}_1/h_1 - [(\mathbf{i}_2/h_2)(\partial f/\partial x_2)] - [(\mathbf{i}_3/h_3)(\partial f/\partial x_3)]}{\{1/h_1^2 + [(1/h_2)(\partial f/\partial x_2)]^2 + [(1/h_3)(\partial f/\partial x_3)]^2\}^{1/2}} \tag{6-71}$$

The velocity vector can be divided into a component normal to the surface and a component tangent to the surface. If the normal velocity is computed as

$$\mathbf{u}_{\text{nor}} = (\mathbf{V} \cdot \mathbf{n})\mathbf{n} \tag{6-72}$$

the small misalignment angle, $\Delta\theta$, representing the orientation of the velocity vector with respect to a surface tangent becomes

$$|\sin(\Delta\theta)| = \frac{|\mathbf{u}_{\text{nor}}|}{|\mathbf{V}|} \tag{6-73}$$

We may write this as

$$\sin(\Delta\theta) = \frac{\mathbf{V} \cdot \mathbf{n}}{|\mathbf{V}|} \tag{6-74}$$

The geometry of this problem is shown in Fig. 6-12. The misalignment angle $\Delta\theta$ is clearly shown. The velocity vector $\mathbf{V}$ represents the velocity computed at the body surface using the integration scheme. If MacCormack's method was used to solve the equations of motion, the velocity vector, $\mathbf{V}$, shown in Fig. 6-12 is the value at the end of the corrector step. Again remember that a forward corrector must be used at the body.

In order to turn the velocity vector through an angle $\Delta\theta$ so that it is parallel to the body, a weak wave is introduced into the flow. If $\Delta\theta$ is positive, an expansion is required. As the flow turns through an angle $\Delta\theta$, the body surface pressure must also

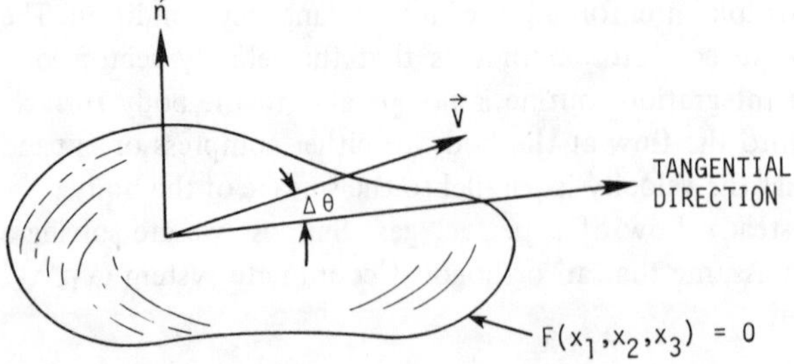

Figure 6-12 Velocity vector orientation on body surface.

change. For weak waves the pressure is related to the flow turning angle by the expression [see NACA Report 1135 (Ames Research Staff, 1953)].

$$\frac{p_2}{p_1} = 1 - \frac{\gamma M^2}{\sqrt{M^2 - 1}}\,\Delta\theta + \gamma M \left[\frac{(\gamma+1)M^4 - 4(M^2-1)}{4(M^2-1)^2}\right]\Delta\theta^2 + \cdots \tag{6-75}$$

In this expression M and p_1 are the Mach number and pressure before turning while p_2 represents the pressure after the turn takes place. With the pressure known from Eq. (6-75), the density change can now be computed. This is one point where Abbett's scheme requires additional information. It is assumed that the value of the surface entropy is known. At least along the streamline which wets the body, the value of p/ρ^γ is known. The new surface pressure (p_2) is used in conjunction with the surface entropy to calculate a new density ρ_2.

The magnitude of the velocity in the tangential direction is computed by use of the steady energy equation. If H is the total enthalpy, the velocity along the body surface is calculated as

$$|\mathbf{V}_2| = \sqrt{2\left(H - \frac{\gamma}{\gamma-1}\frac{p_2}{\rho_2}\right)} \tag{6-76}$$

The velocity components must now be determined. The direction of the new velocity vector along the surface is obtained by subtracting the normal velocity from the original velocity computed using the integration routine. This produces the result

$$\mathbf{V}_T = \mathbf{V} - (\mathbf{V}\cdot\mathbf{n})\mathbf{n} \tag{6-77}$$

and represents the tangential component of the original velocity. It is assumed that the new surface velocity vector is in the same direction. The new velocity $\mathbf{V}_2$ is given by

$$\mathbf{V}_2 = |\mathbf{V}_2|\,\frac{\mathbf{V}_T}{|\mathbf{V}_T|} \tag{6-78}$$

This boundary condition routine is relatively easy to apply and provides excellent results (see Kutler et al., 1973). One of the major problems is that of determining the proper direction for the final velocity vector. Abbett's method assumes this final velocity vector lies in the tangent plane of the body in the direction of the intersection of the tangent plane and the plane formed by the unit normal and the original velocity vector. No out of plane correction is used.

One of the difficult boundary condition problems occurring in shock-capturing calculations is that of enforcing surface tangency when an oblique shock wave strikes a solid boundary. Griffin (1981) reports success replacing the entropy at the next body point $(x + \Delta x, 0)$ by the previous value $(x, 0)$ plus the entropy change between the previous two points above the body surface $(x, \Delta y)$ and $(x - \Delta x, \Delta y)$. In steady supersonic flow problems this procedure gives an estimate of the expected correct surface entropy. According to Griffin, this procedure works very well in establishing the correct surface entropy and providing an appropriate boundary condition procedure for regions with shock-surface intersections.

Kentzer (1970) proposed a scheme of applying surface boundary conditions which essentially utilizes the compatibility equations running from the domain interior toward the boundary along with the surface boundary condition. This approach is similar to that used with the nonconservative SCM method which is discussed in the next section. In this procedure, the surface tangency condition is used in differential form with the appropriate compatibility equation.

6-4 THE SCM METHOD

The split coefficient matrix (SCM) method is a relatively recent addition to the class of finite-difference methods for solving hyperbolic PDE's. The SCM scheme introduced by Chakravarthy (1979) and Chakravarthy et al. (1980) is a nonconservation form of the split flux scheme proposed by Steger (1978). The SCM method utilizes the information on signal propagation provided by the theory of characteristics. Thus, we might expect the results obtained by applying this method to be superior to those obtained using other methods. This is true and the SCM method is recommended for use when hyperbolic PDE's in nonconservation form are solved numerically.

Other methods which use characteristic information have been developed. Gordon (1969) developed a scheme similar to the SCM technique for use in solving hyperbolic systems. Moretti (1978) introduced the "λ" scheme which used the idea of correctly treating the signal propagation directions. In some cases Moretti's λ scheme and the SCM method are identical. However, for multidimensional flows in arbitrary coordinate systems a unique formulation of the required difference equations using the λ scheme does not exist. The SCM scheme precisely formulates the required difference equations for an arbitrary number of dimensions.

The SCM method is easily explained by applying it to the system of two linear equations used in Section 6-2. The system given in Eq. (6-3) was diagonalized and the compatibility relations were written as

$$\begin{aligned} \frac{\partial}{\partial x}(\beta u + v) - \frac{1}{\beta}\frac{\partial}{\partial y}(\beta u + v) &= 0 \\ \frac{\partial}{\partial x}(\beta u - v) + \frac{1}{\beta}\frac{\partial}{\partial y}(\beta u - v) &= 0 \end{aligned} \tag{6-79}$$

As noted in Fig. 6-13, these equations are equivalent to requiring that

$$d(\beta u - v) = 0 \tag{6-80a}$$

on the characteristic

$$x - \beta y = \xi$$

and

$$d(\beta u + v) = 0 \tag{6-80b}$$

on

$$x + \beta y = \eta$$

Suppose we want to generate a simple finite-difference analog of these expressions. Along the positive characteristic (η = constant in Fig. 6-13), we may write

$$(\beta u - v)_D - (\beta u - v)_A = 0 \tag{6-81}$$

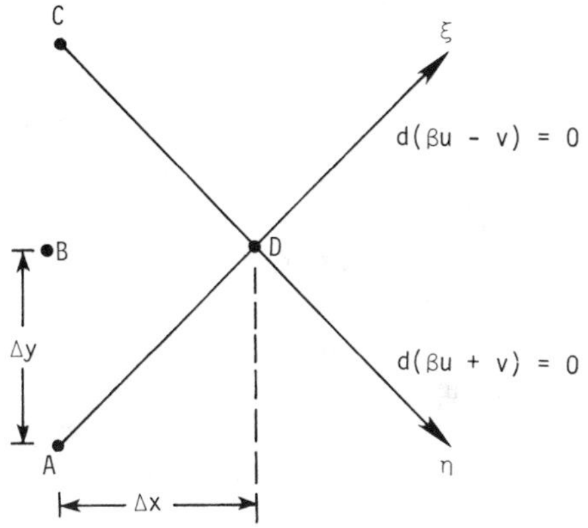

Figure 6-13 Characteristics and grid point locations.

and along $\xi =$ constant

$$(\beta u + v)_D - (\beta u + v)_C = 0 \tag{6-82}$$

These equations can be solved for the unknowns at point D in terms of the known values at A and C. The method of characteristics actually depends upon integrating the compatibility equations [Eqs. (6-80*a*) and (6-80*b*)] along the characteristics. A finite-difference method of characteristics can be obtained by using the partial differential equation equivalent, Eq. (6-79). Suppose we add and subtract the quantity $(\beta u - v)_B$ from Eq. (6-81). This yields

$$\frac{(\beta u - v)_D - (\beta u - v)_B}{\Delta x} - \frac{(\beta u - v)_A - (\beta u - v)_B}{\beta \Delta y} = 0 \tag{6-83}$$

Similarly, Eq. (6-82) may be written

$$\frac{(\beta u + v)_D - (\beta u + v)_B}{\Delta x} - \frac{(\beta u + v)_C - (\beta u + v)_B}{\beta \Delta y} = 0 \tag{6-84}$$

These two relationships are the finite-difference equivalent of Eq. (6-79) and would provide the same solution for the unknowns at point B. The SCM scheme essentially relies on integrating a decoupled form of the compatibility relations. The form used in the SCM scheme is provided by rewriting Eq. (6-79) as

$$\begin{bmatrix} \beta & -1 \\ \beta & 1 \end{bmatrix} \mathbf{w}_x + \begin{bmatrix} 1 & -\dfrac{1}{\beta} \\ 0 & 0 \end{bmatrix} \mathbf{w}_y^+ + \begin{bmatrix} 0 & 0 \\ -1 & -\dfrac{1}{\beta} \end{bmatrix} \mathbf{w}_y^- = 0$$

This may be written

$$\mathbf{w}_x + [A]^+ \mathbf{w}_y^+ + [A]^- \mathbf{w}_y^- = 0 \tag{6-85}$$

where

$$[A]^+ = \frac{1}{2\beta} \begin{bmatrix} 1 & -\dfrac{1}{\beta} \\ -\beta & 1 \end{bmatrix} \qquad [A]^- = \frac{1}{2\beta} \begin{bmatrix} -1 & -\dfrac{1}{\beta} \\ -\beta & -1 \end{bmatrix} \tag{6-86}$$

Referring to Eq. (6-14) we see that

$$[A]^+ + [A]^- = [A] \tag{6-87}$$

The meaning of the notation used in Eq. (6-85) follows from the characteristic equations. The coefficient matrix $[A]$ has been split to account for the correct signal propagation directions when correct one-sided differencing is used. For example, the term $[A]^+\mathbf{w}_y^+$ denotes the contribution from the characteristic which has a positive slope. In this case, a backward difference is used on the derivatives of $\mathbf{w}$ and we are reminded of this by the sign of the eigenvalue as noted. If simple one-sided difference approximations are used for the y derivatives, the split coefficient matrix form, Eq. (6-85), is identical to the finite-difference method of characteristics. If more sophisticated one-sided differences are used, the methods are not identical although the split form still retains the proper information on signal propagation.

Example 6-3 In Example 6-1, we calculated a solution of the Prandtl-Glauert equation for a single period wavy wall. Let us apply the SCM method to compute the solution of this same problem.

Solution: The appropriate finite-difference equations are obtained by writing the two scalar equations [Eq. (6-85)] as

$$\begin{aligned} u_x + \tfrac{1}{2}(u_y - v_y)^+ - \tfrac{1}{2}(u_y + v_y)^- &= 0 \\ v_x + \tfrac{1}{2}(-u_y + v_y)^+ - \tfrac{1}{2}(u_y + v_y)^- &= 0 \end{aligned} \tag{6-88}$$

We now forward difference the x derivatives and apply first-order accurate differences for the plus and minus terms. The first of these difference equations may then be written

$$u_j^{n+1} = u_j^n - \frac{\Delta x}{2\,\Delta y}(u_j^n - u_{j-1}^n - v_j^n + v_{j-1}^n) + \frac{\Delta x}{2\,\Delta y}(u_{j+1}^n - u_j^n + v_{j+1}^n - v_j^n)$$

The second expression is obtained in a similar fashion. The boundary conditions are known and we may integrate the difference equations from the initial data surface by marching in the x direction. We have created a first-order SCM scheme for solving our example problem. The results of the numerical solution of this problem will prove to be very good when compared to the exact result (see Prob. 6-6).

The stability of the explicit form of the SCM method using first-order differencing is determined by the usual CFL condition. A different stability bound is obtained when higher-order one-sided differences are used. This will be discussed in more detail later in this section.

We have presented the method of characteristics and the SCM method for a simple linear problem. This simple problem was used to demonstrate the basic concepts employed in these schemes. Both techniques are applicable to the nonlinear equations of fluid flow as long as they remain hyperbolic. We will now consider the theory for a system of nonlinear hyperbolic PDE's.

The equations which govern the unsteady one-dimensional flow of an inviscid, perfect gas are of the form

$$\frac{\partial \mathbf{w}}{\partial t} + [A]\frac{\partial \mathbf{w}}{\partial x} = 0 \tag{6-89}$$

where $\mathbf{w}$ is a vector of n unknowns and $[A]$ is $n \times n$. The eigenvalues of $[A]$ define the characteristic directions and may be written

$$\frac{dx}{dt} = \lambda_i \qquad i = 1, 2, \ldots, n \tag{6-90}$$

For each eigenvalue λ_i, there is a left eigenvector $\mathbf{L}_i$ which must satisfy

$$\mathbf{L}_i^T([A] - \lambda_i[I]) = 0 \tag{6-91}$$

As in our linear example, we obtain the compatibility equations by multiplying the transpose of $\mathbf{L}_i$ into the original system given by Eq. (6-89).

$$\mathbf{L}_i^T(\mathbf{w}_t + [A]\mathbf{w}_x) = \mathbf{L}_i^T(\mathbf{w}_t + \lambda_i \mathbf{w}_x) = 0 \tag{6-92}$$

This provides the partial differential equations which hold along the characteristics and the SCM scheme is developed from this point. Let $[T]^{-1}$ represent a matrix whose n rows are the n left eigenvectors taken in order. We may then write the compatibility equations in the form

$$[T]^{-1}\mathbf{w}_t + [\Lambda_A][T]^{-1}\mathbf{w}_x = 0 \tag{6-93}$$

where $[\Lambda_A]$ is a diagonal matrix of the eigenvalues of $[A]$. Multiplying by $[T]$ yields

$$\mathbf{w}_t + [T][\Lambda_A][T]^{-1}\mathbf{w}_x = 0 \tag{6-94}$$

We note that the original $[A]$ matrix is now written as

$$[A] = [T][\Lambda_A][T]^{-1} \tag{6-95}$$

If $[\Lambda_A]$ is split into its positive and negative parts,

$$[\Lambda_A] = [\Lambda_A]^+ + [\Lambda_A]^- \tag{6-96}$$

and substituted into Eq. (6-95), we obtain

$$[A] = [T][\Lambda_A]^+[T]^{-1} + [T][\Lambda_A]^-[T]^{-1} \tag{6-97}$$

where $$[A] = [A]^+ + [A]^-$$

The $[A]^+$ and $[A]^-$ matrices are identified with the eigenvalues and the eigenvectors. We again write our equation

$$\mathbf{w}_t + [A]^+\mathbf{w}_x + [A]^-\mathbf{w}_x = 0 \tag{6-98}$$

and the meaning of the ± notation is the same as in our previous example.

Example 6-4 Suppose we wish to develop the split form of the steady two-dimensional Euler equations. This split form can be used to solve for the steady supersonic flow over any two-dimensional surface. The final form of the governing equations for this case is

$$\mathbf{w}_x + [T][\Lambda]^+[T]^{-1}\mathbf{w}_y + [T][\Lambda]^-[T]^{-1}\mathbf{w}_y = 0$$

We have already derived $[T]^{-1}$ and that is given in Eq. (6-19*b*). It remains to compute $[T]$. While some algebra is involved, $[T]$ is easily computed. For problems of this nature, the solution must be evaluated at each step to determine the λ^+ and λ^- values and thus determine the proper spatial differencing. The remaining details of this example are left as an exercise in Prob. 6-7 at the end of this chapter.

An alternative form of the governing equations is favored by Moretti (1971) and others (Salas, 1975; Marconi, 1980). In this form density derivatives in the continuity equation are replaced by derivatives of the pressure by using the equation for the speed of sound. In addition, the entropy equation is used in place of the energy equation. In this setting the governing equations from Section 5-5.4 are

$$\begin{aligned}
\frac{\partial u}{\partial x} + \frac{u}{\gamma}\frac{\partial P}{\partial x} + \frac{\partial v}{\partial y} + v\frac{\partial P}{\partial y} &= 0 \\
u\frac{\partial u}{\partial x} + v\frac{\partial u}{\partial y} + \frac{a^2}{\gamma}\frac{\partial P}{\partial x} &= 0 \\
u\frac{\partial v}{\partial x} + v\frac{\partial v}{\partial y} + \frac{a^2}{\gamma}\frac{\partial P}{\partial y} &= 0 \\
u\frac{\partial s}{\partial x} + v\frac{\partial s}{\partial y} &= 0
\end{aligned} \tag{6-99}$$

The unknowns in this system are the velocity components, u and v; the entropy, s; and P, the natural logarithm of the pressure. It is interesting to note that the entropy equation is decoupled from the other equations when this formulation is used. Consequently, entropy can be directly obtained by integration independent of the other expressions. Moretti (1971) has noted that upwind differencing should always be used on the entropy since the entropy equation is an expression of the constancy of s along a streamline. This requires that the differences employed in the entropy equation for $\partial s/\partial y$ must be forward or backward depending on the sign of v/u. This is consistent with the SCM method. Even if a finite-difference scheme such as MacCormack's is used, upwind differencing on the entropy is advisable.

Additional attention must be given to the appropriate differencing of the derivative terms in the SCM method. In order to include more general ideas, let us consider a nonlinear equation of the form given in Eq. (6-98). If we denote a first-order backward difference by ∇ and the usual forward difference by Δ, a method which is first order in time and space is given by

$$\mathbf{w}_j^{n+1} = \mathbf{w}_j^n - ([A]^+\nabla\mathbf{w}_j^n + [A]^-\Delta\mathbf{w}_j^n)\frac{\Delta t}{\Delta x}$$

The development of second-order methods requires more care. Moretti (1978) suggested a scheme which switches from two-point to three-point differences in the predictor-corrector sequence.

Predictor step:

$$\overline{\mathbf{w}_j^{n+1}} = \mathbf{w}_j^n - \Delta t([A]^+\mathbf{w}_x^- + [A]^-\mathbf{w}_x^+)$$

with the spatial differences given by

$$\mathbf{w}_x^- = \frac{2\mathbf{w}_j^n - 3\mathbf{w}_{j-1}^n + \mathbf{w}_{j-2}^n}{\Delta x}$$
$$\mathbf{w}_x^+ = \frac{\mathbf{w}_{j+1}^n - \mathbf{w}_j^n}{\Delta x} \tag{6-100}$$

Corrector step:

$$\mathbf{w}_j^{n+1} = \mathbf{w}_j^n + \frac{\Delta t}{2}(\mathbf{w}_t^n + \overline{\mathbf{w}_t^{n+1}})$$

where spatial differences in both $\mathbf{w}_t^n$ and $\overline{\mathbf{w}_t^{n+1}}$ are the two-point backward and three-point forward expressions

$$\mathbf{w}_x^- = \frac{\mathbf{w}_j^n - \mathbf{w}_{j-1}^n}{\Delta x}$$
$$\mathbf{w}_x^+ = \frac{-2\mathbf{w}_j^n + 3\mathbf{w}_{j+1}^n - \mathbf{w}_{j+2}^n}{\Delta x} \tag{6-101}$$

with the spatial derivatives in $\overline{\mathbf{w}_t^{n+1}}$ formed using predicted values. The differencing in each step is altered in order to provide the desired second-order result. It is important to notice that second-order accurate explicit methods which use one-sided differences do not have a compact form like the centered schemes. For Eq. (6-98), first-order methods require three points and second-order methods require five points. On the other hand, MacCormack's method is formally second order in space using only three points.

The scheme presented in Eqs. (6-100) and (6-101) has the disadvantage that it does not satisfy the perfect shift condition. This means that at a Courant number of one, the scheme does not exactly follow characteristics for a linear problem. Gabutti (1982) introduced a correction to eliminate this difficulty. Gabutti's scheme requires three steps. For equations of the form given in Eq. (6-99), we have

Predictor step:

$$\overline{\mathbf{w}_j^{n+1}} = \mathbf{w}_j^n - \frac{\Delta t}{\Delta x}([A]^+ \nabla \mathbf{w}_j^n + [A]^- \Delta \mathbf{w}_j^n) \tag{6-102}$$

Intermediate step:

$$\mathbf{w}_t' = [A]^-\left(\frac{2\mathbf{w}_j^n - 3\mathbf{w}_{j+1}^n + \mathbf{w}_{j+2}^n}{\Delta x}\right) - [A]^+\left(\frac{2\mathbf{w}_j^n - 3\mathbf{w}_{j-1}^n + \mathbf{w}_{j-2}^n}{\Delta x}\right) \tag{6-103}$$

Corrector step:

$$\mathbf{w}_j^{n+1} = \mathbf{w}_j^n + \frac{\Delta t}{2} (\mathbf{w}_t' + \overline{\mathbf{w}_t^{n+1}}) \tag{6-104}$$

where $\overline{\mathbf{w}_t^{n+1}}$ is computed in the usual way from the original differential equation using first-order forward and backward differences. It is interesting to note that this scheme is the same as that introduced by Warming and Beam (1975) (see Section 4-1.9) and used by Steger and Warming (1979). The second-order upwind scheme introduced by Beam and Warming uses the predictor in Eq. (6-102) while the corrector is given by

$$\mathbf{w}_j^{n+1} = \mathbf{w}_j^n + \frac{\Delta t}{2} (\mathbf{w}_t^n + \overline{\mathbf{w}_t^{n+1}}) + \frac{\Delta t}{2\,\Delta x} [A]^- \Delta^2 \mathbf{w}_j^n - \frac{\Delta t}{2\,\Delta x} [A]^+ \nabla^2 \mathbf{w}_j^n \tag{6-105}$$

The two additional terms appended to the corrector alter the truncation error in such a way that the scheme is second order but also satisfies the shift condition.

The procedures for the application of boundary conditions for SCM-like methods are reasonably well-defined. At the boundaries of the computational domain, some characteristics point from the interior of the region to the boundary while others point from the exterior to the boundary. We can utilize the compatibility equations along those characteristics which point from the interior toward the boundary. However, information carried along the remaining characteristics can not be used. The corresponding compatibility relations must be replaced by prescribed boundary conditions. These boundary conditions may take the form of the flow tangency condition at a solid boundary or a prescribed pressure or velocity condition at an outflow surface. If we consider supersonic flows with shocks, the shock waves must be fit as a discontinuity. This will be discussed later in this section.

Example 6-5 Suppose we wish to compute the pressure distribution in a one-dimensional nozzle flow starting at stagnation conditions, accelerating through sonic speed in the throat, expanding supersonically, and shocking down to a prescribed subcritical back pressure. A schematic of the nozzle is shown in Fig. 6-14.

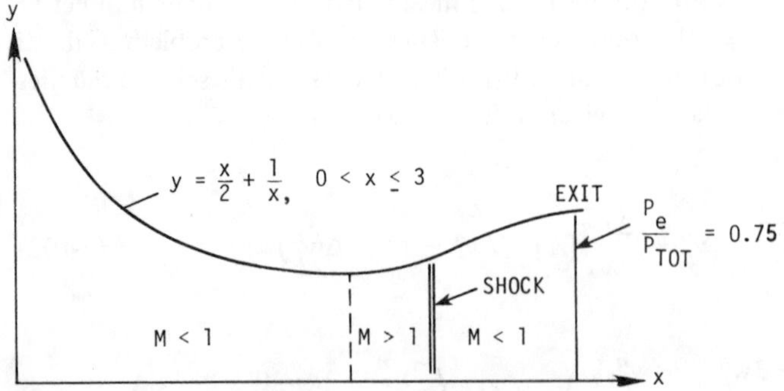

Figure 6-14 One-dimensional nozzle flow.

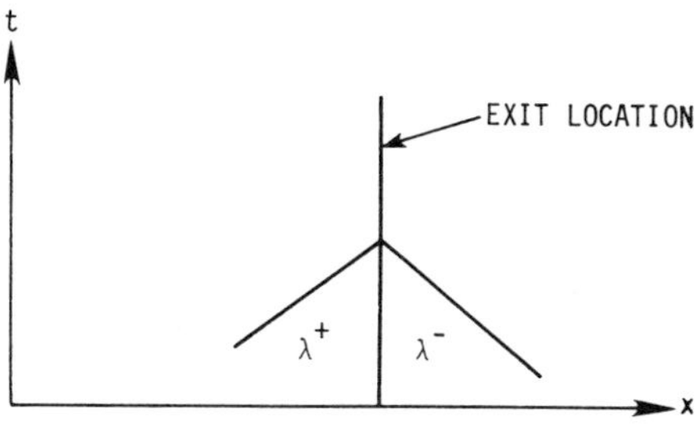

Figure 6-15 Exit plane characteristics.

Solution: The governing equations for one-dimensional flow are

$$\frac{\partial P}{\partial t} + u\,\frac{\partial P}{\partial x} + \gamma\left(\frac{\partial u}{\partial x} + \alpha u\right) = 0$$

$$\frac{\partial u}{\partial t} + u\,\frac{\partial u}{\partial x} + \frac{a^2}{\gamma}\,\frac{\partial P}{\partial x} = 0 \tag{6-106}$$

where

$$\alpha = \frac{1}{\mathrm{A}}\,\frac{\partial A}{\partial x}$$

Since we are using a nonconservative form of the governing equations, we will not correctly capture the shock wave in the nozzle. If we integrate Eq. (6-106), we require that the entropy equation be integrated or that an isentropic flow ($s =$ constant) be assumed. For simplicity, we assume that entropy is a constant for this example. Even though we make this assumption, a unique solution for the nozzle flow is obtained when the equations are integrated in time.

The proper form of the SCM equations is readily obtained from Eq. (6-106) and will not be repeated here. Our major interest is in establishing the boundary conditions to apply at the exit of the nozzle. Figure 6-15 shows the location of the exit plane in (x, t) space along with the characteristics that intersect at the boundary. Since the negative slope characteristic points toward the exit, the compatibility equation must be replaced by the boundary condition $p = 0.75p_{\text{tot}}$ in this example. The compatibility equation along the positive slope characteristic provides useful information and is used. This equation is written

$$a\,\frac{\partial P}{\partial t} + \gamma\,\frac{\partial u}{\partial t} + \lambda^+\left(a\,\frac{\partial P}{\partial x} + \gamma\,\frac{\partial u}{\partial x}\right) + \gamma a \alpha u = 0 \tag{6-107}$$

For a fixed back pressure at the exit, we require $\partial P/\partial t = 0$. When backward differences are used for the x derivatives in Eq. (6-107), the new exit velocity is immediately computed by integration. This result along with the specified exit pressure completely specifies the problem at the exit plane.

The SCM method is applicable to any number of dimensions and an example of the system for a two-dimensional time-dependent flow is instructive. Usually the equations of motion are transformed into computational space. If we use variables s, u, v, P in a Cartesian base system and perform a transformation

$$\begin{aligned} \tau &= t \\ \xi &= \xi(t, x, y) \\ \eta &= \eta(t, x, y) \end{aligned} \tag{6-108}$$

the governing equations in transformed coordinates become

$$\mathbf{w}_\tau + [A]\,\mathbf{w}_\xi + [B]\,\mathbf{w}_\eta + \mathbf{h} = 0 \tag{6-109}$$

Consistent with the notation previously used, we may split both $[A]$ and $[B]$ to obtain

$$\mathbf{w}_\tau + [T][\Lambda_A][T]^{-1}\mathbf{w}_\xi + [S][\Lambda_B][S]^{-1}\mathbf{w}_\eta + \mathbf{h} = 0 \tag{6-110}$$

In this expression, $[\Lambda_A]$ and $[\Lambda_B]$ are diagonal matrices of the eigenvalues of $[A]$ and $[B]$ while the corresponding transposed eigenvector matrices are given by $[T]^{-1}$ and $[S]^{-1}$. For Eq. (6-110)

$$\mathbf{w} = (s, u, v, P)^T \tag{6-111}$$

$$\mathbf{h} = \left(0, 0, 0, \epsilon\gamma\,\frac{v}{y}\right)^T \tag{6-112}$$

where $\epsilon = (1, 0)$ for axisymmetric or two-dimensional flow

$$[A] = \begin{bmatrix} \bar{u} & 0 & 0 & 0 \\ 0 & \bar{u} & 0 & \xi_x \dfrac{a^2}{\gamma} \\ 0 & 0 & \bar{u} & \xi_y \dfrac{a^2}{\gamma} \\ 0 & \gamma\xi_x & \gamma\xi_y & \bar{u} \end{bmatrix} \tag{6-113}$$

$$\bar{u} = \xi_t + u\xi_x + v\xi_y$$

$$[\Lambda_A] = \text{Diagonal}\,(\bar{u}, \bar{u}, \bar{u} + a\sqrt{\xi_x^2 + \xi_y^2}\,, \bar{u} - a\sqrt{\xi_x^2 + \xi_y^2}\,) \tag{6-114}$$

$$[T]^{-1} = \begin{bmatrix} 1 & 0 & 0 & 0 \\ 0 & -k_2 & k_1 & 0 \\ 0 & \dfrac{k_1}{2} & \dfrac{k_2}{2} & \dfrac{a}{2\gamma} \\ 0 & \dfrac{k_1}{2} & \dfrac{k_2}{2} & -\dfrac{a}{2\gamma} \end{bmatrix} \tag{6-115}$$

$$k_1 = \frac{\xi_x}{\sqrt{\xi_x^2 + \xi_y^2}} \qquad k_2 = \frac{\xi_y}{\sqrt{\xi_x^2 + \xi_y^2}}$$

The corresponding matrices for $[B]$, $[\Lambda_B]$, $[S]^{-1}$, $[S]$ are the same with η replacing ξ in the forms given in the elements of $[A]$. Suppose a solid wall exists at $\eta = 0$. Since we are integrating the governing equations in time, we will examine the

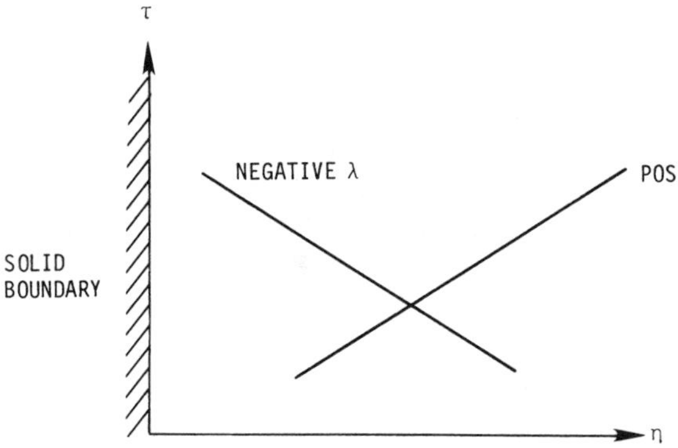

Figure 6-16 Computational (τ, η) space.

characteristic directions in the (τ, η) plane to determine which can be used and which ones must be replaced by boundary conditions. Figure 6-16 shows the (τ, η) plane. Only information carried along the negative slope characteristic (right running) reaches the boundary as time increases. Information normally carried by the other characteristic must be replaced by the wall boundary conditions. The third compatibility equation corresponding to the third eigenvalue given in Eq. (6-114) (with $\bar{u}$ replaced by $\bar{v}$ and ξ replaced by η) cannot be used. The boundary condition for this case is

$$\bar{v} = \eta_t + u\eta_x + v\eta_y = 0 \tag{6-116}$$

We differentiate this expression with respect to time to obtain

$$\eta_x u_\tau + \eta_y v_\tau = 0 \tag{6-117}$$

This expression is used with the three remaining compatibility equations to advance the solution at the boundary. The set of governing equations can be placed in a more formal setting. If we let

$$\mathbf{g} = [T]\,[\Lambda_A]\,[T]^{-1}\mathbf{w}_\xi + \mathbf{h} \tag{6-118}$$

then Eq. (6-110) may be written

$$\mathbf{w}_\tau + [S]\,[\Lambda_B]\,[S]^{-1}\mathbf{w}_\eta + \mathbf{g} = 0 \tag{6-119}$$

The compatibility equations in the (τ, η) plane are

$$[S]^{-1}\mathbf{w}_\tau + [\Lambda_B]\,[S]^{-1}\mathbf{w}_\eta + [S]^{-1}\mathbf{g} = 0 \tag{6-120}$$

The third compatibility equation is now replaced by the differentiated surface tangency condition [Eq. (6-117)] and leads to a system of the form

$$[D]\,\mathbf{w}_\tau + \mathbf{e} + \mathbf{f} = 0 \tag{6-121}$$

where

$$[D] = \begin{bmatrix} s_{11} & s_{12} & s_{13} & s_{14} \\ s_{21} & s_{22} & s_{23} & s_{24} \\ 0 & \eta_x & \eta_y & 0 \\ s_{41} & s_{42} & s_{43} & s_{44} \end{bmatrix}$$

$$\mathbf{e} = \begin{bmatrix} \lambda_1 \sum_{i=1}^{4} s_{1i} w_{i\eta}^{+} \\ \lambda_2 \sum_{i=1}^{4} s_{2i} w_{i\eta}^{+} \\ 0 \\ \lambda_4 \sum_{i=1}^{4} s_{4i} w_{i\eta}^{+} \end{bmatrix}$$

$$\mathbf{f} = \begin{bmatrix} \sum_{i=1}^{4} s_{1i} g_i \\ \sum_{i=1}^{4} s_{2i} g_i \\ 0 \\ \sum_{i=1}^{4} s_{4i} g_i \end{bmatrix}$$

The s_{ij} are elements of $[S]^{-1}$. The explicit expressions for the elements of $\mathbf{w}_\xi$ can now be obtained from Eq. (6-121).

Since we are discussing boundary condition procedures for nonconservative forms of the Euler equations, it is convenient to include shock fitting at this point. Usually we fit the shock as a boundary and compute the shock wave position as part of the solution. In terms of the computational space shown in Fig. 6-16, the shock wave would appear at $\eta = \eta_{\max}$. The process of fitting the shock is a matter of satisfying the Rankine-Hugoniot equations while simultaneously requiring that the solution on the downstream side of the shock be compatible with the rest of the flowfield.

The solution for the flow variables downstream of a shock wave is determined by the freestream conditions, the shock velocity and the shock orientation. If we know the freestream conditions, the initial shock slope and velocity, the shock velocity or shock pressure can be considered as the primary unknown in the shock-fitting procedure. The procedure usually followed is to combine the Rankine-Hugoniot equations with one compatibility equation to provide the expression for shock acceleration or post-shock pressure. For example, once we determine the downstream pressure, we may compute the other downstream flow variables using the equations

$$\mathbf{V}_\infty = \mathbf{i}u_\infty + \mathbf{j}v_\infty$$

$$\mathbf{n}_s = \frac{\mathbf{i}\eta_x + \mathbf{j}\eta_y}{\sqrt{\eta_x^2 + \eta_y^2}}$$

$$u_{\infty n} = |\mathbf{V}_\infty \cdot \mathbf{n}_s|$$

$$M_s = \left\{ \frac{1}{2\gamma} \left[\frac{p_2}{p_\infty} (\gamma + 1) + (\gamma - 1) \right] \right\}^{1/2}$$

$$V_s = a_\infty M_s - u_{\infty n}$$

$$u_{2n} - u_{\infty n} = \frac{2a_\infty (1 - M_\infty^2)}{(\gamma + 1) M_s}$$

$$\rho_2 = \rho_\infty \left(\frac{p_2}{p_\infty} + \frac{\gamma - 1}{\gamma + 1} \right) \left[\frac{1}{1 + (\gamma - 1)(p_2/p_\infty)/(\gamma + 1)} \right]$$

$$\mathbf{V}_2 = \mathbf{V}_\infty + (u_{2n} - u_{\infty n})\mathbf{n}_s [\text{sign}\,(\mathbf{V}_\infty \cdot \mathbf{n}_s)] \tag{6-122}$$

The subscript ∞ refers to freestream conditions while 2 denotes conditions immediately downstream of the shock wave. Subscript s indicates shock surface, and n indicates normal to this surface. Equation (6-122) can be easily derived from the relative velocity expression for the shock motion and the Rankine-Hugoniot equations. Details of this derivation are omitted since they are similar to those given in the previous section where shock fitting was demonstrated in a space-marching problem. Figure 6-17 shows the notation and orientation of the shock wave in physical space. Consistent with the discussion of boundary condition procedures, only one characteristic carries information from the interior to the shock wave. If this characteristic is λ_4, the corresponding compatibility equation is

$$\sum_{i=1}^{4} s_{4i}(w_{i\tau} + \lambda_4 w_{i\eta} + g_i) = 0 \tag{6-123}$$

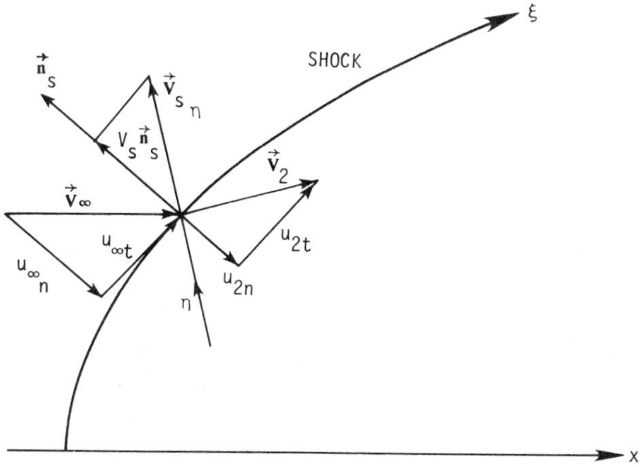

Figure 6-17 Shock geometry.

Since we have the shock wave as one boundary of our domain, we may write

$$\frac{\partial w_i}{\partial \tau} = \frac{\partial w_i}{\partial p} \frac{\partial p}{\partial \tau} \tag{6-124}$$

That is, we explicitly include the dependence of the w_i variables on the shock pressure. The derivative $\partial w_i/\partial p$ can be explicitly evaluated from Eq. (6-122). If we substitute Eq. (6-124) into Eq. (6-123), an expression for the time rate of change of pressure is obtained

$$\frac{\partial p}{\partial \tau} \sum_{i=1}^{4} s_{4i} \frac{\partial w_i}{\partial p} = -\sum_{i=1}^{4} s_{4i}(\lambda_4 w_{i\eta} + g_i) \tag{6-125}$$

The $w_{i\eta}$ derivatives in this expression are evaluated using backward differences which is consistent with the fact that information is being carried along a positive characteristic. The expression given in Eq. (6-125) permits p_τ to be computed, and then time derivatives of the other variables can be obtained using Eq. (6-124). These expressions are then integrated to provide the updated dependent variables. The shock position is updated by integrating the known shock speed. Moretti (1974, 1975) prefers to use the shock speed as the dependent variable. This can be easily accomplished within the above analysis. The dependence of the w_i variables given in Eq. (6-124) is replaced by

$$\frac{\partial w_i}{\partial \tau} = \frac{\partial w_i}{\partial V_s} \frac{\partial V_s}{\partial \tau} \tag{6-126}$$

where we again compute $\partial w_i/\partial V_s$ from the Rankine-Hugoniot equations. Substituting this expression into our compatibility equation yields an equation which may be solved for the shock acceleration

$$\frac{\partial V_s}{\partial \tau} \sum_{i=1}^{4} s_{4i} \frac{\partial w_i}{\partial V_s} = -\sum_{i=1}^{4} s_{4i}(\lambda_4 w_{i\eta} + g_i) \tag{6-127}$$

Once the shock acceleration is known, the velocity and position are obtained by integration in time. The new dependent variables are computed from the Rankine-Hugoniot equations using the new shock velocity.

DeNeef and Moretti (1980) have suggested another approach to shock fitting which they have called a post-correction technique. In this method, the shock wave is assumed to be a boundary in the flow and the equations of motion are integrated behind the shock using appropriate differences. This provides a set of dependent variables based upon integration of the Euler equations carrying information from the interior of the domain. Another estimate is then made for the post shock dependent variables based upon the Rankine-Hugoniot equations. DeNeef and Moretti have shown that the change in shock speed is related to the difference in the dependent variable solutions obtained using the two values noted. Once the change in shock speed is determined, a new shock speed is known and a new shock position can be calculated.

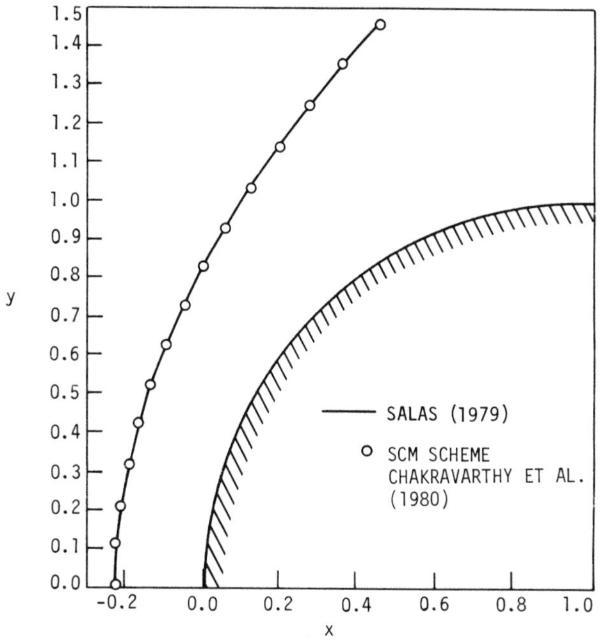

Figure 6-18 Sphere in supersonic flow, $M_\infty = 2.94$.

In closing this section on the SCM method, two examples of flow over bodies are shown in Figs. 6-18 through 6-20. A typical bow shock shape produced by a sphere at a freestream Mach number of 2.94 is shown in Fig. 6-18 while the surface pressure distribution is compared with data provided by Salas (1979) in Fig. 6-19. Figure 6-20 presents shock and sonic line shapes for a 60° cone with a cylindrical afterbody at a freestream Mach number of 1.81. These solutions were obtained using the

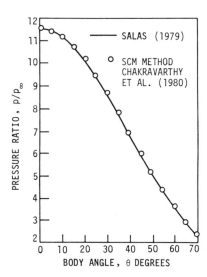

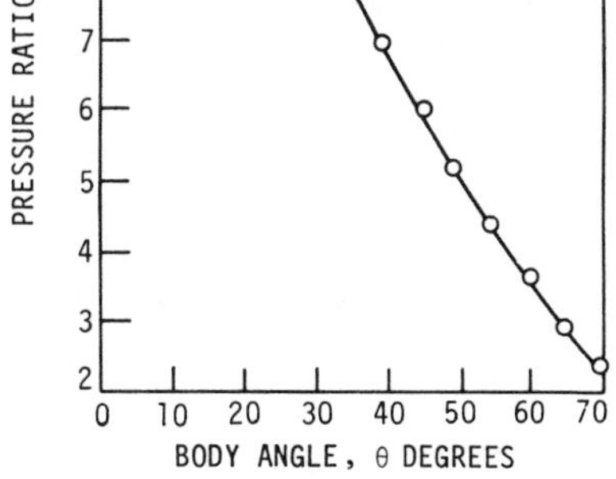

Figure 6-19 Surface pressure comparison for a sphere, $M_\infty = 2.94$.

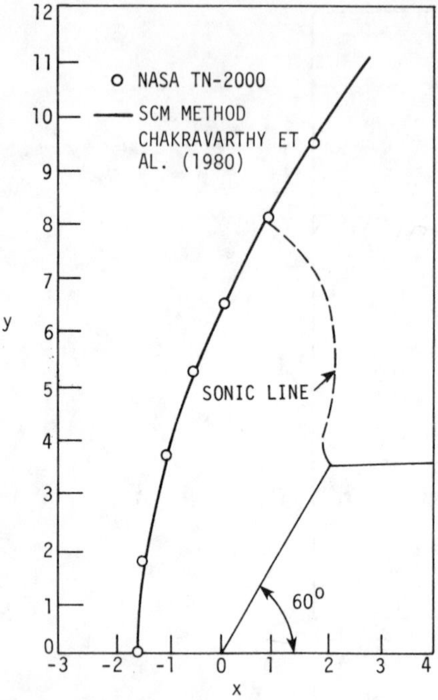

Figure 6-20 Shock and sonic line for a 60° cone with a cylindrical afterbody, $M_\infty = 1.81$.

SCM method and treating the bow shock as a discontinuity. Of methods available for computing the solutions of the Euler equations in nonconservative form, the SCM scheme is an excellent technique to use in conjunction with shock fitting.

6-5 METHODS FOR SOLVING THE POTENTIAL EQUATION

While it is now possible to numerically solve the Euler equations in most cases, a simplified set of equations which can be solved easier at lower cost is desirable. An example might be in preliminary design of supersonic/hypersonic vehicles where many geometric configurations are studied and some sort of design optimization is attempted. For applications of this nature, the cost and computer time required to solve the Euler equations for each configuration of interest is prohibitive.

As is well-known in fluid mechanics, a hierarchy of equations exists based upon the order of the approximation attempted or the assumptions made in the derivation of the governing equations. As we consider reductions from the Euler equations, the next logical step is to consider the solution provided by a full potential equation formulation.

The full potential equation either in conservative or nonconservative form is frequently used for solving transonic flow problems. In developing the full potential equation, the existence of the velocity potential requires that the flow be irrotational. Furthermore, Crocco's equation [Eq. (5-187)] requires that no entropy production occur. Thus no entropy changes are permitted across shocks even in supersonic flows when a full potential formulation is used. At first glance this appears to be a poor assumption. However, experience has shown that the full potential and Euler solutions do not differ significantly if the component of the Mach number normal to the shock is close to one. The entropy production across a weak shock is dependent on the normal Mach number, M_n, and we may write (Liepmann and Roshko, 1957)

$$\frac{\Delta s}{R} \alpha \frac{2\gamma}{\gamma + 1} (M_n^2 - 1)^3 \tag{6-128}$$

This shows that the assumption of no entropy change across a shock is reasonable so long as the normal component of the Mach number is sufficiently close to one. It is important to note that the restriction is on the normal component of the local Mach number and not the freestream Mach number.

If the irrotational flow assumption is a reasonably good one, we expect a solution of the potential equation to yield results nearly as good as the Euler equation solutions even in supersonic and transonic flows with shocks. Difficulties in solving the Euler equations are not completely circumvented by the potential formulation since we retain the nonlinear fluid behavior even with this simplification. We will discuss the application of the potential equation to supersonic and transonic flows in this section. Since we have been studying applications to hyperbolic equations, we begin with supersonic steady flows and follow that by transonic flows.

The full potential equation approximation to the Euler equations can be developed in either a nonconservative or a conservative form. The nonconservative form of the potential equation may be written for two dimensions as [Eq. (5-197)]

$$\left(1 - \frac{u^2}{a^2}\right) \phi_{xx} - \frac{2uv}{a^2} \phi_{xy} + \left(1 - \frac{v^2}{a^2}\right) \phi_{yy} = 0 \tag{6-129}$$

where

$$u = \frac{\partial \phi}{\partial x} \qquad v = \frac{\partial \phi}{\partial y} \tag{6-130}$$

and a is the speed of sound which may be obtained from the energy equation

$$\frac{a^2}{\gamma - 1} + \frac{u^2 + v^2}{2} = H = \text{constant} \tag{6-131}$$

Equation (6-129) is sometimes referred to as the quasi-linear form of the full potential equation. In our discussion of solutions of the Euler equations, the use of the nonconservation form did not produce acceptable results at the shocks. However, in transonic flow this problem does not occur because the shocks are weak. In this section, we will present methods for solving the potential equation in conservation form.

The conservation form of the nondimensional full potential equation in two dimensions may be written

$$\frac{\partial(\rho u)}{\partial x} + \frac{\partial(\rho v)}{\partial y} = 0 \tag{6-132}$$

where the asterisk denoting nondimensional form has been omitted and

$$u = \frac{\partial \phi}{\partial x} \qquad v = \frac{\partial \phi}{\partial y} \tag{6-133}$$

The density is calculated from the energy equation in the form

$$\rho = \left[1 - \frac{\gamma - 1}{2} M_\infty^2 (u^2 + v^2 - 1) \right]^{1/\gamma - 1} \tag{6-134}$$

In this formulation, the density and the velocity components are nondimensionalized by the freestream values. For a supersonic marching problem, we wish to solve Eqs. (6-132)–(6-134) subject to the surface tangency condition written as

$$\frac{\partial \phi}{\partial n} = 0 \tag{6-135}$$

and a prescribed freestream Mach number M_∞. This formulation is certainly less complicated than one using the full Euler equations.

The full potential equation has been used to compute the solution over conical and cambered twisted delta wings by Grossman (1979) and Grossman and Siclari (1980). They used a nonconservative formulation and a transonic relaxation procedure to obtain their solutions. We will present the marching procedure developed by Shankar (1981) and Shankar and Chakravarthy (1981) for the conservative formulation in order to include the unique density linearization idea they have used.

Before the total procedure can be outlined, some discussion of appropriate differencing of the potential equation is necessary. Since we have assumed irrotational flow and used a potential formulation, no dissipative mechanism is present in our system of equations. A result of eliminating changes in entropy is the existence of both expansion and compression shocks as valid solutions using the potential formulation. While this is not a problem for subsonic flow, the existence of expansion shock waves in supersonic flow is unacceptable and must be avoided. This problem is avoided by the use of implicitly or explicitly added artificial viscosity.

Murman and Cole (1971), in a landmark paper treating transonic flow, pointed out that derivatives at each mesh point in the domain of interest must be correctly treated using type-dependent differences. They were particularly interested in solving the transonic small disturbance equation but the same idea is applicable to the full potential equation.

To illustrate type-dependent differencing, consider the nonconservative equation [Eq. (6-129)]. This equation is hyperbolic at points where

$$\frac{u^2 + v^2}{a^2} - 1 > 0$$

and elliptic when

$$\frac{u^2 + v^2}{a^2} - 1 < 0$$

Consider the case when the flow is aligned with the x direction. If the flow is subsonic, the equation is elliptic and central differences are used for the derivatives. If the flow is supersonic, the equation is hyperbolic at the point of interest and the streamwise second derivative is retarded in the upstream direction. The expressions for the finite-difference representation of the second derivatives at point (i, j) become

$$\phi_{xx} = \frac{\phi_{i,j} - 2\phi_{i-1,j} + \phi_{i-2,j}}{(\Delta x)^2}$$

$$\phi_{xy} = \frac{\phi_{i,j+1} - \phi_{i,j-1} - \phi_{i-1,j+1} + \phi_{i-1,j-1}}{2\,\Delta x\,\Delta y} \tag{6-136}$$

$$\phi_{yy} = \frac{\phi_{i,j+1} - 2\phi_{i,j} + \phi_{i,j-1}}{(\Delta y)^2}$$

The grid points used for both supersonic and subsonic points are shown in Fig. 6-21.

The structure of the point clusters shown in Fig. 6-21 illustrates the correct type-dependence for either supersonic or subsonic points. The location of the points used in the finite-difference representation of the steady potential equation shows that it is desirable to use an implicit scheme to compute solutions. If we consider only supersonic flow so that no elliptic points exist in the field, a solution can be obtained using an explicit formulation. This is not advisable if the flow is only slightly supersonic at some field points because the CFL stability criterion prohibits reasonable step sizes. In that case an explicit solution even for purely supersonic flow becomes impractical.

If we examine the truncation error in the finite-difference representation of ϕ_{xx} at hyperbolic points, we find the leading terms to be of the form

$$\Delta x(u^2 - a^2)\phi_{xxx} \tag{6-137}$$

This provides a positive artificial viscosity at all points where $u^2 > a^2$. If the differencing given in Eq. (6-136) is used at an elliptic point, the artificial viscosity becomes negative and a stability problem results. Jameson (1974) pointed out that difficulty arises in those cases where the flow is supersonic, and the x component (u) of the velocity is less than the speed of sound. The problem can be understood by considering a case where the flow is not aligned with the x direction as shown in Fig. 6-22. The proper domain of dependence for all points is not included. One of the y coordinates of a point in the finite-difference molecule lies behind one of the characteristics passing through the point $(i\,\Delta x,\ j\,\Delta y)$. In order to remedy this problem, Jameson

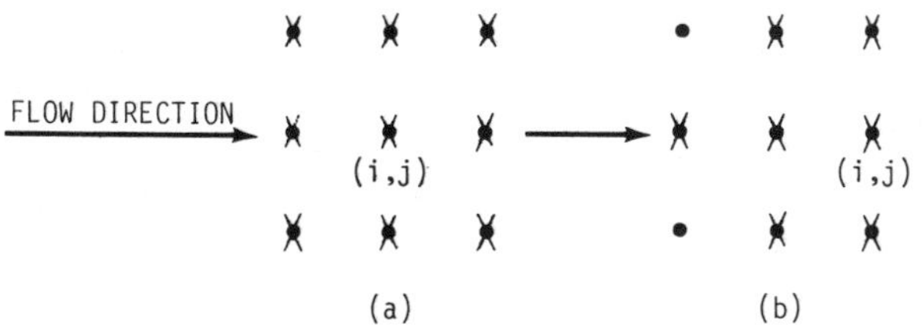

Figure 6-21 Type-dependent differencing. (a) Elliptic point. (b) Hyperbolic point.

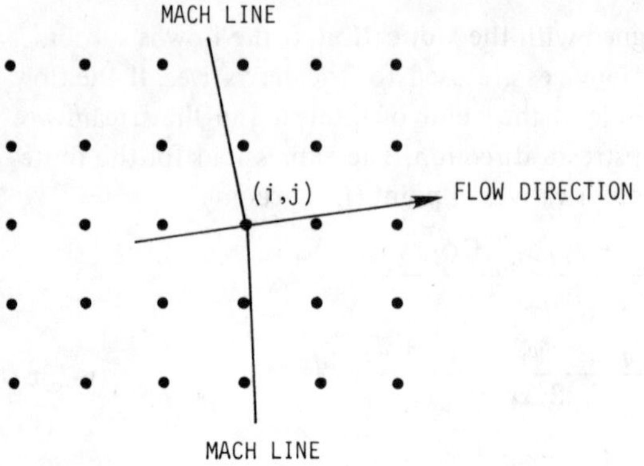

Figure 6-22 Flow with nonaligned mesh system.

introduced his well-known rotated difference scheme. The idea is to write the potential equation in natural coordinates as

$$(a^2 - V^2)\phi_{ss} + a^2\phi_{nn} = 0 \tag{6-138}$$

where s and n are distances along and normal to the streamlines. By applying the chain rule for partial derivatives, the second derivatives may be written in terms of x and y as

$$\begin{aligned} \phi_{ss} &= \frac{1}{V^2}(u^2\phi_{xx} + 2uv\phi_{xy} + v^2\phi_{yy}) \\ \phi_{nn} &= \frac{1}{V^2}(v^2\phi_{xx} - 2uv\phi_{xy} + u^2\phi_{yy}) \end{aligned} \tag{6-139}$$

Both x and y derivative contributions to ϕ_{ss} are lagged or retarded while central differences are used for the ϕ_{nn} term. When the flow is aligned with the grid, the rotated scheme reduces to that given in Eq. (6-136) and produces an artificial viscosity with leading term of the form

$$\left(1 - \frac{a^2}{V^2}\right)(\Delta s u^2 \phi_{sss} + \cdots) \tag{6-140}$$

This provides us with a positive artificial viscosity for all points where the flow is supersonic and we expect shock waves to form only as compressions. While the concept of artificial viscosity is used as a means of explaining the behavior of the solutions of the full potential equation, it should be understood that the same conclusions regarding proper treatment of the various terms can be reached by a careful analysis of the finite-difference equations.

Hafez et al. (1979) applied the idea of artificial compressibility in transonic flows in order to provide artificial viscosity in supersonic regions. This concept was originally introduced by Harten (1978) in attempting to devise better methods of shock capturing in supersonic flows. Holst and Ballhaus (1979) and Holst (1979) used an upwind density bias to provide the necessary artificial viscosity. The method presented below incorporates these ideas and is very useful for solving the full potential equation.

To understand the role of density biasing or artificial compressibility in providing an artificial viscosity, it is instructive to consider the one-dimensional form of the potential equation

$$\frac{\partial}{\partial x}\left(\rho \frac{\partial \phi}{\partial x}\right) = 0 \tag{6-141}$$

This expression may be approximated to second order by writing

$$\nabla(\rho_{i+1/2}\,\Delta\phi_i) = 0 \tag{6-142}$$

where the notation is as previously introduced. For elliptic points, Eq. (6-142) is satisfactory. For hyperbolic points, an artificial viscosity must be added such as that used by Jameson (1975)

$$-\Delta x(\mu\phi_{xx})_x \tag{6-143}$$

where

$$\mu = \min\begin{cases} 0 \\ \rho\left(1 - \dfrac{\phi_x^2}{a^2}\right) \end{cases} \tag{6-144}$$

As previously noted, this explicit addition of artificial viscosity is equivalent to the type-dependent differencing introduced by Murman and Cole (1971). Jameson (1975) has shown that Eq. (6-143) is equivalent to a term with the form

$$-\Delta x(\nu\rho_x\phi_x)_x \tag{6-145}$$

where

$$\nu = \max\begin{cases} 0 \\ 1 - \dfrac{a^2}{\phi_x^2} \end{cases} \tag{6-146}$$

This form is obtained by differentiation of the one-dimensional form of the energy equation. If this artificial viscosity form is incorporated into the potential equation, the finite-difference approximation to Eq. (6-141) becomes

$$\frac{\partial}{\partial x}\left(\rho \frac{\partial \phi}{\partial x}\right) \approx \nabla[\rho_{i+1/2}\,\Delta\phi_i] - \nabla[\nu_i(\rho_{i+1/2} - \rho_{i-1/2})\Delta\phi_i] = 0 \tag{6-147}$$

This expression, due to Holst and Ballhaus (1979), is second-order accurate and centrally differenced in subsonic regions. In supersonic regions, this is a first-order upwind scheme due to the addition of the artificial viscosity. The differencing becomes more strongly biased in the upwind direction as the Mach number increases. In subsonic regions, the density biasing is switched off.

The difference expression given by Eq. (6-147) can also be written

$$\frac{\partial}{\partial x}\left(\rho \frac{\partial \phi}{\partial x}\right) \approx \nabla(\tilde{\rho}_{i+1/2}\,\Delta\phi_i) = 0 \tag{6-148}$$

if the new density is identified by

$$\tilde{\rho}_{i+1/2} = (1 - \nu_i)\rho_{i+1/2} + \nu_i \rho_{i-1/2} \tag{6-149}$$

where the values at the cell midpoint are obtained from the energy equation [Eq. (6-134)]. In this expression for $\rho_{i+1/2}$, only u appears and is evaluated as $(\phi_{i+1} - \phi_i)/\Delta x$. Equations (6-148) and (6-149) show that the effect of adding artificial viscosity is equivalent to using a retarded density. In Jameson's method (1975) the artificial viscosity is explicitly added while in the scheme outlined here, the artificial viscosity is included through the treatment of the density. If the artificial viscosity, ν, is chosen as given in Eq. (6-144), the two techniques give identical results. If $\nu = 0$, the scheme is good only in elliptic regions and is unstable for supersonic flow. However, if ν is set equal to a positive constant, the scheme can be used for both subsonic and supersonic flows. It should be noted that the resulting method is first order and highly dissipative when ν is set equal to a constant. Second-order techniques using density biasing have recently been developed by Steinhoff and Jameson (1981).

Before we apply the artificial compression or density bias approach, the governing equations are transformed into a general computational set (ξ, η) in order to easily apply boundary conditions at prescribed surfaces. If the Viviand (1974) strong conservation transformation is used, Eq. (6-132) becomes

$$\left(\rho \frac{U}{J}\right)_\xi + \left(\rho \frac{V}{J}\right)_\eta = 0 \tag{6-150}$$

where U, V are the contravariant velocity components given by

$$\begin{aligned} U &= A_1 \phi_\xi + A_2 \phi_\eta \\ V &= A_2 \phi_\xi + A_3 \phi_\eta \end{aligned} \tag{6-151}$$

and

$$\begin{gathered} A_1 = \xi_x^2 + \xi_y^2 \qquad A_2 = \xi_x \eta_x + \xi_y \eta_y \qquad A_3 = \eta_x^2 + \eta_y^2 \\ J = \frac{\partial(\xi, \eta)}{\partial(x, y)} = \xi_x \eta_y - \xi_y \eta_x \end{gathered} \tag{6-152}$$

with the density written as

$$\rho = \left[1 - \frac{\gamma - 1}{2} M_\infty^2 (U\phi_\xi + V\phi_\eta - 1)\right]^{1/\gamma - 1} \tag{6-153}$$

The metric coefficients are also required. They are related to the derivatives of the physical coordinates by the expressions

$$\begin{aligned} \xi_x &= J y_\eta \\ \xi_y &= -J x_\eta \\ \eta_x &= -J y_\xi \\ \eta_y &= J x_\xi \end{aligned} \tag{6-154}$$

6-5.1 Methods for Supersonic Flows

In illustrating the procedure used to solve the full potential equation for a supersonic flow, we consider ξ to be the marching direction and assume all information is known at the ith and all previous planes. Our problem is to advance the solution to level $i+1$ and obtain ϕ. In solving the potential equation, we evaluate the first term at $(i+\frac{1}{2}, j)$ and the second term at (i, j) or at $(i+1, j)$ for a fully implicit scheme.

Consider the first term in Eq. (6-150).

$$\frac{\partial}{\partial \xi}\left(\frac{\rho U}{J}\right)$$

In advancing in the ξ direction, we expand ρ in terms of the known values at the ith plane. Since ρ is an explicit function of the velocity components, we write

$$\rho = \rho(\phi_x, \phi_y)$$

and

$$\rho = \rho_i + \Delta\rho \tag{6-155}$$

where

$$\Delta\rho = \frac{\partial \rho}{\partial \phi_x}\Delta\phi_x + \frac{\partial \rho}{\partial \phi_y}\Delta\phi_y$$

Using the density expression given in Eq. (6-153) and the definitions of U and V, the density change is written

$$\Delta\rho = -\frac{\rho_i}{a_i^2}\left(U_i\frac{\partial}{\partial \xi} + V_i\frac{\partial}{\partial \eta}\right)\Delta\phi \tag{6-156}$$

where

$$\Delta\phi = \phi - \phi_i \qquad \Delta\phi_i = \phi_i - \phi_{i-1}$$

Using Eq. (6-155) for the density and Eq. (6-151) for U, the ξ derivative term may be written

$$\frac{\partial}{\partial \xi}\left(\rho\frac{U}{J}\right) = \frac{\partial}{\partial \xi}\left\{\left[\left(\frac{\rho_i A_1}{J} - \frac{\rho_i U_i^2}{a_i^2 J_i}\right)\frac{\Delta\phi}{\Delta\xi} + \frac{\rho_i U_i^2}{a_i^2 J_i}\frac{\Delta\phi_i}{\Delta\xi}\right]\right.$$
$$\left. + \left[\left(\frac{\rho_i A_2}{J} - \frac{\rho_i U_i V_i}{a_i^2 J_i}\right)\frac{\partial(\Delta\phi)}{\partial\eta} + \frac{\rho_i A_2}{J}\frac{\partial \phi_i}{\partial \eta}\right]\right\} \tag{6-157}$$

If a first-order upwind difference is used such as

$$\frac{\partial(\ )}{\partial \xi} = \frac{1}{\Delta\xi}[(\ )_{i+1,j} - (\ )_{i,j}] \tag{6-158}$$

a leading truncation error term is produced which has the form

$$\frac{\rho_i}{J_i a_i^2}\left[1 - \frac{a_i^2 A_1}{U_i^2}\right]U_i^2\phi_{\xi\xi\xi}\,\Delta\xi \tag{6-159}$$

This provides a positive artificial viscosity when

$$\frac{U_i^2}{A_1} > a_i^2 \tag{6-160}$$

This condition [Eq. (6-160)] must be met in order to solve the full potential equation as a marching problem in supersonic flow. If Eq. (6-160) is not satisfied, the artificial viscosity is negative and an instability results. It should be noted that the cross-derivatives appearing in Eq. (6-157) are upwind differenced depending on the signs of their coefficients to maintain diagonal dominance of the implicit scheme. The eta derivative

$$\frac{\partial}{\partial \eta}\left(\frac{\rho V}{J}\right) = \frac{\partial}{\partial \eta}\left[\frac{\rho}{J}(A_2 \phi_\xi + A_3 \phi_\eta)\right]_{i+1,j}$$

is evaluated at $i + 1$ which results in an implicit marching algorithm. If this difference is written in terms of the potential difference, $\Delta\phi$, it becomes

$$\frac{\partial}{\partial \eta}\left(\frac{\rho V}{J}\right) = \left(\frac{\rho_{i+1}^n A_2}{J}\frac{\Delta\phi}{\Delta\xi}\right)_\eta + \left(\frac{\rho_{i+1}^n A_3}{J}\frac{\partial\, \Delta\phi}{\partial \eta}\right)_\eta + \left(\frac{\rho_{i+1}^n A_3}{J}\frac{\partial \phi_i}{\partial \eta}\right)_\eta \tag{6-161}$$

The density ρ_{i+1}^n given in Eq. (6-161) represents the nth iteration at the $i + 1$ grid point. As each step is taken in the ξ direction, we must iterate to determine the new density. This iterative process turns out to consist of only one or two successive passes for most problems. In order to create the correct artificial viscosity, the density is "upwind biased" [see Holst (1979) or Hafez et al. (1979)]. The value of ρ_{i+1}^n in Eq. (6-161) is replaced by

$$\tilde{\rho}_{i+1,j+1/2} = [(1-\nu)\rho_{i+1}^n]_{i+1,j+1/2} + \nu_{i+1,j+1/2}(\rho_{i+1}^n)_{i+1,j+1/2+m} \tag{6-162}$$

where

$$m = -1 \qquad V_{i+1,j+1/2} > 0$$
$$= +1 \qquad V_{i+1,j+1/2} < 0$$

and

$$\nu_{i+1,j+1/2} = \left[1 - \left(\frac{1}{M^n}\right)^2\right]_{i+1,j+s} \tag{6-163}$$

with $s = 0$ for $V_{i+1,j+1/2} > 0$ and $s = 1$ for $V_{i+1,j+1/2} < 0$. The density treatment provided by Eqs. (6-162) and (6-163) produces a positive artificial viscosity for all points where $M_{i+1}^n > 1$ (supersonic). Since we are studying a procedure for solving supersonic flows using a marching procedure, the method fails when the flow becomes subsonic and the artificial viscosity becomes negative.

In advancing the solution to level $i + 1$, boundary conditions are applied implicitly at that level. For any symmetry plane the idea of simple reflection can be used. For example, in a two-dimensional problem when a sweep in the η direction is made, we may implement this at the boundary (assuming we have symmetry boundary conditions at $j = j_{\max-1}$) by setting

$$(\Delta\phi)_{i+1,j_{\max}} = (\Delta\phi)_{i+1,j_{\max-2}} \tag{6-164}$$

The application of the body surface boundary condition is nearly as simple. If a body fitted coordinate system is assumed, the surface boundary condition in this two-dimensional problem is of the form

$$V = A_2\phi_\xi + A_3\phi_\eta = 0$$

We can use a dummy point ($j = 1$) below the body surface ($j = 2$) to implement the surface condition. Let

$$0 = V_{i+1,2} = A_2\left(\frac{\Delta\phi}{\Delta\xi}\right)_{i+1,2} + A_3\,\frac{(\Delta\phi + \phi^n_{i+1})_{i+1,3} - (\Delta\phi + \phi^n_{i+1})_{i+1,1}}{2\,\Delta\eta} \tag{6-165}$$

This expression is used to eliminate the sublayer point ($j = 1$) in the difference equation at the boundary. The surface boundary condition is implicitly included in the calculation, and consequently, no restriction on marching step size occurs because of boundaries. Implicitly including boundary conditions also improves the convergence rate of the density iteration at each marching location.

The structure of the algorithm for computing the solution of this problem results in a tridiagonal system for $\Delta\phi$. This can be verified most easily by considering the system written in the form

$$\left[1 + \frac{C_1}{\beta}\frac{\partial}{\partial\eta} + \frac{1}{\beta}\frac{\partial}{\partial\eta}(C_2) + \frac{1}{\beta}\frac{\partial}{\partial\eta}\left(C_3\,\frac{\partial}{\partial\eta}\right)\right]\Delta\phi = R \tag{6-166}$$

where C_1, C_2, C_3, and β are the coefficients of the differential terms and R is the known right-hand side of the equation. Evaluation of these terms is left as an exercise (see Prob. 6-15). This tridiagonal equation is solved for $\Delta\phi$. Once $\Delta\phi$ is known, the density ρ_{i+1} is calculated. This new density is used to compute a new $\Delta\phi$ in Eq. (6-166) again, and this iterative process is repeated until convergence is attained. In the three-dimensional case, the same procedure is followed except the solution for $\Delta\phi$ is computed using approximate factorization requiring the solution of a tridiagonal system in the η direction and in the ζ direction.

Shankar and Osher (1982) have applied the solution procedure presented here to the full potential equation. Instead of simply linearizing the density, as in Eq. (6-155), the product of the density and the contravariant velocity component in the ξ direction is expanded in the form

$$(\rho U)_{i+1} = (\rho U)_i + \Delta(\rho U)$$

In addition to this improvement, they employed upwind differences based upon the characteristics of the hyperbolic system of equations. This upwind differencing provides a much more concise application of artificial viscosity than the artificial compression or retarded density approach.

Typical results presented by Shankar and Chakravarthy (1981) are shown in Figs. 6-23–6-25. The Euler equation solution for the wedge-flow case is a shock-capturing second-order solution which accounts for the oscillations at the shock. The

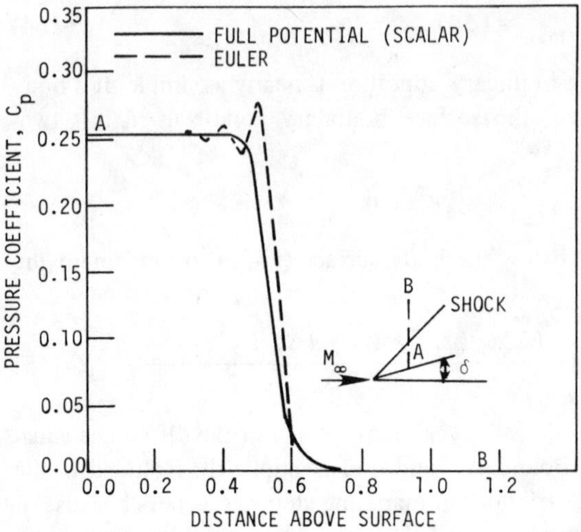

Figure 6-23 Potential equation solution for wedge flow.

results for both the cone and cosine shaped body show very close agreement with the Euler solutions. Not only are the full potential solutions in close agreement with the Euler equation results, but the computer time requirements are as much as an order of magnitude less in some cases. This approach is recommended for use in supersonic problems where the potential equation is valid.

6-5.2 Methods for Transonic Flow

The full potential equation is useful for describing transonic flows when shock wave strength is small. Recent schemes have used the idea of retarded density, and the method of Holst and Ballhaus (1979) is representative of these techniques.

Consider flow over a two-dimensional airfoil at a freestream Mach number M_∞ that may be sufficient to introduce locally supersonic conditions. This flight condition is shown in Fig. 6-26. Inviscid flow over such a configuration is approximated

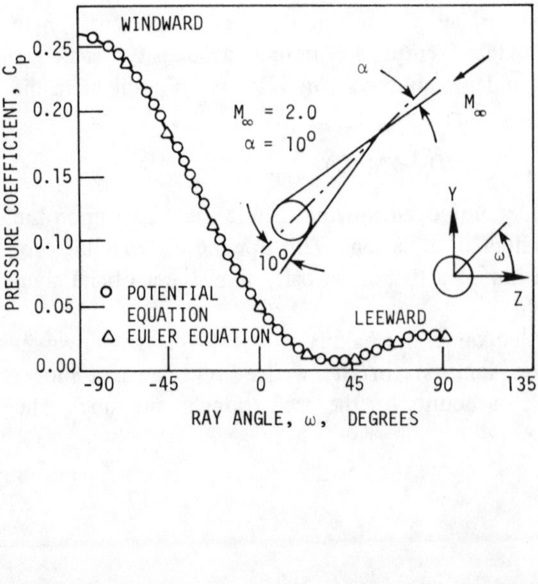

Figure 6-24 Flow over a circular cone.

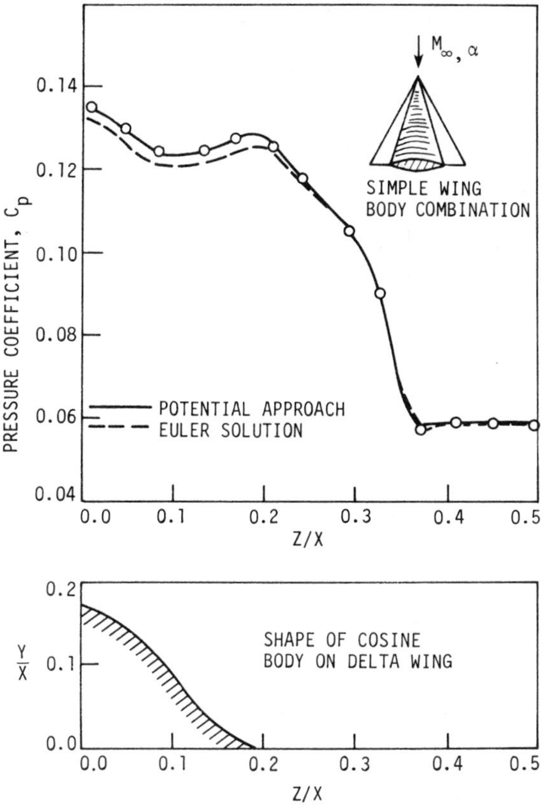

Figure 6-25 Flow over wing body configuration.

by a solution of the full potential equation. In solving for the flowfield, the physical domain is again mapped into computational space (ξ, η) in such a way that the airfoil surface is an η = constant surface as is the outer boundary. The ξ coordinate is defined in such a way that the lower side of the assumed straight trailing slip line is a ξ = constant surface. The ξ coordinate increases around the airfoil to ξ_{max} at the upper side of the vortex sheet. The computational domain is shown in Fig. 6-27. General mappings from physical to computational space will be covered in Chapter 10.

The full potential equation was presented earlier in Eqs. (6-150)-(6-152). Our objective is to solve the full potential equation for the transonic flow over an airfoil as shown in Fig. 6-26. The notation is consistent with the previous development for a supersonic marching problem.

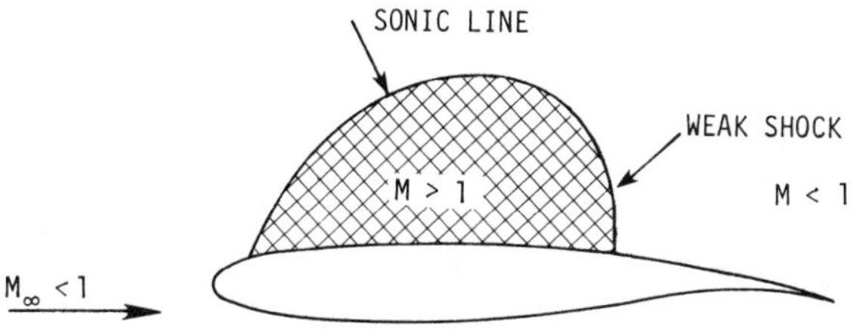

Figure 6-26 Transonic airfoil.

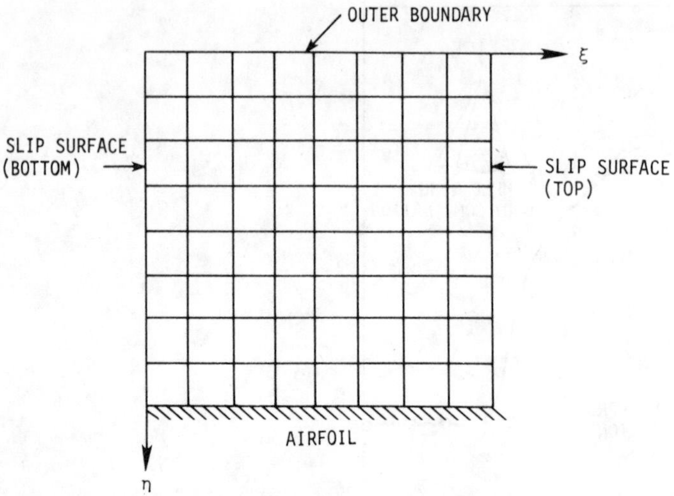

Figure 6-27 Computational domain for transonic airfoil.

A second-order finite-difference approximation to Eq. (6-150) can be written

$$\nabla_\xi \left(\frac{\rho U}{J}\right)_{i+1/2,j} + \nabla_\eta \left(\frac{\rho V}{J}\right)_{i,j+1/2} = 0 \tag{6-167}$$

In this expression the contravariant velocity components are given by

$$U_{i+1/2,j} = A_{1_{i+1/2,j}}(\phi_{i+1,j} - \phi_{i,j}) + \tfrac{1}{4}A_{2_{i+1/2,j}}(\phi_{i+1,j+1} - \phi_{i+1,j-1} + \phi_{i,j+1} - \phi_{i,j-1})$$

$$V_{i,j+1/2} = \tfrac{1}{4}A_{2_{i,j+1/2}}(\phi_{i+1,j+1} - \phi_{i-1,j+1} + \phi_{i+1,j} - \phi_{i-1,j}) + A_{3_{i,j+1/2}}(\phi_{i,j+1} - \phi_{i,j}) \tag{6-168}$$

where the half integer values are obtained by averaging and the A's are given in Eq. (6-152). This is a valid formulation for subsonic flow regions. We now introduce the retarded density and replace Eq. (6-167) with

$$\nabla_\xi \left[\tilde{\rho}_i \left(\frac{U}{J}\right)_{i+1/2,j}\right] + \nabla_\eta \left(\frac{\rho V}{J}\right)_{i,j+1/2} = 0 \tag{6-169}$$

where
$$\tilde{\rho}_i = (1 - \nu_{i+k,j})\rho_{i+1/2,j} + \nu_{i+k,j}\rho_{i+2k-1/2,j} \tag{6-170}$$

with

$$k = \begin{cases} 0 & U_{i+1/2,j} > 0 \\ 1 & U_{i+1/2,j} < 0 \end{cases} \tag{6-171}$$

The value of the artificial viscosity is selected to be

$$\nu = \max\left[0, C_1\left(1 - \frac{1}{M^2}\right)\right]$$

where the constant C_1 is unity for small regions of supersonic flow but must be increased in regions where shock strength is appreciable. This formulation may be used throughout the flow to solve the full potential equation at points which are either elliptic or hyperbolic.

The solution procedure used to solve the set of resulting difference equations may take many forms. Holst and Ballhaus computed solutions for transonic flow using both the conventional successive line over-relaxation (SLOR) technique and approximate factorization (AF) schemes. We will discuss the AF scheme of Holst (called AF2) since SLOR techniques have been presented for other systems of equations.

An AF scheme (and most others) for the full potential equation may be written

$$NC^n + \omega L\phi^n = 0 \tag{6-172}$$

for a relaxation problem governed by a PDE of the form, $L\phi = 0$, where L is a differential operator. The ω is a relaxation parameter, C^n is the correction term $(\phi^{n+1} - \phi^n)$, $L\phi^n$ represents the residual (the PDE is not satisfied by the approximate solution, ϕ^n), and N is an operator which determines the iteration method. In AF schemes, N represents the product of two or more operators.

$$N = N_1 N_2$$

The operators N_1 and N_2 must be selected so their product approximates L, only simple matrix operations are required, and the overall scheme is stable. The AF2 scheme used by Holst uses an N operator of the form

$$\alpha NC^n_{i,j} = -\left[\alpha - \Delta_\eta \left(\frac{\rho A_3}{J}\right)_{i,j-1/2}\right]\left[\alpha \nabla_\eta - \nabla_\xi \tilde{\rho}_i \left(\frac{A_1}{J}\right)_{i+1/2,j} \Delta_\xi\right] C^n_{i,j} \tag{6-173}$$

The α is a free parameter which may be interpreted as $(\Delta t)^{-1}$. A sequence of alphas is used during the calculations in order to reduce both high and low frequency errors in the solution. Holst and Ballhaus (1979) present such a sequence. This sequence can be used for other transonic airfoil calculations and near optimal performance of the AF2 scheme will be obtained.

The AF2 scheme of Eq. (6-173) is implemented in two steps

$$\begin{aligned} \left[\alpha - \Delta_\eta \left(\frac{\rho A_3}{J}\right)_{i,j-1/2}\right] f^n_{i,j} &= \alpha\omega L\phi^n_{i,j} \\ \left[\alpha \nabla_\eta - \nabla_\xi \tilde{\rho}_i \left(\frac{A_1}{J}\right)_{i+1/2,j} \Delta_\xi\right] C^n_{i,j} &= f^n_{i,j} \end{aligned} \tag{6-174}$$

where $f^n_{i,j}$ is an intermediate result required in the numerical scheme. In the first step, $f_{i,j}$ is obtained by solving a bidiagonal system similar to that encountered in the factored split flux scheme in Section 6-3. The solution of a tridiagonal system is required in the second step. For this numerical routine, the sweep direction required is outward away from the airfoil in step 1 and inward for step 2. No limitation on sweep direction due to flow direction exists for this scheme. However if other methods such as SLOR are used to solve the full potential equation, care must be exercised to insure that each sweep is made in the direction of flow in the supersonic regions. This provides the addition of a stabilizing term, $\phi_{\xi t}$, as a result of the direction of sweep.

Jameson (1974) pointed out that stability problems occur near sonic lines when the full potential equation is solved. In order to avoid this difficulty, time-like terms are added to numerical schemes. These time-like terms are of the form $\phi_{\xi t}$ and $\phi_{\eta t}$ and are usually included in the operators in relaxation schemes. In this case, terms of the form

$$\alpha K_1 \nabla_\xi \qquad \alpha K_1 \Delta_\xi$$

are added to the operators in the second step [Eq. (6-174)]. Only the upwind term is used and then usually, only in the supersonic region. The sign is chosen to increase the size of the diagonal term in the second step of Eq. (6-174) insuring diagonal dominance. Hafez et al. (1979) have shown that the addition of these time-like terms can be accomplished by modifying the density. This modification is achieved by adding a ϕ_t term to the square of the velocity in the equation for density [Eq. (6-153)]. This approach provides a stabilizing term which is very close to that suggested by Jameson.

Boundary conditions at the airfoil surface are enforced by using sublayer points and reflecting. On the airfoil, the boundary condition, $V = 0$ is enforced by using

$$\left(\frac{\rho V}{J}\right)_{i,NJ-1/2} = -\left(\frac{\rho V}{J}\right)_{i,NJ+1/2} \tag{6-175}$$

at the sublayer point, where NJ is the surface point. This boundary condition is enforced explicitly in the calculations.

If the airfoil is nonlifting, the outer boundary values of the velocity potential and density are held fixed at their freestream values. For the lifting case, the outer boundary points must have the circulation specified consistent with a vortex solution and updated at the end of each integration cycle. At the end of each step the circulation is computed from the trailing-edge velocity potential jump

$$\Gamma = \phi_{u_{\text{TE}}} - \phi_{l_{\text{TE}}} \tag{6-176}$$

As a new iteration starts, the velocity potential jump is imposed all along the vortex sheet. The correction is differenced across this sheet as

$$\Gamma^{n+1} - \Gamma^n = C_u^n - C_l^n \tag{6-177}$$

with

$$\Gamma^{n+1} = 3(\Gamma^n - \Gamma^{n-1}) + \Gamma^{n-2} \tag{6-178}$$

Results for transonic flow over an NACA 0012 airfoil are shown in Fig. 6-28. This figure compares pressure coefficients computed using the AF2 algorithm with the results of Lock (1970). This method works well as is evidenced by the results obtained and provides a very efficient transonic airfoil routine when coupled with the generalized mapping schemes noted earlier. In addition, this scheme may be extended to three-dimensional configurations. Holst (1980) has extended this approach to three-dimensional wings. This procedure is a straightforward extension of the two-dimensional

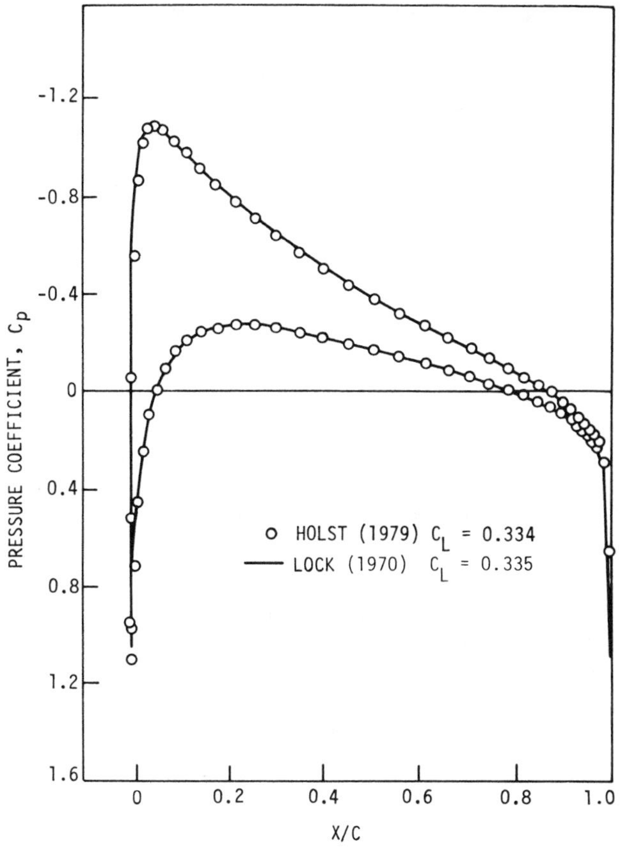

Figure 6-28 Airfoil pressure coefficient comparison NACA 0012, $\alpha = 2°$, $M_\infty = 0.63$.

method. One major point should be noted. In the two-dimensional scheme, using a retarded density in the ξ direction is usually sufficient. If supersonic flow extends to the trailing edge of the airfoil, then upwinding in the η direction is also necessary. For three-dimensional wing calculations, upwinding is used in the chordwise and spanwise directions. If the supersonic region extends to the trailing edge, the same upwinding approach should be used in the normal direction.

Hafez et al. (1979) have pointed out that the artificial viscosity terms which result from the upwind treatment of the density may be viewed as an approximation to the Navier-Stokes equation. Sichel (1963) has derived a viscous transonic small-disturbance equation which is of the form

$$(1 - M^2)\phi_{xx} + \phi_{yy} = -\epsilon\phi_{xxx} \tag{6-179}$$

The viscous term is of the same form as the term introduced through the explicit addition of artificial viscosity in supersonic regions or the use of artificial density or artificial compressibility based upon density modification.

As a final note on the full potential equation, Steinhoff and Jameson (1981) report the occurrence of multiple solutions to the full potential equation in transonic flow. Their results show that in the Mach number range between 0.82 and 0.85 an 11.8% thick Joukowski airfoil exhibits more than one solution for a given condition. This is somewhat disconcerting and further research needs to be done in computing solutions of the Euler equations and the Navier-Stokes equations for similar cases. This would determine if the nonuniqueness was just a property of the full potential equation or a phenomenon to be physically expected. Preliminary results using the Euler equations appear to indicate that this may only be a property of the full potential equation.

6-6 THE TRANSONIC SMALL DISTURBANCE EQUATIONS

The use of the full potential equation for inviscid transonic flows was discussed in the previous section. Results obtained for airfoils and some three-dimensional shapes compare very well with available experimental data. Methods for solving the full potential equation are very efficient and are being used extensively. However, we still find numerous applications where the sophistication provided by the full potential formulation is not required and the accuracy of the solution of the transonic small disturbance equation is sufficient. In addition, a significant advantage accrues in the application of boundary conditions. Boundary conditions for two-dimensional problems are applied on the slit for two dimensions or on the plane for three-dimensional problems. The governing equations are greatly simplified since complex body aligned mappings are unnecessary for the application of boundary conditions. This can result in significant reductions in computer time and storage requirements particularly in three-dimensional problems.

The transonic small disturbance equations may be derived by a systematic expansion procedure. The details of this procedure are given by Cole and Messiter (1957) and Hayes (1966) and provide a means of systematically developing higher-order approximations to the Euler equations. In Chapter 5, the transonic small disturbance equation [Eq. (5-203)] was derived using a perturbation procedure. This may be written in the nondimensional form

$$[K - (\gamma + 1)\phi_x]\phi_{xx} + \phi_{\tilde{y}\tilde{y}} = 0 \tag{6-180}$$

where K is the transonic similarity parameter given by

$$K = \frac{1 - M_\infty^2}{\delta^{2/3}} \tag{6-181}$$

with δ representing the maximum thickness ratio and f is the shape function of an airfoil defined by the expression

$$y = \delta f(x) \tag{6-182}$$

The velocity potential used in Eq. (6-180) is the perturbation velocity potential defined in such a way that the x derivative of ϕ is the perturbation velocity in the x

direction nondimensionalized with respect to the freestream velocity and similarly in the y direction. The scaled coordinate, $\tilde{y}$, is defined by

$$\tilde{y} = \delta^{1/3} y \tag{6-183}$$

Equation (6-180) is formally equivalent to Eq. (5-203) and both are forms of the Guderley-von Kármán transonic small disturbance equations. The similarity form given in Eq. (6-180) is the equation originally treated by Murman and Cole (1971) in calculating the inviscid flow over a nonlifting airfoil. The pressure coefficient is the same as Eq. (5-205) and may be written

$$C_p = -2\phi_x$$

For flows which are not considered transonic, we obtain the Prandtl-Glauert equation for subsonic or supersonic flow. This expression has been used in numerous examples in previous chapters and takes the form

$$(1 - M_\infty^2)\phi_{xx} + \phi_{yy} = 0 \tag{6-184}$$

The main point to remember is that the transonic small disturbance equation is nonlinear and switches from elliptic to hyperbolic in the same manner as the full potential and Euler equations.

In their original paper, Murman and Cole treated the inviscid transonic flow over a nonlifting airfoil and solved the transonic small disturbance equation as given in Eq. (6-180). In addition to the governing PDE, the necessary body surface boundary conditions for zero angle of attack are given by

$$\phi_{\tilde{y}}(x, 0) = f'(x) \tag{6-185}$$

applied in the plane $\tilde{y} = 0$ consistent with the theory. A boundary condition must also be applied at the outer boundary of the computational mesh. For this case, in the far field

$$\phi \simeq \frac{1}{2\pi\sqrt{K}} \frac{Mx}{x^2 + K\tilde{y}^2} \tag{6-186}$$

where

$$M = 2\int_{-1}^{1} f(\xi)\, d\xi + \frac{\gamma + 1}{2} \iint_{-\infty}^{+\infty} d\xi\, d\eta \tag{6-187}$$

and the airfoil is confined to the interval

$$-1 \leqslant x \leqslant 1$$

In the lifting case, the circulation must be imposed and determined by satisfying the Kutta condition on the airfoil. The far field boundary condition in this case takes the form of a vortex with the value of circulation determined by the Kutta condition. For development of the far field boundary condition, the papers by Ludford (1951) and Klunker (1971) are recommended.

Murman and Cole solved Eq. (6-180) for a nonlifting transonic airfoil using line

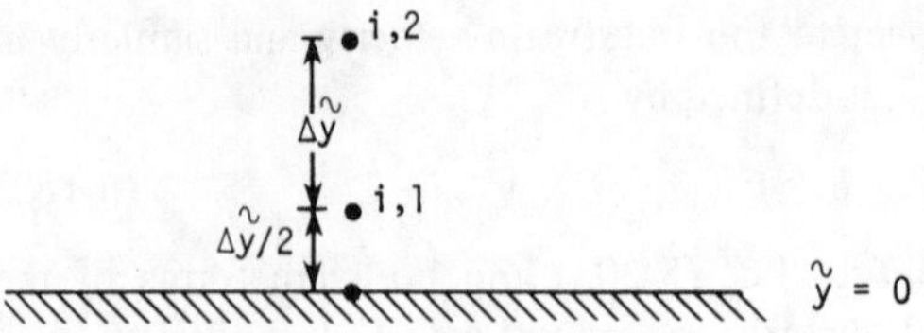

Figure 6-29 Surface boundary point distribution.

relaxation methods. Type-dependent differencing given in Eq. (6-136) was used for hyperbolic regions and central differencing was used in the elliptic regions. The airfoil now appears on the x axis as a line or slit and boundary conditions are applied there. In this case the airfoil lies between two mesh points at the half mesh interval as shown in Fig. 6-29. The boundary condition at $\tilde{y} = 0$ enters as a body slope or derivative of ϕ given in Eq. (6-185). At the $(i, 1)$ point the $\phi_{\tilde{y}\tilde{y}}$ derivative is differenced as

$$\phi_{\tilde{y}\tilde{y}} = \frac{1}{\Delta\tilde{y}}(\phi_{\tilde{y}_{i,3/2}} - \phi_{\tilde{y}_{i,1/2}}) = \frac{1}{\Delta\tilde{y}}\left[\frac{\phi_{i,2} - \phi_{i,1}}{\Delta\tilde{y}} - \phi_{\tilde{y}}(x, 0)\right] \qquad (6\text{-}188)$$

The surface boundary condition explicitly enters the calculation through the $\phi_{\tilde{y}}$ term.

Figure 6-30 shows the pressure distribution for a circular arc airfoil obtained by solving the transonic small disturbance equation. As can be seen, the experimental data of Knechtel (1959) and the computed results compare favorably for both the subcritical and supercritical cases. It is interesting that the shock location and strength for this example agree well with the experimental measurements. The nonconservative equations of small disturbance theory for an inviscid flow underestimate shock strength and produce the same effect as a shock-boundary layer interaction on shock strength and location. Thus the nonconservative form has been popular even though

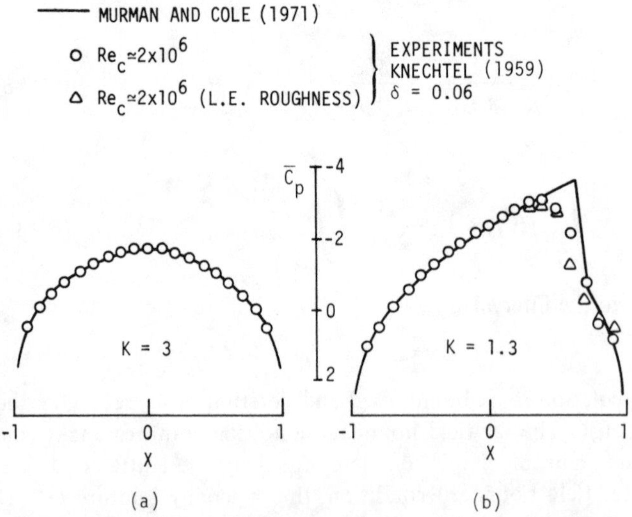

Figure 6-30 Pressure distribution for circular arc airfoil. (a) Subcritical case. (b) Supercritical case.

the conservative form is mathematically most appealing. Numerous applications of the technique, presented by Murman and Cole for solving the transonic small disturbance equation, have been made since it was originally introduced, and many refinements of the basic method have been developed. However, the main point to remember is that a significant simplification over either the Euler equations or the full potential formulation is realized when this approach is used.

The simplified form of the transonic small disturbance equation was used in developing solutions for three-dimensional wings by Bailey and Ballhaus (1972). Their work led to the development of a widely used three-dimensional code for transonic wing analysis. This code has been used extensively in designing improved wings for flight in the transonic speed regime. For those interested in three-dimensional transonic flow over wings, the paper by Bailey and Ballhaus is recommended reading.

Most current work in transonic flow is concentrated on developing full potential or Euler equation solvers. One area where considerable effort is being expended using the transonic small disturbance equation is in the development of design codes. In the design problem, the body pressure is prescribed and the body shape is unknown. For this type of problem a simplified approach offers advantages.

6-7 METHODS FOR SOLVING LAPLACE'S EQUATION

The numerical techniques presented in the previous sections of this chapter were applied to the nonlinear equations governing inviscid fluid flow. Linear PDE's are often used to model both internal and external flows. Examples include Laplace's equation for incompressible, inviscid, irrotational flow and the Prandtl-Glauert equation which is valid in compressible flow if the small perturbation assumptions are satisfied. The methods for solving both of these equations are similar. Finite-difference methods for solving Laplace's equation were presented in Chapter 4 and will not be reviewed here. Instead, the basic idea underlying the use of panel methods will be discussed. These methods have received extensive use in industry.

The advantage of using panel methods is that a solution for the body pressure distribution can be obtained without solving for the flowfield throughout the domain. In this case the problem is reduced to the solution of a system of algebraic equations for source, doublet, or vortex strengths on the boundaries. Using the resulting solution, the body surface pressures can be computed. Panel schemes require the solution of a large system of algebraic equations. For most practical configurations, the storage and speed capability of modern computers has been sufficient. However, judicious selection of the number of surface panels and correct placement is essential in obtaining a good solution for the body surface pressure.

In studying panel methods we will consider the flow of an incompressible, inviscid, irrotational fluid which is governed by a solution of Laplace's equation written in terms of the velocity potential. We require

$$\nabla^2 \phi = 0 \tag{6-189}$$

in the domain of interest and specify either ϕ or $\partial\phi/\partial n$ on the boundary of the domain. For simplicity we restrict our attention to the two-dimensional case although

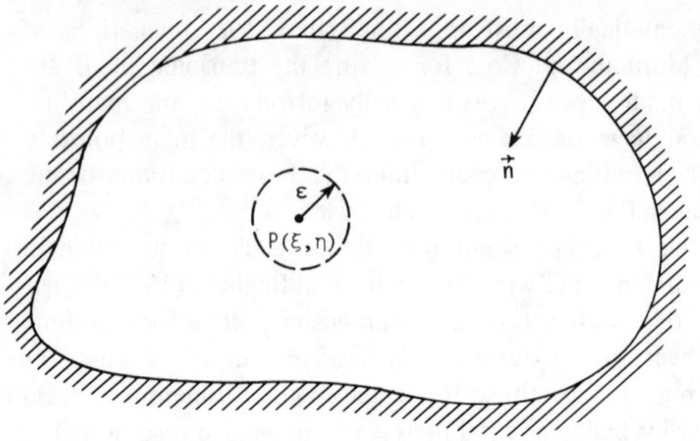

Figure 6-31 Physical domain for Laplace's equation.

the fully three-dimensional problem is conceptually the same. The geometry of the problem under consideration is shown in Fig. 6-31. The basic idea underlying all panel methods is to replace the required solution of Laplace's equation in the domain with a surface integral. This method is developed by the application of Green's second identity to the domain of interest. If u and v are two functions with continuous derivatives through second order (class C^{II}) then Green's second identity may be written

$$\iint_A (u\,\nabla^2 v - v\,\nabla^2 u)\,dA = \int_S (v\,\nabla u - u\,\nabla v)\cdot\mathbf{n}\,ds$$

where $\mathbf{n}$ is the unit normal to the boundary and s is arc length along the boundary. Suppose we choose u to be the potential ϕ and v to be of the form

$$v = \ln(r)$$

where

$$r = \sqrt{(x-\xi)^2 + (y-\eta)^2}$$

We take (ξ, η) as the coordinates of the point P where ϕ is to be determined and (x, y) as the coordinates of the point Q on the boundary where a source is located. In evaluating the integrals in applying Green's identity we must exercise caution as (ξ, η) approach (x, y), that is, when $r \to 0$. In order to avoid this difficulty we think of enclosing the point $P(\xi, \eta)$ with a small circle of radius ϵ and apply Green's identity to the region enclosed by the original boundary (B) and that of the small circle enclosing P. Thus

$$0 = \oint_B (v\,\nabla u - u\,\nabla v)\cdot\mathbf{n}\,ds - \oint_\epsilon (v\,\nabla u - u\,\nabla v)\cdot\mathbf{n}\,ds$$

Consider the second integral with u, v replaced as noted above

$$\oint_\epsilon [\ln(r)\,\nabla\phi - \phi\,\nabla \ln(r)]\cdot\mathbf{n}\,ds$$

On the boundary of the small circle, r equals ϵ and we may write this integral as

$$\ln(\epsilon)\left[\oint \nabla\phi\cdot\mathbf{n}\,ds\right] - \oint \frac{\phi}{r}\,ds$$

By our original hypothesis ϕ is a solution of Laplace's equation, and therefore the first term must vanish (see Prob. 2-7). The second term may be written

$$\frac{1}{\epsilon}\oint_{\epsilon} \phi\,ds$$

which by the mean value property of harmonic functions becomes

$$\frac{1}{\epsilon}\oint \phi\,ds = 2\pi\phi(\xi,\eta)$$

We substitute this result into our original expression to obtain

$$\phi(\xi,\eta) = \frac{1}{2\pi}\oint\left[\ln(r)\frac{\partial\phi}{\partial n} - \phi\frac{\partial \ln(r)}{\partial n}\right]ds \tag{6-190}$$

Thus we have reduced the problem of computing a solution to Laplace's equation in the domain to solving an integral equation over the boundary. The first term represents a Neumann problem where $\partial\phi/\partial n$ is given on the boundary while the second is an example of the classical Dirichlet boundary value problem where ϕ is specified. These integrals correspond to contributions to ϕ of sources and doublets. We could write

$$\phi = \frac{1}{2\pi}\oint\left[\mu\frac{\partial \ln(r)}{\partial n} + \sigma\ln(r)\right]ds \tag{6-191}$$

where we interpret σ as a source distribution and μ as a doublet distribution with axis normal to the bounding surface.

A surface source distribution with density σ per unit length produces a potential at an external point given by

$$\phi = \frac{1}{2\pi}\oint \sigma\ln(r)\,ds \tag{6-192}$$

where the integration is taken over the surface. If we have n surfaces or panels, the total potential at a point P is the sum of the contributions from each panel

$$\phi_i = \sum_{j=1}^{n}\frac{1}{2\pi}\int_j \sigma_j \ln(r_{ij})\,ds_j \tag{6-193}$$

A similar expression can be developed for a doublet distribution.

If a uniform stream is superimposed on the domain which includes the source panels, we include the velocity potential of the freestream and write

$$\phi_i = U_\infty x_i + \sum_{j=1}^{n} \frac{1}{2\pi} \int_j \sigma_j \ln (r_{ij})\, ds_j \tag{6-194}$$

The simplest panel representation to treat numerically is obtained when the source strength of each panel is assumed to be constant. Some advanced methods assume other distributions and the representation of the velocity potential becomes correspondingly more complex. For a constant source strength per panel

$$\phi_i = U_\infty x_i + \sum_{j=1}^{n} \frac{\sigma_j}{2\pi} \int_j \ln (r_{ij})\, ds_j \tag{6-195}$$

The geometry appropriate to the above potential distribution is shown in Fig. 6-32. The problem in using the source panel representation for a given body is to determine the source strengths, σ_j's. This is accomplished by selecting a control point on each panel and requiring that no flow cross the panel. The control point is selected at the midpoint of each panel. We now specify the point P to be at the control point of the ith panel. The boundary condition that no flow goes across the panel at this point is

$$\frac{\partial}{\partial n_i} \phi(x_i, y_i) = 0 \tag{6-196}$$

Since ϕ is the velocity potential, this requires that the normal velocity at the control point of the ith panel vanish. Therefore

$$\sum_{i=1}^{n} \frac{\sigma_j}{2\pi} \int_j \frac{\partial}{\partial n_i} \ln (r_{ij})\, ds_j = -\mathbf{U}_\infty \cdot \mathbf{n}_i \tag{6-197}$$

The dot product is used because the velocity component normal to the surface is required. The velocity induced at the ith control point due to the ith panel is $\sigma_i/2$ and is usually taken out of the above summation. With this convention we may write

$$\frac{\sigma_i}{2\pi} + \sum_{i \neq j}^{n} \frac{\sigma_j}{2\pi} \int \frac{\partial}{\partial n_i} \ln (r_{ij})\, ds_j = -\mathbf{U}_\infty \cdot \mathbf{n}_i \tag{6-198}$$

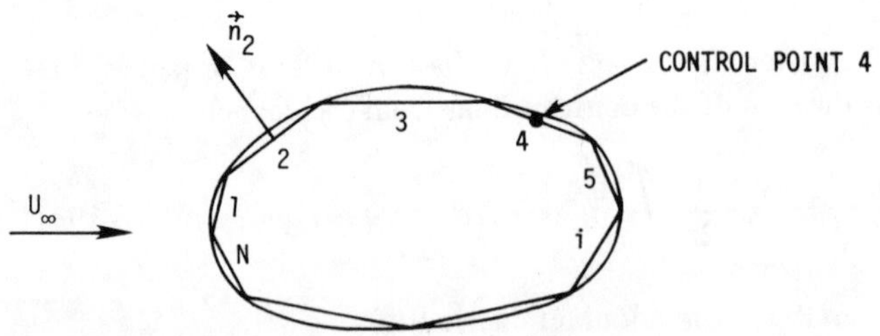

Figure 6-32 Panel representation for general shape.

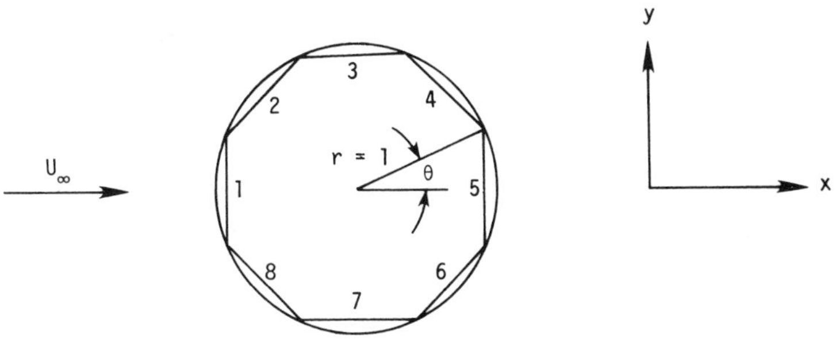

Figure 6-33 Panel representation of cylinder.

Application of this equation to each panel provides n algebraic equations for the n source strengths. Once the σ_j's are computed, the pressure coefficients can be determined. When Eq. (6-198) is used to generate the required panel source strengths, the integrand function is most easily developed by using the vector dot product and may be written

$$\frac{\partial \ln (r_{ij})}{\partial n_i} = \nabla_i \ln (r_{ij}) \cdot \mathbf{n}_i \tag{6-199}$$

An example demonstrating the procedure for generating the required algebraic equations is in order.

Example 6-6 Suppose we wish to solve for the pressure distribution on a cylinder of unit radius in incompressible flow using the method of source panels. The cylinder is to be represented by eight panels and the configuration is shown in Fig. 6-33.

In order to determine the surface pressures, we must calculate the panel strengths required for all eight panels on the cylinder. This is done by solving the system of algebraic equations generated by the application of Eq. (6-198) to each panel. In applying Eq. (6-198) to any panel, the most difficult part is the evaluation of the integral term. In general, it is convenient to view the integral as an influence coefficient and write the system of governing equations in the form

$$[C] \frac{\sigma}{2\pi U_\infty} = -\frac{\mathbf{U}_\infty \cdot \mathbf{n}_i}{U_\infty} \tag{6-200}$$

Using this notation we may write

$$c_{ij} = \begin{cases} \displaystyle\int_j \nabla_i \ln (r_{ij}) \cdot \mathbf{n}_i \, ds_j & i \neq j \\ \pi & i = j \end{cases} \tag{6-201}$$

To demonstrate the application of Eq. (6-201) we elect to compute c_{53} which represents the normal velocity at the control point of panel 5 due to a constant source strength of magnitude $1/U_\infty$ on panel 3. For this case we write the radius as

$$r_{53} = [(x_5 - x_3)^2 + (y_5 - y_3)^2]^{1/2}$$

and

$$\nabla_5 \ln (r_{53}) = \frac{(x_5 - x_3)\mathbf{i} + (y_5 - y_3)\mathbf{j}}{(x_5 - x_3)^2 + (y_5 - y_3)^2} \tag{6-202}$$

The unit normal on panel 5 is just the unit vector in the positive x direction and

$$\nabla_5 \ln (r_{53}) \cdot \mathbf{n}_5 = \frac{x_5 - x_3}{(x_5 - x_3)^2 + (y_5 - y_3)^2}$$

For this integral $x_5 = 0.9239$, $y_5 = 0$, and $y_3 = 0.9239$ while x_3 is a variable on panel 3. This reduces the integral required to the form

$$c_{53} = \int_{-0.3827}^{+0.3827} \frac{0.9239 - x}{x^2 - 1.848x + 1.707}\, dx = 0.4018$$

In this expression the arc length along panel 3 is equal to $x - 0.3827$, therefore $ds_3 = dx$ and we use the x coordinates of the panel end points as the integration limits. Notice that the integration proceeds clockwise around the cylinder which is the positive sense for the domain where a solution of Laplace's equation is required. The $[C]$ matrix is symmetric, i.e., $c_{ij} = c_{ji}$ and the solution for the σ_i's must be such that

$$\sum_{i=1}^{n} \sigma_i = 0$$

This requirement is an obvious result of the requirement that we have a closed body.

A comparison of the eight panel solution with the analytically derived pressure coefficient is shown in Fig. 6-34. Clearly the panel scheme provides a very accurate numerical solution for this case.

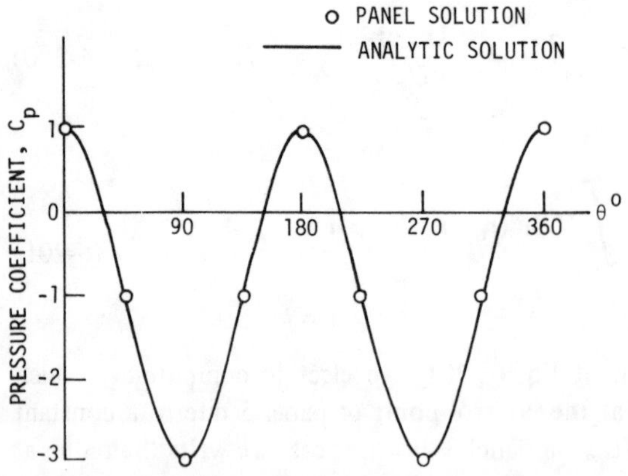

Figure 6-34 Pressure coefficient for a circular cylinder.

We have used the method of source panels in our example to demonstrate the mechanics of applying the technique. We could use doublets or dipoles to construct bodies as well as vortex panels. Clearly we must include circulation if we are concerned with lifting airfoils. This may be done in a number of ways but one technique is to use a vortex panel distribution along the mean camber line to provide circulation and satisfy the Kutta condition at a control point just aft of the trailing edge.

Panel methods represent a powerful set of techniques for solving certain classes of flow problems. They have received extensive use and have resulted in a number of standard codes which are used industry wide. For more details on the development of these schemes, the paper by Hess and Smith (1967) provides basic details while the papers by Rubbert and Saaris (1972) and Johnson and Rubbert (1975) present more advanced ideas.

PROBLEMS

6-1 In Example 6-1, we used a characteristic method to solve for supersonic flow over a wavy wall. Verify the velocity field obtained by solving the Prandtl-Glauert equation [Eq. (6-1)] directly.

6-2 Derive the differential equations of the characteristics of the nonlinear system of equations governing two-dimensional supersonic flow written in rectangular coordinates.

6-3 The differential equations of the characteristics obtained in Prob. 6-2 are written in rectangular coordinates. Transform these results using the streamline angle θ and show that the characteristics are inclined at the local Mach angle, i.e.,

$$\tan(\theta \pm \mu) = \frac{dy}{dx}$$

6-4 Develop the compatibility equations for the nonlinear equations of Prob. 6-2.

6-5 Use the results of Probs. 6-2 and 6-4 and solve Example 6-1 using the method of characteristics for the nonlinear equations.

6-6 Complete the development of the finite-difference equations of Example 6-3 and solve the problem given in Example 6-1 using this SCM formulation.

6-7 Complete the derivation required in Example 6-4. This requires that T be computed. The governing equations are then ready to be numerically integrated using appropriate one-sided differences.

6-8 Develop a code to solve for the supersonic flow over the two-dimensional wedge of Example 6-2. Use a shock-capturing approach and MacCormack's method in solving the steady flow equations.

6-9 Solve the wedge-flow problem of Prob. 6-8 by using the unsteady (time-dependent) approach and the fact that the flow is conical.

6-10 Develop a code to solve the two-dimensional supersonic wedge problem of Prob. 6-8 but fit the shock wave as a discontinuity. Use either conservative or nonconservative form.

6-11 Suppose that a solid boundary lies on a ray ($\theta = \text{const}$) in a two-dimensional flow problem. Use reflection to establish a suitable means of determining the flow variables of the sublayer points. Use rectangular velocity components.

6-12 Develop the appropriate boundary condition procedure using Kentzer's method for the supersonic wedge-flow problem.

6-13 Show that Eq. (6-142) is a second-order representation of the one-dimensional potential equation.

6-14 Show that the retarded density formulation of Eq. (6-148) is equivalent to Eq. (6-147).

6-15 Evaluate the right-hand side of Eq. (6-166).

6-16 Show that Eq. (6-190) is a valid representation for the potential in an incompressible fluid flow.

6-17 Compute c_{43} of Example 6-6.

CHAPTER

SEVEN

NUMERICAL METHODS FOR BOUNDARY-LAYER TYPE EQUATIONS

7-1 INTRODUCTION

It was pointed out in Chapter 5 that the equations that result from the boundary-layer (or thin shear-layer) approximation provide a useful mathematical model for several important flows occurring in engineering applications. Among these are many jet and wake flows, two-dimensional or axisymmetric flows in channels and tubes as well as the classical wall boundary layer. Certain three-dimensional flows can also be economically treated through the boundary-layer approximation. In recent years, methods have been developed to extend the boundary-layer approximation to flows containing small regions of recirculation. Often, a small region exists near the streamwise starting plane of these flows in which the thin-shear layer-approximation is a poor one, but for moderate to large Reynolds numbers, this region is very (and usually negligibly) small.

In this chapter, methods and numerical considerations related to the finite-difference solution of these equations will be presented. The emphasis will be on the application of methods and principles covered earlier in Chapters 3 and 4 rather than on the exposition of a single general difference procedure. Several finite-difference methods for these equations are described in detail elsewhere. Except as an aid in illustrating key principles, those details will not be repeated here.

The history of numerical methods for boundary-layer equations goes back to the 1930's and 1940's. Finite-difference methods in a form very similar to those now in use began emerging in the 1950's (Friedrich and Forstall, 1953; Rouleau and Osterle, 1955). We can think of difference schemes for the boundary-layer equations as being relatively well-developed and tested as compared to methods for some other classes of flows. Despite this, new developments in the numerical treatment of these equations continue to appear regularly.

7-2 BRIEF COMPARISON OF PREDICTION METHODS

Before proceeding with a discussion of finite-difference methods for boundary-layer flows, it is well to remember that over the years useful solutions have been obtained by other methods and for some simple flows, engineering results are available as simple formulas. These results are presented in standard textbooks on fluid mechanics, aerodynamics and heat transfer. The books by Schlichting (1979) and White (1974) are especially valuable references for viscous flows.

Except for a few isolated papers based on similarity methods, the calculation methods for boundary-layer-type problems that appear in the current literature can generally be categorized as (1) integral methods, (2) finite-difference methods, or (3) finite-element methods.

Integral methods can be applied to a wide range of both laminar and turbulent flows and, in fact, any problem which can be solved by a finite-difference method can also be solved by an integral method. Prior to the 1960's integral methods were the primary "advanced" calculation method for solving complex problems in fluid mechanics and heat transfer. Loosely speaking, the method transforms the PDE's into one or more ODE's by integrating out the dependence of one independent variable (usually the normal coordinate) in advance by making assumptions about the general form of the velocity and temperature profiles (often functions of "N" parameters). Many of these procedures can be grouped as weighted residual methods. It can be shown that the solution by the method of weighted residuals approaches the exact solution of the PDE as N becomes very large. Modern versions of integral methods for complex problems make use of digital computers. In practice, it appears that implementing integral methods is not as straightforward (requiring more "intuition" about the problem) as for finite-difference methods. The methods are not as flexible or general as finite-difference methods in that more changes are generally required as boundary or other problem conditions are changed. In recent years the preference of the scientific community has shifted in favor of using finite-difference methods over integral methods for computing the more complex boundary-layer flows. However, integral methods have at least a few strong advocates and can be used to solve important current problems.

Application of finite-element methodology to boundary-layer equations has been observed only relatively recently. Comments on this approach for boundary layers can be found in Chung (1978). The objective of all three of these methods is to transform the problem posed through PDE's to one having an algebraic representation. The methods differ in the procedures used to implement this discretization. In the future, it is likely that hybrid computational schemes will evolve which will tend to preserve the best features of each of these methods.

7-3 FINITE-DIFFERENCE METHODS FOR TWO-DIMENSIONAL OR AXISYMMETRIC STEADY EXTERNAL FLOWS

7-3.1 A Generalized Form of the Equations

The preferred form for the boundary-layer equations will vary from problem to problem. In the case of laminar flows, coordinate transformations are especially useful for maintaining a nearly constant number of grid points across the flow. The energy equation is usually written differently for compressible flow than it is for incompressible flow. In practice it is frequently necessary to extend or alter a difference scheme established for one PDE to accommodate one which is similar but different in some detail. Optimizing the representation often requires a trial and error procedure.

The boundary-layer equations were given in Chapter 5 [Eqs. (5-116)-(5-119)] in physical coordinates. Here, we will utilize the Boussinesq approximation to evaluate the Reynolds shear-stress and heat-flux quantities in terms of a turbulent viscosity μ_T and the turbulent Prandtl number $\Pr_T$. Specifically, we will let

$$-\rho\overline{u'v'} = \mu_T \frac{\partial u}{\partial y}$$

and

$$-\rho c_p \overline{v'T'} = \frac{c_p \mu_T}{\Pr_T} \frac{\partial T}{\partial y}$$

To solve the energy equation numerically using H as the primary thermal variable, it will be helpful to eliminate T in the expression for the Reynolds heat flux by using the definition of total enthalpy, $H = c_p T + u^2/2 + v^2/2$. The $v^2/2$ term can be neglected in keeping with the boundary-layer approximation. These substitutions permit the boundary-layer equations for a steady, compressible, two-dimensional or axisymmetric flow to be written as

x momentum:

$$\rho u \frac{\partial u}{\partial x} + \rho\tilde{v} \frac{\partial u}{\partial y} = \rho_e u_e \frac{du_e}{dx} + \frac{1}{r^m} \frac{\partial}{\partial y}\left[r^m(\mu + \mu_T)\frac{\partial u}{\partial y}\right] \tag{7-1}$$

energy:

$$\rho u \frac{\partial H}{\partial x} + \rho\tilde{v} \frac{\partial H}{\partial y} = \frac{1}{r^m} \frac{\partial}{\partial y}\left(r^m\left\{\left(\frac{\mu}{\Pr} + \frac{\mu_T}{\Pr_T}\right)\frac{\partial H}{\partial y} + \left[\mu\left(1 - \frac{1}{\Pr}\right) + \mu_T\left(1 - \frac{1}{\Pr_T}\right)\right] u \frac{\partial u}{\partial y}\right\}\right) \tag{7-2}$$

continuity:

$$\frac{\partial}{\partial x}(r^m \rho u) + \frac{\partial}{\partial y}(r^m \rho\tilde{v}) = 0 \tag{7-3}$$

state:

$$\rho = \rho(T, p) \tag{7-4}$$

Property relationships are also needed to evaluate μ, k, c_p as a function (usually) of temperature.

As indicated in Chapter 5, m is a flow index equal to unity for axisymmetric flow and equal to zero for two-dimensional flow, and $\tilde{v} = (\bar{\rho}\bar{v} + \overline{\rho' v'})/\bar{\rho}$. When $m = 0$, $r^m = 1$ and the equations are in appropriate form for two-dimensional flows.

The primary dependent variable in the momentum equation is u and it is useful to think of Eq. (7-1) as a "transport" equation for u in which terms representing convection, diffusion and "sources" of u can be recognized. Likewise, the energy equation can be viewed as a transport equation for H with similar categories of terms. This interpretation can also be extended to include the unsteady form of the boundary-layer momentum and energy equations.

Within the transport equation context, both Eqs. (7-1) and (7-2) can usually [an exception may occur with the use of some turbulence models, as in Bradshaw et al. (1967)] be cast into the general form

$$\underbrace{\rho u \frac{\partial \phi}{\partial x} + \rho \tilde{v} \frac{\partial \phi}{\partial y}}_{\text{Convection of } \phi} = \underbrace{\frac{1}{r^m} \frac{\partial}{\partial y}\left(r^m \lambda \frac{\partial \phi}{\partial y}\right)}_{\text{Diffusion of } \phi} + \underbrace{S}_{\text{Source terms}} \tag{7-5}$$

In Eq. (7-5) ϕ is a generalized variable which would be u for the boundary-layer momentum equation and H for the boundary-layer energy equation, λ is a generalized diffusion coefficient and S represents the source terms. Source terms are those terms in the PDE which do not involve a derivative of ϕ. The term $\rho_e u_e \, du_e/dx$ in Eq. (7-1) and the term involving $u \, \partial u/\partial y$ in Eq. (7-2) are examples of source terms. Most of the transport equations for turbulence model parameters given in Chapter 5 also fit the form of Eq. (7-5).

The momentum and energy equations which can be cast into the general form of Eq. (7-5) are parabolic with x as the marching coordinate. By making appropriate assumptions regarding the evaluation of coefficients, it is possible to decouple the finite-difference representation of the equations permitting the momentum, continuity, and energy equations to be marched one step in the x direction independently to provide new values of u_j, H_j, and $\tilde{v}_j$. This strategy is illustrated below:

Equation	*Marched to obtain*
x momentum	u_j^{n+1}
energy	H_j^{n+1}
equation of state + continuity	$\tilde{v}_j^{n+1}$

After each marching step the coefficients in the equations are reevaluated (updated) so that the solutions of the three equations are in fact interdependent–the decoupling is in the algebraic system for one marching step at a time. In some solution schemes, the coupling is maintained so that at each marching step, a larger system of algebraic equations must be solved simultaneously for new values of u_j, H_j, $\tilde{v}_j$. Uncoupling the algebraic system is conceptually the simplest procedure and can usually be made to work satisfactorily for most flow problems.

7-3.2 An Example of a Simple Explicit Procedure

Although the simplest explicit method is no longer widely used for boundary layers due to the restrictive stability constraint associated with it, it will be used here for pedagogical purposes to demonstrate the general solution algorithm for boundary-layer flows (Wu, 1961). Let's consider a two-dimensional laminar incompressible flow without heat transfer. The governing equations in partial differential form are given as Eqs. (5-104) and (5-105).

The difference equations can be written as

x momentum:

$$u_j^n \frac{(u_j^{n+1} - u_j^n)}{\Delta x} + \boxed{v_j^n \frac{(u_{j+1}^n - u_{j-1}^n)}{2\,\Delta y}} = u_e^n \frac{(u_e^{n+1} - u_e^n)}{\Delta x}$$

$$+ \frac{\nu}{(\Delta y)^2}(u_{j+1}^n - 2u_j^n + u_{j-1}^n) + O(\Delta x) + O(\Delta y)^2 \tag{7-6}$$

continuity:

$$\frac{v_j^{n+1} - v_{j-1}^{n+1}}{\Delta y} + \frac{u_j^{n+1} + u_{j-1}^{n+1} - u_j^n - u_{j-1}^n}{2\,\Delta x} = 0 + O(\Delta x) + O(\Delta y)^2 \tag{7-7}$$

For flow over a flat plate (see Fig. 7-1), the computation is usually started by assuming that $u_j^n = u_\infty$ at the leading edge and $v_j^n = 0$. The value v_j^n is required in the explicit algorithm in order to advance the solution to the $n + 1$ level. However, in the formal mathematical formulation of the PDE problem, it is not necessary to specify an initial distribution for v_j^n. A compatible initial distribution can be obtained for v_j^n (Ting, 1965) by first using the continuity equation to eliminate

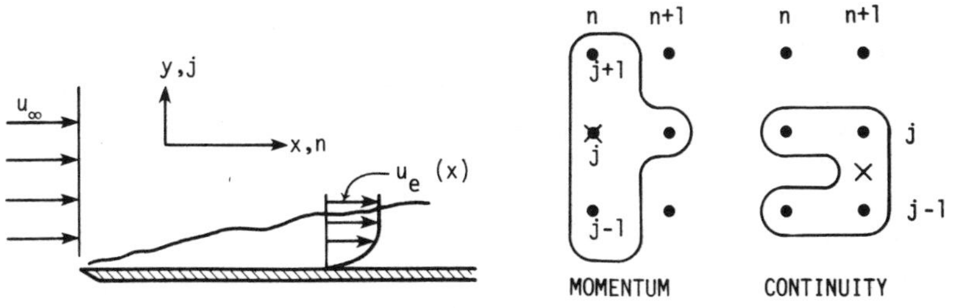

Figure 7-1 Simple explicit procedure.

$\partial u/\partial x$ from the boundary-layer momentum equation. For a laminar, incompressible flow, this gives

$$-u\frac{\partial v}{\partial y}+v\frac{\partial u}{\partial y}=u_e\frac{du_e}{dx}+\nu\frac{\partial^2 u}{\partial y^2}$$

We can observe that

$$-u\frac{\partial v}{\partial y}+v\frac{\partial u}{\partial y}=-u^2\frac{\partial}{\partial y}\left(\frac{v}{u}\right)$$

Thus,
$$\frac{\partial}{\partial y}\left(\frac{v}{u}\right)=-\frac{1}{u^2}\left(u_e\frac{du_e}{dx}+\nu\frac{\partial^2 u}{\partial y^2}\right)$$

and using $v=0$ at $y=0$, we find

$$v(y)=-u\int_0^y\frac{1}{u^2}\left(u_e\frac{du_e}{dx}+\nu\frac{\partial^2 u}{\partial y^2}\right)dy \tag{7-8}$$

For the flat plate problem at hand, we would assume that $u_j^n=u_\infty$ at $x=0$ (the leading edge) except at the wall where $u_1^n=0$. We can use a numerical evaluation of the integral in Eq. (7-8) to obtain an estimate of a compatible initial distribution of v_j^n to use in the explicit difference procedure. Employing the usual central-difference representation for $\partial^2 u/\partial y^2$ about the first point beyond the wall suggests $v_j^n=2\nu/\Delta y$ at all points except at the wall where $v_1^n=0$. In practice, letting $v_j^n=0$ initially throughout is also found to work satisfactorily.

Having initial values for u_j^n, the momentum equation, Eq. (7-6) can be solved for u_j^{n+1} explicitly, usually by starting from the wall and working outward until $u_j^{n+1}/u_e^{n+1}=1-\epsilon\approx 0.9995$; that is, due to the asymptotic boundary condition, we *find* the location of the outer boundary as the solution proceeds. The values of v_j^{n+1} can now be computed from Eq. (7-7), starting with the point next to the lower boundary and computing outward. The difference formulation of the continuity equation and the solution procedure described is equivalent to integrating the continuity equation by the trapezoidal rule for v_j^{n+1}.

The *stability constraint* for this method is:

$$\frac{2\nu\,\Delta x}{u_j^n(\Delta y)^2}\leqslant 1 \qquad\text{and}\qquad \frac{(v_j^n)^2\,\Delta x}{u_j^n\nu}\leqslant 2$$

The second term in the momentum equation has been enclosed by a dashed box for a couple of reasons. First, we should be aware that the presence of this term is mainly responsible for any difference between the stability constraints of Eq. (7-6) and the heat equation, and second, we will suggest an alternative treatment for this term below.

Alternative formulation for explicit method. In order to control the stability of the explicit method by checking only a single inequality, the boxed term in Eq. (7-6), (difference representation of $v\,\partial u/\partial y$) can be expressed as

$$v_j^n \frac{u_j^n - u_{j-1}^n}{\Delta y}$$

when
$$v_j^n > 0$$

and
$$v_j^n \frac{u_{j+1}^n - u_j^n}{\Delta y}$$

when $v_j^n < 0$, whereby the stability constraint becomes

$$\Delta x \leqslant \frac{1}{2\nu/[u_j^n(\Delta y)^2] + |v_j^n|/(u_j^n \Delta y)}$$

The truncation error deteriorates to only $O(\Delta x) + O(\Delta y)$ when this treatment of $v\, \partial u/\partial y$ is used.

Note that the stability constraints for both methods depend upon the local values of u and v. This is typical for equations with variable coefficients. The von Neumann stability analysis has proven to be a reliable guide to stability for boundary-layer equations if the coefficients u and v which appear in the equations are treated as being locally constant. Treatment of the μ_T for turbulent flow in the stability analysis requires further consideration. For some models, μ_T will contain derivatives whose difference representation could contribute to numerical instabilities. In the stability analysis, one can treat μ_T as simply a specified variable property and then by trial and error develop a stable difference representation for μ_T or one can express μ_T in terms of the dependent flow variables and attempt to determine the appropriate stability constraints by the usual methods.

7-3.3 Crank-Nicolson and Fully Implicit Methods

The characteristics of most implicit methods can be visualized by considering the following representation of the compressible laminar boundary-layer equations in physical coordinates on a mesh for which Δy = constant.

momentum:

$$\begin{aligned}
&\frac{[\theta(\rho_j^{n+1}u_j^{n+1}) + (1-\theta)(\rho_j^n u_j^n)]\,(u_j^{n+1} - u_j^n)}{\Delta x} \\
&\quad + \frac{\theta(\rho_j^{n+1}v_j^{n+1})(u_{j+1}^{n+1} - u_{j-1}^{n+1}) + (1-\theta)(\rho_j^n v_j^n)(u_{j+1}^n - u_{j-1}^n)}{2\,\Delta y} \\
&= \frac{[\theta(\rho_e^{n+1}u_e^{n+1}) + (1-\theta)(\rho_e^n u_e^n)]\,(u_e^{n+1} - u_e^n)}{\Delta x} \\
&\quad + \frac{1}{(\Delta y)^2}\{\theta[\mu_{j+1/2}^{n+1}(u_{j+1}^{n+1} - u_j^{n+1}) - \mu_{j-1/2}^{n+1}(u_j^{n+1} - u_{j-1}^{n+1})] \\
&\quad + (1-\theta)[\mu_{j+1/2}^n(u_{j+1}^n - u_j^n) - \mu_{j-1/2}^n(u_j^n - u_{j-1}^n)]\}
\end{aligned} \tag{7-9}$$

In the above, θ is a weighting factor. If:

$\theta = 0$ Method is explicit, most convenient expansion point is (n, j), truncation error is $O(\Delta x) + O(\Delta y)^2$; the von Neumann stability constraint, given previously, presents a severe limitation on the marching step size.

$\theta = \frac{1}{2}$ Crank-Nicolson implicit; the most convenient expansion point is $(n+\frac{1}{2}, j)$; the truncation error is $O(\Delta x)^2 + O(\Delta y)^2$ if coefficients (and properties) are evaluated at $(n+\frac{1}{2}, j)$. No stability constraint arises from the von Neumann analysis but difficulties can arise if diagonal dominance is not maintained for the tridiagonal algorithm (Hirsh and Rudy, 1974).

$\theta = 1$ Fully implicit, expansion point $(n+1, j)$, truncation error $O(\Delta x) + O(\Delta y)^2$ [if properties and coefficients are evaluated at $(n+1, j)$]. No stability constraint by the von Neumann method, but same comment as for $\theta = \frac{1}{2}$ applies for diagonal dominance.

We note that the above scheme becomes implicit if $\theta > 0$ and inherently stable if $\theta \geqslant \frac{1}{2}$. Values of θ between $\frac{1}{2}$ and 1 have been used successfully. The same form of the continuity equation can be used for both the fully implicit and the explicit methods.

continuity:

$$\frac{\rho_j^{n+1} v_j^{n+1} - \rho_{j-1}^{n+1} v_{j-1}^{n+1}}{\Delta y} + \frac{\rho_j^{n+1} u_j^{n+1} - \rho_j^n u_j^n + \rho_{j-1}^{n+1} u_{j-1}^{n+1} - \rho_{j-1}^n u_{j-1}^n}{2\,\Delta x} = 0 \tag{7-10}$$

When $\theta = \frac{1}{2}$, we can consider ρ's and v's in the first term to be at the $n+\frac{1}{2}$ level and rewrite Eq. (7-10) accordingly. This results in a truncation error of $O(\Delta x)^2 + O(\Delta y)^2$ for the continuity equation. Differencing of the energy equation follows the same general pattern as used for the momentum equation. Choosing T as the primary thermal variable as we might for low speed flow, we can write the energy equation as

energy:

$$\rho u c_p \frac{\partial T}{\partial x} + \rho v c_p \frac{\partial T}{\partial y} = \frac{\partial}{\partial y}\left(k \frac{\partial T}{\partial y}\right) + \beta T u \frac{dp}{dx} + \mu \left(\frac{\partial u}{\partial y}\right)^2 \tag{7-11}$$

which, utilizing the θ notation, can be written in difference form as

$$\begin{aligned}
&[\theta(\rho_j^{n+1} u_j^{n+1} c_{pj}^{n+1}) + (1-\theta)(\rho_j^n u_j^n c_{pj}^n)] \frac{T_j^{n+1} - T_j^n}{\Delta x} \\
&\quad + \frac{\theta(\rho_j^{n+1} v_j^{n+1} c_{pj}^{n+1})(T_{j+1}^{n+1} - T_{j-1}^{n+1}) + (1-\theta)(\rho_j^n v_j^n c_{pj}^n)(T_{j+1}^n - T_{j-1}^n)}{2\,\Delta y} \\
&= \frac{1}{(\Delta y)^2} \{\theta [k_{j+1/2}^{n+1}(T_{j+1}^{n+1} - T_j^{n+1}) - k_{j-1/2}^{n+1}(T_j^{n+1} - T_{j-1}^{n+1})] \\
&\quad + (1-\theta)[k_{j+1/2}^n (T_{j+1}^n - T_j^n) - k_{j-1/2}^n (T_j^n - T_{j-1}^n)]\}
\end{aligned} \tag{7-12}$$

$$+ \frac{[\theta(\beta_j^{n+1} T_j^{n+1} u_j^{n+1}) + (1-\theta)(\beta_j^n T_j^n u_j^n)](p_j^{n+1} - p_j^n)}{\Delta x}$$

$$+ \theta \mu_j^{n+1} \left(\frac{u_{j+1}^{n+1} - u_{j-1}^{n+1}}{2\,\Delta y} \right)^2 + (1-\theta)\mu_j^n \left(\frac{u_{j+1}^n - u_{j-1}^n}{2\,\Delta y} \right)^2 \qquad \text{(7-12) (Cont.)}$$

The truncation error for the energy equation is identical to that stated for the momentum equation for $\theta = 0, \frac{1}{2}, 1$.

The fully implicit ($\theta = 1$) scheme can be elevated to formal second-order accuracy by representing streamwise derivatives by three-level ($n-1$, n, $n+1$) second-order accurate differences, such as can be found in Chapter 3. Davis (1963) and Harris (1971) have demonstrated the feasibility of such a procedure.

For any implicit method ($\theta \neq 0$) the finite-difference forms of the momentum and energy equations [Eqs. (7-9) and (7-12)] are algebraically nonlinear in the unknowns due to the appearance of quantities unknown at the $n+1$ level in the coefficients. Linearizing procedures which can and have been utilized are:

1. *Lagging the coefficients*

   The simplest and most common strategy is to linearize the difference equations by evaluating all coefficients at the n level. This is known as "lagging" the coefficients. The procedure provides a consistent representation since, for a general function $\phi(x, y)$, $\phi(x_0 + \Delta x, y_0) = \phi(x_0, y_0) + O(\Delta x)$. This procedure does ensure that the difference scheme is formally no better than first-order accurate in the marching coordinate. Using the generalized form [Eq. (7-5)] for a transport equation, the linearized difference representation obtained by lagging the coefficients can be written

$$\rho_j^n u_j^n \frac{\phi_j^{n+1} - \phi_j^n}{\Delta x} + \frac{\rho_j^n v_j^n}{2\,\Delta y} [\theta(\phi_{j+1}^{n+1} - \phi_{j-1}^{n+1}) + (1-\theta)(\phi_{j+1}^n - \phi_{j-1}^n)] = \frac{1}{(\Delta y)^2}$$

$$\times \{\lambda_{j+1/2}^n [\theta(\phi_{j+1}^{n+1} - \phi_j^{n+1}) + (1-\theta)(\phi_{j+1}^n - \phi_j^n)] - \lambda_{j-1/2}^n [\theta(\phi_j^{n+1} - \phi_{j-1}^{n+1})$$

$$+ (1-\theta)(\phi_j^n - \phi_{j-1}^n)]\} + \theta S_j^{n+1} + (1-\theta) S_j^n \qquad \text{(7-13)}$$

   The three conservation equations in difference form can now be solved in an uncoupled manner. The momentum equation can be solved for u_j^{n+1}, the energy equation for T_j^{n+1}, and an equation of state used to obtain ρ_j^{n+1}. Finally, the continuity equation can be solved for v_j^{n+1}. The matrix of unknowns in each equation (for momentum and energy) is tridiagonal and the Thomas algorithm can be used.

2. *Simple iterative update of coefficients*

   The coefficients can be ultimately evaluated at the $n+1$ level as required in Eqs. (7-9), (7-10), and (7-12) by use of a simple iterative updating procedure. To do this, the coefficients are first evaluated at the n level (lagged) and the system solved for new values of u, T, v at the $n+1$ level. The coefficients can then be updated by utilizing the solution just obtained at the $n+1$ level and the calculation repeated to obtain "better" predictions at $n+1$.

This procedure can be repeated iteratively until changes are small. Usually only two or three iterations are used although Blottner (1975a) points out that up to 19 iterations were required in a sample calculation with the Crank-Nicolson procedure before the solution obtained behaved like a second-order accurate scheme under grid refinement (see Section 3-2). Although the programming changes involved in advancing from the lagged procedure to the simple iterative update are minimal, the use of Newton linearization, to be described next, is more efficient and is recommended for that reason.

3. *Use of Newton linearization to iteratively update coefficients*

Newton linearization (also called quasilinearization) proceeds as follows. Suppose we want to evaluate $(u_j^{n+1})^2$. We let δ_u equal the change in u between two iterative solutions of the difference equations. Thus, $u_j^{n+1} = \hat{u}_j^{n+1} + \delta_u$ where the caret denotes an evaluation of the variable from a previous iteration level. For the first iteration, $\hat{u}_j^{n+1}$ will be taken as the value of the variable from the previous marching station. The δ_u plays the role of Δx used in the Newton-Raphson procedure for finding the root of a transcendental equation. We evaluate $(u_j^{n+1})^2$ as

$$(u_j^{n+1})^2 = (\hat{u}_j^{n+1} + \delta_u)^2 = (\hat{u}_j^{n+1})^2 + 2\delta_u \hat{u}_j^{n+1} + \delta_u^2 \tag{7-14}$$

The right-hand side of Eq. (7-14) is linearized by dropping the term which represents the change raised to the second power, δ_u^2, which is analogous to dropping the term of order $(\Delta x)^2$ in the Newton-Raphson procedure. Our representation of $(u_j^{n+1})^2$ after the linearization is

$$(u_j^{n+1})^2 \simeq (\hat{u}_j^{n+1})^2 + 2\delta_u \hat{u}_j^{n+1} \tag{7-15}$$

in which δ_u is the only unknown. Alternatively, we can use $\delta_u = u_j^{n+1} - \hat{u}_j^{n+1}$ to rewrite Eq. (7-15) as

$$(u_j^{n+1})^2 \simeq 2u_j^{n+1}\hat{u}_j^{n+1} - (\hat{u}_j^{n+1})^2 \tag{7-16}$$

This linearization can be put on a more formal basis by employing a Taylor series expansion as follows. We let $\eta = u_j^{n+1}$, $F(\eta) = \eta^2$, and $\eta_1 = \hat{u}_j^{n+1}$, the value of u_j^{n+1} from a previous iteration level. Expanding about the value at the previous iteration level, we can write

$$F(\eta_1 + \Delta\eta) = F(\eta_1) + F'(\eta_1)\,\Delta\eta + \cdots \tag{7-17}$$

where the series has been truncated after the first derivative term. We note that $F'(\eta_1)\,\Delta\eta = 2\eta_1\,\Delta\eta$. Thus, evaluating Eq. (7-17) in terms of the u_j^{n+1}'s gives the same result as Eq. (7-15).

Both representations, one utilizing the changes, δ_u, and the other eliminating the appearance of δ_u by substitution are found in the literature and are equivalent. We will employ the latter in examples in this chapter. The advantage of the Newton linearization lies in the enhanced convergence rate expected in the iterative updating procedure for the coefficients.

For a more specific example of the use of this procedure we'll consider a fully implicit ($\theta = 1$) example in which the conservation equations are to be solved in an uncoupled manner for an incompressible flow. The most obvious nonlinearity appears in the representation for the $\rho u\, \partial u/\partial x$ term. Applying Newton linearization to the fully implicit finite-difference representation of this term gives

$$\frac{\rho[2\hat{u}_j^{n+1} u_j^{n+1} - (\hat{u}_j^{n+1})^2 - u_j^n u_j^{n+1}]}{\Delta x} \tag{7-18}$$

in which u_j^{n+1} is the only unknown. For the first iteration, $\hat{u}_j^{n+1}$ is evaluated as u_j^n. A slightly different final result is obtained if we apply the linearization procedure to this term in the mathematically equivalent form, $\rho\partial(u^2/2)/\partial x$.

If the conservation equations are to be solved in an uncoupled manner, i.e., one unknown is to be determined independently from each conservation equation, the other nonlinear terms

$$\rho v \frac{\partial u}{\partial y} \quad \text{and} \quad \frac{\partial}{\partial y}\left(\mu \frac{\partial u}{\partial y}\right)$$

are usually evaluated by the simple iterative updating procedure described above.

Evaluating $\rho u\, \partial u/\partial x$ by the Newton linearization as indicated in Eq. (7-18) and using simple updating on other nonlinear terms results in a tridiagonal coefficient matrix which permits use of the Thomas algorithm with no special modifications. The calculation is repeated two or more times at each streamwise location updating variables as indicated.

4. *Newton linearization with coupling*

Several investigators have observed that convergence of the iterations to update coefficients at each streamwise step in the boundary-layer momentum equation can be accelerated greatly by solving the momentum and continuity equations in a coupled manner. Second-order accuracy for the Crank-Nicolson procedure has been observed using only one iteration at each streamwise station when the equations are solved in a coupled manner (Blottner, 1975a). According to Blottner (1975a), coupling was first suggested by R. T. Davis and used by Werle and co-workers (Werle and Bertke, 1972; Werle and Dwoyer, 1972). An example of the coupled procedure for a fully implicit formulation for incompressible, constant property flow follows.

The $u\, \partial u/\partial x$ term is treated as in Eq. (7-18). The $v\, \partial u/\partial y$ term is linearized by using $v_j^{n+1} = \hat{v}_j^{n+1} + \delta_v$ and $u_j^{n+1} = \hat{u}_j^{n+1} + \delta_u$. For the first iteration, $\hat{v}_j^{n+1}$ and $\hat{u}_j^{n+1}$ are most conveniently evaluated as v_j^n and u_j^n, respectively. After terms involving products of δ are dropped, the $v\, \partial u/\partial y$ term becomes

$$\left(v \frac{\partial u}{\partial y}\right)^{n+1} \approx \hat{v}^{n+1}\left(\frac{\partial u}{\partial y}\right)^{n+1} + v^{n+1}\left(\frac{\partial \hat{u}}{\partial y}\right)^{n+1} - \hat{v}^{n+1}\left(\frac{\partial \hat{u}}{\partial y}\right)^{n+1} \tag{7-19}$$

This same result can be developed with the aid of a Taylor series expansion in two variables, v, and $\partial u/\partial y$. The series would be truncated after the first derivative terms.

The continuity and momentum equations are written in difference form as

$$\frac{u_j^{n+1} - u_j^n + u_{j-1}^{n+1} - u_{j-1}^n}{2\,\Delta x} + \frac{v_j^{n+1} - v_{j-1}^{n+1}}{\Delta y} = 0 \tag{7-20}$$

$$\frac{2\hat{u}_j^{n+1} u_j^{n+1} - (\hat{u}_j^{n+1})^2 - u_j^n u_j^{n+1}}{\Delta x} + \frac{\hat{v}_j^{n+1}(u_{j+1}^{n+1} - \hat{u}_{j+1}^{n+1} - u_{j-1}^{n+1} + \hat{u}_{j-1}^{n+1})}{2\,\Delta y}$$
$$+ \frac{v_j^{n+1}(\hat{u}_{j+1}^{n+1} - \hat{u}_{j-1}^{n+1})}{2\,\Delta y} = \frac{\nu}{(\Delta y)^2}(u_{j+1}^{n+1} - 2u_j^{n+1} + u_{j-1}^{n+1}) + \frac{(u_e^{n+1})^2 - u_e^{n+1} u_e^n}{\Delta x} \tag{7-21}$$

To clarify the algebraic formulation of the problem, the momentum equation can be written as

$$B_j u_{j-1}^{n+1} + D_j u_j^{n+1} + A_j u_{j+1}^{n+1} + a_j v_j^{n+1} + b_j v_{j-1}^{n+1} = C_j \tag{7-22}$$

where
$$B_j = -\frac{\hat{v}_j^{n+1}}{2\,\Delta y} - \frac{\nu}{(\Delta y)^2} \qquad D_j = \frac{2\hat{u}_j^{n+1} - u_j^n}{\Delta x} + \frac{2\nu}{(\Delta y)^2}$$

$$A_j = \frac{\hat{v}_j^{n+1}}{2\,\Delta y} - \frac{\nu}{(\Delta y)^2} \qquad a_j = \frac{\hat{u}_{j+1}^{n+1} - \hat{u}_{j-1}^{n+1}}{2\,\Delta y} \qquad b_j = 0$$

$$C_j = \frac{(\hat{u}_j^{n+1})^2}{\Delta x} + \hat{v}_j^{n+1}\frac{\hat{u}_{j+1}^{n+1} - \hat{u}_{j-1}^{n+1}}{2\,\Delta y} + \frac{(u_e^{n+1})^2 - u_e^{n+1} u_e^n}{\Delta x}$$

In this example, b_j could be dropped, since it is equal to zero. We will continue to develop the solution algorithm including b_j because the result will be useful to us for solving other difference equations in this chapter.

For any j value, *four* unknowns (five if $b_j \neq 0$) appear on the left-hand side of Eq. (7-22), u_{j-1}^{n+1}, u_j^{n+1}, u_{j+1}^{n+1}, and v_j^{n+1}. It is obvious that the matrix of coefficients is no longer tridiagonal. However, the continuity equation can be written as

$$v_j^{n+1} = v_{j-1}^{n+1} - e_j(u_{j-1}^{n+1} + u_j^{n+1}) + d_j \tag{7-23}$$

where
$$e_j = \frac{\Delta y}{2\,\Delta x} \qquad d_j = \frac{(u_{j-1}^n + u_j^n)\,\Delta y}{2\,\Delta x}$$

and Eqs. (7-22) and (7-23) together form a coupled system which can be written in "block-tridiagonal" form (see Appendix B) with 2 × 2 blocks. A solution algorithm has been developed (see also Werle et al., 1973, or Blottner, 1975a) for solving this coupled system of equations. In this procedure (often called the modified tridiagonal algorithm), the blocks above the main diagonal are first eliminated. This permits the velocities, u_j^{n+1}, to be calculated from the recursion formula $u_j^{n+1} = E_j u_{j-1}^{n+1} + F_j + G_j v_{j-1}^{n+1}$ after E_j, F_j, G_j, and v_{j-1}^{n+1} are computed as indicated below. At the upper boundary, corresponding to $j = J$, conditions are specified as

$$E_J = 0$$

$$F_J = u_J^{n+1} \text{ (specified boundary value)}$$

$$G_J = 0$$

Then for $j = J-1, J-2, \ldots, 2$ we compute

$$\bar{D}_j = D_j + A_j E_{j+1} - e_j(A_j G_{j+1} + a_j)$$

$$E_j = -\left(\frac{B_j - e_j(A_j G_{j+1} + a_j)}{\bar{D}_j}\right)$$

$$F_j = \frac{C_j - A_j F_{j+1} - d_j(A_j G_{j+1} + a_j)}{\bar{D}_j}$$

$$G_j = -\left(\frac{A_j G_{j+1} + a_j + b_j}{\bar{D}_j}\right)$$

Then the lower boundary conditions are utilized to compute $v_1^{n+1} = 0$, $u_1^{n+1} = 0$ after which the velocities can be computed for $j = 2, \ldots, J$ by utilizing $u_j^{n+1} = E_j u_{j-1}^{n+1} + F_j + G_j v_{j-1}^{n+1}$ and $v_j^{n+1} = v_{j-1}^{n+1} - e_j(u_{j-1}^{n+1} + u_j^{n+1}) + d_j$. The above procedure reduces to the Thomas algorithm (but with elements above the main diagonal being eliminated) for a scalar tridiagonal system whenever a_j, b_j, e_j, and d_j are all set to zero. This system of equations can also be solved by the general algorithm for a block tridiagonal system given in Appendix B. However, the algorithm given above is more efficient because it is specialized to systems exactly of the form given by Eqs. (7-22) and (7-23).

The procedure can be extended to compressible variable property flows readily (see Blottner, 1975a). The energy equation is nearly always solved in an uncoupled manner in this case.

5. *Extrapolating the coefficients*

Values of the coefficients can be obtained at the $n+1$ level by extrapolation based on values already obtained from previous n levels. Formally, the truncation error of this procedure can be made as small as we wish. For example, we can write

$$u_j^{n+1} = u_j^n + \left.\frac{\partial u}{\partial x}\right)_j^n \Delta x_+ + O(\Delta x)^2$$

Approximating $(\partial u/\partial x)_j^n$ by only a first-order accurate representation such as

$$\left.\frac{\partial u}{\partial x}\right)_j^n = \frac{u_j^n - u_j^{n-1}}{\Delta x_-} + O(\Delta x)$$

gives the following representation for u_j^{n+1} which formally has a truncation error of $O(\Delta x)^2$.

$$u_j^{n+1} = u_j^n + \frac{u_j^n - u_j^{n-1}}{\Delta x_-} \Delta x_+ + O[(\Delta x)^2]$$

A similar procedure can be used for other coefficients needed at the $n+1$ level. This approach has been used satisfactorily for boundary-layer flows by Harris (1971).

A recommendation. For many calculations, the linearization introduced by simply lagging the coefficients u and v and the fluid properties (in cases with temperature variations) will cause no serious deterioration of accuracy. Errors associated with linearization of coefficients are simply truncation errors which can be controlled by adjustment of the marching step size. Many investigators have used this procedure satisfactorily. For any problem in which this linearization causes special difficulties, extrapolation of coefficients or Newton linearization with coupling is recommended. The former procedure requires no iterations to update the coefficients and for that reason should be more economical of computation time. In advocating the extrapolation procedure, McDonald (1978) points out that rather than utilizing computation time to iterate only for the sake of reducing the truncation error associated with linearization, overall accuracy might be improved by utilizing the same computation time to reduce the marching step size. This also reduces the truncation errors associated with the derivatives in the marching coordinate. The required level of accuracy varies from problem to problem. Clearly it is desirable to use a method that is consistent so that the numerical errors can be reduced to any level required. Especially for turbulent flow calculations, the uncertainties in the experimental data which are used to guide and verify the calculations and the uncertainties introduced by turbulence modeling add up to several (at least three to five) percent making extreme accuracy in the numerical procedures unrewarding. In this situation, the merits of using a higher-order method (highly accurate in terms of order of truncation error) should be determined on the basis of computer time which can be saved through the use of the coarser grids permitted by the more accurate schemes.

A warning on stability. Implicit schemes are touted as being unconditionally stable (in the von Neumann sense) if $\theta \geqslant \frac{1}{2}$. The Crank-Nicolson scheme just barely satisfies the formal stability requirement in terms of θ and this requirement was based on a heuristic extension of von Neumann's analysis for linear equations to nonlinear ones.

For turbulent flows especially, the Crank-Nicolson procedure has occasionally been found to become unstable. For this reason, the fully implicit scheme has been gaining in popularity. Formal second-order accuracy can be achieved by use of a three-point representation of the streamwise derivative and extrapolation of the coefficients. As an example, for uniform grid spacing, the convective terms

$$u \frac{\partial u}{\partial x} + v \frac{\partial u}{\partial y}$$

can be represented by

$$u \frac{\partial u}{\partial x} + v \frac{\partial u}{\partial y} = \frac{(2u_j^n - u_j^{n-1})(3u_j^{n+1} - 4u_j^n + u_j^{n-1})}{2\,\Delta x} + \frac{(2v_j^n - v_j^{n-1})(u_{j+1}^{n+1} - u_{j-1}^{n+1})}{2\,\Delta y} + O(\Delta x)^2 + O(\Delta y)^2 \quad (7\text{-}24)$$

With a slight increase in algebraic complexity, these representations can be generalized to also provide second-order accurate representations when the mesh increments Δx and Δy are not constant (Harris, 1971).

There is still one very real constraint on the use of the implicit schemes given for boundary-layer flows. Though not detected by the von Neumann stability analysis, a behavior very much characteristic of numerical instability can occur if the choice of grid spacing permits the convective transport (of momentum or energy) to dominate the diffusive transport. Two sources of this difficulty can be identified. First, errors can grow out of hand in the tridiagonal elimination scheme if diagonal dominance is not maintained, that is, in terms of the notation being used for the Thomas algorithm, if $|D_j|$ is not greater than $|B_j| + |A_j|$. This property of elimination methods has long been known, but was discussed in terms of implicit finite-difference schemes only relatively recently by Hirsh and Rudy (1974). Earlier, Patankar and Spalding (1970) noted difficulties in their calculations and referred to their proposed remedy as a "high lateral flux" correction. A second and equally important cause of these unacceptable solutions can be related to a physical implausibility which arises when the choice of grid size permits the algebraic model to be an inaccurate representation for a viscous flow. The same difficulty for the viscous Burgers equation was discussed in Chapter 4. It can be shown that satisfying the conditions required to keep the algebraic representation a physically valid one provides a sufficient condition for diagonal dominance in the elimination scheme.

To illustrate the basis for these difficulties, let's consider the fully implicit procedure applied to the boundary-layer momentum equation for constant property flow with the coefficients lagged. The finite-difference equation can be written as

$$B_j u_{j-1}^{n+1} + D_j u_j^{n+1} + A_j u_{j+1}^{n+1} = C_j \tag{7-25}$$

with

$$B_j = -\frac{v_j^n}{2\,\Delta y} - \frac{\nu}{(\Delta y)^2}$$

$$D_j = \frac{u_j^n}{\Delta x} + \frac{2\nu}{(\Delta y)^2}$$

$$A_j = \frac{v_j^n}{2\,\Delta y} - \frac{\nu}{(\Delta y)^2}$$

$$C_j = \frac{(u_j^n)^2}{\Delta x} + u_e^n \frac{(u_e^{n+1} - u_e^n)}{\Delta x}$$

By reflecting on the implications of Eq. (7-25) as it suggests the predicted behavior of u_j^{n+1} relative to changes in u_{j-1}^{n+1} and u_{j+1}^{n+1}, we would expect A_j and B_j to be both negative to properly imply the expected behavior of a viscous fluid. The expected behavior would be such that a decrease in the velocity of the fluid below or above the point $n+1, j$ would contribute toward a decrease in the velocity at point $n+1, j$ through the effects of viscosity. We should be able to see that such would not be the case if either A_j or B_j would become positive. To keep A_j and B_j negative in value requires

$$\frac{|v_j^n|}{2\,\Delta y} - \frac{\nu}{(\Delta y)^2} < 0$$

or

$$\frac{|v_j^n|\,\Delta y}{\nu} \leqslant 2 \tag{7-26}$$

Equation (7-26) confirms our suspicion that the "correct" representation is one that permits viscous-like behavior in that the inequality can be satisfied for a sufficiently fine mesh which, of course, is achieved at convergence. The term $|v_j^n|\Delta y/\nu$ can be identified as a mesh Reynolds number. Mesh Peclet number, a more general terminology, is also frequently used for this term.

Maintaining the inequality of Eq. (7-26) provides a sufficient (but not the necessary) condition for diagonal dominance of the algebraic system. It appears that keeping the coefficients A_j and B_j negative to provide correct simulation of viscous behavior should be the major concern and that error propagation in the elimination scheme is only a coincidental effect. For this scheme it is probable that we would have to reject the error free solution (if we would obtain it) for $|v_j^n|\Delta y/\nu \gg 2$ on physical grounds. On the other hand, the error propagation in the elimination scheme can be a real effect contributing to difficulties with calculations in some instances.

For some flows, the constraint of Eq. (7-26) tends to require the use of an excessively large number of grid points. This has motivated several investigators to consider ways of altering the difference scheme to eliminate the mesh Reynolds number constraint. Most of the studies on this problem have focused on the more complex Navier-Stokes equations where the motivation for computational economy is stronger. The simplest remedy to the problem of the mesh Reynolds number constraint is to replace the central-difference representation for $v\,\partial u/\partial y$ by an upstream (one-way) difference:

$$v\,\frac{\partial u}{\partial y} \simeq \frac{v_j^n(u_j^{n+1} - u_{j-1}^{n+1})}{\Delta y}$$

when

$$v_j^n > 0$$

and

$$\frac{v_j^n(u_{j+1}^{n+1} - u_j^{n+1})}{\Delta y}$$

when

$$v_j^n < 0$$

The truncation error associated with the upstream (also called "upwind") scheme creates an "artificial viscosity" which tends to enhance viscous-like behavior causing a deterioration in accuracy in some cases.

The question of the most appropriate representation of derivatives for large mesh Reynolds numbers is still hotly debated in the current technical literature and the issue has not been completely and satisfactorily resolved at this time. It is clearly possible to devise upstream weighted schemes having a more favorable truncation error (using two or more upstream grid points) but these can lead to coefficient matrices which are not tridiagonal in form–a distinct disadvantage. Most of the example calculations illustrating the detrimental effects of upstream differencing have been for the Navier-Stokes equations. Less specific information appears to be available for the boundary-layer equations. The tentative conclusion is that the use of upstream differencing for $v\,\partial u/\partial y$ (when mandated by the mesh Reynolds number) is a sufficient solution to the constraint of Eq. (7-26). Use of central differencing for this term is, of course, recommended whenever feasible.

It is common to include logic to switch from one difference representation to another in a computer program. Rather than switching abruptly from the central to the upwind scheme as the mesh Reynolds number exceeds two, the use of a combination (hybrid) of central and upwind schemes is recommended. This concept was originally suggested by Allen and Southwell (1955). Others, apparently not aware of this early work, have more recently proposed similar or identical forms (Spalding, 1972; Raithby and Torrance, 1974). To illustrate this principle for the case when $v_j^n > 0$, we let $R_{\Delta y}$ equal $|v_j^n|\Delta y/\nu$ and R_c equal the desired critical mesh Reynolds number for initiating the hybrid scheme, $R_c \leqslant 2$. Then for $v_j^n > 0$ and $R_{\Delta y} \geqslant R_c$ we represent $v\,\partial u/\partial y$ by

$$v\frac{\partial u}{\partial y} \simeq \underbrace{\left(\frac{R_c}{R_{\Delta y}}\right) v_j^n \frac{(u_{j+1}^{n+1} - u_{j-1}^{n+1})}{2\,\Delta y}}_{\text{Central-difference component}} + \underbrace{\left(1 - \frac{R_c}{R_{\Delta y}}\right) v_j^n \frac{(u_j^{n+1} - u_{j-1}^{n+1})}{\Delta y}}_{\text{Upwind component}} \tag{7-27}$$

The above representation is for $v_j^n > 0$. The upwind representation naturally would be altered when the direction of the "wind" changes; i.e., when $v_j^n < 0$. The appropriate representation in this case is obvious.

We observe that as $R_{\Delta y}$ increases, the weighting shifts toward the upwind representation. As $R_{\Delta y} \to \infty$, the representation is entirely upwind. The hybrid scheme maintains negative values for A_j and B_j in Eq. (7-25) while permitting the maximum utilization of the central-difference representation.

The reader is referred to the work of Raithby (1976), Leonard (1979a, 1979b) and Chow and Tien (1978) for an introduction to the literature on the mesh Reynolds number problem and for some of the more recent proposals for its solution. It is quite likely that the use of the hybrid scheme employing the combination of central and upwind differencing will eventually be replaced by a better remedy for the mesh Reynolds number constraint. At the moment, however, there appears to be no clear consensus either as to the seriousness of the errors introduced by use of the hybrid scheme for purely boundary-layer flows or as to the best alternative procedure to use.

It is interesting that nothing has been noted in the technical literature about the mesh Reynolds number constraint for boundary-layer equations when the equations are solved in a coupled manner as with the Davis coupled scheme discussed in this section or the modified box method to be discussed in Section 7-3.5. When coupling is used, the v in $v\,\partial u/\partial y$ is treated algebraically as an unknown and not merely as a coefficient for the unknown u's. It is possible that the coupling eliminates the "wiggles" and nonphysical behavior observed when central differencing is used for large mesh Reynolds numbers. Whether the smooth solution which results is more accurate than the solution obtained using upwind differencing without coupling is another matter.

Closing comment on C-N and fully implicit methods. The difference schemes presented in this section have been purposely applied to equations in physical coordinates and have been written assuming Δx and Δy were both constant. This has been done primarily to keep the equations as simple as possible as the fundamental characteristics of the schemes were being discussed. As familiarity is gained with the basic concepts involved with differencing the boundary-layer equations, ways of extending schemes to a nonuniform grid will be pointed out.

Computer programs based on implicit methods can be found readily in the literature and are not included here. The well-known Patankar-Spalding method is based on fully implicit procedures and is well-documented in Patankar and Spalding (1970). The STAN5 (Crawford and Kays, 1975) program guide is another source for a complete implicit program listing.

7-3.4 The DuFort-Frankel Method

Another finite-difference procedure which has worked well for both laminar and turbulent boundary layers is an extension of the method proposed by DuFort and Frankel (1953) for the heat equation. The difference representation will be written in a form which will accommodate variable grid spacing. We let $\Delta x_+ = x^{n+1} - x^n$, $\Delta x_- = x^n - x^{n-1}$, $\Delta y_+ = y_{j+1} - y_j$, $\Delta y_- = y_j - y_{j-1}$. The implicit methods of the previous section can be extended in applicability to a nonuniform grid by following a similar procedure.

In presenting the DuFort-Frankel procedure for the momentum and energy equations, the generalized transport partial differential equation, Eq. (7-5) will be employed with the dependent variable ϕ denoting velocity components, a turbulence model parameter, or a thermal variable such as temperature or enthalpy. In the DuFort-Frankel differencing, stability is promoted by eliminating the appearance of ϕ_j^n in the diffusion term through the use of an average of ϕ at the $n+1$ and $n-1$ levels. With unequal spacing, however, Dancey and Pletcher (1974) observed that accuracy was improved by use of a linearly interpolated value of ϕ between $n-1$ and $n+1$ levels instead of a simple average. Here we define the linearly interpolated value as $\overline{\phi_j^n}$ according to $\overline{\phi_j^n} = (\Delta x_+ \phi_j^{n-1} + \Delta x_- \phi_j^{n+1})/(\Delta x_+ + \Delta x_-)$. As before, for turbulent flows it's understood that u and v are time-mean quantities. For a compressible flow $v = \tilde{v}$. For generality we'll let $\bar{\lambda} = \lambda_T + \lambda$ where λ_T is a turbulent diffusion coefficient. The DuFort-Frankel representation of the generalized transport equation becomes

$$\frac{\rho_j^n u_j^n(\phi_j^{n+1} - \phi_j^{n-1})}{\Delta x_+ + \Delta x_-} + \frac{\rho_j^n v_j^n(\phi_{j+1}^n - \phi_{j-1}^n)}{\Delta y_+ + \Delta y_-} = \frac{2}{\Delta y_+ + \Delta y_-}$$

$$\times \left[\frac{\bar{\lambda}_{j+1/2}^n(\phi_{j+1}^n - \bar{\phi}_j^n)}{\Delta y_+} - \frac{\bar{\lambda}_{j-1/2}^n(\bar{\phi}_j^n - \phi_{j-1}^n)}{\Delta y_-}\right] + S_j^n \tag{7-28}$$

In the above, S_j^n denotes the source terms. Examples of source terms which frequently occur include the pressure gradient dp/dx in the x-momentum equation where

$$S_j^n = \frac{p_j^{n+1} - p_j^{n-1}}{\Delta x_+ + \Delta x_-}$$

a viscous dissipation term $\bar{\mu}(\partial u/\partial y)^2$ in the energy equation when T is used as the thermal variable,

$$S_j^n = \bar{\mu}_j^n \left(\frac{u_{j+1}^n - u_{j-1}^n}{\Delta y_+ + \Delta y_-}\right)^2$$

and a dissipation term $C_D \rho (\bar{k})^{3/2}/l$ in the modeled form of the turbulence kinetic energy equation

$$S_j^n = C_D \rho_j^n \left(\frac{\Delta y_+ (\bar{k})_{j-1}^n + \Delta y_- (\bar{k})_{j+1}^n}{\Delta y_+ + \Delta y_-}\right)^{1/2} \left(\frac{\Delta x_+ (\bar{k})_j^{n-1} + \Delta x_- (\bar{k})_j^{n+1}}{\Delta x_+ + \Delta x_-}\right) \Big/ l_j^n$$

Note that this latter representation avoids using the dependent variable $\bar{k}$ at (n, j). This is required by stability (see Malik and Pletcher, 1978) as might be expected in light of the special treatment required for the diffusion term noted above.

We recall (Chapter 4) that the DuFort-Frankel representation is explicit. Although ϕ_j^{n+1} appears in both the left and right sides (within $\bar{\phi}_j^n$) of the equation, the equation can be rearranged to isolate ϕ_j^{n+1} so that we can write ϕ_j^{n+1} = (all known quantities at the n and $n-1$ levels). The formal truncation error for the equation with $\Delta x_+ = \Delta x_-$ and $\Delta y_+ = \Delta y_-$ is $O(\Delta x)^2 + O(\Delta y)^2 + O(\Delta x/\Delta y)^2$. However, the leading term in the truncation error represented by $O(\Delta x/\Delta y)^2$ is actually $(\Delta x/\Delta y)^2$ $(\partial^2 \phi/\partial x^2)$ and $\partial^2 \phi/\partial x^2$ is presumed to be very small for boundary-layer flows. One can show that a deterioration in the formal truncation error is generally expected as the grid spacing becomes unequal although a paper by Blottner (1974) points out exceptions. This deterioration would be observed in all methods presented thus far in this chapter. In practice, the increase in actual error due to the use of unequal spacing may be negligible. In nearly all cases, remedies can be found that will restore the original formal truncation error at the expense of algebraic operations. For example, Hong (1974) demonstrated that the streamwise derivative $\partial\phi/\partial x$ in the DuFort-Frankel method can be written as

$$\frac{(\Delta x_-)^2 \phi_j^{n+1} - (\Delta x_+)^2 \phi_j^{n-1} + (\Delta x_+^2 - \Delta x_-^2)\phi_j^n}{\Delta x_- \, \Delta x_+^2 + \Delta x_+ \, \Delta x_-^2}$$

with second-order accuracy even when $\Delta x_+ \neq \Delta x_-$.

A consistent treatment of the continuity equation is given by

$$\frac{\rho_j^{n+1}v_j^{n+1} - \rho_{j-1}^{n+1}v_{j-1}^{n+1}}{\Delta y_-} + \frac{\rho_j^{n+1}u_j^{n+1} - \rho_j^{n-1}u_j^{n-1} + \rho_{j-1}^{n+1}u_{j-1}^{n+1} - \rho_{j-1}^{n-1}u_{j-1}^{n-1}}{2(\Delta x_+ + \Delta x_-)} = 0 \tag{7-29}$$

with truncation error $O(\Delta x) + O(\Delta y)^2$.

A stability analysis made for Δy = constant (Madni and Pletcher, 1975a, 1975b) suggests that

$$\Delta x_+ \leqslant \frac{\rho_j^n u_j^n \, \Delta y}{|\rho_j^n v_j^n + (\bar{\lambda}_{j-1}^n - \bar{\lambda}_{j+1}^n)/2 \, \Delta y|} \tag{7-30}$$

It would appear that this constraint could also be used to provide a rough guide under variable Δy conditions. In practice this condition has not proven to be especially restrictive on the marching step size probably because v/u is generally very small and the other term in the denominator involves differences in the diffusion coefficient rather than the coefficient itself.

The stability condition was developed from the von Neumann analysis by treating the coefficients locally as constants. It is interesting to note that Eq. (7-30) follows essentially from the CFL condition rather than the diffusion stability limit for the boundary-layer momentum equation. This becomes evident when the diffusion term $\partial/\partial y(\bar{\lambda}\, \partial\phi/\partial y)$ is expanded to two terms and the boundary-layer equation rearranged as

$$\frac{\partial \phi}{\partial x} + \frac{1}{\rho u}\left(\rho v - \frac{\partial \bar{\lambda}}{\partial y}\right)\frac{\partial \phi}{\partial y} = \frac{\bar{\lambda}}{\rho u}\frac{\partial^2 \phi}{\partial y^2} + \frac{S}{\rho u}$$

Now simply applying the CFL condition gives Eq. (7-30).

The boundary-layer calculation begins by utilizing an initial distribution for the ϕ variables. Since the Dufort-Frankel procedure requires information at *two* streamwise levels in order to advance the calculation, some other method must be used to obtain a solution for at least one streamwise station before the DuFort-Frankel scheme can be employed. A simple explicit scheme is most frequently used to provide these starting values. A typical calculation would require the solution to the momentum, continuity, and energy equations. The equations can be solved sequentially starting with the momentum equation in an uncoupled manner. The usual procedure is to solve first for the unknown streamwise velocities from the momentum equation starting with the point nearest the wall and working outward to the outer edge of the boundary layer. The outer edge of the boundary layer is located when the velocity from the solution is within a prescribed tolerance of the velocity specified as the outer boundary condition. The energy equation can be solved in a like manner for the thermal variable. The density at the new station can be evaluated from an equation of state. Finally, the continuity equation is used to obtain the normal component of velocity at the $n + 1$ level starting from the point adjacent to the wall and working outward.

The explicit nature of the DuFort-Frankel procedure is probably its most attractive feature. Those inexperienced in numerical methods are likely to feel more comfortable programming an explicit procedure than they are in applying an implicit scheme. A second significant feature of the scheme is that no additional linearizations, iterations, or assumptions are needed to evaluate coefficients in the equation since these all appear at the n-level where they are known values. Further details on the application of the DuFort-Frankel type schemes to wall boundary layers can be found in Pletcher (1969, 1970, 1971).

7-3.5 The Box and Modified Box Methods

The Keller (1970) box method for parabolic partial differential equations has been adapted to turbulent boundary-layer calculations by Keller and Cebeci (1972) and is described in detail by Cebeci and Smith (1974). The differencing is implicit, formally second order in accuracy, and appears to differ from other second-order accurate implicit procedures in that the formulation assumes that the grid spacing is arbitrary from the beginning. Second derivatives are replaced with first derivatives through the introduction of additional variables (and equations). The scheme is algebraically more complex than most other procedures for the heat equation and was not included in Chapter 4 for that reason.

The starting point for the Keller box scheme will be sketched here so that the point of view can be contrasted to that used by other second-order methods, such as the Crank-Nicolson procedure. In this example we'll revert back to consideration of the heat equation to illustrate the techniques used to represent first and second derivatives by this method. Thus, we consider

$$\frac{\partial u}{\partial t} = \alpha \frac{\partial^2 u}{\partial x^2}$$

but define

$$v = \frac{\partial u}{\partial x}$$

so that we can write the original second-order PDE as a system of two first-order equations:

$$\frac{\partial u}{\partial x} = v \tag{7-31}$$

$$\frac{\partial u}{\partial t} = \alpha \frac{\partial v}{\partial x} \tag{7-32}$$

Now we endeavor to approximate these equations using only central differences making use of the four points at the corners of a "box" about $(n-\frac{1}{2}, j-\frac{1}{2})$ (see Fig. 7-2). The mesh functions whose subscript or superscript contains a $\frac{1}{2}$ are defined as averages as, for example,

$$u_{j-1/2}^n = \frac{u_j^n + u_{j-1}^n}{2}$$

$$v_j^{n-1/2} = \frac{v_j^n + v_j^{n-1}}{2}$$

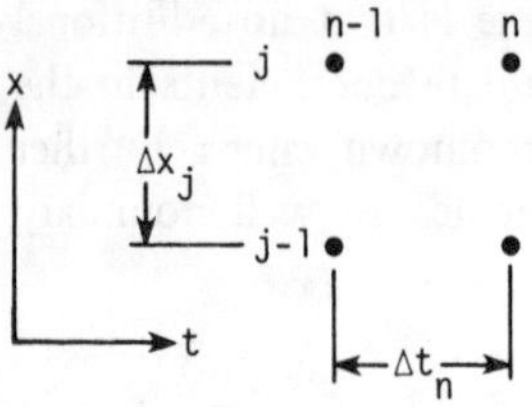

Figure 7-2 Grid for box scheme.

Equations (7-31) and (7-32) are approximated with central differences (see Figs. 7-3 and 7-4) as

$$\frac{u_j^n - u_{j-1}^n}{\Delta x_j} = v_{j-1/2}^n \tag{7-33}$$

and

$$\frac{u_{j-1/2}^n - u_{j-1/2}^{n-1}}{\Delta t_n} = \frac{\alpha(v_j^{n-1/2} - v_{j-1}^{n-1/2})}{\Delta x_j} \tag{7-34}$$

The system of Eqs. (7-33) and (7-34) can be written in block tridiagonal form with 2 X 2 blocks and solved by a block elimination scheme (Keller, 1970). It is also possible to maintain the basic box difference molecule yet combine the difference representations at two adjacent grid points to eliminate one variable and obtain a simple tridiagonal system of equations. The system can be solved efficiently with the standard Thomas algorithm. This revision of the box method which simplifies the final algebraic formulation will be referred to as the modified box method.

Modified box method for the heat equation. As a starting point, we represent Eqs. (7-31) and (7-32) as in the box method but with the marching index $n+1$ used to denote the time level at which the solution is to be determined. This is consistent with the practice used for all other difference schemes for marching problems presented in Chapter 4.

$$\frac{u_j^{n+1} - u_{j-1}^{n+1}}{\Delta x_j} = v_{j-1/2}^{n+1} = \frac{v_j^{n+1} + v_{j-1}^{n+1}}{2} \tag{7-35}$$

$$\frac{u_{j-1/2}^{n+1} - u_{j-1/2}^n}{\Delta t_{n+1}} = \alpha\frac{v_j^{n+1/2} - v_{j-1}^{n+1/2}}{\Delta x_j} = \alpha\frac{v_j^n + v_j^{n+1} - v_{j-1}^n - v_{j-1}^{n+1}}{2\,\Delta x_j} \tag{7-36}$$

As before, mesh functions utilizing a subscript containing a $\frac{1}{2}$ are defined as averages. Equation (7-36) can be rewritten

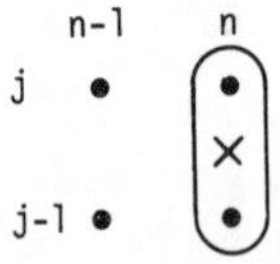

Figure 7-3 Difference molecule for evaluation of $v_{j-1/2}^n$.

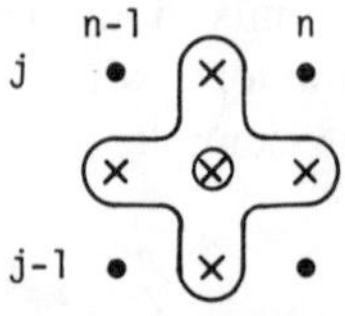

Figure 7-4 Difference molecule for Eq. (7-34).

$$\frac{u_j^{n+1} + u_{j-1}^{n+1}}{\Delta t_{n+1}} = \alpha \frac{v_j^{n+1} - v_{j-1}^{n+1}}{\Delta x_j} + \frac{u_j^n + u_{j-1}^n}{\Delta t_{n+1}} + \alpha \frac{v_j^n - v_{j-1}^n}{\Delta x_j} \tag{7-37}$$

The strategy in the development of the modified box method is to express the v's in terms of u's. v_{j-1}^{n+1} can be eliminated from Eq. (7-37) by a simple substitution using Eq. (7-35). Similarly, v_{j-1}^n can be eliminated through substitution by evaluating Eq. (7-35) for time level n.

This gives

$$\frac{u_j^{n+1} + u_{j-1}^{n+1}}{\Delta t_{n+1}} = 2\alpha \frac{v_j^{n+1}}{\Delta x_j} - 2\alpha \frac{u_j^{n+1} - u_{j-1}^{n+1}}{(\Delta x_j)^2} + \frac{u_j^n + u_{j-1}^n}{\Delta t_{n+1}} + 2\alpha \frac{v_j^n}{\Delta x_j} - 2\alpha \frac{u_j^n - u_{j-1}^n}{(\Delta x_j)^2} \tag{7-38}$$

To eliminate v_j^{n+1} and v_j^n, Eqs. (7-35) and (7-37) can first be rewritten with the j index advanced by one and combined. The result is

$$\frac{u_{j+1}^{n+1} + u_j^{n+1}}{\Delta t_{n+1}} = \frac{-2\alpha v_j^{n+1}}{\Delta x_{j+1}} + \frac{2\alpha(u_{j+1}^{n+1} - u_j^{n+1})}{(\Delta x_{j+1})^2} + \frac{u_{j+1}^n + u_j^n}{\Delta t_{n+1}} + \frac{-2\alpha v_j^n}{\Delta x_{j+1}} + 2\alpha \frac{u_{j+1}^n - u_j^n}{(\Delta x_{j+1})^2} \tag{7-39}$$

The terms v_j^{n+1} and v_j^n can be eliminated by multiplying Eq. (7-38) by Δx_j and Eq. (7-39) by Δx_{j+1} and adding the two products. The result can be written in the tridiagonal format

$$B_j u_{j-1}^{n+1} + D_j u_j^{n+1} + A_j u_{j+1}^{n+1} = C_j \tag{7-40}$$

with

$$B_j = \frac{\Delta x_j}{\Delta t_{n+1}} - \frac{2\alpha}{\Delta x_j} \qquad A_j = \frac{\Delta x_{j+1}}{\Delta t_{n+1}} - \frac{2\alpha}{\Delta x_{j+1}}$$

$$D_j = \frac{\Delta x_j}{\Delta t_{n+1}} + \frac{\Delta x_{j+1}}{\Delta t_{n+1}} + \frac{2\alpha}{\Delta x_j} + \frac{2\alpha}{\Delta x_{j+1}}$$

$$C_j = 2\alpha \frac{u_{j-1}^n - u_j^n}{\Delta x_j} + 2\alpha \frac{u_{j+1}^n - u_j^n}{\Delta x_{j+1}} + (u_j^n + u_{j-1}^n)\frac{\Delta x_j}{\Delta t_{n+1}} + (u_{j+1}^n + u_j^n)\frac{\Delta x_{j+1}}{\Delta t_{n+1}}$$

The above can be simplified somewhat if the spacing in the x direction is uniform. Even then, a few more algebraic operations per time step are required than for the Crank-Nicolson scheme which is also second-order accurate for a uniformly spaced mesh. A conceptual advantage of schemes based on the box difference molecule is that formal second-order accuracy is achieved even when the mesh is nonuniform. The Crank-Nicolson scheme can be extended to cases of nonuniform grid spacing by representing the second derivative term as indicated for Laplace's equation in Eq. (3-98). Formally, the truncation error for that representation is reduced to first order for arbitrary grid spacing. Blottner (1974) has shown that if the variable grid spacing used is one which could be established through a coordinate stretching transformation, then the Crank-Nicolson scheme is also second-order accurate for that variable grid arrangement.

The box method for boundary-layer equations. Keller and Cebeci (1972) have applied the box differencing scheme to the boundary-layer momentum and continuity equations after they had first been transformed to a single third-order PDE using the Mangler and Levy-Lees transformations (see Cebeci and Smith, 1974). The third-order PDE is written as a system of three first-order PDE's using newly defined variables in a manner that parallels the procedure commonly employed in the numerical solution of third-order ordinary differential equations. The box differencing scheme with Newton linearization is then applied to the three first-order PDE's giving rise to a block-tridiagonal system having 3 X 3 blocks which is solved by a block elimination scheme. The corresponding treatment for the energy equation gives rise to a block tridiagonal system with 2 X 2 blocks.

The details of the Cebeci-Keller box method for the boundary-layer equations will not be given here; they can be found in Cebeci and Smith (1974). Instead, we will indicate how a modified box scheme can be developed which only requires the use of the same modified tridiagonal elimination scheme presented in this section for the Davis coupled scheme. From reports in the literature (Blottner, 1975a; Wornam, 1977), the modified box scheme appears to require only on the order of one-half as much computer time as the standard box scheme for the boundary-layer equations.

The momentum and energy equations for a compressible flow can be written in the generalized form given by Eq. (7-5). For rectangular coordinates this becomes

$$\rho u \frac{\partial \phi}{\partial x} + \rho \tilde{v} \frac{\partial \phi}{\partial y} = \frac{\partial}{\partial y}\left(\bar{\lambda} \frac{\partial \phi}{\partial y}\right) + S \tag{7-41}$$

where

$$\bar{\lambda} = \lambda_T + \lambda$$

The continuity equation can be written as

$$\frac{\partial \rho u}{\partial x} + \frac{\partial \rho \tilde{v}}{\partial y} = 0 \tag{7-42}$$

The grid nomenclature is given in Fig. 7-5.

We let

$$\bar{\lambda} \frac{\partial \phi}{\partial y} = q$$

and

$$D = S + \frac{\partial q}{\partial y} - \rho \tilde{v} \frac{\partial \phi}{\partial y}$$

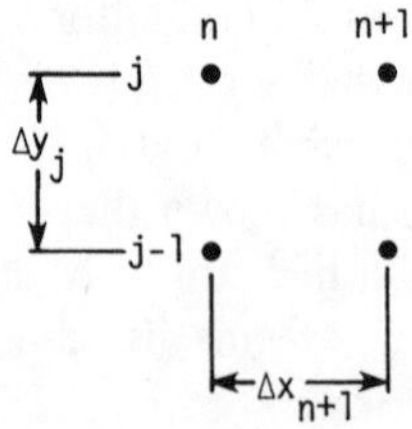

Figure 7-5 Grid arrangement for the modified box scheme.

Equation (7-41) can then be written as

$$\rho u \frac{\partial \phi}{\partial x} = D \tag{7-43}$$

Centering on the box grid gives

$$\frac{(\rho u)_{j-1/2}^{n+1} + (\rho u)_{j-1/2}^{n}}{2} \frac{\phi_{j-1/2}^{n+1} - \phi_{j-1/2}^{n}}{\Delta x_{n+1}} = S_{j-1/2}^{n+1/2} - \frac{(\rho \tilde{v})_{j-1/2}^{n+1/2}(\phi_j^{n+1/2} - \phi_{j-1}^{n+1/2})}{\Delta y_j} + \frac{q_j^{n+1/2} - q_{j-1}^{n+1/2}}{\Delta y_j} \tag{7-44}$$

Utilizing the definition of $q_{j-1/2}^{n+1/2}$,

$$\bar{\lambda}_{j-1/2}^{n+1/2} \frac{\phi_j^{n+1/2} - \phi_{j-1}^{n+1/2}}{\Delta y_j} = \frac{q_j^{n+1/2} + q_{j-1}^{n+1/2}}{2} \tag{7-45}$$

we can eliminate $q_{j-1}^{n+1/2}$ from Eq. (7-44).

This gives

$$\frac{(\rho u)_{j-1/2}^{n+1} + (\rho u)_{j-1/2}^{n}}{2} \frac{\phi_{j-1/2}^{n+1} - \phi_{j-1/2}^{n}}{\Delta x_{n+1}} = S_{j-1/2}^{n+1/2} - (\rho \tilde{v})_{j-1/2}^{n+1/2} \frac{\phi_j^{n+1/2} - \phi_{j-1}^{n+1/2}}{\Delta y_j} + \frac{2q_j^{n+1/2}}{\Delta y_j} - 2\bar{\lambda}_{j-1/2}^{n+1/2} \frac{\phi_j^{n+1/2} - \phi_{j-1}^{n+1/2}}{\Delta y_j^2} \tag{7-46}$$

In a similar manner, the momentum equation can be put into difference form centering about the point $(n+\frac{1}{2}, j+\frac{1}{2})$ and $q_{j+1}^{n+1/2}$ eliminated from that equation using the definition of $q_{j+1/2}^{n+1/2}$.

The result is

$$\frac{(\rho u)_{j+1/2}^{n+1} + (\rho u)_{j+1/2}^{n}}{2} \frac{\phi_{j+1/2}^{n+1} - \phi_{j+1/2}^{n}}{\Delta x_{n+1}} = S_{j+1/2}^{n+1/2} - (\rho \tilde{v})_{j+1/2}^{n+1/2} \frac{\phi_{j+1}^{n+1/2} - \phi_j^{n+1/2}}{\Delta y_{j+1}} + 2\bar{\lambda}_{j+1/2}^{n+1/2} \frac{\phi_{j+1}^{n+1/2} - \phi_j^{n+1/2}}{(\Delta y_{j+1})^2} - \frac{2q_j^{n+1/2}}{\Delta y_{j+1}} \tag{7-47}$$

Equations (7-46) and (7-47) can be combined to eliminate $q_j^{n+1/2}$. This is accomplished by multiplying Eq. (7-46) by Δy_j, and Eq. (7-47) by Δy_{j+1} and adding the two products. After replacing quantities specified for evaluation at grid midpoints by averages from grid points, the results can be written as

$$\begin{aligned}
&\frac{\Delta y_j}{2\,\Delta x_{n+1}} [(\rho u)_j^{n+1} + (\rho u)_{j-1}^{n+1} + (\rho u)_j^{n} + (\rho u)_{j-1}^{n}]\,(\phi_j^{n+1} + \phi_{j-1}^{n+1} - \phi_j^{n} - \phi_{j-1}^{n}) \\
&+ \frac{\Delta y_{j+1}}{2\,\Delta x_{n+1}} [(\rho u)_{j+1}^{n+1} + (\rho u)_j^{n+1} + (\rho u)_{j+1}^{n} + (\rho u)_j^{n}]\,(\phi_{j+1}^{n+1} + \phi_j^{n+1} - \phi_{j+1}^{n} - \phi_j^{n}) \\
&+ \tfrac{1}{2} [(\rho \tilde{v})_j^{n+1} + (\rho \tilde{v})_{j-1}^{n+1} + (\rho \tilde{v})_j^{n} + (\rho \tilde{v})_{j-1}^{n}]\,(\phi_j^{n+1} + \phi_j^{n} - \phi_{j-1}^{n+1} - \phi_{j-1}^{n})
\end{aligned} \tag{7-48}$$

$$+\tfrac{1}{2}\left[(\rho\tilde{v})_{j+1}^{n+1}+(\rho\tilde{v})_{j}^{n+1}+(\rho\tilde{v})_{j+1}^{n}+(\rho\tilde{v})_{j}^{n}\right](\phi_{j+1}^{n+1}+\phi_{j+1}^{n}-\phi_{j}^{n+1}-\phi_{j}^{n})$$

$$=\Delta y_j(S_j^{n+1}+S_{j-1}^{n+1}+S_j^n+S_{j-1}^n)+\Delta y_{j+1}(S_{j+1}^{n+1}+S_j^{n+1}+S_{j+1}^n+S_{j-1}^n)$$

$$+\frac{(\bar{\lambda}_{j+1}^{n+1}+\bar{\lambda}_{j}^{n+1}+\bar{\lambda}_{j+1}^{n}+\bar{\lambda}_{j}^{n})(\phi_{j+1}^{n+1}+\phi_{j+1}^{n}-\phi_{j}^{n+1}-\phi_{j}^{n})}{\Delta y_{j+1}}$$

$$-\frac{(\bar{\lambda}_{j}^{n+1}+\bar{\lambda}_{j-1}^{n+1}+\bar{\lambda}_{j}^{n}+\bar{\lambda}_{j-1}^{n})(\phi_{j}^{n+1}+\phi_{j}^{n}-\phi_{j-1}^{n+1}-\phi_{j-1}^{n})}{\Delta y_j} \qquad \text{(7-48)}$$

(*Cont.*)

Equation (7-48) can be expressed in the tridiagonal format for the unknown ϕ's but, as for all implicit methods, some scheme must be devised for treating the algebraic nonlinearities arising through the coefficients. Conceptually, any of the procedures presented previously in Section 7-3.3 can be employed. The most suitable representation of the continuity equation may depend upon the procedure used to accomplish the linearization of the momentum equation. To date, Newton linearization with coupling (Blottner, 1975a) has been the most commonly used procedure. For this, the continuity equation can be written as

$$\frac{(\rho u)_j^{n+1}+(\rho u)_{j-1}^{n+1}-(\rho u)_j^n-(\rho u)_{j-1}^n}{2\,\Delta x_{n+1}}+\frac{(\rho\tilde{v})_j^{n+1}+(\rho\tilde{v})_j^n-(\rho\tilde{v})_{j-1}^{n+1}-(\rho\tilde{v})_{j-1}^n}{2\,\Delta y_j}=0 \qquad \text{(7-49)}$$

The momentum equation involves $(\rho\tilde{v})_{j+1}^{n+1}$, $(\rho\tilde{v})_j^{n+1}$, and $(\rho\tilde{v})_{j-1}^{n+1}$. To employ the modified tridiagonal elimination scheme, the continuity equation (Eq. 7-49) can be written between the j and $j+1$ levels and $(\rho\tilde{v})_{j+1}^{n+1}$ eliminated from the momentum equation by substitution. After employing Newton linearization in a manner which parallels the procedures illustrated in Section 7-3.3 for the fully implicit Davis coupled method, the coupled momentum and continuity equations can be solved with the modified tridiagonal elimination scheme. The energy equation is usually solved in an uncoupled manner and properties (including the turbulent viscosity) updated iteratively as desired or required by accuracy constraints.

With the box and modified box schemes, the wall shear stress and heat flux are usually determined by evaluating q at the wall ($j=1$). For the modified box scheme this is done after the solutions for ϕ, $\tilde{v}$, and ρ have been determined. An expression for $q_1^{n+1/2}$ can be obtained by writing Eqs. (7-44) and (7-45) for $j=2$ and eliminating $q_2^{n+1/2}$ by a simple substitution.

7-3.6 Other Methods

Exploratory studies of limited scope have indicated that the Barakat and Clark ADE method can be used for solving the boundary-layer equations (R. G. Hindman, 1975, private communication; S. S. Hwang, 1975, private communication). These results indicate that ADE methods are roughly equivalent in accuracy and computation time to the more conventional implicit methods for boundary-layer problems.

Higher-order schemes (up through fourth order) have also been applied to the boundary-layer equations. A critical study of some of these schemes was reported by Wornam (1977). It is worthwhile to note that the accuracy of lower-order methods can also be improved through the use of Richardson extrapolation (Ralston, 1965; Cebeci and Smith, 1974).

It is believed that the most commonly used difference schemes for two-dimensional or axisymmetric boundary layers have been described in this section. No attempt has been made to cover all known methods in detail.

7-3.7 A Comment on Coordinate Transformations for Boundary Layers

The general subject of coordinate transformation has been treated in Chapter 5. In the present chapter the focus has been on the difference schemes themselves and to illustrate these in the simplest possible manner, the equations have been presented in rectangular Cartesian, "physical coordinates." It's well to point out that there may be advantages to employing coordinate transformations or "stretching" on the governing equations prior to formulating the difference equations. Many of the boundary-layer methods appearing in the literature have utilized transformations.

The main objectives of the transformations are generally to obtain a coordinate frame for computation in which the boundary-layer thickness remains as constant as possible and to remove the singularity in the equations at the leading edge or stagnation point. Unfortunately, for complex turbulent flows, the optimum transformation leading to a constant boundary-layer thickness in the transformed plane has not been identified, although the transformation suggested by Carter et al. (1980) shows promise.

The most commonly used transformation makes use of the transverse similarity variable η employed in the Blasius similarity solution to the laminar boundary layer. We will give an example of such a transformation for the constant property laminar boundary layer below. We start with

continuity:

$$\frac{\partial u}{\partial x} + \frac{\partial v}{\partial y} = 0 \tag{7-50}$$

momentum:

$$u\,\frac{\partial u}{\partial x} + v\,\frac{\partial u}{\partial y} = u_e\,\frac{du_e}{dx} + \nu\,\frac{\partial^2 u}{\partial y^2} \tag{7-51}$$

The crucial element of the transformation is the introduction of

$$\eta = \frac{y}{x}\left(\frac{u_e x}{\nu}\right)^{1/2}$$

From this point on several variations are possible but a common procedure is to let $x = x$ (no stretching of x) and $F = u/u_e$. Using the chain rule we note that

$$\left.\frac{\partial}{\partial x}\right)_y = \left.\frac{\partial}{\partial x}\right)_\eta + \left.\frac{\partial \eta}{\partial x}\right)_y \left.\frac{\partial}{\partial \eta}\right)_x = \left.\frac{\partial}{\partial x}\right)_\eta + \left(\frac{\eta}{2u_e}\frac{du_e}{dx} - \frac{\eta}{2x}\right)\left.\frac{\partial}{\partial \eta}\right)_x$$

and
$$\left.\frac{\partial}{\partial y}\right)_x = \left.\frac{\partial x}{\partial y}\right)_x \left.\frac{\partial}{\partial x}\right)_\eta + \left.\frac{\partial \eta}{\partial y}\right)_x \left.\frac{\partial}{\partial \eta}\right)_x = \left(\frac{u_e}{x\nu}\right)^{1/2} \left.\frac{\partial}{\partial \eta}\right)_x$$

Replacing the x and y derivatives in Eqs. (7-50) and (7-51) as indicated and utilizing F, results in the transformed momentum and continuity equations:

momentum:

$$xF\frac{\partial F}{\partial x} + V\frac{\partial F}{\partial \eta} = \beta(1 - F^2) + \frac{\partial^2 F}{\partial \eta^2} \tag{7-52}$$

continuity:

$$x\frac{\partial F}{\partial x} + \frac{\partial V}{\partial \eta} + F\frac{\beta + 1}{2} = 0 \tag{7-53}$$

where
$$V = \frac{\beta - 1}{2} F\eta + \left(\frac{x}{\nu u_e}\right)^{1/2} v$$

and
$$\beta = \frac{x}{u_e}\frac{du_e}{dx}$$

When $x = 0$, the streamwise derivatives vanish from the transformed equations and a system of two ordinary differential equations remain. It is common to solve these equations with a slightly modified version of the marching technique employed for the rest of the flow domain, i.e., for $x > 0$, although special numerical procedures applicable to ODE's could be used.

There is no singular behavior at $x = 0$ in the new coordinate system since the troublesome streamwise derivatives have been eliminated. In fact, for laminar flow over a plate, the solution for $x = 0$ is the well-known Blasius similarity solution. Naturally, for a zero pressure gradient flow, the marching solution for $x > 0$ should reproduce essentially the same solution downstream and the boundary-layer thickness should remain constant. When pressure gradients or wall boundary conditions force a nonsimilar laminar flow solution, the boundary-layer thickness will change somewhat along the flow. For nearly similar laminar flows, we would expect the solution of the transformed equations to provide greater and more uniform accuracy near the leading edge than the solution in physical coordinates because of the tendency of the former procedure to divide the boundary layer into a more nearly constant number of points in the transverse direction. For turbulent flows, we observe the boundary-layer thickness growing along the surface, generally quite significantly, even with the use of the above transformed variables.

For external laminar boundary-layer calculations, the use of transformed coordinates of the similarity type is recommended. For turbulent flows, the advantages of transformations suggested to date are less certain.

7-3.8 Special Considerations for Turbulent Flows

The accurate solution of the boundary-layer equations for turbulent flows using models which evaluate the turbulent viscosity at all points within the flow requires that grid points be located within the viscous sublayer, $y^+ \leqslant 4.0$ for incompressible flow, and perhaps $y^+ \leqslant 1.0$ or 2.0 for flows in which a solution to the energy equation is also being obtained. The use of equal grid spacing for the transverse coordinate would require several thousand grid points across the boundary layer for a typical calculation at moderate Reynolds numbers. This at least provides motivation for considering ways to reduce the number of grid points required to span the boundary layer. The techniques which have been used successfully fall into three categories, use of wall functions, unequal grid spacing, and coordinate transformations.

Use of wall functions. For many turbulent wall boundary layers the inner portion of the flow appears to have a "universal" character captured by the logarithmic "law of the wall" discussed previously (see Fig. 5-7). Basically, this inner region is a zone in which convective transport is relatively unimportant. The law of the wall can be roughly thought of as a solution to the boundary-layer momentum equation using Prandtl's mixing-length turbulence model when convective and pressure gradient terms are unimportant. Corresponding nearly universal behavior has been observed for the temperature distribution for many turbulent flows, and wall functions can be used to provide an inner boundary condition for solutions to the energy equation. Thus, with the wall function approach, the boundary-layer equations are solved using a turbulence model in the outer region on a relatively coarse grid and the near wall region is "patched in" through the use of a form of the law of the wall which, in fact, represents an approximate solution for the near-wall region. In this approach, the law of the wall is usually assumed to be valid in the range $30 < y^+ < 200$ and the first computational point away from wall is located in this interval. Boundary conditions are developed for the dependent variables in the transport equations being solved (u, T, $\bar{k}$, ϵ, etc.) at this point from the wall functions. Many variations in this procedure are possible and details depend upon the turbulence model and difference scheme being used. The procedure has been well-developed for use with the $\bar{k} - \epsilon$ turbulence model and recommended wall functions for u, T, $\bar{k}$, and ϵ can be found in Launder and Spalding (1974).

Like turbulence models themselves, wall functions need modifications to accurately treat effects such as wall blowing and suction, surface roughness, etc. Their use does, however, circumvent the need for many closely spaced points near the wall. It would seem that the use of wall functions may not be necessary or desirable for many boundary-layer flows, but their use becomes more attractive in treating more complex flows perhaps involving a nonboundary layer (elliptic) form of the Reynolds equations where the processes of most interest may be well away from the walls. Over the next few years as faster computers become available and attempts are made to solve the time-dependent Navier-Stokes equations for turbulent flows (without modeling) it is likely that some of the early calculations will employ some form of wall functions to "patch in" the approximate solution over the wall region where the eddies would be the smallest.

Use of unequal grid spacing. Almost without exception, turbulent boundary-layer calculations which have applied the difference scheme right down to the wall have utilized either a variable grid scheme or what is often equivalent, a coordinate transformation. Arbitrary spacing will work. In Pletcher (1969), Δy corresponding to Δy^+ [defined as $\Delta y(\tau_w/\rho)^{1/2}/\nu_w] \simeq 1.0$ was used for several mesh increments nearest the wall and then approximately doubled every few points until Δy^+ reached 100 in the outer part of the flow.

Another commonly used (Cebeci and Smith, 1974) and very workable scheme maintains a constant ratio between two adjacent increments

$$\frac{\Delta y_+}{\Delta y_-} = \frac{\Delta y_{j+1}}{\Delta y_j} = K \tag{7-54}$$

In this constant ratio scheme, each grid spacing is increased by a fixed percentage from the wall outward. This results in a geometric progression in the size of the spacing. K is usually a number between 1.0 and 2.0 for turbulent flows. For the constant ratio scheme it follows that

$$\Delta y_j = K^{j-1}\,\Delta y_1 \qquad \text{and} \qquad y_j = \Delta y_1 \frac{K^{j-1}-1}{K-1} \tag{7-55}$$

The accuracy (and occasionally the stability) of some schemes appears sensitive to the value of K being used. Most methods appear to give satisfactory results for $K \leqslant 1.15$. For a typical calculation using $\Delta y_1^+ \simeq 1.5$, $K = 1.04$ and $y_e^+ \simeq 3000$, Eqs. (7-54) and (7-55) can be used to determine that about 113 grid points in the transverse direction would be required.

As a difference scheme is being generalized to accommodate variable grid spacing, the truncation error should be reevaluated since a deterioration in the formal trunction error is common in these circumstances. For example, the treatment previously recommended for the transverse shear stress derivative is

$$\frac{\partial}{\partial y}\left(\mu \frac{\partial u}{\partial y}\right)_j^n = \frac{2}{\Delta y_+ + \Delta y_-}\left(\mu_{j+1/2}^n \frac{u_{j+1}^n - u_j^n}{\Delta y_+} - \mu_{j-1/2}^n \frac{u_j^n - u_{j-1}^n}{\Delta y_-}\right)$$
$$+ O(\Delta y_+ - \Delta y_-) + O(\Delta y_+ + \Delta y_-)^2$$

which at first appears to be first order in accuracy unless there is a way to show that $O(\Delta y_+ - \Delta y_-) = O(\Delta y)^2$ for a particular scheme. Blottner (1974) has shown that the treatment of derivatives indicated above in his Crank-Nicolson scheme using the constant ratio arrangement for mesh spacing is locally second order in accuracy. To prove this, Blottner interpreted the constant ratio scheme in terms of a coordinate transformation (see below) and verified his findings by calculations which indicated that his scheme behaved as though the truncation errors were second order as the mesh was refined.

Use of coordinate transformations. The general topic of coordinate transformations was treated in Chapter 5. Here we are considering the use of a coordinate transformation for the purpose of providing unequal grid spacing in the physical plane. Transformation 1 of Section 5-6 provides a good example of this concept (see also Fig. 5-8). Such a transformation permits the use of standard equal increment differencing of the governing equations in terms of the transformed coordinates. Thus, the clustering of points near the wall can be achieved without deterioration in the *order* of the truncation error. On the other hand, the equations generally become more complex in terms of the transformed variables and new variable coefficients always appear. The actual *magnitude* of the truncation error will be influenced by the new coefficients.

Example transformations 1 and 2 in Section 5-6 of Chapter 5 are representative of those which can be readily used with the boundary-layer equations.

7-3.9 Example Applications

For laminar flows in which the boundary-layer approximation is valid, finite-difference predictions can easily be made to agree with results of more exact theories to several significant figures. Even with only modest attention to mesh size, agreement to within ±1-2% of some "exact" standard is relatively common. Figure 7-6 compares the velocity profile computed by a DuFort-Frankel type difference scheme (Pletcher, 1971) with the analytical results of van Driest (1952) for laminar flow at a Mach number of 4 and $T_w/T_e = 4$. The temperature profiles are compared for the same flow conditions in Fig. 7-7. The agreement is quite good and typical of what can be expected for laminar boundary-layer flows.

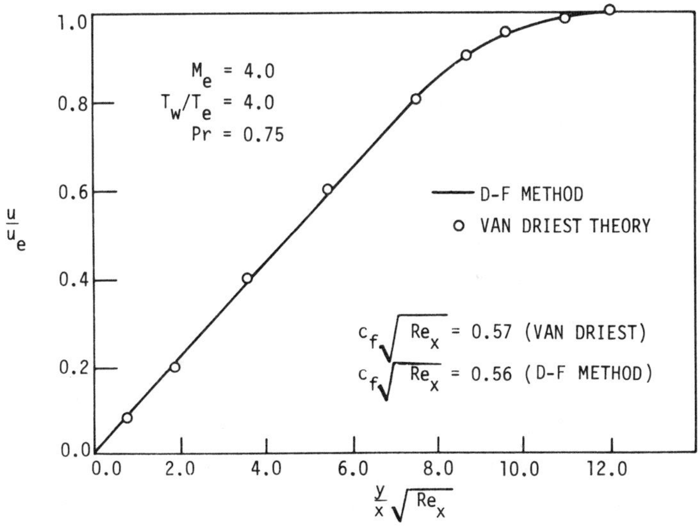

Figure 7-6 Velocity profile comparisons for a laminar compressible boundary layer. Solid line represents predictions from the DuFort-Frankel finite-difference scheme (Pletcher, 1971).

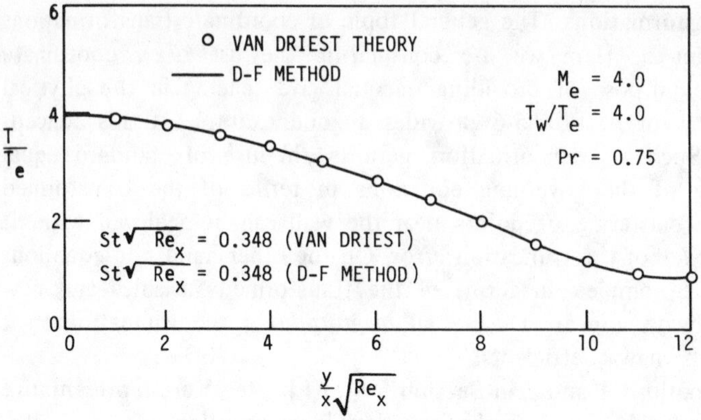

Figure 7-7 Temperature profile comparisons for a laminar compressible boundary layer. Solid line represents predictions from the DuFort-Frankel finite-difference scheme (Pletcher, 1971).

The prediction of turbulent boundary-layer flows is another matter. The issue of turbulence modeling adds complexity and uncertainty to the prediction. Turbulence models can be adjusted to give quite good predictions for a limited class of flows but, when applied to other flows containing conditions not accounted for by the model, poor agreement is often noted. Because of the usual level of uncertainty in both the experimental measurements and turbulence models, agreement to within ±3-4% is generally considered quite good for turbulent flows.

Even a simple algebraic turbulence model can give good predictions over a wide range of Mach numbers for turbulent boundary-layer flows in zero or mild pressure gradients. Figure 7-8 compares the prediction of a Du-Fort-Frankel finite-difference

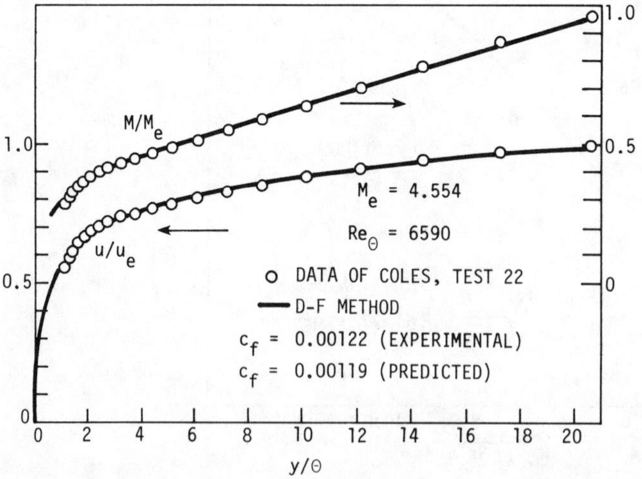

Figure 7-8 Comparison for a compressible flat plate flow measured by Coles (1953). Solid line represents predictions from the DuFort-Frankel finite-difference scheme (Pletcher, 1970).

method with the measurements of Coles (1953) for a turbulent boundary layer on an adiabatic plate at a freestream Mach number of 4.554. The agreement is quite good.

Finite-difference methods easily accommodate step changes in boundary conditions permitting solutions to be obtained for conditions under which simple correlations are especially unreliable. Figure 7-9 compares predictions of an algebraic mixing-length turbulence model used with a DuFort-Frankel type finite-difference procedure with the measurements of Moretti and Kays (1965) for low speed flow over a cooled flat plate with a step change in wall temperature and a favorable pressure gradient. The Stanton number St in Fig. 7-9 is defined as $k(\partial T/\partial y)_w/[\rho_e u_e(H_{aw} - H_w)]$ where H_{aw} is the total enthalpy of the wall under adiabatic conditions.

Examples of cases where predictions of the simplest algebraic turbulence models fail to agree with experimental data abound in the technical literature. Several effects, which are not well-predicted by the simplest models, were cited in Chapter 5. One of these was flow at low Reynolds numbers, especially at supersonic Mach numbers. This low Reynolds number effect is demonstrated in Fig. 7-10 where it can be seen that the point at which the simplest algebraic model (Model A) begins to fail shifts to higher and higher Reynolds numbers as the Mach number of the

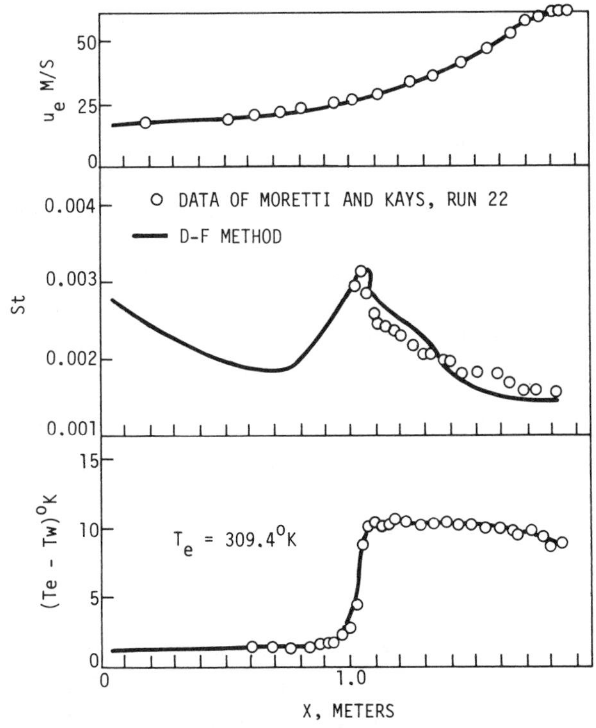

Figure 7-9 Comparison for a cooled flat plate with flow acceleration measured by Moretti and Kays (1965). Solid lines represent predictions from the DuFort-Frankel finite-difference scheme (Pletcher, 1970).

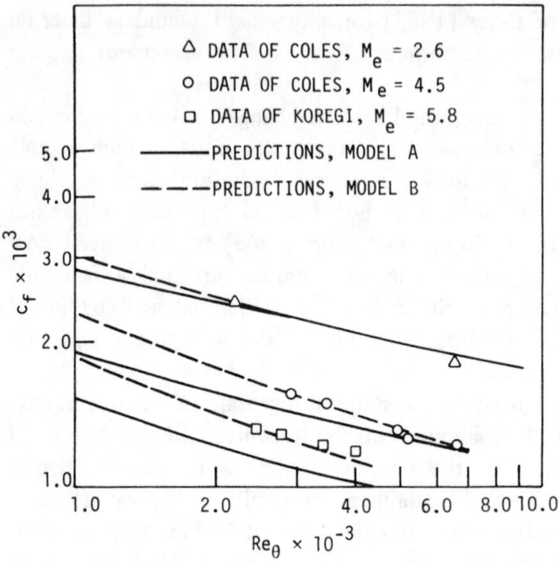

Figure 7-10 Comparison of predicted skin-friction coefficients with the measurements of Coles (1953) and Korkegi (1956) for the compressible turbulent boundary layer on a flat plate at low Reynolds number.

flow increases. Predictions of a model containing the simple modification discussed in Section 5-4.3 for low Reynolds numbers are included in Fig. 7-10 as Model B.

7-3.10 Closure

This section discussed considerations important in the finite-difference solution of the boundary-layer equations for two-dimensional and axisymmetric flows. Several difference schemes have been described. In computational work as in many endeavors, "hands on" participation or "practice" is important. Accordingly, several example problems should be solved using the schemes discussed in order to develop an appreciation of the concepts and issues involved. Just as an engineer could hardly be considered an experimentalist without running an experiment, one likewise should not be considered a computational fluid dynamicist until some computations have been completed.

Which finite-difference scheme is best for the boundary-layer equations? The question is a logical one to be raised at this point, but we need to establish measures by which "best" can be identified. All consistent difference schemes should provide numerical results as accurate as needed with sufficient grid refinement. With ultimate accuracy no longer an issue, the main remaining concerns are computational costs and, to a lesser extent, ease of programming. In the present discussion it is assumed that the user insists on understanding all algebraic operations. We will include the time and effort needed to *understand* a given algorithm as part of the programming effort category. Programming effort then will measure, not the number of statements in a computer program, but the implied algebraic complexity of the steps in the algorithm and the difficulty of following the various steps for the beginner.

A review of the technical literature over the past 10 years suggests that the schemes listed in Table 7-1 have been satisfactorily employed in the solution of the two-dimensional or axisymmetric boundary-layer equations for both laminar and turbulent flow and are recommended based on their well-established performance.

The computation time for a typical calculation for all of the above schemes is expected to be modest (only a few seconds) on present-day computers. More details can be found in the literature cited previously as those methods were discussed. Only a few studies have been reported in which the computer times for several schemes have been compared for the boundary-layer equations. The work of Blottner (1975a) suggests that the Crank-Nicolson scheme with coupling requires about the same time as the modified box scheme for comparable accuracy. Again, for comparable accuracy, Blottner (1975a) found that the box scheme requires two to three times more computer time than the modified box scheme.

For the beginner wishing to establish a general purpose boundary-layer computer program, a reasonable way to start would be with the fully implicit scheme. The scheme is only first-order accurate in the marching direction but second-order accuracy does not appear to be crucial for most boundary-layer calculations. This may be due partly to the fact that the $O(\Delta x)$ term in the truncation error usually includes the second streamwise derivative which is relatively small when the boundary-layer approximation is valid. If second-order accuracy becomes desirable in the streamwise coordinate, it can be achieved with only minor changes through the use of a three-point, second-order representation of the streamwise derivative or by switching to the Crank-Nicolson representation. In increasing order of programming complexity, the logical choices for linearizing the coefficients are lagging, extrapolation, and Newton linearization with coupling. If lagging is adopted as standard, it would be advisable to program one of the latter two more accurate (for the same mesh increment) procedures as an option to provide periodic checks.

7-4 INVERSE METHODS, SEPARATED FLOWS, AND VISCOUS-INVISCID INTERACTION

7-4.1 Introduction

Thus far we have only considered the conventional or "direct" boundary-layer solution methods for the standard equations and boundary conditions given in Section 5-3. An "inverse" calculation method for the boundary-layer equations is a scheme whereby a solution is obtained which satisfies boundary conditions which differ from the standard ones. The usual procedure in an inverse method is to replace the outer boundary condition

$$\lim_{y\to\infty} u(x, y) = u_e(x)$$

by the specification of a displacement thickness or wall shear stress which must be satisfied by the solution. The pressure gradient [or $u_e(x)$] is determined as part of the solution. It should be noted clearly that it is the *boundary conditions* which

Table 7-1 Recommended finite-difference schemes for the boundary-layer equations listed in estimated order of increasing programming effort

1	DuFort-Frankel
2	Fully-implicit (including Patankar-Spalding version)
3	Crank-Nicolson implicit
4	Fully implicit with continuity equation coupling
5	Crank-Nicolson implicit with continuity equation coupling
6	Modified box scheme
7	Box scheme

differ between the conventional direct methods and the inverse methods. It is perhaps more correct to think of the problem specification as being direct or inverse rather than the method. However, we will yield to convention and refer to the solution method as being direct or inverse.

The inverse methods are not merely an alternative way to solve the boundary-layer equations. The successful development of inverse calculation methods has permitted an expansion of the range of usefulness for the boundary-layer approximation.

Clearly, some design applications can be envisioned where it is desirable to calculate the boundary-layer pressure distribution that will accompany a specified distribution of displacement thickness or wall shear stress. This has provided some of the motivation for the development of inverse methods for the boundary-layer equations. Perhaps the most interesting applications of inverse methods have been in connection with separated flows. The computation of separated flows has long been thought to require the solution of the full Navier-Stokes equations. Thus, any suggestion that these flows, which are very important in applications, can be adequately treated with a much simpler mathematical model has been received with great interest. For this reason, the present discussion of inverse methods will emphasize applications to flows containing separated regions. The ability to remove the separation point singularity (Goldstein, 1948) is one of the most unique characteristics of the inverse methods.

7-4.2 Comments on Computing Separated Flows Using the Boundary-Layer Equations

Until fairly recently it was generally thought that the usefulness of the boundary-layer approximation ended as the flow separation point was approached. This was because of the well-known singularity (Goldstein, 1948) of the standard boundary-layer formulation at separation and because the entire boundary-layer approximation is subject to question as the layer thickens and the normal component of velocity becomes somewhat larger (relative to u) than in the usual high Reynolds number flow. It is now known that the inverse formulation is regular at separation (Klineberg and Steger, 1974) and evidence has been accumulating (Williams, 1977;

Kwon and Pletcher, 1979) that the boundary-layer equations provide a useful approximation for flows containing small, confined (bubble) separated regions. In support of the validity of the boundary-layer approximation, it is noted that the formation of a separation bubble normally does not cause the thickness of the viscous region to increase by an order of magnitude; that is, the boundary-layer measure of thinness, $\delta/L \ll 1$ is still met. The "triple-deck theory" of Lighthill (1953) and Stewartson (1974) (see Section 7-4.4) also provides analytical support for the validity of the boundary-layer approximation for large Reynolds number flows containing small separated regions. On the other hand, large local values of $d\delta/dx$ may occur and are expected to induce rather large values of v/u. At best it should be conceded that the boundary-layer model is a weaker approximation for flow containing recirculation even though it may provide estimates of flow parameters accurate enough for many purposes. The full range of applicability of the boundary-layer equations for separated flows is still under study.

Flow separation presents two obstacles to a straightforward space marching solution procedure using conventional boundary conditions with the boundary-layer equations; these are (1) the singularity at separation and (2) the flow reversal which prohibits marching the solution in the direction of the external flow (see Fig. 7-11) unless the convection terms in the equations are altered. When the pressure gradient is fixed near separation by the conventional boundary conditions, the normal component of the velocity and $d\tau_w/dx$ tend toward infinity at the point of separation. A detailed discussion of this singularity can be found in the works of Goldstein (1948) and Brown and Stewartson (1969). This phenomena appears in finite-difference solutions which fix $u_e(x)$ as the tendency for v to increase without limit as the streamwise step size is reduced. This is illustrated in Fig. 7-12 for the Howarth linearly retarded flow (Howarth, 1938). Naturally, a finite v will be obtained for a finite step size, but the solution will not be unique. This singular behavior, which is mathematical rather than physical, can be overcome by the use of an auxiliary pressure interaction relationship with direct methods (Reyhner and Flügge-Lotz, 1968; Napolitano et al., 1978) or by the use of inverse procedures. In this section we will concentrate on the inverse procedures which require no auxiliary relationships to eliminate the singular behavior.

The difficulty with the convective terms can be viewed as follows. We recall that the steady boundary-layer equations are parabolic. For $u > 0$, the solution can

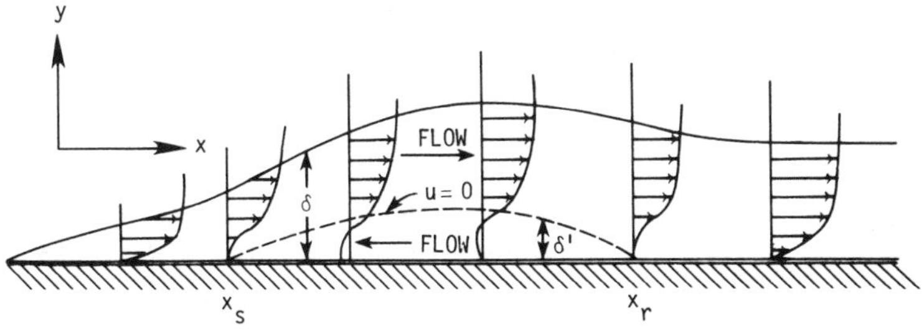

Figure 7-11 Flow containing a separation bubble.

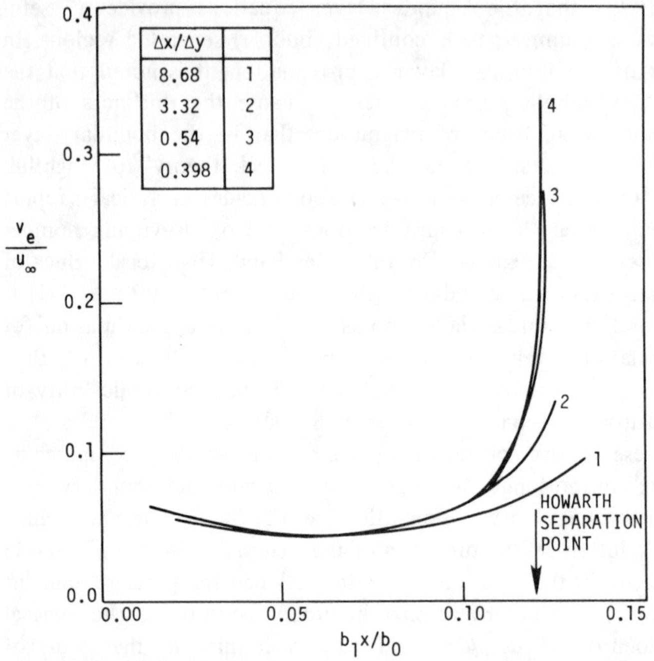

Figure 7-12 Effect of x-grid refinement on v_e for a direct finite-difference boundary-layer calculation (Pletcher and Dancey, 1976) near the separation point for a linearly retarded stream; $u_e = b_0 - b_1 x$, $b_0 = 30.48$ m/s, $b_1 = 300\ \text{s}^{-1}$, $\nu = 1.49 \times 10^{-4}\ \text{m}^2/\text{s}$.

be marched in the positive x direction. Physically, information is carried downstream from the initial plane by the flow. In regions of reversed flow, however, the "downstream" direction is in the negative x direction (Fig. 7-11). Mathematically we observe that when $u < 0$, the boundary-layer momentum equation remains parabolic but the correct marching direction is in the negative x direction.

It would seem, then, that a solution procedure might be devised to overcome the problem associated with the "correct" marching direction by making initial guesses or approximations for the velocities in the reversed flow portion of a flow with a separation bubble, storing these velocities, and correcting them by successive iterative calculation sweeps over the entire flowfield. To do this requires using a difference representation which honors the appropriate marching direction, forward or backward, depending on the direction of flow. To follow this iterative procedure means abandoning the once-through simplicity of the usual boundary-layer approach. Computer storage must also be provided for velocities in and near the region of reversed flow. Such multiple-pass procedures have been employed by Klineberg and Steger (1974), Carter and Wornom (1975) and Cebeci (1976). Some crucial aspects of differencing for multiple-pass procedures will become apparent from the material to be presented in Chapter 8.

Reyhner and Flügge-Lotz (1968) suggested a simpler alternative to the multiple-pass procedure. Noting that the reversed flow velocities are generally quite small for confined regions of recirculation, they suggested that the convective term $u\,\partial u/\partial x$ in the boundary-layer momentum equation be represented in the reversed flow regions by $C|u|\partial u/\partial x$ where C is zero or a small positive constant. This representation has become known as the FLARE approximation and permits the boundary-layer solution to proceed through separated regions by a simple forward marching procedure. It should be clear that the FLARE procedure introduces an additional approximation (or assumption) into the boundary-layer formulation, namely that the $u\,\partial u/\partial x$ term is small relative to other terms in the momentum equation in the region of reversed flow. On the other hand, the FLARE approximation appears to give smooth and plausible solutions for many flows with separation bubbles. Example solutions will be presented in Section 7-4.3. Experimental and computational evidence accumulated to date indicates that for naturally occurring separation bubbles, the u component of velocity in reversed flow regions is indeed fairly small in magnitude, usually less than about 10% of the maximum velocity found in the viscous region.

It should be noted that, although ways of satisfactorily treating the $u\,\partial u/\partial x$ convective term have been presented, it still does not appear possible to obtain a unique, convergent solution of the steady boundary-layer equations alone by a direct marching procedure. Direct calculation procedures reported to date have always employed an interaction relationship whereby the pressure gradient specified becomes dependent upon the displacement thickness (or related parameter) of the viscous regions, usually in a time-dependent manner (Napolitano et al., 1978). This is not necessarily a disadvantage. Viscous-inviscid interaction usually needs to be considered ultimately in obtaining the solution for the complete flowfield containing a separated region, if the boundary-layer equations are used for the viscous regions. Viscous-inviscid interaction will be treated further in Section 7-4.4. On the other hand, we should note that a convergent, unique solution can be obtained for the steady boundary-layer equations alone using inverse methods.

7-4.3 Inverse Finite-Difference Methods

Two procedures will be illustrated. The first is conceptually the simplest and is especially useful for illustrating the concept of the inverse method. It appears to work very well when the flow is attached (no reversed flow region), but gives rise to small controlled oscillations in the skin-friction when reversed flow is present. This oscillatory behavior is overcome by the second method which solves the boundary-layer equations in a coupled manner. The FLARE approximation will be employed in both of these methods. For simplicity, the methods will be illustrated for incompressible flows.

Inverse Method A. The boundary-layer equations are written as follows:

continuity:

$$\frac{\partial u}{\partial x} + \frac{\partial v}{\partial y} = 0 \tag{7-56}$$

momentum:

$$C|u|\frac{\partial u}{\partial x} + v\frac{\partial u}{\partial y} = u_e\frac{du_e}{dx} + \frac{1}{\rho}\frac{\partial \tau}{\partial y} \tag{7-57}$$

In the above, $C = 1.0$ when $u > 0$ and C is small ($\leqslant 0.2$) positive constant when $u \leqslant 0$ and

$$\tau = \mu\frac{\partial u}{\partial y} - \rho\overline{u'v'} = (\mu + \mu_T)\frac{\partial u}{\partial y} \tag{7-58}$$

The above equations are in a form applicable to either laminar or turbulent flow. For laminar flow, the primed velocities and μ_T are zero, and for turbulent flow, the unprimed velocities are time-mean quantities.

The boundary conditions for the inverse procedure are

$$u(x, 0) = v(x, 0) = 0 \tag{7-59}$$

and

$$\int_0^\infty \left(1 - \frac{u}{u_e}\right) dy = \delta^*(x) \tag{7-60}$$

where δ^* is a prescribed function. Alternatively, $\tau_w(x)$ can be specified as a boundary condition. Clearly Eqs. (7-56) and (7-57) can be solved by a direct method utilizing the conventional boundary condition

$$\lim_{y\to\infty} u(x, y) = u_e(x) \tag{7-61}$$

in place of Eq. (7-60) for attached portions of the flow. It is possible to start a boundary-layer calculation in the direct mode and switch to the inverse procedure when desired.

The boundary-layer equations are cast into a fully implicit difference form and the coefficients lagged. Such a difference representation was discussed in Section 7-3 and will not be repeated here.

The inverse treatment of boundary conditions is implemented by varying u_e in successive iterations at each streamwise calculation station until the solution satisfies the specified value of $\delta^*(x)$. In each of these iterations the numerical formulation and implementation of boundary conditions is the same as for a direct method. The displacement thickness is evaluated from the computed velocity distribution by numerical integration (use of either Simpson's rule or the trapezoidal rule is suggested). The appropriate value of u_e needed to satisfy the boundary condition on δ^* (δ^*_{BC}) is determined by considering $\delta^* - \delta^*_{BC}$ to be a function of u_e at each streamwise station, $\delta^* - \delta^*_{BC} = F(u_e)$, and seeking the value of u_e required to establish $F = 0$ by a variable secant (Fröberg, 1969) procedure. In the above, δ^* is the δ^* actually obtained from the solution for a specified value of u_e. Two initial guesses are required for this procedure which usually converges in three or four iterations (Pletcher, 1978).

The variable secant procedure can be thought of as a generalization of Newton's method (also known as the Newton-Raphson method) for finding the root of $F(x)=0$. In Newton's method we expand $F(x)$ in a Taylor series about a reference point x_n:

$$F(x_n + \Delta x) = F(x_n) + F'(x_n)\,\Delta x + \cdots$$

We truncate the series after the first derivative term and compute the value of Δx required to establish $F(x_n + \Delta x) = 0$. For Newton's method this gives

$$x_{n+1} - x_n = \Delta x = -\frac{F(x_n)}{F'(x_n)} \tag{7-62}$$

Thus, starting with an initial guess x_n, an improved approximation, x_{n+1}, can be computed from Eq. (7-62). The process is repeated iteratively until $|(x_{n+1} - x_n)| < \epsilon$.

Newton's method is a simple and effective procedure. Its use does, however, require that $F'(x)$ be evaluated analytically. When this is not possible, the variable secant generalization of the Newton procedure represents a reasonable alternative.

In the variable secant procedure, the derivative is replaced by a secant line approximation through two points

$$F'(x_n) \simeq \frac{F(x_n) - F(x_{n-1})}{x_n - x_{n-1}}$$

After two initial guesses for x, the third approximation to the root is obtained from

$$x_{n+1} = x_n - \frac{F(x_n)}{F(x_n) - F(x_{n-1})}\,(x_n - x_{n-1}) \tag{7-63}$$

In the application of the variable secant method to the inverse boundary-layer calculation, x_n becomes $u_{e,n}$ and $F = \delta^* - \delta^*_{BC}$. The iterative process is illustrated in Fig. 7-13.

When the iterative search for the $u_e(x)$ which provides the specified $\delta^*(x)$ is completed, the solution may be advanced to another streamwise station in the usual manner for parabolic equations. The simplicity of inverse method A is obvious. Apart from small changes to implement the FLARE representation, the difference equations are solved just as for the standard direct method for boundary layers. The method performs reasonably well (Pletcher, 1978; Kwon and Pletcher, 1979) but does predict small oscillations in the wall shear stress when separation is present. These oscillations can be eliminated by suitable coupling of the momentum and continuity equations and are not present in solutions obtained by the method described below.

Inverse Method B. Here we will describe the method developed by Kwon and Pletcher (1981). The overall strategy of this method is to couple all unknowns and the boundary conditions in one simultaneous system of algebraic equations to be solved at each streamwise station. To do so it is convenient to introduce the stream function, ψ. Accordingly,

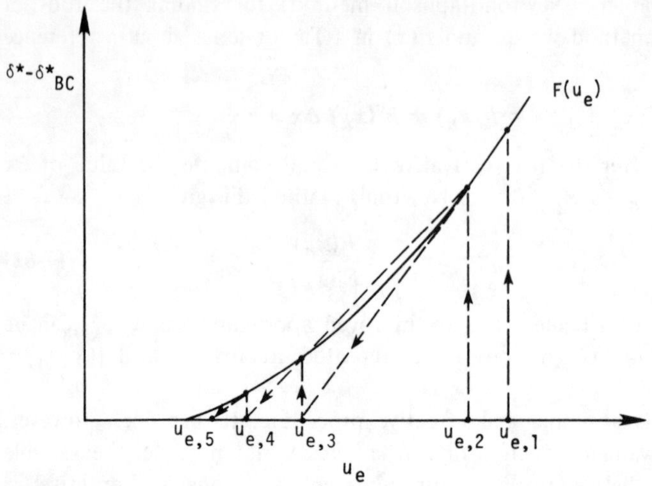

Figure 7-13 Determination of $u_e(x)$ through the use of the variable secant procedure.

$$u = \frac{\partial \psi}{\partial y}$$

and
$$v = -\frac{\partial \psi}{\partial x}$$

The conservation equations for mass and momentum are written as

$$u = \frac{\partial \psi}{\partial y} \tag{7-64}$$

$$Cu\,\frac{\partial u}{\partial x} - \frac{\partial \psi}{\partial x}\,\frac{\partial u}{\partial y} = u_e\,\frac{du_e}{dx} + \frac{1}{\rho}\,\frac{\partial \tau}{\partial y} \tag{7-65}$$

where
$$\tau = \bar{\mu}\,\frac{\partial u}{\partial y}$$

and
$$\bar{\mu} = \mu + \mu_T$$

The boundary conditions are

$$u(x, 0) = \psi(x, 0) = 0 \tag{7-66}$$

and
$$\psi_e = u_e(y_e - \delta^*(x)) \tag{7-67}$$

where $\delta^*(x)$ is a prescribed function. The boundary condition for ψ_e follows from the definition of δ^*:

$$\delta^* = \int_0^\infty \left(1 - \frac{u}{u_e}\right) dy$$

The upper limit of this integral can be replaced by y at the outer edge of the boundary layer, y_e, since the integrand is equal to zero for $y > y_e$. Multiplying by u_e gives

$$u_e \delta^* = u_e y_e - \int_0^{y_e} u\, dy$$

Expressing u in terms of the stream function permits the integral to be evaluated as ψ_e. Rearranging gives Eq. (7-67). When the difference equations below are solved in a direct mode, the outer boundary condition becomes the conventional one given by Eq. (7-61) instead of that specified by Eq. (7-67).

Equations (7-64) and (7-65) are first represented in finite-difference form as

$$\frac{u_j^{n+1} + u_{j-1}^{n+1}}{2} = \frac{\psi_j^{n+1} - \psi_{j-1}^{n+1}}{\Delta y_-} \tag{7-68}$$

$$Cu_j^{n+1} \frac{u_j^{n+1} - u_j^n}{\Delta x} - \frac{\psi_j^{n+1} - \psi_j^n}{\Delta x} \frac{u_{j+1}^{n+1} - u_{j-1}^{n+1}}{\Delta y_+ + \Delta y_-} = \chi^{n+1} + \frac{2}{\rho(\Delta y_+ + \Delta y_-)} \times \left(\bar{\mu}_{j+1/2} \frac{u_{j+1}^{n+1} - u_j^{n+1}}{\Delta y_+} - \bar{\mu}_{j-1/2} \frac{u_j^{n+1} - u_{j-1}^{n+1}}{\Delta y_-} \right) \tag{7-69}$$

In the above

$$C = 1 \quad \text{when} \quad u_j^{n+1} > 0$$

and

$$C = 0 \quad \text{when} \quad u_j^{n+1} < 0$$

$$\chi = -\frac{1}{\rho} \frac{dp}{dx}$$

Newton linearization is next applied to the above nonlinear convective terms following the procedures presented in Section 7-3.3. We let $u_j^{n+1} = \hat{u}_j^{n+1} + \delta_u$ and $\psi_j^{n+1} = \hat{\psi}_j^{n+1} + \delta_\psi$ where the carets indicate provisional values of the variables in an iterative process. The quantities δ_u and δ_ψ are the changes in the variables between two iterative sweeps, i.e., $\delta_\phi = \phi_j^{n+1} - \hat{\phi}_j^{n+1}$ for a general variable ϕ. The resulting difference equations can be written in the form

$$\psi_{j-1}^{n+1} - \psi_j^{n+1} + b_j(u_{j-1}^{n+1} + u_j^{n+1}) = 0 \tag{7-70}$$

$$B_j u_{j-1}^{n+1} + D_j u_j^{n+1} + A_j u_{j+1}^{n+1} + E_j \psi_j^{n+1} = H_j \chi^{n+1} + C_j \tag{7-71}$$

where

$$A_j = -\frac{\hat{\psi}_j^{n+1} - \psi_j^n}{\Delta x(\Delta y_+ + \Delta y_-)} - \frac{2\bar{\mu}_{j+1/2}}{\rho\, \Delta y_+(\Delta y_+ + \Delta y_-)}$$

$$B_j = \frac{\hat{\psi}_j^{n+1} - \psi_j^n}{\Delta x(\Delta y_+ + \Delta y_-)} - \frac{2\bar{\mu}_{j-1/2}}{\rho\, \Delta y_-(\Delta y_+ + \Delta y_-)}$$

$$C_j = \frac{C(\hat{u}_j^{n+1})^2}{\Delta x} - \frac{\hat{\psi}_j^{n+1}(\hat{u}_{j+1}^{n+1} - \hat{u}_{j-1}^{n+1})}{\Delta x(\Delta y_+ + \Delta y_-)}$$

$$D_j = \frac{C(2\hat{u}_j^{n+1} - u_j^n)}{\Delta x} + \frac{2}{\rho(\Delta y_+ + \Delta y_-)} \left(\frac{\bar{\mu}_{j+1/2}}{\Delta y_+} + \frac{\bar{\mu}_{j-1/2}}{\Delta y_-} \right)$$

$$E_j = -\frac{\hat{u}_{j+1}^{n+1} - \hat{u}_{j-1}^{n+1}}{\Delta x(\Delta y_+ + \Delta y_-)}$$

$$H_j = 1$$

$$b_j = \frac{\Delta y_-}{2}$$

The above algebraic formulation is similar to that presented in Section 7-3 in connection with the Davis coupled scheme and solved by the modified Thomas algorithm. Equations (7-70) and (7-71) form a block-tridiagonal system with 2 X 2 blocks and require the simultaneous solution of $2(NJ) - 2$ equations for $2(NJ) - 2$ unknowns at each streamwise marching step. The parameter NJ is the number of grid points across the flow, including boundary points. One difference between this formulation and the algebraic equations arising from the Davis coupled scheme is the appearance of the new term, $H_j\chi^{n+1}$, on the right-hand side of Eq. (7-71). The pressure gradient parameter χ^{n+1} is one of the unknowns in the inverse formulation. The outer boundary conditions are also different. These facts preclude the use of the modified tridiagonal algorithm presented in Section 7-3. However, the blocks below the main diagonal can be eliminated and a recursion formula developed (Kwon and Pletcher, 1981) for the back substitution. Before the back substitution is carried out, however, the parameter χ^{n+1} must be determined by a special procedure to be indicated subsequently.

The unknowns can be computed from

$$u_j^{n+1} = A_j' u_{j+1}^{n+1} + H_j' \chi^{n+1} + C_j' \tag{7-72}$$

$$\psi_j^{n+1} = B_j' u_{j+1}^{n+1} + D_j' \chi^{n+1} + E_j' \tag{7-73}$$

providing the coefficients A_j', H_j', C_j', B_j', D_j', E_j' and the quantities u_{j+1}^{n+1} and χ^{n+1} are known a priori. The coefficients are given by

$$A_j' = -\frac{A_j}{R_1}$$

$$B_j' = A_j' R_2$$

$$C_j' = \frac{C_j - B_j C_{j-1}' - E_j(b_j C_{j-1}' + E_{j-1}')}{R_1}$$

$$D_j' = b_j H_{j-1}' + D_{j-1}' + H_j' R_2$$

$$E_j' = b_j C_{j-1}' + E_{j-1}' + C_j' R_2$$

$$H_j' = \frac{H_j - B_j H_{j-1}' - E_j(b_j H_{j-1}' + D_{j-1}')}{R_1}$$

$$R_1 = D_j + (B_j + E_j b_j) A_{j-1}' + E_j(B_{j-1}' + b_j)$$

$$R_2 = b_j(1 + A_{j-1}') + B_{j-1}'$$

Since the inner ($j = 1$) boundary conditions on u_j^{n+1} and ψ_j^{n+1} are zero, the coefficients $A_1', B_1', C_1', D_1', E_1', H_1'$, are also zero and the coefficients above can be computed started from $j = 2$ and continuing to the outer boundary ($j = NJ$).

The pressure gradient parameter χ^{n+1} is evaluated by simultaneously solving the equations obtained from Eqs. (7-72) and (7-73) by replacing j with $NJ - 1$ and the boundary conditions in the following manner. At $j = NJ - 1$ Eqs. (7-72) and (7.73) become

$$u_{NJ-1}^{n+1} = A_{NJ-1}' u_{NJ}^{n+1} + H_{NJ-1}' \chi^{n+1} + C_{NJ-1}' \tag{7-74}$$

$$\psi_{NJ-1}^{n+1} = B_{NJ-1}' u_{NJ}^{n+1} + D_{NJ-1}' \chi^{n+1} + E_{NJ-1}' \tag{7-75}$$

The boundary conditions are written as

$$\psi_{NJ}^{n+1} = u_{NJ}^{n+1}(y_{NJ} - \delta^{*n+1}) \tag{7-76}$$

and

$$\chi^{n+1} = \frac{1}{\Delta x}\left[(2\hat{u}_{NJ}^{n+1} - u_{NJ}^{n})u_{NJ}^{n+1} - (\hat{u}_{NJ}^{n+1})^2\right] \tag{7-77}$$

Equation (7-68) is written as

$$\psi_{NJ}^{n+1} = \psi_{NJ-1}^{n+1} + \frac{\Delta y_-}{2}(u_{NJ}^{n+1} + u_{NJ-1}^{n+1}) \tag{7-78}$$

Solving Eqs. (7-74)-(7-78) for χ^{n+1} gives

$$\chi^{n+1} = \frac{(F_3/F_1)(2\hat{u}_{NJ}^{n+1} - u_{NJ}^{n}) - (u_{NJ}^{n+1})^2}{\Delta x - (F_2/F_1)(2\hat{u}_{NJ}^{n+1} - u_{NJ}^{n})} \tag{7-79}$$

where

$$F_1 = y_{NJ} - \delta^{*n+1} - B_{NJ-1}' - \frac{\Delta y_-}{2}(1 + A_{NJ-1}')$$

$$F_2 = D_{NJ-1}' + \frac{\Delta y_-}{2} H_{NJ-1}'$$

$$F_3 = E_{NJ-1}' + \frac{\Delta y_-}{2} C_{NJ-1}'$$

Once the pressure gradient parameter χ^{n+1} is determined, the edge velocity u_{NJ}^{n+1} can be calculated using Eqs. (7-74)-(7-78) as

$$u_{NJ}^{n+1} = \frac{F_2}{F_1}\chi^{n+1} + \frac{F_3}{F_1} \tag{7-80}$$

Then, ψ_{NJ}^{n+1} can be computed directly from Eq. (7-76). Now the back substitution process can be initiated using Eqs. (7-72) and (7-73) to compute u_j^{n+1} and ψ_j^{n+1} from the outer edge to the wall. The Newton linearization requires that the system of equations be solved iteratively with $\hat{u}_j^{n+1}$ and $\hat{\psi}_j^{n+1}$ being updated between iterations. The iterative process is continued at each streamwise location until the maximum change, in u's and ψ's between two successive iterations, is less than some predetermined tolerance. The calculation is initiated at each streamwise station by setting $\hat{u}_j^{n+1} = u_j^n$ and $\hat{\psi}_j^{n+1} = \psi_j^n$. In previous applications of this method, only two or three iterations were generally required for the maximum fractional change in the variables, i.e., $\Delta\phi/\phi$, to be reduced to 5×10^{-4}.

Details of other somewhat different coupled inverse boundary-layer finite-difference procedures employing the FLARE approximation can be found in the work of Cebeci (1976) and Carter (1978).

7-4.4 Viscous-Inviscid Interaction

In design it is common to obtain the pressure distribution about aerodynamic bodies from an inviscid flow solution. The inviscid flow solution then provides the edge velocity distribution needed as a boundary condition for solving the boundary-layer equations to obtain the viscous drag on the body. In many cases, the presence of the viscous boundary layer only slightly modifies the flow pattern over the body. It is possible to obtain an improved inviscid flow solution by augmenting the physical thickness of the body by the boundary-layer displacement thickness. The definition of δ^* is such that the new inviscid flow solution properly accounts for the displacement of the inviscid flow caused by the viscous flow near the body. The improved inviscid edge velocity distribution can then be used to obtain yet another viscous flow solution. In principle, this viscous-inviscid interaction procedure can be continued iteratively until changes are small. In practice, however, severe under-relaxation of the changes from one iterative cycle to another is often required for convergence.

Fortunately, for most flows involving an attached boundary layer the changes which arise from accounting for the viscous-inviscid interaction are negligibly small and it suffices for engineering design purposes to compute the inviscid and viscous flows independently (i.e., without considering viscous-inviscid interaction). Flows which separate or contain separation bubbles are a notable exception.

The displacement effect of the separated regions locally alter the pressure distribution in a significant manner. A rapid thickening of the boundary layer under the influence of an adverse pressure gradient even without separation can also alter the pressure distribution locally to the extent that a reasonable flow solution cannot be obtained without accounting for the displacement effect of the viscous flow. Often under such conditions, a boundary-layer calculation obtained using the edge velocity distribution from an inviscid flow solution which neglects the displacement effect will predict separation when the real flow does not separate at all.

It is often possible to confine the region where viscous-inviscid interaction effects are important to the local neighborhood of the "bulge" in the displacement surface. Such a local interaction region is depicted in Fig. 7-14. The inverse boundary procedure described previously in Section 7-4.3 is particularly well-suited for flows in which separation may occur.

The essential elements of a viscous-inviscid interaction calculation procedure are the following:

1. A method for obtaining an improved inviscid flow solution which provides a pressure distribution or edge velocity distribution which accounts for the viscous

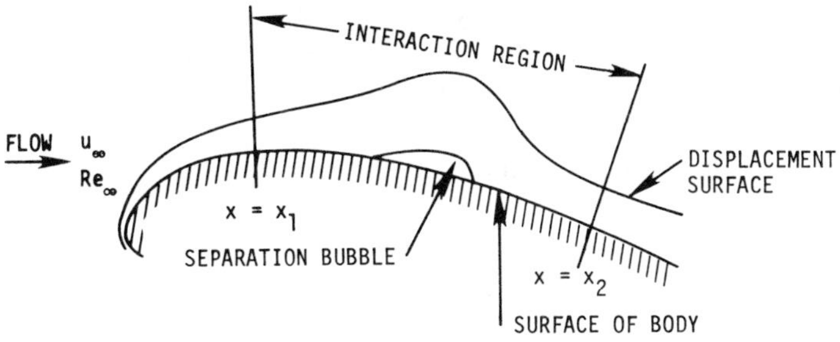

Figure 7-14 Local interaction region on a two-dimensional body.

flow displacement effect. In principle any inviscid flow "solver" could be used but it is also frequently possible to employ a greatly simplified inviscid flow calculation scheme based on a small disturbance approximation.

2. A technique for obtaining a solution to the boundary-layer equations suitable for the problem at hand. For a flow which may separate, an inverse boundary-layer procedure would be appropriate.
3. A procedure for relating the inviscid and viscous flow solutions in a manner which will drive the changes from one iterative cycle to the next toward zero.

Over the years numerous viscous-inviscid interaction schemes have been proposed. It will not be possible to discuss all of these here. Instead, we will summarize a suitable approach for predicting the flow in the neighborhood of a separation bubble on an airfoil in incompressible flow. This configuration is illustrated in Fig. 7-14.

For this case, a good estimate of the effect of the displacement correction for the inviscid flow solution can be obtained by the use of a small disturbance approximation. We let $u_{e,o}$ denote the tangential component of velocity of the inviscid flow over the solid body (neglecting all effects of the viscous flow) and u_c be the velocity on the displacement surface induced only by the sources and sinks distributed on the surface of the body due to the displacement effect of the viscous flow in the interaction region. Then, the x component of velocity of a fluid particle on the displacement surface can be written as

$$u_e = u_{e,o} + u_c \tag{7-81}$$

Following Lighthill (1958) the intensity of the line source or sinks displacing a streamline at the displacement surface of the viscous flow can be evaluated as

$$q = \frac{d(u_e\delta^*)}{dx} \tag{7-82}$$

For small values of δ^*, u_c can be evaluated from the Hilbert integral

$$u_c(x) = \frac{1}{\pi}\int_{-\infty}^{\infty} \frac{d(u_e\delta^*)}{dx'}\,\frac{dx'}{x-x'} \tag{7-83}$$

In the numerical computation of u_c it is usually assumed that strong interaction is limited to the region $x_1 \leqslant x \leqslant x_2$ shown in Fig. 7-14. The intensity of the source or sink caused by the viscous displacement is assumed to approach zero as x approaches $\pm\infty$. Consequently, $d(u_e\delta^*/dx)$ is normally only computed in the region $x_1 < x < x_2$ using the boundary-layer solution. An arbitrary extrapolation of the form (Kwon and Pletcher, 1979)

$$q'(x) = \frac{b}{x^2} \tag{7-84}$$

is often used for the regions $x < x_1$ and $x > x_2$ in order to evaluate the integral in Eq. (7-83). The constant b is chosen to match the q obtained from the boundary-layer solution at x_1 and x_2. Equation (7-83) can now be written as

$$u_c(x) = \frac{1}{\pi}\left[\int_{-\infty}^{x_1} \frac{q'(x')}{x - x'}\,dx' + \int_{x_1}^{x_2} \frac{q(x')}{x - x'}\,dx' + \int_{x_2}^{\infty} \frac{q'(x')}{x - x'}\,dx'\right] \tag{7-85}$$

The first and third integrals can be evaluated analytically. The second integral is evaluated numerically, normally using the trapezoidal rule. The singularity at $x = x'$ can be isolated using the procedure found in Jobe (1974). Some authors have found it possible to evaluate the integral numerically with no special attention given to the singularity as long as $(x - x')$ remained finite [Briley and McDonald (1975)].

The inviscid surface velocity on the solid body (neglecting the boundary layer), $u_{e,o}$, can be obtained by the methods cited in Chapter 6 [as, for example, the Hess and Smith (1967) method], or from experimental data. The Hess and Smith procedure could be used iteratively for all of the inviscid flow calculations. However, the relatively simple small disturbance procedure requires significantly less computer time and has been found to provide sufficient accuracy for incompressible viscous-inviscid interaction calculations of a type which permits the use of the boundary-layer equations for the viscous flow.

The inverse boundary-layer procedures discussed in Section 7-4.3 are quite suitable for computing the viscous portion of the flow which may include separated regions. The iterative updating of the solutions can be effectively carried out by the method successfully demonstrated by Carter (1978) and Kwon and Pletcher (1979).

The interaction calculation proceeds in the following way. First, u_{eo} is obtained for the body of interest and the viscous flow is computed up to the beginning of the interaction region by a conventional direct method. These two solutions do not change. Next, an initial $\delta^*(x)$ distribution is chosen over the region $x_1 < x < x_2$ (see Fig. 7-14). The initial guess is purely arbitrary but should match the $\delta^*(x)$ of the boundary layer computed by the direct method at $x = x_1$. The boundary-layer solution is next obtained by an inverse procedure using this $\delta^*(x)$ as a boundary condition. An edge velocity distribution $u_{e,BL}(x)$ is obtained as an output.

Now the small disturbance inviscid flow procedure, Eq. (7-85), is used to compute the correction to inviscid flow velocity. This establishes a new distribution for edge (surface) velocity distribution $u_{e,\text{inv}}(x)$. The $u_e(x)$ from the two calculations, boundary layer and inviscid, will not agree until convergence has been achieved. The difference between $u_e(x)$ calculated both ways can be used as a potential to calculate an improved distribution for $\delta^*(x)$. To do this formally, one would seek to determine the way in which a change in u_e would influence δ^*. A suitable scheme has been developed for subsonic flows by noting that a response to small excursions in local u_e tends to preserve the volume flow rate per unit width in the boundary layer, i.e., $u_e\delta^* \simeq$ constant. This implies that a local decrease in $u_e(x)$ (associated with a more adverse pressure gradient) causes an increase in $\delta^*(x)$ and a local increase in $u_e(x)$ (associated with a more favorable pressure gradient) causes a decrease in $\delta^*(x)$. This concept is put into practice by computing the appropriate new distribution of δ^* (Carter, 1978) to use for a new pass through the boundary-layer calculation by

$$\delta^*_{k+1} = \delta^*_k \left(\frac{u_{e,BL_k}}{u_{e,\text{inv}_k}}\right) \tag{7-86}$$

where k denotes iteration level. It is important to note that Eq. (7-86) only serves as a basis for correcting δ^* between iterative passes so that no formal justification for its use is required so long as the iterative process converges. At convergence $u_{e,BL} = u_{e,\text{inv}}$; thus, Eq. (7-86) represents an identity thereby having no effect on the final solution. In this sense, the use of Eq. (7-86) is somewhat like the use of an arbitrary over-relaxation factor in the numerical solution of an elliptic equation by SOR. Carter (1978) has given a somewhat more formal justification of Eq. (7-86) based on the von Kármán momentum integral.

The viscous-inviscid interaction calculation is completed by making successive passes first through the inverse boundary-layer scheme, then through the inviscid flow procedure with δ^* being computed by Eq. (7-86) prior to each boundary-layer calculation. When $|u_{e,BL} - u_{e,\text{inv}}|$ is less than a prescribed tolerance, convergence is considered to have been achieved. In some applications of this matching procedure, successive over-relaxation of δ^* in Eq. (7-86) has been observed to speed convergence. An illuminating discussion of several matching procedures can be found in the paper by Wigton and Holt (1981).

Some example predictions (Kwon and Pletcher, 1979) are shown in Figs. 7-15 and 7-16 for the flow in the neighborhood of a transitional separation bubble on a NACA 66_3-018 airfoil. The parameter Tu in the figures is the freestream turbulence level and Re_c is the Reynolds number based on the airfoil chord. Figure 7-15 compares the predicted pressure coefficient with measurements. The dashed line in Fig. 7-15 indicates the pressure coefficient predicted by inviscid flow theory neglecting the presence of the boundary layer. In the neighborhood of the separation bubble centered at $s/c \simeq 0.7$ (s is the distance along the airfoil surface measured from the leading edge and c is the chord) this predicted pressure coefficient is seen to be considerably in error compared to the measurements. The solid line indicates the prediction of a viscous-inviscid interaction procedure which is seen to follow the trend of the measurements fairly closely. Seventeen passes through the viscous-inviscid procedure were required for convergence in this case. Velocity profiles are compared in Fig. 7-16. Reversed flow is evident from the

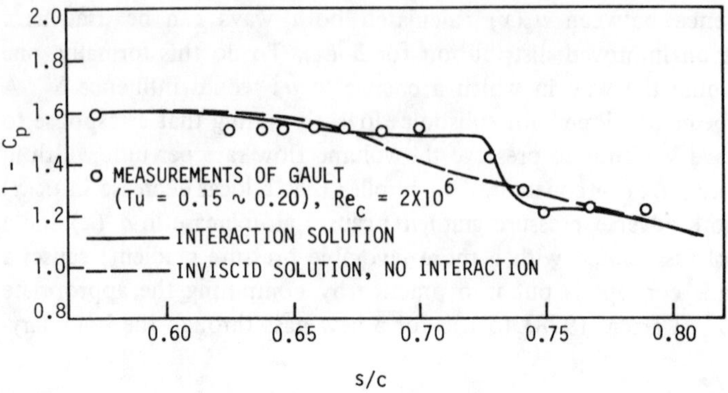

Figure 7-15 Comparison of the predicted pressure distribution with experimental data (Gault, 1955) for a NACA 66_3-018 airfoil at zero angle of incidence.

profiles in the vicinity of $s/c \simeq 0.7$. The predicted results are quite sensitive to the model used for laminar-turbulent transition.

The same general strategy outlined above for viscous-inviscid interaction calculations has also been found to work well for compressible flows including transonic and supersonic applications (Carter, 1981; Werle and Verdon, 1979). When the flow becomes compressible, the boundary-layer form of the energy equation is solved in the viscous flow region usually with the use of the FLARE approximation. The solution procedure used for the inviscid flow normally varies as the flow regime changes. A relaxation solution of the full potential equation for inviscid flow as used by Carter (1981) in his transonic viscous-inviscid interaction calculations. For

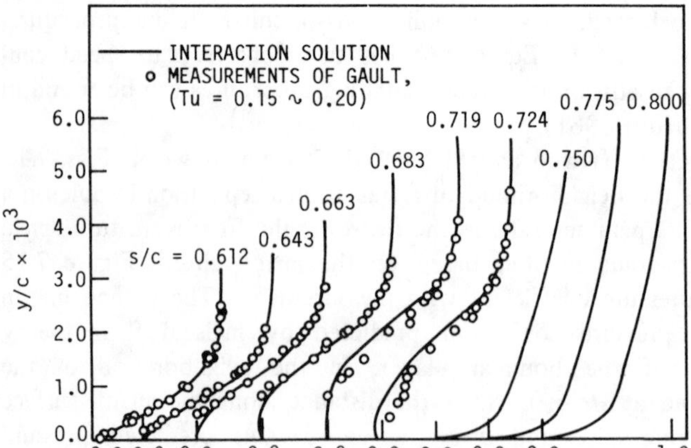

Figure 7-16 Comparison of the predicted mean velocity profiles with experimental data (Gault, 1955) for NACA 66_3-018 airfoil at zero angle of attack.

fully supersonic streams, the concept of a small disturbance approximation (linearized theory) is again useful and the component of the pressure gradient attributed to viscous displacement can be related to the second derivative of the boundary-layer displacement thickness in a very simple manner. The exact form of the appropriate pressure gradient relationship varies somewhat with the application considered. The reader is referred to the work of Werle and Vatsa (1974) and Burggraf et al. (1979) for specific examples. Despite the fact that the pressure gradient depends upon local quantities in the case of a supersonic external stream, a downstream condition must be imposed (usually it is on δ^*) in order to obtain a unique solution.

Various time-dependent interaction schemes have also been successfully applied to both subsonic and supersonic flows (Briley and McDonald, 1975; Werle and Vatsa, 1974).

Mention is often made of "triple-deck theory" or "triple-deck structure" in connection with viscous-inviscid interactions. It is natural to wonder if this theory introduces something which ought to be taken into account by those applying finite-difference methods to viscous-inviscid interaction problems. The theory itself is based on a multistructured asymptotic expansion valid as $\mathrm{Re} \to \infty$ for laminar flow in the neighborhood of a perturbation to a boundary-layer flow such as would occur due to small separated regions or near the trailing edge of a flat plate. We will primarily concentrate on the application of triple-deck theory to the small separation problem.

Several individuals have contributed to the theory. Some of the early concepts were introduced by Lighthill (1958). Stewartson and co-workers have made several contributions. An excellent review of developments in the theory up through 1974 is given by Stewartson (1974).

The theory is applicable if the streamwise length of the disturbance is relatively short. Thus, the theory would be applicable to small separation bubbles, but not to catastrophic separation. The length of perturbation region where the triple-deck analysis would be applicable is of order $\mathrm{Re}^{-3/8}$ where Re is the Reynolds number based on the origin of the boundary layer. The "decks" in the theory are flow regions measured normal to the wall. The thickness of the lower deck is of order $\mathrm{Re}^{-5/8}$. The flow in this thin lower region has very little inertia so that it responds quite readily to disturbances transmitted by the pressure gradient. The thickness of the middle (main) deck is of order $\mathrm{Re}^{-1/2}$. The flow in this region is essentially a streamwise continuation of the upstream boundary-layer flow and is predominantly rotational and inviscid. All flow quantities in this region are only perturbed slightly from those in a conventional noninteracting boundary layer. The disturbances being transmitted by the lower deck displace the main deck boundary layer outward. The upper deck is of order $\mathrm{Re}^{-3/8}$ in thickness. The upper deck flow is the perturbed part of the inviscid, irrotational flow.

Triple-deck theory provides the equations and boundary conditions needed to match the solutions in each of the three regions. The results are only valid for laminar flows where $\mathrm{Re} \to \infty$ so in a sense are of limited practical value. These equations are frequently solved numerically using viscous-inviscid interaction procedures (Jobe and Burggraf, 1974).

To the computational fluid dynamicist, the most important ideas and conclusions that come from the development of the triple-deck theory to date are:

1. The equations which result from the triple-deck theory as applied to flows containing small perturbations (such as small closed separated regions and trailing edge flows) contain no terms which are not present in the boundary-layer viscous-inviscid interaction model. This tends to confirm that the boundary-layer viscous-inviscid interaction model is correct in the limit as $\mathrm{Re} \to \infty$. Normal pressure gradients are neglected in the triple-deck theory when applied to the class of flows being considered here.
2. Triple-deck theory identifies length scales which can prove useful in finite-difference computations for laminar flows. The theory predicts that the lower deck is of order $\mathrm{Re}^{-5/8}$ in thickness. Although this conclusion is only strictly valid in the limit as $\mathrm{Re} \to \infty$, it would appear prudent to use a mesh near the wall sufficiently fine as to resolve this lower deck region where pressure variations can have a fairly drastic effect on the flow. The importance of honoring this scaling is confirmed by the finite-difference study made by Burggraf et al. (1979).
3. The theory provides clear evidence that the supersonic separation problem is boundary value in nature requiring a downstream boundary condition in order to select a unique solution from the branching solutions which might otherwise be obtained. This requirement is not immediately obvious in the supersonic case because the boundary-layer equations themselves are parabolic and according to linearized theory, the pressure depends only on the *local* slope of the displacement body. The downstream boundary condition is usually invoked as a prescribed value of the displacement thickness.

The paper by Burggraf et al. (1979) is helpful in clarifying the differences between the application of boundary-layer viscous-inviscid interaction schemes and the numerical solution of the triple-deck equations. At very large Reynolds numbers (10^9) the boundary-layer viscous-inviscid interaction calculation agreed very well with the triple-deck results for separating supersonic flow past a compression ramp. As the Reynolds number decreased, the predictions of the boundary-layer viscous-inviscid interaction procedure and the triple-deck results differed very noticeably.

7-5 METHODS FOR INTERNAL FLOWS

7-5.1 Introduction

The thin-shear-layer equations provide a reasonably accurate mathematical model for two-dimensional and axisymmetric internal flows. These include the developing flow in straight tubes and in the annulus formed between two concentric straight tubes. In addition, the flow in the central portion of a large aspect ratio straight rectangular channel ("parallel plate duct") is often found to be reasonably two dimensional. These flow configurations are illustrated in Fig. 7-17. The flow cross-sectional area does not change with axial distance in these standard geometries. The boundary-layer model also provides a good approximation for some internal flows in channels having abrupt expansions in cross-sectional area which cause regions of flow reversal. These new areas of possible applicability of the thin-shear-layer equations will be discussed further in Section 7-5.2.

The finite-difference approach is particularly useful in analyzing the flow from the inlet to the region of fully developed flow. The flow is said to be hydrodynamically fully developed when the velocity distribution is no longer changing with the axial distance along the flow passage. The hydrodynamic fully developed idealization is generally only realized for flows in which fluid property variations in the main flow direction are negligible. The thermal development of the flow is also of interest and can be predicted for the class of flow mentioned above by solving

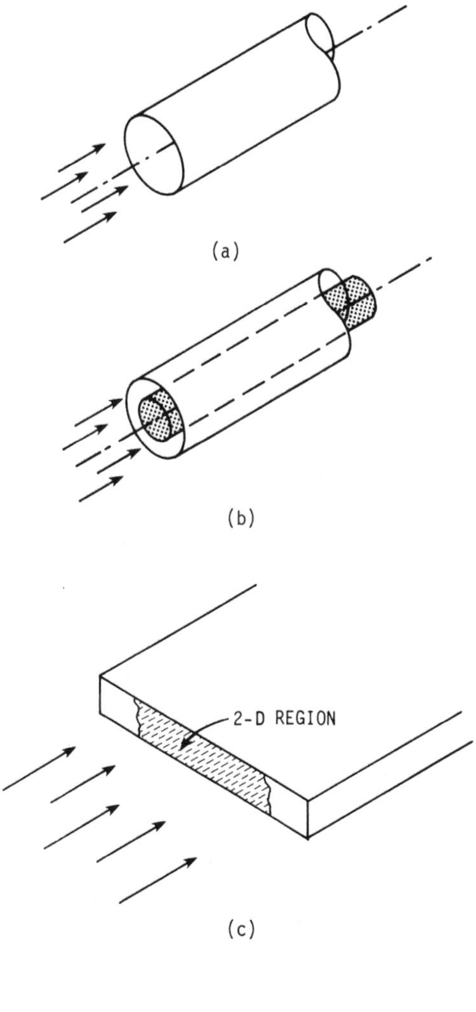

Figure 7-17 Internal flow configurations in which the thin-shear-layer equations are applicable. (a) Circular duct. (b) Annular passage. (c) Large aspect ratio rectangular channel.

the thin-shear-layer form of the energy equation simultaneously with the momentum and continuity equations. Under constant property assumptions and with either constant wall temperature or uniform wall heat-flux thermal boundary conditions, it is possible for the nondimensional temperature distribution to become independent of the axial direction. The book by Shah and London (1978) provides an excellent discussion of the thermal aspects of internal flows.

The finite-difference approach is of less interest for treating fully-developed flow since the governing PDE's reduce to ODE's under these conditions. Laminar fully-developed flow in a tube is the well-known Hagen-Poiseuille flow (White, 1974). If a relatively simple algebraic turbulence model is used, even turbulent fully-developed flow can be treated by numerical methods appropriate for ODE's. With heat transfer present, it becomes more likely that the property variations will prevent the flow from reaching a fully-developed state.

The characteristic Reynolds number for internal flows makes use of the channel hydraulic diameter, D_H, as the characteristic length. The hydraulic diameter is evaluated as $4A/P$ where A is the flow cross-sectional area and P is the wetted perimeter. For circular ducts, D_H reduces to the duct diameter.

We expect to find a small region near the channel inlet where the boundary-layer approximation is poor. This corresponds to the low Reynolds number leading edge region in external flows. For channel Reynolds numbers greater than about 75, this region is negligibly small. Comparisons of various numerical models for very low Reynolds number channel inlet flows can be found in the work of McDonald et al. (1972) and Chilukuri and Pletcher (1980).

7-5.2 Computational Strategies for Internal Flows

It is very important to observe that for steady channel flow, the flux of mass across any plane perpendicular to the channel axis is constant in the absence of wall blowing or suction. Since an initial velocity and temperature distribution must be given as part of the problem specification for the parabolic equations, the mass flow rate can also be considered as specified. If wall blowing or suction occurs, the normal component of velocity at the walls would be required as part of the boundary conditions for the boundary-layer equations; hence, the changing mass flow rate through the channel can be computed from the problem specifications. For simplicity, we will assume that no flow passes through the channel walls in the discussion to follow. However, computational procedures can easily be modified to account for these effects. This additional information about the global or overall mass flow in the channel permits a constraint to be placed on the solution from which the pressure gradient can be determined. In a sense, this mass flow constraint serves the same purpose as the simple relationship between $u_e(x)$ and dp/dx which can be obtained from the steady Euler momentum equation for external flows. In the usual treatment for external flows, the flow outside the boundary layer is assumed to be inviscid and at the outer edge we specialize the Euler equation to $dp/dx = -\rho u_e \, du_e/dx$. Thus, for external flows we usually think of the pressure gradient being specified. What we mean is that dp/dx is either given or easily calculated from $u_e(x)$. When we apply the boundary-layer equations alone to calculate steady internal flows, no information is available from an inviscid flow solution and "outer" boundary conditions on u are established from geometric considerations. In general, viscous effects may be important throughout the flow so that the Euler momentum equation cannot be used in any manner to obtain the pressure gradient. Instead, the global mass flow constraint is used. Thus, in steady internal flows, *the pressure gradient is determined from the solution* (with the help of the global mass flow constraint) rather than being "specified" as for external flows. This is the primary difference between the numerical treatment of internal and external flows.

The thin-shear-layer equations can be written in a form applicable to two-dimensional internal flows as follows:

momentum:

$$\rho u \frac{\partial u}{\partial x} + \rho \tilde{v} \frac{\partial u}{\partial y} = -\frac{dp}{dx} + \frac{1}{r^m} \frac{\partial}{\partial y} (r^m \tau) \tag{7-87}$$

energy:

$$\rho u c_p \frac{\partial T}{\partial x} + \rho \tilde{v} c_p \frac{\partial T}{\partial y} = \frac{1}{r^m} \frac{\partial}{\partial y} (-r^m q_y) + \beta T u \frac{dp}{dx} + \tau \frac{\partial u}{\partial y} \tag{7-88}$$

mass:

$$\frac{\partial}{\partial x} (\rho u r^m) + \frac{\partial}{\partial y} (\rho \tilde{v} r^m) = 0 \tag{7-89}$$

global mass:

$$\dot{m} = \int_A \rho u \, dA = \text{constant} \tag{7-90}$$

In the above, A is the cross-sectional area perpendicular to the channel axis. In addition, an equation of state is normally used to relate density to temperature and pressure. When $m = 0$, the above equations are applicable to two-dimensional flows and when $m = 1$ they apply to axisymmetric flows. For turbulence models utilizing the Boussinesq assumption, we find

$$\tau = \mu \frac{\partial u}{\partial y} - \rho \overline{u'v'} = (\mu + \mu_T) \frac{\partial u}{\partial y} \tag{7-91}$$

and

$$q_y = -k \frac{\partial T}{\partial y} + \rho c_p \overline{v'T'} = \left(-k + \frac{c_p \mu_T}{\Pr_T}\right) \frac{\partial T}{\partial y} \tag{7-92}$$

The governing equations reduce to a form applicable to laminar flows whenever the fluctuating (Reynolds) terms above are equal to zero.

The wall boundary conditions remain the same as for external flows. For flows in straight tubes and parallel plate channels, a symmetry line or plane exists and outer boundary conditions of the form

$$\left.\frac{\partial u}{\partial y}\right)_{r=0} = \left.\frac{\partial T}{\partial y}\right)_{r=0} = 0 \tag{7-93}$$

are used. For tube flow, the shear-stress and heat-flux terms in Eqs. (7-87) and (7-88) are singular at $r = 0$. A correct representation can be found from an application of L'Hospital's rule from which we find

$$\lim_{r\to 0} \frac{1}{r}\frac{\partial}{\partial y}\left(\mu r \frac{\partial \phi}{\partial y}\right) = 2\frac{\partial}{\partial y}\left(\mu \frac{\partial \phi}{\partial y}\right)$$

Except for the treatment of the pressure gradient, the differencing of the governing equations proceeds just as for external boundary-layer flow. The pressure gradient is treated as an unknown in the internal flow case, its value to be determined with the aid of the global mass flow constraint, as indicated previously. This can be done in several ways.

When explicit difference schemes are used, the pressure gradient can be determined as follows. The finite-difference form of the momentum equation can be written in the form

$$u_j^{n+1} = Q_j^n + \frac{dp}{dx} R_j^n \tag{7-94}$$

where Q_j^n and R_j^n contain quantities which are all known. Equation (7-94) is then multiplied by the density $\hat{\rho}_j^{n+1}$ and the resulting equation integrated numerically over the channel cross-section by Simpson's or the trapezoidal rule. This gives

$$\int_A \hat{\rho}_j^{n+1} u_j^{n+1}\, dA = \dot{m} = \int_A \hat{\rho}_j^{n+1} Q_j^n\, dA + \frac{dp}{dx}\int_A \hat{\rho}_j^{n+1} R_j^n\, dA \tag{7-95}$$

The density $\hat{\rho}_j^{n+1}$ is not known a priori at the $n + 1$ level at the time the pressure gradient is being determined. The caret is being used to indicate the provisional nature of this one variable. Very good results have been obtained by simply letting $\hat{\rho}_j^{n+1} = \rho_j^n$. In fact, this has been the most common procedure. An alternative is to evaluate $\hat{\rho}_j^{n+1}$ from ρ_j^n and ρ_j^{n-1} by a second-order accurate extrapolation. Since $\dot{m}$ is specified by the problem initial conditions and the integrals in Eq. (7-95) contain all known quantities, dp/dx can be determined as

$$\frac{dp}{dx} = \frac{\dot{m} - \int_A \hat{\rho}_j^{n+1} Q_j^n\, dA}{\int_A \hat{\rho}_j^{n+1} R_j^n\, dA} \tag{7-96}$$

Once dp/dx has been evaluated, the finite-difference form of the momentum, continuity, and energy equations can be solved just as for external flows. The most widely used explicit scheme for internal flows appears to be of the DuFort-Frankel type. The DuFort-Frankel scheme was given for the thin-shear-layer equations in Section 7-3.4. A typical comparison between the predictions of the DuFort-Frankel scheme and experimental measurements of Barbin and Jones (1963) is shown in Fig.

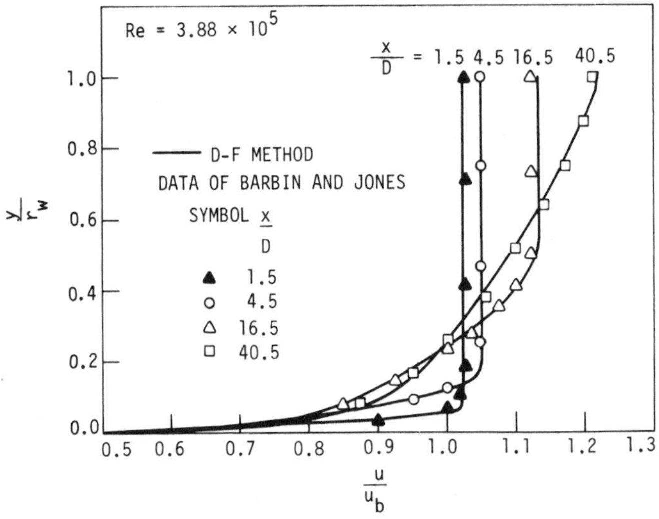

Figure 7-18 Comparison of predicted and measured turbulent velocity profiles in the entrance region of a pipe (Nelson and Pletcher, 1974).

7-18 for the turbulent flow of air in a tube. In the figure, u_b denotes the bulk velocity in the tube and r_w is the radius of the tube. Even very near the inlet $(x/D = 1.5)$, the predictions are seen to be in good agreement with the measurements. A simple algebraic turbulence model was used in the predictions.

The internal flow problem is conceptually very similar to the inverse boundary-layer problem discussed in Section 7-4 for external flows. This is most evident when implicit difference schemes are used. For internal flows the "correct" pressure gradient must be determined which will give velocities which satisfy the global mass flow constraint. This corresponds to adjusting the pressure gradient (or edge velocity) until the velocity distribution satisfies the specified displacement thickness in inverse methods for external flows. Several different procedures have been used with implicit methods to determine the pressure gradient. A number of these are briefly discussed below.

1. *Variable secant iteration.* The pressure gradient can be varied iteratively at each streamwise location until the global mass flow constraint is met (Briley, 1974) by employing the variable secant procedure discussed in Section 7-4.3 in connection with imposing the δ^* boundary condition for the inverse boundary-layer method. For fixed coefficients, velocities vary linearly with the pressure gradient so that convergence is usually obtained with three iterations.
2. *Lagging the pressure adjustment.* Patankar and Spalding (1970) have pointed out that iterating at each streamwise station is uneconomical and have suggested that a value for the pressure gradient be guessed to advance the solution and then let the knowledge of any resulting error in mass flow rate guide our choice of pressure gradient for the *next* step. That is, in analogy with the way an

automobile is steered, adjustments are made to correct the course without going back to retrace the path. This was the common-sense approach included in the early versions of the Patankar-Spalding finite-difference method for confined flows. Although the common-sense aspect of this logic cannot be denied, the algorithm appears a bit too approximate by present-day standards and is not recommended. In concert with the trend toward lower computer costs observed over the past decade there has been a cost-equalizing trend toward the use of algorithms which are potentially more accurate.

3. *Newton's method.* Recently Raithby and Schneider (1979) have proposed a scheme suitable for incompressible flows which requires one-third less effort than the minimum (three iterations) variable secant calculation. The scheme assumes that the coefficients in the difference equations will remain constant, i.e., no form of updating is employed as the pressure gradient is adjusted so that the global mass flow constraint is satisfied. The idea is that once an initial guess for dp/dx is made and a provisional solution obtained for the difference equations, a correction can be obtained by employing a form of Newton's method. With "frozen" coefficients, the velocities will vary linearly with the pressure gradient and it follows that one Newton-type correction should provide the correct pressure gradient. To illustrate, we will let $S = dp/dx$. We make an initial guess for $dp/dx = (dp/dx)^*$ and calculate provisional velocities $(u_j^{n+1})^*$ and a provisional mass flow rate $\dot{m}^*$. Due to the linearity of the momentum equation with frozen coefficients, we observe from an application of Newton's method (see Section 7-4.2) that the correct velocity at each point would be

$$u_j^{n+1} = (u_j^{n+1})^* + \frac{\partial u_j^{n+1}}{\partial S} \Delta S \tag{7-97}$$

where ΔS is the change in the pressure gradient required to satisfy the global mass flow constraint. We define $u_{p,j}^{n+1} = \partial u_j^{n+1}/\partial S$. The difference equations are actually differentiated with respect to the pressure gradient (S) to obtain difference equations for $u_{p,j}^{n+1}$ which are tridiagonal in form. The coefficients for the unknowns in these equations will be the same as for the original implicit difference equations. The Thomas algorithm is used to solve the system of algebraic equations for $u_{p,j}^{n+1}$. The boundary conditions on $u_{p,j}^{n+1}$ must be consistent with the velocity boundary conditions. On boundaries where the velocity is specified, $u_{p,j}^{n+1} = 0$, whereas on boundaries where the velocity gradient is specified, $\partial u_{p,j}^{n+1}/\partial n = 0$ (n normal to boundary). The solution for $u_{p,j}^{n+1}$ is then used to compute ΔS by noting that $u_{p,j}^{n+1}\, \Delta S$ is the correction in velocity at each point required to satisfy the global mass flow constraint. Thus we can write

$$\dot{m} - \dot{m}^* = \Delta S \int_A \rho u_{p,j}^{n+1}\, dA \tag{7-98}$$

where the integral is evaluated by numerical means. The $\dot{m}$ in Eq. (7-98) is the known value specified by the initial conditions. The required value of ΔS is determined from Eq. (7-98). The correct values of velocity u_j^{n+1} can then be determined from Eq. (7-97). The continuity equation is then used to determine v_j^{n+1}. The computational effort of this procedure is roughly equivalent to two iterations of the method employing the variable secant procedure.

4. *Treating the pressure gradient as a dependent variable.* In all of the procedures listed above, the pressure gradient is treated as a known quantity whenever the simultaneous algebraic equations for the new velocities are solved. The standard Thomas algorithm can be used for methods 1-3 above. Here we consider schemes in which the pressure gradient is treated as an unknown in the algebraic formulation. The coefficient matrix is no longer tridiagonal. Early methods of this type (Hornbeck, 1963) tended to employ conventional Gaussian elimination. More recent procedures (Blottner, 1977; Cebeci and Chang, 1978; and Kwon and Pletcher, 1981) have used more efficient block elimination procedures. The method of Kwon and Pletcher (1981) is a modification of inverse method B presented in Section 7-4.3. The procedure employs the FLARE approximation to permit the calculation of separated regions in internal flows. The changes which must be made in inverse method B in order to treat internal incompressible flows in a two-dimensional channel will now be described. The flow is assumed to be symmetric about the channel centerline located at $y = H/2$ where y is measured from the channel wall. The channel height is H. Equations (7-64) and (7-65) apply. The outer boundary conditions become

$$\left.\frac{\partial u}{\partial y}\right)_{y=H/2} = 0 \qquad \psi\left(x, \frac{H}{2}\right) = \frac{\dot{m}}{2\rho} \tag{7-99}$$

where $\dot{m}$ is the mass flow rate per unit width for a two-dimensional channel. The difference equations, Eqs. (7-68)-(7-71), are applicable and χ^{n+1} represents the unknown pressure gradient

$$-\frac{1}{\rho}\frac{dp}{dx}$$

as before. The procedures for internal and external flow differ in the way in which χ^{n+1} and u_{NJ}^n are determined from the outer boundary conditions, which are different in the two cases. We express $\partial u/\partial y)_{H/2}$ in terms of a one-sided second-order accurate difference representation,

$$\left(\frac{\partial u}{\partial y}\right)_{NJ}^{n+1} \simeq \frac{u_{NJ}^{n+1}}{2}\left(\frac{4}{\Delta y_-} - \frac{1}{\Delta y_{--}}\right) - \frac{2u_{NJ-1}^{n+1}}{\Delta y_-} + \frac{u_{NJ-2}^{n+1}}{2\,\Delta y_{--}} \tag{7-100}$$

where
$$\Delta y_- = y_{NJ} - y_{NJ-1}$$

and
$$\Delta y_{--} = y_{NJ-1} - y_{NJ-2}$$

The outer boundary conditions, Eq. (7-99) can now be written as

$$u_{NJ}^{n+1} = c_1 u_{NJ-1}^{n+1} - c_2 u_{NJ-2}^{n+1} \tag{7-101}$$

and
$$\psi_{NJ}^{n+1} = \frac{\dot{m}}{2\rho} \tag{7-102}$$

where
$$c_1 = \frac{4}{4-K}$$
and
$$c_2 = \frac{K}{4-K}$$
and
$$K = \frac{\Delta y_-}{\Delta y_{--}}$$

Equations (7-101) and (7-102) are to be solved with Eqs. (7-74), (7-75), and (7-78). However, one additional relationship is needed since five unknowns appear (u_{NJ}^{n+1}, u_{NJ-1}^{n+1}, u_{NJ-2}^{n+1}, ψ_{NJ-1}^{n+1}, χ^{n+1}) and only four independent relationships among them have been identified thus far [Eqs. (7-101), (7-74), (7-75), and (7-78)]. The additional equation can be obtained by specializing Eq. (7-72) for u_{NJ-2}^{n+1} as

$$u_{NJ-2}^{n+1} = A'_{NJ-2}u_{NJ-1}^{n+1} + H'_{NJ-2}\chi^{n+1} + C'_{NJ-2} \tag{7-103}$$

This system of equations can be solved for χ^{n+1} by defining

$$\alpha_1 = 1 - A'_{NJ-1}(c_1 - c_2A'_{NJ-2})$$
$$\alpha_2 = (c_1 - c_2A'_{NJ-2})H'_{NJ-1} - c_2H'_{NJ-2}$$
$$\alpha_3 = (c_1 - c_2A'_{NJ-2})C'_{NJ-1} - c_2C'_{NJ-2}$$
$$\alpha_4 = 1 + \frac{2}{\Delta y_-}B'_{NJ-1} + A'_{NJ-1}$$
$$\alpha_5 = -\left(H'_{NJ-1} + \frac{2}{\Delta y_-}D'_{NJ-1}\right)$$
$$\alpha_6 = \frac{\dot{m}}{\rho\,\Delta y_-} - \frac{2}{\Delta y_-}E'_{NJ-1} - C'_{NJ-1}$$

then
$$\chi^{n+1} = \frac{\alpha_1\alpha_6 - \alpha_3\alpha_4}{\alpha_2\alpha_4 - \alpha_1\alpha_5} \tag{7-104}$$

The axial component of velocity at the line of symmetry can be found from

$$u_{NJ}^{n+1} = \frac{\alpha_2}{\alpha_1}\chi^{n+1} + \frac{\alpha_3}{\alpha_1} \tag{7-105}$$

At this point the back substitution process can be initiated using Eqs. (7-72) and (7-73) to compute u_j^{n+1} and ψ_j^{n+1} from the outer boundary to the wall. The remaining portions of the algorithm are as discussed in Section 7-4.3. The only differences between inverse method B and the related procedure for internal flows is due to the minor differences in the boundary conditions for the two cases. This requires that slightly different algebraic procedures be used to evaluate χ^{n+1} and u_{NJ}^{n+1} prior to the back substitution step in the block tridiagonal solution procedure.

An interesting application of this method has been made to laminar channel flows having a sudden symmetric expansion which creates a region of recirculation downstream of the expansion. The general flow pattern of such a flow is illustrated in Fig. 7-19. The predictions were obtained by the boundary-layer method described above utilizing a fully developed velocity profile at the step. Re_h is the Reynolds number based on step height and H_1/H_2 is the ratio of the channel height before and after the expansion. Such flows have customarily been predicted by solving the full Navier-Stokes equations.

Figure 7-20 compares velocity profiles predicted by the same method with experimental data and Navier-Stokes solutions for a symmetric sudden expansion flow. In the figure, $u_{i,\max}$ denotes the maximum velocity just upstream of the expansion (step), H_0 is the channel height downstream of the expansion, and y_{CL} is the distance from the wall to the channel centerline. The symbols h, H_1, H_2 are as defined previously. The method based on boundary-layer equations requires an order of magnitude less computer time than required for the solution of the Navier-Stokes equations. The boundary-layer method provides good agreement with experimental data in this case.

In Fig. 7-21 the predicted reattachment length is compared with results from Navier-Stokes solutions. The Reynolds number, Re_{Hi}, is based on the height of the channel upstream of the expansion. The boundary-layer predictions generally appear to be in good agreement with the Navier-Stokes results except for Reynolds numbers less than about 20 where the trends indicated by the Navier-Stokes solutions and the boundary-layer solutions appear to diverge.

7-5.3 Concluding Remarks

Details of several difference schemes suitable for the thin-shear-layer equations were presented in Section 7-3 as they applied to ordinary external boundary-layer flows.

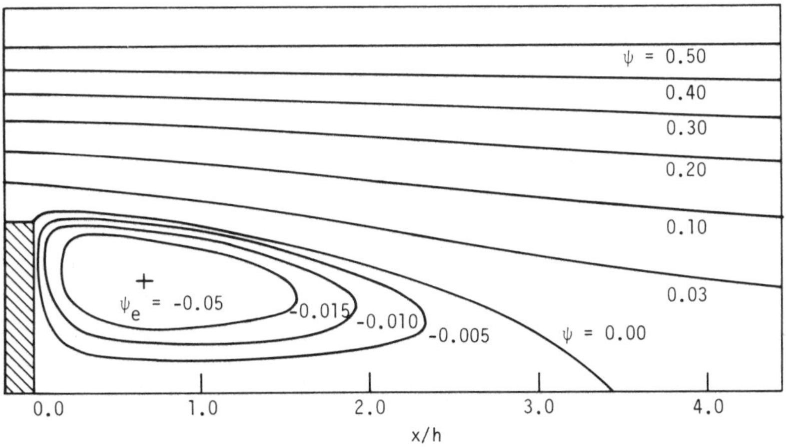

Figure 7-19 Streamline contours predicted from boundary-layer equations (Kwon and Pletcher, 1981) for a laminar flow in a channel with a symmetric sudden expansion, $\mathrm{Re}_h = 50$, $H_1/H_2 = 0.5$.

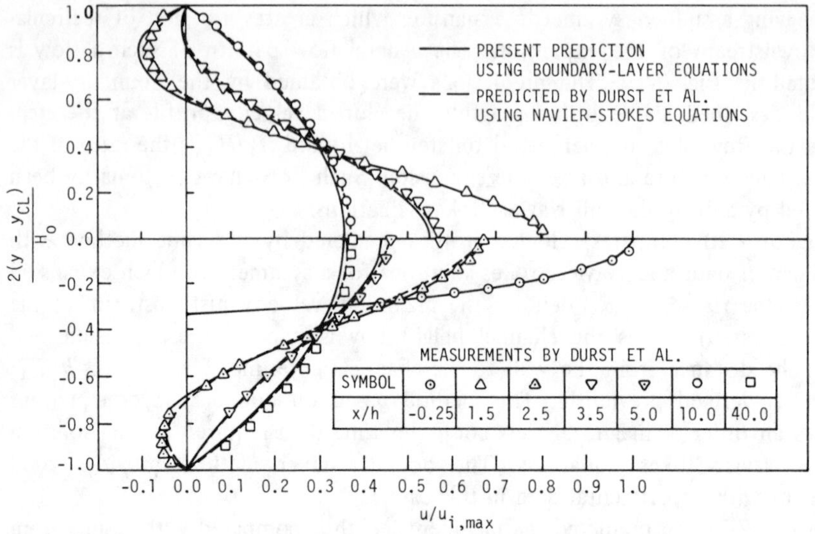

Figure 7-20 Velocity profiles for a laminar flow in a channel with a symmetric sudden expansion, Re_h (based on u_{max}) = 56, $H_1/H_2 = \frac{1}{3}$ (Kwon and Pletcher, 1981).

As methods for confined flows have been considered in the present section, numerical details which remain the same as for external flows have not been repeated. However, an attempt has been made to clearly point out and emphasize those details which change and are unique to internal flows.

Coverage in this section has been limited to flows in straight channels. Blottner

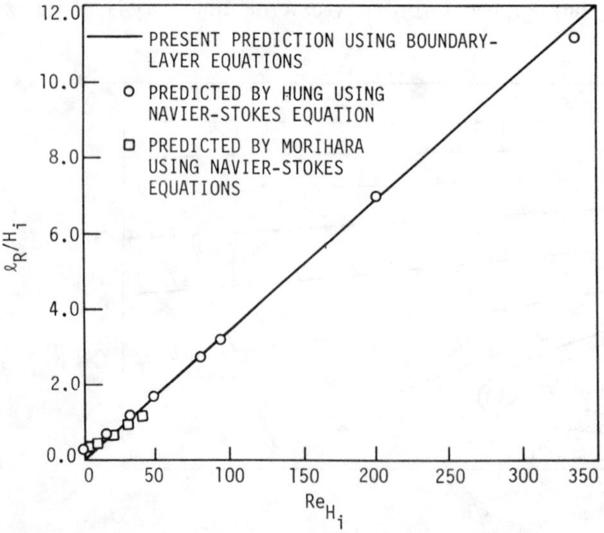

Figure 7-21 Prediction of the reattachment length for laminar flows in a channel with a symmetric sudden expansion, $h_1/h_2 = 0.5$ (Kwon and Pletcher, 1981).

(1977) demonstrated that the thin-shear-layer approximation (also known as the "slender channel" approximation) can be extended to curved two-dimensional channels with varying channel height. The equations solved are the boundary-layer equations with longitudinal curvature (Van Dyke, 1969). The normal pressure gradient induced by the channel curvature is accounted for through

$$\frac{\partial p}{\partial n} = \frac{\kappa \rho u^2}{1 + \kappa n} = 0$$

where n is the coordinate normal to the channel centerline and κ is the curvature of the centerline.

Viscous-inviscid interaction schemes can be applied to internal flows in which an inviscid core region can be identified. The interaction effect is expected to be negligibly small except very near the inlet for low Reynolds number flows and under conditions in which the channel cross-sectional area changes abruptly. Viscous-inviscid interaction permits information to be transmitted upstream and can give improved predictions for flows in which the pressure field at a point is expected to be influenced by conditions farther downstream. The incompressible inviscid flow in channels is conveniently determined through a numerical solution of Laplace's equation for the stream function. Inverse method B discussed in Section 7-4.3 can be used for the viscous portion of the flow. Such a combination has been employed interactively to predict the flow over a rearward facing step in a channel.

7-6 APPLICATION TO FREE-SHEAR FLOWS

The thin-shear-layer equations provide a fairly accurate mathematical model for a number of free-shear flows. These include the plane or axisymmetric jet discharging to a quiescent or co-flowing ambient, the planar mixing layer, and simple wake flows. The majority of free-shear layers encountered in engineering applications are turbulent. To date, turbulence models for free-shear flows have not exhibited nearly the degree of generality as those used for wall boundary layers. It is still a major challenge to find models which can provide accurate predictions for the development of both the planar and axisymmetric jet without requiring adjustments in the model parameters.

A complete treatise on the subject of the finite-difference prediction of free-shear flows might devote 60% of its content to turbulence modeling, 25% to coverage of the physics of various categories of free-shear flows, and 15% to numerical procedures. The numerical procedures, which are our main concern here, are the least troublesome aspect of the problem of obtaining accurate predictions for turbulent free-shear flows.

The round jet has been studied extensively both experimentally and analytically and provides a representative example of a free-shear flow. The thin-shear-layer equations will provide a good mathematical model for the round jet following a straight trajectory if pressure in the interior of the jet can be assumed to be equal to that of the surrounding medium. This requires that the surface tension of the jet be negligible and that the jet be fully expanded, i.e., the pressure at the discharge plane equals the pressure in the surrounding medium. A subsonic jet discharging from a tube can always be considered as fully expanded. For the jet cross section to remain round and the trajectory to remain straight, it is necessary that no forces act on the jet in the lateral direction. This requires that the medium into which the jet is injected be at rest or flowing in the same direction as the discharging jet (co-flowing) and that body forces (such as buoyancy) be negligible. Under these conditions, the form of the thin-shear-layer equations given by Eqs. (5-116)-(5-119) are applicable. These equations are specialized further below for the steady incompressible flow of a round jet in the absence of a pressure gradient:

continuity:

$$\frac{\partial(yu)}{\partial x} + \frac{\partial(yv)}{\partial y} = 0 \tag{7-106}$$

momentum:

$$u\frac{\partial u}{\partial x} + v\frac{\partial u}{\partial y} = \frac{1}{y}\frac{\partial}{\partial y}\left[y\left(\nu\frac{\partial u}{\partial y} - \overline{u'v'}\right)\right] \tag{7-107}$$

Numerically, the primary difference between the wall boundary layer and the round jet is in the specification of the boundary conditions. Figure 7-22 illustrates the round jet flow configuration. Due to the symmetry which exists about the jet centerline, the appropriate boundary conditions at $y = 0$ are $(\partial u/\partial y)_{y=0} = 0$ and $v(x, 0) = 0$. The outer boundary condition is identical to that for a wall boundary layer,

$$\lim_{y\to\infty} u(x, y) = u_e$$

Initial conditions are also needed for the finite-difference calculation. For turbulent

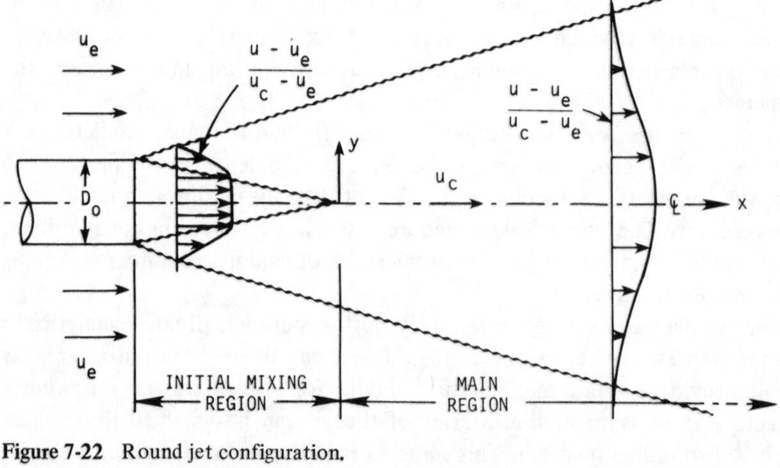

Figure 7-22 Round jet configuration.

jets especially, the initial streamwise velocity distribution is usually taken as a uniform stream at the discharge velocity, u_0. Naturally, this cannot be completely correct in that the velocities in a small region must exhibit the retarding effects of the tube walls. On the other hand, the boundary-layer equations are not expected to provide an extremely accurate solution very near the discharge plane, i.e., for x/D_0 less than about one where D_0 is the diameter of the jet at discharge. Using a uniform velocity distribution at discharge for the turbulent jet appears to provide fairly accurate results for $x/D_0 > 1$ which includes the region of most interest in engineering applications. Some finite-difference schemes applied to the jet problem in the Cartesian coordinate system will also require an initial distribution for v. As was mentioned in Section 7-3 for wall boundary layers, this is a requirement of the numerical procedure and not a requirement in the mathematical specification of the problem. When an initial distribution for v is required, using $v(0, y) = 0$ is recommended. Taking several very small streamwise steps near the starting plane helps confine the effects of the starting singularity (which is due to the very large values of $\partial u/\partial x$ associated with the vanishingly small initial mixing zone) to a small region. This starting singularity is similar to that observed at the leading edge of a flat plate for the wall boundary layer when the equations are solved in the Cartesian coordinate system.

For a turbulent jet discharging to a quiescent ambient, the initial mixing region, indicated in Fig. 7-22, extends to an x/D of about 5. For jets discharging to a co-flowing stream, the initial mixing region is even longer. This initial mixing region is characterized by the fact that the fluid at the centerline is moving at the jet discharge velocity. Beyond the initial mixing region, the velocities throughout the jet are influenced by the ambient stream velocity, u_e. The growth properties of the jet differ in the two regions, initial and main, and when algebraic turbulence models are used, it is expected that somewhat different models (or values for the constants in the models) will be required in the two regions.

Most of the finite-difference schemes discussed in Section 7-3 have been observed to work well for jets. Several methods are described in the *Proceedings of the Langley Working Conference on Free Turbulent Shear Flows* (NASA, 1972). This reference should provide a good starting point for obtaining a background on the special problems associated with achieving accurate predictions for several types of turbulent free-shear flows. Many numerical details can also be found in the works of Hornbeck (1973), Madni and Pletcher (1975a, 1975b, 1977a), and Hwang and Pletcher (1978). This latter work gives the difference equations used in evaluating the fully implicit, Crank-Nicolson implicit, DuFort-Frankel, Larkin ADE, Saul'yev ADE, and Barakat and Clark ADE methods for the round jet. A useful evaluation of available experimental data on uniform density turbulent free-shear layers has been provided by Rodi (1975).

The boundary-layer form of the energy equation is also applicable to free-shear flows. For a heated jet discharging vertically into a quiescent ambient, with or without thermal stratification, the trajectory of the jet is straight and no special difficulties arise with the boundary-layer model. For the heated jet discharging at other angles or discharging at any angle with a crossflow, the jet is expected to follow a curved trajectory. Such flows have been treated by solving the fully three-dimensional Navier-Stokes equations by Patankar et al. (1977) and others and by more approximate parabolic finite-difference models that assume the flow remains axisymmetric (Madni and Pletcher, 1977b; Hwang and Pletcher, 1978). In these latter axisymmetric models, the momentum equation in the transverse direction is treated in a lumped manner which yields an ordinary differential equation for the angle between the tangent to the trajectory of the jet centerline and the horizontal direction. This approach requires only slightly more computational effort than solving the axisymmetric boundary-layer equations and gives surprisingly good agreement with experimental measurements, especially with regards to the trajectory of the jet.

No finite-difference algorithms will be provided in this section since the procedures discussed in Section 7-3 can be adapted to free-shear flows in a straightforward manner. However, one numerical anomaly which sometimes occurs in the prediction of jets discharging to an ambient at rest is worth mentioning. For this case, some schemes are unable to correctly predict u to asymptotically approach the freestream velocity of zero. The problem is thought to be related to the treatment of the coefficients of the convective terms and the procedures used to locate the outer boundary. The difficulty is most evident if the coefficients are lagged. It is commonplace to overcome this problem in a practical manner by letting u_e be a small positive velocity on the order of 1–3% of the jet centerline velocity. Reports in the literature claim that this approximation does not seriously degrade the accuracy of the calculations. Hornbeck (1973) shows that a satisfactory solution for $u_e = 0$ can be obtained with implicit methods by iteratively updating the coefficients.

7-7 THREE-DIMENSIONAL BOUNDARY LAYERS

7-7.1 Introduction

The majority of the flows that occur in engineering applications are three dimensional. In this section we will consider finite-difference methods for those three-dimensional flows which are "thin" (i.e., with large velocity gradients) in only one coordinate direction. Such flows are sometimes referred to as "boundary sheets." Many flows occurring in applications are of this type. These are predominately external flows. Examples include much of the viscous portion of the flow over wings and general aerodynamic bodies.

An example of a three-dimensional boundary-layer flow is illustrated in Fig. 7-23. The presence of the cylinder alters the pressure field causing the inviscid flow streamlines to turn as indicated qualitatively in the figure. In accordance with the equations of motion, a component of the pressure gradient (responsible for the turning) is directed away from the center of curvature of the inviscid flow streamlines. Because the viscous layer is thin, this pressure gradient does not change in the direction normal to the surface. As a result, the velocity vector rotates

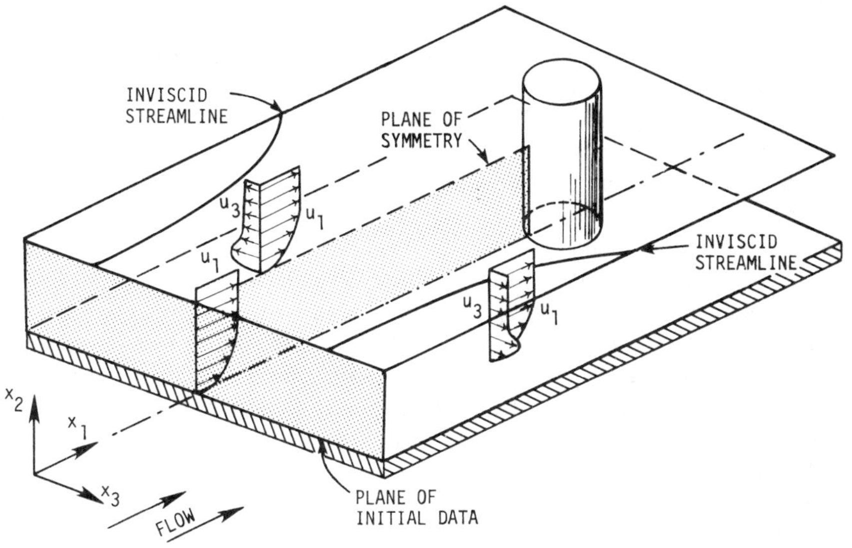

Figure 7-23 An example of subsonic three-dimensional boundary-layer flow.

toward the center of curvature of the inviscid streamlines as we move down within the boundary layer. This occurs because the pressure gradient remains fixed but the inertia of the fluid decreases as we move nearer the wall. This requires that the radius of streamline curvature decrease as we move in the normal direction toward the wall in the boundary layer. Thus, the crossflow component of velocity will generally reach a maximum at some point within the boundary layer, as indicated in Fig. 7-23. This pressure induced "crossflow" is referred to as a secondary flow in some applications and is responsible for such phenomena as the transport of sand toward the inside bank of a curved riverbed and the migration of tea leaves toward the center (near the bottom) of a stirred cup of tea.

Another interesting example of three-dimensional boundary-layer flow occurs on bodies of revolution at incidence. Such flows on a prolate spheroid, for example, have been studied extensively by several investigators including Wang (1974, 1975), Blottner and Ellis (1973), Patel and Choi (1979), and Cebeci et al. (1979a).

The three-dimensional boundary-layer equations are not applicable to flows near the intersection of two surfaces (for example, near wing-body junctions and corners in channels) because stress gradients in two directions are important in those regions. Other reduced forms of the Navier-Stokes equations can be used to treat the flow near corners. These will be discussed in Chapter 8.

The subject of three-dimensional boundary layers will not be covered in great detail here. Instead, we will outline the general numerical strategy required for the solution of the problem posed by these equations, making use of the material developed in earlier sections for the two-dimensional boundary-layer equations. Several new considerations arise with the three-dimensional problem and these will be emphasized.

7-7.2 The Equations

The three-dimensional boundary-layer equations are presented in Chapter 5 in Cartesian coordinates [Eqs. (5-120)-(5-123)] and in body intrinsic orthogonal curvilinear coordinates [Eqs. (5-124)-(5-128)]. For certain special conditions (the laminar supersonic flow over a cone at incidence being one of them), the number of independent variables can be reduced from three to two. Special cases of this sort will not be discussed here.

The Cartesian coordinate system can be used for flows over developable surfaces (those which can be formed by bending a plane without stretching or shrinking) including, of course, the special case of a flat surface. Curvilinear systems are required for flows over more general bodies. A few studies have been made using a curvilinear coordinate system coinciding with the inviscid streamlines (see Cebeci et al., 1973). However, most three-dimensional boundary-layer computations have been made with coordinate systems related to the geometry of the surface. Even with a body-oriented coordinate system, choices remain in the selection of the coordinate axes. Blottner (1975b) provides a review of the coordinate systems which have been used for three-dimensional flows.

The three-dimensional boundary-layer equations presented in Chapter 5 are singular at the origin of the x_1 coordinate. This singularity is of the same type as found at the leading edge of a flat plate in two-dimensional flow (see Section 7-3.7). Several investigators have satisfactorily used the equations in this form both for flows in the Cartesian coordinate system (Klinksiek and Pierce, 1973) and for more complex flows over axisymmetric bodies (Wang, 1972; Patel and Choi, 1979). These authors have generally used a separate procedure to generate a satisfactory stagnation point solution before utilizing their three-dimensional solution scheme.

It is also quite common to eliminate the singular behavior of the equations by use of a suitable dependent variable transformation. No single transformation has proven optimum for all flows. Blottner (1975b) discusses several which have been used for specific problems. An example transformation will be presented here which will remove the singularity at the origin of x_1 and will permit the stagnation point profiles to be obtained from the solution of the ordinary differential equations that remain when $x_1 = 0$. The transformed equations will permit the solution to be advanced from the stagnation point in a smooth and systematic manner. For laminar flows, the boundary layer will tend to have a nearly uniform thickness in the transformed coordinates.

We first note that the boundary-layer equations revert to the Euler equations at the outer edge of the boundary layer where the viscous terms vanish and $\partial u_1/\partial x_2$ and $\partial u_3/\partial x_2 \to 0$. This permits the components of the pressure gradient in Eqs. (5-126) and (5-127) to be written as

$$-\frac{1}{h_1}\frac{\partial p}{\partial x_1} = \frac{\rho u_{1,e}}{h_1}\frac{\partial u_{1,e}}{\partial x_1} + \frac{\rho u_{3,e}}{h_3}\frac{\partial u_{1,e}}{\partial x_3} + u_{1,e}u_{3,e}K_1 - \rho u_{3,e}^2 K_3 \tag{7-108}$$

$$-\frac{1}{h_3}\frac{\partial p}{\partial x_3} = \frac{\rho u_{1,e}}{h_1}\frac{\partial u_{3,e}}{\partial x_1} + \frac{\rho u_{3,e}}{h_3}\frac{\partial u_{3,e}}{\partial x_3} - u_{1,e}^2 K_1 + \rho u_{1,e}u_{3,e}K_3 \tag{7-109}$$

where $u_{1,e}(x_1, x_3)$ and $u_{3,e}(x_1, x_3)$ are given by the inviscid flow solution over the body. The subscript e denotes quantities evaluated at the outer edge of the boundary layer.

Let us assume that the Reynolds stresses for a turbulent flow will be evaluated by a viscosity model. That is, let

$$-\rho\overline{u_1' u_2'} = \mu_T \frac{\partial u_1}{\partial x_2} \qquad -\rho\overline{u_3' u_2'} = \mu_T \frac{\partial u_3}{\partial x_2}$$

$$-\rho c_p \overline{u_2' T'} = k_T \frac{\partial T}{\partial x_2} \qquad \frac{\mu_T c_p}{k_T} = \Pr_T$$

and
$$\bar{\mu} = \mu_T + \mu$$

The model for μ_T may be simple or complex; no assumption about the complexity of μ_T is made at this time. The equations remain valid for laminar flow where $\bar{\mu} = \mu$.

It is convenient to introduce nondimensional variables for velocity components defined by

$$F = \frac{u_1}{u_{1,e}} \qquad G = \frac{u_3}{W_e} \qquad I = \frac{H}{H_e}$$

where W_e will be designated later as either $u_{1,e}$ or $u_{3,e}$.

We now let $x = x_1$, $z = x_3$, and

$$\eta = \left(\frac{u_{1,e}}{x(\rho\mu)_e}\right)^{1/2} \int_0^{x_2} \rho\, dx_2$$

Using the chain rule for differentiation, derivatives with respect to the original independent variables can be replaced according to

$$\frac{\partial}{\partial x_1} = \frac{\partial}{\partial x} + \frac{\partial \eta}{\partial x}\frac{\partial}{\partial \eta}$$

$$\frac{\partial}{\partial x_2} = \frac{\partial \eta}{\partial x_2}\frac{\partial}{\partial \eta} = \left[\frac{u_{1,e}}{x(\rho\mu)_e}\right]^{1/2} \rho \frac{\partial}{\partial \eta}$$

$$\frac{\partial}{\partial x_3} = \frac{\partial}{\partial z} + \frac{\partial \eta}{\partial z}\frac{\partial}{\partial \eta}$$

Making the indicated substitutions permits Eqs. (5-125)-(5-128) to be written as

continuity:

$$\frac{x}{h_1 h_3}\frac{\partial(h_3 F)}{\partial x} + \frac{F}{2h_1}(1 + \beta_1) + \frac{\partial V}{\partial \eta} + \frac{1}{h_1 h_3 [(\rho\mu)_e u_{1,e}/x]^{1/2}} \times \frac{\partial}{\partial z}\left\{\frac{h_1 W_e G}{u_{1,e}} [x u_{1,e}(\rho\mu)_e]^{1/2}\right\} = 0 \tag{7-110}$$

x momentum:

$$\frac{xF}{h_1}\frac{\partial F}{\partial x} + V\frac{\partial F}{\partial \eta} + \underbrace{\frac{xG}{h_3}\frac{\partial F}{\partial z}}_{(1)} + \underbrace{FGxK_1}_{(2)} - \underbrace{xG^2K_3}_{(3)} = \beta_1(\theta - \underbrace{F^2}_{(4)}) + \beta_2\left(\frac{\theta u_{3,e}}{u_{1,e}} - \underbrace{FG}_{(5)}\right)$$

$$+ \theta\left(\frac{xu_{3,e}K_1}{u_{1,e}} - \frac{xu_{3,e}^2K_3}{u_{1,e}^2}\right) + \frac{\partial}{\partial \eta}\left(\frac{\rho\bar{\mu}}{(\rho\mu)_e}\frac{\partial F}{\partial \eta}\right) \qquad (7\text{-}111)$$

z momentum:

$$\frac{xF}{h_1}\frac{\partial G}{\partial x} + V\frac{\partial G}{\partial \eta} + \frac{xG}{h_3}\frac{\partial G}{\partial z} + xFGK_3 - xF^2K_1 = \theta\left(\beta_3 + \beta_4 + \frac{xK_3u_{3,e}}{u_{1,e}} - xK_1\right)$$

$$- \beta_5 GF - \beta_6 G^2 + \frac{\partial}{\partial \eta}\left(\frac{\rho\bar{\mu}}{(\rho\mu)_e}\frac{\partial G}{\partial \eta}\right) \qquad (7\text{-}112)$$

energy:

$$\frac{xF}{h_1}\frac{\partial I}{\partial x} + V\frac{\partial I}{\partial \eta} + \frac{xGW_e}{h_3u_{1,e}}\frac{\partial I}{\partial z} = -\beta_7 FI - \frac{\beta_8 GW_eI}{u_{1,e}} + \frac{\partial}{\partial \eta}\left\{\left(\frac{\mu}{\Pr} + \frac{\mu_T}{\Pr_T}\right)\frac{\rho}{(\rho\mu)_e}\frac{\partial I}{\partial \eta}\right.$$

$$+ \frac{\rho u_{1,e}^2 F}{H_e(\rho\mu)_e}\left[\mu\left(1 - \frac{1}{\Pr}\right) + \mu_T\left(1 - \frac{1}{\Pr_T}\right)\right]\frac{\partial F}{\partial \eta}$$

$$\left. + \frac{\rho W_e^2 G}{H_e(\rho\mu)_e}\left[\mu\left(1 - \frac{1}{\Pr}\right) + \mu_T\left(1 - \frac{1}{\Pr_T}\right)\right]\frac{\partial G}{\partial \eta}\right\} \qquad (7\text{-}113)$$

where

$$V = \rho\tilde{u}_2\left[\frac{x}{u_{1,e}(\rho\mu)_e}\right]^{1/2} + \frac{xF}{h_1}\frac{\partial \eta}{\partial x} + \frac{xGW_e}{h_3u_{1,e}}\frac{\partial \eta}{\partial z}$$

$$\theta = \frac{\rho_e}{\rho} \qquad \beta_1 = \frac{x}{h_1u_{1,e}}\frac{\partial u_{1,e}}{\partial x}$$

$$\beta_2 = \frac{x}{h_3u_{1,e}}\frac{\partial u_{1,e}}{\partial z} \qquad \beta_3 = \frac{x}{h_1W_e}\frac{\partial u_{3,e}}{\partial x}$$

$$\beta_4 = \frac{xu_{3,e}}{W_eh_3u_{1,e}}\frac{\partial u_{3,e}}{\partial z} \qquad \beta_5 = \frac{x}{W_eh_1}\frac{\partial W_e}{\partial x}$$

The metrics and geodesic curvatures of the surface coordinate lines are as defined in Chapter 5.

$$\beta_6 = \frac{x}{h_3u_{1,e}}\frac{\partial W_e}{\partial z} \qquad \beta_7 = \frac{x}{h_1H_e}\frac{\partial H_e}{\partial x} \qquad \beta_8 = \frac{x}{h_3H_e}\frac{\partial H_e}{\partial z}$$

An equation of state, $\rho = \rho(p, T)$, is needed to close the system of equations for a compressible flow. Several terms in the equation above have been numbered for future reference.

The usual boundary conditions are

$\eta = 0$

$$V = F = G = 0 \qquad I = I(x, 0, z)$$

or

$$\left.\frac{\partial I}{\partial \eta}\right)_{\eta=0} = Q(x, 0, z)$$

$\eta \to \infty$

$$F = G = 1, \qquad G = \frac{u_{3,e}}{W_e}$$

where $Q(x, 0, z)$ is a specified function related to the wall heat flux. In addition, initial distributions of F, G, and I must be provided. Distributions of $u_{1,e}$, $u_{3,e}$, and H_e are also required.

The question of initial conditions requires careful consideration. Examination of the three-dimensional boundary-layer equations in the original orthogonal curvilinear coordinates prior to the transformation indicated above (or for that matter, in Cartesian coordinates) indicates that the role of the x_1 and x_3 coordinates are interchangeable; that is, the equations are symmetric with respect to the interchange of x_1 and x_3 coordinates. As long as both u_1 and u_3 are positive, no single coordinate direction emerges as the obvious "marching" direction from considering the equations alone. Since first derivatives of u_1, u_3, and H appear with respect to both x_1 and x_3, it is expected that initial data should be provided in two intersecting planes to permit marching the dependent variables in both the x_1 and x_3 directions. The correct (permissible) marching direction is dictated by the zone of dependence principle which will be discussed later. For now we will proceed under the assumption that it is possible to march the solution in either the x_1 or x_3 directions and that initial data is needed on two intersecting planes. It is generally easy to determine a "main" flow direction from considerations of the body geometry and the direction of the oncoming stream. In defining η above, we have already assumed that the x or x_1 coordinates are in this main flow direction and the x_3 or z coordinates are in the crossflow direction. We will first discuss the determination of initial distributions of F, G, and I in the z, η plane which will provide information appropriate for marching the solution in the x direction.

If the origin of the x coordinate is taken at the stagnation point (or line in some flows) the momentum and energy equations reduce to ordinary differential equations which can be solved with the continuity equation to provide the necessary initial conditions in one plane. For the flow illustrated in Fig. 7-23, the appropriate form of the equations is obtained by simply neglecting all terms multiplied by x (which becomes equal to zero). This starting condition is similar to the flow at the leading edge of a sharp flat plate. On blunt bodies having a true stagnation point, $u_{1,e}$ and $u_{3,e}$ are known (Howarth, 1951) to vary linearly with x in the stagnation region. Thus, some of the terms which vanish for the sharp leading edge starting condition now have a nonzero limiting value as $x \to 0$ for blunt bodies. Blottner and Ellis (1973) discuss the stagnation point formulation in detail for incompressible flow.

In most three-dimensional boundary-layer flows, it is possible to compute initial distributions of F, G, and I (and u_1, u_3, and H when the untransformed curvilinear system is used) on a second intersecting plane by solving the PDE's on a *plane of symmetry*. The formulation of the plane of symmetry problem will be discussed below, but first, it is well to mention that in a few problems it is not possible to identify a plane of symmetry. An example of this is the sharp spinning cone considered by Dwyer (1971) and Dwyer and Sanders (1975). Controversy has erupted over the question as to whether this flow can be treated as an initial value problem with the boundary-layer equations (Lin and Rubin, 1973a). It appears that use of difference schemes which lag the representation of the crossflow derivatives (Dwyer and Sanders, 1975; Kitchens et al., 1975) permits the solution to be marched away from a single initial data plane into those regions not forbidden by the zone of dependence principle. Such difference representations as well as the zone of dependence principle will be discussed in Section 7-7.3.

The plane of symmetry is indicated in Fig. 7-23 for the flow over a flat plate with an attached cylinder. Flows over nonspinning bodies of revolution at incidence typically have both a windward and a leeward plane of symmetry, the former being most commonly used to develop the required second plane of initial data. Along the plane of symmetry,

$$G = \frac{\partial F}{\partial z} = \frac{\partial V}{\partial z} = \frac{\partial^2 G}{\partial z^2} = 0 \tag{7-114}$$

The inviscid flow and fluid properties are also symmetric about the plane of symmetry. Using Eq. (7-114), the x-momentum and energy equations reduce to two-dimensional form. The problem remains three dimensional, however, because the cross-derivative term in the continuity equation does *not* vanish on the plane of symmetry. Expanding out the cross-derivative term in Eq. (7-110) and invoking the symmetry conditions, Eq. (7-114), permits the continuity equation to be written as

$$\frac{x}{h_1 h_3}\frac{\partial(h_3 F)}{\partial x} + \frac{F}{2h_1}(1+\beta_1) + \frac{\partial V}{\partial \eta} + \frac{xG_z}{h_3 u_{1,e}}\frac{\partial u_{3,e}}{\partial z} = 0 \tag{7-115}$$

where

$$G_z = \frac{\partial u_3/\partial z}{\partial u_{3,e}/\partial z}$$

The z-momentum equation in the form given by Eq. (7-112) provides no useful information because $G = 0$ everywhere in the plane of symmetry. However, differentiating the z-momentum equation with respect to z and again invoking the symmetry conditions provides an equation which can be solved for the required values of G_z:

$$\frac{xF}{h_1}\frac{\partial G_z}{\partial x} + V\frac{\partial G_z}{\partial \eta} + xFG_zK_3 = \beta_a(\theta - FG_z) + \beta_{10}(\theta - G_z^2) + x\theta K_3 \tag{7-116}$$

$$+\frac{\partial}{\partial \eta}\left[\frac{\rho\bar{\mu}}{(\rho\mu)_e}\frac{\partial G_z}{\partial \eta}\right] \qquad \text{(7-116)} \; (Cont.)$$

Defining $W_{e,z} = \partial u_{3,e}/\partial z$, we can express the parameters β_9 and β_{10} as

$$\beta_9 = \frac{x}{h_1 W_{e,z}} \frac{\partial W_{e,z}}{\partial x}, \qquad \beta_{10} = \frac{x W_{e,z}}{h_3 u_{1,e}}$$

$W_{e,z}$ is to be obtained from the inviscid flow solution. Equation (7-116) for G_z has the same general form as the original z-momentum equation and can be solved by marching in the x direction along the plane of symmetry.

The arbitrary parameter W_e used in nondimensionalizing the crossflow velocity component is chosen to avoid singular behavior. We take $W_e = u_{3,e}$ at the stagnation point and along the plane of symmetry and $W_e = u_{1,e}$ elsewhere.

7-7.3 Comments on Solution Methods for Three-Dimensional Flows

The three-dimensional boundary-layer problem involves several complicating features and considerations not present in the two-dimensional flows considered thus far. The inviscid flow solution required to provide the pressure gradient input for the boundary-layer solution is often considerably more difficult to obtain than for two-dimensional flows. Generation of the metrics and other information needed to establish the curvilinear body-oriented coordinate system can also be a significant task for complex bodies. Turbulence models need to be extended to provide representation for the new apparent stress. In addition, the following features require special attention in the difference formulation: (1) implementation of the zone of dependence principle, and (2) representation of the crossflow convective derivatives in a manner to permit a stable solution for both positive and negative crossflow velocity components.

The three-dimensional boundary-layer equations have a hyperbolic character in the x-z plane and the mathematical constraint which results is very much like the CFL condition discussed in connection with the wave equation. Major contributions to the formulation and interpretation of the zone of dependence principle for three-dimensional boundary layers can be found in the work of Raetz (1957), Der and Raetz (1962), Wang (1971), and Kitchens et al. (1975). The principle actually addresses both a zone of dependence and a zone of influence and is sometimes just referred to as the "influence" principle. The dependence part of the principle is the most relevant to the proper establishment of difference schemes and so has been given emphasis here.

If we consider the point labeled P in Fig. 7-24 within a three-dimensional boundary layer, the influence principle states that the influence of the solution at P is transferred instantaneously by diffusion to all points in the viscous flow on a line (labeled A-B in Fig. 7-24) normal to the surface passing through P and by convection downstream along all streamlines through that point. Normals to the body surface form the characteristic surfaces and the speed of propagation is

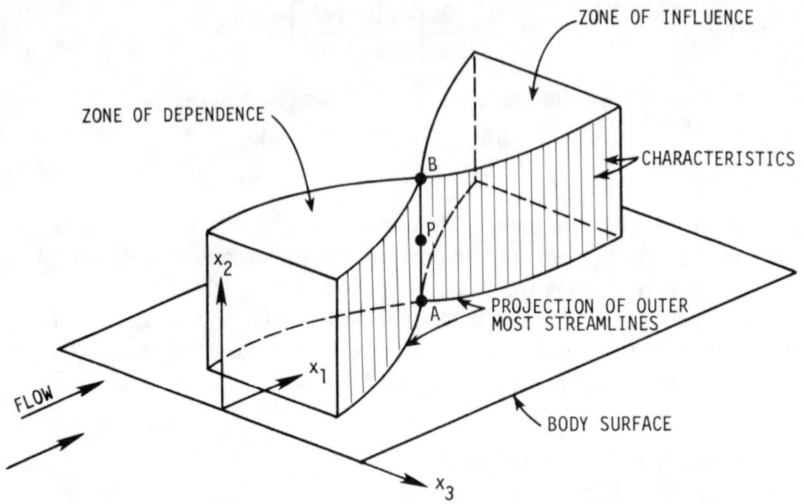

Figure 7-24 Zones of dependence and influence in three-dimensional boundary layers.

infinite in that direction. Disturbances anywhere along A-B are felt instantaneously along the whole line A-B and are carried downstream by *all* streamlines passing through A-B. The positions of the two outermost streamlines through A-B and extending downstream define the lateral extent of the wedge-shaped *zone of influence* for points on A-B. Events along A-B can influence the flow within the region bounded by the characteristics (lines normal to the wall) through these outermost streamlines. Typically, one outermost streamline is the limiting streamline at the wall and the other is the inviscid flow streamlines. The flow along A-B is obviously influenced by the flow upstream and a *zone of dependence* is defined by the characteristic lines passing through the two outermost streamlines extending upstream. Events at all points within this wedge-shaped region upstream can influence events along A-B. The "outermost" streamlines are those having the maximum and minimum angular displacement from the constant x_3 (or z) surface passing through A-B. The zone of dependence then designates the minimum amount of initial data which must be supplied to determine the solution along A-B. These concepts apply to the PDE's. It is important for the difference molecule used along A-B not to exclude information in the zone of dependence; that is, the zone of dependence implied by the difference representation must be at least as large as the zone identified with the PDE's. This has also been shown previously for hyperbolic PDE's and identified as the CFL condition. The exact quantitative statement of the zone of dependence principle depends upon the difference molecule employed. For example, using a scheme which represents $\partial G/\partial z$ centrally (with Δz constant) at the n marching level as the solution is advanced to the $n+1$ level, the zone of dependence principle would require that

$$F > 0 \qquad \left|\frac{h_1}{h_3}\frac{\Delta x G}{\Delta z F}\right| \leqslant 1 \tag{7-117}$$

Equation (7-117) indicates that the local angle made by the streamlines with the plane of constant z must be contained within the angle whose tangent is given by the mesh parameter $h_3\ \Delta z/(h_1\ \Delta x)$. We would like Eq. (7-117) to be satisfied at a given x level with Δx being the increment back upstream. It would be unprofitable to iterate simply to establish the allowable step size, so the usual procedure is to utilize the most recently calculated values of G and F to establish the new step size using a safety factor to allow for anticipated changes in G and F over a Δx increment. In using Eq. (7-117) to establish the maximum allowable marching step increment, the inequality should be checked at each internal point at a given x level before Δx is established for the next step. With the use of certain difference schemes for flows in which G does not change sign, the zone of dependence constraint is automatically satisfied as will be illustrated below for the three-dimensional Crank-Nicolson scheme.

Stability is also a concern in three-dimensional boundary-layer calculations. The presence of the additional convective derivative in the momentum equation generally influences the stability properties of the difference scheme. The stability constraint of a scheme is very likely to change as it is extended from two- to three-dimensional flow. The concept of stability is separate from the concept of the zone of dependence. This point is demonstrated very well in the work of Kitchens et al. (1975). For some schemes, the constraint imposed by the zone of dependence principle will coincide with the stability constraint determined by the usual von Neumann analysis, but not always. Kitchens et al. (1975) show that for four difference schemes investigated, errors tend to grow whenever the zone of dependence principle was violated, but that for some schemes, the solutions remained very smooth and "stable" in appearance even though the errors were very significant. In other schemes, violation of the zone of dependence principle may trigger unstable behavior characterized by large oscillations even when such behavior is not predicted by a stability analysis. It is even possible to devise inherently unstable schemes which satisfy the zone of dependence constraint.

A few common difference schemes for three-dimensional boundary layers will be briefly described. In the following discussion, the indices n, j, k will be associated with the coordinate directions x_1, x_2, x_3 (or x, η, z). The solution is being advanced from the nth marching plane to the $n+1$ plane. The solution at the $n+1$ level will start at $k=1$ (usually on the plane of symmetry) where the equations will be solved for the unknowns for all values of j. That is, fixing n and k, we obtain the solution along a line normal to the wall. Then, the k index is advanced by one and the solution is obtained for another "column" of points along the surface normal. Thus, the marching (or "sweeping") at the $n+1$ level is in the crossflow direction. In difference representations below, the unknowns will be variables at the $n+1, k$ levels.

Crank-Nicolson scheme. The three-dimensional extension of the Crank-Nicolson scheme has been used by several investigators. Its use is restricted to flows in which the crossflow component of velocity does not change sign due to zone of dependence and stability considerations. The difference molecule is centered at

$n+\frac{1}{2}, j,k-\frac{1}{2}$. Figure 7-25(a) illustrates the molecule as we look down on the flow (i.e., only points in the x-z plane are shown). The shaded area indicates the approximate maximum zone of dependence permitted by the molecule. The circled point indicates the location of the unknowns and the x indicates the center of the molecule. The presence of negative crossflow components of velocity causes this scheme to violate the zone of dependence principle because no information is contained in the molecule which would reflect flow conditions in the negative coordinate direction as the solution is advanced to the $n+1, k$ level. On the other hand, the zone of dependence principle imposes no restriction on the size of Δx as long as $G \geqslant 0$ since the molecule spans all possible flow angles for which $F>0$, $G \geqslant 0$.

More than one variation of the Crank-Nicolson scheme has been proposed. In the most frequently used version, terms of the form $\partial/\partial\eta(a\,\partial\phi/\partial\eta)$ are differenced as for the two-dimensional Crank-Nicolson scheme but *averaged* between $k-1$ and k. Likewise, $\partial\phi/\partial x$ and $\partial\phi/\partial\eta$ terms are represented as for the two-dimensional Crank-Nicolson scheme but averaged over k and $k-1$. Derivatives in the crossflow direction [as for example in the term labeled (1) in Eq. (7-111)] are represented by

$$\left.\frac{\partial\phi}{\partial z}\right)_{j,k-1/2}^{n+1/2} \simeq \frac{\phi_{j,k}^{n+1}+\phi_{j,k}^{n}-\phi_{j,k-1}^{n+1}-\phi_{j,k-1}^{n}}{2\,\Delta z}$$

For flows over curved surfaces for which the curvature parameters K_1 and K_3 are nonzero, new terms of the form represented by terms labeled (2) and (3) in Eq. (7-111) must be represented. Similar terms appear in the untransformed equations in orthogonal curvilinear coordinates given in Chapter 5. Terms of this general form, not involving derivatives of the dependent variables, are considered source terms according to the definition given in Section 7-3.1. The terms labeled (4) and (5) in Eq. (7-111) are two additional source terms which arise due to the introduction of

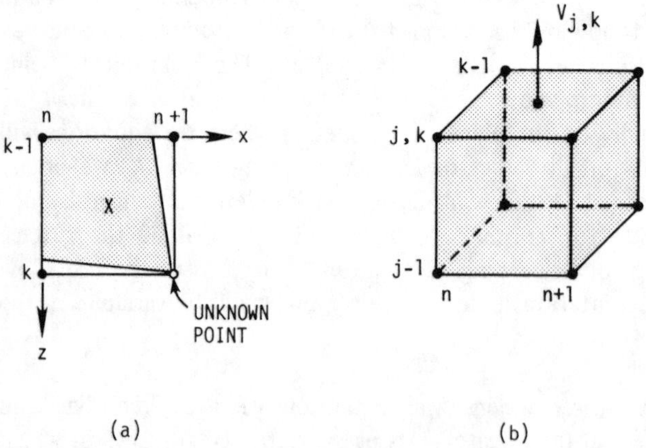

Figure 7-25 The Crank-Nicolson scheme. (a) The difference molecule projected on the x-z plane. (b) The control volume for the continuity equation.

the F and G variables. These terms, (2)-(5) in Eq. (7-111), require linearization in the difference representation as do the convective terms. Any of the linearization techniques suggested in Section 7-3.3 can be used, although coupling of the equations is not commonly used at present. The source terms are represented at the center of the difference molecule $(n+\frac{1}{2},\ j,k-\frac{1}{2})$ by appropriate averages of variables at neighboring grid points. As an example, the term labeled (2) in Eq. (7-111) can be represented as

$$(FGxK_1)^{n+1}_{j,k-1/2} \simeq x^{n+1/2}K^{n+1/2}_{1_{k-1/2}}(F^n_{j,k}+F^n_{j,k-1}+F^{n+1}_{j,k}+F^{n+1}_{j,k-1})$$
$$\times(G^n_{j,k}+G^n_{j,k-1}+G^{n+1}_{j,k-1}+\hat{G}^{n+1}_{j,k})/16 \qquad (7\text{-}118)$$

The only quantity treated algebraically as an unknown in Eq. (7-118) is $F^{n+1}_{j,k}$. The linearization is implemented by treating $\hat{G}^{n+1}_{j,k}$ algebraically as a known. The value of $\hat{G}^{n+1}_{j,k}$ can be determined by extrapolation, updated iteratively, or simply lagged, although lagging is not often used for the three-dimensional boundary-layer equations. Considerable flexibility exists in the way in which the various terms can be linearized. Other source terms appear on the right-hand side of Eq. (7-111), but these do not require linearization. The algebraic formulation for each momentum equation results in a simultaneous system of equations for the unknowns along the $n+1$, k column of points. The coefficient matrix is tridiagonal so that the Thomas algorithm can be used.

Most current procedures for the three-dimensional boundary-layer equations solve the continuity equation separately for $V^{n+1}_{j,k}$ after F and G have been determined from the solution to the momentum equations. The difference representation for the continuity equation is usually established by considering a control volume centered about $(n+\frac{1}{2},\ j-\frac{1}{2},\ k-\frac{1}{2})$. Such a control volume is illustrated in Fig. 7-25(b). For F and G, the average value of these quantities over a face of the control volume is established by taking the average of the quantities of the four corners of the face. Values of V are only needed at locations $n+\frac{1}{2},\ j,k-\frac{1}{2}$ in the momentum equations. Thus, computational effort is normally saved by simply letting the value of V determined from the continuity equation be the value at the center of the x-z planes of the control volume. For computer storage, the V physically considered to be located at $n+\frac{1}{2}, j,k-\frac{1}{2}$ is usually assigned the subscript $n+1$, j,k. The location of $V^{n+1}_{j,k}$ is indicated in Fig. 7-25(b) where the labeling usually used for computer storage is employed. Grid schemes in which the dependent variables are evaluated at different locations in the computational domain are usually referred to as "staggered" grids. In this staggered grid, all variables except V are evaluated at regular grid points. Further examples of staggered grids will arise in Chapter 8. The Crank-Nicolson scheme has the potential of being formally second-order accurate {(truncation error $0[(\Delta x)^2,\ (\Delta\eta)^2,\ (\Delta z)^2]$}. The truncation error may be less favorable depending on how linearizations and unequal mesh sizes are handled.

Krause zig-zag scheme. The Krause (1969) scheme has been widely used for flows in which the crossflow velocity component changes sign. The difference molecule is centered at $n + \frac{1}{2}$, j,k and its projection on the x-z plane is given in Fig. 7-26(a). The shaded area again denotes the approximate maximum zone of dependence permitted by the molecule. We note that the molecule includes information in both z directions from point $n+1, j,k$ so that, within limits, crossflow in both directions is permitted as long as the flow direction remains within the zone of dependence of the molecule. As for the Crank-Nicolson scheme, we observe that no mesh size constraint occurs when $F > 0$, $G \geqslant 0$. However, a constraint is observed when crossflow in the negative z direction occurs. The zone of dependence and the stability constraint for the Krause scheme can be stated as

$$F > 0 \qquad \frac{\Delta x h_1 G}{\Delta z h_3 F} \geqslant -1$$

It should be noted that the permitted flow direction can be altered by changing the aspect ratio, $\Delta z/\Delta x$, of the molecule.

The Krause difference representation is somewhat simpler algebraically than the Crank-Nicolson scheme, primarily because most difference representations are only averaged between $n+1$ and n, but *not* between two k levels. For the Krause scheme, terms of the form $\partial/\partial\eta(a\,\partial\phi/\partial\eta)$ and $\partial\phi/\partial x$ are differenced in the same manner as in the two-dimensional Crank-Nicolson scheme. Derivatives in the crossflow direction in the momentum equations are differenced using points in the zig-zag pattern denoted by the dashed lines in Fig. 7-26(a). For equal Δz increments this representation can be written

$$\left.\frac{\partial\phi}{\partial z}\right)_{j,k}^{n+1/2} \simeq \frac{\phi_{j,k+1}^{n} - \phi_{j,k}^{n} + \phi_{j,k}^{n+1} - \phi_{j,k-1}^{n+1}}{2\,\Delta z} \tag{7-119}$$

Since the sweep in the z direction is from columns $(n+1, k-1)$ to $(n+1, k)$, $\phi_{j,k}^{n+1}$ is the only unknown in Eq. (7-119). The problem of linearization of the algebraic representations is much the same as for the Crank-Nicolson scheme except

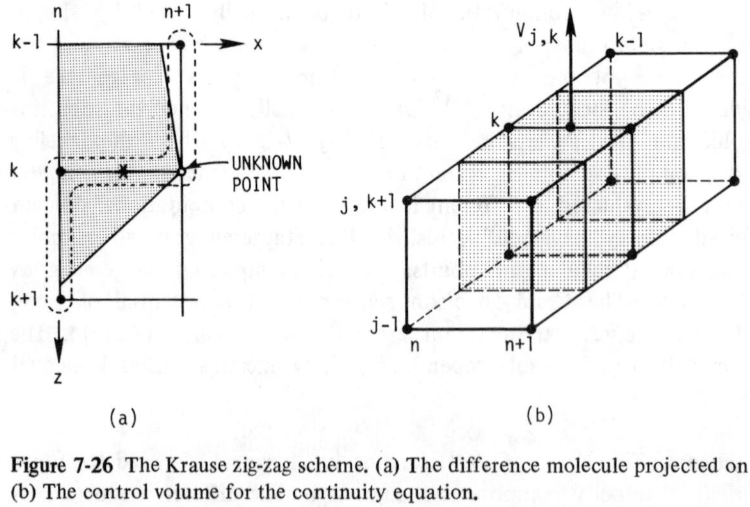

Figure 7-26 The Krause zig-zag scheme. (a) The difference molecule projected on the x-z plane. (b) The control volume for the continuity equation.

that the molecule is more compact because quantities generally only need to be averaged between two grid points instead of four. For example, the term labeled (2) in Eq. (7-111) can be represented as

$$(FGxK_1)_{j,k}^{n+1/2} = \frac{x^{n+1/2} K_{1k}^{n+1/2} (F_{j,k}^{n} + F_{j,k}^{n+1})(G_{j,k}^{n} + \hat{G}_{j,k}^{n+1})}{4} \tag{7-120}$$

A tridiagonal system of algebraic equations results from the Krause formulation which can be solved by the Thomas algorithm.

The difference equation for the continuity equation is established by considering a control volume centered about $(n + \frac{1}{2}, j - \frac{1}{2}, k)$ as indicated in Fig. 7-26(b). The average value of F on a η-z face of the control volume can be determined from averaging only in the η direction since the middle of the plane coincides with a k-level. A zig-zag (or diagonal) average is used to represent G on a x-η plane. To illustrate this for equal Δz increments, we would represent a term of the form $\partial(aG)/\partial z$ in the Krause continuity equation as

$$\left.\frac{\partial(aG)}{\partial z}\right)_{j-1/2,k}^{n+1/2} \simeq \{[(aG)_{j,k+1}^{n} + (aG)_{j-1,k+1}^{n} + (aG)_{j,k}^{n+1} + (aG)_{j-1,k}^{n+1}] - [(aG)_{j,k}^{n} + (aG)_{j-1,k}^{n} + (aG)_{j,k-1}^{n+1} + (aG)_{j-1,k-1}^{n+1}]\}/4\,\Delta z \tag{7-121}$$

The V determined from the Krause continuity equation is located at the center of the upper x-z plane of the control volume (at $n + \frac{1}{2}$, j,k, but usually stored as $n + 1$, j,k). The storage index is the one indicated in the labeling of Fig. 7-26(b). The truncation error for the Krause scheme is the same as for the Crank-Nicolson scheme. Further details on the Crank-Nicolson and Krause schemes can be found in Blottner and Ellis (1973).

Some variations. Two variations on the Krause scheme which have proven to be suitable for both positive and negative crossflow velocity components will be mentioned briefly. Wang (1973) has developed a second-order accurate two-step method which eliminates the need to linearize terms in the momentum equations. As with all multilevel methods, initial data must be provided at two marching levels. This is usually accomplished through the use of some other scheme for one or more steps. The projection of the two-step molecule on the x-z plane is shown in Fig. 7-27. The shaded area again indicates the approximate zone of dependence permitted by the molecule. Known data on the $n - 1$ and n levels are used to advance the solution. The method is implicit and centered at the point (n, j, k). Derivatives in the x and z direction are represented centrally about (n, j, k). Derivatives of the form

$$\frac{\partial}{\partial \eta}\left(a \frac{\partial \phi}{\partial \eta}\right)$$

are represented at $(n + 1,\ j,k)$ and $(n - 1,\ j,k)$ and averaged. The zone of dependence constraint is given by

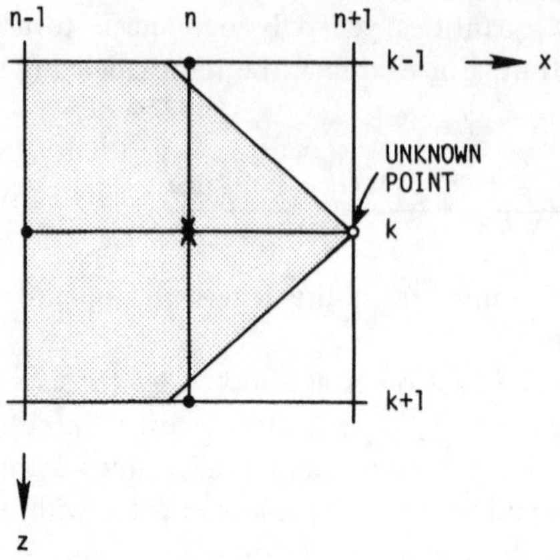

Figure 7-27 The two-step scheme.

$$F > 0 \qquad \left| \frac{\Delta x h_1 G}{\Delta z h_3 F} \right| \leqslant 1$$

No formal stability constraint is observed so long as $F > 0$.

Kitchens et al. (1975) compared the properties of four schemes for three-dimensional boundary layers and found that their scheme D had quite favorable error growth and stability properties. In addition, the results seemed relatively insensitive to violations in the zone of dependence constraint. The projection of this difference molecule on the *x-z* plane is shown in Fig. 7-28. The shaded area indicates the approximate zone of dependence for the method. The method is implicit. Derivatives in the x direction are represented centrally about $(n + \frac{1}{2}, j, k)$ with one very unique twist which converts an otherwise unstable scheme into a stable one. In the representation for $\partial\phi/\partial x$, values needed at n, j, k are replaced by the average of $\phi^n_{j,k+1}$ and $\phi^n_{j,k-1}$. Thus, for equal increments we would use

$$\frac{\partial \phi}{\partial x} \simeq \frac{\phi^{n+1}_{j,k} - 0.5(\phi^n_{j,k+1} + \phi^n_{j,k-1})}{\Delta x}$$

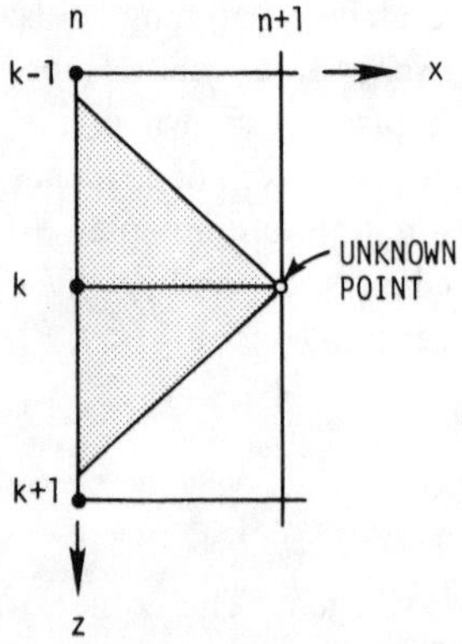

Figure 7-28 Scheme D (Kitchens et al., 1975).

Derivatives in the crossflow direction are represented centrally about (n, j, k). Derivatives of the form $\partial/\partial\eta(a\,\partial\phi/\partial\eta)$ are represented at $(n+1, j, k)$ and (n, j, k) and averaged. The truncation error stated by Kitchens et al. (1975) is $O[\Delta x, (\Delta z)^2/\Delta x, (\Delta\eta)^2, (\Delta z)^2]$. The zone of dependence constraint for this method is the same as for the two-step method. In this case, the stability restriction is the same as the zone of dependence constraint.

7.7-4 Example Calculations

Here we briefly present some example computational results for the sample three-dimensional flow illustrated in Fig. 7-23. The results were obtained by application of the Krause scheme to Eqs. (7-110) to (7-112) for an incompressible laminar flow. The Crank-Nicolson scheme was used at the last z station to permit the calculation to end without requiring information from the $k+1$ level. Computed results for this flow have been reported in the literature by several investigators [see, for example, Cebeci (1975)]. For this flow, the inviscid velocity distribution is given by

$$u_{1,e} = u_\infty\left(1 + a^2\,\frac{\gamma_2}{\gamma_1}\right), \qquad u_{3,e} = -2u_\infty a^2\,\frac{\gamma_3}{\gamma_1^2}$$

where u_∞ is the reference freestream velocity and $\gamma_1 = (x-x_0)^2 + z^2$, $\gamma_2 = -(x - x_0)^2 + z^2$, and $\gamma_3 = (x-x_0)z$. The parameter x_0 is the distance of the cylinder axis from the leading edge, a is the cylinder radius, and x and z denote the distance measured from the leading edge and plane of symmetry, respectively. It is also useful to know $\partial u_{3,e}/\partial z$ along the plane of symmetry:

$$\left.\frac{\partial u_{3,e}}{\partial z}\right)_{z=0} = \frac{-2u_\infty a^2}{(x-x_0)^3}$$

Calculations were made for $u_\infty = 30.5$ m/s, $a = 0.061$ m, $x_0 = 0.457$ m using $\Delta x = 0.0061$ m, $\Delta\eta = 0.28$, and $\Delta z = 0.0061$ m. Typical velocity profiles for this flow are shown in Fig. 7-29. In particular we note that the crossflow velocity component reaches a maximum within the inner one-third of the boundary layer. The variation in the flow angle (in x-z plane) with distance from the wall is shown in Fig. 7-30(a). The maximum skewing is observed to occur close to the wall. The velocity vector is seen to rotate through an angle of about 13° along the surface normal. This corresponds to the included angle made by the zone of dependence at this location (see Fig. 7-24). The variation in the skin-friction coefficient is shown in Fig. 7-30(b). The presence of the cylinder causes the flow on the plane of symmetry to separate at $x \simeq 0.26$ m. Conventional boundary-layer calculation methods can proceed no farther along the plane of symmetry because both the x and z components of velocity have vanished. It will be interesting to see in the future whether extensions of inverse boundary-layer methods will prove useful in advancing the solution through singular points in three-dimensional flows.

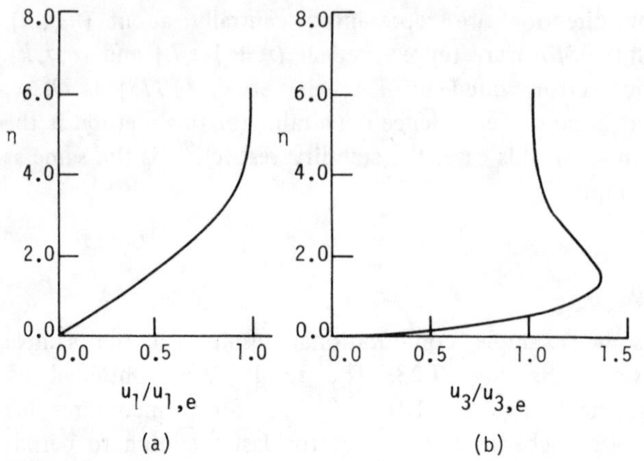

Figure 7-29 Velocity profiles at $x = 0.219$ m, $z = 0.079$ m, for an example three-dimensional flow over a flat plate with an attached cylinder. (a) Streamwise velocity distribution. (b) Crossflow velocity distribution.

7-7.5 Concluding Remarks

Only a few representative difference schemes for three-dimensional boundary layers have been discussed in this chapter. Many other useful procedures have been suggested. Several of these are discussed by Wang (1974), Kitchens et al. (1975), and Blottner (1975b). Cebeci (1975) has extended the box scheme to three-dimensional flows and recently (Cebeci et al., 1979a) have implemented a zig-zag feature which permits the calculation of three-dimensional flows in which the crossflow velocity component changes sign. No single scheme has emerged to date as being superior for all flows. Several investigators have found the need to employ

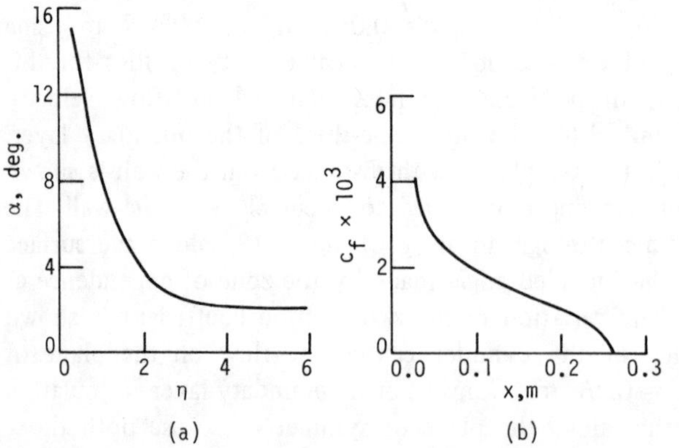

Figure 7-30 Example of three-dimensional boundary-layer flow over a flat plate with an attached cylinder. (a) Variation of flow angle (measured from x-η plane) along the surface normal at $x = 0.219$ m, $z = 0.079$ m. (b) Variation of skin-friction coefficient along plane of symmetry.

more than one scheme in order to cover all regions efficiently in some flows. The Krause zig-zag scheme is recommended as a reasonable starting point for the development of a three-dimensional boundary-layer finite-difference procedure. After the Krause procedure is well in hand, the user should be encouraged to explore the possible advantages offered by the several variations which have been suggested.

Turbulence modeling is certainly an important concern in three-dimensional flows. Most three-dimensional turbulent calculations to date have assumed that the turbulent viscosity is a scalar and have employed the extension of Prandtl's mixing-length formula given in Chapter 5 [see Eq. (5-131a)]. Several recent investigations have explored the use of a "nonisotropic" modification in the outer portion of the flow (McLean and Randall, 1979; Lin et al., 1981). Recent measurements tend to support the view that in the outer portion of the flow the apparent viscosity in a Boussinesq evaluation of the stress in the crossflow direction may be substantially less (by a factor ~0.4–0.7) than the viscosity for the apparent stress in the streamwise direction. Further research on turbulence modeling for three-dimensional flows would seem desirable.

The most successful application of three-dimensional boundary-layer theory in recent years has probably been for flows over wings. Reasonably refined computer programs have been documented for this application (Cebeci et al., 1977; McLean and Randall, 1979). It is possible to employ viscous-inviscid interaction in three-dimensional flows although the computation of the displacement surface is somewhat more complex. Such viscous-inviscid interaction calculations have been carried out for wing geometries (McLean and Randall, 1979). The simple Cauchy integral small disturbance correction [Eq. (7-83)] is not applicable in three dimensions and the inviscid flow is generally recomputed in its entirety for each iterative pass. However, rather than recomputing the inviscid flow for the body modified by the displacement surface, it is frequently advantageous to maintain the same body in the inviscid calculation and represent the effects of the viscous flow by a distribution of sources and sinks (Lighthill, 1958). It was actually Lighthill's source-sink concept, which was used in the representation in Section 7-4.4, that reduced to the Cauchy integral formulation of Eq. (7-83) in two-dimensional incompressible flow when the viscous disturbance was small. In the full inviscid potential flow solution, the sources and sinks (related to space derivatives of the displacement thicknesses) are represented as normal velocity boundary conditions (blowing or suction) at the body surface. This formulation offers an advantage when direct methods are used to solve the elliptic partial differential equations for the subsonic flow in that the influence matrix and its inverse need not be recomputed for each interaction iteration. The viscous-inviscid interaction calculations reported to date for fully three-dimensional flows have employed the usual direct method for solving the boundary-layer equations. Little is known about the applicability of the inverse boundary-layer approach for fully three-dimensional flows.

Several papers and reports in the literature of a review or general nature should prove useful in obtaining a broad view of the status of predictions in three-dimensional boundary layers. The list includes Wang (1974, 1975), Bushnell et al. (1976), Blottner (1975b), and Kitchens et al. (1975).

7-8 UNSTEADY BOUNDARY LAYERS

It is frequently desirable to predict unsteady boundary-layer behavior, especially in the design of flight vehicles. Numerical aspects of this problem are reasonably well understood; however, challenging aspects of turbulence modeling remain. We will limit our discussion to two-dimensional unsteady boundary layers although many of the concepts carry over to the three-dimensional case.

The unsteady two-dimensional boundary-layer equations appear as Eqs. (5-116)–(5-118) in Chapter 5. They differ from their steady flow counterparts only through the appearance of the term $\rho \partial u/\partial t$ in the momentum equation and $\partial \rho/\partial t$ in the continuity equation. The unsteady equations are also parabolic but with time as the marching parameter. Values of u, v, H, and fluid properties must be stored at grid points throughout the flow domain. Initial values of u, v, and H must be specified for all x and y. Boundary conditions may vary with time. The usual boundary conditions are:

1. At $x = x_0$, $u(t, x_0, y)$, and $H(t, x_0, y)$ are prescribed for all y and t.
2. At $y = 0$, $u(t, x, 0) = v(t, x, 0) = 0$.
3. $\lim_{y\to\infty} u(t, x, y) = u_e(t, x)$.

The main objective is to develop computational procedures which will provide accurate and stable solutions when flow reversal ($u < 0$) occurs. In this respect, the unsteady two-dimensional boundary-layer problem is similar to the three-dimensional steady problem where the concern was to identify methods which would permit flow reversal in the crossflow direction. When flow reversal occurs in the unsteady problem, it is crucial to employ a difference representation which permits upstream influence. This principle has not been formulated in terms of a zone of dependence concept for the two-dimensional unsteady boundary-layer equations, but to ignore the possibility of information being convected in the flow direction is physically implausible. Furthermore, the steady boundary-layer equations are parabolic in the x direction which again requires that information move in the direction of the x component of velocity, otherwise it would not be possible to achieve the correct steady-state solution from the transient formulation.

An adaptation of the zig-zag representation introduced by Krause for the three-dimensional boundary-layer equations has been frequently used to represent $\partial u/\partial x$ when flow reversal is present. This representation is illustrated in Fig. 7-31. Using the mesh notation introduced in the figure, the zig-zag representation of the streamwise derivative for equal Δx increments is

$$\frac{\partial u}{\partial x} \simeq \frac{u_{i,j}^{n+1} - u_{i-1,j}^{n+1} + u_{i+1,j}^{n} - u_{i,j}^{n}}{2\,\Delta x} \tag{7-122}$$

The j index is associated with the normal coordinate. The representation of Eq. (7-122) can be used with a difference scheme centered at $n + \frac{1}{2}, i, j$ which can be

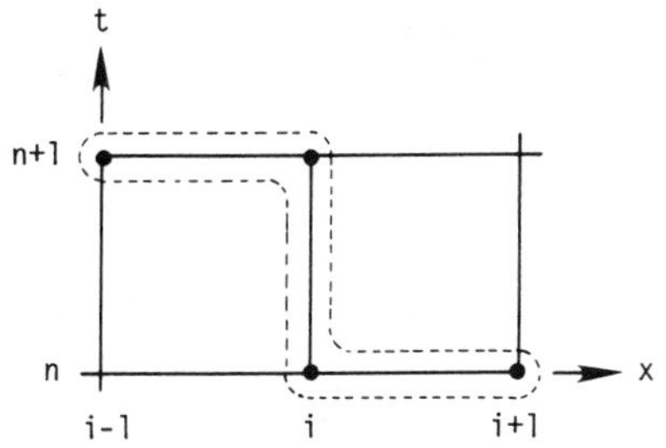

Figure 7-31 The zig-zag representation for streamwise derivatives in unsteady flows.

thought of as the unsteady two-dimensional boundary-layer version of the Krause scheme for three-dimensional boundary layers.

Concepts from the zig-zag box scheme for three-dimensional boundary layers have also been used to develop a zig-zag, box representation for unsteady boundary layers. This scheme also appears applicable when flow reversal is present (Cebeci et al., 1979a).

Other types of upwind differencing have been found to work satisfactorily with flow reversal by Telionis et al. (1973) and Murphy and Prenter (1981). The method of Murphy and Prenter (1981) also utilizes a fourth-order accurate discretization in the normal direction.

Blottner (1975) provides a helpful review of computational work on the unsteady boundary-layer equations. Other useful references include Telionis et al. (1973), Tsahalis and Telionis (1974), Telionis and Tsahalis (1976), Cebeci et al. (1979b), Phillips and Ackerberg (1973), and Murphy and Prenter (1981).

PROBLEMS

7-1 Verify the stability constraints given in Section 7-3.2 for the two versions of the simple explicit procedure for the boundary-layer equations.

7-2 The term $(\partial u/\partial y)^2$ needs to be evaluated at the $n+1$ marching level where u is an unknown. The marching coordinate is x, and y denotes the normal distance from the wall in a viscous flow problem. Utilize Newton linearization to obtain a difference representation for $(\partial u/\partial y)^2$ which could be used iteratively with the Thomas algorithm and which would be linear in the unknowns at each application of the algorithm.

7-3 Verify that Eqs. (7-20) and (7-21) give rise to a block tridiagonal system of algebraic equations. Identify the elements in the blocks.

7-4 Verify Eq. (7-24).

7-5 Generalize Eq. (7-24) to provide a second-order accurate representation when the mesh increments, Δx and Δy, are not constant.

7-6 Consider the following proposed implicit representation for the boundary-layer momentum equation.

$$u_j^n \frac{u_j^{n+1} - u_j^n}{\Delta x} + v_j^n \frac{u_j^{n+1} - u_{j-1}^{n+1}}{\Delta y} = \frac{\nu}{(\Delta y)^2} (u_{j+1}^{n+1} - 2u_j^{n+1} + u_{j-1}^{n+1})$$

Would you expect to find any mesh Reynolds number restrictions on the use of this representation when employing the Thomas algorithm for u and $v > 0$? Substantiate your answer.

7-7 Work Prob. 7-6 for the difference equation which results when the second term in the equation is replaced by

$$v_j^n \frac{u_{j+1}^{n+1} - u_j^{n+1}}{\Delta y}$$

7-8 Work out the hybrid representation for $v\, \partial u/\partial y$ in a form similar to Eq. (7-27) but for $v_j^n < 0$.

7-9 Verify the stability constraint given by Eq. (7-30).

7-10 Carry out the steps indicated in the development of the modified box scheme for the heat equation. Verify Eqs. (7-38)–(7-40).

7-11 Establish that the algebraic system represented by Eqs. (7-48) and (7-49) is block tridiagonal with 2 × 2 blocks. Verify that it fits the format required by the modified tridiagonal elimination scheme given in Section 7-3.3 for solving the momentum and continuity equations in a coupled manner.

7-12 Use the modified box scheme [Eq. (7-37)] to solve Prob. 4-25.

7-13 Write a computer program using an implicit method (either fully implicit, Crank-Nicolson, or the modified box scheme) to solve the incompressible laminar boundary-layer equations for flat plate flow in both physical (scheme A) and transformed (scheme B) [Eqs. (7-52) and (7-53)] coordinates. Linearize the difference equations either by lagging or extrapolating the coefficients v and u. Solve the momentum and continuity equations in an uncoupled manner. Use the tridiagonal elimination scheme to solve the system of simultaneous equations. Use $\Delta\eta = 0.3$ for scheme B and $\rho u_\infty \Delta y/\mu = 60$ for scheme A. For scheme A the boundary layer will grow with increasing x so that it will be necessary to add points to the computational domain as the calculation progresses. It will be possible to increase the marching step size in proportion to the boundary-layer thickness. It is suggested that the first step be established as $\Delta x = \rho u_\infty (\Delta y)^2/2\mu$ for scheme A.

Compare schemes A and B for ease of programming and accuracy. Consider the similarity solution tabulated in Schlichting (1979) as an "exact" solution for purposes of comparison. Calculate

$$c_f = \frac{\mu(\partial u/\partial y)_w}{\rho u_e^2/2}$$

from the solution. Determine $(\partial u/\partial y)_w$ by fitting a second-degree polynomial through the solution near the wall. Limit the downstream extent of the calculation to 75 streamwise steps. Investigate the sensitivity of methods to the streamwise step size. Perform the calculations for $\Delta x = 1\delta, 2\delta, 4\delta$. For scheme B, study the influence of the starting procedure on accuracy by first performing a streamwise calculation sweep by using $v = 0$ at $x = 0$ in the momentum equation and then repeating the calculation determining v at $x = 0$ iteratively through the use of the continuity equation.

7-14 Work Prob. 7-13 with the following changes. Select an implicit scheme and choose either physical or transformed coordinates in which to express the boundary-layer equations. Scheme A linearizes the coefficients by lagging and scheme B implements the linearization through the Newton procedure with coupling of the continuity equation.

7-15 Work Prob. 7-13 using either physical or transformed coordinates. Let scheme A be an implicit scheme of your choice and scheme B be an explicit procedure such as DuFort-Frankel, hopscotch or ADE.

7-16 Modify a difference scheme used in working Probs. 7-13 through 7-15 to permit the calculation of a boundary-layer flow in a pressure gradient. Verify your difference scheme by comparing the predicted velocity profiles with the results from the similarity solutions to the

Faulkner-Skan equations (see Schlichting, 1979) for a potential flow given by $u_e(x) = u_1 x^m$ where u_1 and m are constants and x is the streamwise coordinate. Make your comparisons for $m = \frac{1}{3}$ and -0.0654. You may choose any convenient value for u_1.

7-17 Modify a difference scheme used in working Probs. 7-13 through 7-15 to permit the calculation of a boundary-layer flow with blowing or suction. Verify your difference scheme by comparing the predicted velocity profiles with the results obtained by Hartnett and Eckert (1957) for blowing and suction distributions given by

$$\frac{v_w(x)}{u_\infty} \sqrt{\text{Re}_x} = 0.25 \text{ and } -2.5$$

7-18 Develop a finite-difference scheme for compressible laminar boundary-layer flow. Solve the energy equation in an uncoupled manner. Use the computer program to predict the skin-friction coefficient and Stanton number distributions for the flow of air over a flat plate at $M_e = 4$ and $T_w/T_\infty = 2$. Use the Sutherland equation [Eq. (5-40)] to evaluate the fluid viscosity as a function of temperature. Assume constant values of Pr and c_p ($\text{Pr} = 0.75$, $c_p = 1 \times 10^3$ J/kg K). Compare your predictions with the analytical results of van Driest (1952); (heat transfer results can be found in Kays and Crawford, 1980).

7-19 Modify a difference scheme used in working Probs. 7-13 through 7-15 to permit calculation of an incompressible turbulent boundary layer on a flat plate. Use an algebraic turbulence model from Chapter 5. Use $u_\infty = 33$ m/s and $\nu = 1.51 \times 10^{-5}$ m²/s. Compare your velocity profiles in law of the wall coordinates with Fig. 5-7. Compare your predicted values of c_f with the measurements of Wieghardt and Tillmann (1951) tabulated below:

x, m	c_f
0.087	0.00534
0.187	0.00424
0.287	0.00386
0.387	0.00364
0.487	0.00345
0.637	0.00337
0.787	0.00317
0.937	0.00317
1.087	0.00308

7-20 Verify Eq. (7-79).

7-21 Work out the details of the terms Q_j^n and R_j^n in Eq. (7-94) for the DuFort-Frankel scheme for internal flows.

7-22 Verify Eq. (7-97) for a fully implicit method.

7-23 Verify Eq. (7-100).

7-24 Verify Eq. (7-104).

7-25 Derive Eq. (7-116).

7-26 Specialize Eqs. (7-110)–(7-112) for an incompressible three-dimensional laminar flow in the Cartesian coordinate system. Write out the Crank-Nicolson representation for the equations. Explain your scheme for linearizing the algebraic equations.

7-27 Work Prob. 7-26 for the Krause zig-zag scheme.

7-28 Choose a suitable implicit finite-difference scheme to solve the three-dimensional laminar boundary-layer equations on the plane of symmetry for the example flow described in Section 7-7.4. Compare your predicted skin-friction coefficients with the results in Cebeci (1975) and/or Fig. 7-30(b).

7-29 Solve the example flow of Section 7-7.4 using the Crank-Nicolson scheme. Use the grid described in the example. Compare your results with those given in Cebeci (1975).

7-30 Write out a Krause-type difference scheme for the two-dimensional, unsteady, incompressible boundary-layer equations.

CHAPTER

EIGHT

NUMERICAL METHODS FOR THE "PARABOLIZED" NAVIER-STOKES EQUATIONS

8-1 INTRODUCTION

The boundary-layer equations can be utilized to solve many viscous flow problems as discussed in Chap. 7. There are, however, a number of very important viscous flow problems which cannot be solved by using the boundary-layer equations. In these problems, the boundary-layer assumptions are not valid. For example, if the inviscid flow is fully merged with the viscous flow, the two flows cannot be solved independent of each other as required by boundary-layer theory. As a result, it becomes necessary to solve a set of equations which are valid in both the inviscid and viscous flow regions.

Examples of viscous flowfields where the boundary-layer equations are not the appropriate governing equations are shown in Fig. 8-1. The hypersonic rarefied flow near the sharp leading edge of a flat plate (Fig. 8-1a) is a classical example of a viscous flowfield which cannot be solved by the boundary-layer equations. In fact, very near the leading edge, the flow is not a continuum, so that the Navier-Stokes equations are invalid. In the merged layer region where the flow can first be considered a continuum, the shock layer and the viscous layer are fully merged and indistinguishable from each other. Further downstream, the shock layer coalesces into a discontinuity and a distinct inviscid layer develops between the shock wave and the viscous layer. This is the beginning of the interaction region which is further divided into the strong- and weak-interaction regions. The weak-interaction region eventually evolves into the classical Prandtl boundary-layer flow further downstream. Obviously, the boundary-layer equations cannot be used in the merged layer region because the viscous layer and the shock layer are completely merged. At the beginning of the strong-interaction region, the viscous flow cannot be solved independent of the inviscid flow because of

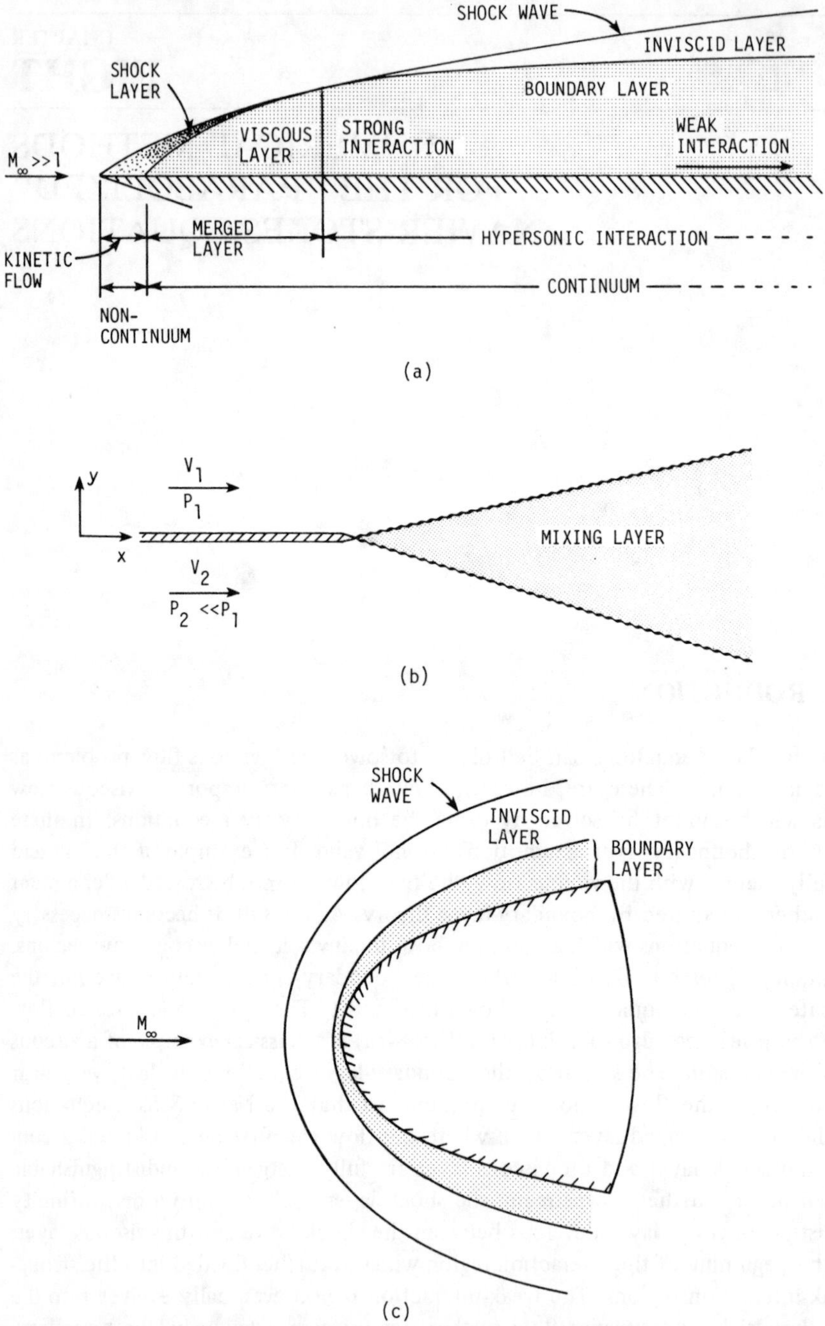

Figure 8-1 Examples of flowfields where the boundary-layer equations are not applicable. (a) Leading edge of a flat plate in a hypersonic rarefied flow. (b) Mixing layer with a strong transverse pressure gradient. (c) Blunt body in a supersonic flow at high altitude.

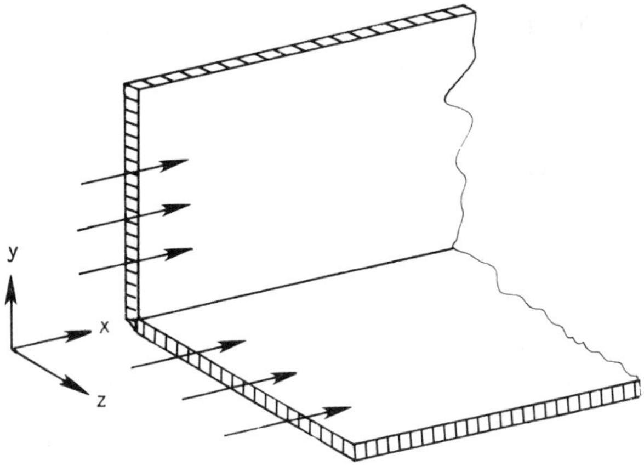

(d)

Figure 8-1 Examples of flowfields where the boundary-layer equations are not applicable (Cont.). (d) Flow along a streamwise corner.

the strong interaction. In the weak-interaction region, it is possible to solve the inviscid and viscous portions of the flow separately but this must be done in an iterative fashion as discussed in Chap. 7. That is, the boundary-layer equations can be computed initially using approximate edge conditions. With the computed displacement thickness, the inviscid portion of the flowfield can then be determined. This provides new edge conditions for the recomputation of the boundary layer. This procedure can be repeated until the solution for the entire flowfield does not change between iterations. Unless the interaction is very weak, it has been observed that this iterative procedure is often inferior to solving a set of equations which is valid in both the inviscid and viscous flow regions (Davis and Rubin, 1980).

Figure 8-1*b* illustrates a mixing layer problem for which the boundary-layer (thin-shear-layer) equations are not applicable. Across the mixing layer a strong normal pressure gradient exists. Consequently, the usual boundary-layer (thin-shear-layer) equations which contain the normal momentum equation

$$\frac{\partial p}{\partial y} = 0 \tag{8-1}$$

are not valid. In this case, a more complete normal momentum equation is required. Another example of a flowfield where the boundary-layer equations may not be applicable is the supersonic flow around a blunt body at high altitude as seen in Fig. 8-1*c*. In the region between the shock wave and the body (i.e., the shock layer) there exists a strong interaction between the boundary layer and the inviscid flow region. As a result, sets of equations which are valid in both the inviscid and viscous regions are normally used to compute this type of flowfield.

The flow along the corner formed by two intersecting surfaces, illustrated in Fig. 8-1*d*, provides a final example of a flow for which the boundary-layer equations are not applicable. As pointed out in Chap. 7, the boundary-layer equations only include viscous derivatives with respect to a single "normal" coordinate direction. Very near the corner, viscous derivatives with respect to *both* "normal" directions will be important. Such a flow configuration occurs often in applications, as for example, near wing-body junctures and in rectangular channels.

The complete Navier-Stokes equations are an obvious set of equations which can be used to solve the flowfields in Fig. 8-1 as well as all other viscous flowfields for which the boundary-layer equations are not applicable. In some cases they are the only equations which apply. Unfortunately, the Navier-Stokes equations are very difficult to solve in their complete form. In general, a very large amount of computer time and storage is necessary to obtain a solution with these equations. This is particularly true for the compressible Navier-Stokes equations which are a mixed set of elliptic-parabolic equations for a steady flow and a mixed set of hyperbolic-parabolic equations for an unsteady flow. The time-dependent solution procedure is normally used when a steady flowfield is computed. That is, the unsteady Navier-Stokes equations are integrated in time until a steady-state solution is achieved. Thus, for a three-dimensional flowfield a four-dimensional (3 space, 1 time) problem must be solved when the compressible Navier-Stokes equations are employed. Methods for solving the complete Navier-Stokes equations are discussed in Chap. 9.

Fortunately, for many of the viscous flow problems where the boundary-layer equations are not applicable, it is possible to solve a reduced set of equations that fall between the complete Navier-Stokes equations and the boundary-layer equations in terms of complexity. These reduced equations belong to a class of equations which are often referred to as the "thin-layer" or "parabolized" Navier-Stokes equations. There are several sets of equations that fall within this class. Some of the names of the equations are:

1. "Thin-layer" Navier-Stokes equations
2. "Parabolized" Navier-Stokes equations
3. "Partially parabolized" Navier-Stokes equations
4. Viscous shock-layer equations
5. "Conical" Navier-Stokes equations

The sets of equations in this class are characterized by the fact that they are applicable to both inviscid and viscous flow regions. In addition, the equations all contain a nonzero normal pressure gradient. This is a necessary requirement if viscous and inviscid regions are to be solved simultaneously.

There are two very important advantages that result when these equations are used instead of the complete Navier-Stokes equations. First of all, there are fewer terms in the equations which leads to some reduction in the required computation time. Secondly, and by far the most important advantage, is the fact that for a steady flow most of the equations in this class are a mixed set of hyperbolic-parabolic equations in the streamwise direction (provided that certain conditions are met). In other words the Navier-Stokes equations are "parabolized" in the streamwise direction. As a consequence, the equations can be solved using a boundary-layer type of marching technique so that a typical problem is reduced from four dimensions to three spatial dimensions. A substantial reduction in computation time and storage is thus achieved. In this chapter we will discuss the derivation of the equations in the thin-layer Navier-Stokes class and present a number of methods for solving them.

8-2 THIN-LAYER NAVIER-STOKES EQUATIONS

The unsteady boundary-layer equations can be formally derived from the complete Navier-Stokes equations by neglecting terms on the order of $1/(\mathrm{Re}_L)^{1/2}$ and smaller. As a consequence of this order of magnitude analysis, all viscous terms containing derivatives parallel to the body surface are dropped since they are substantially smaller than viscous terms containing derivatives normal to the wall. In addition, the normal momentum equation is reduced to a simple equation [i.e., Eq. (8-1) for a Cartesian coordinate system] which indicates that the normal pressure gradient is negligible. In the thin-layer approximation to the Navier-Stokes equations, the viscous terms containing derivatives in the directions parallel to the body surface are again neglected in the unsteady Navier-Stokes equations, but all other terms in the momentum equations are retained. One of the principal advantages of retaining the terms which are normally neglected in boundary-layer theory is that separated and reverse flow regions can be computed in a straightforward manner. Also, flows which contain a large normal pressure gradient, such as the ones shown in Fig. 8-1, can be readily computed.

The concept of the thin-layer approximation also arises from a detailed examination of typical high Reynolds number computations involving the complete Navier-Stokes equations (Baldwin and Lomax, 1978). In these computations, a substantial fraction of the available computer storage and time is expended in resolving the normal gradients in the boundary layer since a highly stretched grid is required. As a result, the gradients parallel to the body surface are usually not resolved in an adequate manner even though the corresponding viscous terms are retained in the computations. Hence, for many Navier-Stokes computations it makes sense to drop those terms that are not being adequately resolved provided that they are reasonably small. This naturally leads to the use of the thin-layer Navier-Stokes equations.

Upon simplifying the complete Navier-Stokes equations using the thin-layer approximation for the flow geometry shown in Fig. 8-2, the thin-layer Navier-Stokes equations in Cartesian coordinates become:

continuity:

$$\frac{\partial \rho}{\partial t} + \frac{\partial \rho u}{\partial x} + \frac{\partial \rho v}{\partial y} + \frac{\partial \rho w}{\partial z} = 0 \tag{8-2}$$

x momentum:

$$\frac{\partial \rho u}{\partial t} + \frac{\partial}{\partial x}(p + \rho u^2) + \frac{\partial}{\partial y}\left(\rho u v - \mu \frac{\partial u}{\partial y}\right) + \frac{\partial}{\partial z}(\rho u w) = 0 \tag{8-3}$$

y momentum:

$$\frac{\partial \rho v}{\partial t} + \frac{\partial}{\partial x}(\rho u v) + \frac{\partial}{\partial y}\left(p + \rho v^2 - \frac{4}{3}\mu \frac{\partial v}{\partial y}\right) + \frac{\partial}{\partial z}(\rho v w) = 0 \tag{8-4}$$

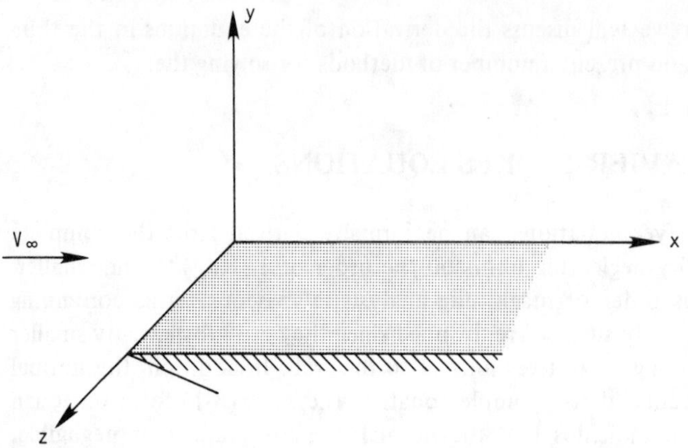

Figure 8-2 Flow over a flat plate.

z momentum:

$$\frac{\partial \rho w}{\partial t} + \frac{\partial}{\partial x}(\rho u w) + \frac{\partial}{\partial y}\left(\rho v w - \mu \frac{\partial w}{\partial y}\right) + \frac{\partial}{\partial z}(p + \rho w^2) = 0 \tag{8-5}$$

energy:

$$\frac{\partial E_t}{\partial t} + \frac{\partial}{\partial x}(E_t u + pu) + \frac{\partial}{\partial y}\left(E_t v + pv - \mu u \frac{\partial u}{\partial y} - \frac{4}{3}\mu v \frac{\partial v}{\partial y} - \mu w \frac{\partial w}{\partial y} - k \frac{\partial T}{\partial y}\right)$$
$$+ \frac{\partial}{\partial z}(E_t w + pw) = 0 \tag{8-6}$$

These equations are written for a laminar flow, but they can be readily modified to apply to a turbulent flow using the techniques of Section 5-4.

For more complicated body geometries it becomes necessary to map the body surface into a transformed coordinate surface in order to apply the thin-layer approximation. Suppose we apply the general transformation given by

$$\begin{aligned} \xi &= \xi(x, y, z, t) \\ \eta &= \eta(x, y, z, t) \\ \zeta &= \zeta(x, y, z, t) \\ t &= t \end{aligned} \tag{8-7}$$

to the complete Navier-Stokes equations (see Section 5-6.2) and let the body surface be defined as $\eta = 0$, as seen in Fig. 8-3. The transformed Navier-Stokes equations in strong conservation-law form become

$$\left(\frac{\mathbf{U}}{J}\right)_t + \left(\frac{\mathbf{U}\xi_t + \mathbf{E}\xi_x + \mathbf{F}\xi_y + \mathbf{G}\xi_z}{J}\right)_\xi + \left(\frac{\mathbf{U}\eta_t + \mathbf{E}\eta_x + \mathbf{F}\eta_y + \mathbf{G}\eta_z}{J}\right)_\eta$$
$$+ \left(\frac{\mathbf{U}\zeta_t + \mathbf{E}\zeta_x + \mathbf{F}\zeta_y + \mathbf{G}\zeta_z}{J}\right)_\zeta = 0 \tag{8-8}$$

where J is the Jacobian of the transformation and $\mathbf{U}$, $\mathbf{E}$, $\mathbf{F}$, and $\mathbf{G}$ are defined by Eqs. (5-44). We now apply the thin-layer approximation to the transformed Navier-Stokes equations. This approximation allows us to drop all viscous terms containing partial derivatives with respect to ξ and ζ. The resulting thin-layer equations may be written as (Pulliam and Steger, 1978):

$$\frac{\partial \mathbf{U}_2}{\partial t} + \frac{\partial \mathbf{E}_2}{\partial \xi} + \frac{\partial \mathbf{F}_2}{\partial \eta} + \frac{\partial \mathbf{G}_2}{\partial \zeta} = \frac{\partial \mathbf{S}_2}{\partial \eta} \tag{8-9}$$

where

$$\mathbf{U}_2 = \frac{\mathbf{U}}{J}$$

$$\mathbf{E}_2 = \frac{1}{J}\begin{bmatrix} \rho U \\ \rho u U + \xi_x p \\ \rho v U + \xi_y p \\ \rho w U + \xi_z p \\ (E_t + p)U - \xi_t p \end{bmatrix}$$

$$\mathbf{F}_2 = \frac{1}{J}\begin{bmatrix} \rho V \\ \rho u V + \eta_x p \\ \rho v V + \eta_y p \\ \rho w V + \eta_z p \\ (E_t + p)V - \eta_t p \end{bmatrix} \tag{8-10}$$

$$\mathbf{G}_2 = \frac{1}{J}\begin{bmatrix} \rho W \\ \rho u W + \zeta_x p \\ \rho v W + \zeta_y p \\ \rho w W + \zeta_z p \\ (E_t + p)W - \zeta_t p \end{bmatrix}$$

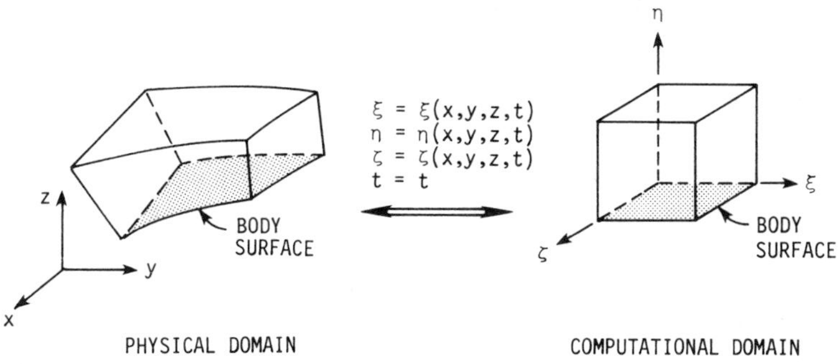

Figure 8-3 Generalized transformation.

and all the viscous terms are contained in

$$\mathbf{S}_2 = \frac{1}{J}\begin{bmatrix} 0 \\ \mu(\eta_x^2 + \eta_y^2 + \eta_z^2)u_\eta + \dfrac{\mu}{3}(\eta_x u_\eta + \eta_y v_\eta + \eta_z w_\eta)\eta_x \\ \mu(\eta_x^2 + \eta_y^2 + \eta_z^2)v_\eta + \dfrac{\mu}{3}(\eta_x u_\eta + \eta_y v_\eta + \eta_z w_\eta)\eta_y \\ \mu(\eta_x^2 + \eta_y^2 + \eta_z^2)w_\eta + \dfrac{\mu}{3}(\eta_x u_\eta + \eta_y v_\eta + \eta_z w_\eta)\eta_z \\ (\eta_x^2 + \eta_y^2 + \eta_z^2)\left[\dfrac{\mu}{2}(u^2 + v^2 + w^2)_\eta + kT_\eta\right] \\ + \dfrac{\mu}{3}(\eta_x u + \eta_y v + \eta_z w)(\eta_x u_\eta + \eta_y v_\eta + \eta_z w_\eta) \end{bmatrix} \tag{8-11}$$

For compactness, Eqs. (8-10) are written in terms of the contravariant velocity components (U, V, W) which are defined by

$$\begin{aligned} U &= \xi_t + \xi_x u + \xi_y v + \xi_z w \\ V &= \eta_t + \eta_x u + \eta_y v + \eta_z w \\ W &= \zeta_t + \zeta_x u + \zeta_y v + \zeta_z w \end{aligned} \tag{8-12}$$

The contravariant velocity components U, V, W are in directions normal to constant ξ, η, ζ surfaces, respectively.

Although the thin-layer Navier-Stokes equations are considerably less complicated than the complete Navier-Stokes equations, a substantial amount of computer effort is still required to solve these equations. The thin-layer Navier-Stokes equations are a mixed set of hyperbolic-parabolic PDE's in time. As a consequence, the "time-dependent" approach can be applied in an identical manner to the procedure normally used to solve the compressible Navier-Stokes equations. Thus, we will postpone our discussion of finite-difference methods for solving the thin-layer Navier-Stokes equations until Chap. 9 where the methods for solving the complete Navier-Stokes equations are discussed.

8-3 "PARABOLIZED" NAVIER-STOKES EQUATIONS

The "parabolized" Navier-Stokes (PNS) equations have recently gained popularity because they can be used to predict complex three-dimensional, steady, supersonic, viscous flowfields in an efficient manner. This efficiency is achieved because the equations can be solved using a space-marching finite-difference technique as opposed to the time-marching technique which is normally employed for the complete Navier-Stokes equations. As a result, the computational effort required to solve the PNS equations for an entire supersonic flowfield is similar to the effort required to solve either the inviscid portion of the flowfield using the Euler equations or the viscous portion of the flowfield using the boundary-layer equations. Furthermore, since the PNS equations are valid in both the inviscid and viscous portions of the flowfield, the interaction between these regions of the flowfield is automatically taken into account.

The term "parabolized" Navier-Stokes equations is somewhat of a misnomer since the equations are actually a mixed set of hyperbolic-parabolic equations provided that certain conditions are met. These conditions include the requirements that the inviscid outer region of the flow be supersonic and the streamwise velocity component be everywhere positive. Note that the last requirement excludes streamwise flow separation but crossflow separation is permitted. An additional constraint is caused by the presence of the streamwise pressure gradient in the streamwise momentum equation. If this term is included everywhere in the flowfield, then upstream influence can occur in the subsonic portion of the boundary layer and a space-marching method of solution is not well-posed. This leads to exponentially growing solutions which are often called *departure solutions.* Several techniques have been proposed to circumvent this difficulty and they will be discussed shortly.

8-3.1 Derivation of PNS Equations

The derivation of the PNS equations from the complete Navier-Stokes equations is, in general, not as rigorous as the derivation of the boundary-layer equations. Because of this, slightly different versions of the PNS equations have appeared in the literature. These versions differ in some cases because of the type of flow problem being considered. However, in all cases the normal pressure gradient term is retained and the second derivative terms with respect to the streamwise direction are omitted.

One of the earliest studies involving the use of the PNS equations was by Rudman and Rubin (1968). In their study, the hypersonic laminar flow near the leading edge of a flat plate (see Fig. 8-1a) was computed using a set of PNS equations. Rudman and Rubin derived their PNS equations from the complete Navier-Stokes equations using a series expansion technique. This method for reducing the complexity of the Navier-Stokes equations is an alternative to the order of magnitude analysis used in Chap. 5 to derive the boundary-layer equations. In the series expansion method, the flow variables are first nondimensionalized with respect to local reference conditions in order to estimate the magnitude of the various terms in the Navier-Stokes equations. The flow variables are then expanded in an appropriate series. Rudman and Rubin assumed the following form

$$
\begin{aligned}
u &= V_\infty(u_0^* + \epsilon u_1^* + \cdots)\\
v &= V_\infty \delta^*(v_0^* + \epsilon v_1^* + \cdots)\\
p &= p_\infty p_{\text{ref}}^*(p_0^* + \epsilon p_1^* + \cdots)\\
\rho &= \rho_\infty \rho_{\text{ref}}^*(\rho_0^* + \epsilon \rho_1^* + \cdots)\\
T &= T_\infty T_{\text{ref}}^*(T_0^* + \epsilon T_1^* + \cdots)\\
\mu &= \mu_\infty \mu_{\text{ref}}^*(\mu_0^* + \epsilon \mu_1^* + \cdots)\\
x &= x^* L \qquad y = y^*\delta \qquad \delta = \delta^* L
\end{aligned}
\tag{8-13}
$$

where a term with a subscript ref represents the local reference value of a flow variable nondimensionalized with respect to the freestream value, L is the characteristic length in the x direction, and δ is the characteristic length in the y direction. The first term in the series expansion (denoted with a subscript zero) is used to obtain the zeroth-order solution while both the first and second terms are needed to obtain the first-order solution. The relative magnitude of the coefficient ϵ is determined later in the analysis. For the relatively thin disturbed region shown in Fig. 8-1*a*, the gradients normal to the surface are much greater than the gradients parallel to the surface and δ^* can be assumed to be small.

When the expansions are substituted into the two-dimensional, steady, Navier-Stokes equations, the following nondimensional equations result (for convenience the subscript 0 has been dropped)

continuity:

$$\frac{\partial \rho^* u^*}{\partial x^*} + \frac{\partial \rho^* v^*}{\partial y^*} = O(\epsilon) \tag{8-14}$$

x momentum:

$$\rho^* u^* \frac{\partial u^*}{\partial x^*} + \rho^* v^* \frac{\partial u^*}{\partial y^*} = -\Delta^2 \frac{\partial p^*}{\partial x^*} + \frac{1}{(\delta^*)^2 \mathrm{Re_{ref}}} \frac{\partial}{\partial y^*}\left(\mu^* \frac{\partial u^*}{\partial y^*}\right) + O[\epsilon, (\mathrm{Re_{ref}})^{-1}] \tag{8-15}$$

y momentum:

$$\rho^* u^* \frac{\partial v^*}{\partial x^*} + \rho^* v^* \frac{\partial v^*}{\partial y^*} = -\left(\frac{\Delta}{\delta^*}\right)^2 \frac{\partial p^*}{\partial y^*} + \frac{1}{(\delta^*)^2 \mathrm{Re_{ref}}}\left[\frac{4}{3}\frac{\partial}{\partial y^*}\left(\mu^* \frac{\partial v^*}{\partial y^*}\right) + \frac{\partial}{\partial x^*}\left(\mu^* \frac{\partial u^*}{\partial y^*}\right) - \frac{2}{3}\frac{\partial}{\partial y^*}\left(\mu^* \frac{\partial u^*}{\partial x^*}\right)\right] + O[\epsilon, (\mathrm{Re_{ref}})^{-1}] \tag{8-16}$$

energy:

$$\Delta^2\left[\rho^* u^* \frac{\partial T^*}{\partial x^*} + \rho^* v^* \frac{\partial T^*}{\partial y^*} + (\gamma-1)p^*\left(\frac{\partial u^*}{\partial x^*} + \frac{\partial v^*}{\partial y^*}\right) - \frac{\gamma}{\mathrm{Pr}} \frac{1}{(\delta^*)^2 \mathrm{Re_{ref}}} \frac{\partial}{\partial y^*}\left(\mu^* \frac{\partial T^*}{\partial y^*}\right)\right] = \frac{\gamma-1}{(\delta^*)\mathrm{Re_{ref}}}\Bigg\{\mu^*\left(\frac{\partial u^*}{\partial y^*}\right)^2 + (\delta^*)^2\left[\frac{4}{3}\mu^*\left(\frac{\partial u^*}{\partial x^*}\right)^2 + \frac{4}{3}\mu^*\left(\frac{\partial v^*}{\partial y^*}\right)^2 - \frac{4}{3}\mu^* \frac{\partial u^*}{\partial x^*}\frac{\partial v^*}{\partial y^*} + 2\mu^* \frac{\partial v^*}{\partial x^*}\frac{\partial u^*}{\partial y^*}\right] + \epsilon\left[\mu_1^*\left(\frac{\partial u^*}{\partial y^*}\right)^2 + 2\mu^* \frac{\partial u^*}{\partial y^*}\frac{\partial u_1^*}{\partial y^*}\right]\Bigg\} + O\left[\Delta^2(\mathrm{Re_{ref}})^{-1}, (\delta^*)^2(\mathrm{Re_{ref}})^{-1}, \frac{\epsilon^2(\mathrm{Re_{ref}})^{-1}}{(\delta^*)^2}\right] \tag{8-17}$$

In the above equations, $\mathrm{Re}_{\mathrm{ref}} = (\rho_\infty V_\infty L/\mu_\infty)(\rho^*_{\mathrm{ref}}/\mu^*_{\mathrm{ref}})$, $\Delta^2 = T^*_{\mathrm{ref}}/M^2_\infty \gamma$, and a perfect gas is assumed.

The next step in the process is to determine which terms can be neglected in comparison to other terms in Eqs. (8-14)-(8-17). In order to do this, we need to obtain estimates for the magnitudes of $\mathrm{Re}_{\mathrm{ref}}$, Δ^2, and $(\Delta/\delta^*)^2$ in the various regions of the flowfield. From our previous discussions on boundary layers we know that for a thin viscous layer, $\mathrm{Re}_{\mathrm{ref}}$ is of order $1/(\delta^*)^2$. Also, near the edge of the viscous layer Δ^2 is proportional to $(M^2_\infty)^{-1}$ since $T^*_{\mathrm{ref}} = 1$ in this region. From compressible boundary-layer theory (Schlichting, 1968) it is known that Δ^2 can achieve a maximum value on the order of $(\gamma - 1)/2A\gamma$ where A varies between $\mathrm{Pr}^{-1/2}$ for an adiabatic wall to about 4 in the cold wall limit. Hence, for most cases we can assume $\Delta^2 \ll 1$ provided that $M_\infty \geqslant 5$. Rudman and Rubin (1968) have shown that $(\Delta/\delta^*)^2$ is of order unity in the merged layer region. Further downstream in the strong-interaction region, they have shown that $(\Delta/\delta^*)^2$ is very large near the wall but decreases in value to order unity at the edge of the boundary layer. Using the above information for the relative magnitudes of $\mathrm{Re}_{\mathrm{ref}}$, Δ^2, and $(\Delta/\delta^*)^2$ in the various regions of the flowfield, we can now simplify Eqs. (8-14)-(8-17). For the set of equations valid to zeroth-order ($M_\infty \geqslant 5$), we can neglect terms of order $(\delta^*)^2$, Δ^2, and ϵ; but we must retain terms of order $(\Delta/\delta^*)^2$. As a result, the continuity equation and the y-momentum equation cannot be reduced further. On the other hand, the x-momentum equation is simplified since the streamwise pressure gradient term can be dropped and the energy equation reduces to

$$\frac{\partial u^*}{\partial y^*} = 0 \tag{8-18}$$

If we combine Eq. (8-18) with the x-momentum equation, we find that

$$u^* = \text{constant} = 1 \tag{8-19}$$

or

$$u = V_\infty$$

Obviously this is a trivial result (applicable only in the freestream), and we are forced to retain higher-order terms [i.e., $(\delta^*)^2$, Δ^2, and ϵ] in order to obtain a meaningful energy equation. Note that we can eliminate many of the higher-order terms by employing Eq. (8-19). The final forms of the zeroth-order equations in dimensional form become

continuity:

$$\frac{\partial \rho u}{\partial x} + \frac{\partial \rho v}{\partial y} = 0 \tag{8-20}$$

x momentum:

$$\rho u \frac{\partial u}{\partial x} + \rho v \frac{\partial u}{\partial y} = \frac{\partial}{\partial y}\left(\mu \frac{\partial u}{\partial y}\right) \tag{8-21}$$

y momentum:

$$\rho u \frac{\partial v}{\partial x} + \rho v \frac{\partial v}{\partial y} = -\frac{\partial p}{\partial y} + \frac{4}{3}\frac{\partial}{\partial y}\left(\mu \frac{\partial v}{\partial y}\right) + \frac{\partial}{\partial x}\left(\mu \frac{\partial u}{\partial y}\right) - \frac{2}{3}\frac{\partial}{\partial y}\left(\mu \frac{\partial u}{\partial x}\right) \quad (8\text{-}22)$$

energy:

$$\rho u c_v \frac{\partial T}{\partial x} + \rho v c_v \frac{\partial T}{\partial y} = -p\left(\frac{\partial u}{\partial x} + \frac{\partial v}{\partial y}\right) + \frac{\partial}{\partial y}\left(k \frac{\partial T}{\partial y}\right) + \mu\left(\frac{\partial u}{\partial y}\right)^2 + \frac{4}{3}\mu\left(\frac{\partial v}{\partial y}\right)^2 \quad (8\text{-}23)$$

The zeroth-order equations are valid for leading edge flowfields when $M_\infty \geqslant 5$ while the first-order equations are applicable when $M_\infty \geqslant 2$. The zeroth-order equations were derived by neglecting terms of order $(\delta^*)^2$, Δ^2, and ϵ. Since ϵ is the coefficient of the first-order terms, its order is given by the largest of $(\delta^*)^2$ and Δ^2. Rudman and Rubin have shown that in order for $(\delta^*)^2$ to be very small (i.e. $\leqslant 0.05$) the zeroth-order equations are not valid upstream of the point at which

$$\frac{\chi_\infty}{M_\infty^2} \cong 2$$

where χ_∞ is the strong-interaction parameter defined by

$$\chi_\infty = \left(\frac{\mu_{\text{wall}} T_\infty}{\mu_\infty T_{\text{wall}}}\right)^{1/2} (M_\infty^3 \operatorname{Re}_{x\infty})^{-1/2}$$

Consequently, an initial starting solution is required for the present leading edge problem. The same is true for all other problems which are solved using the PNS equations. For the present problem it is permissible to employ an approximate starting solution located very close to the leading edge because it will have a small effect on the flowfield further downstream. This is because only a small amount of mass flow passes between the plate and the shock layer edge at this initial station as compared with the mass flow passing between the plate and the shock wave at a station further downstream. For other problems, however, the initial starting solution will have a definite effect on the downstream flowfield and in many cases the starting solution must be determined accurately.

The set of PNS equations derived by Rudman and Rubin do not contain a streamwise pressure gradient term so that there can be no upstream influence through the subsonic portion of the boundary layer. As a result the equations behave in a strictly "parabolic" manner in the boundary-layer region. Because of this, Davis and Rubin (1980) refer to these equations as the parabolic Navier-Stokes equations instead of the "parabolized" Navier-Stokes equations. They use the latter name to refer to the sets of equations which do contain a streamwise pressure gradient term.

The PNS equations derived by Rudman and Rubin have been used to solve leading edge flows about both two- and three-dimensional geometries including flat plates, rectangular corners, cones, and wing tips (see Lin and Rubin, 1973b, for references). The three-dimensional equations are derived in a similar manner to the two-dimensional equations. The coordinates x, y, z are first nondimensionalized using L, δ_y, and δ_z, respectively. The velocities u, v, w are nondimensionalized using V_∞, $V_\infty \delta_y^*$, and

$V_\infty \delta_z^*$, respectively, where $\delta_y^* = \delta_y/L$ and $\delta_z^* = \delta_z/L$. Terms of order $(\delta_z^*)^2$, $(\delta_y^*)^2$, $\delta_y^* \delta_z^*$, etc., are assumed small. After substituting the series expansions into the Navier-Stokes equations and neglecting higher-order terms, the three-dimensional zeroth-order equations become

continuity:

$$\frac{\partial \rho u}{\partial x} + \frac{\partial \rho v}{\partial y} + \frac{\partial \rho w}{\partial z} = 0 \tag{8-24}$$

x momentum:

$$\rho u \frac{\partial u}{\partial x} + \rho v \frac{\partial u}{\partial y} + \rho w \frac{\partial u}{\partial z} = \frac{\partial}{\partial y}\left(\mu \frac{\partial u}{\partial y}\right) + \frac{\partial}{\partial z}\left(\mu \frac{\partial u}{\partial z}\right) \tag{8-25}$$

y momentum:

$$\begin{aligned} \rho u \frac{\partial v}{\partial x} + \rho v \frac{\partial v}{\partial y} + \rho w \frac{\partial v}{\partial z} = &-\frac{\partial p}{\partial y} + \frac{4}{3}\frac{\partial}{\partial y}\left(\mu \frac{\partial v}{\partial y}\right) + \frac{\partial}{\partial z}\left(\mu \frac{\partial v}{\partial z}\right) + \frac{\partial}{\partial x}\left(\mu \frac{\partial u}{\partial y}\right) \\ &- \frac{2}{3}\frac{\partial}{\partial y}\left(\mu \frac{\partial u}{\partial x} + \mu \frac{\partial w}{\partial z}\right) + \frac{\partial}{\partial z}\left(\mu \frac{\partial w}{\partial y}\right) \end{aligned} \tag{8-26}$$

z momentum:

$$\begin{aligned} \rho u \frac{\partial w}{\partial x} + \rho v \frac{\partial w}{\partial y} + \rho w \frac{\partial w}{\partial z} = &-\frac{\partial p}{\partial z} + \frac{4}{3}\frac{\partial}{\partial z}\left(\mu \frac{\partial w}{\partial z}\right) + \frac{\partial}{\partial y}\left(\mu \frac{\partial w}{\partial y}\right) + \frac{\partial}{\partial x}\left(\mu \frac{\partial u}{\partial z}\right) \\ &- \frac{2}{3}\frac{\partial}{\partial z}\left(\mu \frac{\partial v}{\partial y} + \mu \frac{\partial u}{\partial x}\right) + \frac{\partial}{\partial y}\left(\mu \frac{\partial v}{\partial z}\right) \end{aligned} \tag{8-27}$$

energy:

$$\begin{aligned} \rho u c_v \frac{\partial T}{\partial x} + \rho v c_v \frac{\partial T}{\partial y} + \rho w c_v \frac{\partial T}{\partial z} = &-p\left(\frac{\partial u}{\partial x} + \frac{\partial v}{\partial y} + \frac{\partial w}{\partial z}\right) + \frac{\partial}{\partial y}\left(k \frac{\partial T}{\partial y}\right) \\ &+ \frac{\partial}{\partial z}\left(k \frac{\partial T}{\partial z}\right) + \mu\left[\left(\frac{\partial u}{\partial y}\right)^2 + \left(\frac{\partial u}{\partial z}\right)^2 + \left(\frac{\partial w}{\partial y} + \frac{\partial v}{\partial z}\right)^2\right] \\ &+ \frac{4}{3}\mu\left[\left(\frac{\partial v}{\partial y}\right)^2 + \left(\frac{\partial w}{\partial z}\right)^2 - \frac{\partial v}{\partial y}\frac{\partial w}{\partial z}\right] \end{aligned} \tag{8-28}$$

A set of PNS equations very similar to those of Rudman and Rubin were derived independently by Cheng et al. (1970). The equations of Cheng et al. included a streamwise pressure gradient term. Probably the most common form of the PNS equations (Lubard and Helliwell, 1973, 1974) is obtained by assuming that the streamwise viscous derivative terms (including heat flux terms) are negligible compared to the normal and transverse viscous derivative terms. In other words, the streamwise viscous derivative terms are assumed to be of $O(1)$ while the normal and transverse viscous derivative terms are of $O(\mathrm{Re}_L^{1/2})$. Hence, these PNS equations are derived by simply dropping all viscous terms containing partial derivatives with respect to the streamwise direction from the steady Navier-Stokes equations. The resulting set of equations for a Cartesian coordinate system (x is the streamwise direction) is given by

continuity:

$$\frac{\partial \rho u}{\partial x} + \frac{\partial \rho v}{\partial y} + \frac{\partial \rho w}{\partial z} = 0 \tag{8-29}$$

x momentum:

$$\rho u \frac{\partial u}{\partial x} + \rho v \frac{\partial u}{\partial y} + \rho w \frac{\partial u}{\partial z} = -\frac{\partial p}{\partial x} + \frac{\partial}{\partial y}\left(\mu \frac{\partial u}{\partial y}\right) + \frac{\partial}{\partial z}\left(\mu \frac{\partial u}{\partial z}\right) \tag{8-30}$$

y momentum:

$$\rho u \frac{\partial v}{\partial x} + \rho v \frac{\partial v}{\partial y} + \rho w \frac{\partial v}{\partial z} = -\frac{\partial p}{\partial y} + \frac{4}{3}\frac{\partial}{\partial y}\left(\mu \frac{\partial v}{\partial y}\right) + \frac{\partial}{\partial z}\left(\mu \frac{\partial v}{\partial z}\right)$$
$$+ \frac{\partial}{\partial z}\left(\mu \frac{\partial w}{\partial y}\right) - \frac{2}{3}\frac{\partial}{\partial y}\left(\mu \frac{\partial w}{\partial z}\right) \tag{8-31}$$

z momentum:

$$\rho u \frac{\partial w}{\partial x} + \rho v \frac{\partial w}{\partial y} + \rho w \frac{\partial w}{\partial z} = -\frac{\partial p}{\partial z} + \frac{4}{3}\frac{\partial}{\partial z}\left(\mu \frac{\partial w}{\partial z}\right) + \frac{\partial}{\partial y}\left(\mu \frac{\partial w}{\partial y}\right)$$
$$+ \frac{\partial}{\partial y}\left(\mu \frac{\partial v}{\partial z}\right) - \frac{2}{3}\frac{\partial}{\partial z}\left(\mu \frac{\partial v}{\partial y}\right) \tag{8-32}$$

energy:

$$\rho u c_v \frac{\partial T}{\partial x} + \rho v c_v \frac{\partial T}{\partial y} + \rho w c_v \frac{\partial T}{\partial z} = -p\left(\frac{\partial u}{\partial x} + \frac{\partial v}{\partial y} + \frac{\partial w}{\partial z}\right) + \frac{\partial}{\partial y}\left(k \frac{\partial T}{\partial y}\right)$$
$$+ \frac{\partial}{\partial z}\left(k \frac{\partial T}{\partial z}\right) + \mu\left[\left(\frac{\partial u}{\partial y}\right)^2 + \left(\frac{\partial u}{\partial z}\right)^2 + \left(\frac{\partial w}{\partial y} + \frac{\partial v}{\partial z}\right)^2\right]$$
$$+ \frac{4}{3}\mu\left[\left(\frac{\partial v}{\partial y}\right)^2 + \left(\frac{\partial w}{\partial z}\right)^2 - \frac{\partial v}{\partial y}\frac{\partial w}{\partial z}\right] \tag{8-33}$$

It is interesting to compare this set of PNS equations with the equations of Rudman and Rubin [Eqs. (8-24)-(8-28)]. We note that the continuity and energy equations are identical but the momentum equations are different. In particular, the present x-momentum equation contains the streamwise pressure gradient term as discussed previously.

We now wish to express the PNS equations in terms of a generalized coordinate system. For the generalized transformation described in Section 5-6.2, the complete Navier-Stokes equations can be written as

$$\frac{\partial}{\partial t}\left(\frac{\mathbf{U}}{J}\right)+\frac{\partial}{\partial \xi}\left\{\frac{1}{J}\left[\xi_x(\mathbf{E}_i-\mathbf{E}_v)+\xi_y(\mathbf{F}_i-\mathbf{F}_v)+\xi_z(\mathbf{G}_i-\mathbf{G}_v)\right]\right\}$$
$$+\frac{\partial}{\partial \eta}\left\{\frac{1}{J}\left[\eta_x(\mathbf{E}_i-\mathbf{E}_v)+\eta_y(\mathbf{F}_i-\mathbf{F}_v)+\eta_z(\mathbf{G}_i-\mathbf{G}_v)\right]\right\}$$
$$+\frac{\partial}{\partial \zeta}\left\{\frac{1}{J}\left[\zeta_x(\mathbf{E}_i-\mathbf{E}_v)+\zeta_y(\mathbf{F}_i-\mathbf{F}_v)+\zeta_z(\mathbf{G}_i-\mathbf{G}_v)\right]\right\}=0 \qquad (8\text{-}34)$$

where

$$\mathbf{U}=\begin{bmatrix}\rho\\ \rho u\\ \rho v\\ \rho w\\ E_t\end{bmatrix}$$

$$\mathbf{E}_i=\begin{bmatrix}\rho u\\ \rho u^2+p\\ \rho uv\\ \rho uw\\ (E_t+p)u\end{bmatrix} \qquad \mathbf{E}_v=\begin{bmatrix}0\\ \tau_{xx}\\ \tau_{xy}\\ \tau_{xz}\\ u\tau_{xx}+v\tau_{xy}+w\tau_{xz}-q_x\end{bmatrix}$$

$$\mathbf{F}_i=\begin{bmatrix}\rho v\\ \rho uv\\ \rho v^2+p\\ \rho vw\\ (E_t+p)v\end{bmatrix} \qquad \mathbf{F}_v=\begin{bmatrix}0\\ \tau_{xy}\\ \tau_{yy}\\ \tau_{yz}\\ u\tau_{xy}+v\tau_{yy}+w\tau_{yz}-q_y\end{bmatrix} \qquad (8\text{-}35)$$

$$\mathbf{G}_i=\begin{bmatrix}\rho w\\ \rho uw\\ \rho vw\\ \rho w^2+p\\ (E_t+p)w\end{bmatrix} \qquad \mathbf{G}_v=\begin{bmatrix}0\\ \tau_{xz}\\ \tau_{yz}\\ \tau_{zz}\\ u\tau_{xz}+v\tau_{yz}+w\tau_{zz}-q_z\end{bmatrix}$$

and

$$
\begin{aligned}
E_t &= \rho\left(e + \frac{u^2 + v^2 + w^2}{2}\right) \\
\tau_{xx} &= \tfrac{2}{3}\mu[2(\xi_x u_\xi + \eta_x u_\eta + \zeta_x u_\zeta) - (\xi_y v_\xi + \eta_y v_\eta + \zeta_y v_\zeta) \\
&\quad - (\xi_z w_\xi + \eta_z w_\eta + \zeta_z w_\zeta)] \\
\tau_{yy} &= \tfrac{2}{3}\mu[2(\xi_y v_\xi + \eta_y v_\eta + \zeta_y v_\zeta) - (\xi_x u_\xi + \eta_x u_\eta + \zeta_x u_\zeta) \\
&\quad - (\xi_z w_\xi + \eta_z w_\eta + \zeta_z w_\zeta)] \\
\tau_{zz} &= \tfrac{2}{3}\mu[2(\xi_z w_\xi + \eta_z w_\eta + \zeta_z w_\zeta) - (\xi_x u_\xi + \eta_x u_\eta + \zeta_x u_\zeta) \\
&\quad - (\xi_y v_\xi + \eta_y v_\eta + \zeta_y v_\zeta)] \\
\tau_{xy} &= \mu(\xi_y u_\xi + \eta_y u_\eta + \zeta_y u_\zeta + \xi_x v_\xi + \eta_x v_\eta + \zeta_x v_\zeta) \\
\tau_{xz} &= \mu(\xi_z u_\xi + \eta_z u_\eta + \zeta_z u_\zeta + \xi_x w_\xi + \eta_x w_\eta + \zeta_x w_\zeta) \\
\tau_{yz} &= \mu(\xi_z v_\xi + \eta_z v_\eta + \zeta_z v_\zeta + \xi_y w_\xi + \eta_y w_\eta + \zeta_y w_\zeta) \\
q_x &= -k(\xi_x T_\xi + \eta_x T_\eta + \zeta_x T_\zeta) \\
q_y &= -k(\xi_y T_\xi + \eta_y T_\eta + \zeta_y T_\zeta) \\
q_z &= -k(\xi_z T_\xi + \eta_z T_\eta + \zeta_z T_\zeta)
\end{aligned}
\tag{8-36}
$$

Note that the usual **E**, **F**, and **G** vectors have been split into an inviscid part (subscript i) and a viscous part (subscript v). The reason for doing this will become evident later when we describe numerical procedures for solving the PNS equations. The PNS equations in generalized coordinates can now be obtained by simply dropping the unsteady terms and the viscous terms containing partial derivatives with respect to the streamwise direction ξ. The resulting equations become

$$\frac{\partial \mathbf{E}_3}{\partial \xi} + \frac{\partial \mathbf{F}_3}{\partial \eta} + \frac{\partial \mathbf{G}_3}{\partial \zeta} = 0 \tag{8-37}$$

where

$$
\begin{aligned}
\mathbf{E}_3 &= \frac{1}{J}(\xi_x \mathbf{E}_i + \xi_y \mathbf{F}_i + \xi_z \mathbf{G}_i) \\
\mathbf{F}_3 &= \frac{1}{J}[\eta_x(\mathbf{E}_i - \mathbf{E}_v') + \eta_y(\mathbf{F}_i - \mathbf{F}_v') + \eta_z(\mathbf{G}_i - \mathbf{G}_v')] \\
\mathbf{G}_3 &= \frac{1}{J}[\zeta_x(\mathbf{E}_i - \mathbf{E}_v') + \zeta_y(\mathbf{F}_i - \mathbf{F}_v') + \zeta_z(\mathbf{G}_i - \mathbf{G}_v')]
\end{aligned}
\tag{8-38}
$$

and the prime is used to indicate that terms containing partial derivatives with respect to ξ have been omitted. Hence, the shear stress and heat flux terms in Eqs. (8-35) reduce to

$$\begin{aligned}
\tau'_{xx} &= \tfrac{2}{3}\mu[2(\eta_x u_\eta + \zeta_x u_\zeta) - (\eta_y v_\eta + \zeta_y v_\zeta) - (\eta_z w_\eta + \zeta_z w_\zeta)] \\
\tau'_{yy} &= \tfrac{2}{3}\mu[2(\eta_y v_\eta + \zeta_y v_\zeta) - (\eta_x u_\eta + \zeta_x u_\zeta) - (\eta_z w_\eta + \zeta_z w_\zeta)] \\
\tau'_{zz} &= \tfrac{2}{3}\mu[2(\eta_z w_\eta + \zeta_z w_\zeta) - (\eta_x u_\eta + \zeta_x u_\zeta) - (\eta_y v_\eta + \zeta_y v_\zeta)] \\
\tau'_{xy} &= \mu(\eta_y u_\eta + \zeta_y u_\zeta + \eta_x v_\eta + \zeta_x v_\zeta) \\
\tau'_{xz} &= \mu(\eta_z u_\eta + \zeta_z u_\zeta + \eta_x w_\eta + \zeta_x w_\zeta) \\
\tau'_{yz} &= \mu(\eta_z v_\eta + \zeta_z v_\zeta + \eta_y w_\eta + \zeta_y w_\zeta) \\
q'_x &= -k(\eta_x T_\eta + \zeta_x T_\zeta) \\
q'_y &= -k(\eta_y T_\eta + \zeta_y T_\zeta) \\
q'_z &= -k(\eta_z T_\eta + \zeta_z T_\zeta)
\end{aligned} \tag{8-39}$$

For many applications (Schiff and Steger, 1979), the thin-layer approximation can also be applied to the PNS equations. With this additional assumption, the resulting equations are simply the steady form of the thin-layer Navier-Stokes equations. For the generalized transformation described previously, these equations can be written as

$$\frac{\partial \mathbf{E}_2}{\partial \xi} + \frac{\partial \mathbf{F}_2}{\partial \eta} + \frac{\partial \mathbf{G}_2}{\partial \zeta} = \frac{\partial \mathbf{S}_2}{\partial \eta} \tag{8-40}$$

where $\mathbf{E}_2$, $\mathbf{F}_2$, $\mathbf{G}_2$, and $\mathbf{S}_2$ are defined by Eqs. (8-10) and (8-11).

8-3.2 Streamwise Pressure Gradient

The presence of the streamwise pressure gradient term in the streamwise momentum equation permits information to be propagated upstream through subsonic portions of the flowfield such as a boundary layer. As a consequence, a space-marching method of solution is not well-posed, and in many cases exponentially growing solutions (departure solutions) are encountered. These departure solutions are characterized by either a separation-like increase in wall pressure or an expansion-like decrease in wall pressure. A similar behavior (Lighthill, 1953) is observed for the boundary-layer equations when the streamwise pressure gradient is not prescribed. The one difference, however, is that in the case of the PNS equations, the normal momentum equation allows a pressure interaction to occur between the critical subsonic boundary-layer region and the inviscid outer region.

In order to better understand why departure solutions occur, let us examine the influence of the streamwise pressure gradient term on the mathematical nature of the PNS equations. For simplicity, let us consider the two-dimensional PNS equations and assume a perfect gas with constant viscosity. With these assumptions, Eqs. (8-29)–(8-33) can be reduced to the following vector representation

$$\frac{\partial \mathbf{E}}{\partial x} + \frac{\partial \mathbf{F}}{\partial y} = \frac{\partial \mathbf{F}_v}{\partial y} \tag{8-41}$$

where

$$
\mathbf{E} = \begin{bmatrix} \rho u \\ \rho u^2 + \omega p \\ \rho u v \\ \left[\dfrac{\gamma}{\gamma - 1} p + \dfrac{\rho}{2}(u^2 + v^2)\right] u \end{bmatrix}
$$

$$
\mathbf{F} = \begin{bmatrix} \rho v \\ \rho u v \\ \rho v^2 + p \\ \left[\dfrac{\gamma}{\gamma - 1} p + \dfrac{\rho}{2}(u^2 + v^2)\right] v \end{bmatrix} \tag{8-42}
$$

$$
\mathbf{F}_v = \mu \begin{bmatrix} 0 \\ u_y \\ \dfrac{4}{3} v_y \\ u u_y + \dfrac{4}{3} v v_y + \dfrac{k}{\mu} T_y \end{bmatrix}
$$

Note that in these equations a parameter ω has been inserted in front of the streamwise pressure gradient term in the x-momentum equation. Thus if ω is set equal to zero, the streamwise pressure gradient term is omitted. On the other hand, if ω is set equal to one, the term is retained completely.

If we first consider the inviscid limit ($\mu \to 0$), Eq. (8-41) reduces to the Euler equation

$$
\frac{\partial \mathbf{E}}{\partial x} + \frac{\partial \mathbf{F}}{\partial y} = 0 \tag{8-43}
$$

which is equivalent to

$$
[A_1]\mathbf{Q}_x + [B_1]\mathbf{Q}_y = 0 \tag{8-44}
$$

where

$$
[A_1] = \begin{bmatrix} u & \rho & 0 & 0 \\ 0 & \rho u & 0 & \omega \\ 0 & 0 & \rho u & 0 \\ 0 & \rho u^2 + \dfrac{\gamma p}{\gamma - 1} & \rho u v & \dfrac{\gamma u}{\gamma - 1} \end{bmatrix} \qquad \mathbf{Q} = \begin{bmatrix} \rho \\ u \\ v \\ p \end{bmatrix} \tag{8-45}
$$

$$[B_1] = \begin{bmatrix} v & 0 & \rho & 0 \\ 0 & \rho v & 0 & 0 \\ 0 & 0 & \rho v & 1 \\ 0 & \rho u v & \rho v^2 + \dfrac{\gamma p}{\gamma - 1} & \dfrac{\gamma v}{\gamma - 1} \end{bmatrix}$$

These equations are hyperbolic in x provided that the eigenvalues of $[A_1]^{-1}[B]$ are real (see Section 2-5). The eigenvalues are

$$\lambda_{1,2} = \frac{v}{u}$$
$$\lambda_{3,4} = \frac{-b \pm \sqrt{b^2 - 4ac}}{2a} \tag{8-46}$$

where

$$a = [\gamma - \omega(\gamma - 1)]u^2 - \omega a^2$$
$$b = -uv[1 + \gamma - \omega(\gamma - 1)]$$
$$c = v^2 - a^2$$

and a is the speed of sound. If the streamwise pressure gradient is retained completely (i.e., $\omega = 1$), it is easy to show that the eigenvalues are all real provided that

$$u^2 + v^2 \geqslant a^2$$

or

$$M \geqslant 1$$

This is the usual requirement which must be satisfied if the Euler equations are to be integrated using a space-marching technique. However, if only a fraction of the streamwise pressure gradient is retained (i.e., $0 \leqslant \omega \leqslant 1$), the eigenvalues will remain real even in subsonic regions provided that

$$\omega \leqslant \frac{\gamma M_x^2}{1 + (\gamma - 1)M_x^2} \tag{8-47}$$

where $M_x = u/a$. This condition on the streamwise pressure gradient is derived by assuming the normal component of velocity (v) is much smaller than the streamwise component (u).

We next consider the viscous limit by ignoring terms in Eq. (8-41) containing first derivatives with respect to y. The resulting equations can be written as

$$[A_2]\mathbf{Q}_x = [B_2]\mathbf{Q}_{yy} \tag{8-48}$$

where

$$[A_2] = \begin{bmatrix} u & \rho & 0 & 0 \\ u^2 & 2\rho u & 0 & \omega \\ uv & \rho v & \rho u & 0 \\ \dfrac{u(u^2 + v^2)}{2} & \dfrac{\gamma p}{\gamma - 1} + \dfrac{\rho(3u^2 + v^2)}{2} & \rho u v & \dfrac{\gamma u}{\gamma - 1} \end{bmatrix} \tag{8-49}$$

$$[B_2] = \mu \begin{bmatrix} 0 & 0 & 0 & 0 \\ 0 & 1 & 0 & 0 \\ 0 & 0 & \frac{4}{3} & 0 \\ \frac{-\gamma p}{(\gamma-1)\rho^2 \Pr} & u & \frac{4}{3}v & \frac{\gamma}{(\gamma-1)\rho \Pr} \end{bmatrix}$$

These equations are parabolic in the positive x direction if the eigenvalues of $[A_2]^{-1}[B_2]$ are real and positive (see Section 2-5). The eigenvalues must be positive in order for a positive viscosity to produce damping in the streamwise direction. The eigenvalues can be found from the following polynomial (assuming $u \neq 0$)

$$\lambda\left(\frac{\rho u}{\mu}\lambda - \frac{4}{3}\right)\left(\left(\frac{\rho u}{\mu}\lambda\right)^2 \left\{M_x^2[\gamma - \omega(\gamma-1)] - \omega\right\} + \left(\frac{\rho u}{\mu}\lambda\right)\left\{\left[\omega(\gamma-1) - \gamma\left(\frac{1+\Pr}{\Pr}\right)\right] M_x^2 + \frac{\omega}{\Pr}\right\} + \frac{\gamma M_x^2}{\Pr}\right) = 0 \quad (8\text{-}50)$$

Vigneron et al. (1978a) have shown that the eigenvalues determined from this equation will be real and positive if

$$u > 0 \qquad (8\text{-}51)$$

and

$$\omega < \frac{\gamma M_x^2}{1 + (\gamma - 1)M_x^2} \qquad (8\text{-}52)$$

Equation (8-51) prohibits reverse flows while Eq. (8-52) places a restriction on the streamwise pressure gradient term in an identical manner to that given previously by Eq. (8-47). From this, we can conclude that the instability caused by the presence of the streamwise pressure gradient term in the PNS equations is actually an inviscid phenomenon.

Note that the right-hand side of Eq. (8-52), denoted by $f(M_x)$, is a function of the local streamwise Mach number (M_x) which becomes equal to 1 when $M_x = 1$ and is greater than 1 when $M_x > 1$ (see Fig. 8-4). Hence, the streamwise pressure gradient term can be included fully when $M_x > 1$. However, when $M_x < 1$ only a fraction of this term (i.e., $\omega\, \partial p/\partial x$) can be retained if the eigenvalues are to remain real and positive. Note also that ω approaches zero close to a wall where $M_x = 0$. Thus we see that space-marched solutions of the PNS equations are subject to instabilities (departure solutions) when the streamwise pressure gradient term is retained fully in the subsonic portion of the boundary layer since an "elliptic-like" behavior is introduced. A number of different techniques have been proposed to circumvent this difficulty and they will now be discussed.

The obvious technique is to drop completely the streamwise pressure gradient term in subsonic regions. This will produce a stable marching scheme but will introduce

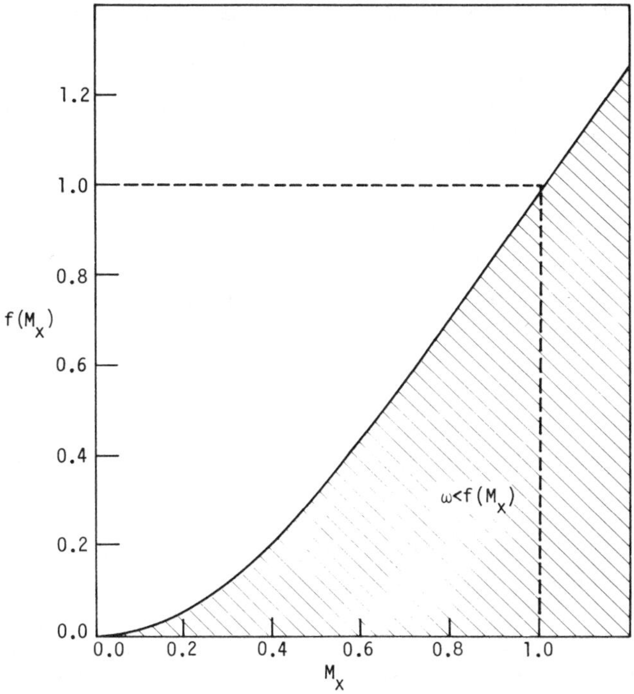

Figure 8-4 Constraint on streamwise pressure gradient term in subsonic regions.

errors in flowfields with large streamwise pressure gradients. It should be noted, however, that streamwise pressure variations will still exist in the numerical solution being evaluated through the y-momentum equation and the energy equation. An alternative procedure is to specify the variation of the streamwise pressure gradient. Obviously, setting the pressure gradient equal to zero is just one of many ways that this can be done. If the streamwise pressure gradient is specified, we can remove this term from matrices $[A_1]$ and $[A_2]$ in Eqs. (8-44) and (8-48) and treat it as a source term in the eigenvalue analyses. As a consequence, the streamwise pressure gradient will not affect the mathematical character of the equations. For the solution of the boundary-layer equations, the streamwise pressure gradient is usually known either from the external inviscid flow, or, for the case of internal flows, from the conservation of mass law. Unfortunately, for the flowfields normally computed with the PNS equations, the streamwise pressure gradient is not known a priori but must be computed as part of the solution.

In several studies, the streamwise pressure gradient term has been retained in the subsonic viscous region by employing a backward-difference formula which can be computed using information from the previous marching step. For example, when the solution at the $i + 1$ station is computed, $\partial p/\partial x$ can be evaluated from

$$\frac{\partial p}{\partial x} \cong \frac{p_i - p_{i-1}}{\Delta x} \tag{8-53}$$

which is a first-order backward-difference expression. Lubard and Helliwell (1973) have studied the stability (departure behavior) of using a backward-difference formula for the streamwise pressure gradient term in both the momentum and energy equations. They applied a simple implicit differencing scheme to the PNS equations and used a Fourier stability analysis to show that an instability will occur if

$$\Delta x < (\Delta x)_{\min} \tag{8-54}$$

This stability condition is highly unusual since we normally find from a Fourier stability analysis that an instability occurs when Δx is greater than some $(\Delta x)_{\max}$. When this analysis is applied to the two-dimensional PNS equations given by Eqs. (8-48)–(8-49), $(\Delta x)_{\min}$ is given by

$$(\Delta x)_{\min} = \frac{\frac{1}{4}(\rho u/\mu)[(1/M_x^2) - 1](\Delta y)^2}{\gamma \sin^2(\beta/2)} \tag{8-55}$$

where β is the wave number ($k_m \Delta y$). Lubard and Helliwell have also shown that if the streamwise pressure gradient term is differenced implicitly, like the rest of the terms in the PNS equations when simple implicit differencing is applied, the minimum allowable step size $(\Delta x)_{\min}$ is doubled. In order to explain these unusual stability conditions, Rubin (1981) has observed that $(\Delta x)_{\min}$ appears to represent the extent of the upstream elliptic interaction. If $(\Delta x) > (\Delta x)_{\min}$ the interaction is overstepped and a forward marching procedure is stable. On the other hand, if $(\Delta x) < (\Delta x)_{\min}$, the numerical solution attempts to represent the elliptic interaction and this leads to departure solutions since upstream effects are not permitted by a forward-marched solution. Rubin and Lin (1980) have shown that the extent of the elliptic interaction region is of the order of the thickness of the subsonic region. Thus, if the subsonic region is relatively large, the minimum allowable Δx may be too large to permit accurate (or stable) calculations.

Another method which has been used to treat the streamwise pressure gradient term is called the "sublayer approximation" technique. This method was originally proposed by Rubin and Lin (1971) and later applied to the PNS equations by Schiff and Steger (1979). In the "sublayer approximation" technique, the pressure gradient term in the subsonic viscous region is calculated at a supersonic point outside of the sublayer region. This approximation is based on the fact that for a thin subsonic viscous layer, $\partial p/\partial y$ is negligible. Since the pressure gradient is specified in the subsonic region, it would appear that this technique would lead to stable space-marched solutions. However, it has been observed by Schiff and Steger that departure solutions still exist for some cases. This may be due to the pressure interaction between the supersonic and subsonic regions which is permitted by the normal momentum equation and the energy equation.

A novel technique for handling the streamwise pressure gradient term was proposed by Vigneron et al. (1978a). In this approach, a fraction of the pressure-gradient term $\omega(\partial p/\partial x)$ in the streamwise momentum equation is retained in the subsonic viscous region and the remainder $(1-\omega)(\partial p/\partial x)$ is either omitted or is evaluated explicitly using a backward-difference formula or the "sublayer approximation" technique. For this approach, Eq. (8-41) is rewritten as

$$\frac{\partial \mathbf{E}}{\partial x} + \frac{\partial \mathbf{P}}{\partial x} + \frac{\partial \mathbf{F}}{\partial y} = \frac{\partial \mathbf{F}_v}{\partial y} \tag{8-56}$$

where

$$\mathbf{P} = \begin{bmatrix} 0 \\ (1-\omega)p \\ 0 \\ 0 \end{bmatrix} \tag{8-57}$$

and **E**, **F**, and $\mathbf{F}_v$ are defined in Eqs. (8-42). The parameter ω is computed using Eq. (8-47) with a safety factor σ applied:

$$\omega = \frac{\sigma \gamma M_x^2}{1 + (\gamma - 1) M_x^2} \tag{8-58}$$

Vigneron et al. (1978b) have used a Fourier stability analysis to study the "departure behavior" of this technique. They applied the simple implicit (Euler implicit) scheme to Eq. (8-56), with $\partial \mathbf{F}/\partial y$ omitted, and used a backward difference for $\partial \mathbf{P}/\partial x$. As expected, they found that if the explicit pressure gradient term $\partial \mathbf{P}/\partial x$ is omitted, this technique will always lead to a stable space-marched solution since the equations remain hyperbolic-parabolic. However, if this term is retained, an instability results if Δx is less than some $(\Delta x)_{\min}$. For $\omega = 0$, it was found that $(\Delta x)_{\min}$ is given by Eq. (8-55) which confirms the previous findings of Lubard and Helliwell. Other techniques for treating the streamwise pressure gradient term include those proposed by Lin and Rubin (1979), Buggeln et al. (1980), and Yanenko et al. (1980).

For many flow problems, the upstream elliptic effects are relatively small and the techniques described above will successfully prevent departure solutions while permitting an accurate solution to be computed with a single marching sweep through the flowfield. For other problems where the upstream influence is significant (due to separation, wakes, shocks, etc.), the above techniques prove to be inadequate. Either a departure solution results or the inconsistency introduced into the PNS equations to prevent the departure solution will lead to large errors. For these cases, a *global pressure relaxation procedure* (Rubin and Lin, 1980) can be used. In this procedure, the entire pressure distribution is initially specified in order to determine the pressure gradient at each point. The initial pressure distribution can be obtained by either setting the streamwise pressure gradient equal to zero, by using the "Vigneron" technique with $\partial \mathbf{P}/\partial x = 0$, or by taking a sufficiently large Δx. With the pressure gradient known, the PNS equations can be solved in a stable manner using a space-marching finite-difference technique provided that the pressure gradient term is differenced in an appropriate manner. The resulting solution will contain a new pressure distribution which can be used to determine the pressure gradient for the next sweep of the flowfield. This iteration procedure is continued until the solution converges. In order for the elliptic character of the flowfield to be properly modeled, the pressure gradient term must introduce downstream contributions. This can be accomplished by applying a forward difference to the pressure gradient term. That is, when the solution at the $i + 1$ station is computed, the streamwise pressure gradient term is differenced as

$$\frac{\partial p}{\partial x} \cong \frac{p_{i+2} - p_{i+1}}{\Delta x} \tag{8-59}$$

This type of differencing is only possible when the global pressure relaxation procedure is used since p_{i+2} is normally unknown. Rubin and Lin have studied the stability of using a forward difference for $\partial p/\partial x$ and have shown that it is unconditionally stable. However, the stability is marginal when the subsonic region becomes very large.

The global pressure relaxation procedure shows great promise for solving problems where the upstream influence is significant. It should be remembered, however, that this procedure requires substantially more computer time than a typical PNS calculation which employs one sweep. In some cases, the computation time will approach that required for the solution of the complete Navier-Stokes equations. Hence, for these problems, the PNS equations may not offer any advantages over the complete Navier-Stokes equations.

8-3.3 Numerical Solution of PNS Equations

As discussed previously, the PNS equations are a mixed set of hyperbolic-parabolic equations in the streamwise direction provided that the following conditions are satisfied:

1. Inviscid flow is supersonic.
2. Streamwise velocity component is everywhere greater than zero.
3. Streamwise pressure gradient term in streamwise momentum equation is either omitted or the "departure behavior" is suppressed using one of the techniques described in the last section.

If these conditions are met, the PNS equations can be solved using finite-difference methods similar to those employed for the parabolic boundary-layer equations. Thus, the solution can be marched downstream in a stable manner from an initial data surface to the desired final station.

Some of the earliest solutions of the PNS equations were obtained using an explicit finite-difference technique. This was done more for convenience than efficiency since we have demonstrated in Chap. 7 that implicit methods are much more efficient for equations of this type. In later studies, the PNS equations have been solved using a variety of implicit algorithms. Nardo and Cresci (1971) employed the Peaceman-Rachford ADI scheme while Rubin and Lin (1972) and Lubard and Helliwell (1973) used similar iterative-implicit schemes. Rubin and Lin's predictor-corrector multiple iteration scheme is described in Section 4-5.10 where it is applied to the 3-D linear Burgers equation

$$u_x + cu_y + du_z = \mu(u_{yy} + u_{zz}) \tag{8-60}$$

The 3-D linear Burgers equation is a useful model equation for the PNS equations, but, of course, it does not represent the nonlinear character of these equations. Thus, when the predictor-corrector multiple iteration method is applied to the PNS equations, nonlinear terms such as $(u_{i+1,j,k}^{m+1})^2$ appear where m is the iteration level, $x = i\,\Delta x$, $y = j\,\Delta y$, and $z = k\,\Delta z$. These nonlinear terms are linearized using a Newton-Raphson procedure (see Section 7-3.3). That is, if $f = f(x_1, x_2, \ldots, x_l)$ is a nonlinear term, then

$$f^{m+1} = f^m + \sum_{k=1}^{l} \left(\frac{\partial f}{\partial x_k}\right)^m (x_k^{m+1} - x_k^m) \tag{8-61}$$

where x_k denotes the dependent variables. Applying this formula to the nonlinear term $(u_{i+1,j,k}^{m+1})^2$ gives

$$(u_{i+1,j,k}^{m+1})^2 = 2u_{i+1,j,k}^{m+1} u_{i+1,j,k}^m - (u_{i+1,j,k}^m)^2 \tag{8-62}$$

After all the nonlinear terms are linearized in this manner, the resulting set of algebraic equations (at iteration level $m + 1$) can be solved using an efficient block tridiagonal solver. The iteration is continued until the solution converges at the $i + 1$ station. This method is implicit in the y direction, where the gradients are largest, but is explicit in the z direction (see Section 4-5.10) which leads to the following stability condition when applied to the 3-D PNS equations:

$$\Delta x \leqslant \Delta z \left| \frac{w}{u} \right| \tag{8-63}$$

Until recently, the PNS equations have been solved using iterative, implicit finite-difference schemes such as the one described above. Vigneron et al. (1978a) were the first to employ a more efficient noniterative, implicit, approximate-factorization, finite-difference scheme to solve the PNS equations. Their algorithm is adapted from the class of ADI schemes developed by Lindemuth and Killeen (1973), McDonald and Briley (1975), and Beam and Warming (1978) to solve time-dependent equations such as the Navier-Stokes equations. In order to explain this algorithm, let us apply it to the 3-D PNS equations written in Cartesian coordinates (x is the streamwise direction) for a perfect gas. In this case, the generalized coordinates become

$$\begin{aligned} \xi &= x \\ \eta &= y \\ \zeta &= z \end{aligned} \tag{8-64}$$

and Eqs. (8-37)–(8-38) reduce to

$$\frac{\partial \mathbf{E}}{\partial x} + \frac{\partial \mathbf{F}}{\partial y} + \frac{\partial \mathbf{G}}{\partial z} = 0 \tag{8-65}$$

where

$$\begin{aligned} \mathbf{E} &= \mathbf{E}_i \\ \mathbf{F} &= \mathbf{F}_i - \mathbf{F}_v \\ \mathbf{G} &= \mathbf{G}_i - \mathbf{G}_v \end{aligned} \tag{8-66}$$

The vectors $\mathbf{E}_i$, $\mathbf{F}_i$, $\mathbf{G}_i$, $\mathbf{F}_v$, and $\mathbf{G}_v$ are given by Eqs. (8-35) and contain the following "parabolized" shear stress and heat flux terms

$$
\begin{aligned}
\tau_{xx} &= \tfrac{2}{3}\mu(-v_y - w_z) \\
\tau_{yy} &= \tfrac{2}{3}\mu(2v_y - w_z) \\
\tau_{zz} &= \tfrac{2}{3}\mu(2w_z - v_y) \\
\tau_{xy} &= \mu u_y \\
\tau_{xz} &= \mu u_z \\
\tau_{yz} &= \mu(v_z + w_y) \\
q_x &= 0 \\
q_y &= -kT_y \\
q_z &= -kT_z
\end{aligned}
\tag{8-67}
$$

In order to use the "Vigneron" technique for handling the streamwise pressure gradient, $\mathbf{E}$ can be replaced by $\mathbf{E}' + \mathbf{P}$, so that, Eq. (8-65) becomes

$$
\frac{\partial \mathbf{E}'}{\partial x} + \frac{\partial \mathbf{P}}{\partial x} + \frac{\partial \mathbf{F}}{\partial y} + \frac{\partial \mathbf{G}}{\partial z} = 0 \tag{8-68}
$$

where $\mathbf{E}'$ and $\mathbf{P}$ are given by

$$
\mathbf{E}' = \begin{bmatrix} \rho u \\ \rho u^2 + \omega p \\ \rho u v \\ \rho u w \\ (E_t + p)u \end{bmatrix} \qquad \mathbf{P} = \begin{bmatrix} 0 \\ (1-\omega)p \\ 0 \\ 0 \\ 0 \end{bmatrix} \tag{8-69}
$$

The solution of Eq. (8-65) is marched in x using the following difference formula suggested by Beam and Warming (1978)

$$
\Delta^i \mathbf{E} = \frac{\theta_1 \Delta x}{1+\theta_2} \frac{\partial}{\partial x}(\Delta^i \mathbf{E}) + \frac{\Delta x}{1+\theta_2} \frac{\partial}{\partial x}(\mathbf{E}^i) + \frac{\theta_2}{1+\theta_2} \Delta^{i-1}\mathbf{E} + O\left[\left(\theta_1 - \frac{1}{2} - \theta_2\right)(\Delta x)^2 + (\Delta x)^3\right] \tag{8-70}
$$

where

$$
\Delta^i \mathbf{E} = \mathbf{E}^{i+1} - \mathbf{E}^i \tag{8-71}
$$

and $x = i\Delta x$. This general difference formula, with the appropriate choice of the parameters θ_1 and θ_2, reproduces many of the standard difference schemes as seen in Table 8-1. For the PNS equations, either the first-order Euler implicit scheme ($\theta_1 = 1$, $\theta_2 = 0$) or the second-order, three-point backward scheme ($\theta_1 = 1$, $\theta_2 = \frac{1}{2}$) are normally used. As shown by Beam and Warming, the second-order trapezoidal

Table 8-1 Finite-difference schemes contained in Eq. (8-70)

θ_1	θ_2	Scheme	Truncation error in Eq. (8-70)
0	0	Euler explicit	$O[(\Delta x)^2]$
0	$-\frac{1}{2}$	Leapfrog (explicit)	$O[(\Delta x)^3]$
$\frac{1}{2}$	0	Trapezoidal (implicit)	$O[(\Delta x)^3]$
1	0	Euler implicit	$O[(\Delta x)^2]$
1	$\frac{1}{2}$	Three-point-backward (implicit)	$O[(\Delta x)^3]$

differencing scheme ($\theta_1 = \frac{1}{2}, \theta_2 = 0$) will lead to unstable calculations when applied to parabolic equations. Note that the truncation error in Table 8-1 is for $\Delta^i \mathbf{E}$. When $\partial \mathbf{E}/\partial x$ is replaced by $\Delta^i \mathbf{E}/\Delta x$ in the numerical scheme, the truncation error is divided by Δx.

Substituting Eq. (8-65) into Eq. (8-70) yields

$$\Delta^i \mathbf{E} = -\frac{\theta_1 \Delta x}{1+\theta_2}\left[\frac{\partial}{\partial y}(\Delta^i \mathbf{F}) + \frac{\partial}{\partial z}(\Delta^i \mathbf{G})\right] - \frac{\Delta x}{1+\theta_2}\left[\frac{\partial}{\partial y}(\mathbf{F}^i) + \frac{\partial}{\partial z}(\mathbf{G}^i)\right] + \frac{\theta_2}{1+\theta_2}\Delta^{i-1}\mathbf{E} \tag{8-72}$$

with the truncation error term omitted. This difference formula is in the so-called "delta" form as discussed in Section 4-4.7. The "delta" terms $\Delta^i \mathbf{E}$, $\Delta^i \mathbf{F}$, and $\Delta^i \mathbf{G}$ which can be written as

$$\begin{aligned} \Delta^i \mathbf{E} &= \Delta^i \mathbf{E}' + \Delta^i \mathbf{P} \\ \Delta^i \mathbf{F} &= \Delta^i \mathbf{F}_i - \Delta^i \mathbf{F}_v \\ \Delta^i \mathbf{G} &= \Delta^i \mathbf{G}_i - \Delta^i \mathbf{G}_v \end{aligned} \tag{8-73}$$

are linearized using truncated Taylor series expansions. In order to linearize the inviscid "delta" terms $\Delta^i \mathbf{E}'$, $\Delta^i \mathbf{F}_i$, and $\Delta^i \mathbf{G}_i$, we make use of the fact that $\mathbf{E}'$, $\mathbf{F}_i$, and $\mathbf{G}_i$ are functions only of the $\mathbf{U}$ vector

$$\mathbf{U} = \begin{bmatrix} \rho \\ \rho u \\ \rho v \\ \rho w \\ E_t \end{bmatrix} = \begin{bmatrix} U_1 \\ U_2 \\ U_3 \\ U_4 \\ U_5 \end{bmatrix} \tag{8-74}$$

For example, $\mathbf{F}_i$ can be expressed as

$$\mathbf{F}_i = \begin{bmatrix} U_3 \\ \dfrac{U_2 U_3}{U_1} \\ \dfrac{U_3^2}{U_1} + (\gamma - 1)\left(U_5 - \dfrac{U_2^2 + U_3^2 + U_4^2}{2U_1}\right) \\ \dfrac{U_3 U_4}{U_1} \\ \left[U_5 + (\gamma - 1)\left(U_5 - \dfrac{U_2^2 + U_3^2 + U_4^2}{2U_1}\right)\right]\dfrac{U_3}{U_1} \end{bmatrix} \tag{8-75}$$

As a consequence, we can readily expand $\mathbf{E}'$, $\mathbf{F}_i$, and $\mathbf{G}_i$ as

$$\begin{aligned} (\mathbf{E}')^{i+1} &= (\mathbf{E}')^i + \left(\frac{\partial \mathbf{E}'}{\partial \mathbf{U}}\right)^i \Delta^i \mathbf{U} + O[(\Delta x)^2] \\ (\mathbf{F}_i)^{i+1} &= (\mathbf{F}_i)^i + \left(\frac{\partial \mathbf{F}_i}{\partial \mathbf{U}}\right)^i \Delta^i \mathbf{U} + O[(\Delta x)^2] \\ (\mathbf{G}_i)^{i+1} &= (\mathbf{G}_i)^i + \left(\frac{\partial \mathbf{G}_i}{\partial \mathbf{U}}\right)^i \Delta^i \mathbf{U} + O[(\Delta x)^2] \end{aligned} \tag{8-76}$$

or

$$\begin{aligned} \Delta^i \mathbf{E}' &= [Q]^i \Delta^i \mathbf{U} + O[(\Delta x)^2] \\ \Delta^i \mathbf{F}_i &= [R]^i \Delta^i \mathbf{U} + O[(\Delta x)^2] \\ \Delta^i \mathbf{G}_i &= [S]^i \Delta^i \mathbf{U} + O[(\Delta x)^2] \end{aligned} \tag{8-77}$$

where $[Q]$, $[R]$, and $[S]$ are the Jacobian matrices $\partial \mathbf{E}'/\partial \mathbf{U}$, $\partial \mathbf{F}_i/\partial \mathbf{U}$, and $\partial \mathbf{G}_i/\partial \mathbf{U}$ given by

$$\frac{\partial \mathbf{E}'}{\partial \mathbf{U}} = \begin{bmatrix} 0 & 1 & 0 & 0 & 0 \\ \dfrac{\omega(\gamma-1)-2}{2}u^2 + \dfrac{\omega(\gamma-1)}{2}(v^2+w^2) & [2-\omega(\gamma-1)]u & -\omega(\gamma-1)v & -\omega(\gamma-1)w & \omega(\gamma-1) \\ -uv & v & u & 0 & 0 \\ -uw & w & 0 & u & 0 \\ \left[-\dfrac{\gamma E_t}{\rho} + (\gamma-1)(u^2+v^2+w^2)\right]u & \dfrac{\gamma E_t}{\rho} - (\gamma-1)\dfrac{3u^2+v^2+w^2}{2} & -(\gamma-1)uv & -(\gamma-1)uw & \gamma u \end{bmatrix} \tag{8-78}$$

$$\frac{\partial \mathbf{F}_i}{\partial \mathbf{U}} = \begin{bmatrix} 0 & 0 & 1 & 0 & 0 \\ -uv & v & u & 0 & 0 \\ \frac{\gamma-1}{2}(u^2+w^2)+\frac{\gamma-3}{2}v^2 & -(\gamma-1)u & (3-\gamma)v & -(\gamma-1)w & \gamma-1 \\ -vw & 0 & w & v & 0 \\ \left[-\frac{\gamma E_t}{\rho}+(\gamma-1)(u^2+v^2+w^2)\right]v & -(\gamma-1)uv & \frac{\gamma E_t}{\rho}-\frac{\gamma-1}{2}(u^2+3v^2+w^2) & -(\gamma-1)vw & \gamma v \end{bmatrix} \tag{8-79}$$

$$\frac{\partial \mathbf{G}_i}{\partial \mathbf{U}} = \begin{bmatrix} 0 & 0 & 0 & 1 & 0 \\ -uw & w & 0 & u & 0 \\ -vw & 0 & w & v & 0 \\ \frac{\gamma-1}{2}(u^2+v^2)+\frac{\gamma-3}{2}w^2 & -(\gamma-1)u & -(\gamma-1)v & (3-\gamma)w & \gamma-1 \\ \left[-\frac{\gamma E_t}{\rho}+(\gamma-1)(u^2+v^2+w^2)\right]w & -(\gamma-1)uw & -(\gamma-1)vw & \frac{\gamma E_t}{\rho}-\frac{\gamma-1}{2}(u^2+v^2+3w^2) & \gamma w \end{bmatrix} \tag{8-80}$$

The expression for the Jacobian $\partial \mathbf{E}'/\partial \mathbf{U}$ is derived by assuming ω to be locally independent of $\mathbf{U}$.

The viscous "delta" terms can be linearized using a method suggested by Steger (1977). In order to apply this linearization method, the coefficients of viscosity (μ) and thermal conductivity (k) are assumed to be locally independent of $\mathbf{U}$ and the cross-derivative viscous terms are neglected. As a result of these assumptions, elements of $\mathbf{F}_v$ and $\mathbf{G}_v$ have the general form

$$\begin{aligned} f_k &= \alpha_k \frac{\partial}{\partial y}(\beta_k) \\ g_k &= \alpha_k \frac{\partial}{\partial z}(\beta_k) \end{aligned} \tag{8-81}$$

where α_k is independent of $\mathbf{U}$ and β_k is a function of $\mathbf{U}$. These elements are linearized in the following manner

$$\begin{aligned} f^{i+1} &= f^i + \alpha_k^i \frac{\partial}{\partial y}\left[\sum_{l=1}^{5}\left(\frac{\partial \beta_k}{\partial U_l}\right)^i \Delta^i U_l\right] + O[(\Delta x)^2] \\ g^{i+1} &= g^i + \alpha_k^i \frac{\partial}{\partial z}\left[\sum_{l=1}^{5}\left(\frac{\partial \beta_k}{\partial U_l}\right)^i \Delta^i U_l\right] + O[(\Delta x)^2] \end{aligned} \tag{8-82}$$

so that we can write

$$\begin{aligned} \Delta^i \mathbf{F}_v &= [V]^i \Delta^i U + O[(\Delta x)^2] \\ \Delta^i \mathbf{G}_v &= [W]^i \Delta^i \mathbf{U} + O[(\Delta x)^2] \end{aligned} \tag{8-83}$$

where $[V]$ and $[W]$ are the Jacobian matrices $\partial \mathbf{F}_v/\partial \mathbf{U}$ and $\partial \mathbf{G}_v/\partial \mathbf{U}$ given by

$$\frac{\partial \mathbf{F}_v}{\partial \mathbf{U}} = \mu \left[\begin{array}{c:c:c:c:c} 0 & 0 & 0 & 0 & 0 \\ \hdashline -\partial_y\left(\frac{u}{\rho}\right) & \partial_y\left(\frac{1}{\rho}\right) & 0 & 0 & 0 \\ \hdashline -\frac{4}{3}\,\partial_y\left(\frac{v}{\rho}\right) & 0 & \frac{4}{3}\,\partial_y\left(\frac{1}{\rho}\right) & 0 & 0 \\ \hdashline -\partial_y\left(\frac{w}{\rho}\right) & 0 & 0 & \partial_y\left(\frac{1}{\rho}\right) & 0 \\ \hdashline \begin{array}{c} -\partial_y\left(\frac{u^2}{\rho}\right) - \frac{4}{3}\,\partial_y\left(\frac{v^2}{\rho}\right) - \partial_y\left(\frac{w^2}{\rho}\right) \\ -\frac{\gamma}{\Pr}\,\partial_y\left[\frac{p}{(\gamma-1)\rho^2} - \frac{u^2+v^2+w^2}{2\rho}\right] \end{array} & \left(1-\frac{\gamma}{\Pr}\right)\partial_y\left(\frac{u}{\rho}\right) & \left(\frac{4}{3}-\frac{\gamma}{\Pr}\right)\partial_y\left(\frac{v}{\rho}\right) & \left(1-\frac{\gamma}{\Pr}\right)\partial_y\left(\frac{w}{\rho}\right) & \frac{\gamma}{\Pr}\,\partial_y\left(\frac{1}{\rho}\right) \end{array} \right] \tag{8-84}$$

$$\frac{\partial \mathbf{G}_v}{\partial \mathbf{U}} = \mu \left[\begin{array}{c:c:c:c:c} 0 & 0 & 0 & 0 & 0 \\ \hdashline -\partial_z\left(\frac{u}{\rho}\right) & \partial_z\left(\frac{1}{\rho}\right) & 0 & 0 & 0 \\ \hdashline -\partial_z\left(\frac{v}{\rho}\right) & 0 & \partial_z\left(\frac{1}{\rho}\right) & 0 & 0 \\ \hdashline -\frac{4}{3}\,\partial_z\left(\frac{w}{\rho}\right) & 0 & 0 & \frac{4}{3}\,\partial_z\left(\frac{1}{\rho}\right) & 0 \\ \hdashline \begin{array}{c} -\partial_z\left(\frac{u^2}{\rho}\right) - \partial_z\left(\frac{v^2}{\rho}\right) - \frac{4}{3}\,\partial_z\left(\frac{w^2}{\rho}\right) \\ -\frac{\gamma}{\Pr}\,\partial_z\left[\frac{p}{(\gamma-1)\rho^2} - \frac{u^2+v^2+w^2}{2\rho}\right] \end{array} & \left(1-\frac{\gamma}{\Pr}\right)\partial_z\left(\frac{u}{\rho}\right) & \left(1-\frac{\gamma}{\Pr}\right)\partial_z\left(\frac{v}{\rho}\right) & \left(\frac{4}{3}-\frac{\gamma}{\Pr}\right)\partial_z\left(\frac{w}{\rho}\right) & \frac{\gamma}{\Pr}\,\partial_z\left(\frac{1}{\rho}\right) \end{array} \right] \tag{8-85}$$

In these Jacobian matrices, ∂_y and ∂_z represent the partial derivatives $\partial/\partial y$ and $\partial/\partial z$.

We now substitute Eqs. (8-73), (8-77), and (8-83) into Eq. (8-72) to obtain

$$\left\{\left(\frac{\partial \mathbf{E}'}{\partial \mathbf{U}}\right)^i + \frac{\theta_1 \Delta x}{1+\theta_2}\left[\frac{\partial}{\partial y}\left(\frac{\partial \mathbf{F}_i}{\partial \mathbf{U}} - \frac{\partial \mathbf{F}_v}{\partial \mathbf{U}}\right) + \frac{\partial}{\partial z}\left(\frac{\partial \mathbf{G}_i}{\partial \mathbf{U}} - \frac{\partial \mathbf{G}_v}{\partial \mathbf{U}}\right)\right]^i\right\} \Delta^i \mathbf{U}$$

$$= -\frac{\Delta x}{1+\theta_2}\left[\frac{\partial}{\partial y}(\mathbf{F}^i) + \frac{\partial}{\partial z}(\mathbf{G}^i)\right] + \frac{\theta_2}{1+\theta_2}\Delta^{i-1}\mathbf{E} - \Delta^i \mathbf{P} \tag{8-86}$$

where the expression

$$\left[\frac{\partial}{\partial y}\left(\frac{\partial \mathbf{F}_i}{\partial \mathbf{U}} - \frac{\partial \mathbf{F}_v}{\partial \mathbf{U}}\right)\right] \Delta^i \mathbf{U}$$

implies

$$\frac{\partial}{\partial y}\left[\left(\frac{\partial \mathbf{F}_i}{\partial \mathbf{U}} - \frac{\partial \mathbf{F}_v}{\partial \mathbf{U}}\right) \Delta^i \mathbf{U}\right]$$

and the partial derivatives appearing in $\partial \mathbf{F}_v/\partial \mathbf{U}$ and $\partial \mathbf{G}_v/\partial \mathbf{U}$ are to be applied to all terms on their right including $\Delta^i \mathbf{U}$. Note that in Eq. (8-86), all the implicit terms have been placed on the left-hand side of the equation while all the explicit terms appear on the right-hand side. Included in the right-hand side of the equation is the explicit pressure gradient term $\Delta^i \mathbf{P}$ which can be evaluated using a suitable backward-difference formula. A first-order backward-difference formula which is consistent with the Euler implicit scheme is given by

$$\Delta^i \mathbf{P} = \Delta^{i-1}\mathbf{P} + O(\Delta x)^2 \tag{8-87}$$

whereas for second-order accurate schemes the following second-order backward-difference formula can be used

$$\Delta^i \mathbf{P} = 2\Delta^{i-1}\mathbf{P} - \Delta^{i-2}\mathbf{P} + O[(\Delta x)^3] \tag{8-88}$$

The left-hand side of Eq. (8-86) is approximately factored in the following manner

$$\left\{\left[\left(\frac{\partial \mathbf{E}'}{\partial \mathbf{U}}\right)^i + \frac{\theta_1 \Delta x}{1+\theta_2}\frac{\partial}{\partial z}\left(\frac{\partial \mathbf{G}_i}{\partial \mathbf{U}} - \frac{\partial \mathbf{G}_v}{\partial \mathbf{U}}\right)^i\right]\left[\left(\frac{\partial \mathbf{E}'}{\partial \mathbf{U}}\right)^i\right]^{-1}\right.$$

$$\left.\times\left[\left(\frac{\partial \mathbf{E}'}{\partial \mathbf{U}}\right)^i + \frac{\theta_1 \Delta x}{1+\theta_2}\frac{\partial}{\partial y}\left(\frac{\partial \mathbf{F}_i}{\partial \mathbf{U}} - \frac{\partial \mathbf{F}_v}{\partial \mathbf{U}}\right)^i\right]\right\} \Delta^i \mathbf{U} = \mathbf{RHS}\ [\text{Eq. (8-86)}] \tag{8-89}$$

The order of accuracy of this factored expression can be determined by multiplying out the factored terms and comparing the result with the left-hand side of Eq. (8-86). Upon doing this, we obtain

$$\left\{\left(\frac{\partial \mathbf{E}'}{\partial \mathbf{U}}\right)^i + \frac{\theta_1 \Delta x}{1+\theta_2}\left[\frac{\partial}{\partial y}\left(\frac{\partial \mathbf{F}_i}{\partial \mathbf{U}} - \frac{\partial \mathbf{F}_v}{\partial \mathbf{U}}\right) + \frac{\partial}{\partial z}\left(\frac{\partial \mathbf{G}_i}{\partial \mathbf{U}} - \frac{\partial \mathbf{G}_v}{\partial \mathbf{U}}\right)\right]^i\right.$$

$$\left. + \left(\frac{\theta_1 \Delta x}{1+\theta_2}\right)^2 \frac{\partial}{\partial z}\left(\frac{\partial \mathbf{G}_i}{\partial \mathbf{U}} - \frac{\partial \mathbf{G}_v}{\partial \mathbf{U}}\right)^i \left[\left(\frac{\partial \mathbf{E}'}{\partial \mathbf{U}}\right)^i\right]^{-1} \frac{\partial}{\partial y}\left(\frac{\partial \mathbf{F}_i}{\partial \mathbf{U}} - \frac{\partial \mathbf{F}_v}{\partial \mathbf{U}}\right)^i\right\} \Delta^i \mathbf{U}$$

$$= \textbf{RHS} \text{ [Eq. (8-86)]} \tag{8-90}$$

so that

$$\textbf{LHS} \text{ [Eq. 8-89)]} = \textbf{LHS} \text{ [Eq. (8-86)]} + O[(\Delta x)^2] \tag{8-91}$$

As a consequence, the formal accuracy of the finite-difference algorithm is not affected by the approximate factorization.

The partial derivatives $\partial/\partial y$ and $\partial/\partial z$ in Eq. (8-89) are approximated with second-order accurate central differences. For example, the inviscid term

$$\frac{\partial}{\partial y}\left(\frac{\partial \mathbf{F}_i}{\partial \mathbf{U}}\right)^i \Delta^i \mathbf{U}$$

is differenced as

$$\frac{[(\partial \mathbf{F}_i/\partial \mathbf{U})^i \Delta^i \mathbf{U}]_{j+1} - [(\partial \mathbf{F}_i/\partial \mathbf{U})^i \Delta^i \mathbf{U}]_{j-1}}{2\Delta y} \tag{8-92}$$

and each element of the viscous term

$$\frac{\partial}{\partial y}\left(\frac{\partial \mathbf{F}_v}{\partial \mathbf{U}}\right)^i \Delta^i \mathbf{U}$$

which has the general form

$$\frac{\partial}{\partial y}\left[\alpha \frac{\partial}{\partial y}(\beta \Delta^i U_l)\right]$$

is differenced as

$$\frac{\{\alpha[\partial(\beta\Delta^i U_l)/\partial y]\}_{j+1/2} - \{\alpha[\partial(\beta\Delta^i U_l)/\partial y]\}_{j-1/2}}{\Delta y}$$

$$\simeq \frac{\alpha_{j+1/2}[(\beta\Delta^i U_l)_{j+1} - (\beta\Delta^i U_l)_j] - \alpha_{j-1/2}[(\beta\Delta^i U_l)_j - (\beta\Delta^i U_l)_{j-1}]}{(\Delta y)^2}$$

$$\simeq \frac{(\alpha_j + \alpha_{j+1})[(\beta\Delta^i U_l)_{j+1} - (\beta\Delta^i U_l)_j] - (\alpha_j + \alpha_{j-1})[(\beta\Delta^i U_l)_j - (\beta\Delta^i U_l)_{j-1}]}{2(\Delta y)^2} \tag{8-93}$$

The algorithm given by Eq. (8-89) is implemented in the following manner:

Step 1:

$$\left[\left(\frac{\partial \mathbf{E}'}{\partial \mathbf{U}}\right)^i + \frac{\theta_1 \Delta x}{1+\theta_2}\frac{\partial}{\partial z}\left(\frac{\partial \mathbf{G}_i}{\partial \mathbf{U}} - \frac{\partial \mathbf{G}_v}{\partial \mathbf{U}}\right)^i\right] \Delta^i \mathbf{U}_1 = \textbf{RHS} \text{ [Eq. (8-86)]} \tag{8-94}$$

Step 2:

$$\Delta^i \mathbf{U}_2 = \left(\frac{\partial \mathbf{E}'}{\partial \mathbf{U}}\right)^i \Delta^i \mathbf{U}_1 \tag{8-95}$$

Step 3:

$$\left[\left(\frac{\partial \mathbf{E}'}{\partial \mathbf{U}}\right)^i + \frac{\theta_1 \Delta x}{1+\theta_2}\frac{\partial}{\partial y}\left(\frac{\partial \mathbf{F}_i}{\partial \mathbf{U}} - \frac{\partial \mathbf{F}_v}{\partial \mathbf{U}}\right)^i\right]\Delta^i \mathbf{U} = \Delta^i \mathbf{U}_2 \tag{8-96}$$

Step 4:

$$\mathbf{U}^{i+1} = \mathbf{U}^i + \Delta^i \mathbf{U} \tag{8-97}$$

In Step 1, $\Delta^i \mathbf{U}_1$ represents the vector quantity

$$\left[\left(\frac{\partial \mathbf{E}'}{\partial \mathbf{U}}\right)^i\right]^{-1}\left[\left(\frac{\partial \mathbf{E}'}{\partial \mathbf{U}}\right)^i + \frac{\theta_1 \Delta x}{1+\theta_2}\frac{\partial}{\partial y}\left(\frac{\partial \mathbf{F}_i}{\partial \mathbf{U}} - \frac{\partial \mathbf{F}_v}{\partial \mathbf{U}}\right)^i\right]\Delta^i \mathbf{U}$$

which is determined by solving the system of equations given by Eq. (8-94). This system of equations has the following block tridiagonal structure

$$\begin{bmatrix} [B_1] & [C_1] & & & & 0 \\ [A_2] & [B_2] & [C_2] & & & \\ & [A_3] & [B_3] & [C_3] & & \\ & & \ddots & \ddots & \ddots & \\ & & & [A_{K-1}] & [B_{K-1}] & [C_{K-1}] \\ 0 & & & & [A_K] & [B_K] \end{bmatrix} \begin{bmatrix} [\Delta^i U_1]_1 \\ [\Delta^i U_1]_2 \\ [\Delta^i U_1]_3 \\ \vdots \\ [\Delta^i U_1]_{K-1} \\ [\Delta^i U_1]_K \end{bmatrix} = \begin{bmatrix} [\text{RHS}]_1 \\ [\text{RHS}]_2 \\ [\text{RHS}]_3 \\ \vdots \\ [\text{RHS}]_{K-1} \\ [\text{RHS}]_K \end{bmatrix} \tag{8-98}$$

where $[A]$, $[B]$, and $[C]$ are 5×5 matrices and $[\Delta^i \mathbf{U}_1]$ and [RHS] are column matrices whose elements are the components of the vectors $\Delta^i \mathbf{U}_1$ and **RHS** [Eq. (8-86)]. This system of equations can be solved using the block tridiagonal solver given in Appendix B. Once $\Delta^i \mathbf{U}_1$ is determined, it is multiplied by $(\partial \mathbf{E}'/\partial \mathbf{U})^i$ in Step 2. As a result of this multiplication, the inverse matrix $[(\partial \mathbf{E}'/\partial \mathbf{U})^i]^{-1}$ does not have to be determined in the solution process. In step 3, the block tridiagonal system of equations in the y direction is solved. Finally, in Step 4, the vector of unknowns at station $i+1$ (i.e., $\mathbf{U}^{i+1}$) is determined by simply adding $\Delta^i \mathbf{U}$ to the vector of unknowns at station i. The primitive variables can then be obtained from $\mathbf{U}^{i+1}$ in the following manner:

$$
\begin{aligned}
\rho^{i+1} &= U_1^{i+1} \\
u^{i+1} &= \frac{U_2^{i+1}}{U_1^{i+1}} \\
v^{i+1} &= \frac{U_3^{i+1}}{U_1^{i+1}} \\
w^{i+1} &= \frac{U_4^{i+1}}{U_1^{i+1}} \\
e^{i+1} &= \frac{U_5^{i+1}}{U_1^{i+1}} - \frac{(u^{i+1})^2 + (v^{i+1})^2 + (w^{i+1})^2}{2}
\end{aligned}
\tag{8-99}
$$

For algorithms of the present type, it is often necessary to add smoothing in order to suppress high frequency oscillations. This can easily be accomplished by adding a fourth-order explicit dissipation term of the form

$$
-\epsilon_e \left[(\Delta y)^4 \frac{\partial^4}{\partial y^4} (\mathbf{U}^i) + (\Delta z)^4 \frac{\partial^4}{\partial z^4} (\mathbf{U}^i) \right]
\tag{8-100}
$$

to the right-hand side of Eq. (8-86). Since this is a fourth-order term it does not affect the formal accuracy of the algorithm. The negative sign is required in front of the fourth-derivatives in order to produce positive damping [see Eq. (4-21)]. The smoothing coefficient ϵ_e should be less than approximately $\frac{1}{16}$ for stability. The fourth-derivative terms are evaluated using the following finite-difference approximations

$$
\begin{aligned}
(\Delta y)^4 \frac{\partial^4}{\partial y^4} (\mathbf{U}^i) &\simeq \mathbf{U}_{j+2,k}^i - 4\mathbf{U}_{j+1,k}^i + 6\mathbf{U}_{j,k}^i - 4\mathbf{U}_{j-1,k}^i + \mathbf{U}_{j-2,k}^i \\
(\Delta z)^4 \frac{\partial^4}{\partial z^4} (\mathbf{U}^i) &\simeq \mathbf{U}_{j,k+2}^i - 4\mathbf{U}_{j,k+1}^i + 6\mathbf{U}_{j,k}^i - 4\mathbf{U}_{j,k-1}^i + \mathbf{U}_{j,k-2}^i
\end{aligned}
\tag{8-101}
$$

A noniterative implicit algorithm similar to the one described above has been developed by Schiff and Steger (1979). In the Schiff and Steger algorithm, as well as the one developed by Vigneron et al. (1978a), the solution is advanced using computational planes (i.e. solution surfaces) normal to the body axis. Most body shapes can be treated in this manner. However, for bodies with large surface slopes, the axial component of velocity in the inviscid part of the flowfield may become subsonic which prevents the computation from proceeding further. To alleviate this difficulty, Tannehill et al. (1982) have applied the numerical scheme (described previously) to the PNS equations written in general nonorthogonal coordinates, Eqs. (8-37)–(8-39). As a result, the orientation of each solution surface (ξ = constant) is left arbitrary so that the most appropriate orientation can be selected for a given problem. In general, the optimum orientation occurs when the solution surface is nearly perpendicular to the local flow direction. In a similar manner, Helliwell et al. (1980) have incorporated a nonorthogonal coordinate system into the Lubard-Helliwell method to permit a more optimum orientation of the computational planes.

Other implicit algorithms for solving the PNS equations have been developed by McDonald and Briley (1975) and Briley and McDonald (1980) who utilize a consistently split linearized block implicit (LBI) scheme and by Li (1981) who uses an iterative, factored implicit scheme. The LBI scheme of McDonald and Briley has a linearized block implicit structure which is identical to the structure of the "delta" form of the Beam-Warming scheme.

8-4 PARABOLIZED AND PARTIALLY PARABOLIZED NAVIER-STOKES PROCEDURES FOR SUBSONIC FLOWS

Previous sections in this chapter have dealt with flows which were predominantly supersonic. In this section we will discuss two computational strategies which are particularly useful for subsonic flows. For both, the starting point is a form of the PNS equations. The approaches differ in the way that the pressure is treated.

8-4.1 Parabolic Procedures for 3-D Confined Flows

This approach is applicable to internal flows in which a predominate flow direction can be identified. The velocity component in this primary flow direction must be greater than zero; i.e., flow reversal in the primary direction is not permitted. No restrictions are placed on the velocity components in the cross-flow direction. As with all forms of the PNS equations, diffusion in the streamwise direction is neglected.

Unless further steps are taken, the PNS equations will permit transmission of influences in the streamwise direction through the pressure field for subsonic flows as discussed in Section 8-3.2. In the present approach, this elliptic behavior is suppressed in the streamwise direction by utilizing an approximation first suggested by Gosman and Spalding (1971). The computational strategy will be illustrated by considering flow through a straight rectangular channel. This permits use of the conservation equations in the Cartesian coordinate system. The same concepts are applicable to curved channels of constant cross-sectional area but a different coordinate system must be used. The 3-D parabolic model has been extended to more general geometries by Briley and McDonald (1979).

The channel axis is in the x direction. Thus, the y and z coordinates span planes perpendicular to the primary flow direction. The equations will be written in a form applicable to either laminar or turbulent flow. The variables are understood to represent time-mean quantities. This is the same convention as employed in Chap. 7. In developing the parabolized form of the Reynolds equations, diffusion in the streamwise direction by both molecular and turbulent mechanisms will be neglected. Furthermore, since only subsonic applications are to be considered, it will be assumed that $\overline{\rho' u'}/\bar{\rho}\bar{u}$, $\overline{\rho' v'}/\bar{\rho}\bar{v}$, and $\overline{\rho' w'}/\bar{\rho}\bar{w}$ are small so that the difference between conventional and mass-weighted variables can be neglected. Terms involving pressure fluctuations in the energy equation will also be neglected. The symbol τ will denote the effective stress due to both molecular and turbulent mechanisms. Similarly, the symbol q will denote heat flux quantities from both molecular and turbulent mechanisms. Apart from the pressure gradient terms, which will be discussed below, the equations for the 3-D parabolic procedure follow from Eqs. (5-68), (5-73), and (5-84) after the simplifying assumptions given above are invoked:

continuity:

$$\frac{\partial \rho u}{\partial x} + \frac{\partial \rho v}{\partial y} + \frac{\partial \rho w}{\partial z} = 0 \tag{8-102a}$$

$$\int_A \rho u \, dA = \text{constant} \quad \text{(global)} \tag{8-102b}$$

x momentum:

$$\rho u \frac{\partial u}{\partial x} + \rho v \frac{\partial u}{\partial y} + \rho w \frac{\partial u}{\partial z} = -\frac{d\hat{p}}{dx} + \frac{\partial \tau_{xy}}{\partial y} + \frac{\partial \tau_{xz}}{\partial z} \tag{8-103}$$

y momentum:

$$\rho u \frac{\partial v}{\partial x} + \rho v \frac{\partial v}{\partial y} + \rho w \frac{\partial v}{\partial z} = -\frac{\partial p}{\partial y} + \frac{\partial \tau_{yy}}{\partial y} + \frac{\partial \tau_{yz}}{\partial z} \tag{8-104}$$

z momentum:

$$\rho u \frac{\partial w}{\partial x} + \rho v \frac{\partial w}{\partial y} + \rho w \frac{\partial w}{\partial z} = -\frac{\partial p}{\partial z} + \frac{\partial \tau_{zy}}{\partial y} + \frac{\partial \tau_{zz}}{\partial z} \tag{8-105}$$

energy:

$$\rho u c_p \frac{\partial T}{\partial x} + \rho v c_p \frac{\partial T}{\partial y} + \rho w c_p \frac{\partial T}{\partial z} = \frac{\partial}{\partial y}(-q_y) + \frac{\partial}{\partial z}(-q_z) + \beta T u \frac{d\hat{p}}{\partial x} + \tau_{xy} \frac{\partial u}{\partial y} + \tau_{xz} \frac{\partial u}{\partial z} \tag{8-106}$$

state:

$$\rho = \rho(p, T) \tag{8-107}$$

In the pressure approximation of Gosman and Spalding (1971), a pressure $\hat{p}$ is defined for use in the x-momentum equation which is assumed to vary *only in the* x *direction.* The pressure $\hat{p}$ will be determined with the aid of the global mass flow constraint much as for 2-D or axisymmetric channel flows computed through the thin-shear-layer equations. On the other hand, the p employed in the y- and z-momentum equations is permitted to vary across the channel cross section. The static pressure in the channel is assumed to be the sum of $\hat{p}$ and p.

The physical assumption in this decoupling procedure is that the pressure variations across the channel are so small that they would have a negligible effect if included in the streamwise momentum equation. Thus, cross-plane pressure variations have been neglected in the streamwise momentum equation. On the other hand, these small pressure variations are included in the momentum equations in the y and z directions since they play an important role in the distribution of the generally small components of velocity in the directions normal to the channel walls. The determination of $\hat{p}$ requires no information from downstream; $\hat{p}$ is a function of x only and can be uniquely determined at each cross section by employing the global mass flow constraint in combination with the momentum equations. This permits a "once through" calculation of the flow in a parabolic manner. On the other hand, since p varies with both y and z, the equations are elliptic (for subsonic flow) in the y-z plane. In fact, a Poisson equation can be developed for $p(y, z)$ in the cross plane from the y- and z-momentum equations. The overall calculation scheme then requires the use of procedures for elliptic equations in each cross plane, but the solution can be advanced in the x direction in a parabolic manner.

Using the Boussinesq approximation, the stresses (using summation notation) in the above equations can be evaluated from

$$\tau_{ij} = (\mu + \mu_T)\left(\frac{\partial u_i}{\partial x_j} + \frac{\partial u_j}{\partial x_i} - \frac{2}{3}\delta_{ij}\frac{\partial u_k}{\partial x_k}\right) - \frac{2}{3}\bar{\rho}\bar{k}\delta_{ij} \tag{8-108}$$

With similar modeling assumptions, the heat flux quantities are normally represented by

$$q_y = -\left(k + \frac{\mu_T c_p}{\text{Pr}_T}\right)\frac{\partial T}{\partial y}$$

$$q_z = -\left(k + \frac{\mu_T c_p}{\text{Pr}_T}\right)\frac{\partial T}{\partial z}$$

Further simplifications to Eq. (8-108) are often found in specific applications including the fully incompressible representation given by $\tau_{ij} = (\mu + \mu_T)\,\partial u_i/\partial x_j$. Suitable turbulence modeling for μ_T and Pr_T must be employed to close the system of equations. The usual boundary conditions for channel flow apply.

The most commonly used solution strategy will be outlined briefly. We note that for a specified pressure field, the momentum and energy equations would be entirely parabolic and the solution could be marched in the primary flow direction using the x-momentum equation to obtain u, the y-momentum equation to obtain v, and the z-momentum equation to obtain w. The energy equation provides T and the density is obtained from the equation of state. For all but exactly the correct cross-plane pressure distribution, the velocity components will not satisfy the continuity equation. This, of course, is the crux of the problem—the momentum, energy, and state equations are a natural combination to use to advance the solution for the velocity components and density. The way in which the continuity and momentum equations can be used to determine the correct pressure distribution is less obvious. Workable procedures have been devised for correcting the pressure field and these will be discussed below. The computational strategy of solving the conservation equations in an uncoupled manner, sequentially for one variable at a time, is known as the *segregated approach.*

In principle, the complete set of equations could be advanced simultaneously by a direct solution method and then corrected for the nonlinear coupling by an iterative procedure. At the present time, however, the most efficient direct solvers (Buneman, 1969; Schwartztrauber and Sweet, 1977; Bank, 1977) are only applicable to a special class of difference equations and boundary conditions which severely limits their applicability to the present problem. Other direct solution techniques are not economically feasible at present. On the other hand, advances have been made in block iterative methods for solving systems of algebraic equations of the type appearing in the present problem. Using these strongly implicit procedures (Stone, 1968; Schneider and Zedan, 1981; Rubin and Khosla, 1981), it may be possible to develop more efficient algorithms which can maintain coupling between the pressure and velocity variables in the 3-D parabolic equations. These strongly implicit procedures for the 3-D parabolic equations are in an early stage of development at this time.

Most of the solutions reported in the literature for the 3-D parabolic equations have followed the general segregated strategy outlined by Patankar and Spalding (1972) as the SIMPLE (Semi-Implicit Method for Pressure-Linked Equations) procedure. Some notable improvements in some of the solution steps have been suggested recently and these will be mentioned below. The Patankar and Spalding (1972) approach in turn draws heavily upon the earlier work of Harlow and Welch (1965), Amsden and Harlow (1970), and Chorin (1968). The segregated strategy proceeds as follows. The superscript $n + 1$ refers to the present streamwise station.

1. Employing suitable linearization for coefficients in Eq. (8-103), the pressure $\hat{p}^{n+1}$ can be determined much as for two-dimensional and axisymmetric channel flows solved by means of the boundary-layer equations (see Section 7-5), by making use of the global conservation of mass constraint. Then $u_{j,k}^{n+1}$ can be determined from the finite-difference solution of Eq. (8-103). The energy equation can be solved for $T_{j,k}^{n+1}$ and the equation of state used to determine $\rho_{j,k}^{n+1}$. An ADI scheme works very well for solving the momentum and energy equations.
2. Using an assumed pressure distribution in Eqs. (8-104) and (8-105), provisional values of v and w can be determined from a marching solution (an ADI scheme is recommended here too) to these momentum equations just as for the x-momentum equation.
3. These provisional solutions for v and w in the cross plane will not generally satisfy the difference form of the continuity equation. By applying the continuity equation to the provisional solutions for the velocity components, mass sources (or sinks) can be computed at each grid point. We now seek a means for adjusting the pressure field in the cross plane so as to eliminate the mass sources. It is in the computation of the velocity and pressure corrections that the 3-D parabolic methods differ the most. Several investigators including Briley (1974), Ghia et al. (1977b), and Ghia and Sokhey (1977) have followed the suggestion of Chorin (1968) and assumed that the corrective flow in the cross plane is irrotational, being driven by a pressure-like potential in such a manner as to annihilate the mass sources. A Poisson equation can be developed for this potential from the continuity equation. Using p subscripts to denote provisional velocities and c subscripts to denote corrective quantities, we demand that

$$\frac{\partial \rho u}{\partial x} + \frac{\partial}{\partial y}[\rho(v_p + v_c)] + \frac{\partial}{\partial z}[\rho(w_p + w_c)] = 0 \qquad (8\text{-}109)$$

The streamwise derivative term and derivatives of the provisional velocities are known at the time the corrections are sought and can be incorporated into a single source term S_ϕ. Thus, we can define a potential function $\hat{\phi}$ by $\rho v_c = \partial\hat{\phi}/\partial y$, $\rho w_c = \partial\hat{\phi}/\partial z$, and write Eq. (8-109) as

$$\frac{\partial^2 \hat{\phi}}{\partial y^2} + \frac{\partial^2 \hat{\phi}}{\partial z^2} = S_\phi \tag{8-110}$$

The required velocity corrections can then be computed from the $\hat{\phi}$ distribution resulting from the numerical solution of the Poisson equation in the cross plane. This approach preserves the vorticity of the original v_p and w_p velocity fields.

The original Patankar and Spalding proposal assumed that the velocity corrections were driven by pressure corrections in accordance with a very approximate form of the momentum equations in which the streamwise convective terms were equated to the pressure terms. This can be indicated symbolically by

$$\rho u \frac{\partial v_c}{\partial x} = -\frac{\partial p'}{\partial y} \tag{8-111}$$

$$\rho u \frac{\partial w_c}{\partial x} = -\frac{\partial p'}{\partial z} \tag{8-112}$$

In the above, p' can be viewed merely as a potential function (much like $\hat{\phi}$) used to generate velocity corrections which satisfy the continuity equation. In some schemes [as in the original Patankar and Spalding (1972) proposal] p' is viewed as an actual correction to be added to the provisional values of pressure. Since the velocity corrections can be assumed to be zero at the previous streamwise station, Eqs. (8-111) and (8-112) can be interpreted as

$$v_c = -A \frac{\partial p'}{\partial y} \tag{8-113}$$

$$w_c = -B \frac{\partial p'}{\partial z} \tag{8-114}$$

where A and B are coefficients which involve ρ, u, and Δx. The derivatives of p' are, of course, eventually to be represented on the finite-difference grid. The similarity between Eqs. (8-113) and (8-114) and the representation given earlier for the velocity corrections in terms of the potential $\hat{\phi}$ should be noted. Equations (8-113) and (8-114) can now be used in the continuity equation to develop a Poisson equation of the form

$$\frac{\partial^2 p'}{\partial y^2} + \frac{\partial^2 p'}{\partial z_2} = S_{p'} \tag{8-115}$$

The required velocity corrections can then be computed from the numerical solution of Eq. (8-115) using Eqs. (8-113) and (8-114). This approach is known as the p' procedure for obtaining velocity corrections. Improvements on this procedure have been suggested which attempt to employ a more complete form of the momentum equation in relating velocity corrections to p'. The paper by Raithby and Schneider (1979) describes several variations of the p' approach.

The next step is the pressure update. The velocity corrections just obtained have not been required to satisfy a complete momentum equation. It is now necessary to take steps to develop the improved pressure field in the cross plane which, when used in the complete momentum equations, will produce velocities which satisfy the continuity equation. Several procedures have been used. The corrected velocities can be employed in the difference form of the momentum equations to provide expressions for the pressure gradients which would be consistent with new velocities. We denote these symbolically by

$$\frac{\partial p}{\partial y} = F_1 \tag{8-116}$$

$$\frac{\partial p}{\partial z} = F_2 \tag{8-117}$$

One estimate of the "best" revised pressure field can be obtained by solving the Poisson equation which is developed from Eqs. (8-116) and (8-117):

$$\frac{\partial^2 p}{\partial y^2} + \frac{\partial^2 p}{\partial z^2} = \frac{\partial F_1}{\partial y} + \frac{\partial F_2}{\partial z} = S_p \tag{8-118}$$

The right-hand side of Eq. (8-118) is evaluated from the difference form of the momentum equations using the corrected velocities and is treated as a source term. Patankar (1980) has suggested a slightly different formulation which also results in a Poisson equation to be solved for the updated pressure (the SIMPLER algorithm). SIMPLER stands for SIMPLE Revised. In all of these solutions of the Poisson equation, care must be taken in establishing the numerical representation of the boundary conditions. The differencing and solution strategy must ensure that the Gauss divergence theorem (see Section 3-3.7) is satisfied. A more detailed example of the boundary treatment for the Poisson equation for pressure will be given in Section 8-4.3.

Raithby and Schneider (1979) have proposed a scheme for updating the pressure which does not require the solution of a second Poisson equation. They refer to this as the procedure for Pressure Update from Multiple Path INtegration (PUMPIN). The idea is that the pressure change from grid point to grid point can be computed from integrating Eqs. (8-116) and (8-117) again using the corrected velocities in the momentum equations to evaluate F_1 and F_2. For exactly the correct velocities v and w, the pressure change computed by this procedure between any two points within the cross plane would be independent of path. If the velocities v and w are not exactly correct (they will only be correct as convergence is achieved), then each different path between two points will lead to a different result. We can fix one point as a reference and compute pressures at other points in the cross plane by averaging the pressures obtained by integrating over several different paths between the reference point and the grid point of interest. Raithby and Schneider (1979) reported good success at averaging pressures over only two paths, namely from the reference point to the point of interest (*a*) along constant y and then constant z, and (*b*) along constant z and then constant y.

The pressure can also be updated very simply by accepting the p' obtained in the velocity correction procedure of Patankar and Spalding (1972) [see Eq. (8-115)] as the correction to be added to the pressure.

5. Because the momentum and continuity equations have not been satisfied simultaneously in the procedures just described, steps (2) through (4) are normally repeated iteratively in sequence at each cross plane before the solution is advanced to the next marching station. Under-relaxation is commonly used for both the velocity and pressure corrections. That is, in moving from step (3) to (4), only a fraction of the computed velocity corrections may be added to the provisional v and w velocities. The fraction will vary from method to method. Likewise, it is common to only adjust the pressure by a fraction of the computed pressure correction before moving to step (2). Time-dependent forms of the governing equations are sometimes used to carry out this iterative process. Because steps (2) through (4) are to be repeated iteratively, it is common practice to terminate the intermediate Poisson equation solutions for velocity and pressure corrections (especially the latter) short of full convergence in early iterative passes through steps (2) through (4). The objective is to obtain an *improvement* in the pressure field with each iterative pass through steps (2) through (4). Until overall convergence is approached, there is little point in obtaining the best possible pressure field based on the wrong velocity distribution. The iterative sweeps through steps (2) through (4) are terminated when a pressure field has been established which will yield solutions to the momentum equations that satisfy the continuity equation within a specified tolerance; i.e., velocity corrections are no longer required.
6. After convergence to the specified degree is achieved, steps (1) through (5) are repeated for the next streamwise station.

Raithby and Schneider (1979) have reported on a comparative study of several of the methods described above for achieving the velocity and pressure corrections. The number of iterations through steps (2)-(5) above, required for convergence, was taken as the primary measure of merit. The computation time required for the various algorithms would be of interest, but was not reported. Fixing the method for the pressure update, they observed that all of the methods given above for achieving the velocity corrections worked satisfactorily. There was very little difference between them in terms of the required number of iterations.

When the method for obtaining velocity corrections was fixed and several different methods for obtaining the pressure update were compared, the p' method of Patankar and Spalding (1972) was observed to require notably more iterations for convergence than the other methods evaluated. The methods utilizing a Poisson equation and the PUMPIN procedure required only about half as many iterations as the p' method. The PUMPIN method required the fewest iterations by a slim margin. Use of the p' (Patankar and Spalding, 1972) method would not be recommended on the basis of the Raithby and Schneider (1979) study. This conclusion is confirmed by Patankar's (1980) recommendation that his SIMPLER algorithm which employs the Poisson equation formulation be used instead of the older p' method for updating the pressure. It is possible that the p' method may appear more competitive with the other methods when computation time rather than number of iterations is taken as the measure of merit.

Several investigators have reported calculations based on the 3-D parabolic model. These include the work of Patankar and Spalding (1972), Caretto et al. (1972), Briley

(1974), Ghia et al. (1977b), Ghia and Sokhey (1977), and Patankar et al. (1974). For flows through channels of varying cross-sectional area, suggestions have been made to include an inviscid flow pressure (determined a priori) in the analysis to partially account for elliptic influences in the primary flow direction. Both regular and staggered grids have been used. The concepts of the mathematical model appear well established. Further improvements in the algorithm are likely, particularly through the use of strongly implicit procedures which permit the equations to be solved in a coupled rather than segregated manner. The properties of the 3-D parabolic procedure for flows in which the velocities approach sonic values are not well known.

8-4.2 Parabolic Procedures for 3-D Free-Shear and Other Flows

The 3-D parabolic procedure discussed in the last section is not restricted entirely to confined flows. The essential feature of the model was the decoupling of the pressure gradient terms in the primary and cross-flow directions. For confined flows, the pressure gradient in the primary flow direction was determined with the aid of the global conservation of mass constraint. The main elements of the procedure can be used for other types of three-dimensional flows if the pressure gradient in the primary flow direction can be neglected or prescribed in advance. One such application occurs in the discharge of a subsonic free jet from a rectangular-shaped nozzle into a co-flowing or quiescent ambient. The shape of such a jet gradually changes in the streamwise direction, eventually becoming round in cross section. For such flows it is reasonable to neglect the streamwise pressure gradient. Small pressure variations in the cross plane must still be considered, as for 3-D confined flows. McGuirk and Rodi (1977) and Hwang and Pletcher (1978) have computed such flows by the 3-D parabolic procedure by setting $d\hat{p}/dx = 0$. An example of the 3-D parabolic procedure applied to free surface flows is given by Raithby and Schneider (1980).

8-4.3 The Partially Parabolized Model

The partially parabolized (PPNS) model for subsonic flows utilizes equations which again are conceptually identical to the PNS equations. That is, streamwise diffusion is the only physical process neglected and terms representing this diffusion are dropped from the Navier-Stokes equations. To date, most applications of this model have been for incompressible flows so that the remaining stress terms usually have appeared in a form somewhat simplified from that given in Section 8-3.2. In PNS applications having a supersonic stream, downstream influences are suppressed by one of the techniques described in Section 8-3.2. In the partially parabolized model, these elliptic effects transmitted by the pressure field are actually computed. Thus, the flow is only *partially* parabolized. The elliptic behavior associated with the pressure field remains. This requires that the solution be marched in the streamwise direction in an iterative manner.

The equations for the partially parabolized model are as given in Eqs. (8-102)–(8-107) with $d\hat{p}/dx$ replaced by $\partial p/\partial x$. The primary flow direction is aligned with the x-coordinate axis. The model was first suggested by Pratap and Spalding (1976). Other partially parabolized procedures have been proposed by Dodge (1977), Moore and Moore (1979), and Chilukuri and Pletcher (1980).

The scheme was originally thought to be restricted to flows in which flow reversal in the primary direction does not occur. For these flows, three-dimensional storage is only required for the pressure (and the source term in the Poisson equation for pressure if the Poisson equation formulation is used) and not for the velocity components. This is the main computational advantage of the PPNS procedure compared to procedures for the full Navier-Stokes equations. Madavan and Pletcher (1982) have recently demonstrated that the PPNS model can be extended to two-dimensional applications in which reversal occurs in the component of velocity in the primary direction. This procedure requires that computer storage also be used for velocity components in and near the regions of primary flow reversal. Dodge (1977) also implies that his PPNS method could be extended to applications in which streamwise flow reversal occurs.

We will briefly describe how the PPNS strategy of Chilukuri and Pletcher (1980) can be applied to a steady, incompressible two-dimensional laminar flow. The improvements suggested by Madavan and Pletcher (1982) will be included. For such a flow, the PPNS equations can be written

continuity:

$$\frac{\partial u}{\partial x} + \frac{\partial v}{\partial y} = 0 \tag{8-119}$$

x momentum:

$$u \frac{\partial u}{\partial x} + v \frac{\partial u}{\partial y} = -\frac{1}{\rho} \frac{\partial p}{\partial x} + \nu \frac{\partial^2 u}{\partial y^2} \tag{8-120}$$

y momentum:

$$u \frac{\partial v}{\partial x} + v \frac{\partial v}{\partial y} = -\frac{1}{\rho} \frac{\partial p}{\partial y} + \nu \frac{\partial^2 v}{\partial y^2} \tag{8-121}$$

Staggered grids are frequently used for flows treated in orthogonal coordinate systems. The staggered grid was first used by Harlow and Welch (1965). We will use it in the present two-dimensional example of the PPNS procedure.

The idea is to define a different grid for each velocity component. This is illustrated in Fig. 8-5. To avoid confusion, only the grid location for the scalar variables (pressure and the velocity correction potential $\hat{\phi}$, in this example) are denoted by solid point symbols in the figure. Velocity components are calculated for "points" or locations which are on the faces of a control volume which could be drawn around the pressure points. The velocity components are located midway between pressure points which means that for an unequal grid, the pressure points are not necessarily in the geometric center of such a control volume. The locations of the velocity components are indicated by arrows in Fig. 8-5. Vertical arrows denote locations for v and horizontal arrows indicate the locations for u. It is convenient to refer to the variables with a single set of grid indices, despite the fact that the variables are actually

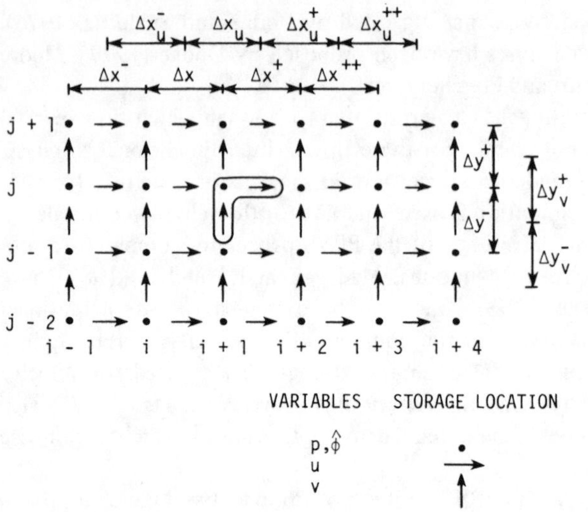

Figure 8-5 Grid spacing definitions and spatial location of variables on a staggered grid.

defined at different locations. Thus, the designation $(i + 1, j)$ identifies a cluster of three distinct spatial locations as indicated by the boomerang shaped enclosure in Fig. 8-5. In the staggered grid, $v_{i+1,j}^{n+1}$ is below $p_{i+1,j}$, and $u_{i+1,j}$ is to the right of $p_{i+1,j}$.

The staggered grid permits the divergence of the velocity field to be represented with second-order accuracy (for equally spaced grid points) at the solid grid points using velocity components at adjacent locations. Such a configuration ensures that the difference representation for this divergence has the conservative property. Also, the pressure difference between adjacent grid points becomes the natural driving force for velocity components located midway between the points. That is, a simple forward-difference representation for pressure derivatives is "central" relative to the location of velocity components. This permits the development of a Poisson equation for pressure which automatically satisfies the Gauss divergence theorem so long as care is taken in the treatment of the boundary conditions. Such boundary conditions are also more easily handled on the staggered grid. Patankar (1980) provides an excellent and more detailed discussion of the advantages of using a staggered grid for problems such as the present one.

The computational boundaries are most conveniently located along grid lines where components of velocity normal to the boundaries are located. This is illustrated in Fig. 8-6 for a lower boundary. Fictitious points are located outside of the physical boundary as necessary for imposing suitable boundary conditions. As an example, we will suppose that it is desired to impose no-slip boundary conditions at the lower boundary illustrated in Fig. 8-6. The v-component of velocity is located at the physical boundary and it is easy to simply specify $v_{i+1,1} = 0$. The treatment for the u-component is not so obvious since no u grid points are located on this boundary. Numerous possibilities exist. The main requirement is that the boundary formulation used must imply that the tangential component of velocity is zero at the location of the physical boundary. This can be achieved by developing a special difference form of the conservation equations for the control volume at the boundary, or by constraining the solution near the boundary such that an extrapolation to the boundary would satisfy the no-slip condition. A third and often used procedure is to employ a fictitious velocity point below the boundary with the constraint that $(u_{i+1,1} + u_{i+1,2})/2 = 0$. This is similar to the reflection technique for enforcing boundary conditions for inviscid flow which was discussed in Chap. 6. The velocity at the fictitious point would then be used as required in the momentum equations in the interior. Values of the potential function used to correct the velocities are also often obtained from points outside the physical boundaries. Values of pressure are not needed from such points for boundaries on which velocity boundary conditions are specified in the formulation to be described below. More details on the treatment of boundary conditions on a staggered grid can be found in the work of Amsden and Harlow (1970).

Several choices exist for the representation of the convective derivatives in the momentum equations. The scheme illustrated below makes use of three-point second-order accurate upwind representations for convective terms of the form $u\,\partial\phi/\partial x$. A hybrid scheme (see Section 7-3.3) will be used for terms of the form $v\,\partial\phi/\partial y$. These representations are linearized by extrapolating the coefficients based on values at the two adjacent upstream stations. When streamwise flow reversal is present, the direction of the "wind" changes and this is taken into account in the representation used for the streamwise derivatives and in the extrapolation direction for the coefficients.

In the ensuing discussion, the following notation is adopted. The superscript $n+1$ denotes the current marching sweep; the subscript $i+1$ denotes the current streamwise step for which the solution is sought; and the subscript j denotes the grid points in the y direction. For the forward going flow, the following representation is used to obtain the extrapolated value of the coefficient $u_{i+1,j}^{n+1}$:

$$\hat{u}_{i+1,j}^{n+1} = \left(1 + \frac{\Delta x_u}{\Delta x_u^-}\right) u_{i,j}^{n+1} - \frac{\Delta x_u}{\Delta x_u^-}\, u_{i-1,j}^{n+1}$$

The caret indicates that $\hat{u}_{i+1,j}^{n+1}$ is a known quantity, determined by extrapolation. An extrapolation for $\hat{v}_{i+1,j}^{n+1}$ is obtained in a like manner using appropriate streamwise mesh increments. If $\hat{u}_{i+1,j}^{n+1}$ in the above expression becomes negative, the flow at $(i+1, j)$ is assumed to be reversed. When this occurs, the recommended procedure is to simply represent $\hat{u}_{i+1,j}^{n+1}$ by $u_{i+1,j}^{n}$ and $\hat{v}_{i+1,j}^{n+1}$ by $v_{i+1,j}^{n}$, making use of velocities

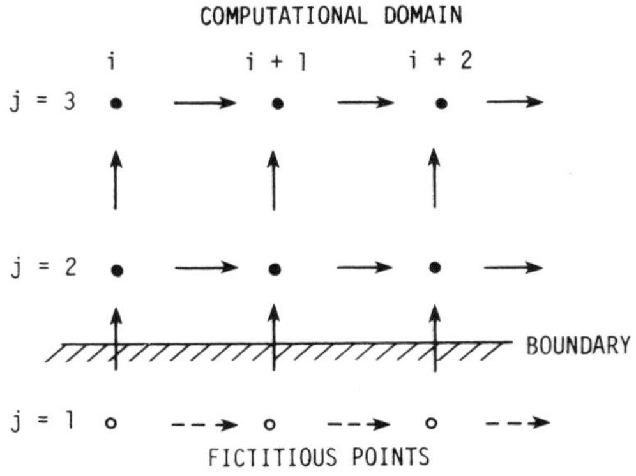

Figure 8-6 The staggered finite-difference grid near a boundary.

from the previous iteration which are stored for points in and near regions of reversed flow. As an alternative, extrapolation can also be used according to the expression

$$\hat{u}_{i+1,j}^{n+1} = \left(1 + \frac{\Delta x_u^+}{\Delta x_u^{++}}\right) u_{i+2,j}^n - \frac{\Delta x_u^+}{\Delta x_u^{++}} u_{i+3,j}^n$$

An expression for $\hat{v}_{i+1,j}^{n+1}$ can also be obtained by extrapolation in regions of reversed flow. The streamwise convective derivatives are then represented as follows. For the forward going flow,

$$\left(u \frac{\partial u}{\partial x}\right)_{i+1,j}^{n+1} \simeq \hat{u}_{i+1,j}^{n+1} \left(\frac{\Delta x_u^- + 2\,\Delta x_u}{\Delta x_u(\Delta x_u^- + \Delta x_u)} u_{i+1,j}^{n+1} - \frac{\Delta x_u^- + \Delta x_u}{\Delta x_u^- \,\Delta x_u} u_{i,j}^{n+1} + \frac{\Delta x_u}{\Delta x_u^-(\Delta x_u^- + \Delta x_u)} u_{i-1,j}^{n+1}\right) \tag{8-122}$$

and for the reversed flow region,

$$\left(u \frac{\partial u}{\partial x}\right)_{i+1,j}^{n+1} \simeq -\hat{u}_{i+1,j}^{n+1} \left(\frac{\Delta x_u^{++} + 2\Delta x_u^+}{\Delta x_u(\Delta x_u^{++} + \Delta x_u^+)} u_{i+1,j}^{n+1} + \frac{\Delta x_u^{++} + \Delta x_u^+}{\Delta x_u^{++}\,\Delta x_u^+} u_{i+2,j}^n - \frac{\Delta x_u^+}{\Delta x_u^{++}(\Delta x_u^{++} + \Delta x_u^+)} u_{i+3,j}^n\right) \tag{8-123}$$

The term $v\,\partial u/\partial y$ is represented by a hybrid scheme as follows:

$$\begin{aligned}\left(v \frac{\partial u}{\partial y}\right)_{i+1,j}^{n+1} \simeq &\left[\hat{v}_{i+1,j}^{n+1}(u_{i+1,j}^{n+1} - u_{i+1,j-1}^{n+1}) \frac{\Delta y^+}{\Delta y^+ + \Delta y^-}\right. \\ &\left. + \hat{v}_{i+1,j+1}^{n+1} \frac{u_{i+1,j+1}^{n+1} - u_{i+1,j}^{n+1}}{\Delta y^+} \frac{\Delta y^-}{\Delta y^+ + \Delta y^-}\right] W \\ &+ \hat{v}_{i+1,j}^{n+1} \frac{u_{i+1,j}^{n+1} - u_{i+1,j-1}^{n+1}}{\Delta y^-} (1 - W)\,A \\ &+ \hat{v}_{i+1,j+1}^{n+1} \frac{u_{i+1,j+1}^{n+1} - u_{i+1,j}^{n+1}}{\Delta y^+} (1 - W)\,B\end{aligned} \tag{8-124}$$

The magnitudes of W, A, and B are determined as follows. Defining

$$R_m^+ = \frac{\hat{v}_{i+1,j+1}^{n+1}\,\Delta y^-}{\nu}$$

$$R_m^- = \frac{\hat{v}_{i+1,j}^{n+1}\,\Delta y^+}{\nu}$$

R_c = critical mesh Reynolds number = 1.9 (see Section 7-3.3)

When $R_m^+ > R_c$:

$$W = \frac{R_c}{R_m^+} \qquad A = 1 \qquad B = 0$$

When $R_m^+ < -R_c$:

$$W = \frac{R_c}{R_m^-} \qquad A = 0 \qquad B = 1$$

When $R_m^- < R_c < R_m^+$:

$$W = 1 \qquad A = 0 \qquad B = 0$$

This scheme is thus a weighted average of central and upwind differences for larger mesh Reynolds numbers and degenerates to central differencing for small mesh Reynolds numbers.

The second derivative term is represented by

$$\left(\frac{\partial^2 u}{\partial y^2}\right)_{i+1,j}^{n+1} \simeq \frac{2}{\Delta y^+ + \Delta y^-}\left(\frac{u_{i+1,j+1}^{n+1} - u_{i+1,j}^{n+1}}{\Delta y^+} - \frac{u_{i+1,j}^{n+1} - u_{i+1,j-1}^{n+1}}{\Delta y^-}\right) \tag{8-125}$$

The pressure derivative in the streamwise momentum equation is represented by

$$\left.\frac{\partial p}{\partial x}\right)_{i+1,j}^{n} \simeq \frac{p_{i+2,j}^n - p_{i+1,j}^n}{\Delta x^+} \tag{8-126}$$

The differencing of the pressure gradient term ensures that $u_{i+1,j}^{n+1}$ is influenced by the pressure downstream.

The y-momentum equation is differenced in a similar manner. Because of the staggered grid being used, $v_{i+1,j}^{n+1}$ is not located at the same point in the flow as $u_{i+1,j}^{n+1}$. The evaluation of the coefficients in the difference representation of the y-momentum equation should reflect this. For example, in representing the term $u\,\partial v/\partial x$, the coefficient should be formed using the average of u at two j levels. The pressure derivative utilizes pressure values on both sides of $v_{i+1,j}^{n+1}$,

$$\left.\frac{\partial p}{\partial y}\right)_{i+1,j}^{n} \simeq \frac{p_{i+1,j}^n - p_{i+1,j-1}^n}{\Delta y^-} \tag{8-127}$$

As the momentum equations are solved, the best current estimate of the pressure field is used. Additional details on how this pressure field is determined will be discussed below. With the pressure fixed, the momentum equations are parabolic and are solved in a segregated manner, the x-momentum equation for $u_{i+1,j}^{n+1}$ and the y-momentum equation for $v_{i+1,j}^{n+1}$. The system of algebraic equations for the unknowns at the $i+1$ level is tridiagonal and can be solved by employing the Thomas algorithm. As was observed for the 3-D parabolic procedure, the solution for the velocities will not satisfy the continuity equation until the correct pressure field is determined. Thus, the velocities obtained from the solutions for the momentum equations are provisional. It is assumed that velocity corrections are driven by a potential, $\hat{\phi}$ in such a manner that the continuity equation is satisfied by the corrected velocities. This requires that

$$\frac{\partial(u_p + u_c)}{\partial x} + \frac{\partial(v_p + v_c)}{\partial y} = 0 \tag{8-128}$$

where u_c and v_c are velocity corrections and u_p and v_p are the provisional velocities obtained from the solution to the momentum equations at marching level $i+1$. Defining a potential function $\hat{\phi}$ by

$$u_c = \frac{\partial \hat{\phi}}{\partial x} \qquad v_c = \frac{\partial \hat{\phi}}{\partial y} \tag{8-129}$$

we obtain

$$\frac{\partial^2 \hat{\phi}}{\partial x^2} + \frac{\partial^2 \hat{\phi}}{\partial y^2} = -\frac{\partial u_p}{\partial x} - \frac{\partial v_p}{\partial y} = S_\phi \tag{8-130}$$

In difference form this becomes

$$\frac{1}{\Delta x_u}\left(\frac{\hat{\phi}_{i+2,j} - \hat{\phi}_{i+1,j}}{\Delta x^+} - \frac{\hat{\phi}_{i+1,j} - \hat{\phi}_{i,j}}{\Delta x}\right)$$
$$+ \frac{1}{\Delta y_v^+}\left(\frac{\hat{\phi}_{i+1,j+1} - \hat{\phi}_{i+1,j}}{\Delta y^+} - \frac{\hat{\phi}_{i+1,j} - \hat{\phi}_{i+1,j-1}}{\Delta y^-}\right)$$
$$= -2\,\frac{(u_p)_{i+1,j} - (u_p)_{i,j}}{\Delta x + \Delta x^+} - 2\,\frac{(v_p)_{i+1,j+1} - (v_p)_{i+1,j}}{\Delta y^+ + \Delta y^-} = (S_\phi)_{i+1,j} \tag{8-131}$$

Such an algebraic equation can be written for each $\hat{\phi}$ grid point across the flow; $j = 2, 3, \ldots, NJ$ where $j = 2$ is the first $\hat{\phi}$ grid point above the lower boundary and $j = NJ$ denotes the $\hat{\phi}$ grid point just below the upper boundary. This results in a tridiagonal system of equations for the unknown $\hat{\phi}_{i+1,j}$'s if $\hat{\phi}_{i,j}$ and $\hat{\phi}_{i+2,j}$ are known. The assumptions made to evaluate $\hat{\phi}_{i,j}$ and $\hat{\phi}_{i+2,j}$ are

$$(a) \quad \hat{\phi}_{i,j} = \hat{\phi}_{i+1,j}$$

This implies that no corrective flow is present from the ith station where conservation of mass has already been established.

$$(b) \quad \hat{\phi}_{i+2,j} = 0$$

This implies that $(v_c)_{i+2,j}$ is zero which must be the case when convergence is achieved. Any other assumption regarding $\hat{\phi}_{i+2,j}$ would appear to be inconsistent with convergence. The boundary conditions used when solving the tridiagonal system of equations to determine $\hat{\phi}_{i+1,j}$ are chosen to be consistent with the prescribed velocity boundary conditions. For example, if velocities are prescribed along the top and bottom boundaries, v_c would be zero along these boundaries and the conditions used would be $\hat{\phi}_{i+1,1} = \hat{\phi}_{i+1,2}$ and $\hat{\phi}_{i+1,NJ} = \hat{\phi}_{i+1,NJ+1}$.

After the $\hat{\phi}_{i+1,j}$'s are determined, velocity corrections are evaluated from the finite-difference representation of Eq. (8-129), namely

$$(u_c)_{i+1,j} = -\frac{\phi_{i+1,j}}{\Delta x^+}$$

and

$$(v_c)_{i+1,j} = \frac{\phi_{i+1,j} - \phi_{i+1,j-1}}{\Delta y^-}$$

The corrected velocities now satisfy continuity at each grid point at the $i + 1$ marching level but unfortunately, until convergence, these velocities do not satisfy the momentum equations exactly.

The pressure is updated between marching sweeps by solving a Poisson equation for pressure using the method of SOR by points. The Poisson equation is formed from the difference representation of the momentum equations. That is, we can write

$$\frac{\partial p}{\partial x} = -\rho\left(u\frac{\partial u}{\partial x} + v\frac{\partial u}{\partial y} - \nu\frac{\partial^2 u}{\partial y^2}\right) = G1$$

$$\frac{\partial p}{\partial y} = -\rho\left(u\frac{\partial v}{\partial x} + v\frac{\partial v}{\partial y} - \nu\frac{\partial^2 v}{\partial y^2}\right) = G2$$

When the above equations are differenced, the G's are considered to be located midway between the pressure points used in representing the pressure derivatives on the left-hand side. Thus, $G1$ "points" are coincident with u locations, and $G2$ "points" are coincident with v locations. Then

$$\frac{\partial^2 p}{\partial x^2} + \frac{\partial^2 p}{\partial y^2} = \frac{\partial G1}{\partial x} + \frac{\partial G2}{\partial y} = S_p \tag{8-132}$$

$G1$ and $G2$ are evaluated by using the *corrected* velocities which satisfy the continuity equation. The use of corrected velocities contributes to the development of a pressure field which will ultimately force the solutions to the momentum equations to conserve mass locally. The S_p terms are evaluated and stored as the marching integration sweep of the momentum equations proceeds. Normally, one SOR sweep of the pressure field is made during this marching procedure. It is easy to update the pressure by one line relaxation before advancing the velocity solution to the next i level. Several more SOR passes are made at the conclusion of the marching sweep. Over-relaxation factors of 1.7 have been successfully used, but the source term, S_p, is typically under-relaxed by a factor ranging from 0.2 to 0.65, the smaller factor being used for the earliest marching sweeps.

The boundary conditions for the Poisson equation for pressure are all Neumann as derived from the momentum equations. The divergence theorem requires that

$$\iint S_p \, dx \, dy = \int \frac{\partial p}{\partial n} \, dC$$

where C represents the boundary of the flow domain and $\partial p/\partial n$ is the magnitude of the Neumann boundary condition. The finite-difference equivalent of this constraint must be satisfied before the solution procedure for the Poisson equation will converge. With the staggered grid, this constraint can be satisfied by relating the boundary point pressures to the pressures in the interior through the specified derivative boundary conditions by an equation which is implicit with respect to iteration levels in the method of SOR by points. This step eliminates all dependence on the boundary pressures themselves (Miyakoda, 1962) when solving the Poisson equation for pressure. As long as the difference representation for S_p has the conservative property, the iterative procedure will converge. This boundary treatment is illustrated by writing Eq. (8-132) in difference form for a p point just inside the lower boundary

$$\frac{1}{\Delta x_u}\left(\frac{p_{i+2,2}^{k}-p_{i+1,2}^{k+1}}{\Delta x^+}-\frac{p_{i+1,2}^{k+1}-p_{i,2}^{k+1}}{\Delta x}\right)$$
$$+\frac{1}{\Delta y_v^+}\left(\frac{p_{i+1,3}^{k+1}-p_{i+1,2}^{k+1}}{\Delta y^+}-\frac{p_{i+1,2}^{k+1}-p_{i+1,1}^{k+1}}{\Delta y^-}\right)$$
$$=\frac{G1_{i+1,2}-G1_{i,2}}{\Delta x_u}+\frac{G2_{i,3}-G2_{i,2}}{\Delta y_v^+} \tag{8-133}$$

In the above, k refers to the iteration level in the SOR procedure for the Poisson equation and $k + 1$ denotes the current level. The boundary condition on the Poisson equation at the lower boundary is taken as

$$\left(\frac{\partial p}{\partial y}\right)_{\text{bdy}} = G2$$

That is, the boundary pressure derivative is evaluated from the momentum equation. In difference form this becomes

$$\frac{p_{i+1,2}^{k+1}-p_{i+1,1}^{k+1}}{\Delta y^-} = G2_{i+1,2} \tag{8-134}$$

where the pressures have been written implicitly at the present iteration level. The pressure below the lower boundary, $p_{i+1,1}^{k+1}$, can now be eliminated from the Poisson equation by substituting Eq. (8-134) into Eq. (8-133). This gives

$$\frac{1}{\Delta x_u}\left(\frac{p_{i+2,2}^{k}-p_{i+1,2}^{k+1}}{\Delta x^+}-\frac{p_{i+1,2}^{k+1}-p_{i,2}^{k+1}}{\Delta x}\right)$$
$$+\frac{1}{\Delta y_v^+}\left(\frac{p_{i+1,3}^{k+1}-p_{i+1,2}^{k+1}}{\Delta y^+}\right)=\frac{G1_{i+1,2}-G1_{i,2}}{\Delta x_u}+\frac{G2_{i,3}}{\Delta y_v^+} \tag{8-135}$$

An examination of the representation for S_p substantiates that the constraint imposed by the divergence theorm is satisfied by this procedure. An evaluation of $\iint S_p\,dx\,dy$ leaves only terms involving $G1$ and $G2$ along the boundaries due to cancellation of all other G's. These boundary $G1$ and $G2$'s are exactly equal to $\int (\partial p/\partial n)\,dC$ when the boundary conditions are expressed in terms of the G's as illustrated by Eqs. (8-134) and (8-135).

The steps in the PPNS solution procedure are summarized below.

1. The momentum equations are solved for tentative velocity profiles at the $i + 1$ station using an estimated pressure field. For the first streamwise sweep, this pressure field can be obtained by (*a*) assuming that $\partial p/\partial x = -\rho u_e(du_e/dx)$ and $\partial p/\partial y = 0$ or (*b*) assuming that $\partial p/\partial y = 0$ and using a secant procedure (see Section 7-4.3) to determine the value of $\partial p/\partial x$ which will conserve mass globally across the flow much as is done when solving internal flows with boundary-layer equations. For sweeps beyond the first, a block adjustment can be added (or subtracted) to the downstream pressure through use of the secant procedure at each i station to ensure that mass is conserved globally across the flow. This forces the algebraic sum of the mass sources across the flow to be zero and appears

to speed convergence in some cases. Also for the first streamwise sweep only, the FLARE approximation (see Section 7-4.2) is used to advance the solution through any regions of reversed flow.

2. The velocities are corrected to satisfy continuity locally using the potential function $\hat{\phi}$ as indicated above.
3. The pressure at $i+1$ is now updated by one SOR pass across the flow at the $i+1$ level. This is optional at this point as all pressures are further improved at the end of the marching sweep.
4. Steps 1-3 are repeated for all streamwise stations until the downstream boundary is reached.
5. At the conclusion of the marching sweep, the pressures throughout the flow are updated by several passes through the Poisson equation. This completes one global iteration. The next marching sweep then starts at the inflow boundary using the revised pressure field. The process continues until the velocity corrections become negligible; i.e., the pressure field obtained permits solutions to the momentum equations which also satisfy the continuity equation.

Sample computational results from the PPNS procedure are shown in Figs. 8-7 and 8-8. Chilukuri and Pletcher (1980) found that solutions to the PPNS equations for two-dimensional laminar channel inlet flows agreed well with solutions to the full Navier-Stokes equations for channel Reynolds numbers as low as 10. Velocity profiles

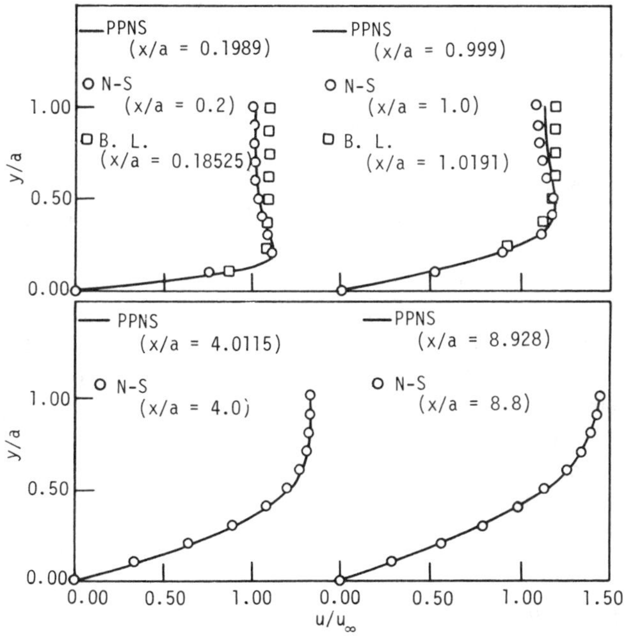

Figure 8-7 Comparison of velocity profiles predicted by the PPNS procedure (Chilukuri and Pletcher, 1980) with the Navier-Stokes solutions of McDonald et al. (1972) and with boundary-layer solutions obtained using the method of Nelson and Pletcher (1974), Re = 75.

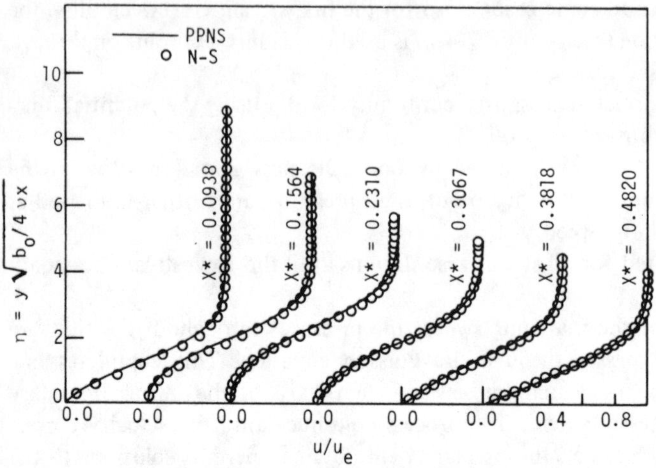

Figure 8-8 Comparison of velocity profiles predicted by the PPNS procedure (Madavan and Pletcher, 1982) with the Navier-Stokes solutions of Briley (1971) for a laminar separating and reattaching flow; $x^* = b_1 x/b_0$, $b_0 = 30.48$ m/s, $b_1 = 300\ \mathrm{s}^{-1}$.

predicted by the PPNS procedure are compared with the Navier-Stokes solutions obtained by McDonald et al. (1972) for a channel Reynolds number ($\mathrm{Re} = u_\infty a/\nu$, a = channel half-width) equal to 75 in Fig. 8-7. For reference, solutions to the boundary-layer equations are also shown in the figure. The boundary-layer solutions fail to exhibit the velocity overshoots characteristic of solutions to the PPNS and NS equations. The PPNS scheme employed 32 grid points in the streamwise direction and 18 across the flow. The sum of the magnitudes of the mass sources at any streamwise station were reduced to less than 1% of the channel mass flow rate in 7 streamwise marching sweeps.

PPNS results obtained by Madavan and Pletcher (1982) for a separated external flow are compared with numerical solutions to the Navier-Stokes equations obtained by Briley (1971) in Fig. 8-8. The flow separates under the influence of a linearly decelerating external stream. At a point downstream of separation, the external stream velocity becomes constant causing the flow to reattach. Reversed flow exists over approximately one-third of the streamwise extent of the computational domain. In the PPNS calculation, 35 grid points were employed in the streamwise direction and 32 across the flow. Sixteen streamwise sweeps were required to reduce the sum of the magnitudes of the mass sources at any streamwise station to less than 1% of the mass flow rate. The computation was continued for a total of 43 streamwise sweeps at which time the sum of the magnitudes of the mass sources at any streamwise station was less than 0.05%.

8-5 VISCOUS SHOCK-LAYER EQUATIONS

The viscous shock-layer (VSL) equations are a more approximate set of equations than the PNS equations. In terms of complexity, they fall between the PNS equations and the boundary-layer equations. The major advantage of the viscous shock-layer equations is that they remain hyperbolic-parabolic in both the streamwise and crossflow directions. Thus, the VSL equations can be solved using a marching procedure in both directions very similar to techniques employed for the three-dimensional boundary-layer equations. This is in contrast to the PNS equations which must be solved simultaneously over the entire crossflow plane. As a consequence, the VSL equations can be solved (in most cases) with less computer time than the PNS equations. An additional advantage of the VSL equations is that they can be used to compute the viscous flow in the subsonic blunt nose region where the PNS equations are not applicable. Thus, for bodies with blunt noses, the VSL equations can be solved to provide a starting solution for a subsequent PNS computation. The major disadvantage of the VSL equations is that they cannot be used to compute flowfields with crossflow separations. This is a direct result of the fact that the VSL equations are not elliptic in the crossflow plane.

The concept of using a set of equations such as the VSL equations to solve for the high Mach number flow past a blunt body had its origins in the work of Cheng (1963) and Davis and Flügge-Lotz (1964). As mentioned previously, the solution of a set of equations like the VSL equations avoids the need to explicitly determine the second-order boundary-layer effects of vorticity and displacement thickness. Furthermore, the difficulty encountered in matching the viscous and inviscid solutions, when the boundary layer is significantly merged with the outer inviscid flow, is eliminated.

Of all the early studies involving the use of the VSL equations, the method of Davis (1970) was the most successful. He solved the axisymmetric VSL equations in order to determine the hypersonic, laminar flow over a hyperboloid. The VSL equations used by Davis are derived by first nondimensionalizing the Navier-Stokes equations with variables which are of order one in the boundary layer for large Reynolds numbers. In a similar manner, another set of equations is formed by nondimensionalizing the Navier-Stokes equations with variables which are of order one in the inviscid region of the flowfield. In both sets of equations, terms up to second order in ϵ are retained

$$\epsilon = \left[\frac{\mu_{\text{ref}}}{\rho_\infty V_\infty r_{\text{nose}}}\right]^{1/2} \tag{8-136}$$

where μ_{ref} is the coefficient of viscosity evaluated at the reference temperature

$$T_{\text{ref}} = \frac{V_\infty^2}{c_{p_\infty}} \tag{8-137}$$

The two sets of equations are then compared and combined into a single set of equations which is valid to second order from the body to the shock. For a 2-D ($m = 0$) or axisymmetric ($m = 1$) body intrinsic coordinate system (see Fig. 5-3), the VSL equations in nondimensional form become

continuity:

$$\frac{\partial}{\partial \xi^*}[(r^* + \eta^* \cos \phi)^m \rho^* u^*] + \frac{\partial}{\partial \eta^*}[(1 + K^*\eta^*)(r^* + \eta^* \cos \phi)^m \rho^* v^*] = 0 \tag{8-138}$$

ξ momentum:

$$\rho^*\left[\frac{u^*}{1+K^*\eta^*}\frac{\partial u^*}{\partial \xi^*}+v^*\frac{\partial u^*}{\partial \eta^*}+\frac{K^*u^*v^*}{1+K^*\eta^*}\right]+\frac{1}{1+K^*\eta^*}\frac{\partial p^*}{\partial \xi^*}$$

$$=\frac{\epsilon^2}{(1+K^*\eta^*)^2(r^*+\eta^*\cos\phi)^m}\frac{\partial}{\partial \eta^*}[(1+K^*\eta^*)^2(r^*+\eta^*\cos\phi)^m\tau^*] \quad (8\text{-}139)$$

where

$$\tau^*=\mu^*\left(\frac{\partial u^*}{\partial \eta^*}-\frac{K^*u^*}{1+K^*\eta^*}\right)$$

η momentum:

$$\rho^*\left[\frac{u^*}{1+K^*\eta^*}\frac{\partial v^*}{\partial \xi^*}+v^*\frac{\partial v^*}{\partial \eta^*}-\frac{K^*(u^*)^2}{1+K^*\eta^*}\right]+\frac{\partial p^*}{\partial \eta^*}=0 \quad (8\text{-}140)$$

energy:

$$\rho^*\left(\frac{u^*}{1+K^*\eta^*}\frac{\partial T^*}{\partial \xi^*}+v^*\frac{\partial T^*}{\partial \eta^*}\right)-\frac{u^*}{1+K^*\eta^*}\frac{\partial p^*}{\partial \xi^*}-v^*\frac{\partial p^*}{\partial \eta^*}=\frac{\epsilon^2(\tau^*)^2}{\mu^*}$$

$$+\frac{\epsilon^2}{(1+K^*\eta^*)(r^*+\eta^*\cos\phi)^m}\frac{\partial}{\partial \eta^*}\left[(1+K^*\eta^*)(r^*+\eta^*\cos\phi)^m\frac{\mu^*}{\Pr}\frac{\partial T^*}{\partial \eta^*}\right]$$

$$(8\text{-}141)$$

These equations have been nondimensionalized in the following manner:

$$\xi^*=\frac{\xi}{r_{\text{nose}}} \qquad \eta^*=\frac{\eta}{r_{\text{nose}}} \qquad r^*=\frac{r}{r_{\text{nose}}} \qquad K^*=\frac{K}{r_{\text{nose}}}$$

$$u^*=\frac{u}{V_\infty} \qquad v^*=\frac{v}{V_\infty} \qquad T^*=\frac{T}{T_{\text{ref}}} \quad (8\text{-}142)$$

$$p^*=\frac{p}{\rho_\infty V_\infty^2} \qquad \rho^*=\frac{\rho}{\rho_\infty} \qquad \mu^*=\frac{\mu}{\mu_{\text{ref}}}$$

By assuming a thin shock layer, the normal momentum equation reduces to:

η momentum (thin shock-layer approximation):

$$\frac{\partial p^*}{\partial \eta^*}=\frac{K^*\rho^*(u^*)^2}{1+K^*\eta^*} \quad (8\text{-}143)$$

The above equations can be readily converted to a 2-D Cartesian coordinate system by setting

$$m=0$$

$$K^*=0$$

$$x^*=\xi^* \quad (8\text{-}144)$$

$$y^*=\eta^*$$

The resulting VSL equations in Cartesian coordinates can then be compared directly with the PNS equations given previously by Eqs. (8-29)-(8-33). This comparison shows that the continuity and x-momentum equations are the same but the y-momentum and energy equations in the VSL set of equations are simpler than the corresponding PNS equations.

In the original solution technique of Davis, the VSL equations were normalized with variable values behind the shock. This was done to permit the same grid in the normal direction to be used over the entire body. An initial global solution was obtained by utilizing the thin shock-layer assumption. This assumption makes the VSL equations totally parabolic and permits the use of standard boundary-layer solution algorithms. Subsequent global iterations retained the complete normal momentum equation. Also, for the first global iteration the shock was assumed to be concentric with the body. This assumption was possible because only hyperboloid body shapes were considered due to the difficulties associated with curvature discontinuities in body shapes such as sphere-cones. The shock angles for the second iteration were determined from the shock-layer thicknesses computed during the first iteration.

The marching procedure was initiated from an approximate stagnation streamline solution. This stagnation streamline solution was obtained from the VSL equations which reduce to ordinary differential equations along $\xi = 0$. The solution at each subsequent ξ station was obtained by solving the VSL equations individually in the following order:

1. energy
2. ξ momentum
3. continuity
4. η momentum

The original method of Davis was not entirely satisfactory because of several limitations. First of all, the method was restricted to analytic body shapes such as hyperboloids. This difficulty was circumvented by Miner and Lewis (1975) who computed the flow around a sphere-cone body. They started with an initial shock shape from an inviscid blunt body solution and used a transition function near the sphere-cone juncture in order to obtain a smooth distribution of curvature. Later, Srivastava et al. (1978) overcame this same difficulty by applying special difference formulas to the jump conditions across surface discontinuities.

Another difficulty associated with the original Davis method was the poor convergence of the shock shape when the shock layer became thick. This problem was resolved by Srivastava et al. (1978, 1979) who noted that the relaxation process associated with the shock shape was similar to the interaction between displacement thickness and the outer inviscid flow in supersonic interacting boundary-layer theory. As a result of this observation, they were able to solve the shock shape divergence problem by adapting the ADI method of Werle and Vatsa (1974) for interacting boundary layers. Another problem with the original Davis method was that it was not able to solve the flow far downstream on slender bodies. This difficulty was traced to the fact that the VSL equations were being solved in an uncoupled manner. In particular, the two first-order equations (continuity and normal momentum) introduced instabilities which grew in the streamwise direction. By solving the continuity and normal momentum equations in a coupled fashion, Waskiewicz et al. (1978) were able to eliminate this stability problem. In a similar manner, Hosny et al. (1978) overcame the problem by completely coupling the VSL equations through a quasi-linearization technique.

With the above difficulties eliminated, the application of the VSL equations to more complicated problems has become possible. Murray and Lewis (1978) have applied the VSL equations to solve the flow around general three-dimensional body shapes at angle of attack. Their code has been used to successfully compute a large number of different flowfields. More recently, Lewis and associates have included turbulence modeling (Szema and Lewis, 1980) and real gas effects (Thareja et al., 1982; Swaminathan et al., 1983) in their computations.

8-6 "CONICAL" NAVIER-STOKES EQUATIONS

The conical flow assumption for inviscid flows makes use of the fact that a significant length scale is missing in the conical direction for a flowfield surrounded by conical boundaries. As a result, no variations in flow properties in the radial direction can occur and a three-dimensional inviscid flow problem is reduced to a two-dimensional problem. This leads to a self-similar solution which is the same for all constant radius surfaces but scales linearly with the radius. The concept of conical flow is strictly valid only for inviscid flows. However, the viscous portions of the same flowfields have been observed in experiments to be strongly dominated by the outer inviscid conical flow. For these flowfields, Anderson (1973) suggested that a quick estimate of the heat transfer and skin friction could be obtained by solving the unsteady Navier-Stokes equations in a time-dependent fashion on the unit sphere with all derivatives in the radial direction set equal to zero. Thus, the Navier-Stokes equations are solved subject to a local conical approximation. We will refer to the equations solved in this manner as the "conical" Navier-Stokes equations. The local Reynolds number is determined by the radial position where the solution is computed. As a result, the solution is not self-similar in the sense of inviscid conical flow but is scaled through the local Reynolds number which remains in the resulting set of equations.

The "conical" Navier-Stokes equations were originally used by McRae (1976) to compute the laminar flow over a cone at high angle of attack. Since then, Vigneron et al. (1978a) and Bluford (1978) have computed the laminar flow over a delta wing and Tannehill and Anderson (1980) have computed the flow in a 3-D axial corner. Also, McRae and Hussaini (1978) have employed an eddy viscosity model in conjunction with the "conical" Navier-Stokes equations to compute the turbulent flow over a cone at high angle of attack. In all of the above cases (except one where the inviscid flow was not completely conical) the computed inviscid and viscous structure of the flowfield agrees surprisingly well with the available experimental data.

The "conical" Navier-Stokes equations have also proved quite useful in computing starting solutions for PNS calculations of flows over conical (or pointed) body shapes. Schiff and Steger (1979) have incorporated a marching step-back method in their PNS algorithm which is equivalent to solving the "conical" Navier-Stokes equations using the time-dependent approach described above. In their method, the flow variables are initially set equal to their freestream values and the equations are marched from $x = x_0$ to $x = x_0 + \Delta x$ using the same implicit scheme as used to solve the PNS equations but with $\partial p/\partial x = 0$. After each marching step, the solution is scaled back to $x = x_0$. The computation is repeated until no change in flow variables occurs.

The "conical" Navier-Stokes equations are derived from the complete Navier-Stokes equations

$$\frac{\partial \mathbf{U}^*}{\partial t^*} + \frac{\partial \mathbf{E}^*}{\partial x^*} + \frac{\partial \mathbf{F}^*}{\partial y^*} + \frac{\partial \mathbf{G}^*}{\partial z^*} = 0 \tag{8-145}$$

where $\mathbf{U}^*$, $\mathbf{E}^*$, $\mathbf{F}^*$, and $\mathbf{G}^*$ are the nondimensional vectors defined by Eqs. (5-46). The following conical transformation

$$\begin{aligned} \alpha &= [(x^*)^2 + (y^*)^2 + (z^*)^2]^{1/2} \\ \beta &= \frac{y^*}{x^*} \\ \gamma &= \frac{z^*}{x^*} \\ \tau &= t^* \end{aligned} \tag{8-146}$$

is initially applied to these equations. The resulting transformed equations can be written in the following strong conservation-law form

$$\frac{\partial}{\partial \tau}\left(\frac{\alpha^2}{\lambda^3}\mathbf{U}^*\right) + \frac{\partial}{\partial \alpha}\left[\frac{\alpha^2}{\lambda^4}(\mathbf{E}^* + \beta\mathbf{F}^* + \gamma\mathbf{G}^*)\right] + \frac{\partial}{\partial \beta}\left[\frac{\alpha}{\lambda^2}(-\beta\mathbf{E}^* + \mathbf{F}^*)\right] + \frac{\partial}{\partial \gamma}\left[\frac{\alpha}{\lambda^2}(-\gamma\mathbf{E}^* + \mathbf{G}^*)\right] = 0 \tag{8-147}$$

where $$\lambda = (1 + \beta^2 + \gamma^2)^{1/2}$$

The assumption of local conical self-similarity requires that

$$\frac{\partial \mathbf{E}^*}{\partial \alpha} = 0 \qquad \frac{\partial \mathbf{F}^*}{\partial \alpha} = 0 \qquad \frac{\partial \mathbf{G}^*}{\partial \alpha} = 0 \tag{8-148}$$

which reduces Eq. (8-147) to

$$\frac{\partial}{\partial \tau}\left(\frac{\alpha^2}{\lambda^3}\mathbf{U}^*\right) + \frac{2\alpha}{\lambda^4}(\mathbf{E}^* + \beta\mathbf{F}^* + \gamma\mathbf{G}^*) + \frac{\partial}{\partial \beta}\left[\frac{\alpha}{\lambda^2}(-\beta\mathbf{E}^* + \mathbf{F}^*)\right] + \frac{\partial}{\partial \gamma}\left[\frac{\alpha}{\lambda^2}(-\gamma\mathbf{E}^* + \mathbf{G}^*)\right] = 0 \tag{8-149}$$

The solution is computed on a spherical surface whose nondimensional radius ($r^* = r/L$) is equal to 1. On this computational surface, $\alpha = 1$ since

$$r^* = [(x^*)^2 + (y^*)^2 + (z^*)^2]^{1/2} = \alpha$$

As a result, Eq. (8-149) can be rewritten as

$$\frac{\partial \mathbf{U}_4}{\partial \tau} + \frac{\partial \mathbf{F}_4}{\partial \beta} + \frac{\partial \mathbf{G}_4}{\partial \gamma} + \mathbf{H}_4 = 0 \tag{8-150}$$

where

$$\begin{aligned} \mathbf{U}_4 &= \frac{\mathbf{U}^*}{\lambda^3} \\ \mathbf{F}_4 &= \frac{-\beta \mathbf{E}^* + \mathbf{F}^*}{\lambda^2} \\ \mathbf{G}_4 &= \frac{-\gamma \mathbf{E}^* + \mathbf{G}^*}{\lambda^2} \\ \mathbf{H}_4 &= \frac{2(\mathbf{E}^* + \beta \mathbf{F}^* + \gamma \mathbf{G}^*)}{\lambda^4} \end{aligned} \tag{8-151}$$

The partial derivatives appearing in the viscous terms of $\mathbf{E}^*$, $\mathbf{F}^*$, and $\mathbf{G}^*$ are readily transformed using

$$\begin{aligned} \frac{\partial}{\partial x^*} &= -\beta\lambda \frac{\partial}{\partial \beta} - \gamma\lambda \frac{\partial}{\partial \gamma} \\ \frac{\partial}{\partial y^*} &= \lambda \frac{\partial}{\partial \beta} \\ \frac{\partial}{\partial z^*} &= \lambda \frac{\partial}{\partial \gamma} \end{aligned} \tag{8-152}$$

Thus, the shear-stress and heat-flux terms, given in Eqs. (5-47), become

$$\begin{aligned} \tau_{xx}^* &= \frac{2\mu^*}{3\,\mathrm{Re}_L} (-2\beta\lambda u_\beta^* - 2\gamma\lambda u_\gamma^* - \lambda v_\beta^* - \lambda w_\gamma^*) \\ \tau_{yy}^* &= \frac{2\mu^*}{3\,\mathrm{Re}_L} (2\lambda v_\beta^* + \beta\lambda u_\beta^* + \gamma\lambda u_\gamma^* - \lambda w_\gamma^*) \\ \tau_{zz}^* &= \frac{2\mu^*}{3\,\mathrm{Re}_L} (2\lambda w_\gamma^* + \beta\lambda u_\beta^* + \gamma\lambda u_\gamma^* - \lambda v_\beta^*) \\ \tau_{xy}^* &= \frac{\mu^*}{\mathrm{Re}_L} (\lambda u_\beta^* - \beta\lambda v_\beta^* - \gamma\lambda v_\gamma^*) \\ \tau_{xz}^* &= \frac{\mu^*}{\mathrm{Re}_L} (\lambda u_\gamma^* - \beta\lambda w_\beta^* - \gamma\lambda w_\gamma^*) \\ \tau_{yz}^* &= \frac{\mu^*}{\mathrm{Re}_L} (\lambda v_\gamma^* + \lambda w_\beta^*) \end{aligned} \tag{8-153}$$

$$q_x^* = \frac{\mu^*}{(\gamma - 1)M_\infty^2 \operatorname{Re}_L \operatorname{Pr}} (-\beta\lambda T_\beta^* - \gamma\lambda T_\gamma^*)$$

$$q_y^* = \frac{\mu^*}{(\gamma - 1)M_\infty^2 \operatorname{Re}_L \operatorname{Pr}} \lambda T_\beta^*$$

$$q_z^* = \frac{\mu^*}{(\gamma - 1)M_\infty^2 \operatorname{Re}_L \operatorname{Pr}} \lambda T_\gamma^* \tag{8-153 (Cont.)}$$

Note that the Reynolds number Re_L remains in the expressions for shear stress and heat flux. This Reynolds number is evaluated using

$$\operatorname{Re}_L = \frac{\rho_\infty V_\infty L}{\mu_\infty} \tag{8-154}$$

where L is the radius of the spherical surface where the solution is computed. As a consequence, solutions of the "conical" Navier-Stokes equations depend directly on the position $(r = L)$ where they are computed. This is different from inviscid solutions which are independent of r, and thus, truly conical.

The "conical" Navier-Stokes equations can be solved using the same "time-dependent" algorithms that are applied in Chap. 9 to the 2-D compressible Navier-Stokes equations. Thus, we will postpone our discussion of numerical schemes for the "conical" Navier-Stokes equations until then. In closing, it should be remembered that the "conical" Navier-Stokes equations are a very approximate form of the complete Navier-Stokes equations, and as such, they should not be used for flow problems where a high degree of accuracy is required.

PROBLEMS

8-1 Verify Eq. (8-8).

8-2 Derive Eqs. (8-9), (8-10), and (8-11).

8-3 Reduce the thin-layer equations in Cartesian coordinates to the set of boundary equations which are valid at a no-slip wall ($y = 0$). Assume the wall is held at a constant temperature of T_w.

8-4 Reduce the thin-layer equations written in the transformed coordinate system [Eqs. (8-9), (8-10), and (8-11)] to the set of boundary equations which are valid at a no-slip wall ($\eta = 0$). Assume the wall is held at a constant temperature of T_w.

8-5 Obtain Eq. (8-15) from Eq. (5-19).

8-6 Obtain Eq. (8-16) from Eq. (5-19).

8-7 Obtain Eq. (8-17) from Eq. (5-31).

8-8 Obtain Eq. (8-23) from Eq. (8-17).

8-9 Derive the compressible, laminar, boundary-layer equations starting with Eqs. (8-14)–(8-17). Note that in the boundary-layer region, $\Delta^2 \sim O(1)$ and $(\Delta/\delta^*)^2 \gg 1$.

8-10 Apply the thin-layer approximation to Eqs. (8-37)–(8-39) and show that they are equivalent to Eqs. (8-40), (8-10), and (8-11).

8-11 Verify that Eq. (8-44) is equivalent to Eq. (8-43).

8-12 Show that the eigenvalues of Eq. (8-44) are given by Eq. (8-46).
Hint: $|\lambda[I] - [A_1]^{-1}[B_1]| = |[A_1]^{-1}|\,|\lambda[A_1] - [B_1]|$.

8-13 Derive Eq. (8-47).

8-14 Verify Eqs. (8-48)–(8-49).

8-15 Derive Eq. (8-50).

8-16 For the flow conditions,

$$M_x = 0.6$$

$$\frac{\text{Re}}{L} = \frac{\rho u}{\mu} = \frac{1000}{m}$$

$$\gamma = 1.4$$

$$\text{Pr} = 0.72$$

solve Eq. (8-50) and show that all the roots will be real and positive if $\omega = 0.4$ which satisfies Eq. (8-52).

8-17 Repeat Prob. 8-16 with $\omega = 0.5$ and show that at least one root of Eq. (8-50) will *not* be real and positive.

8-18 If all the eigenvalues of Eq. (8-50) are real, show that these eigenvalues are positive provided that the conditions given by Eqs. (8-51) and (8-52) are satisfied.

8-19 Place an ω in front of the streamwise pressure gradient term in both the streamwise momentum equation and the energy equation and evaluate the condition which must be satisfied in order for Eq. (8-44) to remain hyperbolic if $\omega < 1$. You may assume that $v \ll u$.

8-20 Linearize the following terms using Eq. (8-61):

(*a*) $u_{i+1,j,k}^{m+1} v_{i+1,j,k}^{m+1}$

(*b*) $(u_{i+1,j,k}^{m+1})^2 v_{i+1,j,k}^{m+1}$

(*c*) $(u_{i+1,j,k}^{m+1})^3 v_{i+1,j,k}^{m+1}$

(*d*) $u_{i+1,j,k}^{m+1} v_{i+1,j,k}^{m+1} w_{i+1,j,k}^{m+1}$

(*e*) $u_{i+1,j,k}^{m+1} (v_{i+1,j,k}^{m+1})^2 w_{i+1,j,k}^{m+1}$

8-21 Derive the expression for the Jacobian $\partial \mathbf{E}^*/\partial \mathbf{U}$ given by Eq. (8-78).

8-22 Derive the expression for the Jacobian $\partial \mathbf{F}/\partial \mathbf{U}$ given by Eq. (8-79).

8-23 Derive the expression for the Jacobian $\partial \mathbf{G}/\partial \mathbf{U}$ given by Eq. (8-80).

8-24 If ω can be approximated by

$$\omega \simeq \gamma M_x^2$$

derive the expression for the Jacobian $\partial \mathbf{E}^*/\partial \mathbf{U}$ which results when ω is no longer assumed independent of $\mathbf{U}$.

8-25 Derive the expression for the Jacobian $\partial \mathbf{F}_v/\partial \mathbf{U}$ given by Eq. (8-84).

8-26 Derive the expression for the Jacobian $\partial \mathbf{G}_v/\partial \mathbf{U}$ given by Eq. (8-85).

8-27 The elements of the matrix $[C]_k$ in Eq. (8-98) can be represented by $(c_{lm})_k$ where $l = 1, 2, \ldots, 5$ and $m = 1, 2, \ldots, 5$. Determine the element $(c_{24})_k$.

8-28 Determine the element $(c_{32})_k$ in Prob. 8-27.

8-29 Determine the element $(c_{43})_k$ in Prob. 8-27.

8-30 The elements of the matrix $[B]_k$ in Eq. (8-98) can be represented by $(b_{lm})_k$ where $l = 1, 2, \ldots, 5$ and $m = 1, 2, \ldots, 5$. Determine the element $(b_{24})_k$.

8-31 Determine the element $(b_{43})_k$ in Prob. 8-30.

8-32 Determine the elements $(a_{33})_k$, $(b_{33})_k$, and $(c_{33})_k$ of the matrices $[A]_k$, $[B]_k$, and $[C]_k$ in Eq. (8-98).

8-33 Apply the difference formula given in Eq. (8-70) to the 2-D PNS equation

$$\frac{\partial \mathbf{E}^*}{\partial x} + \frac{\partial \mathbf{P}}{\partial x} + \frac{\partial \mathbf{F}}{\partial y} = 0$$

and develop a solution algorithm like that given by Eqs. (8-94)–(8-97) for the 3-D PNS equation.

8-34 Work out the details for a velocity correction procedure for the 3-D parabolic procedure for an incompressible flow in a rectangular channel. Use both the $\hat{\phi}$ potential and p' methods. Employ a staggered grid.

8-35 Write the y-momentum equation in finite-difference form for the PPNS model following the strategy outlined in Section 8-4.3 for the x-momentum equation.

8-36 Prove that the formulation described for the Poisson equation for pressure in the PPNS model satisfies the constraint

$$\iint S_p \, dx \, dy = \int \frac{\partial p}{\partial n} \, dC$$

8-37 Suggest a way that the PPNS procedure might be extended to three-dimensional flows.

8-38 Explain how the velocity boundary conditions can be implemented for a boundary which is a line of symmetry (such as the centerline of a two-dimensional channel) when a staggered grid is used for the PPNS momentum equations. Explain in terms of the Thomas algorithm.

8-39 Apply the transformation

$$\alpha = x^*$$

$$\beta = \frac{y^*}{x^*}$$

$$\gamma = \frac{z^*}{x^*}$$

$$\tau = t^*$$

to Eq. (8-145) and derive the "conical" Navier-Stokes equations which can be used to compute a solution at the station $x = L$ where $\alpha = x^* = 1$.

CHAPTER

NINE

NUMERICAL METHODS FOR THE NAVIER-STOKES EQUATIONS

9-1 INTRODUCTION

For certain viscous flow problems, it is not possible to obtain an accurate solution using the simplified flow equations discussed in Chaps. 6, 7, and 8. Examples of such flow problems include shock-boundary layer interactions, leading edge flows, certain wake flows, and other flows which involve strong viscous-inviscid interactions with large separated flow regions. For these cases, it becomes necessary to solve the complete set of Navier-Stokes (N-S) equations (or the Reynolds averaged form of these equations). Unfortunately, these equations are very complex and require a substantial amount of computer time in order to obtain a solution. However, if the flow is incompressible, the equations can be simplified considerably and the required computer time is decreased accordingly.

The unsteady compressible N-S equations are a mixed set of hyperbolic-parabolic equations, while the unsteady incompressible N-S equations are a mixed set of elliptic-parabolic equations. As a consequence, different numerical techniques must be used to solve the N-S equations in the compressible and incompressible flow regimes. These techniques will be discussed separately in this chapter beginning with the techniques for solving the compressible N-S equations.

9-2 COMPRESSIBLE N–S EQUATIONS

The compressible N-S equations in Cartesian coordinates without body forces or external heat addition can be written (see Section 5-1.5) as

$$\frac{\partial \mathbf{U}}{\partial t} + \frac{\partial \mathbf{E}}{\partial x} + \frac{\partial \mathbf{F}}{\partial y} + \frac{\partial \mathbf{G}}{\partial z} = 0 \tag{9-1}$$

where **U**, **E**, **F**, and **G** are vectors given by

$$\mathbf{U} = \begin{bmatrix} \rho \\ \rho u \\ \rho v \\ \rho w \\ E_t \end{bmatrix} \tag{9-2}$$

$$\mathbf{E} = \begin{bmatrix} \rho u \\ \rho u^2 + p - \tau_{xx} \\ \rho uv - \tau_{xy} \\ \rho uw - \tau_{xz} \\ (E_t + p)u - u\tau_{xx} - v\tau_{xy} - w\tau_{xz} + q_x \end{bmatrix} \tag{9-3}$$

$$\mathbf{F} = \begin{bmatrix} \rho v \\ \rho uv - \tau_{xy} \\ \rho v^2 + p - \tau_{yy} \\ \rho vw - \tau_{yz} \\ (E_t + p)v - u\tau_{xy} - v\tau_{yy} - w\tau_{yz} + q_y \end{bmatrix} \tag{9-4}$$

$$\mathbf{G} = \begin{bmatrix} \rho w \\ \rho uw - \tau_{xz} \\ \rho vw - \tau_{yz} \\ \rho w^2 + p - \tau_{zz} \\ (E_t + p)w - u\tau_{xz} - v\tau_{yz} - w\tau_{zz} + q_z \end{bmatrix} \tag{9-5}$$

and the components of the shear-stress tensor and heat-flux vector are given by

$$\tau_{xx} = \frac{2}{3}\mu\left(2\frac{\partial u}{\partial x} - \frac{\partial v}{\partial y} - \frac{\partial w}{\partial z}\right)$$

$$\tau_{yy} = \frac{2}{3}\mu\left(2\frac{\partial v}{\partial y} - \frac{\partial u}{\partial x} - \frac{\partial w}{\partial z}\right)$$

$$\tau_{zz} = \frac{2}{3}\mu\left(2\frac{\partial w}{\partial z} - \frac{\partial u}{\partial x} - \frac{\partial v}{\partial y}\right) \tag{9-6}$$

$$\tau_{xy} = \mu\left(\frac{\partial u}{\partial y} + \frac{\partial v}{\partial x}\right) = \tau_{yx}$$

$$\tau_{xz} = \mu\left(\frac{\partial w}{\partial x} + \frac{\partial u}{\partial z}\right) = \tau_{zx}$$

$$\tau_{yz} = \mu\left(\frac{\partial v}{\partial z} + \frac{\partial w}{\partial y}\right) = \tau_{zy}$$

$$q_x = -k\frac{\partial T}{\partial x}$$

$$q_y = -k\frac{\partial T}{\partial y}$$

$$q_z = -k\frac{\partial T}{\partial z} \qquad (9\text{-}6)\ (Cont.)$$

These equations can be expressed in terms of a generalized orthogonal curvilinear coordinate system (x_1, x_2, x_3) using the formulas in Section 5-1.7. In addition, the compressible N-S equations can be written in terms of a generalized nonorthogonal curvilinear coordinate system (ξ, η, ζ) using the general transformation described in Section 5-6.2:

$$\begin{aligned} \xi &= \xi(x, y, z) \\ \eta &= \eta(x, y, z) \\ \zeta &= \zeta(x, y, z) \end{aligned} \qquad (9\text{-}7)$$

The transformed equations are given in Chap. 8 as Eqs. (8-34)–(8-36).

The thin-layer approximation to the compressible N-S equations was discussed in Section 8-2. This approximation allows one to drop a number of terms from the complete N-S equations. However, the mathematical character of the resulting equations is identical to that of the original N-S equations, and as a result, the two sets of equations are normally solved in the same manner. The thin-layer N-S equations are given in Chap. 8 for a Cartesian coordinate system [Eqs. (8-2)–(8-6)] and for a general nonorthogonal coordinate system [Eqs. (8-9)–(8-12)].

For turbulent flows, it is convenient to use the Reynolds-averaged equations instead of the N-S equations. Employing the Boussinesq approximation (see Section 5-4.2), the N-S equations can be changed to a modeled form of the Reynolds-averaged equations by replacing the coefficient of viscosity μ with

$$\mu + \mu_T$$

and by also replacing the coefficient of thermal conductivity k with

$$k + k_T$$

where μ_T is the eddy viscosity and k_T is the turbulent thermal conductivity. The turbulent thermal conductivity can be expressed in terms of the eddy viscosity using the turbulent Prandtl number Pr_T

$$k_T = \frac{c_p \mu_T}{\text{Pr}_T} \tag{9-8}$$

Techniques for determining μ_t are described in detail in Section 5-4.

As mentioned previously, the unsteady compressible N-S equations are a mixed set of hyperbolic-parabolic equations in time. If the unsteady terms are dropped from these equations, the resulting equations become a mixed set of hyperbolic-elliptic equations which are difficult to solve because of the differences in numerical techniques required for hyperbolic and elliptic type equations. As a consequence, nearly all successful solutions of the compressible N-S equations have employed the unsteady form of the equations. The steady-state solution is obtained by marching the solution in time until convergence is achieved. This procedure is called the *time-dependent approach* and is the method that will be discussed in this chapter for solving the compressible N-S equations.

Both explicit and implicit finite-difference schemes have been used with the time-dependent approach to solve the compressible N-S equations. Nearly all of these methods are second-order accurate in "space" and are either first- or second-order accurate in "time." If an accurate time evolution of the flow is required, the numerical scheme should at least be second-order accurate in "time." On the other hand, if only the steady-state solution is desired, it is often advantageous to employ a scheme which is not time accurate since the steady-state solution may be achieved with fewer time steps. Because of the added complexity, only a handful of third-order (or higher) methods have appeared in the literature to solve the compressible N-S equations. Many feel that a second-order method is the optimum choice since higher-order accuracy is at the expense of substantially more computer time. For a complete review of nearly all papers which report solutions to the compressible N-S equations prior to 1976, the reader is urged to consult the excellent survey paper of Peyret and Viviand (1975). We will now begin our detailed discussion of methods for solving the compressible N-S equations.

9-2.1 Explicit MacCormack Method

When the original MacCormack scheme (1969) is applied to the compressible N-S equations given by Eq. (9-1) the following algorithm results:

Predictor:

$$\overline{\mathbf{U}_{i,j,k}^{n+1}} = \mathbf{U}_{i,j,k}^{n} - \frac{\Delta t}{\Delta x}(\mathbf{E}_{i+1,j,k}^{n} - \mathbf{E}_{i,j,k}^{n}) - \frac{\Delta t}{\Delta y}(\mathbf{F}_{i,j+1,k}^{n} - \mathbf{F}_{i,j,k}^{n})$$
$$- \frac{\Delta t}{\Delta z}(\mathbf{G}_{i,j,k+1}^{n} - \mathbf{G}_{i,j,k}^{n}) \tag{9-9}$$

Corrector:

$$\mathbf{U}_{i,j,k}^{n+1} = \frac{1}{2}\left[\mathbf{U}_{i,j,k}^{n} + \overline{\mathbf{U}_{i,j,k}^{n+1}} - \frac{\Delta t}{\Delta x}(\overline{\mathbf{E}_{i,j,k}^{n+1}} - \overline{\mathbf{E}_{i-1,j,k}^{n+1}}) - \frac{\Delta t}{\Delta y}(\overline{\mathbf{F}_{i,j,k}^{n+1}} - \overline{\mathbf{F}_{i,j-1,k}^{n+1}})\right.$$
$$\left. - \frac{\Delta t}{\Delta z}(\overline{\mathbf{G}_{i,j,k}^{n+1}} - \overline{\mathbf{G}_{i,j,k-1}^{n+1}})\right] \tag{9-10}$$

where $x = i\,\Delta x$, $y = j\,\Delta y$, and $z = k\,\Delta z$. This explicit scheme is second-order accurate in both space and time. In the present form of this scheme, forward differences are used for all spatial derivatives in the predictor step while backward differences are used in the corrector step. The forward and backward differencing can be alternated between predictor and corrector steps as well as between the three spatial derivatives in a sequential fashion. This eliminates any bias due to the one-sided differencing. An example of a suitable sequence is given in Table 9-1.

The derivatives appearing in the viscous terms of **E**, **F**, and **G** must be differenced correctly in order to maintain second-order accuracy. This is accomplished in the following manner. The x derivative terms appearing in **E** are differenced in the opposite direction to that used for $\partial\mathbf{E}/\partial x$ while the y derivatives and the z derivatives are approximated with central differences. Likewise, the y derivative terms appearing in **F** and the z derivative terms appearing in **G** are differenced in the opposite direction to that used for $\partial\mathbf{F}/\partial y$ and $\partial\mathbf{G}/\partial z$, respectively, while the cross-derivative terms in **F** and **G** are approximated with central differences. For example, consider the following term in **F** which corresponds to the x-momentum equation

$$F_2 = \rho uv - \mu \frac{\partial u}{\partial y} - \mu \frac{\partial v}{\partial x} \tag{9-11}$$

In the predictor step, given by Eq. (9-9), this term in $\mathbf{F}^n_{i,j,k}$ is differenced as

$$(F_2)^n_{i,j,k} = (\rho uv)^n_{i,j,k} - \mu^n_{i,j,k} \frac{u^n_{i,j,k} - u^n_{i,j-1,k}}{\Delta y} - \mu^n_{i,j,k} \frac{v^n_{i+1,j,k} - v^n_{i-1,j,k}}{2\,\Delta x} \tag{9-12}$$

while in the corrector step, given by Eq. (9-10), this term in $\mathbf{F}^{\overline{n+1}}_{i,j-1,k}$ is differenced as

$$(F_2)^{\overline{n+1}}_{i,j-1,k} = (\rho uv)^{\overline{n+1}}_{i,j-1,k} - \mu^{\overline{n+1}}_{i,j-1,k} \frac{u^{\overline{n+1}}_{i,j,k} - u^{\overline{n+1}}_{i,j-1,k}}{\Delta y}$$
$$- \mu^{\overline{n+1}}_{i,j-1,k} \frac{v^{\overline{n+1}}_{i+1,j-1,k} - v^{\overline{n+1}}_{i-1,j-1,k}}{2\,\Delta x} \tag{9-13}$$

Table 9-1 Differencing sequence for MacCormack scheme[a]

	Predictor			Corrector		
Step	x derivative	y derivative	z derivative	x derivative	y derivative	z derivative
1	F	F	F	B	B	B
2	B	B	F	F	F	B
3	F	F	B	B	B	F
4	B	F	B	F	B	F
5	F	B	F	B	F	B
6	B	F	F	F	B	B
7	F	B	B	B	F	F
8	B	B	B	F	F	F
9	F	F	F	B	B	B
.	.	.	.	.	.	.
.	.	.	.	.	.	.

[a]F, forward difference; B, backward difference.

Because of the complexity of the compressible N-S equations, it is not possible to obtain a closed-form stability expression for the MacCormack scheme applied to these equations. However, the following empirical formula (Tannehill et al., 1975) can normally be used

$$\Delta t \leqslant \frac{\sigma(\Delta t)_{\mathrm{CFL}}}{1 + 2/\mathrm{Re}_\Delta} \tag{9-14}$$

where σ is the safety factor ($\simeq 0.9$), $(\Delta t)_{\mathrm{CFL}}$ is the inviscid CFL condition (MacCormack, 1971)

$$(\Delta t)_{\mathrm{CFL}} \leqslant \left(\frac{|u|}{\Delta x} + \frac{|v|}{\Delta y} + \frac{|w|}{\Delta z} + a \sqrt{\frac{1}{(\Delta x)^2} + \frac{1}{(\Delta y)^2} + \frac{1}{(\Delta z)^2}} \right)^{-1} \tag{9-15}$$

Re_Δ is the minimum mesh Reynolds number given by

$$\mathrm{Re}_\Delta = \min\left(\mathrm{Re}_{\Delta x}, \mathrm{Re}_{\Delta y}, \mathrm{Re}_{\Delta z}\right) \tag{9-16}$$

where

$$\begin{aligned} \mathrm{Re}_{\Delta x} &= \frac{\rho |u| \Delta x}{\mu} \\ \mathrm{Re}_{\Delta y} &= \frac{\rho |v| \Delta y}{\mu} \\ \mathrm{Re}_{\Delta z} &= \frac{\rho |w| \Delta z}{\mu} \end{aligned} \tag{9-17}$$

and a is the local speed of sound

$$a = \sqrt{\frac{\gamma p}{\rho}}$$

Before each step, Δt can be computed for each grid point using Eq. (9-14). The smallest of these Δt's is then used to advance the solution over the entire mesh. If only the steady-state solution is desired, Li (1973) has suggested that the solution at each point be advanced using the maximum possible Δt, as computed from Eq. (9-14), in order to accelerate the convergence of the solution. Along these same lines, the over-relaxation procedure described in Section 4-5.6 can be used to reduce the number of steps required for convergence.

After each predictor or corrector step, the primitive variables (ρ, u, v, w, e, p, T) can be found by "decoding" the $\mathbf{U}$ vector

$$\mathbf{U} = \begin{bmatrix} \rho \\ \rho u \\ \rho v \\ \rho w \\ E_t \end{bmatrix} = \begin{bmatrix} U_1 \\ U_2 \\ U_3 \\ U_4 \\ U_5 \end{bmatrix} \tag{9-18}$$

in the following manner

$$
\begin{aligned}
\rho &= U_1 \\
u &= \frac{U_2}{U_1} \\
v &= \frac{U_3}{U_1} \\
w &= \frac{U_4}{U_1} \\
e &= \frac{U_5}{U_1} - \frac{u^2 + v^2 + w^2}{2} \\
p &= p(\rho, e) \\
T &= T(\rho, e)
\end{aligned}
\tag{9-19}
$$

MacCormack (1971) modified his original method by incorporating time splitting into the scheme. This revised method, which was applied to the viscous Burgers equation in Section 4-5.8, "splits" the original MacCormack scheme into a sequence of one-dimensional operations. As a result, the stability condition is based on a one-dimensional scheme which is less restrictive than the original three-dimensional scheme. Thus, it becomes possible to advance the solution in each direction with the maximum possible time step. This is particularly advantageous if the allowable time steps (Δt_x, Δt_y, Δt_z) are much different because of large differences in the mesh spacings (Δx, Δy, Δz). In order to apply this algorithm to Eq. (9-1), we define the one-dimensional difference operators $L_x(\Delta t_x)$, $L_y(\Delta t_y)$, and $L_z(\Delta t_z)$ in the following manner. The $L_x(\Delta t_x)$ operator applied to $\mathbf{U}^*_{i,j,k}$

$$\mathbf{U}^{**}_{i,j,k} = L_x(\Delta t_x)\mathbf{U}^*_{i,j,k} \tag{9-20}$$

is equivalent to the two-step formula

$$
\begin{aligned}
\overline{U^{**}_{i,j,k}} &= \mathbf{U}^*_{i,j,k} - \frac{\Delta t_x}{\Delta x}(\mathbf{E}^*_{i+1,j,k} - \mathbf{E}^*_{i,j,k}) \\
\mathbf{U}^{**}_{i,j,k} &= \frac{1}{2}\left[\mathbf{U}^*_{i,j,k} + \overline{\mathbf{U}^{**}_{i,j,k}} - \frac{\Delta t_x}{\Delta x}(\overline{\mathbf{E}^{**}_{i,j,k}} - \overline{\mathbf{E}^{**}_{i-1,j,k}})\right]
\end{aligned}
\tag{9-21}
$$

These expressions make use of the dummy time indices * and **. The $L_y(\Delta t_y)$ and $L_z(\Delta t_z)$ operators are defined in a similar manner. That is, the $L_y(\Delta t_y)$ operator applied to $\mathbf{U}^*_{i,j,k}$

$$\mathbf{U}^{**}_{i,j,k} = L_y(\Delta t_y)\mathbf{U}^*_{i,j,k} \tag{9-22}$$

is equivalent to

$$
\begin{aligned}
\overline{\mathbf{U}^{**}_{i,j,k}} &= \mathbf{U}^*_{i,j,k} - \frac{\Delta t_y}{\Delta y}(\mathbf{F}^*_{i,j+1,k} - \mathbf{F}^*_{i,j,k}) \\
\mathbf{U}^{**}_{i,j,k} &= \frac{1}{2}\left[\mathbf{U}^*_{i,j,k} + \overline{\mathbf{U}^{**}_{i,j,k}} - \frac{\Delta t_y}{\Delta y}(\overline{\mathbf{F}^{**}_{i,j,k}} - \overline{\mathbf{F}^{**}_{i,j-1,k}})\right]
\end{aligned}
\tag{9-23}
$$

and the $L_z(\Delta t_z)$ operator applied to $\mathbf{U}^*_{i,j,k}$

$$\mathbf{U}^{**}_{i,j,k} = L_z(\Delta t_z)\mathbf{U}^*_{i,j,k} \tag{9-24}$$

is equivalent to

$$\begin{aligned}\overline{\mathbf{U}^{**}_{i,j,k}} &= \mathbf{U}^*_{i,j,k} - \frac{\Delta t_z}{\Delta z}(\mathbf{G}^*_{i,j,k+1} - \mathbf{G}^*_{i,j,k})\\ \mathbf{U}^{**}_{i,j,k} &= \frac{1}{2}\left[\mathbf{U}^*_{i,j,k} + \overline{\mathbf{U}^{**}_{i,j,k}} - \frac{\Delta t_z}{\Delta z}(\overline{\mathbf{G}^{**}_{i,j,k}} - \overline{\mathbf{G}^{**}_{i,j,k-1}})\right]\end{aligned} \tag{9-25}$$

As mentioned in Section 4-5.8, a sequence of operators is consistent if the sums of the time steps for each of the operators are equal and is second-order accurate if the sequence is symmetric. A sequence which satisfies these criteria and is applicable to Eq. (9-1) is given by

$$\mathbf{U}^{n+2}_{i,j,k} = L_x(\Delta t_x)L_y(\Delta t_y)L_z(\Delta t_z)L_z(\Delta t_z)L_y(\Delta t_y)L_x(\Delta t_x)\mathbf{U}^n_{i,j,k} \tag{9-26}$$

Another sequence which satisfies these criteria, and is applicable when $\Delta y \ll \min(\Delta x, \Delta z)$, is given by

$$\mathbf{U}^{n+2}_{i,j,k} = L_x(\Delta t_x)\left[L_y\left(\frac{\Delta t_y}{m}\right)\right]^m L_z(\Delta t_z)L_z(\Delta t_z)\left[L_y\left(\frac{\Delta t_y}{m}\right)\right]^m L_x(\Delta t_x)\mathbf{U}^n_{i,j,k} \tag{9-27}$$

where m is an integer.

The algorithms resulting from a sequence of operators such as Eqs. (9-26) and (9-27) are stable if the time step size in the argument of each does not exceed the maximum allowed for that operator. Since it is not possible to analyze the stability of each operator applied to the complete N-S equations, a one-dimensional form of the empirical stability formula, given by Eq. (9-14), can be used for each operator

$$\begin{aligned}\Delta t_x &\leqslant \frac{\sigma\,\Delta x}{(|u| + a)(1 + 2/\mathrm{Re}_{\Delta x})}\\ \Delta t_y &\leqslant \frac{\sigma\,\Delta y}{(|v| + a)(1 + 2/\mathrm{Re}_{\Delta y})}\\ \Delta t_z &\leqslant \frac{\sigma\,\Delta z}{(|w| + a)(1 + 2/\mathrm{Re}_{\Delta z})}\end{aligned} \tag{9-28}$$

where σ is the safety factor and a is the local speed of sound.

Computations involving the compressible N-S equations sometimes "blow up" because of numerical oscillations. These oscillations are the result of inadequate mesh refinement in regions of large gradients. In many cases, it is impractical to refine the mesh in these regions, particularly if they are far removed from the region of interest. For such situations, MacCormack and Baldwin (1975) have devised a "product" fourth-order smoothing scheme which is an alternative to the fourth-order type of smoothing given by Eq. (8-100). In the MacCormack type of smoothing, dissipation terms are added to the $L_x(\Delta t_x)$ operator in the following manner

$$\overline{\mathbf{U}_{i,j,k}^{**}} = \mathbf{U}_{i,j,k}^{*} - \frac{\Delta t_x}{\Delta x}(\mathbf{E}_{i+1,j,k}^{*} + \mathbf{S}_{i+1,j,k}^{*} - \mathbf{E}_{i,j,k}^{*} - \mathbf{S}_{i,j,k}^{*})$$

$$\mathbf{U}_{i,j,k}^{**} = \frac{1}{2}\left[\mathbf{U}_{i,j,k}^{*} + \overline{\mathbf{U}_{i,j,k}^{**}} - \frac{\Delta t_x}{\Delta x}(\overline{\mathbf{E}_{i,j,k}^{**}} + \overline{\mathbf{S}_{i,j,k}^{**}} - \overline{\mathbf{E}_{i-1,j,k}^{**}} - \overline{\mathbf{S}_{i-1,j,k}^{**}})\right] \tag{9-29}$$

where

$$\mathbf{S}_{i,j,k}^{*} = \epsilon_e\left[(|u_{i,j,k}^{*}| + a_{i,j,k}^{*})\frac{|\delta_x^2 p_{i,j,k}^{*}|}{(p_{i+1,j,k}^{*} + 2p_{i,j,k}^{*} + p_{i-1,j,k}^{*})}(\mathbf{U}_{i,j,k}^{*} - \mathbf{U}_{i-1,j,k}^{*})\right]$$

$$\overline{\mathbf{S}_{i,j,k}^{**}} = \epsilon_e\left[(|\overline{u_{i,j,k}^{**}}| + \overline{a_{i,j,k}^{**}})\frac{|\delta_x^2 \overline{p_{i,j,k}^{**}}|}{(\overline{p_{i+1,j,k}^{**}} + 2\overline{p_{i,j,k}^{**}} + \overline{p_{i-1,j,k}^{**}})}(\overline{\mathbf{U}_{i+1,j,k}^{**}} - \overline{\mathbf{U}_{i,j,k}^{**}})\right] \tag{9-30}$$

and $0 \leqslant \epsilon_e \leqslant 0.5$ for stability. Thus, an artificial viscosity term of the form

$$\epsilon_e(\Delta x)^4 \frac{\partial}{\partial x}\left[\frac{|u| + a}{4p}\left|\frac{\partial^2 p}{\partial x^2}\right|\frac{\partial \mathbf{U}}{\partial x}\right] \tag{9-31}$$

has been added to the N-S equations. This smoothing term has a very small magnitude except in regions of pressure oscillations where the truncation error is already producing erroneous results.

The explicit MacCormack algorithm is a suitable method for solving both steady and unsteady flows at moderate to low Reynolds numbers. However, it is not a satisfactory method for solving high Reynolds numbers flows where the viscous regions become very thin. For these flows, the mesh must be highly refined in order to accurately resolve the viscous regions. This leads to small time steps and subsequently long computer times if an explicit scheme such as the MacCormack method is used. In order to explain this further, let us consider the two-dimensional flow over a flat plate at high Reynolds number. In this case a very fine mesh is required near the flat plate in order to resolve the boundary layer, but a coarser grid can be used in the inviscid portion of the flowfield as illustrated in Fig. 9-1. In the coarse grid region, the MacCormack time-split scheme can be applied in the following manner

$$\mathbf{U}_{i,j}^{n+1} = L_x\left(\frac{\Delta t}{2}\right)L_y(\Delta t)L_x\left(\frac{\Delta t}{2}\right)\mathbf{U}_{i,j}^{n} \tag{9-32}$$

where

$$\Delta t \leqslant \min\,(2\,\Delta t_x, \Delta t_y)_{\text{coarse mesh}} \tag{9-33}$$

In the fine grid region, the following sequence of operators can be used

$$\mathbf{U}_{i,j}^{n+1} = \left[L_y\left(\frac{\Delta t}{2m}\right)L_x\left(\frac{\Delta t}{m}\right)L_y\left(\frac{\Delta t}{2m}\right)\right]^m \mathbf{U}_{i,j} \tag{9-34}$$

where m is the smallest integer such that

$$\frac{\Delta t}{m} \leqslant \min\,(\Delta t_x, 2\,\Delta t_y)_{\text{fine mesh}} \tag{9-35}$$

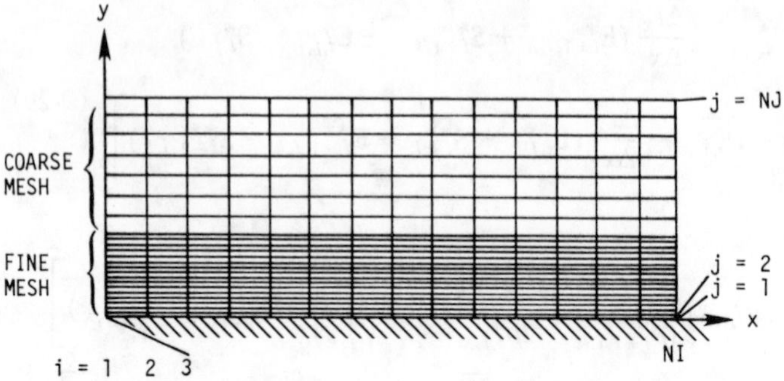

Figure 9-1 Mesh for high Reynolds number flow over a flat plate.

For high Reynolds numbers, the fine grid region becomes very thin requiring Δy to be very small. This causes Δt_y in the L_y operator to be very small and the integer m to be very large. Consequently, a substantial amount of calculation time is required in the fine grid region. To overcome this difficulty, MacCormack (1976) developed a hybrid version of his scheme which is known as the MacCormack rapid solver method. This hybrid method is part explicit and part implicit. For the flat plate problem described above, the rapid solver method is implemented by replacing the $L_y(\Delta t/2m)$ operator in Eq. (9-34) with

$$L_{yH}\left(\frac{\Delta t}{2m}\right)L_{yP}\left(\frac{\Delta t}{2m}\right)$$

where the L_{yH} operator is applied to the inviscid (hyperbolic) portion of the N-S equations, i.e.,

$$\frac{\partial \mathbf{U}}{\partial t}+\frac{\partial \mathbf{F}_H}{\partial y}=0 \tag{9-36}$$

with $\mathbf{F}_H$ defined as

$$\mathbf{F}_H=\begin{bmatrix}\rho v\\ \rho uv\\ \rho v^2+p\\ (E_t+p)v\end{bmatrix} \tag{9-37}$$

The L_{Hp} operator is applied to the viscous (parabolic) portion of the N-S equations

$$\frac{\partial \mathbf{U}}{\partial t}+\frac{\partial \mathbf{F}_P}{\partial y}=0 \tag{9-38}$$

where $\mathbf{F}_P=\mathbf{F}-\mathbf{F}_H$. The L_{yH} operator solves Eq. (9-36) using either the method of characteristics or the original MacCormack scheme (Li, 1977; Shang, 1977). The L_{yp} operator solves Eq. (9-38) using an implicit scheme such as the Crank-Nicolson or Laasonen schemes. Thus, it is possible to solve Eqs. (9-36) and (9-38) using a time step which is not limited by the viscous stability constraint. The rapid solver method has proved to be from 10 to 100 times faster than the time-split scheme for high Reynolds number flows. However, because of its complexity it is rather difficult to program on a computer. More recently, MacCormack (1981) has developed an implicit version of his original scheme. This will be described in Section 9-2.4.

9-2.2 Other Explicit Methods

In addition to the MacCormack scheme, other explicit methods which can be used to solve the compressible N-S equations include the following:

1. Hopscotch method (Section 4-2.12)
2. Leapfrog/DuFort-Frankel method (Section 4-5.2)
3. Brailovskaya method (Section 4-5.3)
4. Allen-Cheng method (Section 4-5.4)
5. Lax-Wendroff method (Section 4-5.5)

These methods were discussed in earlier sections (as indicated) where they were applied to either the heat equation or the viscous Burgers equation. When these methods are applied to the more complicated compressible N-S equations, certain difficulties can arise as we have seen before. For example, the mixed derivative terms create a problem for the hopscotch method. If these terms are differenced in the usual manner by applying Eq. (3-51), the hopscotch method is no longer explicit since a matrix inversion is required. This problem can be circumvented by lagging the mixed derivative terms (i.e., evaluating them at the previous time level).

All of the above methods, except the Lax-Wendroff scheme, are first-order accurate in time so that they cannot be used to accurately compute the time evolution of a flowfield. In addition, all of the methods have a stability restriction which limits the maximum time step. However, the stability conditions for the hopscotch and Allen-Cheng methods are independent of the viscosity which gives them a distinct advantage over the other methods. The allowable time step for the hopscotch method is given by the inviscid CFL condition, which for a 2-D problem becomes

$$(\Delta t)_{\text{CFL}} \leqslant \frac{\Delta x}{|u| + |v| + 2\sqrt{a}} \tag{9-39}$$

if $\Delta x = \Delta y$. An important advantage of the Brailovskaya method is that the viscous terms need to be computed only once during the two-step procedure. Additional explicit methods for solving the compressible N-S equations can be found in the survey paper of Peyret and Viviand (1975).

9-2.3 Beam-Warming Scheme

The Beam-Warming finite-difference scheme (Beam and Warming, 1978) for solving the compressible N-S equations belongs to the same class of ADI schemes developed by Lindemuth and Killeen (1973) and McDonald and Briley (1975). Under certain conditions, these schemes can be shown to be equivalent. The Briley-McDonald scheme was discussed previously in Section 4-5.7 where it was applied to the viscous Burgers equation.

For simplicity, we will apply the Beam-Warming finite-difference scheme to the 2-D compressible N-S equations which can be written in the following vector form

$$\frac{\partial \mathbf{U}}{\partial t}+\frac{\partial \mathbf{E(U)}}{\partial x}+\frac{\partial \mathbf{F(U)}}{\partial y}=\frac{\partial \mathbf{V}_1(\mathbf{U},\mathbf{U}_x)}{\partial x}+\frac{\partial \mathbf{V}_2(\mathbf{U},\mathbf{U}_y)}{\partial x}+\frac{\partial \mathbf{W}_1(\mathbf{U},\mathbf{U}_x)}{\partial y}+\frac{\partial \mathbf{W}_2(\mathbf{U},\mathbf{U}_y)}{\partial y} \tag{9-40}$$

where

$$\mathbf{U}=\begin{bmatrix}\rho\\ \rho u\\ \rho v\\ E_t\end{bmatrix} \qquad \mathbf{E(U)}=\begin{bmatrix}\rho u\\ \rho u^2+p\\ \rho uv\\ (E_t+p)u\end{bmatrix} \qquad \mathbf{F(U)}=\begin{bmatrix}\rho v\\ \rho uv\\ \rho v^2+p\\ (E_t+p)v\end{bmatrix}$$

$$\mathbf{V}_1+\mathbf{V}_2=\begin{bmatrix}0\\ \frac{2}{3}\mu(2u_x-v_y)\\ \mu(u_y+v_x)\\ \mu v(u_y+v_x)+\frac{2}{3}\mu u(2u_x-v_y)+kT_x\end{bmatrix} \tag{9-41}$$

$$\mathbf{W}_1+\mathbf{W}_2=\begin{bmatrix}0\\ \mu(u_y+v_x)\\ \frac{2}{3}\mu(2v_y-u_x)\\ \mu u(u_y+v_x)+\frac{2}{3}\mu v(2v_y-u_x)+kT_y\end{bmatrix}$$

In the Beam-Warming scheme, the solution is marched in time using the following difference formula

$$\Delta^n\mathbf{U}=\frac{\theta_1\Delta t}{1+\theta_2}\frac{\partial}{\partial t}(\Delta^n\mathbf{U})+\frac{\Delta t}{1+\theta_2}\frac{\partial}{\partial t}(\mathbf{U}^n)+\frac{\theta_2}{1+\theta_2}\Delta^{n-1}\mathbf{U}$$
$$+O\left[\left(\theta_1-\frac{1}{2}-\theta_2\right)(\Delta t)^2+(\Delta t)^3\right] \tag{9-42}$$

where $\Delta^n\mathbf{U}=\mathbf{U}^{n+1}-\mathbf{U}^n$. This general difference formula, with the appropriate choice of the parameters θ_1 and θ_2, represents many of the standard difference schemes as we have seen previously in Section 8-3.3. For the compressible N-S equations, either the Euler implicit scheme ($\theta_1=1, \theta_2=0$), which is first-order accurate in time, or the three-point backward implicit scheme ($\theta_1=1$, $\theta_2=\frac{1}{2}$), which is second-order accurate in time, is normally used.

After substituting Eq. (9-40) into Eq. (9-42), we obtain

$$\Delta^n \mathbf{U} = \frac{\theta_1 \Delta t}{1+\theta_2}\left[\frac{\partial}{\partial x}(-\Delta^n \mathbf{E} + \Delta^n \mathbf{V}_1 + \Delta^n \mathbf{V}_2) + \frac{\partial}{\partial y}(-\Delta^n \mathbf{F} + \Delta^n \mathbf{W}_1 + \Delta^n \mathbf{W}_2)\right]$$
$$+ \frac{\Delta t}{1+\theta_2}\left[\frac{\partial}{\partial x}(-\mathbf{E}^n + \mathbf{V}_1^n + \mathbf{V}_2^n) + \frac{\partial}{\partial y}(-\mathbf{F}^n + \mathbf{W}_1^n + \mathbf{W}_2^n)\right]$$
$$+ \frac{\theta_2}{1+\theta_2}\Delta^{n-1}\mathbf{U} + O\left[\left(\theta_1 - \frac{1}{2} - \theta_2\right)(\Delta t)^2 + (\Delta t)^3\right] \quad (9\text{-}43)$$

This difference formula is in the so-called delta form which we have discussed previously. The delta terms are linearized using truncated Taylor series expansions. For example, $\Delta^n \mathbf{E}$ is linearized using

$$\mathbf{E}^{n+1} = \mathbf{E}^n + \left(\frac{\partial \mathbf{E}}{\partial \mathbf{U}}\right)^n (\mathbf{U}^{n+1} - \mathbf{U}^n) + O[(\Delta t)^2] \quad (9\text{-}44)$$

which can be rewritten as

$$\Delta^n \mathbf{E} = [A]^n \Delta^n \mathbf{U} + O[(\Delta t)^2] \quad (9\text{-}45)$$

where $[A]$ is the Jacobian matrix $\partial \mathbf{E}/\partial \mathbf{U}$ given by

$$[A] = -\left[\begin{array}{c|c|c|c} 0 & -1 & 0 & 0 \\ \frac{3-\gamma}{2}u^2 + \frac{1-\gamma}{2}v^2 & (\gamma-3)u & (\gamma-1)v & (1-\gamma) \\ uv & -v & -u & 0 \\ \frac{\gamma E_t u}{\rho} + (1-\gamma)u(u^2+v^2) & -\frac{\gamma E_t}{\rho} + \frac{\gamma-1}{2}(3u^2+v^2) & (\gamma-1)uv & -\gamma u \end{array}\right] \quad (9\text{-}46)$$

and γ is the ratio of specific heats. This Jacobian matrix is derived assuming a perfect gas. In a like manner, $\Delta^n \mathbf{F}$ can be linearized as

$$\Delta^n \mathbf{F} = [B]^n \Delta^n \mathbf{U} + O[(\Delta t)^2] \quad (9\text{-}47)$$

where $[B]$ is the Jacobian matrix $\partial \mathbf{F}/\partial \mathbf{U}$ given by

$$[B] = -\left[\begin{array}{c|c|c|c} 0 & 0 & -1 & 0 \\ uv & -v & -u & 0 \\ \frac{3-\gamma}{2}v^2 + \frac{1-\gamma}{2}u^2 & (\gamma-1)u & (\gamma-3)v & 1-\gamma \\ \frac{\gamma E_t v}{\rho} + (1-\gamma)v(u^2+v^2) & (\gamma-1)uv & -\frac{\gamma E_t}{\rho} + \frac{\gamma-1}{2}(3v^2+u^2) & -\gamma v \end{array}\right] \quad (9\text{-}48)$$

The viscous delta term $\Delta^n \mathbf{V}_1(\mathbf{U}, \mathbf{U}_x)$ is linearized by writing

$$\begin{aligned}\Delta^n \mathbf{V}_1 &= \left(\frac{\partial \mathbf{V}_1}{\partial \mathbf{U}}\right)^n \Delta^n \mathbf{U} + \left(\frac{\partial \mathbf{V}_1}{\partial \mathbf{U}_x}\right)^n \Delta^n \mathbf{U}_x + O[(\Delta t)^2] \\ &= [P]^n \Delta^n \mathbf{U} + [R]^n \Delta^n \mathbf{U}_x + O[(\Delta t)^2] \\ &= ([P] - [R_x])^n \Delta^n \mathbf{U} + \frac{\partial}{\partial x}([R]^n \Delta^n \mathbf{U}) + O[(\Delta t)^2] \end{aligned} \tag{9-49}$$

where $[P]$ is the Jacobian $\partial \mathbf{V}_1/\partial \mathbf{U}$, $[R]$ is the Jacobian $\partial \mathbf{V}_1/\partial \mathbf{U}_x$, and $[R_x] = \partial[R]/\partial x$. These matrices can be written as

$$[P] - [R_x] = -\frac{1}{\rho}\left[\begin{array}{c|c|c|c} 0 & 0 & 0 & 0 \\ -u\left(\frac{4}{3}\mu\right)_x & \left(\frac{4}{3}\mu\right)_x & 0 & 0 \\ -v\mu_x & 0 & \mu_x & 0 \\ -u^2\left(\frac{4}{3}\mu\right)_x - v^2\mu_x & u\left(\frac{4}{3}\mu\right)_x & v\mu_x & 0 \end{array}\right] \tag{9-50}$$

$$[R] = \frac{1}{\rho}\left[\begin{array}{c|c|c|c} 0 & 0 & 0 & 0 \\ -\frac{4}{3}\mu u & \frac{4}{3}\mu & 0 & 0 \\ -\mu v & 0 & \mu & 0 \\ -\left(\frac{4}{3}\mu - \frac{k}{c_v}\right)u^2 - \left(\mu - \frac{k}{c_v}\right)v^2 - \frac{k}{c_v}\frac{E_t}{\rho} & \left(\frac{4}{3}\mu - \frac{k}{c_v}\right)u & \left(\mu - \frac{k}{c_v}\right)v & \frac{k}{c_v} \end{array}\right] \tag{9-51}$$

The matrix for $[P] - [R_x]$ is obtained by assuming that μ and k are locally independent of $\mathbf{U}$. In a like manner, $\Delta^n \mathbf{W}_2(\mathbf{U}, \mathbf{U}_y)$ is linearized as

$$\Delta^n \mathbf{W}_2 = ([Q] - [S_y])^n \Delta^n \mathbf{U} + \frac{\partial}{\partial y}([S]^n \Delta^n \mathbf{U}) + O[(\Delta t)^2] \tag{9-52}$$

where

$$[Q] - [S_y] = -\frac{1}{\rho}\left[\begin{array}{c|c|c|c} 0 & 0 & 0 & 0 \\ -u\mu_y & \mu_y & 0 & 0 \\ -v\left(\frac{4}{3}\mu\right)_y & 0 & \left(\frac{4}{3}\mu\right)_y & 0 \\ -v^2\left(\frac{4}{3}\mu\right)_y - u^2\mu_y & u\mu_y & v\left(\frac{4}{3}\mu\right)_y & 0 \end{array}\right] \tag{9-53}$$

and

$$[S] = \frac{1}{\rho}\left[\begin{array}{c|c|c|c} 0 & 0 & 0 & 0 \\ -\mu u & \mu & 0 & 0 \\ -\frac{4}{3}\mu v & 0 & \frac{4}{3}\mu & 0 \\ -\left(\frac{4}{3}\mu - \frac{k}{c_v}\right)v^2 - \left(\mu - \frac{k}{c_v}\right)u^2 - \frac{k}{c_v}\frac{E_t}{\rho} & \left(\mu - \frac{k}{c_v}\right)u & \left(\frac{4}{3}\mu - \frac{k}{c_v}\right)v & \frac{k}{c_v} \end{array}\right] \tag{9-54}$$

The cross-derivative terms can be evaluated explicitly without loss of accuracy by noting that

$$\begin{aligned} \Delta^n \mathbf{V}_2 &= \Delta^{n-1}\mathbf{V}_2 + O[(\Delta t)^2] \\ \Delta^n \mathbf{W}_1^n &= \Delta^{n-1}\mathbf{W}_1 + O[(\Delta t)^2] \end{aligned} \tag{9-55}$$

for a uniform time step Δt. By evaluating the cross-derivative terms in this manner, the block tridiagonal form of the final equations is maintained. The Steger method (Steger, 1977) for linearizing viscous terms, which was described in Section 8-3.3, can be used in place of the linearizations given by Eqs. (9-49) and (9-52). The Steger form of linearization is particularly useful when coordinate transformations have been applied to the N-S equations.

Substituting Eqs. (9-45), (9-47), (9-49), (9-52), and (9-55) into Eq. (9-43) yields

$$\begin{aligned} &\Bigg\{[I] + \frac{\theta_1 \Delta t}{1+\theta_2}\left[\frac{\partial}{\partial x}([A] - [P] + [R_x])^n - \frac{\partial^2}{\partial x^2}[R]^n \right.\\ &\left. + \frac{\partial}{\partial y}([B] - [Q] + [S_y])^n - \frac{\partial^2}{\partial y^2}[S]^n\right]\Bigg\}\Delta^n\mathbf{U} \\ &= \frac{\Delta t}{1+\theta_2}\left[\frac{\partial}{\partial x}(-\mathbf{E} + \mathbf{V}_1 + \mathbf{V}_2)^n + \frac{\partial}{\partial y}(-\mathbf{F} + \mathbf{W}_1 + \mathbf{W}_2)^n\right] \\ &+ \frac{\theta_1 \Delta t}{1+\theta_2}\left[\frac{\partial}{\partial x}(\Delta^{n-1}\mathbf{V}_2) + \frac{\partial}{\partial y}(\Delta^{n-1}\mathbf{W}_1)\right] + \frac{\theta_2}{1+\theta_2}\Delta^{n-1}\mathbf{U} \\ &+ O\left[\left(\theta_1 - \frac{1}{2} - \theta_2\right)(\Delta t)^2, (\Delta t)^3\right] \end{aligned} \tag{9-56}$$

where $[I]$ is the unity matrix. In Eq. (9-56), expressions such as

$$\left[\frac{\partial}{\partial x}([A] - [P] + [R_x])^n\right]\Delta^n\mathbf{U}$$

are equivalent to

$$\frac{\partial}{\partial x}[([A]-[P]+[R_x])^n\,\Delta^n\mathbf{U}]$$

The left-hand side of Eq. (9-56) is factored in the following manner

$$\left\{[I]+\frac{\theta_1\Delta t}{1+\theta_2}\left[\frac{\partial}{\partial x}([A]-[P]+[R_x])^n-\frac{\partial^2}{\partial x^2}[R]^n\right]\right\}$$
$$\times\left\{[I]+\frac{\theta_1\Delta t}{1+\theta_2}\left[\frac{\partial}{\partial y}([B]-[Q]+[S_y])^n-\frac{\partial^2}{\partial y^2}[S]^n\right]\right\}\Delta^n\mathbf{U}$$
$$=\text{LHS [Eq. (9-56)]}+O[(\Delta t)^3] \tag{9-57}$$

and the final form of the Beam-Warming algorithm becomes

$$\text{LHS [Eq. (9-57)]} = \text{RHS [Eq. (9-56)]} \tag{9-58}$$

The partial derivatives in the algorithm are evaluated using second-order accurate central differences.

The Beam-Warming algorithm is implemented in the following manner:

Step 1:

$$\left\{[I]+\frac{\theta_1\Delta t}{1+\theta_2}\left[\frac{\partial}{\partial x}([A]-[P]+[R_x])^n-\frac{\partial^2}{\partial x^2}[R]^n\right]\right\}\Delta^n\mathbf{U}_1=\text{RHS [Eq. (9-56)]} \tag{9-59}$$

Step 2:

$$\left\{[I]+\frac{\theta_1\Delta t}{1+\theta_2}\left[\frac{\partial}{\partial y}([B]-[Q]+[S_y])^n-\frac{\partial^2}{\partial y^2}[S]^n\right]\right\}\Delta^n\mathbf{U}=\Delta^n\mathbf{U}_1 \tag{9-60}$$

Step 3:

$$\mathbf{U}^{n+1}=\mathbf{U}^n+\Delta^n\mathbf{U} \tag{9-61}$$

In Step 1, $\Delta^n\mathbf{U}_1$ represents the remaining terms on the left-hand side of Eq. (9-57). Equations (9-59) and (9-60) represent systems of equations which have the same block tridiagonal structure as shown in Eq. (8-98), except that for the 2-D compressible N-S equations the blocks are 4×4 matrices.

Warming and Beam (1977) have studied the stability of their algorithm by applying it to both the 2-D wave equation

$$u_t+c_1u_x+c_2u_y=0 \tag{9-62}$$

and the diffusive equation

$$u_t=au_{xx}+bu_{xy}+cu_{yy} \tag{9-63}$$

The latter equation is parabolic if $b^2<4ac$ and $(a, c)>0$. They found that the algorithm is unconditionally stable when applied to Eq. (9-62) provided that $\theta_2>0$. When applied to Eq. (9-63), the algorithm is unconditionally stable provided that $\theta_2 \geqslant 0.385$. Note that neither the leapfrog scheme ($\theta_1 = 0$, $\theta_2 = -\frac{1}{2}$) nor the trapezoidal scheme ($\theta_1 = \frac{1}{2}$, $\theta_2 = 0$) is unconditionally stable when applied to Eq. (9-63). However, the three-point backward scheme ($\theta_1 = 1$, $\theta_2 = \frac{1}{2}$) is unconditionally stable and can be used when second-order temporal accuracy is desired.

In order to successfully initiate a computation with approximate starting data as well as to suppress high frequency oscillations in the solution, it is often necessary to add damping to the Beam-Warming algorithm. This can be accomplished by adding a fourth-order explicit dissipation term of the form given by Eq. (8-100) to the right-hand side of Eq. (9-56). In addition, if only the steady-state solution is of interest, a second-order implicit smoothing term can also be added to the left-hand side of Eq. (9-56). This latter smoothing term can be second-order since it has no effect on the steady-state solution where $\Delta^n \mathbf{U} = 0$. After the smoothing terms are added, the final differenced form of the algorithm becomes

Step 1:

$$\left\{[I] + \frac{\theta_1 \Delta t}{1+\theta_2}\left[\bar{\delta}_x([A] - [P] + [R_x])^n - \delta_x^2 [R]^n - \epsilon_i \delta_x^2\right]\right\} \Delta^n \mathbf{U}_1$$

$$= \text{RHS [Eq. (9-56)]} - \epsilon_e(\delta_x^4 + \delta_y^4)\mathbf{U}^n \tag{9-64}$$

Step 2:

$$\left\{[I] + \frac{\theta_1 \Delta t}{1+\theta_2}\left[\bar{\delta}_y([B] - [Q] + [S_y])^n - \delta_y^2 [S]^n - \epsilon_i \delta_y^2\right]\right\} \Delta^n \mathbf{U} = \Delta^n \mathbf{U}_1 \tag{9-65}$$

Step 3:

$$\mathbf{U}^{n+1} = \mathbf{U}^n + \Delta^n \mathbf{U} \tag{9-66}$$

where $\bar{\delta}$, δ^2, and δ^4 are the usual central-difference operators and ϵ_e and ϵ_i are the coefficients of the explicit and implicit smoothing terms, respectively. Using a Fourier stability analysis, it can be shown that the coefficient of the explicit smoothing term must be in the range

$$0 \leqslant \epsilon_e \leqslant \frac{1 + 2\theta_2}{8(1+\theta_2)} \tag{9-67}$$

to ensure stability.

Désidéri et al. (1978) have investigated the possibility of maximizing the rate of convergence of the time-dependent solution by using the proper ratio of the coefficients of the smoothing terms. They found that when the Beam-Warming scheme (with Euler implicit differencing) is applied to the Euler equations, the rate of convergence is optimized when

$$\frac{\epsilon_i}{\epsilon_e} = 2 \tag{9-68}$$

Beam and Warming have pointed out that their algorithm can be simplified considerably if μ is assumed locally constant. In this case, $(\mu_x, \mu_y) = 0$ and Eqs. (9-50) and (9-53) reduce to

$$[P] - [R_x] = 0$$

$$[Q] - [S_y] = 0$$

If only the steady-state solution is desired, Tannehill et al. (1978) have suggested that all the viscous terms on the left-hand side of the algorithm (i.e., $[P]$, $[R_x]$, $[R]$, $[Q]$, $[S_y]$, $[S]$) can be set equal to zero provided that implicit smoothing ($\epsilon_i > 0$) is retained. This takes advantage of the fact that the left-hand side of Eq. (9-57) approaches zero as the steady-state solution is approached. With this simplification, the complexity of the Beam-Warming algorithm is greatly reduced, particularly if a non-Cartesian coordinate system is employed. It is believed that this simplifying technique can be used in all moderate to high Reynolds number computations since tests confirm that the convergence rate is not affected for these cases. To reduce computation time further, Chaussee and Pulliam (1981) have transformed the coupled set of thin-layer N-S equations into an uncoupled diagonal form.

9-2.4 Implicit MacCormack Method

MacCormack (1981) has developed an implicit analogue of his explicit finite-difference method. This new method consists of two stages. The first stage uses the original MacCormack scheme while the second stage employs an implicit scheme to eliminate any stability restrictions. The resulting matrix equations are either upper or lower block bidiagonal equations which can be solved in an easier fashion than the usual tridiagonal systems. In order to explain the implicit MacCormack scheme we will first apply it to the linear Burgers equation

$$u_t = -cu_x + \mu u_{xx} \qquad c, \mu > 0 \tag{9-69}$$

The original explicit MacCormack scheme applied to Eq. (9-69), see Section 4-5.6, results in the following predictor-corrector equations:

Predictor:

$$(\Delta u_i^n)_{\text{explicit}} = \frac{-c\,\Delta t}{\Delta x}(u_{i+1}^n - u_i^n) + \frac{\mu\,\Delta t}{(\Delta x)^2}(u_{i+1}^n - 2u_i^n + u_{i-1}^n) \tag{9-70}$$

$$(\overline{u_i^{n+1}})_{\text{explicit}} = u_i^n + (\Delta u_i^n)_{\text{explicit}} \tag{9-71}$$

Corrector:

$$(\Delta \overline{u_i^{n+1}})_{\text{explicit}} = \frac{-c\,\Delta t}{\Delta x}(\overline{u_i^{n+1}} - \overline{u_{i-1}^{n+1}}) + \frac{\mu\,\Delta t}{(\Delta x)^2}(\overline{u_{i+1}^{n+1}} - 2\overline{u_i^{n+1}} + \overline{u_{i-1}^{n+1}}) \tag{9-72}$$

$$(u_i^{n+1})_{\text{explicit}} = \tfrac{1}{2}[u_i^n + (\overline{u_i^{n+1}})_{\text{explicit}} + (\Delta \overline{u_i^{n+1}})_{\text{explicit}}] \tag{9-73}$$

These equations are written in "delta" form with

$$\Delta u_i^n = u_i^{n+1} - u_i^n \tag{9-74}$$

and $x = i\,\Delta x$. The subscript (explicit) is used to denote that the present quantities are computed using the explicit MacCormack scheme. The implicit MacCormack method is the implicit analog of Eqs. (9-70)-(9-74) and it is given by

Predictor:

$$\left(1 + \frac{\lambda\,\Delta t}{\Delta x}\right) \Delta \overline{u_i^{n+1}} = (\Delta u_i^n)_{\text{explicit}} + \frac{\lambda\,\Delta t}{\Delta x}\, \Delta \overline{u_{i+1}^{n+1}} \tag{9-75}$$

$$\overline{u_i^{n+1}} = u_i^n + \Delta \overline{u_i^{n+1}} \tag{9-76}$$

Corrector:

$$\left(1 + \frac{\lambda\,\Delta t}{\Delta x}\right) \Delta u_i^{n+1} = (\Delta \overline{u_i^{n+1}})_{\text{explicit}} + \frac{\lambda\,\Delta t}{\Delta x}\, \Delta u_{i-1}^{n+1} \tag{9-77}$$

$$u_i^{n+1} = \tfrac{1}{2}(u_i^n + \overline{u_i^{n+1}} + \Delta u_i^{n+1}) \tag{9-78}$$

where $(\Delta u_i^n)_{\text{explicit}}$ and $(\Delta \overline{u_i^{n+1}})_{\text{explicit}}$ are determined from Eqs. (9-70) and (9-72), respectively, and λ is chosen so that

$$\lambda \geqslant \max \left[\left(c + \frac{2\mu}{\Delta x} - \frac{\Delta x}{\Delta t}\right),\ 0.0\right] \tag{9-79}$$

This method is unconditionally stable and is second-order accurate in both space and time, provided that $\mu\,\Delta t/(\Delta x)^2$ is bounded as Δt and Δx approach zero. The second-order accuracy can be easily shown since the net result of the terms added to the original second-order MacCormack scheme, to obtain Eqs. (9-75)-(9-78), is of third order. That is, Eq. (9-75) can be written as

$$\Delta \overline{u_i^{n+1}} = (\Delta u_i^n)_{\text{explicit}} + \frac{\lambda\,\Delta t}{\Delta x}\,(\Delta \overline{u_{i+1}^{n+1}} - \Delta \overline{u_i^{n+1}})$$

$$= (\Delta u_i^n)_{\text{explicit}} + \lambda(\Delta t)^2 \frac{\partial}{\partial x}\left(\frac{\partial u}{\partial t}\right) + O[(\Delta t)^3] \tag{9-80}$$

and similarly Eq. (9-77) can be written as

$$\Delta u_i^{n+1} = (\Delta \overline{u_i^{n+1}})_{\text{explicit}} - \lambda(\Delta t)^2 \frac{\partial}{\partial x}\left(\frac{\partial u}{\partial t}\right) + O[(\Delta t)^3] \tag{9-81}$$

Substituting Eqs. (9-76), (9-80), and (9-81) into Eq. (9-78), we obtain

$$u_i^{n+1} = \tfrac{1}{2}[2u_i^n + (\Delta u_i^n)_{\text{explicit}} + (\Delta \overline{u_i^{n+1}})_{\text{explicit}}] + O[(\Delta t)^3] \tag{9-82}$$

or

$$u_i^{n+1} = \tfrac{1}{2}[u_i^n + (\overline{u_i^{n+1}})_{\text{explicit}} + (\Delta \overline{u_i^{n+1}})_{\text{explicit}}] + O[(\Delta t)^3] \tag{9-83}$$

Thus, we have shown that

$$[\text{Eq. (9-78)}] = [\text{Eq. (9-73)}] + O[(\Delta t)^3] \tag{9-84}$$

Equations (9-75) and (9-77) represent bidiagonal sets of algebraic equations which can be readily solved using a single sweep through the mesh. For example, Eq. (9-75) can be written as

$$\Delta u_i^{\overline{n+1}} = \frac{(\Delta u_i^n)_{\text{explicit}} + (\lambda\, \Delta t/\Delta x)\, \Delta u_{i+1}^{\overline{n+1}}}{1 + \lambda\, \Delta t/\Delta x} \tag{9-85}$$

so that if the solution is swept from the right boundary ($i = NI$), where u is known, to the left boundary ($i = 1$) we can directly determine $\Delta u_i^{\overline{n+1}}$. This sweeping procedure is reminiscent of those used in the ADE methods of Section 4-2.10.

The parameter λ is chosen by examining the stability limit on the original explicit MacCormack method which is given approximately by

$$(\Delta t)_{\text{explicit}} \leqslant \frac{\Delta x}{c + 2\mu/\Delta x} \tag{9-86}$$

If Δt is less than $(\Delta t)_{\text{explicit}}$, then

$$c + \frac{2\mu}{\Delta x} - \frac{\Delta x}{\Delta t} \leqslant 0 \tag{9-87}$$

and λ is set equal to zero by Eq. (9-79). In this case, the implicit procedures of the second stage are not needed for stability and the implicit MacCormack method reduces to the original explicit MacCormack method. On the other hand, if $\Delta t > (\Delta t)_{\text{explicit}}$ then λ is chosen so that

$$\lambda \geqslant c + \frac{2\mu}{\Delta x} - \frac{\Delta x}{\Delta t} \tag{9-88}$$

When the implicit MacCormack scheme is applied to the 2-D compressible N-S equations

$$\frac{\partial \mathbf{U}}{\partial t} + \frac{\partial \mathbf{E}}{\partial x} + \frac{\partial \mathbf{F}}{\partial y} = 0 \tag{9-89}$$

the following algorithm results

Predictor:

$$(\Delta \mathbf{U}_{i,j}^n)_{\text{explicit}} = -\Delta t \left(\frac{\Delta_x \mathbf{E}_{i,j}^n}{\Delta x} + \frac{\Delta_y \mathbf{F}_{i,j}^n}{\Delta y} \right) \tag{9-90}$$

$$\left([I] - \frac{\Delta t\, \Delta_x [A']}{\Delta x} \right) \left([I] - \frac{\Delta t\, \Delta_y [B']}{\Delta y} \right) \Delta \mathbf{U}_{i,j}^{\overline{n+1}} = (\Delta \mathbf{U}_{i,j}^n)_{\text{explicit}} \tag{9-91}$$

$$\mathbf{U}_{i,j}^{\overline{n+1}} = \mathbf{U}_{i,j}^n + \Delta \mathbf{U}_{i,j}^{\overline{n+1}} \tag{9-92}$$

Corrector:

$$(\Delta \mathbf{U}_{i,j}^{\overline{n+1}})_{\text{explicit}} = -\Delta t \left(\frac{\nabla_x \mathbf{E}_{i,j}^{\overline{n+1}}}{\Delta x} + \frac{\nabla_y \mathbf{F}_{i,j}^{\overline{n+1}}}{\Delta y} \right) \tag{9-93}$$

$$\left([I] + \frac{\Delta t\, \nabla_x [A']}{\Delta x} \right) \left([I] + \frac{\Delta t\, \nabla_y [B']}{\Delta y} \right) \Delta \mathbf{U}_{i,j}^{n+1} = (\Delta \mathbf{U}_{i,j}^{\overline{n+1}})_{\text{explicit}} \tag{9-94}$$

$$\mathbf{U}_{i,j}^{n+1} = \tfrac{1}{2}(\mathbf{U}_{i,j}^n + \mathbf{U}_{i,j}^{\overline{n+1}} + \Delta \mathbf{U}_{i,j}^{n+1}) \tag{9-95}$$

In the above equations, Δ_x, Δ_y, ∇_x, and ∇_y represent the usual forward and backward spatial differences and Δ represents the forward time difference Δ_t. Also, expressions such as

$$\left([I] - \frac{\Delta t \, \Delta_y [B']}{\Delta y}\right) \Delta \overline{\mathbf{U}_{i,j}^{n+1}}$$

are equivalent to

$$\Delta \overline{\mathbf{U}_{i,j}^{n+1}} - \frac{\Delta t}{\Delta y} \Delta_y ([B'] \, \Delta \overline{\mathbf{U}_{i,j}^{n+1}})$$

The viscous terms in $\mathbf{E}$ and $\mathbf{F}$ are differenced in the same manner as the explicit MacCormack scheme. The matrices $[A']$ and $[B']$ have positive eigenvalues and are related to the Jacobian matrices $[A] = \partial \mathbf{E}/\partial \mathbf{U}$ and $[B] = \partial \mathbf{F}/\partial \mathbf{U}$ as discussed in the next paragraph.

If the viscous terms are ignored and a perfect gas is assumed, the Jacobian matrices $[A]$ and $[B]$ can be diagonalized as

$$\begin{aligned} [A] &= [S_x]^{-1} [\Lambda_A] [S_x] \\ [B] &= [S_y]^{-1} [\Lambda_B] [S_y] \end{aligned} \tag{9-96}$$

where

$$[S_x] = \begin{bmatrix} 1 & 0 & 0 & -\dfrac{1}{a^2} \\ 1 & \rho a & 0 & 1 \\ 0 & 0 & 1 & 0 \\ 0 & -\rho a & 0 & 1 \end{bmatrix} \times \begin{bmatrix} 1 & 0 & 0 & 0 \\ -\dfrac{u}{\rho} & \dfrac{1}{\rho} & 0 & 0 \\ -\dfrac{v}{\rho} & 0 & \dfrac{1}{\rho} & 0 \\ \alpha\beta & -u\beta & -v\beta & \beta \end{bmatrix} \tag{9-97}$$

$$[S_y] = \begin{bmatrix} 1 & 0 & 0 & -\dfrac{1}{a^2} \\ 0 & 1 & 0 & 0 \\ 0 & 0 & \rho a & 1 \\ 0 & 0 & -\rho a & 1 \end{bmatrix} \times \begin{bmatrix} 1 & 0 & 0 & 0 \\ -\dfrac{u}{\rho} & \dfrac{1}{\rho} & 0 & 0 \\ -\dfrac{v}{\rho} & 0 & \dfrac{1}{\rho} & 0 \\ \alpha\beta & -u\beta & -v\beta & \beta \end{bmatrix} \tag{9-98}$$

$$[\Lambda_A] = \begin{bmatrix} u & 0 & 0 & 0 \\ 0 & u+a & 0 & 0 \\ 0 & 0 & u & 0 \\ 0 & 0 & 0 & u-a \end{bmatrix} \tag{9-99}$$

$$[\Lambda_B] = \begin{bmatrix} v & 0 & 0 & 0 \\ 0 & v & 0 & 0 \\ 0 & 0 & v+a & 0 \\ 0 & 0 & 0 & v-a \end{bmatrix} \tag{9-100}$$

and $\alpha = (u^2 + v^2)/2$, $\beta = \gamma - 1$, and a is the speed of sound, $\sqrt{\gamma p/\rho}$. The matrices $[A']$ and $[B']$ differ from $[A]$ and $[B]$ in that all of the eigenvalues are positive and viscous effects are approximately included. These matrices are defined by

$$
\begin{aligned}
[A'] &= [S_x]^{-1}[D_A][S_x] \\
[B'] &= [S_y]^{-1}[D_B][S_y]
\end{aligned}
\tag{9-101}
$$

where

$$
[D_A] = \begin{bmatrix} d_{A_1} & 0 & 0 & 0 \\ 0 & d_{A_2} & 0 & 0 \\ 0 & 0 & d_{A_3} & 0 \\ 0 & 0 & 0 & d_{A_4} \end{bmatrix}
\tag{9-102}
$$

$$
[D_B] = \begin{bmatrix} d_{B_1} & 0 & 0 & 0 \\ 0 & d_{B_2} & 0 & 0 \\ 0 & 0 & d_{B_3} & 0 \\ 0 & 0 & 0 & d_{B_4} \end{bmatrix}
\tag{9-103}
$$

$$
\begin{aligned}
d_{A_1} &= \max\left[\left(|u| + \frac{2\nu}{\rho\,\Delta x} - \frac{1}{2}\frac{\Delta x}{\Delta t}\right), 0.0\right] \\
d_{A_2} &= \max\left[\left(|u+a| + \frac{2\nu}{\rho\,\Delta x} - \frac{1}{2}\frac{\Delta x}{\Delta t}\right), 0.0\right] \\
d_{A_3} &= \max\left[\left(|u| + \frac{2\nu}{\rho\,\Delta x} - \frac{1}{2}\frac{\Delta x}{\Delta t}\right), 0.0\right] \\
d_{A_4} &= \max\left[\left(|u-a| + \frac{2\nu}{\rho\,\Delta x} - \frac{1}{2}\frac{\Delta x}{\Delta t}\right), 0.0\right] \\
d_{B_1} &= \max\left[\left(|v| + \frac{2\nu}{\rho\,\Delta y} - \frac{1}{2}\frac{\Delta y}{\Delta t}\right), 0.0\right] \\
d_{B_2} &= \max\left[\left(|v| + \frac{2\nu}{\rho\,\Delta y} - \frac{1}{2}\frac{\Delta y}{\Delta t}\right), 0.0\right] \\
d_{B_3} &= \max\left[\left(|v+a| + \frac{2\nu}{\rho\,\Delta y} - \frac{1}{2}\frac{\Delta y}{\Delta t}\right), 0.0\right] \\
d_{B_4} &= \max\left[\left(|v-a| + \frac{2\nu}{\rho\,\Delta y} - \frac{1}{2}\frac{\Delta y}{\Delta t}\right), 0.0\right]
\end{aligned}
\tag{9-104}
$$

and

$$\nu = \max\left(\tfrac{4}{3}\mu, k\right)$$

If Δt satisfies the explicit stability conditions

$$\Delta t \leqslant \frac{1}{2}\left[\frac{|u| + a}{\Delta x} + \frac{2\nu}{\rho(\Delta x)^2}\right]^{-1}$$
$$\Delta t \leqslant \frac{1}{2}\left[\frac{|v| + a}{\Delta y} + \frac{2\nu}{\rho(\Delta y)^2}\right]^{-1} \tag{9-105}$$

in a given region of the flowfield, then d_A and d_B are set equal to zero by Eqs. (9-104) and the implicit MacCormack scheme reduces to the original explicit MacCormack scheme. On the other hand, if Δt exceeds the explicit limitations, then the implicit stage of the implicit MacCormack scheme is required for stability. The resulting difference equations are either upper or lower block bidiagonal equations which can be readily solved. For example, Eq. (9-91) can be written as

$$\left([I] + \frac{\Delta t}{\Delta x}[A']^n_{i,j}\,\Delta\mathbf{U}^*_{i,j}\right) = (\Delta\mathbf{U}^n_{i,j})_{\text{explicit}} + \frac{\Delta t}{\Delta x}[A']^n_{i+1,j}\,\Delta\mathbf{U}^*_{i+1,j} \tag{9-106}$$

where $\Delta\mathbf{U}^*_{i,j}$ denotes

$$\left([I] - \frac{\Delta t\,\Delta_y[B']}{\Delta y}\right)\Delta\mathbf{U}^{\overline{n+1}}_{i,j}$$

Equation (9-106) is an upper block bidiagonal equation which can be solved for each j by sweeping in the decreasing i direction. Once $\Delta\mathbf{U}^*_{i,j}$ is determined for all (i, j), then $\Delta\mathbf{U}^{\overline{n+1}}_{i,j}$ is obtained from

$$\left([I] + \frac{\Delta t}{\Delta y}[B']^n_{i,j}\right)\Delta\mathbf{U}^{\overline{n+1}}_{i,j} = \Delta\mathbf{U}^*_{i,j} + \frac{\Delta t}{\Delta y}[B']^n_{i,j+1}\,\Delta\mathbf{U}^{\overline{n+1}}_{i,j+1} \tag{9-107}$$

which is also an upper block bidiagonal equation. It can be solved for each i by sweeping in the decreasing j direction. In order to illustrate how Eq. (9-107) is solved at (i, j), we rewrite it as

$$\left([\mathrm{I}] + \frac{\Delta t}{\Delta y}[S_y]^{-1}[D_B][S_y]\right)\Delta\mathbf{U}^{\overline{n+1}}_{i,j} = \mathbf{W} \tag{9-108}$$

where $\mathbf{W}$ represents the right-hand side of Eq. (9-107) and $[B']^n_{i,j}$ has been replaced by $[S_y]^{-1}[D_B][S_y]$. Equation (9-108) is equivalent to

$$\left([S_y] + \frac{\Delta t}{\Delta y}[D_B][S_y]\right)\Delta\mathbf{U}^{\overline{n+1}}_{i,j} = [S_y]\mathbf{W} = \mathbf{X} \tag{9-109}$$

or

$$\left([I] + \frac{\Delta t}{\Delta y}[D_B]\right)[S_y]\,\Delta\mathbf{U}^{\overline{n+1}}_{i,j} = \mathbf{X} \tag{9-110}$$

Hence,

$$[S_y]\,\Delta \mathbf{U}_{i,j}^{n+1} = \left([I] + \frac{\Delta t}{\Delta y}[D_B]\right)^{-1}\mathbf{X} = \mathbf{Y} \tag{9-111}$$

which can be solved as

$$\Delta \overline{\mathbf{U}_{i,j}^{n+1}} = [S_y]^{-1}\mathbf{Y} \tag{9-112}$$

This solution procedure for Eq. (9-107) can be summarized as follows:

1. $\mathbf{W} = \text{RHS}$ [Eq. (9-107)]
2. $\mathbf{X} = [S_y]\mathbf{W}$
3. $\mathbf{Y} = ([I] + (\Delta t/\Delta y)[D_B])^{-1}\mathbf{X}$
4. $\Delta \overline{\mathbf{U}_{i,j}^{n+1}} = [S_y]^{-1}\mathbf{Y}$

Note that the matrix inversion in Step 3 is trivial since the matrix is diagonal. Also, $[S_y]^{-1}$, required in Step 4, can be readily derived from Eq. (9-98). The term, $(\Delta t/\Delta y)[B']_{i,j}^{n}\,\Delta \mathbf{U}_{i,j}^{n+1}$, in $\mathbf{W}$ can now be determined for the next grid point $(i, j-1)$ in the sweeping procedure using

5. $\dfrac{\Delta t}{\Delta y}[B']_{i,j}^{n}\,\Delta \overline{\mathbf{U}_{i,j}^{n+1}} = \mathbf{W} - \Delta \overline{\mathbf{U}_{i,j}^{n+1}}$

During the initial part of some calculations, it may be necessary to increase ν in order to prevent instabilities caused by the large transients.

The implicit MacCormack scheme applied to the compressible N-S equations is second-order accurate in both space and time provided that $\nu\,\Delta t/\rho(\Delta x)^2$ and $\nu\,\Delta t/\rho(\Delta y)^2$ remain bounded as Δx, Δy, and Δt approach zero. The main advantage of this scheme is that only block bidiagonal equations need to be solved instead of the usual block tridiagonal equations. A disadvantage of this scheme is due to the difficulties encountered in applying non-Dirichlet boundary conditions.

9-3 INCOMPRESSIBLE N–S EQUATIONS

The incompressible N-S equations can be derived from the compressible N-S equations by assuming an incompressible flow ($M = 0$, $a = \infty$). Hence, the incompressible N-S equations are a subset of the compressible N-S equations and one might wonder why these equations are treated separately. That is, why not use the compressible N-S equations to compute an incompressible flow? The main reason for not doing this is that an exorbitant amount of computer time would be required. This is not only due to the additional complexity of the compressible N-S equations but is also the result of a time-step limitation. In order to explain the time-step limitation, we note that all the explicit methods for solving the compressible N-S equations, as well as the implicit MacCormack scheme, are limited to a time step which is less than that given by the CFL condition

$$\Delta t \leqslant \frac{1}{(|u|/\Delta x) + (|v|/\Delta y) + a\sqrt{[1/(\Delta x)^2] + [1/(\Delta y)^2]}} \tag{9-113}$$

From this condition, we observe that Δt approaches zero as the speed of sound a approaches its incompressible value of infinity. As a result, an infinite amount of computer time would be required to compute a truly incompressible flow in this manner. Implicit methods such as the Beam-Warming scheme will permit a larger Δt, but the maximum value is normally less than 5-10 times that given by Eq. (9-113) because truncation errors become unacceptable. Thus, even if an implicit scheme is used, it is not practical to compute a truly incompressible flow using the compressible N-S equations. We will now begin our discussion of methods for solving the incompressible N-S equations.

The incompressible N-S equations for a constant property flow without body forces or external heat addition are given by (see Chap. 5):

continuity:

$$\nabla \cdot \mathbf{V} = 0 \tag{9-114}$$

momentum:

$$\rho \frac{D\mathbf{V}}{Dt} = -\nabla p + \mu \nabla^2 \mathbf{V} \tag{9-115}$$

energy:

$$\rho c_v \frac{DT}{Dt} = k \nabla^2 T + \Phi \tag{9-116}$$

These equations (1 vector, 2 scalar) are a mixed set of elliptic-parabolic equations which contain the unknowns ($\mathbf{V}$, p, T). Note that the temperature appears only in the energy equation so that we can uncouple this equation from the continuity and momentum equations. For many applications the temperature changes are either insignificant or unimportant and it is not necessary to solve the energy equation. However, if we wish to find the temperature distribution, this can be easily accomplished since the unsteady energy equation is a parabolic PDE provided that $\mathbf{V}$ is already computed. With this in mind, we will focus our attention on methods for solving the continuity and momentum equations during the remainder of this chapter.

The 2-D incompressible N-S equations written in Cartesian coordinates (without the energy equation) are

continuity:

$$\frac{\partial u}{\partial x} + \frac{\partial v}{\partial y} = 0 \tag{9-117}$$

x momentum:

$$\frac{\partial u}{\partial t} + u \frac{\partial u}{\partial x} + v \frac{\partial u}{\partial y} = -\frac{1}{\rho} \frac{\partial p}{\partial x} + \nu \left(\frac{\partial^2 u}{\partial x^2} + \frac{\partial^2 u}{\partial y^2} \right) \tag{9-118}$$

y momentum:

$$\frac{\partial v}{\partial t} + u \frac{\partial v}{\partial x} + v \frac{\partial v}{\partial y} = -\frac{1}{\rho} \frac{\partial p}{\partial y} + \nu \left(\frac{\partial^2 v}{\partial x^2} + \frac{\partial^2 v}{\partial y^2} \right) \tag{9-119}$$

where ν is the kinematic viscosity μ/ρ. These equations are written in the so-called *primitive-variable form* where p, u, v are the *primitive variables.* One of the popular techniques for solving the incompressible N-S equations involves replacing the primitive variables with the vorticity (ζ) and stream function (ψ). We will discuss this technique in Section 9-3.1. The alternate technique is to solve Eqs. (9-117)-(9-119) in their present form. This technique is referred to as the primitive-variable approach and will be discussed in Section 9-3.2.

9-3.1 Vorticity-Stream Function Approach

The vorticity-stream function approach is one of the most popular methods for solving the 2-D incompressible N-S equations. In this approach, a change of variables is made which replaces the velocity components with the vorticity (ζ) and the stream function (ψ). The vorticity vector ($\boldsymbol{\zeta}$) was defined in Chap. 5 as

$$\boldsymbol{\zeta} = \nabla \times \mathbf{V} \tag{9-120}$$

The magnitude of the vorticity vector

$$\zeta = |\boldsymbol{\zeta}| = |\nabla \times \mathbf{V}| \tag{9-121}$$

can be written as

$$\zeta = \frac{\partial v}{\partial x} - \frac{\partial u}{\partial y} \tag{9-122}$$

for a 2-D Cartesian coordinate system. Also in this coordinate system, the stream function ψ is defined by the equations

$$\begin{aligned} \frac{\partial \psi}{\partial y} &= u \\ \frac{\partial \psi}{\partial x} &= -v \end{aligned} \tag{9-123}$$

Using these new dependent variables, the two momentum equations [Eqs. (9-118) and (9-119)] can be combined (thereby eliminating pressure) to give

$$\frac{\partial \zeta}{\partial t} + u\frac{\partial \zeta}{\partial x} + v\frac{\partial \zeta}{\partial y} = \nu\left(\frac{\partial^2 \zeta}{\partial x^2} + \frac{\partial^2 \zeta}{\partial y^2}\right) \tag{9-124}$$

or

$$\frac{D\zeta}{Dt} = \nu\,\nabla^2 \zeta \tag{9-125}$$

This parabolic PDE is called the *vorticity transport equation.* The one-dimensional form of this equation

$$\frac{\partial \zeta}{\partial t} + u\frac{\partial \zeta}{\partial x} = \nu\frac{\partial^2 \zeta}{\partial x^2} \tag{9-126}$$

is the 1-D advection-diffusion equation which is often used as a model equation. In addition, the nonlinear Burgers equation can be used to model the vorticity transport equation. In fact, the numerical techniques described in Section 4-5 to solve the nonlinear Burgers equation can be directly applied to the vorticity transport equation.

An additional equation involving the new dependent variables ζ and ψ can be obtained by substituting Eq. (9-123) into Eq. (9-122) which gives

$$\frac{\partial^2 \psi}{\partial x^2} + \frac{\partial^2 \psi}{\partial y^2} = -\zeta \tag{9-127}$$

or

$$\nabla^2 \psi = -\zeta \tag{9-128}$$

This elliptic PDE is the *Poisson equation.* Methods for solving equations of this type were discussed in Section 4-3.

As a result of the change of variables, we have been able to separate the mixed elliptic-parabolic 2-D incompressible N-S equations into one parabolic equation (the vorticity transport equation) and one elliptic equation (the Poisson equation). These equations are normally solved using a time-marching procedure which is described by the following steps:

1. Specify initial values for ζ and ψ at time $t = 0$.
2. Solve the vorticity transport equation for ζ at each interior grid point at time $t + \Delta t$.
3. Iterate for new ψ values at all points by solving the Poisson equation using new ζ's at interior points.
4. Find the velocity components from $u = \psi_y$ and $v = -\psi_x$.
5. Determine values of ζ on the boundaries using ψ and ζ values at interior points.
6. Return to Step 2 if the solution is not converged.

Upon completing these steps, the velocity components are determined at each grid point. In order to determine the pressure at each grid point, it is necessary to solve an additional equation which is referred to as the *Poisson equation for pressure.* This equation is derived by differentiating Eq. (9-118) with respect to x,

$$\frac{\partial}{\partial t}\left(\frac{\partial u}{\partial x}\right) + \left(\frac{\partial u}{\partial x}\right)^2 + u\frac{\partial^2 u}{\partial x^2} + \frac{\partial v}{\partial x}\frac{\partial u}{\partial y} + v\frac{\partial^2 u}{\partial x\,\partial y} = -\frac{1}{\rho}\frac{\partial^2 p}{\partial x^2} + \nu\frac{\partial}{\partial x}(\nabla^2 u) \tag{9-129}$$

differentiating Eq. (9-119) with respect to y,

$$\frac{\partial}{\partial t}\left(\frac{\partial v}{\partial y}\right) + \left(\frac{\partial v}{\partial y}\right)^2 + v\frac{\partial^2 v}{\partial y^2} + \frac{\partial u}{\partial y}\frac{\partial v}{\partial x} + u\frac{\partial^2 v}{\partial x\,\partial y} = -\frac{1}{\rho}\frac{\partial^2 p}{\partial y^2} + \nu\frac{\partial}{\partial y}(\nabla^2 v) \tag{9-130}$$

and adding the results to obtain

$$\frac{\partial}{\partial t}\left(\frac{\partial u}{\partial x} + \frac{\partial v}{\partial y}\right) + \left(\frac{\partial u}{\partial x}\right)^2 + \left(\frac{\partial v}{\partial y}\right)^2 + 2\left(\frac{\partial v}{\partial x}\right)\left(\frac{\partial u}{\partial y}\right) + u\left(\frac{\partial^2 u}{\partial x^2} + \frac{\partial^2 v}{\partial x\,\partial y}\right)$$
$$+ v\left(\frac{\partial^2 u}{\partial x\,\partial y} + \frac{\partial^2 v}{\partial y^2}\right) = -\frac{1}{\rho}\nabla^2 p + \nu\left[\frac{\partial}{\partial x}(\nabla^2 u) + \frac{\partial}{\partial y}(\nabla^2 v)\right] \tag{9-131}$$

Using the continuity equation, Eq. (9-131) can be reduced to

$$\nabla^2 p = 2\rho \left(\frac{\partial u}{\partial x}\frac{\partial v}{\partial y} - \frac{\partial u}{\partial y}\frac{\partial v}{\partial x} \right) \tag{9-132}$$

In terms of the stream function this equation can be rewritten as

$$\nabla^2 p = S \tag{9-133}$$

where

$$S = 2\rho \left[\left(\frac{\partial^2 \psi}{\partial x^2} \right) \left(\frac{\partial^2 \psi}{\partial y^2} \right) - \left(\frac{\partial^2 \psi}{\partial x\, \partial y} \right)^2 \right] \tag{9-134}$$

Thus, we have obtained a Poisson equation for pressure which is analogous to Eq. (9-128). In fact, all the methods discussed in Section 4-3 for solving Eq. (9-128) will also apply to Eq. (9-133) if S is differenced in an appropriate manner. A suitable second-order difference representation is given by

$$S_{i,j} = 2\rho_{i,j} \left[\left(\frac{\psi_{i+1,j} - 2\psi_{i,j} + \psi_{i-1,j}}{(\Delta x)^2} \right) \left(\frac{\psi_{i,j+1} - 2\psi_{i,j} + \psi_{i,j-1}}{(\Delta y)^2} \right) - \left(\frac{\psi_{i+1,j+1} - \psi_{i+1,j-1} - \psi_{i-1,j+1} + \psi_{i-1,j-1}}{4\,\Delta x\,\Delta y} \right)^2 \right] \tag{9-135}$$

For a steady flow problem, the Poisson equation for pressure is only solved once, i.e., after the steady-state values of ζ and ψ have been computed. If only the wall pressures are desired, it is not necessary to solve the Poisson equation over the entire flowfield. Instead, a simpler equation can be solved for the wall pressures. This equation is obtained by applying the tangential momentum equation to the fluid adjacent to the wall surface. For a wall located at $y = 0$ in a Cartesian coordinate system (see Fig. 9-2), the steady, tangential momentum equation (x momentum equation) reduces to

$$\left.\frac{\partial p}{\partial x}\right)_{\text{wall}} = \mu \left.\frac{\partial^2 u}{\partial y^2}\right)_{\text{wall}} \tag{9-136}$$

or

$$\left.\frac{\partial p}{\partial x}\right)_{\text{wall}} = -\mu \left.\frac{\partial \zeta}{\partial y}\right)_{\text{wall}} \tag{9-137}$$

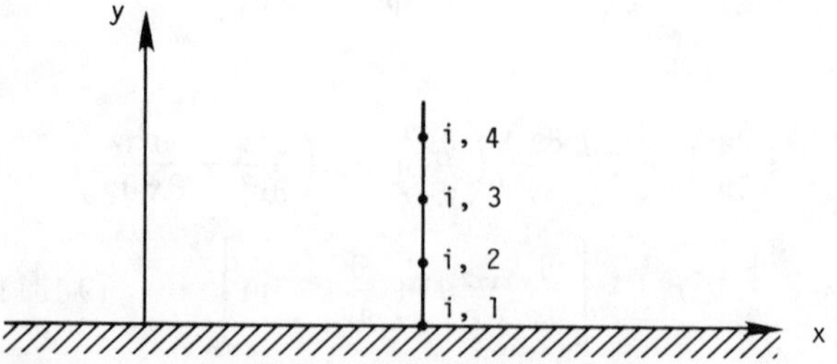

Figure 9-2 Grid points normal to a flat plate at $y = 0$.

which can be differenced as

$$\frac{p_{i+1,1} - p_{i-1,1}}{2\,\Delta x} = -\mu\left(\frac{-3\zeta_{i,1} + 4\zeta_{i,2} - \zeta_{i,3}}{2\,\Delta y}\right) \tag{9-138}$$

In order to apply Eq. (9-138), the pressure must be known for at least one point on the wall surface. The pressure at the adjacent point can be determined using a first-order, one-sided difference expression for $\partial p/\partial x$ in Eq. (9-137). Thereafter, Eq. (9-138) can be used to find the pressure at all other wall points. For a body intrinsic coordinate system, Eq. (9-137) becomes

$$\left.\frac{\partial p}{\partial s}\right)_{\text{wall}} = -\mu \left.\frac{\partial \zeta}{\partial n}\right)_{\text{wall}} \tag{9-139}$$

where s is measured along the body surface and n is normal to it.

The time-marching procedure described earlier for solving the vorticity transport equation and the Poisson equation requires that appropriate expressions for ψ and ζ be specified at the boundaries. The specification of these boundary conditions is extremely important since it directly affects the stability and accuracy of the solution. Let us examine the application of boundary conditions on a wall located at $y = 0$. At the wall surface, ψ is a constant which is usually set equal to zero. In order to find ζ at the wall surface, we expand ψ using a Taylor series about the wall point $(i, 1)$

$$\psi_{i,2} = \psi_{i,1} + \left.\frac{\partial \psi}{\partial y}\right)_{i,1} \Delta y + \frac{1}{2}\left.\frac{\partial^2 \psi}{\partial y^2}\right)_{i,1} (\Delta y)^2 + \cdots \tag{9-140}$$

Since

$$\begin{aligned} \left.\frac{\partial \psi}{\partial y}\right)_{i,1} &= u_{i,1} = 0 \\ \left.\frac{\partial^2 \psi}{\partial y^2}\right)_{i,1} &= \left.\frac{\partial u}{\partial y}\right)_{i,1} \end{aligned} \tag{9-141}$$

and

$$\zeta_{i,1} = \overbrace{\left.\frac{\partial v}{\partial x}\right)_{i,1}}^{0} - \left.\frac{\partial u}{\partial y}\right)_{i,1} = -\left.\frac{\partial^2 \psi}{\partial y^2}\right)_{i,1} \tag{9-142}$$

we can rewrite Eq. (9-140) as

$$\psi_{i,2} = \psi_{i,1} - \tfrac{1}{2}\zeta_{i,1}(\Delta y)^2 + O[(\Delta y)^3]$$

or

$$\zeta_{i,1} = \frac{2(\psi_{i,1} - \psi_{i,2})}{(\Delta y)^2} + O(\Delta y) \tag{9-143}$$

This first-order expression for $\zeta_{i,1}$ often gives better results than higher-order expressions which are susceptible to instabilities at higher Reynolds numbers. For example, the following second-order expression, which was first used by Jensen (1959), leads to unstable calculations at moderate to high Reynolds numbers.

$$\zeta_{i,1} = \frac{7\psi_{i,1} - 8\psi_{i,2} + \psi_{i,3}}{2(\Delta y)^2} + O[(\Delta y)^2] \tag{9-144}$$

Briley (1970) explained the instability by noting that the polynomial expression for ψ, assumed in the derivation of Eq. (9-144), is inconsistent with the evaluation of $u = \partial\psi/\partial y$ at $(i, 2)$ using a central difference. By evaluating u at $(i, 2)$ using the following expression, which is consistent with Eq. (9-144),

$$u_{i,2} = \left.\frac{\partial\psi}{\partial y}\right)_{i,2} = \frac{-5\psi_{i,1} + 4\psi_{i,2} + \psi_{i,3}}{4\,\Delta y} + O[(\Delta y)^2] \tag{9-145}$$

Briley found his computations to be stable even at high Reynolds numbers.

A classical problem which has wall boundaries surrounding the entire computational region is the driven cavity problem illustrated in Fig. 9-3. In this problem, the incompressible viscous flow in the cavity is driven by the uniform translation of the upper surface (lid). The boundary conditions for this problem are indicated in Fig. 9-3. The driven cavity problem is an excellent test case for comparing methods which solve the incompressible N-S equations. A standard test condition of $\mathrm{Re}_l = 100$ is frequently chosen in these comparisons, where

$$\mathrm{Re}_l = \frac{Ul}{\nu} \tag{9-146}$$

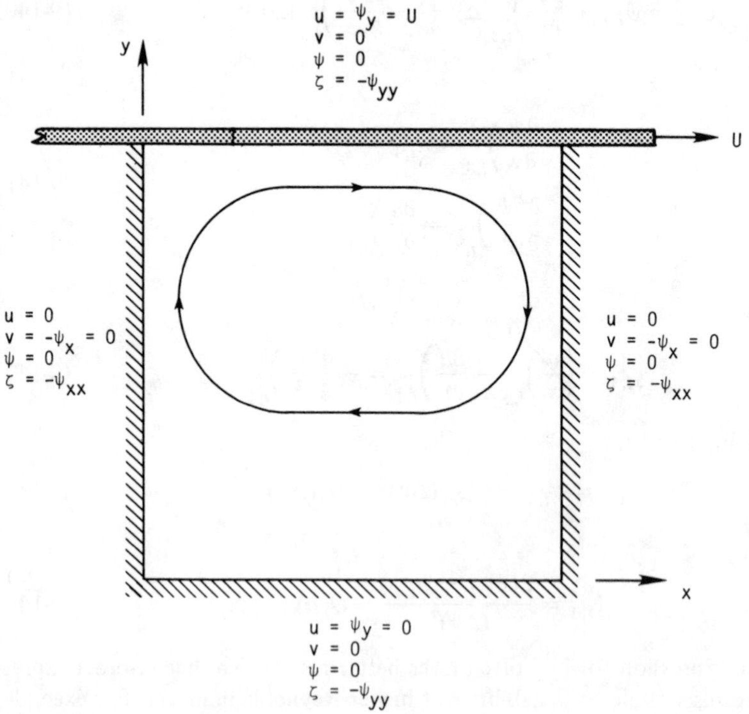

Figure 9-3 Driven cavity problem.

and l is the width of the cavity. Detailed computational results for the driven cavity problem can be found in Burggraf (1966), Bozeman and Dalton (1973), and Rubin and Harris (1975), while experimental data is available in Mills (1965) and Pan and Acrivos (1967).

The specification of appropriate values for ζ and ψ at other types of boundaries such as symmetry lines, upper surfaces, inflow and outflow planes, slip lines, etc. is extremely important and care must be taken to ensure that the physics of the problem is correctly modeled. An excellent discussion on how to treat these various boundaries can be found in Roache (1972).

An alternate way of solving the incompressible N-S equations written in the vorticity-stream function formulation, involves using the steady form of the vorticity transport equation

$$u \frac{\partial \zeta}{\partial x} + v \frac{\partial \zeta}{\partial y} = \nu \nabla^2 \zeta \tag{9-147}$$

This equation is elliptic and can be solved using methods similar to those employed for the Poisson equation. This approach has been successfully used by several investigators, but it appears to be susceptible to instabilities. For this reason, the transient approach is recommended over this steady-state method.

The extension of the vorticity-stream function approach to three-dimensional problems is complicated by the fact that a stream function does not exist for a truly three-dimensional flow. However, the following vector potential (not to be confused with velocity potential)

$$\boldsymbol{\psi} = \psi_x \mathbf{i} + \psi_y \mathbf{j} + \psi_z \mathbf{k} \tag{9-148}$$

does exist, as shown by Aziz and Hellums (1967), which satisfies the continuity equation

$$\nabla \cdot \mathbf{V} = 0 \tag{9-149}$$

so that

$$\mathbf{V} = \nabla \times \boldsymbol{\psi} \tag{9-150}$$

and

$$u = \frac{\partial \psi_z}{\partial y} - \frac{\partial \psi_y}{\partial z}$$

$$v = -\frac{\partial \psi_z}{\partial x} + \frac{\partial \psi_x}{\partial z}$$

$$w = \frac{\partial \psi_y}{\partial x} - \frac{\partial \psi_x}{\partial y}$$

After inserting Eq. (9-150) into Eq. (9-120), we obtain

$$\nabla \times (\nabla \times \boldsymbol{\psi}) = \boldsymbol{\zeta} \tag{9-151}$$

Since the vector potential can be arbitrarily chosen to satisfy

$$\nabla \cdot \boldsymbol{\psi} = 0$$

we can simplify Eq. (9-151) to yield

$$\nabla^2 \psi = -\zeta \tag{9-152}$$

This vector Poisson equation represents three scalar Poisson equations which must be solved after each time step. Likewise, the vorticity transport equation for a 3-D problem is a vector equation which must be separated into three scalar parabolic equations

$$\begin{aligned}
\frac{\partial \zeta_x}{\partial t} + u\frac{\partial \zeta_x}{\partial x} + v\frac{\partial \zeta_x}{\partial y} + w\frac{\partial \zeta_x}{\partial z} - \zeta_x\frac{\partial u}{\partial x} - \zeta_y\frac{\partial u}{\partial y} - \zeta_z\frac{\partial u}{\partial z} &= \nu\nabla^2\zeta_x \\
\frac{\partial \zeta_y}{\partial t} + u\frac{\partial \zeta_y}{\partial x} + v\frac{\partial \zeta_y}{\partial y} + w\frac{\partial \zeta_y}{\partial z} - \zeta_x\frac{\partial v}{\partial x} - \zeta_y\frac{\partial v}{\partial y} - \zeta_z\frac{\partial v}{\partial z} &= \nu\nabla^2\zeta_y \qquad (9\text{-}153) \\
\frac{\partial \zeta_z}{\partial t} + u\frac{\partial \zeta_z}{\partial x} + v\frac{\partial \zeta_z}{\partial y} + w\frac{\partial \zeta_z}{\partial z} - \zeta_x\frac{\partial w}{\partial x} - \zeta_y\frac{\partial w}{\partial y} - \zeta_z\frac{\partial w}{\partial z} &= \nu\nabla^2\zeta_z
\end{aligned}$$

to find the three components $(\zeta_x, \zeta_y, \zeta_z)$ of the vorticity vector. Thus, we are forced to solve three parabolic and three elliptic PDE's at each time level. As a result, it does not appear that the vorticity-stream function approach offers any advantage over the primitive-variable approach when solving a 3-D problem.

Before moving on to a discussion of the primitive-variable approach, we will briefly describe an approach which can be considered a hybrid of the stream function-vorticity approach and the primitive-variable approach. In this hybrid approach, the dependent variables are the vorticity components $(\zeta_x, \zeta_y, \zeta_z)$ and the velocity components (u, v, w). The vorticity components are obtained by solving Eq. (9-153) and the velocity components are determined from

$$\nabla^2 \mathbf{V} = -\nabla \times \boldsymbol{\zeta} \tag{9-154}$$

This vector equation is derived by taking the vector cross-product of the del operator with the definition of the vorticity vector and then simplifying the resulting expression

$$\nabla \times (\nabla \times \mathbf{V}) = \nabla \times \boldsymbol{\zeta}$$

using the appropriate vector identity. Agarwal (1981) states that this hybrid approach avoids the necessity of using a staggered-grid arrangement which is required in some primitive-variable approaches. Also, the application of boundary conditions is easier in the hybrid approach than in the vector potential approach described previously.

9-3.2 Primitive-Variable Approach

The vorticity-stream function approach for solving the incompressible N-S equations loses its attractiveness when applied to a three-dimensional flow because a single scalar stream function does not exist in this case (as discussed in the last section). Consequently, the incompressible N-S equations are normally solved in their primitive-variable form (u, v, w, p) for a three-dimensional problem. The incompressible N-S equations written in nondimensional, primitive-variable form for a Cartesian coordinate system are given by continuity:

$$\frac{\partial u^*}{\partial x^*} + \frac{\partial v^*}{\partial y^*} + \frac{\partial w^*}{\partial z^*} = 0 \tag{9-155}$$

x momentum:

$$\frac{\partial u^*}{\partial t^*} + u^* \frac{\partial u^*}{\partial x^*} + v^* \frac{\partial u^*}{\partial y^*} + w^* \frac{\partial u^*}{\partial z^*} = -\frac{\partial p^*}{\partial x^*} + \frac{1}{\mathrm{Re}_L}\left(\frac{\partial^2 u^*}{\partial x^{*2}} + \frac{\partial^2 u^*}{\partial y^{*2}} + \frac{\partial^2 u^*}{\partial z^{*2}}\right) \tag{9-156}$$

y momentum:

$$\frac{\partial v^*}{\partial t^*} + u^* \frac{\partial v^*}{\partial x^*} + v^* \frac{\partial v^*}{\partial y^*} + w^* \frac{\partial v^*}{\partial z^*} = -\frac{\partial p^*}{\partial y^*} + \frac{1}{\mathrm{Re}_L}\left(\frac{\partial^2 v^*}{\partial x^{*2}} + \frac{\partial^2 v^*}{\partial y^{*2}} + \frac{\partial^2 v^*}{\partial z^{*2}}\right) \tag{9-157}$$

z momentum:

$$\frac{\partial w^*}{\partial t^*} + u^* \frac{\partial w^*}{\partial x^*} + v^* \frac{\partial w^*}{\partial y^*} + w^* \frac{\partial w^*}{\partial z^*} = -\frac{\partial p^*}{\partial z^*} + \frac{1}{\mathrm{Re}_L}\left(\frac{\partial^2 w^*}{\partial x^{*2}} + \frac{\partial^2 w^*}{\partial y^{*2}} + \frac{\partial^2 w^*}{\partial z^{*2}}\right) \tag{9-158}$$

These equations are nondimensionalized using

$$\begin{aligned} u^* &= \frac{u}{V_\infty} \qquad x^* = \frac{x}{L} \qquad p^* = \frac{p}{\rho_\infty V_\infty^2} \\ v^* &= \frac{v}{V_\infty} \qquad y^* = \frac{y}{L} \qquad t^* = \frac{tV_\infty}{L} \\ w^* &= \frac{w}{V_\infty} \qquad z^* = \frac{z}{L} \qquad \mathrm{Re}_L = \frac{V_\infty L}{\nu_\infty} \end{aligned} \tag{9-159}$$

One of the early techniques proposed for solving the incompressible N-S equations in primitive-variable form is the artificial compressibility method of Chorin (1967). In this method, the continuity equation is modified to include an artificial compressibility term which vanishes when the steady-state solution is reached. With the addition of this term to the continuity equation, the resulting N-S equations are a mixed set of hyperbolic-parabolic equations which can be solved using a standard time-dependent approach. In order to explain this method, let us apply it to Eqs. (9-155)–(9-158). The continuity equation is replaced by

$$\frac{\partial \tilde{\rho}^*}{\partial \tilde{t}^*} + \frac{\partial u^*}{\partial x^*} + \frac{\partial v^*}{\partial y^*} + \frac{\partial w^*}{\partial z^*} = 0 \tag{9-160}$$

where $\tilde{\rho}^*$ is an artificial density and $\tilde{t}^*$ is a fictitious time which is analogous to real time in a compressible flow. The artificial density is related to the pressure by the artificial equation of state

$$p^* = \frac{\tilde{\rho}^*}{\beta} \tag{9-161}$$

where β is the artificial compressibility factor to be determined later. Note that at steady state the solution is independent of $\tilde{\rho}^*$ and $\tilde{t}^*$ since $\partial\tilde{\rho}^*/\partial\tilde{t}^* \to 0$. After replacing t^* with $\tilde{t}^*$ in Eqs. (9-156)–(9-158) and substituting Eq. (9-161) into Eq. (9-160), we can apply a suitable finite-difference technique to the resulting equations and march the solution in $\tilde{t}^*$ to obtain a final steady-state incompressible solution. Obviously, this technique is applicable only to steady flow problems since it is not time accurate.

In order to facilitate the application of a finite-difference scheme, Eqs. (9-155)–(9-158) and Eqs. (9-160)–(9-161) can be combined into the following vector form

$$\frac{\partial \mathbf{u}^*}{\partial \tilde{t}^*} + \frac{\partial \mathbf{e}^*}{\partial x^*} + \frac{\partial \mathbf{f}^*}{\partial y^*} + \frac{\partial \mathbf{g}^*}{\partial z^*} = \frac{1}{\text{Re}_L}\left(\frac{\partial^2}{\partial x^{*2}} + \frac{\partial^2}{\partial y^{*2}} + \frac{\partial^2}{\partial z^{*2}}\right)[D]\,\mathbf{u}^* \qquad (9\text{-}162)$$

where

$$\mathbf{u}^* = \begin{bmatrix} p^* \\ u^* \\ v^* \\ w^* \end{bmatrix} \qquad \mathbf{e}^* = \begin{bmatrix} \dfrac{u^*}{\beta} \\ p^* + (u^*)^2 \\ u^*v^* \\ u^*w^* \end{bmatrix}$$

$$\mathbf{f}^* = \begin{bmatrix} \dfrac{v^*}{\beta} \\ u^*v^* \\ p^* + (v^*)^2 \\ v^*w^* \end{bmatrix} \qquad \mathbf{g}^* = \begin{bmatrix} \dfrac{w^*}{\beta} \\ u^*w^* \\ v^*w^* \\ p^* + (w^*)^2 \end{bmatrix} \qquad (9\text{-}163)$$

$$[D] = \begin{bmatrix} 0 & 0 & 0 & 0 \\ 0 & 1 & 0 & 0 \\ 0 & 0 & 1 & 0 \\ 0 & 0 & 0 & 1 \end{bmatrix}$$

In the original paper of Chorin, the leapfrog/DuFort-Frankel finite-difference scheme (see Section 4-5.2) was applied to the governing equations. Chorin derived the following stability condition for this explicit scheme

$$\Delta\tilde{t}^* \leqslant \frac{2\beta^{1/2}\,\Delta^*_{\min}}{N^{1/2}(1+\sqrt{5})} \qquad (9\text{-}164)$$

where N is the number of space dimensions and $\Delta^*_{\min}$ is the minimum of $(\Delta x^*, \Delta y^*, \Delta z^*)$. An additional relationship between $\Delta\tilde{t}^*$ and β can be obtained by noting that the artificial equation of state implies the existence of an artificial sound speed ($\tilde{a}^*$) given by

$$\tilde{a}^* = \frac{1}{\beta^{1/2}} \tag{9-165}$$

Since the maximum artificial Mach number ($\tilde{M}_{max}$) based on this artificial sound speed is required to be less than unity, the following additional condition is obtained

$$\tilde{M}_{max} = \frac{V^*_{max}}{\tilde{a}^*} = \beta^{1/2} V^*_{max} < 1 \tag{9-166}$$

where V^*_{max} is the maximum value of V^* given by

$$V^* = \sqrt{(u^*)^2 + (v^*)^2 + (w^*)^2} \tag{9-167}$$

Thus, the two parameters $\Delta\tilde{t}^*$ and β must be assigned values which satisfy Eqs. (9-164) and (9-166). It is possible to increase the rate of convergence by selecting optimum values of $\Delta\tilde{t}^*$ and β, but this has to be done on a trial and error basis for each problem. In most cases, a value of $\tilde{M}_{max}$ around 0.5 will produce satisfactory results.

In general, an implicit finite-difference scheme is recommended over an explicit scheme to solve Eq. (9-162). Steger and Kutler (1976) have applied the implicit Beam-Warming approximate factorization algorithm (see Section 9-2.3) to Eq. (9-162) in order to compute incompressible vortex wakes. They found that if β is too small, large errors can be introduced into the solution as a result of the approximate factorization.

The artificial compressibility method, just described, is one technique for solving the incompressible N-S equations in primitive-variable form. Probably the most common primitive-variable approach, however, involves using a Poisson equation for pressure in place of the continuity equation. This is done to separate the majority of the "pressure effects" into a single equation so that the elliptic nature of the flow can be suitably modeled. The Poisson equation for pressure is derived in the same manner as Eq. (9-131). In nondimensional form this equation can be written as

$$\nabla^2 p^* = S_p^* - \frac{\partial D^*}{\partial t^*} \tag{9-168}$$

where

$$\begin{aligned} S_p^* = \frac{d}{dx^*}&\left[-(u^*u_x^* + v^*u_y^* + w^*u_z^*) + \frac{1}{\mathrm{Re}_L}(u_{xx}^* + u_{yy}^* + u_{zz}^*)\right] \\ &+ \frac{d}{dy^*}\left[-(u^*v_x^* + v^*v_y^* + w^*v_z^*) + \frac{1}{\mathrm{Re}_L}(v_{xx}^* + v_{yy}^* + v_{zz}^*)\right] \\ &+ \frac{d}{dz^*}\left[-(u^*w_x{}^* + v^*w_y^* + w^*w_z^*) + \frac{1}{\mathrm{Re}_L}(w_{xx}^* + w_{yy}^* + w_{zz}^*)\right] \end{aligned}$$

and D^* is the local dilatation term given by

$$D^* = u_x^* + v_y^* + w_z^*$$

and terms such as u_x^* denote $\partial u^*/\partial x^*$. The derivative of the local dilatation term is purposely not set to zero in order to account for the differences between an intermediate solution and the final converged solution of the Poisson equation. Equation (9-168) was first used by Harlow and Welch (1965) and Welch et al. (1966) in conjunction with their Marker-and-Cell (MAC) method for solving the incompressible N-S equations.

In the approach of Ghia and co-workers (Ghia et al., 1977, 1979, 1981), an ADI scheme is applied to the momentum equations [Eqs. (9-156)-(9-158)] and a SOR method is then used to solve the Poisson pressure equation. To start the procedure, the pressure gradient terms in the momentum equations are given approximate values. After the velocities are computed using the momentum equations, the pressures are found from the Poisson equation. Using these pressures, the pressure gradient terms can be evaluated in the momentum equations. The momentum equations are then solved again to find new velocities. This procedure is repeated until the solution converges.

The flowfields computed by Ghia et al. include the flow in a driven cavity and the flow in a channel. In both of these cases, the boundary conditions for the Poisson equation for pressure consist of the normal pressure gradient ($\partial p/\partial n$) evaluated from the appropriate momentum equation. Thus, the solution of a Neumann problem is required to determine the pressure. This solution must satisfy the following integral constraint as required by the divergence theorem

$$\iint_A \nabla^2 p \, dA = \oint_C \frac{\partial p}{\partial n} \, ds \tag{9-169}$$

where C is the closed boundary of the solution domain of area A and s is a differential length along C. During the transient stages, Eq. (9-169) will not be satisfied exactly. In order to account for this inconsistency, the source term S_p^* in Eq. (9-168) can be modified at every grid point by the amount $\Delta S_p^*/A^*$ where ΔS_p^* is given by

$$\Delta S_p^* = \iint_{A^*} S_p^* \, dA^* - \oint_C \frac{\partial p^*}{\partial n^*} \, ds^* \tag{9-170}$$

and A^*, n^*, and s^* are nondimensionalized quantities. Briley (1974) and Ghia et al. have successfully applied this integral constraint in finite-difference form to obtain a solution of the required Neumann problem.

In order to illustrate the calculation of the normal pressure gradient at a boundary, let us consider a wall located at $y = 0$ as seen in Fig. 9-4. Note that a fictitious row of grid points for pressure has been added below the wall surface in this non-staggered grid. At the wall surface, the y momentum equation [Eq. (9-157)] reduces to

$$\left.\frac{\partial p^*}{\partial y^*}\right)_{i,1} = \frac{1}{\mathrm{Re}_L} \left.\frac{\partial^2 v^*}{\partial y^{*2}}\right)_{i,1} \tag{9-171}$$

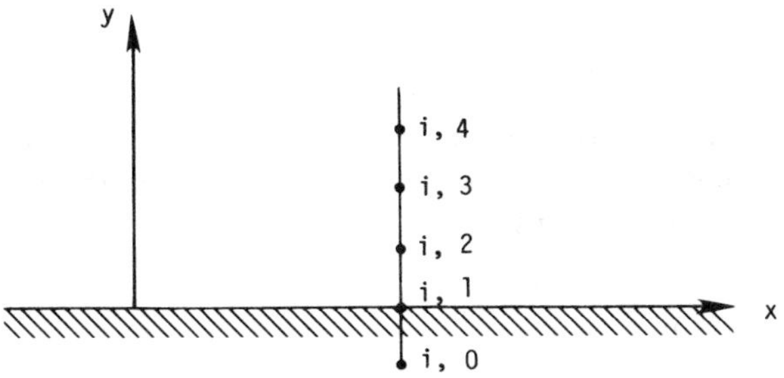

Figure 9-4 Grid points for determination of pressure boundary condition.

This equation can be differenced using the familiar second-order accurate central-difference expressions

$$\frac{p^*_{i,2} - p^*_{i,0}}{2\,\Delta y^*} = \frac{1}{\mathrm{Re}_L}\left(\frac{v^*_{i,2} - \overset{0}{\cancel{2v^*_{i,1}}} + v^*_{i,0}}{(\Delta y^*)^2}\right) \tag{9-172}$$

where $v^*_{i,0}$ is the value of v^* at the fictitious point. An expression for $v^*_{i,0}$ can be determined from the continuity equation which reduces to

$$\left.\frac{\partial v^*}{\partial y^*}\right)_{i,1} = 0 \tag{9-173}$$

at the wall. Using a third-order accurate finite-difference expression for this reduced form of the continuity equation

$$\left.\frac{\partial v^*}{\partial y^*}\right)_{i,1} = \frac{-2v^*_{i,0} - \overset{0}{\cancel{3v^*_{i,1}}} + 6v^*_{i,2} - v^*_{i,3}}{6(\Delta y^*)} + O[(\Delta y^*)^3] = 0 \tag{9-174}$$

allows us to compute $v^*_{i,0}$ and retain second-order accuracy in Eq. (9-172). Similar techniques can be used to find the pressure gradient at other boundaries in order to solve the Poisson pressure equation.

The SIMPLE (Semi-Implicit Method for Pressure-Linked Equations) procedure of Patankar and Spalding (1972), which was described in Section 8-4.1 for the solution of the subsonic PNS equations, can also be applied to the incompressible N-S equations [see Caretto et al. (1972) and Patankar (1975, 1981)]. This procedure is based on a cyclic series of guess-and-correct operations to solve the governing equations. The velocity components are first calculated from the momentum equations using a guessed pressure field. The pressures and velocities are then corrected so as to satisfy continuity. This process continues until the solution converges.

In this procedure, the actual pressure p is written as

$$p = p_0 + p' \tag{9-175}$$

where p_0 is the estimated (or intermediate) value of pressure and p' is the pressure correction. Likewise, the actual velocity components (for a two-dimensional flow) are written as

$$
\begin{aligned}
u &= u_0 + u' \\
v &= v_0 + v'
\end{aligned}
\tag{9-176}
$$

where u_0, v_0 are the estimated (or intermediate) values of velocity and u', v' are the velocity corrections. The pressure corrections are related to the velocity corrections by approximate forms of the momentum equations:

$$
\begin{aligned}
\rho \frac{\partial u'}{\partial t} &= -\frac{\partial p'}{\partial x} \\
\rho \frac{\partial v'}{\partial t} &= -\frac{\partial p'}{\partial y}
\end{aligned}
\tag{9-177}
$$

Since the velocity corrections can be assumed to be zero at the previous iteration step, the above equations can be written as

$$
\begin{aligned}
u' &= -A \frac{\partial p'}{\partial x} \\
v' &= -A \frac{\partial p'}{\partial y}
\end{aligned}
\tag{9-178}
$$

where A is a fictitious time increment divided by density. After combining Eqs. (9-176) and (9-178) and substituting the result into the continuity equation, we obtain

$$
\cancelto{0}{\left(\frac{\partial u}{\partial x} + \frac{\partial v}{\partial y}\right)} - \left(\frac{\partial u_0}{\partial x} + \frac{\partial v_0}{\partial y}\right) + A\left(\frac{\partial^2 p'}{\partial x^2} + \frac{\partial^2 p'}{\partial y^2}\right) = 0 \tag{9-179}
$$

or

$$
\nabla^2 p' = \frac{1}{A}(\nabla \cdot \mathbf{V}_0) \tag{9-180}
$$

where $\mathbf{V}_0$ is the estimated velocity vector. This Poisson equation can be solved for the pressure correction. Note that if the estimated velocity vector satisfies continuity at every point, then the pressure correction is zero at every point. In the actual SIMPLE algorithm, an equivalent differenced form of Eq. (9-180) is used as shown by Raithby and Schneider (1979).

The SIMPLE procedure can now be described by the following steps:

1. Guess the pressure (p_0) at each grid point.
2. Solve the momentum equations to find the velocity components (u_0, v_0). A staggered grid in conjunction with a block iterative method is recommended by Patankar and Spalding.
3. Solve the pressure-correction equation [i.e., Eq. (9-180)] to find p' at each grid point.
4. Correct the pressure and velocity using Eqs. (9-175) and (9-178):

$$
p = p_0 + p'
$$

$$u = u_0 - \frac{A}{2\,\Delta x}\,(p'_{i+1,j} - p'_{i-1,j})$$

$$v = v_0 - \frac{A}{2\,\Delta y}\,(p'_{i,j+1} - p'_{i,j-1})$$

5. Replace the previous intermediate values of pressure and velocity (p_0, u_0, v_0) with the new corrected values (p, u, v) and return to Step 2. Repeat this process until the solution converges.

The SIMPLE procedure has been used successfully to solve a number of incompressible flow problems. However, in certain cases it is found that the rate of convergence is not satisfactory. This is due to the fact that the pressure correction equation tends to overestimate the value of p' even though the corresponding velocity corrections are reasonable. Because of this, Eq. (9-175) is often replaced with

$$p = p_0 + \alpha_p p'$$

where α_p is an under-relaxation constant. For the same reason, under-relaxation is also employed in the solution of the momentum equations. In the present formulation, under-relaxation can be accommodated by varying the parameter A in Eqs. (9-178) and (9-180).

Since it is not possible to readily determine the optimum under-relaxation parameters, the SIMPLE procedure has been revised to improve the rate of convergence (Patankar, 1981). This new procedure is called SIMPLER (SIMPLE revised). In SIMPLER, the velocity corrections are computed in the same manner as in SIMPLE, but a complete Poisson equation for pressure is used to compute pressure. Also, the velocity field is guessed initially instead of the pressure field. In most problems it is easier to estimate a reasonable velocity field rather than a pressure field. Since the pressures computed in SIMPLER are closer to the correct pressures, the need for under-relaxation is greatly reduced and a converged solution is obtained with fewer iterations. In most cases, a reduction in computer time of 30-50% is obtained despite the fact that SIMPLER requires about 30% more computational effort per iteration than SIMPLE.

PROBLEMS

9-1 Show how all the terms in the two-dimensional y momentum equation are differenced when the explicit MacCormack method is applied to the compressible N-S equations.

9-2 Repeat Prob. 9-1 for the 2-D energy equation.

9-3 Apply the explicit MacCormack scheme to the N-S equations written in cylindrical coordinates (see Section 5-1.7) and show how all the terms in the r momentum equation are differenced.

9-4 Apply the Allen-Cheng method instead of the explicit MacCormack method in Prob. 9-1.

9-5 Derive the Jacobian matrix $[A]$ given by Eq. (9-46).

9-6 Derive the Jacobian matrix $[B]$ given by Eq. (9-48).

9-7 Derive the Jacobian matrix $[R]$ given by Eq. (9-51).

9-8 Derive the Jacobian matrix $[S]$ given by Eq. (9-54).

9-9 Derive the matrix $[P] - [R_x]$ given by Eq. (9-50).

9-10 Derive the matrix $[Q] - [S_y]$ given by Eq. (9-53).

9-11 Determine the amplification factor G for the explicit MacCormack scheme applied to the linearized Burgers equation. Does Eq. (9-86) satisfy $|G| \leq 1$ for all values of β when $\nu = \frac{1}{2}$ and $r = \frac{1}{4}$?

9-12 Repeat Prob. 9-11 for $\nu = 1$ and $r = \frac{1}{2}$.

9-13 Use the implicit MacCormack method to solve the linearized Burgers equation for the initial condition

$$u(x, 0) = 0 \qquad 0 \leq x \leq 1$$

and the boundary conditions

$$u(0, t) = 100$$

$$u(1, t) = 0$$

on a 21 grid point mesh. Find the steady-state solution for the conditions

$$r = 0.5$$

$$\nu = 0.5$$

and compare the numerical solution with the exact solution.

9-14 Derive the Jacobian matrix $[A]$ in Eq. (9-96) and show that it is equivalent to $[S_x]^{-1} \times [\Lambda_A][S_x]$.

9-15 Derive the Jacobian matrix $[B]$ in Eq. (9-96) and show that it is equivalent to $[S_y]^{-1} \times [\Lambda_B][S_y]$.

9-16 Obtain Eq. (9-124).

9-17 Solve the square driven cavity problem for $Re_l = 50$. Use the FTCS method to solve the vorticity transport equation and the SOR method to solve the Poisson equation. Employ a first-order evaluation of the vorticity at the wall and use an 8 × 8 grid.

9-18 Repeat Prob. 9-17 for $Re_l = 100$ and a 15 × 15 grid.

9-19 Derive the vorticity transport equations for a 3-D Cartesian coordinate system.

9-20 Use the artificial compressibility method to solve the square driven cavity problem for $Re_l = 100$. Apply the leapfrog/DuFort-Frankel finite-difference scheme to the governing equations on a 15 × 15 grid. Determine pressure at the wall using a suitable finite-difference representation of the normal momentum equation applied at the wall.

CHAPTER

TEN

GRID GENERATION

10-1 INTRODUCTION

The solution of a system of partial differential equations can be greatly simplified by a well-constructed grid. It is also true that a grid which is not well suited to the problem can lead to an unsatisfactory result. In some applications, improper choice of grid point locations can lead to an apparent instability or lack of convergence. One of the central problems in computing numerical solutions to partial differential equations is that of grid generation.

Early work using finite-difference methods was restricted to problems where suitable coordinate systems could be selected in order to solve the governing equations in that base system. As experience in computing complex flowfields was gained, general mappings were employed to transform the physical plane into a computational domain. Numerous advantages accrue when this procedure is followed. For example, the body surface can be selected as a boundary in the computational plane permitting easy application of surface boundary conditions. In general, transformations are used which lead to a uniformly spaced grid in the computational plane while points in physical space may be unequally spaced. This situation is shown in Fig. 10-1. When this procedure is used, it is necessary to include the metrics of the mapping in the differential equation. This can be understood better through a simple example.

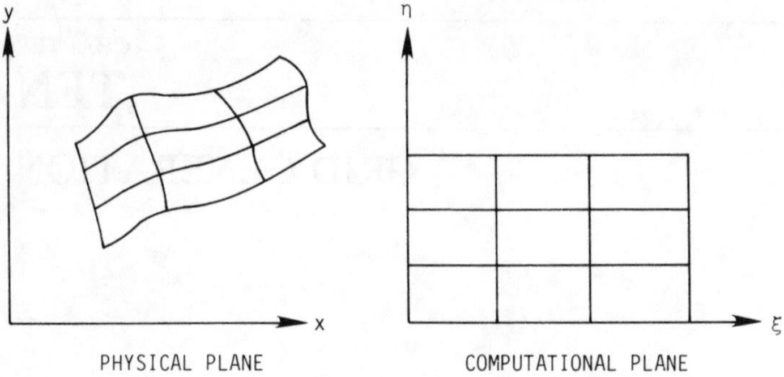

Figure 10-1 Mapping to computational space.

Example 10-1 Suppose we desire a solution of the simple equation

$$\frac{\partial u}{\partial x} + c\,\frac{\partial u}{\partial y} + y = 0$$

on a given domain subject to appropriate boundary and initial data. Since numerical solutions are usually computed in a computational domain, the transformation relating physical and computational space may be written as

$$\xi = \xi(x, y), \qquad \eta = \eta(x, y)$$

The original partial differential equation is transformed from physical coordinates (x, y) to computational coordinates (ξ, η) by applying the chain rule for partial derivatives.

$$\begin{aligned}\frac{\partial u}{\partial x} &= \frac{\partial u}{\partial \xi}\,\xi_x + \frac{\partial u}{\partial \eta}\,\eta_x \\ \frac{\partial u}{\partial y} &= \frac{\partial u}{\partial \xi}\,\xi_y + \frac{\partial u}{\partial \eta}\,\eta_y\end{aligned} \tag{10-1}$$

The original PDE becomes

$$(\xi_x + c\xi_y)\,\frac{\partial u}{\partial \xi} + (\eta_x + c\eta_y)\,\frac{\partial u}{\partial \eta} + y(\xi, \eta) = 0 \tag{10-2}$$

This equation is solved on a uniform grid in the computational plane. It is clear that we need to determine the relationship between the physical and computational planes. This relationship establishes the metrics of the transformation which are the ξ_x, ξ_y, η_x, and η_y terms in the partial differential equation to be solved.

The problem of grid generation is that of determining the mapping which takes the grid points from the physical domain (D) to the computational domain (CD). Several requirements must be placed on such mappings. A partial list can be stated as follows:

1. The mapping must be one to one.
2. The grid lines should be smooth to provide continuous transformation derivatives.
3. Grid points should be closely spaced in D where large numerical errors are expected.
4. Excessive grid skewness should be avoided. It has been shown that grid skewness sometimes exaggerates truncation errors (Raithby, 1976).

The generation of suitable grids in one dimension is reasonably simple. There are many functions (or other methods) which can be used to generate a satisfactory grid. In addition, the problem of complex boundaries does not occur in one-dimensional problems. For these reasons most work on grid generation has been done in two dimensions. Numerous examples in two dimensions will be presented in this chapter. Grid generation in three dimensions is very complicated and few methods are available which provide satisfactory results.

Grid generation techniques can be roughly classified into three categories.

1. Complex variable methods
2. Algebraic methods
3. Differential equation techniques

Complex variable techniques have the advantage that the transformations used are analytic or partially analytic as opposed to those methods that are entirely numerical. Unfortunately, complex variable methods are restricted to two dimensions. For this reason, the technique has limited applicability and will not be covered here. For details of the application of complex variable methods, work by Churchill (1948), Moretti (1979), and Davis (1979) should be consulted. Algebraic and differential equation techniques can be used on complicated three-dimensional problems. Of the methods available for generating computational grids these two schemes show the most promise for continued development and use in conjunction with finite-difference methods. The application of these techniques will be studied in this chapter and a number of examples showing generated grids will be presented.

10-2 ALGEBRAIC METHODS

In Chapter 5, we used algebraic expressions to cluster grid points near solid boundaries to provide adequate resolution of the viscous boundary layer. In another example, a domain normalizing transformation was used in order to align the grid lines with the body and shock wave in physical space. These are examples of simple algebraic mappings. To generate computational grids using this technique, known functions are used in one, two, or three dimensions to take arbitrarily shaped physical regions into a rectangular computational domain. Although the computational domain is not required to be rectangular, the usual procedure uses a rectangular region for simplicity.

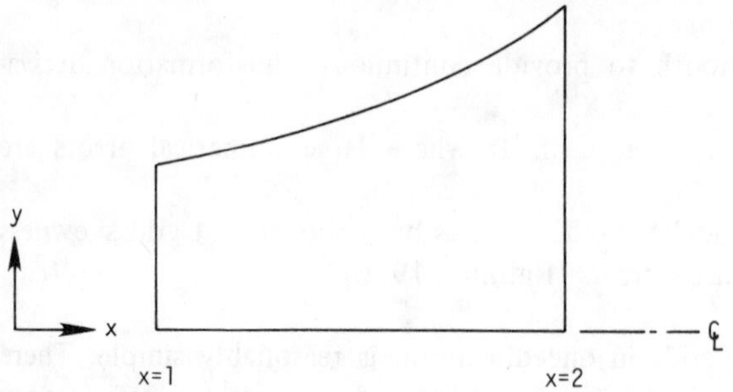

Figure 10-2 Nozzle geometry.

The simplest procedure available which may be used to produce a boundary fitted computational mesh is the normalizing transformation discussed in Section 5-6.

Suppose a body fitted mesh is desired to solve for the flow in a diverging nozzle. The geometry of the nozzle is shown in Fig. 10-2 and the describing function for the nozzle is given as

$$y = x^2, \qquad 1.0 \leqslant x \leqslant 2.0 \tag{10-3}$$

In this example, a computational grid can easily be generated by choosing equally spaced increments in the x direction and using uniform division in the y direction. This may be described as

$$\begin{aligned} \xi &= x \\ \eta &= \frac{y}{y_{\max}} \end{aligned} \tag{10-4}$$

where $y_{\max}$ denotes the nozzle boundary equation. In this case the values of x and y for a given ξ and η are easily recovered. The mesh generated in the physical domain is shown in Fig. 10-3.

Care must be exercised when the metrics of the transformation are derived. In particular, the η_x derivative of Eq. (10-4) is written

$$\eta_x = -\frac{y}{y_{\max}^2}\frac{dy_{\max}}{dx} = -\frac{2\eta}{\xi} \tag{10-5}$$

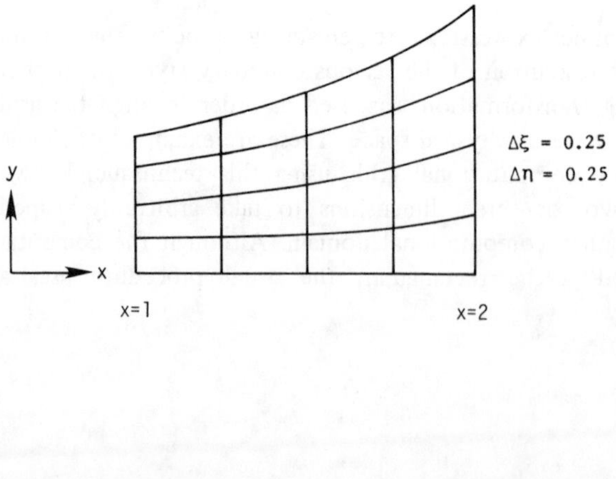

Figure 10-3 Computational mesh in physical space.

and $$\eta_y = \frac{1}{y_{\max}} = \frac{1}{\xi^2} \tag{10-6}$$

In the example just completed, the transformation was analytic and the point distribution was obtained through the given mapping. The same normalizing transformation could have been constructed by assigning points in the physical plane along constant ξ and constant η lines and numerically computing the metrics by using second-order central differences. This has the advantage of permitting assignment of grid points in the physical plane where desired. The disadvantage is that all metrics must be determined using numerical techniques. In this case the transformation would be numerical and not algebraic.

If numerical methods are used to generate the required transformation, the terms x_ξ, x_η, y_ξ, and y_η are determined using finite differences. The quantities ξ_x, ξ_y, η_x, and η_y appear in the differential equation which must be solved. These quantities are obtained from the expressions

$$\begin{aligned} \xi_x &= \frac{y_\eta}{J} \\ \xi_y &= -\frac{x_\eta}{J} \\ \eta_x &= -\frac{y_\xi}{J} \\ \eta_y &= \frac{x_\xi}{J} \\ J &= x_\xi y_\eta - y_\xi x_\eta \end{aligned} \tag{10-7}$$

More details will be presented in the section treating mappings governed by differential equations.

Example 10-2 Compare the metrics for the simple normalizing transformation just discussed by computing them analytically and using a finite-difference approximation.

We select the point (1.75, 2.2969) in the nozzle of Fig. 10-3 to compare the metrics. From Eq. (10-5), the analytic evaluation is

$$\eta_x = -\frac{2(0.75)}{1.75} = -0.85714$$

The numerical calculation is performed by using Eq. (10-7). The Jacobian is evaluated first as

$$J = \cancelto{1}{x_\xi} y_\eta - \cancelto{0}{y_\xi} x_\eta = \frac{3.0625 - 1.53125}{2(0.25)} = 3.06250$$

Next, the y_ξ term is computed as

$$y_\xi = \frac{3 - 1.6875}{0.5} = 2.6250$$

Thus
$$\eta_x = -\frac{2.6250}{3.0625} = -0.85714$$

In this example, the metric computed by analytic and numerical methods give equally good results. Of course this is not true for many problems.

Example 10-3 The trapezoidal region shown in Fig. 10-4 is mapped into a corresponding rectangular region by the equations given by

$$x = \left(\frac{1 + \xi}{2}\right)\left(\frac{3 - \eta}{2}\right)$$
$$y = \frac{\eta + 1}{2} \tag{10-8}$$

In this example the physical domain is mapped into a rectangular region centered at the origin. This demonstrates the use of a normalizing transformation in one direction along with a simple translation. It is clear that any quadrilateral physical domain can be transformed into a rectangle in computational space by the use of a normalizing transformation.

Very complex algebraic functions can be used to generate appropriate grid systems. Smith and Weigel (1980) have developed a flexible method for directly providing grids. In this method, the physical coordinates are rectangular and two disconnected boundaries are mapped into the computational plane. Let these two disconnected boundaries in the physical plane be written as

$$x_{B1} = x_1(\xi)$$
$$y_{B1} = y_1(\xi) \tag{10-9}$$

and
$$x_{B2} = x_2(\xi)$$
$$y_{B2} = y_2(\xi)$$

The range on ξ in the computational plane is

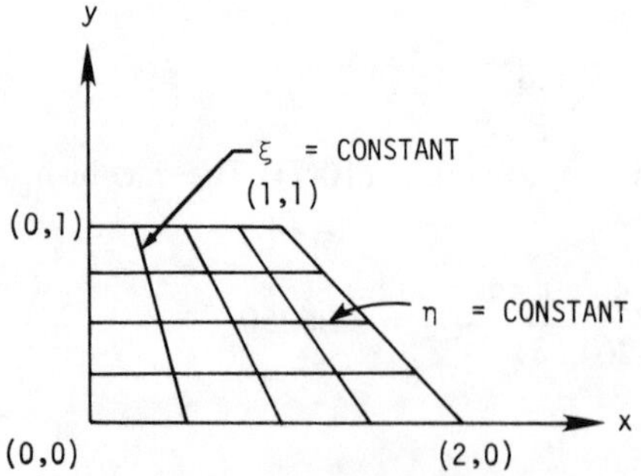

Figure 10-4 Trapezoid to rectangle mapping.

$$0 \leqslant \xi \leqslant 1$$

and the transformation is defined so that at $\eta = 0$

$$x_{B1} = x_1(\xi) = x(\xi, 0)$$
$$y_{B1} = y_1(\xi) = y(\xi, 0)$$

and at $\eta = 1$

$$x_{B2} = x_2(\xi) = x(\xi, 1)$$
$$y_{B2} = y_2(\xi) = y(\xi, 1) \tag{10-10}$$

A function defined on $0 \leqslant \eta \leqslant 1$ with parameters on the two boundaries completes the algebraic relation. This is chosen to be of the form

$$x = x(\xi, \eta) = F\left(x_1, \frac{dx_1}{d\eta}, \ldots, x_2, \frac{dx_2}{d\eta}, \ldots\right)$$
$$y = y(\xi, \eta) = G\left(y_1, \frac{dy_1}{d\eta}, \ldots, y_2, \frac{dy_2}{d\eta}, \ldots\right) \tag{10-11}$$

Smith and Weigel suggest the use of either linear or cubic polynomials. If we choose the linear function then

$$x = x_1(\xi)(1 - \eta) + x_2(\xi)\eta$$
$$y = y_1(\xi)(1 - \eta) + y_2(\xi)\eta \tag{10-12}$$

Example 10-4 To demonstrate the use of this approach, suppose we wish to map the trapezoid defined by the equations

$$x = 0$$
$$x = 1$$
$$y = 0$$
$$y = 1 + x$$

into the computational plane. In this case the upper and lower boundaries may be written

$$x_{B1} = x_1(\xi) = \xi$$
$$y_{B1} = y_1(\xi) = 0$$
$$x_{B2} = x_2(\xi) = \xi$$
$$y_{B2} = y_2(\xi) = 1 + \xi$$

This produces the mapping required in Eq. (10-12) and is of the form

$$x = \xi$$
$$y = (1 + \xi)\eta \tag{10-13}$$

This parameterization produces the simple normalizing transformation discussed earlier in this section. In this example, both the right and left boundaries are also correctly mapped. This is coincidental and will not occur in more general problems. A different point distribution can be obtained by choosing a nonlinear function for the boundary parameterization. For example, if

$$x_1 = \xi^2$$

$$x_2 = \xi^2$$

then

$$x = \xi^2$$

$$y = \eta(1 + \xi^2)$$

If cubic polynomials are used the form of the transformation is altered to

$$\begin{aligned} x &= x_1(\xi)f_1(\eta) + x_2(\xi)f_2(\eta) + \frac{dx_1}{d\eta}(\xi)f_3(\eta) + \frac{dx_2}{d\eta}(\xi)f_4(\eta) \\ y &= y_1(\xi)f_1(\eta) + y_2(\xi)f_2(\eta) + \frac{dy_1}{d\eta}(\xi)f_3(\eta) + \frac{dy_2}{d\eta}(\xi)f_4(\eta) \end{aligned} \tag{10-14}$$

where

$$f_1(\eta) = 2\eta^3 - 3\eta^2 + 1$$

$$f_2(\eta) = -2\eta^3 + 3\eta^2$$

$$f_3(\eta) = \eta^3 - 2\eta^2 + \eta$$

$$f_4(\eta) = \eta^3 - \eta^2$$

The derivatives of the boundary in the physical plane provide even more flexibility in the mapping. For instance, orthogonality at the boundary can be forced in the physical plane [see Kowalski (1980) for details].

In most problems, the boundaries are not analytic functions but are simply prescribed as a set of data points. In this case, the boundary must be approximated by a curve fitting procedure to employ algebraic mappings. Eiseman and Smith (1980) discuss possible methods of accomplishing this and particularly recommend tension splines. Tension splines are suggested because higher-order approximations including cubic splines tend to produce wiggles in the boundary. The tension parameter in the tension spline allows control of this phenomenon.

The two-boundary or two-surface method presented in this section is only one type of algebraic grid generation scheme. A number of other techniques have been used such as the multisurface method of Eiseman (1979). This scheme is similar to the two-surface approach but also defines a grid structure on any number of intermediate control surfaces. Other methods such as transfinite interpolation used by Gordon and Hall (1973) have recently received attention. This method is described in detail by Rizzi and Eriksson (1981) and is similar to the two-surface technique when both coordinate positions and derivatives are specified on the boundaries. The main advantages of using algebraic mappings are that they are direct and the metrics of the transformation can be analytically computed. In addition, they can be applied to three-dimensional problems in a straightforward way. However, a certain amount of ingenuity is required to produce a grid with points properly positioned.

10-3 DIFFERENTIAL EQUATION METHODS

In the previous section algebraic methods were presented for generating suitable grids. Any procedure that results in an acceptable grid is a valid one. One of the most highly developed techniques for generating acceptable grids is the differential equation method. If a differential equation is used to generate a grid, we can exploit the properties of the solution of the grid generating equation in producing the mesh. Laplace's and Poisson's equations have been extensively used for this purpose.

The choice of Laplace's equation can be better understood by considering the solution of a steady heat conduction problem in two dimensions with Dirichlet boundary conditions. The solution of this problem produces isotherms which are smooth (C^{II} continuous) and are nonintersecting. The number of isotherms in a given region can be increased by adding a source term. If the isotherms were used as grid lines, they would be smooth, nonintersecting, and could be densely packed in any region by control of the source term.

Thompson et al. (1974) have worked extensively on using elliptic PDE's to generate grids. This procedure is similar to that used by Winslow (1966) and transforms the physical plane into the computational plane where the mapping is controlled by a Poisson equation. This mapping is constructed by specifying the desired grid points (x, y) on the boundary of the physical domain. The distribution of points on the interior is then determined by solving

$$\begin{aligned} \xi_{xx} + \xi_{yy} &= P(\xi, \eta) \\ \eta_{xx} + \eta_{yy} &= Q(\xi, \eta) \end{aligned} \tag{10-15}$$

where (ξ, η) represent the coordinates in the computational domain and P and Q are terms which control the point spacing on the interior of D. Equations (10-15) are then transformed to computational space by interchanging the roles of the independent and dependent variables. This yields a system of two elliptic equations of the form

$$\begin{aligned} \alpha x_{\xi\xi} - 2\beta x_{\xi\eta} + \gamma x_{\eta\eta} &= -J^2(Px_\xi + Qx_\eta) \\ \alpha y_{\xi\xi} - 2\beta y_{\xi\eta} + \gamma y_{\eta\eta} &= -J^2(Py_\xi + Qy_\eta) \end{aligned} \tag{10-16}$$

where

$$\alpha = x_\eta^2 + y_\eta^2$$

$$\beta = x_\xi x_\eta + y_\xi y_\eta$$

$$\gamma = x_\xi^2 + y_\xi^2$$

$$J = \frac{\partial(x, y)}{\partial(\xi, \eta)} = x_\xi y_\eta - x_\eta y_\xi$$

This system of equations is solved on a uniformly-spaced grid in the computational plane. This provides the (x, y) coordinates of each point in physical space. For simply connected regions, Dirichlet boundary conditions can be used at all boundary points. The advantages of using this technique to generate a computational mesh are many. The resulting grid is smooth, the transformation is one to one and complex boundaries are easily treated. Of course there are some disadvantages. Specification of P and Q is not an easy task, grid point control on the interior is difficult to achieve and boundaries may be changing with time. In the latter case, the grid must be computed after each time step. This can consume large amounts of computer time.

A simple example demonstrating the application of the Thompson scheme is shown in Fig. 10-5. The region between two concentric circles is mapped into the computational domain and the resulting constant ξ and η surfaces in physical space are shown. The inner circle is of radius r_0 and the outer circle is of radius r_1. For this problem, the circle is cut at $\theta = 0$ and mapped into the region between 1 and $\xi_{\max}$ and 1 and $\eta_{\max}$ in computational space. In this problem the mapping is determined by a solution of Laplace's equation

$$\nabla^2 \xi = 0$$

$$\nabla^2 \eta = 0$$

subject to boundary conditions

$$r = r_0 \qquad \eta = 1$$

$$r = r_1 \qquad \eta = \eta_{\max}$$

and

$$\theta = 0 \qquad \xi = 1$$

$$\theta = 2\pi \qquad \xi = \xi_{\max}$$

The solution is of the form

$$x = R \cos \phi$$

$$y = R \sin \phi$$

where

$$R = r_0 \left(\frac{r_1}{r_0} \right)^{(\eta - 1)/(\eta_{\max} - 1)}$$

$$\phi = \left(\frac{\xi - 1}{\xi_{\max} - 1} \right) 2\pi$$

This solution is interesting in that a uniform grid in the physical domain is not achieved in this case. The distribution in the radial direction is a series of concentric circles. To obtain the mapping with a series of uniformly spaced concentric circles, $P = 0$ and $Q = 1/\eta$ (see Prob. 10-9).

As previously noted, one of the difficulties with this scheme is point control on the interior of the domain. This requires that methods for developing P and Q be devised in order to obtain the desired point distribution. Middlecoff and Thomas (1979) have developed a method which provides approximate control of point

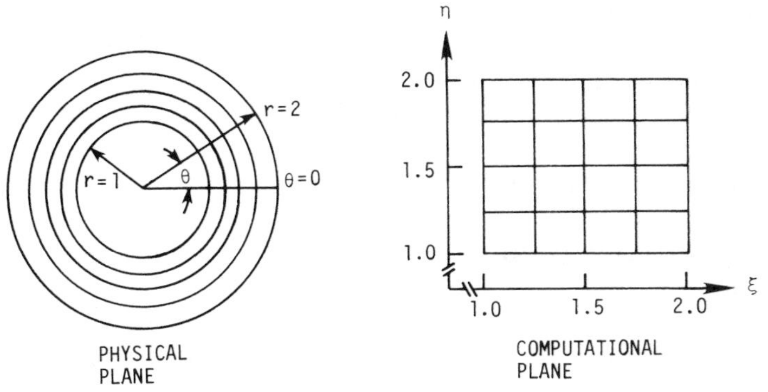

Figure 10-5 Application of Thompson scheme.

spacing by evaluating P and Q according to the desired point distribution on the boundary.

In order to demonstrate this idea, we suppose that a solution of Eq. (10-15) is required subject to Dirichlet boundary conditions. We elect to write P and Q in the form

$$\begin{aligned} P &= \phi(\xi, \eta)(\xi_x^2 + \xi_y^2) \\ Q &= \psi(\xi, \eta)(\eta_x^2 + \eta_y^2) \end{aligned} \tag{10-17}$$

where ϕ and ψ will be specified later through the boundary conditions. With this convention, our original system [Eq. (10-15)] may be written

$$\begin{aligned} \alpha(x_{\xi\xi} + \phi x_\xi) - 2\beta x_{\xi\eta} + \gamma(x_{\eta\eta} + \psi x_\eta) &= 0 \\ \alpha(y_{\xi\xi} + \phi y_\xi) - 2\beta y_{\xi\eta} + \gamma(y_{\eta\eta} + \psi y_\eta) &= 0 \end{aligned} \tag{10-18}$$

On the boundaries, the ϕ and ψ functions are determined by setting the quantities in parentheses equal to zero. For example, along $\xi =$ constant boundaries, Middlecoff and Thomas require

$$\begin{aligned} x_{\eta\eta} + \psi x_\eta &= 0 \\ y_{\eta\eta} + \psi y_\eta &= 0 \end{aligned} \tag{10-19}$$

and along $\eta =$ constant boundaries

$$\begin{aligned} x_{\xi\xi} + \phi x_\xi &= 0 \\ y_{\xi\xi} + \phi y_\xi &= 0 \end{aligned} \tag{10-20}$$

Since x and y are known at all boundary points, the ϕ and ψ functions can be determined by using central differences for the required derivatives. It should be noted that ϕ and ψ are determined from one of the two expressions in each of the sets given by Eqs. (10-19) and (10-20). In general, if ϕ is determined from one of the expressions in Eq. (10-20), the other will not be zero for that choice of ϕ. The same can be said of ψ as determined from Eq. (10-19). Using this approach, the two generating functions are determined on the boundary and ϕ and ψ on the interior are obtained by simply extrapolating ϕ and ψ onto the interior mesh points. This method provides a means of control for the interior point distribution based upon requirements at the boundary. Numerous schemes for point control have been proposed. For a summary of some of the more successful ideas, the papers by Thompson et al. (1975) and Thompson (1980) are recommended.

Other types of partial differential equations can be used to generate grids. In a recent paper, Steger and Sorenson (1980) described a method using a system of hyperbolic equations to generate a mesh around a configuration where the outer boundary is not specified in advance. In this application, the body forms the inner boundary and the hyperbolic system is marched outward in the hyperbolic direction which requires an open domain. Steger and Sorenson propose an arc length orthogonality scheme and a volume orthogonality scheme. Only the latter will be presented in detail here.

In a two-dimensional problem, the Jacobian of the transformation controls the magnification of area elements between the physical and computational planes. If we imagine that mesh spacing in computational space is given by $\Delta\xi = \Delta\eta = 1$ then the area elements are also one unit in size. The quantity

$$x_\xi y_\eta - y_\xi x_\eta = J \tag{10-21}$$

then represents the area in physical space for a given area element in computational space. If J is specified as a function of position then Eq. (10-21) can be used as a single equation specifying grid control in the physical plane. A second equation is obtained by requiring that the grid lines be orthogonal at the boundary in physical space. Along a boundary where $\xi(x, y) = \text{constant}$ we may write

$$d\xi = 0 = \xi_x\, dx + \xi_y\, dy$$

or

$$\left.\frac{dy}{dx}\right)_{\xi=\text{constant}} = -\frac{\xi_x}{\xi_y} = \frac{y_\eta}{x_\eta} \tag{10-22}$$

Along an $\eta = \text{constant}$ surface

$$\left.\frac{dy}{dx}\right)_{\eta=\text{constant}} = -\frac{\eta_x}{\eta_y} = \frac{y_\xi}{x_\xi} \tag{10-23}$$

If we require that ξ and η surfaces be perpendicular, the slopes must be negative reciprocals. This requirement becomes

$$x_\xi x_\eta + y_\xi y_\eta = 0 \tag{10-24}$$

The system given by Eqs. (10-21) and (10-24) is linearized by expanding about a known state denoted by the tilde notation. Using this convention we may linearize one of the terms in Eq. (10-24) as

$$\begin{aligned} x_\xi y_\eta &= (\tilde{x} + x - \tilde{x})_\xi (\tilde{y} + y - \tilde{y})_\eta \\ &= \tilde{x}_\xi \tilde{y}_\eta + \tilde{y}_\eta (x_\xi - \tilde{x}_\xi) + \tilde{x}_\xi (y_\eta - \tilde{y}_\eta) + O(\Delta^2) \\ &= \tilde{y}_\eta x_\xi + \tilde{x}_\xi y_\eta - \tilde{x}_\xi \tilde{y}_\eta + O(\Delta^2) \end{aligned} \tag{10-25}$$

If the other terms are linearized in a similar manner, we obtain

$$[A]\mathbf{w}_\xi + [B]\mathbf{w}_\eta = \mathbf{f} \tag{10-26}$$

where

$$\mathbf{w} = \begin{bmatrix} x \\ y \end{bmatrix}$$

$$[A] = \begin{bmatrix} \tilde{x}_\eta & \tilde{y}_\eta \\ \tilde{y}_\eta & -\tilde{y}_\eta \end{bmatrix}, \quad [B] = \begin{bmatrix} \tilde{x}_\xi & \tilde{y}_\xi \\ -\tilde{y}_\xi & \tilde{x}_\xi \end{bmatrix}, \quad \mathbf{f} = \begin{bmatrix} 0 \\ J + \tilde{J} \end{bmatrix} \tag{10-27}$$

The eigenvalues of $[B]^{-1}[A]$ must be real if the system is hyperbolic in the η direction. These eigenvalues are

$$\lambda_{1,2} = \pm \sqrt{\frac{\tilde{x}_\eta^2 + \tilde{y}_\eta^2}{\tilde{x}_\xi^2 + \tilde{y}_\xi^2}} \tag{10-28}$$

This shows that Eq. (10-26) is hyperbolic in the η direction and can be marched in η so long as $\tilde{x}_\xi^2 + \tilde{y}_\xi^2 \neq 0$.

The procedure to use in generating a grid with this scheme is to assume the body is the $\eta = 0$ surface and specify the distribution of points along the body. Next the quantity J in Eq. (10-21) is required. Steger and Sorenson suggest that J be determined by laying out a straight line with length equal to that of the body surface, (l), and lay out the body point distribution on this line. Next a line parallel to the first is drawn at an $\eta =$ constant surface as desired. Once this is done, the quantity J is easily determined by estimating the area elements of the grid. This procedure is illustrated in Fig. 10-6. The system of governing equations given by Eq. (10-27) is now solved using any standard method for solving systems of hyperbolic PDE's.

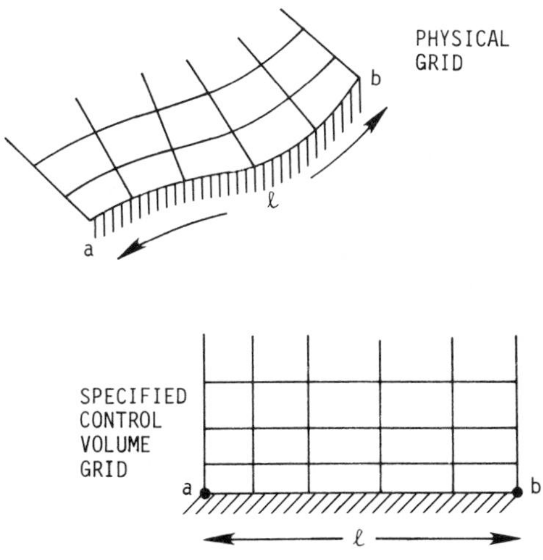

Figure 10-6 Area element computation.

Since we specify J in this scheme, a smoothly varying grid is obtained if J is well chosen. However, poor selection of the J variation leads to possible "shocks" or discontinuous propagation of this information through the mesh. It is also true that discontinuous boundary data are propagated in the mesh. On the other hand, the mesh is orthogonal and is generated very rapidly. Figure 10-7 shows the grid generated about a typical airfoil shape. In this case, points have been clustered near the body in order to permit resolution of the viscous boundary layer.

From the discussion of techniques presented in this section, it is clear that an unlimited number of schemes for generating grids can be developed. Any system which yields a suitable grid is acceptable. All the methods presented thus far in this chapter require that the grid points be determined initially before any calculations have been attempted in solving the PDE's governing the fluid flow. In the next section, we will present some ideas for mesh generation where the grid point locations evolve as part of the solution of the problem.

10-4 ADAPTIVE GRIDS

Techniques for generating computational grids as a prelude to numerically solving a PDE were presented in the previous section. One problem in solving a PDE with a fixed grid is that points are distributed in the physical domain before details of the solution are known. As a consequence, the grid may not be the best one for the particular problem.

The term "best" is subject to interpretation. In many problems, the main interest is in constructing a grid which can move and adjust as a domain changes shape. An example is the supersonic blunt body problem. The shock wave is usually fit as a boundary and this boundary moves in time when a time asymptotic solution

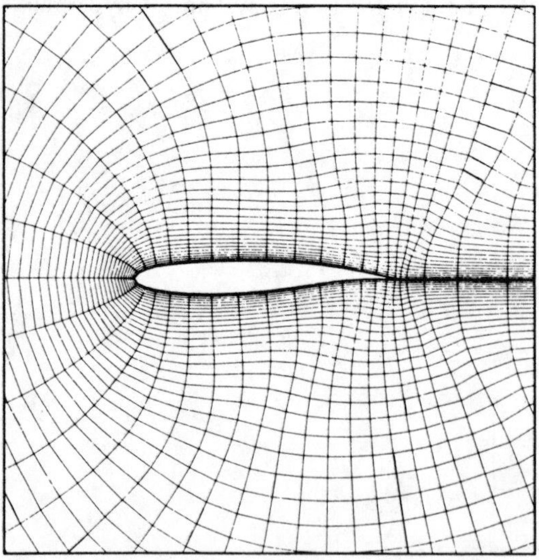

Figure 10-7 Grid for an airfoil configuration.

of the governing equations is desired. In this case the motion of points on the interior can be scaled to the motion of the boundary and acceptable results are obtained. For many problems, this approach for developing a grid which gives an acceptable solution is sufficient. In other examples, we may wish to move the grid points in the domain to provide adequate resolution in the flowfield for either fixed or moving boundaries. This offers the advantage of moving the grid points to regions where large gradients occur without a priori knowledge of the solution. Such regions are often the result of widely varying length scales in the flowfield. Of course, a suitable technique for adjusting the mesh must be devised.

A better strategy is to adjust the mesh to reduce the error in the solution. In this way, a fixed number of grid points can be adjusted to provide a "best" solution using the error measure selected. This usually eliminates the question of resolution because regions where high resolution is required are those where large errors are incurred when a fixed mesh is used. In this section we will give examples of adaptive mesh schemes which are aimed at the questions of resolution and error reduction.

When a technique for generating a solution-adaptive grid is desired, there are two viewpoints that must be considered. In order to understand the two viewpoints we again consider the simple one-dimensional wave equation

$$\frac{\partial u}{\partial t} + c\,\frac{\partial u}{\partial x} = 0 \tag{10-29}$$

If computational space is denoted by (τ, ξ) this equation may be written

$$\frac{\partial u}{\partial \tau} + (\xi_t + c\xi_x)\,\frac{\partial u}{\partial \xi} = 0 \tag{10-30}$$

where the transformation is given by

$$\begin{aligned} \tau &= t \\ \xi &= \xi(t, x) \end{aligned} \tag{10-31}$$

The terms ξ_t and ξ_x in Eq. (10-30) provide the coupling between the physical and the transformed plane. These terms are used to determine the transformation relating the two domains of interest. If we solve for x_τ and x_ξ we obtain

$$\begin{aligned} x_\xi &= \frac{1}{\xi_x} \\ x_\tau &= -\frac{\xi_t}{\xi_x} \end{aligned} \tag{10-32}$$

The quantities x_τ and x_ξ represent the grid point speed and spacing respectively in physical space.

One technique for generating an adaptive grid, used by Dwyer et al. (1979, 1980) and by Klopfer and McRae (1981a), is to position the points after each integration step or after a selected number of steps have been taken. Using this technique, the x's or coordinates of the points in physical space can be placed where they are needed using resolution or any criteria desired. Since x_ξ is then known, ξ_x is also known and x_τ is obtained by using a backward-difference approximation. The term ξ_t is then computed and the quantities required for continued integration of Eq. (10-30) are known. When this approach is used the grid point speed is lagged in the governing equation.

The other way to view the adaptive grid problem is to postulate a law governing the grid speeds. This law can use resolution, error, or any criteria as a means of generating grid speeds. This expression is conveniently used in the computational plane by prescribing ξ_t. At any time, the term ξ_t is known and the grid point speed in physical space, x_τ, is obtained from Eq. (10-32). Since x_τ is known, this can be integrated with the governing PDE to yield the new point locations. The advantage of this technique is that the grid point locations and speeds are time accurate in that no time lag is inherent in the construction of these terms.

Many techniques for constructing adaptive grids have been used and these approaches appear to be quite different. However, closer scrutiny of the basic ideas shows that the fundamental concept employed in a large percentage of these schemes is the same. In this section, variational, equidistribution, and grid speed methods are briefly explored and some results obtained using adaptive grids are presented.

10-4.1 A Variational Method

Brackbill and Saltzman (1980) and Brackbill (1982) have developed a new technique for constructing an adaptive grid using a variational approach. In their scheme, a function which contains a measure of grid smoothness, orthogonality, and volume variation is minimized using variational principles. The smoothness of the transformation is represented by the integral

$$I_s = \int_D [(\nabla\xi)^2 + (\nabla\eta)^2]\, dV \tag{10-33}$$

A measure of orthogonality is provided by

$$I_0 = \int_D (\nabla\xi \cdot \nabla\eta)^2 J^3\, dV \tag{10-34}$$

and the volume measure is given as

$$I_v = \int_D wJ\, dV \tag{10-35}$$

where w is a given weighting function.

The transformation relating D and CD is determined by minimizing a linear combination of the above three integrals. This linear combination with coefficient multipliers λ_v and λ_0 is written

$$I = I_s + \lambda_v I_v + \lambda_0 I_0 \tag{10-36}$$

In order to minimize I, the Euler-Lagrange equations must be formed, Weinstock (1952). As an example, the smoothness measure, Eq. (10-33), may be written

$$I_s = \iint \left(\frac{x_\xi^2 + x_\eta^2 + y_\xi^2 + y_\eta^2}{J} \right) d\xi \, d\eta \tag{10-37}$$

when the variables are interchanged and the integration is performed in computational space. If we construct the Euler-Lagrange equations corresponding to I_s, they are of the form

$$\begin{aligned} \left(\frac{\partial}{\partial x} - \frac{\partial}{\partial \xi}\frac{\partial}{\partial x_\xi} - \frac{\partial}{\partial \eta}\frac{\partial}{\partial x_\eta} \right) \left(\frac{x_\xi^2 + x_\eta^2 + y_\xi^2 + y_\eta^2}{J} \right) &= 0 \\ \left(\frac{\partial}{\partial y} - \frac{\partial}{\partial \xi}\frac{\partial}{\partial y_\xi} - \frac{\partial}{\partial \eta}\frac{\partial}{\partial y_\eta} \right) \left(\frac{x_\xi^2 + x_\eta^2 + y_\xi^2 + y_\eta^2}{J} \right) &= 0 \end{aligned} \tag{10-38}$$

If the differentiation is performed, these expressions may be written

$$\begin{aligned} A(\alpha x_{\xi\xi} - 2\beta x_{\xi\eta} + \gamma x_{\eta\eta}) - B(\alpha y_{\xi\xi} - 2\beta y_{\xi\eta} + \gamma y_{\eta\eta}) &= 0 \\ -B(\alpha x_{\xi\xi} - 2\beta x_{\xi\eta} + \gamma x_{\eta\eta}) + C(\alpha y_{\xi\xi} - 2\beta y_{\xi\eta} + \gamma y_{\eta\eta}) &= 0 \end{aligned} \tag{10-39}$$

The coefficients A, B, C, α, β, and γ are functions of the metrics and their evaluation is left as an exercise (see Prob. 10-16). If

$$B^2 - AC \neq 0$$

these equations may be written as

$$\begin{aligned} \alpha x_{\xi\xi} - 2\beta x_{\xi\eta} + \gamma x_{\eta\eta} &= 0 \\ \alpha y_{\xi\xi} - 2\beta y_{\xi\eta} + \gamma y_{\eta\eta} &= 0 \end{aligned} \tag{10-40}$$

This is the form of the original mapping given by Winslow and is also the basic system of equations for Thompson's work. If I as defined in Eq. (10-36) is minimized, each of the integrals, I_v and I_0, contribute terms to a significantly more complicated set of Euler-Lagrange equations than those given in Eq. (10-40).

The use of a variational approach provides a solid mathematical basis for the grid but also entails additional effort in solving more PDE's. The Euler-Lagrange equations must be solved in addition to those governing the fluid motion. In the example shown here, the adaptive grid is constructed by implementing a new mesh after each iteration or time step and computing the grid speed by using a backward difference. The variational approach clearly offers a powerful method for constructing computational grids. The disadvantage is that a considerable effort must be expended in solving the equations which govern the grid generation. If a linear combination of the integrals of Eq. (10-36) is used, the λ's must also be selected. However, some remarkable results have been obtained with the proper choice of these coefficients.

10-4.2 Equidistribution Schemes

Many applications of adaptive grids require grid motion along one coordinate. For this reason, we consider minimizing the one-dimensional form of the functional, I_v, defined in Eq. (10-35)

$$I_v = \int_D w x_\xi \, dx \tag{10-41}$$

The Euler-Lagrange equation for this integral may be written

$$\left(\frac{\partial}{\partial \xi} - \frac{d}{dx}\frac{\partial}{\partial \xi_x}\right)\left[\frac{w(x)}{\xi_x}\right] = 0 \tag{10-42}$$

Brackbill and Saltzman have shown that a first integral may be directly written in the form

$$w x_\xi^2 = C_1$$

or

$$\sqrt{w}\, x_\xi = w_1 x_\xi = C_1 \tag{10-43}$$

This solution states that the product of the mesh spacing and the weight function, w_1, should remain fixed in physical space. An integral of Eq. (10-43) can be computed to obtain either the physical or computational coordinate. Suppose $x = 0$ when $\xi = 0$ and $x = x_{max}$ when $\xi = \xi_{max}$. If Eq. (10-43) is integrated to obtain the computational coordinate,

$$\xi = \xi_{max} \frac{\int_0^x w_1 \, dx}{\int_0^{x_{max}} w_1 \, dx} \tag{10-44}$$

while the expression

$$x = x_{max} \frac{\int_0^\xi 1/w_1 \, d\xi}{\int_0^{\xi_{max}} 1/w_1 \, d\xi} \tag{10-45}$$

is obtained when the physical coordinate is computed.

Equation (10-44) is the grid adaption law used by Dwyer et al. (1979, 1980). Many of the applications of this law have been in combustion and heat or mass transfer and the results obtained have been very good. The form of the weight function, w_1, was selected to be a linear combination of the derivatives of some dependent variable of interest. If the static temperature is used, w_1 is written

$$w_1 = 1 + a\left|\frac{\partial T}{\partial x}\right| + b\left|\frac{\partial^2 T}{\partial x^2}\right| \tag{10-46}$$

The physical coordinate position given by Eq. (10-45) is the same as that used by Gnoffo (1980). Gnoffo applied this law for grid motion along one coordinate direction to compute solutions of the Navier-Stokes equations. In this case only first derivatives of the dependent variable were used in the weight function.

White (1982) has applied the equidistribution idea to the solution of one-dimensional problems. In his work, the equidistribution law states that the product of arc length and a weight function is held constant. This is also the concept employed in applying either Eq. (10-44) or Eq. (10-45) along one coordinate direction in a two-dimensional problem. However, in White's example, the arc length was taken to be along the solution surface of the dependent variable being monitored. The weight function was selected to be of the form

$$w_1 = 1 + a|\kappa| \tag{10-47}$$

where κ is the curvature of the solution. This approach provides clustering automatically in regions with high gradients and the point density in regions of high curvature can be controlled through the clustering constant, a.

10-4.3 Grid Speed Schemes

In a recent paper, Hindman and Spencer (1983) have developed a grid speed scheme which also incorporates the idea of equidistribution. Since differential equation techniques are the most popular for generating grids, it is reasonable to differentiate Eq. (10-43) to obtain the second-order differential equation satisfied by the grid. This expression may be written

$$x_{\xi\xi} + \frac{w_{1_\xi} x_\xi}{w_1} = 0 \tag{10-48}$$

This is the steady grid equation which may be solved at any time level to determine the grid point distribution. In Hindman and Spencer's approach, the grid speeds are established by differentiating the steady grid equation and solving the resulting equation for x_τ. In this example, the grid speeds are given by a solution of

$$x_{\tau_{\xi\xi}} + \frac{w_{1_\xi}}{w_1} x_{\tau_\xi} = -\frac{x_\xi}{w_1}\left(w_{1_{\xi\tau}} - \frac{w_{1_\xi} w_{1_\tau}}{w_1}\right) \tag{10-49}$$

One way of advancing the grid to the next time level is simply to integrate the grid speeds. The steady grid equation serves only as a starting point to arrive at Eq. (10-49). However, when an adaptive grid is used which causes points to cluster on the interior, the grid tends to relax. A better approach is to solve Eq. (10-49) for the grid speeds and use these values when the partial differential equations governing the physical process are integrated. The grid speeds are then integrated to obtain the approximate mesh position. Next, the steady grid equation [Eq. (10-48)] is solved using the approximate values as an initial estimate. This procedure provides assurance that the grid speeds are correct and the correct steady grid equation is satisfied. The main problem with this procedure is that of establishing a suitable estimate for the time derivative of the weight function, w_{1_τ}. This is most easily accomplished by a numerical approximation. In all cases except that of a simple scalar equation, analytic expressions for this term are very difficult to obtain.

In an earlier paper, Hindman et al. (1979) used a similar technique in developing a modular approach for solving the Euler equations. Different regions of the flow were solved in separate domains coupled through the boundaries. These boundaries could either be permeable or impermeable. In this application, grid point motion was induced only through boundary motion. The grid point speeds were obtained by differentiating the Thompson grid generation equations [Eq. (10-16)] with respect to time. This produces a system of PDE's of the form

$$[s]\,\mathbf{w}_\tau = \mathbf{r} \tag{10-50}$$

where the $\mathbf{w}$ vector is given by

$$\mathbf{w} = \begin{bmatrix} x \\ y \end{bmatrix} \tag{10-51}$$

and the $[s]$ matrix and the $\mathbf{r}$ vector are the coefficient matrix and grid driving function, respectively. The new grid point locations for each iteration or time level were obtained by integrating the grid speeds. This approach worked very well for the cases studied. However, it should be noted that no interior clustering was employed since both P and Q were zero.

Rai and Anderson (1980, 1982) have developed a technique where the grid speeds are governed by estimates of the local errors in the numerical solution. This can be achieved by requiring that points in large error regions attract while points in low error regions repel other points. It is also reasonable to assume that the larger the separation distance between any two points, the smaller the effect they have on each other. If we again denote the physical coordinates by (x, t) and the computational coordinates by (ξ, τ), the grid speed equation can be written in the form

$$\begin{aligned} (\xi_t)_i &= k\left(\sum_{j=i+1}^{n} \frac{|e|_j - |e|_{av}}{r_{ij}^n} - \sum_{j=1}^{i-1} \frac{|e|_j - |e|_{av}}{r_{ij}^n}\right) \\ (x_\tau)_i &= \frac{-(\xi_t)_i}{(\xi_x)_i} \end{aligned} \tag{10-52}$$

In this grid speed equation, e represents some measure of the local solution error and the subscript av means averaged over all mesh points. The constant k is arbitrary because no physical law relates grid speed and error, and r_{ij}^n is the distance between point i and j raised to the power n. One of the main difficulties with this approach is establishing an appropriate error measure. Rai and Anderson suggest using the truncation error terms in the modified equation. They have also used gradients instead of local errors to provide better resolution. Due to the general form of the grid speed given in Eq. (10-52), any reasonable choice for e may be used to control grid motion. Equation (10-52) should be recognized as a scheme for equally distributing error on the grid. As such, the predicted grid speed can be interpreted as a residual in the numerical solution for a grid satisfying some equidistribution law.

While one-dimensional examples in grid generation do not provide insight into the difficulties of treating higher-order problems, they do show the effect produced when an adaptive grid is used. The time-dependent viscous Burgers equation

$$u_t + uu_x = \mu u_{xx} \tag{10-53}$$

with initial conditions

$$u(0, x) = \begin{cases} 1, & x = 0 \\ 0, & 0 < x \leqslant 1 \end{cases}$$

and boundary conditions

$$u(t, 0) = 1$$
$$u(t, 1) = 0$$

was solved using gradients for e in Eq. (10-52). The results obtained by numerically solving Burgers' equation may be compared with the exact solution

$$u = \bar{u} \tanh\left[\frac{\text{Re}}{2}(1 - x)\right]$$

where $\bar{u}$ is obtained from

$$\frac{\bar{u} - 1}{\bar{u} + 1} = e^{-\bar{u}\,\text{Re}}$$

and

$$\text{Re} = \frac{1}{\mu}$$

This solution approaches a steady state in which the slope at the right end of the interval becomes steeper as the Reynolds number increases. Figures 10-8 and 10-9

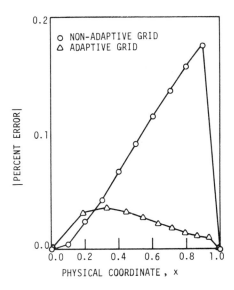

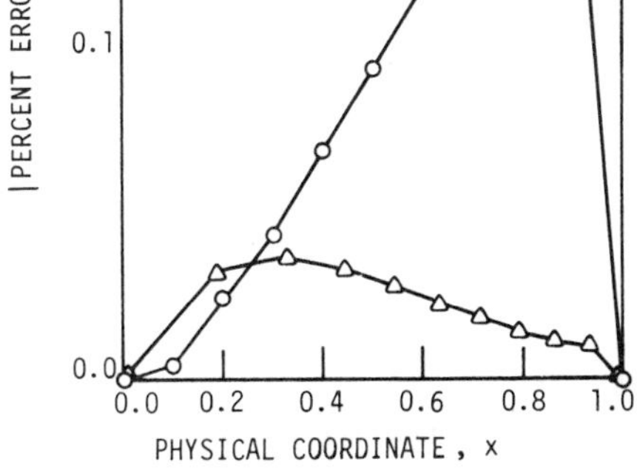

Figure 10-8 Comparison of errors, Re = 2.

show results at two different Reynolds numbers. It is apparent that the error in the solution is reduced for both cases. It should be noted that the solution is smoothed before the gradients are formed. This provides better control for point motion. If the solution is used with no smoothing, very "noisy" grid results are obtained.

The grid speed method is easily adapted to two dimensions. If we use the form given in Eq. (10-52) for ξ_t, then a similar expression for η_t can be derived where the e terms depend only upon the derivatives in the ξ or η directions. In this case η_t is assumed to depend only upon the η derivatives of the dependent variables. For a transformation of the form

$$t = \tau$$
$$\xi = \xi(x, y, t)$$
$$\eta = \eta(x, y, t)$$

we obtain

$$x_\tau = \frac{\xi_y \eta_t - \eta_y \xi_t}{J}$$
$$y_\tau = \frac{\eta_x \xi_t - \xi_x \eta_t}{J} \tag{10-54}$$
$$J = \xi_x \eta_y - \eta_x \xi_y$$

Equations (10-54) permit the grid speeds in physical space to be computed. Figure 10-10 shows an adaptive grid for supersonic inviscid flow over a cylinder. This problem was solved using both a fixed and an adaptive grid. Gradient information was used to drive the grid and a typical comparison is presented in Fig. 10-11. Since the exact solution for this problem is unknown, a calculation based upon a 19 X 19

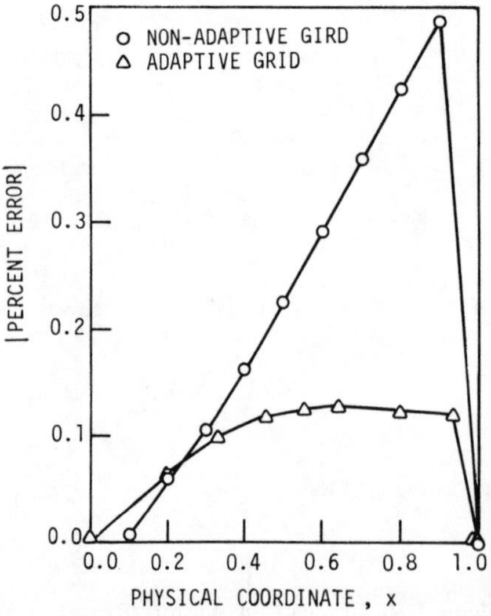

Figure 10-9 Comparison of errors, Re = 3.

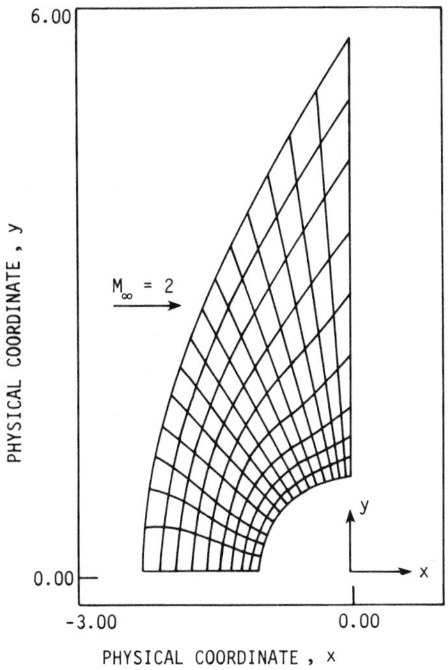

Figure 10-10 Converged adaptive grid for a cylinder.

point fixed grid is used as a basis for comparison. Again we see that errors are reduced even in this case where the adaptive grid is not greatly different from the fixed mesh. Based upon this example, it is apparent that this adaptive grid scheme is applicable to three-dimensional flows. However, another point speed equation is required to generate a full 3-D grid.

An appropriate measure of error for driving the grid is sometimes difficult to obtain. As noted earlier, the most appropriate measure seems to be the first error term of the modified PDE's. Klopfer and McRae (1981b) have derived the first truncation error terms of the modified equations obtained by applying a finite-difference method to the Euler equations. They have used this as an error measure

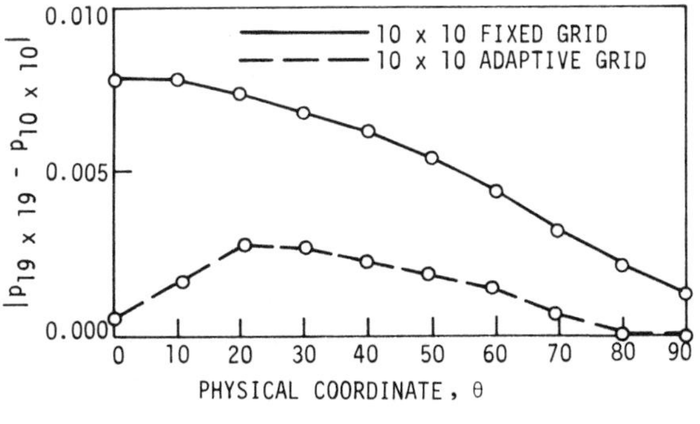

Figure 10-11 Pressure comparison for cylinder problem.

for point redistribution in one-dimensional examples. This results in more accurate error representations in some cases and may prove useful even in multidimensional problems. The error term of interest in the modified PDE is always the first one. The order of the derivative associated with this term depends upon the finite-difference method being employed. For example, if a second-order scheme is applied to a first-order equation, the first error term is a third derivative term and the estimate of local error should be proportional to third derivatives. As a result, the grid for a given problem will be different and depend upon the numerical method selected. However, the grid will still be error reducing.

Another problem of interest in which an adaptive grid scheme is useful is that of shock alignment. Suppose we seek a time asymptotic solution to the two-dimensional Euler equations. The Rankine-Hugoniot conditions for the steady flow problem form the required relations of the flow variables across any discontinuity and the weak solution requirement on the PDE provides the mathematical relation between the variables and the shock slope. If the time asymptotic form of the equations is

$$\frac{\partial \mathbf{E}}{\partial x} + \frac{\partial \mathbf{F}}{\partial y} = 0 \tag{10-55}$$

the jump condition across the shock wave is given by (Section 4-4)

$$[\mathbf{E}] \cos \alpha_1 + [\mathbf{F}] \cos \alpha_2 = 0 \tag{10-56}$$

where $\cos \alpha_1$ and $\cos \alpha_2$ are the direction cosines between the unit normal to the shock and the x and y axes. If oscillations are to be eliminated, the shock should align in such a way that $\cos \alpha_2 = 0$. If this occurs, we see that

$$[\mathbf{E}] = 0 \tag{10-57}$$

The **E** vector is then continuous across the shock and it is appropriate to apply finite differences. This can be achieved using finite-difference methods that align the mesh with the shock wave. Since we only require that $\cos \alpha_2 = 0$, it is necessary to align only one coordinate with the shock, MacCormack and Paullay (1972).

The technique used by Rai and Anderson (1981) for aligning one coordinate with the shock is based upon generating grid speeds which produce an effective rotation of line segments connecting grid points parallel to the level surfaces of a pertinent flow variable. Consider the point distribution in computational space shown in Fig. 10-12. Suppose that point 0 represents the midpoint of the line segment connecting points A and B. Let h represent any physical variable such as the pressure.

Using this technique, the speed of any point C is written as

$$(\xi_c)_t = \frac{K|h_\xi||h_\eta|(-1)^k}{r_{oc}^n} \tag{10-58}$$

where $|h_\xi|$ and $|h_\eta|$ are the absolute values of the gradients of h along and normal to line AB, K and n are constants, r_{oc} is the distance from point 0 to point C, and k is given by

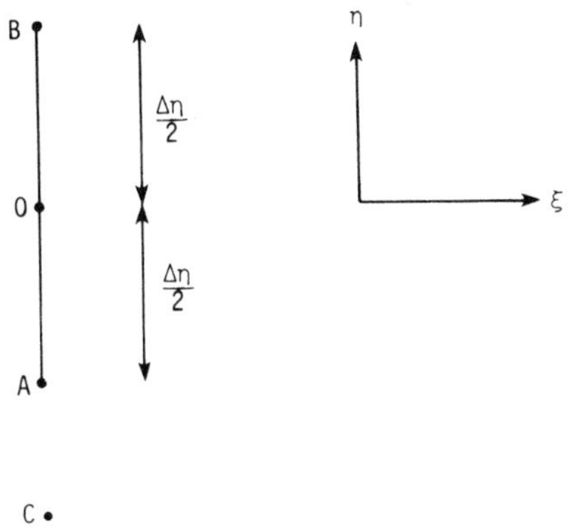

Figure 10-12 Point distribution for shock alignment.

$$k = \begin{cases} 1, & \text{if} \quad \text{sgn}\left(\dfrac{h_\xi}{h_\eta}\right) \text{sgn}\,(\eta_0 - \eta_c) < 0 \\ 2, & \text{if} \quad \text{sgn}\left(\dfrac{h_\xi}{h_\eta}\right) \text{sgn}\,(\eta_0 - \eta_c) > 0 \end{cases} \tag{10-59}$$

where sgn means sign of the argument. If we imagine that h_ξ, h_η are both positive and point C is below point A, the grid speed $(\xi_c)_t$ is also positive indicating a clustering toward the high gradient region. The total grid point speed is obtained by summing the contributions from all line segments in the field. In this manner, a line segment rotates toward the regions of large gradient, but motion ceases when a line segment is parallel to a constant h surface since in that case either h_ξ or $h_\eta = 0$. This permits local coordinate alignment with high gradient regions.

An example demonstrating this technique is shown in Figs. 10-13 and 10-14. Figure 10-13 shows a two-dimensional duct with a surface equation given by

$$y = 0.25 + (y_{\text{exit}} - 0.25)x^2$$

An inlet Mach number of 1.5 was assumed and a back pressure was selected to provide a normal shock at $x = 0.5$ using one-dimensional theory. The two-dimensional time-dependent Euler equations in conservative form were solved to obtain the

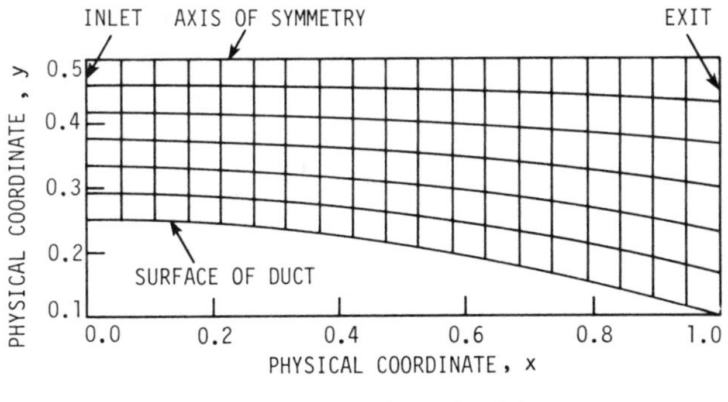

Figure 10-13 Fixed grid for a two-dimensional duct.

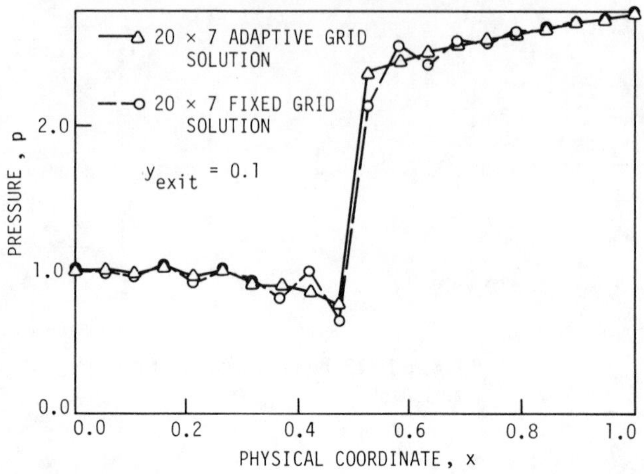

Figure 10-14 Pressure profiles along the axis of symmetry for duct problem.

flowfield. Figure 10-14 shows the static pressure variation along the duct centerline when a fixed grid is used and also when an adaptive grid is used. The oscillations are typical of a second-order shock-capturing technique used with a fixed grid. However, the oscillations do not occur in the steady solution when the shock aligning grid is used.

An additional example consisting of a uniform supersonic freestream with a straight oblique shock can also be used to demonstrate the shock aligning property of the scheme. This is shown in Fig. 10-15. Since this is a fully supersonic problem we experience no difficulty with outflow boundary conditions and solve the time-dependent Euler equations in conservation-law form on the grid shown. The converged grid is presented in Fig. 10-16. The local alignment of the shock and grid is dramatically demonstrated in this figure. The pressure profile at $y = 0$ is shown in Fig. 10-17 for both the fixed and adaptive grids. Again the absence of oscillations in the shock solution using the adaptive grid is striking. The approach of using shock aligning grids clearly is an advantage when shock-capturing schemes are used.

In this section we have introduced the idea of solution adaptive grids. Solutions to PDE's obtained using numerical methods in conjunction with adaptive grids show significant improvements in either accuracy or resolution. In addition to the methods suggested, numerous other approaches can be taken. The ideas used in

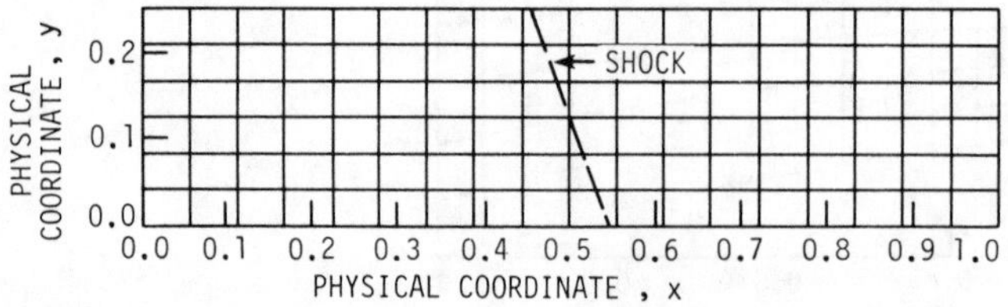

Figure 10-15 Fixed grid for the straight oblique shock problem.

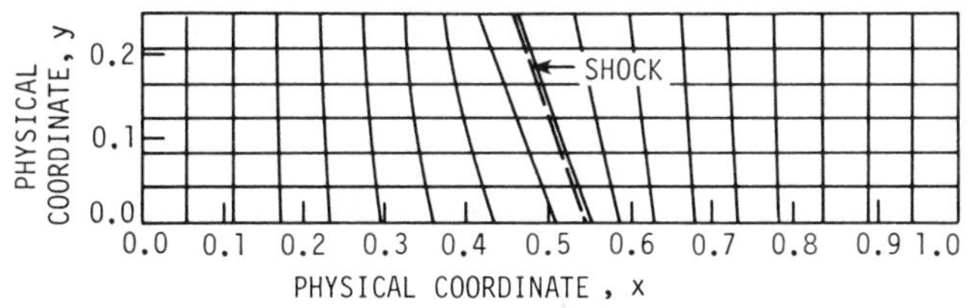

Figure 10-16 Converged grid for the straight oblique shock problem.

construction of adaptive grid techniques are limited only by one's imagination, and any scheme that works in the sense of providing a better solution is a good one.

10-5 OTHER CONSIDERATIONS

We conclude this chapter on grid generation by noting that a grid which is well designed for a particular problem must be appropriately coupled to the governing partial differential equations. This coupling is provided through the transformation metrics as they appear in the PDE system.

An example of proper treatment of the geometry is given by careful examination of the Viviand strong conservation-law form of the governing equations given in Eq. (5-240). Suppose the original system was a rectangular system before the transformation to (τ, ξ, η, ζ) space and that we consider an expanded form of the inviscid equations for simplicity. In a two-dimensional setting, Eq. (5-240) may be written

$$\frac{\partial}{\partial \tau}\left(\frac{\mathbf{U}}{J}\right) + \frac{\partial}{\partial \xi}\left[(y_\tau x_\eta - x_\tau y_\eta)\mathbf{U} + y_\eta \mathbf{E} - x_\eta \mathbf{F}\right]$$
$$+ \frac{\partial}{\partial \eta}\left[(x_\tau y_\xi - y_\tau x_\xi)\mathbf{U} - y_\xi \mathbf{E} + x_\xi \mathbf{F}\right] = 0 \qquad (10\text{-}60)$$

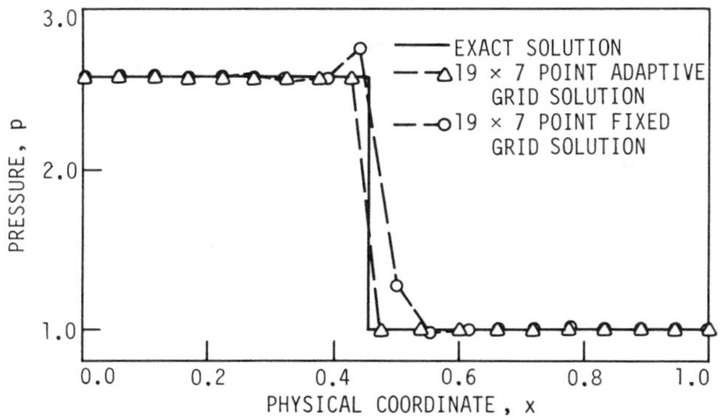

Figure 10-17 Pressure profiles at $y = 0$ for the straight oblique shock problem.

We now replace the differential operator with a finite-difference operator Γ and apply Eq. (10-60) in finite-difference form to a flow which is uniform everywhere. This leads to the requirement that

$$\Gamma_\tau\left(\frac{1}{J}\right) + \Gamma_\xi(y_\tau x_\eta - x_\tau y_\eta)^{(1)} + \Gamma_\eta(x_\tau y_\xi - y_\tau x_\xi)^{(1)} = 0 \tag{10-61}$$

$$\Gamma_\xi(y_\eta^{(2)}) - \Gamma_\eta(y_\xi^{(2)}) = 0 \tag{10-62}$$

$$\Gamma_\xi(x_\eta^{(2)}) - \Gamma_\eta(x_\xi^{(2)}) = 0 \tag{10-63}$$

The subscript on the finite-difference operator indicates the appropriate time or space direction and the superscripts (1) and (2) identify two different numerical representations of the same quantity.

Equations (10-62) and (10-63) are identically satisfied if the same finite-difference scheme is used as is employed in Eq. (10-61). However, Eq. (10-61) is a numerical form of an identity called the geometric conservation law by Thomas and Lombard (1978). The differential equation equivalent of this expression

$$\frac{\partial}{\partial \tau}\left(\frac{1}{J}\right) + \frac{\partial}{\partial \xi}(y_\tau x_\eta - x_\tau y_\eta) + \frac{\partial}{\partial \eta}(x_\tau y_\xi - y_\tau x_\xi) = 0$$

must be differenced exactly the same way as the flow equations given in Eq. (10-61). Obviously this equation provides no useful information if the grid is not changing with time.

Equations (10-61)-(10-63) are a result of the form of Eq. (10-60). If we had not used the Viviand strong conservation form but had used the weak conservation form, a different requirement on the metrics would have resulted. If the non-conservative form of the governing equations is used, no special geometry differencing requirements arise as a consequence of geometry.

The two sets of metrics identified by the superscripts (1) and (2) need further discussion. The metrics identified by (2) must be calculated in a manner consistent with the constraints given in Eqs. (10-62) and (10-63). The calculation of the (1) metrics is still a free choice. Hindman et al. (1981) have shown that the correct way to compute the (1) metrics is dictated by the accuracy of the integrated Jacobian from Eq. (10-61) compared to the actual values from the mapping.

The example presented above shows that the form of the governing equations can impose additional constraints on the manner in which the geometry is treated. Hindman (1981) suggests that the expanded or chain rule form of the governing equations be used even for shock-capturing applications. The expanded form does not incorporate the metrics in the flux terms and as a consequence no special geometric constraints arise. This example should serve as a warning that extreme caution is necessary in solving any system where the grid generator and the original partial differential equations are coupled.

PROBLEMS

10-1 Verify the equations for the transformation metrics given in Eq. (10-7).

10-2 Suppose that a physical domain is defined on the interval $0 \leqslant x \leqslant 1$ with an upper boundary given by

$$y_{\text{upper}} = 1 + 0.2 \sin \pi x$$

and a lower boundary given by

$$y_{\text{lower}} = 0.1 \cos \pi x$$

Devise a transformation which provides a uniform distribution of mesh points between the upper and lower boundaries. Use a simple normalizing transformation.

10-3 In Prob. 10-2 the interval was defined by two $x =$ constant lines. If the left boundary is defined as

$$y_{\text{L}} = 10x$$

and the right boundary is defined by

$$y_{\text{R}} = 4(x - 1)$$

with the same upper and lower boundaries, determine a normalizing transformation to provide equal grid spacing in the physical plane. Why does this become so much more complicated than the transformation of Prob. 10-2?

10-4 Work Prob. 10-2 using the algebraic method demonstrated in Example 10-4. Use linear functions to verify your results and then use cubic functions.

10-5 Work Prob. 10-3 using linear functions with the method given in Example 10-4.

10-6 Suppose that you are required to solve a system of partial differential equations in (t, x, y) on the rectangular domain

$$0 \leqslant x \leqslant 1$$

$$0 \leqslant y \leqslant 1$$

A surface $F(t, x, y) = 0$ is to be tracked similar to a shock and computed as part of the solution. Devise a transformation which converts the physical plane into two rectangular computational domains joined at the boundary $F(t, x, y) = 0$. Assume that the surface is smooth and always intersects the left and right boundaries in physical space.

10-7 Verify the transformation given in Eq. (10-14) and the associated f_i functions.

10-8 The Thompson scheme for generating grids is based upon Eq. (10-15). Derive the computational domain equations given in Eq. (10-16).

10-9 Show that the mapping governed by the differential equations

$$\nabla^2 \xi = 0$$

$$\nabla^2 \eta = \frac{1}{\eta}$$

maps uniformly spaced circles in physical space into a uniform rectangular grid in the computational plane.

10-10 Show that a solution of the Cauchy-Riemann equations is a solution of Laplace's equation but the reverse is not necessarily true.

10-11 Repeat Prob. 10-3 and use the Thompson technique to obtain the mapping using the method of Middlecoff and Thomas [Eq. (10-19) and Eq. (10-20)] to effectively determine P and Q. Discuss your result and point out any difficulties encountered in establishing your choice in selecting ϕ and ψ.

10-12 The one-dimensional Burgers equation is given in Eq. (10-53) and the steady analytic solution for $0 \leqslant x \leqslant 1$ is also presented. Use the adaptive grid scheme proposed by Dwyer and solve this problem numerically. Use any reasonable criteria for establishing mesh spacing in the physical domain. Use a second-order finite-difference method.

10-13 Repeat Prob. 10-12 using the method of Rai and Anderson. Use gradient information to drive the grid. Repeat this problem using third derivative information.

10-14 Verify the grid speed for two dimensions given in Eq. (10-54).

10-15 Show that the requirements of Eqs. (10-61)–(10-63) are correct.

10-16 Complete the differentiation indicated in Eq. (10-38), and determine the coefficients identified in Eq. (10-39).

10-17 Construct the Euler-Lagrange equations which result when a mesh is obtained using a minimization of the orthogonality measure given by Eq. (10-34).

APPENDIX

A

SUBROUTINE FOR SOLVING A TRIDIAGONAL SYSTEM OF EQUATIONS

Subroutine SY solves a tridiagonal system of equations following the Thomas algorithm described in Chapter 4. To use the subroutine, the equations must be of the form

$$\begin{bmatrix} D_{IL} & A_{IL} & & \\ B_I & D_I & A_I & \\ & \ddots & \ddots & \ddots \\ & & B_{IU} & D_{IU} \end{bmatrix} \begin{bmatrix} U_{IL} \\ U_I \\ \vdots \\ U_{IU} \end{bmatrix} = \begin{bmatrix} C_{IL} \\ C_I \\ \vdots \\ C_{IU} \end{bmatrix} \tag{A-1}$$

The call statement for subroutine SY is of the form

$$\text{CALL SY}(IL, IU, B, D, A, C)$$

B, D, A, and C are the array names for the singly subscripted real variables $B(I)$, $D(I)$, $A(I)$, $C(I)$. IL and IU are unsubscripted integer variables. The arrays must be defined for subscripts ranging from IL to IU according to

B, Coefficient behind (to the left of) the main diagonal
D, Coefficient on the main diagonal
A, Coefficient ahead (to the right of) the main diagonal
C, Element in the constant vector

The equations in the system are ordered according to the value of the subscript. The variable IL corresponds to the subscript of the first equation in the system and IU

corresponds to the subscript of the last equation in the system. The number of equations in the system is $IU - IL + 1$. *The solution vector, U, is returned to the calling program in the* C *array.* That is, the constant vector **C** is overwritten in the subroutine with the solution. The *D* array is also altered by the subroutine. *A* and *B* remain unchanged.

LISTING OF SUBROUTINE SY

```
C...
      SUBROUTINE SY(IL,IU,BB,DD,AA,CC)
      DIMENSION AA(1),BB(1),CC(1),DD(1)
C...
C...SUBROUTINE SY SOLVES TRIDIAGONAL SYSTEM BY ELIMINATION
C...IL = SUBSCRIPT OF FIRST EQUATION
C...IU = SUBSCRIPT OF LAST EQUATION
C...BB = COEFFICIENT BEHIND DIAGONAL
C...DD = COEFFICIENT ON DIAGONAL
C...AA = COEFFICIENT AHEAD OF DIAGONAL
C...CC = ELEMENT OF CONSTANT VECTOR
C...
C...ESTABLISH UPPER TRIANGULAR MATRIX
C...
      LP = IL+1
      DO 10 I = LP,IU
      R = BB(I)/DD(I-1)
      DD(I) = DD(I)-R*AA(I-1)
   10 CC(I) = CC(I)-R*CC(I-1)
C...
C...BACK SUBSTITUTION
C...
      CC(IU) = CC(IU)/DD(IU)
      DO 20 I = LP,IU
      J = IU-I+IL
   20 CC(J) = (CC(J)-AA(J)*CC(J+1))/DD(J)
C...
C...SOLUTION STORED IN CC
C...
      RETURN
      END
```

APPENDIX

B

SUBROUTINES FOR SOLVING BLOCK TRIDIAGONAL SYSTEMS OF EQUATIONS

The subroutines described here for solving block tridiagonal systems of equations were provided by Sukumar R. Chakravarthy of Rockwell International Science Center. Subroutine NBTRIP solves a block tridiagonal system of equations of the form

$$\begin{bmatrix} B_{IL} & C_{IL} & & & \\ A_I & B_I & C_I & & \\ & \ddots & \ddots & \ddots & \\ & & & A_{IU} & B_{IU} \end{bmatrix} \begin{bmatrix} X_{IL} \\ X_I \\ | \\ X_{IU} \end{bmatrix} = \begin{bmatrix} D_{IL} \\ D_I \\ | \\ D_{IU} \end{bmatrix} \tag{B-1}$$

Subroutine PBTRIP solves a periodic block tridiagonal system of equations in the form

$$\begin{bmatrix} B_{IL} & C_{IL} & & & A_{IL} \\ A_I & B_I & C_I & & \\ & \ddots & \ddots & \ddots & \\ C_{IU} & & & A_{IU} & B_{IU} \end{bmatrix} \begin{bmatrix} X_{IL} \\ X_I \\ | \\ X_{IU} \end{bmatrix} = \begin{bmatrix} D_{IL} \\ D_I \\ | \\ D_{IU} \end{bmatrix} \tag{B-2}$$

The block matrices A, B, and C are $N \times N$ matrices at every point I with N being any integer greater than 1. Note that for $N = 1$, the Thomas algorithm of Appendix A can be employed. The right-hand side vector D has length N at each point I. The total number of I points at which the matrices are defined (denoted by NI) is given by

$$NI = (IU - IL + 1) \tag{B-3}$$

The matrices A, B, and C are dimensioned as

$$A(N, N, NI)$$

$$B(N, N, NI)$$

$$C(N, N, NI)$$

while the vector D is dimensioned as

$$D(N, NI)$$

The call statement for subroutine NBTRIP is

CALL NBTRIP($A, B, C, D, IL, IU, ORDER$)

with arguments defined by

A, Subdiagonal block matrix
B, Diagonal block matrix
C, Superdiagonal block matrix
D, Right-hand side vector
IL, Lower value of I for which matrices are defined
IU, Upper value of I for which matrices are defined
$ORDER$, N (order can be any integer greater than 1)

The solution (X) is returned to the calling program by overwriting the D vector with the X vector. The calling statement for subroutine PBTRIP is

CALL PBTRIP($A, B, C, D, IL, IU, ORDER$)

with the same arguments as subroutine NBTRIP. However, if $ORDER$ is greater than 5 a dimension statement must be changed in this subroutine (see listing of subroutine).

Subroutines NBTRIP and PBTRIP employ no pivoting strategy in their elimination schemes. It should be noted that a specialized subroutine for solving a block tridiagonal system of equations can be written for each value of N which will be faster than the general subroutines given here.

LISTING OF SUBROUTINE NBTRIP

```
C...
C...SUBROUTINE TO SOLVE NON-PERIODIC BLOCK TRIDIAGONAL
C...SYSTEM OF EQUATIONS WITHOUT PIVOTING STRATEGY
C...WITH THE DIMENSIONS OF THE BLOCK MATRICES BEING
C...N x N (N IS ANY NUMBER GREATER THAN 1).
C...
      SUBROUTINE NBTRIP(A,B,C,D,IL,IU,ORDER)
C...
      DIMENSION A(1),B(1),C(1),D(1)
      INTEGER ORDER,ORDSQ
C...
C...A = SUB DIAGONAL MATRIX
C...B =     DIAGONAL MATRIX
C...C = SUP DIAGONAL MATRIX
C...D = RIGHT HAND SIDE VECTOR
```

```
C...IL = LOWER VALUE OF INDEX FOR WHICH MATRICES ARE DEFINED
C...IU = UPPER VALUE OF INDEX FOR WHICH MATRICES ARE DEFINED
C...     (SOLUTION IS SOUGHT FOR BTRI(A,B,C)*X = D
C...     FOR INDICES OF X BETWEEN IL AND IU (INCLUSIVE).
C...     SOLUTION WRITTEN IN D VECTOR (ORIGINAL CONTENTS
C...     ARE OVERWRITTEN)).
C...ORDER = ORDER OF A,B,C MATRICES AND LENGTH OF D VECTOR
C...     AT EACH POINT DENOTED BY INDEX I
C...     (ORDER CAN BE ANY INTEGER GREATER THAN 1).
C...
C...THE MATRICES AND VECTORS ARE STORED IN SINGLE SUBSCRIPT FORM
C...
      ORDSQ = ORDER**2
C...
C...FORWARD ELIMINATION
C...
      I = IL
      I0MAT = 1+(I-1)*ORDSQ
      I0VEC = 1+(I-1)*ORDER
      CALL LUDECO(B(I0MAT),ORDER)
      CALL LUSOLV(B(I0MAT),D(I0VEC),D(I0VEC),ORDER)
      DO 100 J=1,ORDER
      I0MATJ = I0MAT+(J-1)*ORDER
      CALL LUSOLV(B(I0MAT),C(I0MATJ),C(I0MATJ),ORDER)
  100 CONTINUE
  200 CONTINUE
      I = I+1
      I0MAT = 1+(I-1)*ORDSQ
      I0VEC = 1+(I-1)*ORDER
      I1MAT = I0MAT-ORDSQ
      I1VEC = I0VEC-ORDER
      CALL MULPUT(A(I0MAT),D(I1VEC),D(I0VEC),ORDER)
      DO 300 J=1,ORDER
      I0MATJ = I0MAT+(J-1)*ORDER
      I1MATJ = I1MAT+(J-1)*ORDER
      CALL MULPUT(A(I0MAT),C(I1MATJ),B(I0MATJ),ORDER)
  300 CONTINUE
      CALL LUDECO(B(I0MAT),ORDER)
      CALL LUSOLV(B(I0MAT),D(I0VEC),D(I0VEC),ORDER)
      IF(I.EQ.IU) GO TO 500
      DO 400 J=1,ORDER
      I0MATJ = I0MAT+(J-1)*ORDER
      CALL LUSOLV(B(I0MAT),C(I0MATJ),C(I0MATJ),ORDER)
  400 CONTINUE
      GO TO 200
  500 CONTINUE
C...
C...BACK SUBSTITUTION
C...
      I = IU
  600 CONTINUE
      I = I-1
      I0MAT = 1+(I-1)*ORDSQ
      I0VEC = 1+(I-1)*ORDER
      I1VEC = I0VEC+ORDER
      CALL MULPUT(C(I0MAT),D(I1VEC),D(I0VEC),ORDER)
      IF (I.GT.IL) GO TO 600
C...
      RETURN
      END
```

LISTING OF SUBROUTINE PBTRIP

```
C...
C...SUBROUTINE TO SOLVE PERIODIC BLOCK TRIDIAGONAL
C...SYSTEM OF EQUATIONS WITHOUT PIVOTING STRATEGY.
C...EACH BLOCK MATRIX MAY BE OF DIMENSION N WITH
C...N ANY NUMBER GREATER THAN 1.
C...
      SUBROUTINE PBTRIP(A,B,C,D,IL,IU,ORDER)
      DIMENSION A(1),B(1),C(1),D(1)
      DIMENSION AD(25),CD(25)
      INTEGER ORDER,ORDSQ
C...
C...A = SUB DIAGONAL MATRIX
C...B =     DIAGONAL MATRIX
C...C = SUP DIAGONAL MATRIX
C...D = RIGHT HAND SIDE VECTOR
C...IL = LOWER VALUE OF INDEX FOR WHICH MATRICES ARE DEFINED
C...IU = UPPER VALUE OF INDEX FOR WHICH MATRICES ARE DEFINED
C...     (SOLUTION IS SOUGHT FOR BTRI(A,B,C)*X = D
C...     FOR INDICES OF X BETWEEEN IL AND IU (INCLUSIVE).
C...     SOLUTION WRITTEN IN D VECTOR (ORIGINAL CONTENTS
C...     ARE OVERWRITTEN)).
C...ORDER = ORDER OF A,B,C MATRICES AND LENGTH OF D VECTOR
C...     AT EACH POINT DENOTED BY INDEX I
C...     (ORDER CAN BE ANY INTEGER GREATER THAN 1)
C...     (ARRAYS AD AND CD MUST BE AT LEAST OF LENGTH ORDER**2)
C...     (CURRENT LENGTH OF 25 ANTICIPATES MAXIMUM ORDER OF 5).
C...
      IS = IL+1
      IE = IU-1
      ORDSQ = ORDER**2
      IUMAT = 1+(IU-1)*ORDSQ
      IUVEC = 1+(IU-1)*ORDER
      IEMAT = 1+(IE-1)*ORDSQ
      IEVEC = 1+(IE-1)*ORDER
C...
C...FORWARD ELIMINATION
C...
      I = IL
      IOMAT = 1+(I-1)*ORDSQ
      IOVEC = 1+(I-1)*ORDER
      CALL LUDECO(B(IOMAT),ORDER)
      CALL LUSOLV(B(IOMAT),D(IOVEC),D(IOVEC),ORDER)
      DO 10 J=1,ORDER
      IOMATJ = IOMAT+(J-1)*ORDER
      CALL LUSOLV(B(IOMAT),C(IOMATJ),C(IOMATJ),ORDER)
      CALL LUSOLV(B(IOMAT),A(IOMATJ),A(IOMATJ),ORDER)
   10 CONTINUE
C...
      DO 200 I = IS,IE
      IOMAT = 1+(I-1)*ORDSQ
      IOVEC = 1+(I-1)*ORDER
      I1MAT = IOMAT-ORDSQ
      I1VEC = IOVEC-ORDER
      DO 20 J=1,ORDSQ
```

```
      IOMATJ = J-1+IOMAT
      IUMATJ = J-1+IUMAT
      AD(J) = A(IOMATJ)
      CD(J) = C(IUMATJ)
      A(IOMATJ) = 0.0
      C(IUMATJ) = 0.0
   20 CONTINUE
      CALL MULPUT(AD,D(I1VEC),D(IOVEC),ORDER)
      DO 22 J=1,ORDER
      IOMATJ = IOMAT+(J-1)*ORDER
      I1MATJ = I1MAT+(J-1)*ORDER
      CALL MULPUT(AD,C(I1MATJ),B(IOMATJ),ORDER)
      CALL MULPUT(AD,A(I1MATJ),A(IOMATJ),ORDER)
   22 CONTINUE
      CALL LUDECO(B(IOMAT),ORDER)
      CALL LUSOLV(B(IOMAT),D(IOVEC),D(IOVEC),ORDER)
      DO 24 J=1,ORDER
      IOMATJ = IOMAT+(J-1)*ORDER
      CALL LUSOLV(B(IOMAT),C(IOMATJ),C(IOMATJ),ORDER)
      CALL LUSOLV(B(IOMAT),A(IOMATJ),A(IOMATJ),ORDER)
   24 CONTINUE
      CALL MULPUT(CD,D(I1VEC),D(IUVEC),ORDER)
      DO 26 J=1,ORDER
      IUMATJ = IUMAT+(J-1)*ORDER
      I1MATJ = I1MAT+(J-1)*ORDER
      CALL MULPUT(CD,A(I1MATJ),B(IUMATJ),ORDER)
      CALL MULPUT(CD,C(I1MATJ),C(IUMATJ),ORDER)
   26 CONTINUE
  200 CONTINUE
C...
      DO 30 J=1,ORDSQ
      IUMATJ = J-1+IUMAT
      AD(J) = A(IUMATJ)+C(IUMATJ)
   30 CONTINUE
      CALL MULPUT(AD,D(IEVEC),D(IUVEC),ORDER)
      DO 32 J=1,ORDER
      IUMATJ = IUMAT+(J-1)*ORDER
      IEMATJ = IEMAT+(J-1)*ORDER
      CALL MULPUT(AD,C(IEMATJ),B(IUMATJ),ORDER)
      CALL MULPUT(AD,A(IEMATJ),B(IUMATJ),ORDER)
   32 CONTINUE
      CALL LUDECO(B(IUMAT),ORDER)
      CALL LUSOLV(B(IUMAT),D(IUVEC),D(IUVEC),ORDER)
C...
C...BACK SUBSTITUTION
C...
      DO 40 IBAC = IL,IE
      I = IE-IBAC+IL
      IOMAT = 1+(I-1)*ORDSQ
      IOVEC = 1+(I-1)*ORDER
      I1VEC = IOVEC+ORDER
      CALL MULPUT(A(IOMAT),D(IUVEC),D(IOVEC),ORDER)
      CALL MULPUT(C(IOMAT),D(I1VEC),D(IOVEC),ORDER)
   40 CONTINUE
C...
      RETURN
      END
```

```
C...
C...SUBROUTINE TO CALCULATE L-U DECOMPOSITION
C...OF A GIVEN MATRIX A AND STORE RESULT IN A
C...(NO PIVOTING STRATEGY IS EMPLOYED)
C...
      SUBROUTINE LUDECO(A,ORDER)
C...
      DIMENSION A(ORDER,1)
      INTEGER ORDER
C...
      DO 8 JC=2,ORDER
    8 A(1,JC) = A(1,JC)/A(1,1)
      JRJC = 1
   10 CONTINUE
      JRJC = JRJC+1
      JRJCM1 = JRJC-1
      JRJCP1 = JRJC+1
      DO 14 JR=JRJC,ORDER
      SUM = A(JR,JRJC)
      DO 12 JM=1,JRJCM1
   12 SUM = SUM-A(JR,JM)*A(JM,JRJC)
   14 A(JR,JRJC) = SUM
      IF (JRJC.EQ.ORDER) RETURN
      DO 18 JC = JRJCP1,ORDER
      SUM = A(JRJC,JC)
      DO 16 JM=1,JRJCM1
   16 SUM = SUM-A(JRJC,JM)*A(JM,JC)
   18 A(JRJC,JC) = SUM/A(JRJC,JRJC)
      GO TO 10
      END
```

```
C...
C...SUBROUTINE TO MULTIPLY A VECTOR B BY A MATRIX A,
C...SUBTRACT RESULT FROM ANOTHER VECTOR C AND STORE
C...RESULT IN C.  THUS VECTOR C IS OVERWRITTEN.
C...
      SUBROUTINE MULPUT(A,B,C,ORDER)
C...
      DIMENSION A(1),B(1),C(1)
      INTEGER ORDER
C...
      DO 200 JR=1,ORDER
      SUM = 0.0
      DO 100 JC=1,ORDER
      IA = JR+(JC-1)*ORDER
  100 SUM = SUM+A(IA)*B(JC)
  200 C(JR) = C(JR)-SUM
C...
      RETURN
      END
```

```
C...
C...SUBROUTINE TO SOLVE LINEAR ALGEBRAIC SYSTEM OF
C...EQUATIONS A*C=B AND STORE RESULTS IN VECTOR C.
C...MATRIX A IS INPUT IN L-U DECOMPOSITION FORM.
C...(NO PIVOTING STRATEGY HAS BEEN EMPLOYED TO
C...COMPUTE THE L-U DECOMPOSITION OF THE MATRIX A).
C...
      SUBROUTINE LUSOLV(A,B,C,ORDER)
C...
      DIMENSION A(ORDER,1),B(1),C(1)
      INTEGER ORDER
C...
C...FIRST L(INV)*B
C...
      C(1) = C(1)/A(1,1)
      DO 14 JR=2,ORDER
      JRM1 = JR-1
      SUM = B(JR)
      DO 12 JM=1,JRM1
   12 SUM = SUM-A(JR,JM)*C(JM)
   14 C(JR) = SUM/A(JR,JR)
C...
C...NEXT U(INV) OF L(INV)*B
C...
      DO 18 JRJR=2,ORDER
      JR = ORDER-JRJR+1
      JRP1 = JR+1
      SUM = C(JR)
      DO 16 JMJM = JRP1,ORDER
      JM = ORDER-JMJM+JRP1
   16 SUM = SUM-A(JR,JM)*C(JM)
   18 C(JR) = SUM
C...
      RETURN
      END
```

APPENDIX

C

THE MODIFIED STRONGLY IMPLICIT PROCEDURE

This appendix describes the Modified Strongly Implicit (MSI) procedure (Schneider and Zedan, 1981) for solving a class of elliptic PDE's. The overall strategy of this procedure was described in Chapter 4. This appendix supplies further details. Schneider and Zedan (1981) presented the procedure as a means for solving the algebraic equations arising from the finite-difference representation of the elliptic equation

$$\frac{\partial}{\partial x}\left(k_x \frac{\partial u}{\partial x}\right) + \frac{\partial}{\partial y}\left(k_y \frac{\partial u}{\partial y}\right) = q(x,y) \tag{C-1}$$

which governs two-dimensional steady-state heat conduction when u is temperature. In the above, k_x and k_y are thermal conductivities for heat flow in the x and y directions, respectively, and $q(x, y)$ is a source term accounting for possible heat generation. It should be clear that a wide variety of problems are governed by equations of the form given by Eq. (C-1). With $k_x = k_y =$ constant and $q(x, y) \neq 0$, Eq. (C-1) becomes the Poisson equation. With $k_x = k_y =$ constant and $q(x, y) = 0$, Eq. (C-1) reduces to the Laplace equation. Only numerical examples for the solution to the Laplace equation were presented in Schneider and Zedan (1981). Examples presented employed Dirichlet, Neumann, and Robins (convective) boundary conditions.

The algorithm is developed to handle a nine-point finite-difference representation of Eq. (C-1) and treats the five-point representation as a special case. A nine-point [see Eq. (4-114)] representation of Eq. (C-1) can be written in the general form

$$\begin{aligned} A^1_{i,j} u_{i,j+1} + A^2_{i,j} u_{i+1,j+1} + A^3_{i,j} u_{i+1,j} + A^4_{i,j} u_{i+1,j-1} + A^5_{i,j} u_{i,j-1} + A^6_{i,j} u_{i-1,j-1} \\ + A^7_{i,j} u_{i-1,j} + A^8_{i,j} u_{i-1,j+1} + A^9_{i,j} u_{i,j} = q_{i,j} \end{aligned} \tag{C-2}$$

The i, j subscript refers to location within the grid network rather than the matrix row-column designation. Note that superscripts are used to identify the coefficients in the difference equation written for the general point (i, j). The five-point representation becomes a special case in which

$$A_{i,j}^2 = A_{i,j}^4 = A_{i,j}^6 = A_{i,j}^8 = 0$$

The equations can be written in the form

$$[A]\,\mathbf{u} = \mathbf{C} \tag{C-3}$$

where the coefficient matrix has the form

$[A] =$

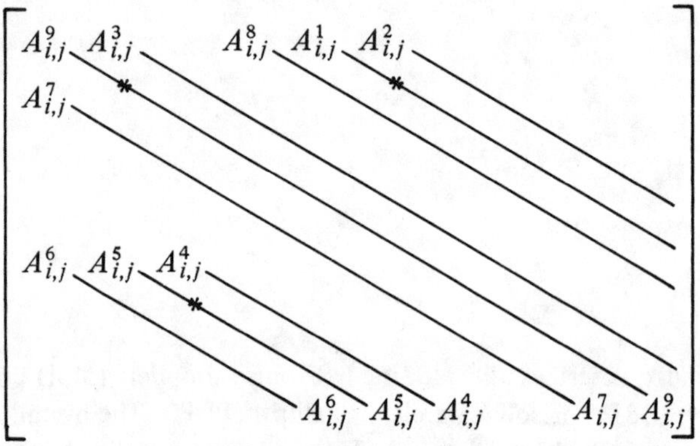

For reference, the diagonals corresponding to grid points having the same value for the i index (same grid column) are identified by an asterisk. We now construct a matrix

$$[B] = [A + P]$$

such that $[B]$ can be decomposed into upper and lower triangular matrices, $[L]$ and $[U]$. We require that the original nine coefficients ($A_{i,j}^1$ through $A_{i,j}^9$) remain unchanged as $[A + P]$ is constructed. The $[L]$ and $[U]$ matrices have the form

$[L] =$

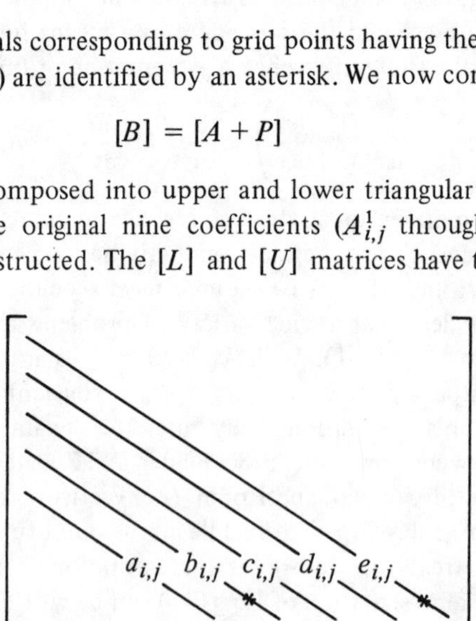

$[U] =$ 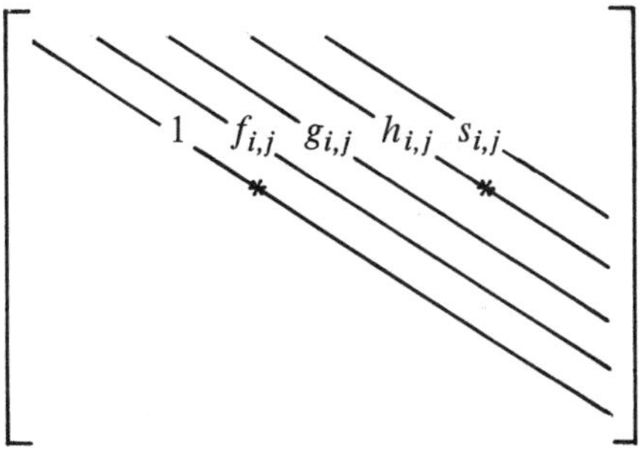

Again the asterisk is used to identify diagonals corresponding to grid points having the same value for the i index.

The equations to be used to determine the coefficients of $[L]$ and $[U]$ such that the original nine coefficients in $[A]$ remain unchanged in $[B]$ are

$$a_{i,j} = A_{i,j}^6 \tag{C-3a}$$

$$a_{i,j} f_{i-1,j-1} + b_{i,j} = A_{i,j}^5 \tag{C-3b}$$

$$b_{i,j} f_{i,j-1} + C_{i,j} = A_{i,j}^4 \tag{C-3c}$$

$$a_{i,j} h_{i-1,j-1} + b_{i,j} g_{i,j-1} + d_{i,j} = A_{i,j}^7 \tag{C-3d}$$

$$a_{i,j} s_{i-1,j-1} + b_{i,j} h_{i,j-1} + c_{i,j} g_{i+1,j-1} + d_{i,j} f_{i-1,j} + e_{i,j} = A_{i,j}^9 \tag{C-3e}$$

$$b_{i,j} s_{i,j-1} + c_{i,j} h_{i+1,j-1} + e_{i,j} f_{i,j} = A_{i,j}^3 \tag{C-3f}$$

$$d_{i,j} h_{i-1,j} + e_{i,j} g_{i,j} = A_{i,j}^8 \tag{C-3g}$$

$$d_{i,j} s_{i-1,j} + e_{i,j} h_{i,j} = A_{i,j}^1 \tag{C-3h}$$

$$e_{i,j} s_{i,j} = A_{i,j}^2 \tag{C-3i}$$

The modified coefficient matrix $[B] = [A + P]$ has the form

$[B] =$ 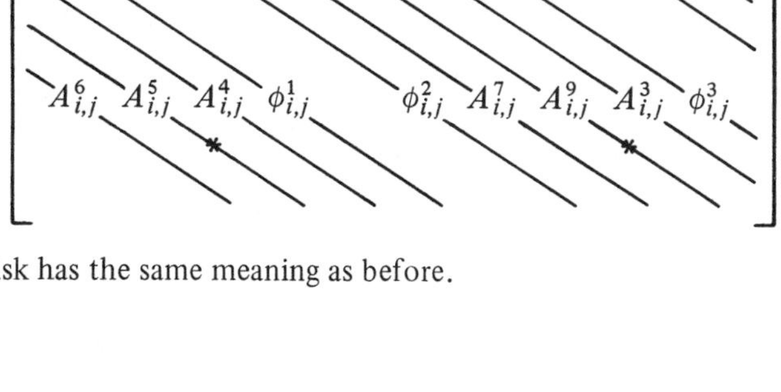

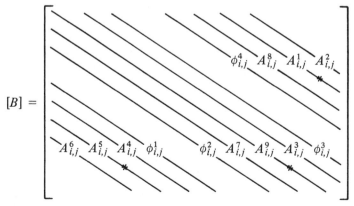

where the asterisk has the same meaning as before.

The elements in $[B]$ denoted by $\phi^1_{i,j}, \phi^2_{i,j}, \phi^3_{i,j}$, and $\phi^4_{i,j}$ are determined from

$$\phi^1_{i,j} = c_{i,j} f_{i+1,j-1} \tag{C-4a}$$

$$\phi^2_{i,j} = a_{i,j} g_{i-1,j-1} \tag{C-4b}$$

$$\phi^3_{i,j} = c_{i,j} s_{i+1,j-1} \tag{C-4c}$$

$$\phi^4_{i,j} = d_{i,j} g_{i-1,j} \tag{C-4d}$$

The numerical molecule associated with the modified matrix [B] is shown schematically in Fig. C-1.

Schneider and Zedan (1981) employed Taylor-series expansions to obtain values of $u_{i-2,j}, u_{i+2,j}, u_{i+2,j-1}$, and $u_{i-2,j+1}$ in terms of u's in the original nine-point molecule to partially cancel the influence of the additional ($\phi_{i,j}$) terms in the $[B]$ matrix. These are

$$u_{i-2,j} = -u_{i,j} + 2u_{i-1,j} \tag{C-5a}$$

$$u_{i+2,j} = -u_{i,j} + 2u_{i+1,j} \tag{C-5b}$$

$$u_{i+2,j-1} = -2u_{i,j} + 2u_{i+1,j} + u_{i,j-1} \tag{C-5c}$$

$$u_{i-2,j+1} = -2u_{i,j} + 2u_{i-1,j} + u_{i,j+1} \tag{C-5d}$$

Other "extrapolation" schemes for obtaining values outside the original molecule may work equally well. The use of such approximations affects only the approach to convergence of the iterative sequence and not the final converged solution.

An iterative parameter α is employed to implement partial cancellation of the influence of the $\phi_{i,j}$ terms appearing in $[B]$. This is done by using a modified representation for the nine-point scheme in the form

$$\begin{aligned} &A^5_{i,j} u_{i,j-1} + A^7_{i,j} u_{i-1,j} + A^9_{i,j} u_{i,j} + A^1_{i,j} u_{i,j+1} + A^3_{i,j} u_{i+1,j} \\ &\quad + A^6_{i,j} u_{i-1,j-1} + A^8_{i,j} u_{i-1,j+1} + A^2_{i,j} u_{i+1,j+1} \\ &\quad + A^4_{i,j} u_{i+1,j-1} + \phi^1_{i,j}[u_{i+2,j-1} - \alpha(-2u_{i,j} + 2u_{i+1,j} + u_{i,j-1})] \\ &\quad + \phi^2_{i,j}[u_{i-2,j} - \alpha(-u_{i,j} + 2u_{i-1,j})] + \phi^3_{i,j}[u_{i+2,j} - \alpha(-u_{i,j} + 2u_{i+1,j})] \\ &\quad + \phi^4\,[u_{i-2,j+1} - \alpha(-2u_{i,j} + 2u_{i-1,j} + u_{i,j+1})] = q_{i,j} \end{aligned} \tag{C-6}$$

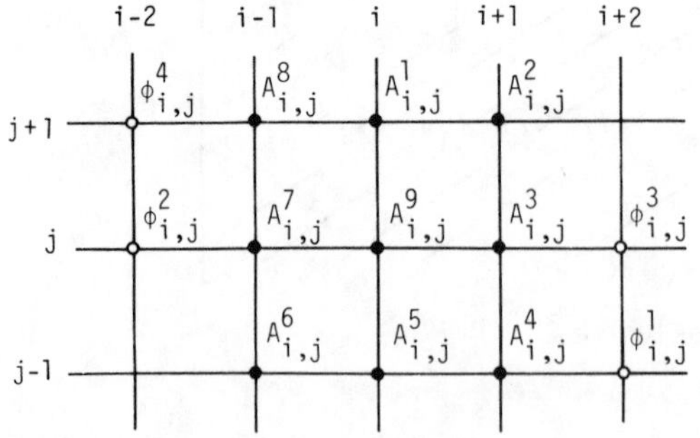

Figure C-1 The numerical molecule for the MSI procedure for a nine-point formulation, points labeled $A^2_{i,j}, A^4_{i,j}, A^6_{i,j}, A^8_{i,j}, \phi^2_{i,j}, \phi^3_{i,j}$ are eliminated when a five-point formulation is used.

Equations (C-3) and (C-4) are modified to include the partial cancellation indicated in Eq. (C-6) and rearranged to permit the explicit evaluation of the elements of $[L]$ and $[U]$:

$$a_{i,j} = A^6_{i,j} \tag{C-7a}$$

$$b_{i,j} = \frac{A^5_{i,j} - a_{i,j} f_{i-1,j-1} - \alpha A^4_{i,j} f_{i+1,j-1}}{1 - \alpha f_{i,j-1} f_{i+1,j-1}} \tag{C-7b}$$

$$c_{i,j} = A^4_{i,j} - b_{i,j} f_{i,j-1} \tag{C-7c}$$

$$d_{i,j} = \frac{A^7_{i,j} - a_{i,j} h_{i-1,j-1} - b_{i,j} g_{i,j-1} - 2\alpha a_{i,j} g_{i-1,j-1}}{1 + 2\alpha g_{i-1,j}} \tag{C-7d}$$

$$e_{i,j} = A^9_{i,j} - a_{i,j} s_{i-1,j-1} - b_{i,j} h_{i,j-1} - c_{i,j} g_{i+1,j-1} - d_{i,j} f_{i-1,j} + \alpha(2\phi^1_{i,j} + \phi^2_{i,j} + \phi^3_{i,j} + 2\phi^4_{i,j}) \tag{C-7e}$$

$$f_{i,j} = \frac{A^3_{i,j} - b_{i,j} s_{i,j-1} - c_{i,j} h_{i+1,j-1} - 2\alpha(\phi^1_{i,j} + \phi^3_{i,j})}{e_{i,j}} \tag{C-7f}$$

$$g_{i,j} = \frac{A^8_{i,j} - d_{i,j} h_{i-1,j}}{e_{i,j}} \tag{C-7g}$$

$$h_{i,j} = \frac{A^1_{i,j} - d_{i,j} s_{i-1,j} - \alpha\phi^4_{i,j}}{e_{i,j}} \tag{C-7h}$$

$$s_{i,j} = \frac{A^2_{i,j}}{e_{i,j}} \tag{C-7i}$$

The $\phi_{i,j}$'s appearing in the above are evaluated as indicated in Eqs. (C-4) using the values of a, b, c, d, f, g, and s obtained from Eqs. (C-7). Note that the $\phi_{i,j}$'s are needed in Eqs. (C-7) and should be evaluated as soon as the evaluation of $d_{i,j}$ is complete. The results obtained by Schneider and Zedan (1981) indicate that the MSI procedure is not extremely sensitive to the choice of α. Values of α between 0.3 and 0.6 worked well in their calculations.

It is important to observe that when the MSI procedure is used for the five-point difference representation,

$$A^2_{i,j} = A^4_{i,j} = A^6_{i,j} = A^8_{i,j} = 0 \tag{C-8}$$

and, as a result,

$$a_{i,j} = s_{i,j} = \phi^2_{i,j} = \phi^3_{i,j} = 0 \tag{C-9}$$

The iterative sequence is developed as follows. Adding $[P]\mathbf{u}$ to both sides of Eq. (C-3) gives

$$[A + P]\mathbf{u} = \mathbf{C} + [P]\mathbf{u} \tag{C-10}$$

We evaluate the unknowns on the right-hand side at the n iteration level to write

$$[A + P]\mathbf{u}^{n+1} = \mathbf{C} + [P]\mathbf{u}^n \tag{C-11}$$

Decomposing $[A+P]$ into the $[L]$ and $[U]$ matrices gives

$$[L]\,[U]\,\mathbf{u}^{n+1} = \mathbf{C} + [P]\,\mathbf{u}^{n} \tag{C-12}$$

Defining an intermediate vector $\mathbf{V}^{n+1}$ by

$$\mathbf{V}^{n+1} = [U]\,\mathbf{u}^{n+1} \tag{C-13}$$

we can employ the two-step process

Step 1: $$[L]\,\mathbf{V}^{n+1} = \mathbf{C} + [P]\,\mathbf{u}^{n} \tag{C-14a}$$

Step 2: $$[U]\,\mathbf{u}^{n+1} = \mathbf{V}^{n+1} \tag{C-14b}$$

The elements of $[P]$ are simply the $\phi^1, \phi^2, \phi^3, \phi^4$ (only ϕ^1 and ϕ^4 when the five-point scheme is used) values determined from Eqs. (C-4).

Alternatively, we can define a difference vector

$$\boldsymbol{\delta}^{n+1} = \mathbf{u}^{n+1} - \mathbf{u}^{n} \tag{C-15}$$

and a residual vector

$$\mathbf{R}^{n} = \mathbf{C} - [A]\,\mathbf{u}^{n} \tag{C-16}$$

so that Eq. (C-11) becomes

$$[A+P]\,\boldsymbol{\delta}^{n+1} = \mathbf{R}^{n} \tag{C-17}$$

Replacing $[A+P]$ by the [L] [U] product gives

$$[L]\,[U]\,\boldsymbol{\delta}^{n+1} = \mathbf{R}^{n}$$

Defining an intermediate vector $\mathbf{W}^{n+1}$ by

$$\mathbf{W}^{n+1} = [U]\,\boldsymbol{\delta}^{n+1} \tag{C-18}$$

the solution procedure can again be written as a two-step process:

Step 1: $$[L]\,\mathbf{W}^{n+1} = \mathbf{R}^{n} \tag{C-19a}$$

Step 2: $$[U]\,\boldsymbol{\delta}^{n+1} = \mathbf{W}^{n+1} \tag{C-19b}$$

The processes represented by Eqs. (C-14) and (C-19) consist of a forward substitution to determine $\mathbf{V}^{n+1}$ or $\mathbf{W}^{n+1}$ followed by a backward substitution to obtain $\mathbf{u}^{n+1}$ or $\boldsymbol{\delta}^{n+1}$. The coefficients remain unchanged for the iterative process. The right-hand side of the Step 1 equation is then updated and the process is repeated.

NOMENCLATURE

a	speed of sound
$\mathbf{A}$	area vector
c	wave speed
c_f	skin-friction coefficient
c_p	specific heat at constant pressure
c_v	specific heat at constant volume
$\mathbf{dr}$	differential length along a line
e	internal energy per unit mass
E_t	total energy per unit volume, [$= \rho(e + V^2/2)$ if only internal energy and kinetic energy are included]
$\mathbf{f}$	body force per unit mass
f_x, f_y, f_z	components of body force per unit mass in a Cartesian system
F	denotes function, nondimensional velocity variable
$\mathbf{g}$	gravity vector
G	amplification factor
G	nondimensional velocity variable
h	height
h	enthalpy per unit mass ($= e + p/\rho$)
h_1, h_2, h_3	scale factors in an orthogonal curvilinear coordinate system
H	total enthalpy ($= h + V^2/2$)
$\mathbf{i}_1, \mathbf{i}_2, \mathbf{i}_3$	unit vectors in a generalized curvilinear coordinate system
$\mathbf{i}, \mathbf{j}, \mathbf{k}$	unit vectors in a Cartesian coordinate system
I	nondimensional enthalpy variable
J	Jacobian

k	coefficient of thermal conductivity
$\bar{k}$	kinetic energy of turbulence
k_m	wave number
k_T	turbulent thermal conductivity
K	local body curvature
K	$\Delta y_+/\Delta y_-$
l	mixing length
l_ϵ	dissipation length
$l_{1,2}$	scalar components of $\mathbf{L}$
L	reference length
$\mathbf{L}$	eigenvector
$\dot{m}$	mass flow rate
M	Mach number
M_x	local streamwise Mach number
n	normal distance
$\mathbf{n}$	unit normal
p	pressure
Pr	Prandtl number
Pr_T	turbulent Prandtl number
q	intensity of line source or sink
$\mathbf{q}$	heat flux vector
Q	external heat addition per unit volume
r	$\alpha \Delta t/(\Delta x)^2$
r	radius, radial distance
R	gas constant
Re	Reynolds number
Re_L	freestream Reynolds number based on length L, $(= \rho_\infty V_\infty L/\mu_\infty)$
$\mathrm{Re}_{\Delta x}$	mesh Reynolds number ($= c\,\Delta x/\mu$ for Burgers equation)
s	entropy per unit mass
s	arc length
S	source term
t	time
T	temperature
u^+	nondimensional velocity used in turbulent flow
u, v, w	velocity components in a Cartesian system
u_1, u_2, u_3	velocity components in a generalized coordinate system
u_r, u_θ, u_z	velocity components in a cylindrical coordinate system
u_r, u_θ, u_ϕ	velocity components in a spherical coordinate system
U, V, W	contravariant velocity components
U_∞	freestream velocity in x direction
$\mathbf{V}$	velocity vector
V	magnitude of velocity vector
$\mathbf{w}$	primitive variable vector
x, y, z	Cartesian coordinates
x_1, x_2, x_3	generalized curvilinear coordinates

y^+	nondimensional distance used in turbulent flow
α	thermal diffusivity
α, β, γ	conical coordinates
β	artificial compressibility factor
β	stretching parameter
β	$k_m \Delta x$
β	$\sqrt{M^2 - 1}$
β	pressure gradient parameter, $[= (x/u_e)\, du_e/dx]$
β_x	$k_m \Delta x$
β_y	$k_m \Delta y$
γ	ratio of specific heats
Γ	finite-difference operator, circulation
δ	characteristic length in y direction
δ	boundary-layer thickness
δ	central-difference operator defined by Eq. (3-14)
δ_u	represents change in u between two iterations
$\bar{\delta}$	central-difference operator defined by Eq. (3-13)
$\hat{\delta}$	central-difference operator defined by Eqs. (4-100)
δ^*	displacement thickness
δ_{ij}	Kronecker delta
Δ	forward-difference operator defined by Eq. (3-9)
Δx_+	$x_{j+1} - x_j$
Δx_-	$x_j - x_{j-1}$
Δy_+	$y_{j+1} - y_j$
Δy_-	$y_j - y_{j-1}$
$\Delta^n(\ )$	$(\ )^{n+1} - (\ )^n$
η	nondimensional distance variable
ϵ	turbulence dissipation rate
ϵ_i	coefficient of implicit smoothing term
ϵ_e	coefficient of explicit smoothing term
ζ	vorticity $(= \nabla \times \mathbf{V})$
ζ	magnitude of vorticity vector
θ	angle measured in circumferential direction
θ	parameter controlling type of difference scheme
θ	momentum thickness
θ_1, θ_2	parameters controlling type of difference scheme
κ	coefficient of bulk viscosity
κ	von Kármán constant
λ	eigenvalue
λ	generalized diffusion coefficient
$\bar{\lambda}$	$\lambda_T + \lambda$
μ	viscous coefficient in Burgers' equation
μ	coefficient of viscosity
μ	averaging operator defined by Eq. (3-16)
$\bar{\mu}$	$\mu + \mu_T$

μ'	second coefficient of viscosity
μ_T	eddy viscosity
ξ, η, ζ	transformed coordinates
π	3.14159 . . .
Π_{ij}	stress tensor
ρ	density
ρ	artificial density
σ	shock angle
τ	computational time
τ	shear stress
τ_{ij}	viscous stress tensor
ν	kinematic viscosity ($= \mu/\rho$)
ν	$c\,\Delta t/\Delta x$
ϕ	velocity potential
ϕ	angle in spherical coordinate system
ϕ	phase angle
ϕ	generalized variable
ϕ, ψ	boundary point clustering function
Φ	dissipation function
χ	strong-interaction parameter
χ	pressure gradient parameter, $[= (-1/\rho)\, dp/dx]$
ψ	stream function
ψ	vector potential
ω	fraction of streamwise pressure gradient term
$\omega, \bar{\omega}$	overrelaxation parameters
∇	backward-difference operator defined by Eq. (3-11)
∇	vector differential operator
∇^2	Laplacian operator ($= \nabla \cdot \nabla$)

SUBSCRIPTS

B	boundary value
CFL	Courant-Friedrichs-Lewy condition
e	exact value
i	inner
i	inviscid term
i, j, k	grid locations in x, y, z directions
l	lower
inv	denotes inviscid value
lam	laminar-like in form
min	minimum
max	maximum
n	normal or normal component
nose	nose value

o	intermediate (or estimated) value
o	initial value
o	outer
ref	reference conditions
s	shock value
stag	stagnation value
t	tangential or tangential component
t	thermal
T	turbulent
turb	turbulent quantity
u	upper
v	viscous term
wall	wall value
x	partial differentiation with respect to x
y	partial differentiation with respect to y
z	partial differentiation with respect to z
x, y, z	components in x, y, z directions
1	conditions in front of shock
2	conditions behind shock
∞	freestream value

SUPERSCRIPTS

i	index in marching direction
m	iteration level
n	time level
$*$	dummy time index
$**$	dummy time index
$*$	denotes a nondimensional quantity
$\hat{}$	denotes value of variable from previous iteration
$\sim$	denotes mass-averaged variables [see Eq. (5-64)]
$'$	denotes fluctuation in turbulent flow, conventionally-averaged variables
$'$	perturbation quantity
$'$	correction term
$''$	denotes fluctuation in turbulent flow, mass-averaged variables
$-$	denotes time-averaged quantity

REFERENCES

Abbett, M. J. (1973). Boundary Condition Calculation Procedures for Inviscid Supersonic Flow Fields, *Proc. AIAA Computational Fluid Dynamics Conference*, Palm Springs, California, pp. 153-172.

Adams Jr., J. C. and Hodge, B. K. (1977). The Calculation of Compressible, Transitional, Turbulent, and Relaminarizational Boundary Layers Over Smooth and Rough Surfaces Using an Extended Mixing Length Hypothesis, AIAA Paper 77-682, Albuquerque, New Mexico.

Agarwal, R. K. (1981). A Third-Order-Accurate Upwind Scheme for Navier-Stokes Solutions in Three Dimensions, *Proc. ASME/AIAA Conference on Computers in Flow Predictions and Fluid Dynamics Experiments*, Washington, D.C., pp. 73-82.

Allen, D. N. de G. (1954). *Relaxation Methods*, McGraw-Hill, New York.

Allen, D. and Southwell, R. V. (1955). Relaxation Methods Applied to Determine the Motion, in Two Dimensions, of a Viscous Fluid Past a Fixed Cylinder, *Q. J. Mech. Appl. Math.*, vol. 8, pp. 129-145.

Allen, J. S. and Cheng, S. I. (1970). Numerical Solutions of the Compressible Navier-Stokes Equations for the Laminar Near Wake, *Phys. Fluids*, vol. 13, pp. 37-52.

Ames Research Staff (1953). Equations, Tables, and Charts for Compressible Flow, NACA Report 1135.

Ames, W. F. (1977). *Numerical Methods for Partial Differential Equations*, 2d ed., Academic, New York.

Amsden, A. A. and Harlow, F. H. (1970). The SMAC Method: A Numerical Technique for Calculating Incompressible Fluid Flows, Los Alamos Scientific Laboratory Report LA-4370, Los Alamos, New Mexico.

Anderson, J. D. (1982). *Modern Compressible Flow*, McGraw-Hill, New York.

Aziz, K. and Hellums, J. D. (1967). Numerical Solution of theThree-dimensional Equations of Motion for Laminar Natural Convection, *Phys. Fluids*, vol. 10, pp. 314-324.

Bailey, F. R. and Ballhaus, W. F. (1972). Relaxation Methods for Transonic Flow about Wing-Cylinder Combinations and Lifting Swept Wings, Proc. Third Int. Conf. Num. Methods Fluid Mech., *Lecture Notes in Physics*, vol. 19, Springer-Verlag, New York, pp. 2-9.

Baker, R. J. and Launder, B. E. (1974). The Turbulent Boundary Layer with Foreign Gas Injection: II–Predictions and Measurements in Severe Streamwise Pressure Gradients, *Int. J. Heat Mass Transfer*, vol. 17, pp. 293-306.

Baldwin, B. S. and Lomax, H. (1978). Thin Layer Approximation and Algebraic Model for Separated Turbulent Flows, AIAA Paper 78-257, Huntsville, Alabama.

Bank, R. E. (1977). Marching Algorithms for Elliptic Boundary Value Problems: II–The Variable Coefficient Case, *SIAM J. Numer. Anal.*, vol. 5, pp. 950–970.

Barakat, H. Z. and Clark, J. A. (1966). On the Solution of the Diffusion Equations by Numerical Methods, *J. Heat Transfer*, vol. 87–88, pp. 421–427.

Barbin, A. R. and Jones, J. B. (1963). Turbulent Flow in the Inlet Region of a Smooth Pipe, *Trans. ASME, J. Basic Eng.*, vol. 85, pp. 29–34.

Beam, R. M. and Warming, R. F. (1976). An Implicit Finite-Difference Algorithm for Hyperbolic Systems in Conservation Law Form, *J. Comp. Phys.*, vol. 22, pp. 87–110.

Beam, R. M. and Warming, R. F. (1978). An Implicit Factored Scheme for the Compressible Navier-Stokes Equations, *AIAA J.*, vol. 16, pp. 393–401.

Beckwith, I. E. and Gallagher, J. J. (1961). Local Heat Transfer and Recovery Temperatures on a Yawed Cylinder at a Mach Number of 4.15 and High Reynolds Numbers, NASA TR R-104.

Benton, E. R. and Platzman, G. W. (1972). A Table of Solutions of the One-dimensional Burgers Equation, *Q. Appl. Math.*, vol. 30, pp. 195–212.

Birch, S. F. (1976). A Critical Reynolds Number Hypothesis and Its Relation to Phenomenological Turbulence Models, *Proc. 1976 Heat Transfer and Fluid Mechanics Institute*, Stanford University Press, Stanford, California, pp. 152–164.

Birkhoff, G., Varga, R. S., and Young, D. (1962). Alternating Direction Implicit Methods, *Advances in Computers*, vol. 3, Academic, New York, pp. 189–273.

Blottner, F. G. (1974). Variable Grid Scheme Applied to Turbulent Boundary Layers, *Comput. Methods Appl. Mech. Eng.*, vol. 4, pp. 179–194.

Blottner, F. G. (1975a). Investigation of Some Finite-Difference Techniques for Solving the Boundary Layer Equations, *Comput. Methods Appl. Mech. Eng.*, vol. 6, pp. 1–30.

Blottner, F. G. (1975b). Computational Techniques for Boundary Layers, *AGARD Lecture Series No. 73 on Computational Methods for Inviscid and Viscous Two- and Three-dimensional Flowfields*, pp. (3-1)–(3-51).

Blottner, F. G. (1977). Numerical Solution of Slender Channel Laminar Flows, *Comput. Methods Appl. Mech. Eng.*, vol. 11, pp. 319–339.

Blottner, F. G. and Ellis, M. A. (1973). Finite-Difference Solution of the Incompressible Three-dimensional Boundary Layer Equations for a Blunt Body, *Comput. Fluids*, vol. 1, Pergamon, Oxford, pp. 133–158.

Bluford, G. S. (1978). Navier-Stokes Solution of Supersonic and Hypersonic Flow Field Around Delta Wings, AIAA Paper 78-1136, Seattle, Washington.

Boussinesq, J. (1877). Essai Sur La Théorie Des Eaux Courantes, *Mem. Présentés Acad. Sci.*, vol. 23, Paris, p. 46.

Bozeman, J. D. and Dalton, C. (1973). Numerical Study of Viscous Flow in a Cavity, *J. Comp. Phys.*, vol. 12, pp. 348–363.

Brackbill, J. U. (1982). Coordinate System Control: Adaptive Meshes, *Numerical Grid Generation, Proceedings of a Symposium on the Numerical Generation of Curvilinear Coordinate Systems and their Use in the Numerical Solution of Partial Differential Equations* (J. F. Thompson, ed.), Elsevier, New York, pp. 277–294.

Brackbill, J. U. and Saltzman, J. (1980). An Adaptive Computation Mesh for the Solution of Singular Perturbation Problems, *Numerical Grid Generation Techniques*, NASA Conference Publication 2166, pp. 193–196.

Bradshaw, P., Ferriss, D. H., and Altwell, N. D. (1967). Calculation of Boundary Layer Development Using the Turbulent Energy Equation, *J. Fluid Mech.*, vol. 28, pp. 593–616.

Bradshaw, P., Dean, R. B., and McEligot, D. M. (1973). Calculation of Interacting Turbulent Shear Layers: Duct Flow, *J. Fluids Eng.*, vol. 95, pp. 214–219.

Brailovskaya, I. (1965). A Difference Scheme for Numerical Solution of the Two-dimensional Nonstationary Navier-Stokes Equations for a Compressible Gas, *Sov. Phys. Dokl.*, vol. 10, pp. 107–110.

Briley, W. R. (1970). A Numerical Study of Laminar Separation Bubbles using the Navier-Stokes Equations, United Aircraft Research Laboratories, Report J110614-1, East Hartford, Connecticut.

Briley, W. R. (1971). A Numerical Study of Laminar Separation Bubbles using the Navier-Stokes Equations, *J. Fluid Mech.*, vol. 47, pp. 713-736.

Briley, W. R. (1974). Numerical Method for Predicting Three-dimensional Steady Viscous Flow in Ducts, *J. Comp. Phys.*, vol. 14, pp. 8-28.

Briley, W. R. and McDonald, H. (1973). Solution of the Three-dimensional Compressible Navier-Stokes Equations by an Implicit Technique, Proc. Fourth Int. Conf. Num. Methods Fluid Dyn., Boulder Colorado, *Lecture Notes in Physics*, vol. 35, Springer-Verlag, New York, pp. 105-110.

Briley, W. R. and McDonald, H. (1975). Numerical Prediction of Incompressible Separation Bubbles, *J. Fluid Mech.*, vol. 69, pp. 631-656.

Briley, W. R. and McDonald, H. (1979). Analysis and Computation of Viscous Subsonic Primary and Secondary Flows, AIAA Paper 79-1453, Williamsburg, Virginia.

Briley, W. R. and McDonald, H. (1980). On the Structure and Use of Linearized Block Implicit Schemes, *J. Comp. Phys.*, vol. 34, pp. 54-73.

Brown, S. N. and Stewartson, K. (1969). Laminar Separation, *Annu. Rev. Fluid Mech.*, vol. 1, Annual Reviews, Inc., Palo Alto, California, pp. 45-72.

Buggeln, R. C., McDonald, H., Kreskovsky, J. P., and Levy, R. (1980). Computation of Three-dimensional Viscous Supersonic Flow in Inlets, AIAA Paper 80-0194, Pasadena, California.

Buneman, O. (1969). A Compact Non-Iterative Poisson Solver, Institute for Plasma Research SUIPR Report 294, Stanford University, California.

Burgers, J. M. (1948). A Mathematical Model Illustrating the Theory of Turbulence, *Adv. Appl. Mech.*, vol. 1, pp. 171-199.

Burggraf, O. R. (1966). Analytical and Numerical Studies of the Structure of Steady Separated Flows, *J. Fluid Mech.*, vol. 24, pp. 113-151.

Burggraf, O. R., Werle, M. J., Rizzetta, D., and Vatsa, V. N. (1979). Effect of Reynolds Number on Laminar Separation of a Supersonic Stream, *AIAA J.*, vol. 17, pp. 336-343.

Burstein, S. Z. and Mirin, A. A. (1970). Third Order Difference Methods for Hyperbolic Equations, *J. Comp. Phys.*, vol. 5, pp. 547-571.

Bushnell, D. M., Cary Jr., A. M., and Holley, B. B. (1975). Mixing Length in Low Reynolds Number Compressible Turbulent Boundary Layers, *AIAA J.*, vol. 13, pp. 1119-1121.

Bushnell, D. M., Cary Jr., A. M., and Harris, J. E. (1976). Calculation Methods for Compressible Turbulent Boundary Layers, von Kármán Institute for Fluid Dynamics, *Lecture Series 86 on Compressible Turbulent Boundary Layers*, vol. 2, Rhode Saint Genese, Belgium.

Buzbee, B. L., Golub, G. H., and Nielson, C. W. (1970). On Direct Methods for Solving Poisson's Equations, *SIAM J. Numer. Anal.*, vol. 7, pp. 627-656.

Caretto, L. S., Gosman, A. D., Patankar, S. V., and Spalding, D. (1972). Two Calculation Procedures for Steady, Three-dimensional Flows with Recirculation, Proc. Third Int. Conf. Num. Methods Fluid Mech., *Lecture Notes in Physics*, vol. 19, Springer-Verlag, New York, pp. 60-68.

Carter, J. E. (1971). Numerical Solutions of the Supersonic, Laminar Flow over a Two-dimensional Compression Corner, Ph.D. thesis, Virginia Polytechnic Institute and State University, Blacksburg.

Carter, J. E. (1978). A New Boundary-Layer Interaction Technique for Separated Flows, NASA TM-78690.

Carter, J. E. (1981). Viscous-Inviscid Interaction Analysis of Transonic Turbulent Separated Flow, AIAA Paper 81-1241, Palo Alto, California.

Carter, J. E. and Wornom, S. F. (1975). Forward Marching Procedure for Separated Boundary Layer Flows, *AIAA J.*, vol. 13, pp. 1101-1103.

Carter, J. E., Edwards, D. E., and Werle, M. J. (1980). A New Coordinate Transformation for Turbulent Boundary Layer Flows, *Numerical Grid Generation Techniques*, NASA Conference Publication 2166, pp. 197-212.

Cebeci, T. (1975). Calculation of Three-dimensional Boundary Layers–II. Three-dimensional Flows in Cartesian Coordinates, *AIAA J.*, vol. 13, pp. 1056-1064.

Cebeci, T. (1976). Separated Flows and Their Representation by Boundary Layer Equations, Report ONR-CR215-234-2, Office of Naval Research, Arlington, Virginia.

Cebeci, T. and Chang, K. C. (1978). A General Method for Calculating Momentum and Heat Transfer in Laminar and Turbulent Duct Flows, *Numer. Heat Transfer*, vol. 1, pp. 39–68.

Cebeci, T. and Smith, A. M. O. (1974). *Analysis of Turbulent Boundary Layers*, Academic, New York.

Cebeci, T., Kaups, K., Mosinskis, G. J., and Rehn, J. A. (1973). Some Problems of the Calculation of Three-dimensional Boundary-Layer Flows on General Configurations, NASA CR-2285.

Cebeci, T., Kaups, K., and Ramsey, J. A. (1977). A General Method for Calculating Three-dimensional Compressible Laminar and Turbulent Boundary Layers on Arbitrary Wings, NASA CR-2777.

Cebeci, T., Khattab, A. A., and Stewartson, K. (1979a). Prediction of Three-dimensional Laminar and Turbulent Boundary Layers on Bodies of Revolution at High Angles of Attack, *Proc. Second Symposium on Turbulent Shear Flows*, Imperial College, London, pp. 15.8–15.13.

Cebeci, T., Carr, L. W., and Bradshaw, P. (1979b). Prediction of Unsteady Turbulent Boundary Layers with Flow Reversal, *Proc. Second Symposium on Turbulent Shear Flows*, Imperial College, London, pp. 14.23–14.28.

Chakravarthy, S. R. (1979). The Split-Coefficient Matrix Method for Hyperbolic Systems of Gasdynamic Equations, Ph.D. dissertation, Department of Aerospace Engineering, Iowa State University, Ames.

Chakravarthy, S. R., Anderson, D. A., and Salas, M. D. (1980). The Split-Coefficient Matrix Method for Hyperbolic Systems of Gasdynamic Equations, AIAA Paper 80-0268, Pasadena, California.

Chambers, T. L. and Wilcox, D. C. (1976). A Critical Examination of Two-Equation Turbulence Closure Models, AIAA Paper 76-352, San Diego, California.

Chan, Y. Y. (1972). Compressible Turbulent Boundary Layer Computations Based on an Extended Mixing Length Approach, *Canadian Aeronautics and Space Institute Transactions*, vol. 5, pp. 21–27.

Chapman, A. J. (1974). *Heat Transfer*, 3d ed., Macmillan, New York.

Chapman, D. R. (1975). Introductory Remarks, NASA SP-347, pp. 4–7.

Chapman, D. R. (1979). Computational Aerodynamics Development and Outlook, *AIAA J.*, vol. 17, pp. 1293–1313.

Chaussee, D. S. and Pulliam, T. H. (1981). Two-dimensional Inlet Simulation Using a Diagonal Implicit Algorithm, *AIAA J.*, vol. 19, pp. 153–159.

Cheng, H. K. (1963). The Blunt-Body Problem in Hypersonic Flow at Low Reynolds Number, Cornell Aeronautical Laboratory, AF-1285-A-10, Buffalo, New York.

Cheng, H. K., Chen, S. Y., Mobley, R., and Huber, C. R. (1970). The Viscous Hypersonic Slender-Body Problem: A Numerical Approach Based on a System of Composite Equations, The Rand Corporation, RM-6193-PR, Santa Monica, California.

Chilukuri, R. and Pletcher, R. H. (1980). Numerical Solutions to the Partially Parabolized Navier-Stokes Equations for Developing Flow in a Channel, *Numer. Heat Transfer*, vol. 3, pp. 169–188.

Chorin, A. J. (1967). A Numerical Method for Solving Incompressible Viscous Flow Problems, *J. Comp. Phys.*, vol. 2, pp. 12–26.

Chorin, A. J. (1968). Numerical Solution of the Navier-Stokes Equations, *Math. Comput.*, vol. 22, pp. 745–762.

Chow, L. C. and Tien, C. L. (1978). An Examination of Four Differencing Schemes for Some Elliptic-Type Convection Equations, *Numer. Heat Transfer*, vol. 1, pp. 87–100.

Chung, T. J. (1978). *Finite Element Analysis in Fluid Dynamics*, McGraw-Hill, New York.

Churchill, R. V. (1941). *Fourier Series and Boundary Value Problems*, McGraw-Hill, New York.

Churchill, R. V. (1948). *Introduction to Complex Variables*, McGraw-Hill, New York.

Churchill, R. V. (1960). *Introduction to Complex Variables and Applications*, 2d ed., McGraw-Hill, New York.

Churchill, S. W. (1974). *The Interpretation and Use of Rate Data: The Rate Concept*, Hemisphere, Washington, D.C.

Cole, J. D. and Messiter, A. F. (1957). Expansion Procedures and Similarity Laws for Transonic Flow, *Z. Angew. Math. Phys.*, vol. 8, pp. 1-25.

Coles, D. E. (1953). Measurements in the Boundary Layer on a Smooth Flat Plate in Supersonic Flow, III. Measurements in a Flat Plate Boundary Layer at the Jet Propulsion Laboratory, Jet Propulsion Laboratory Report 20-71, California Institute of Technology, Pasadena, California.

Courant, R. and Friedrichs, K. O. (1948). *Supersonic Flow and Shock Waves*, Interscience Publishers, New York.

Courant, R., Friedrichs, K. O., and Lewy, H. (1928). Über die Partiellen Differenzengleichungen der Mathematischen Physik, *Mathematische Annalen*, vol. 100, pp. 32-74. (Translated to: On the Partial Difference Equations of Mathematical Physics, *IBM J. Res. Dev.*, vol. 11, pp. 215-234, 1967.)

Crank, J. and Nicolson, P. (1947). A Practical Method for Numerical Evaluation of Solutions of Partial Differential Equations of the Heat-Conduction Type, *Proc. Cambridge Philos. Soc.*, vol. 43, pp. 50-67.

Crawford, M. E. and Kays, W. M. (1975). STAN5–A Program for Numerical Computation of Two-dimensional Internal/External Boundary Layer Flows, Report No. HMT-23, Thermosciences Division, Department of Mechanical Engineering, Stanford University, California.

Crowley, W. P. (1967). Second-Order Numerical Advection, *J. Comp. Phys.*, vol. 1, pp. 471-484.

Daly, B. J. and Harlow, F. H. (1970). Transport Equations in Turbulence, *Phys. Fluids*, vol. 13, pp. 2634-2649.

Dancey, C. L. and Pletcher, R. H. (1974). A Boundary Layer Finite Difference Method for Calculating Through the Separation Point and Into the Region of Recirculation in Incompressible Laminar Flow, Engineering Research Institute Technical Report 74103/HTL-2, Iowa State University, Ames.

Davis, R. T. (1963). Laminar Compressible Flow Past Axisymmetric Blunt Bodies (Results of a Second Order Theory), Ph.D. dissertation, Stanford University, California.

Davis, R. T. (1970). Numerical Solution of the Hypersonic Viscous Shock-Layer Equations, *AIAA J.*, vol. 8, pp. 843-851.

Davis, R. T. (1979). Numerical Methods for Coordinate Generation Based on Schwarz-Christoffel Transformations, AIAA Paper 79-1463, Williamsburg, Virginia.

Davis, R. T. and Flügge-Lotz, I. (1964). Second-Order Boundary-Layer Effects in Hypersonic Flow Past Axisymmetric Blunt Bodies, *J. Fluid Mech.*, vol. 20, pp. 593-623.

Davis, R. T. and Rubin, S. G. (1980). Non-Navier-Stokes Viscous Flow Computations, *Comput. Fluids*, vol. 8, pp. 101-131.

Daywitt, J. E. and Anderson, D. A. (1974). Analysis of a Time-Dependent Finite-Difference Technique for Shock Interaction and Blunt-Body Flows, Engineering Research Institute Technical Report 74074, Iowa State University, Ames.

Deardorff, J. W. (1970). A Numerical Study of Three-dimensional Turbulent Channel Flow at Large Reynolds Numbers, *J. Fluid Mech.*, vol. 41, pp. 453-480.

De Neef, T. and Moretti, G. (1980). Shock Fitting For Everybody, *Comput. Fluids*, vol. 8, pp. 327-334.

Der Jr., J. and Raetz, G. S. (1962). Solution of General Three-dimensional Laminar Boundary Layer Problems by an Exact Numerical Method, Institute of the Aerospace Sciences, Paper 62-70, New York.

Désidéri, J.-A. and Tannehill, J. C. (1977a). Over-Relaxation Applied to the MacCormack Finite-Difference Scheme, *J. Comp. Phys.*, vol. 23, pp. 313-326.

Désidéri, J.-A. and Tannehill, J. C. (1977b). Time-Accuracy of the Over-Relaxed MacCormack Finite-Difference Scheme, ERI Report 77251, Iowa State University, Ames.

Désidéri, J.-A., Steger, J. L., and Tannehill, J. C. (1978). On Improving the Iterative Convergence Properties of an Implicit Approximate-Factorization Finite Difference Algorithm, NASA Technical Memorandum 78495.

Dodge, P. R. (1977). Numerical Method for 2D and 3D Viscous Flows, *AIAA J.*, vol. 15, pp. 961-965.

Donaldson, C. duP. (1972). Calculation of Turbulent Shear Flows for Atmospheric and Vortex Motions, *AIAA J.*, vol. 10, pp. 4–12.

Donaldson, C. duP. and Rosenbaum, H. (1968). Calculation of Turbulent Shear Flows Through Closure of the Reynolds Equations by Invariant Modeling, Aero. Res. Assoc. of Princeton Report 127.

Dorrance, W. H. (1962). *Viscous Hypersonic Flow*, McGraw-Hill, New York.

Douglas Jr., J. (1955). On the Numerical Integration of $\partial^2 u/\partial x^2 + \partial^2 u/\partial y^2 = \partial u/\partial t$ by Implicit Methods, *J. Soc. Ind. Appl. Math.*, vol. 3, pp. 42–65.

Douglas, J. and Gunn, J. E. (1964). A General Formulation of Alternating Direction Methods–Part I. Parabolic and Hyperbolic Problems, *Numerische Mathematik*, vol. 6, pp. 428–453.

Douglas, J. and Rachford, H. H. (1956). On the Numerical Solution of Heat Conduction Problems in Two and Three Space Variables, *Trans. Amer. Math. Soc.*, vol. 82, pp. 421–439.

DuFort, E. C. and Frankel, S. P. (1953). Stability Conditions in the Numerical Treatment of Parabolic Differential Equations, *Mathematical Tables and Other Aids to Computation*, vol. 7, pp. 135–152.

Dwyer, H. A. (1971). Hypersonic Boundary Layer Studies on a Spinning Sharp Cone at Angle of Attack, AIAA Paper 71-57, New York, New York.

Dwyer, H. A. and Sanders, B. R. (1975). A Physically Optimum Difference Scheme for Three-dimensional Boundary Layers, Proc. Fourth Int. Conf. Num. Methods Fluid Dyn., *Lecture Notes in Physics*, vol. 35, Springer-Verlag, New York, pp. 144–150.

Dwyer, H. A., Kee, R. J., and Sanders, B. R. (1979). An Adaptive Grid Method for Problems in Fluid Mechanics and Heat Transfer, AIAA Paper 79-1464, Williamsburg, Virginia.

Dwyer, H. A. Raiszadeh, F., and Otey, G. (1980). A Study of Reactive Diffusion Problems with Stiff Integrators and Adaptive Grids, Proc. Seventh Int. Conf. Num. Methods Fluid Dyn., *Lecture Notes in Physics*, vol. 141, Springer-Verlag, New York, pp. 170–175.

Eiseman, P. (1979). A Multi-Surface Method of Coordinate Generation, *J. Comp. Phys.*, vol. 33, pp. 118–150.

Eiseman, P. K. and Smith, R. E. (1980). Mesh Generation Using Algebraic Techniques, *Numerical Grid Generation Techniques*, NASA Conference Publication 2166, pp. 73–120.

Emery, A. F. and Gessner, F. B. (1976). The Numerical Prediction of the Turbulent Flow and Heat Transfer in the Entrance Region of a Parallel Plate Duct, *J. Heat Transfer*, vol. 98, pp. 594–600.

Evans, M. E. and Harlow, F. H. (1957). The Particle-in-Cell Method for Hydrodynamic Calculations, Los Alamos Scientific Laboratory Report LA-2139, Los Alamos, New Mexico.

Favre, A. (1965). Équations des Gaz Turbulents Compressibles: 1. Formes Générales, *Journal de Mécanique*, vol. 4, pp. 361–390.

Fike, C. T. (1970). *PL/1 for Scientific Programmers*, Prentice-Hall, Englewood Cliffs, New Jersey.

Forsythe, G. E. and Wasow, W. (1960). *Finite Difference Methods for Partial Differential Equations*, Wiley, New York.

Frankel, S. P. (1950). Convergence Rates of Iterative Treatments of Partial Differential Equations, *Mathematical Tables and Other Aids to Computation*, vol. 4, pp. 65–75.

Friedrich, C. M. and Forstall Jr., W. (1953). A Numerical Method for Computing the Diffusion Rate of Coaxial Jets, *Proceedings of the Third Midwestern Conference on Fluid Mechanics*, Univ. of Minnesota Inst. of Technol., Minneapolis, Minnesota, pp. 635–649.

Fröberg, C. (1969). *Introduction to Numerical Analysis*, 2d ed., Addison-Wesley, Reading, Massachusetts, pp. 21–28.

Fromm, J. E. (1968). A Method for Reducing Dispersion in Convective Difference Schemes, *J. Comp. Phys.*, vol. 3, pp. 176–189.

Gabutti, B. (1982). On Two Upwind Finite-Difference Schemes for Hyperbolic Equations in Non-conservative Form, *Comput. Fluids* (in press).

Garabedian, P. R. (1964). *Partial Differential Equations*, Wiley New York.

Gardner, W. D. (1982). The Independent Inventor, *Datamation*, vol. 28, pp. 12–22.

Gary, J. (1962). Numerical Computation of Hydrodynamic Flows Which Contain a Shock, Courant Institute of Mathematical Sciences Report NYO 9603, New York University.

Gary, J. (1969). The Numerical Solution of Partial Differential Equations, National Center for Atmospheric Research, NCAR Manuscript 69-54, Boulder, Colorado.

Gault, D. E. (1955). An Experimental Investigation of Regions of Separated Laminar Flow, NACA TN-3505.

Ghia, K. N. and Sokhey, J. S. (1977). Laminar Incompressible Viscous Flow in Curved Ducts of Regular Cross-Sections, *J. Fluids Eng.*, vol. 99, pp. 640–648.

Ghia, K. N., Hankey Jr., W. L., and Hodge, J. K. (1977). Study of Incompressible Navier-Stokes Equations in Primitive Variables Using Implicit Numerical Technique, AIAA Paper 77-648, Albuquerque, New Mexico.

Ghia, K. N., Hankey Jr., W. L., and Hodge, J. K. (1979). Use of Primitive Variables in the Solution of Incompressible Navier-Stokes Equations, *AIAA J.*, vol. 17, pp. 298-301.

Ghia, U., Ghia, K. N., and Struderus, C. J. (1977). Three-dimensional Laminar Incompressible Flow in Straight Polar Ducts, *Comput. Fluids*, vol. 5, pp. 205–218.

Ghia, U., Ghia, K. N., Rubin, S. G., and Khosla, P. K. (1981). Study of Incompressible Flow Separation Using Primitive Variables, *Comput. Fluids*, vol. 9, pp. 123–142.

Gnoffo, P. A. (1980). Complete Supersonic Flowfields Over Blunt Bodies in a Generalized Orthogonal Coordinate System, NASA TM 81784.

Godunov, S. K. (1959). Finite-Difference Method for Numerical Computation of Discontinuous Solutions of the Equations of Fluid Dynamics, *Mat. Sb.*, vol. 47, pp. 271–306.

Goldstein, S. (1948). On Laminar Boundary Layer Flow near a Position of Separation, *Q. J. Mech. Appl. Math.*, vol. 1, pp. 43–69.

Gordon, P. (1969). The Diagonal Form of Quasi-Linear Hyperbolic Systems as a Basis for Difference Equations, General Electric Company Final Report, Naval Ordnance Laboratory Contract No. N60921-7164, pp. II.D-1, II.D-22.

Gordon, W. and Hall, C. (1973). Construction of Curvilinear Coordinate Systems and Application to Mesh Generation, *International Journal for Numerical Methods in Engineering*, vol. 7, pp. 461–477.

Gosman, A. D. and Spalding, D. B. (1971). The Prediction of Confined Three-dimensional Boundary Layers, *Salford Symposium on Internal Flows*, Paper 19, Inst. Mech. Engrs., London.

Greenspan, D. (1961). *Introduction to Partial Differential Equations*, McGraw-Hill, New York.

Grossman, B. (1979). Numerical Procedure for the Computation of Irrotational Conical Flows, *AIAA J.*, vol. 17, pp. 828-837.

Grossman, B. and Siclari, M. J. (1980). The Nonlinear Supersonic Potential Flow over Delta Wings, AIAA Paper 80-0269, Pasadena, California.

Hadamard, J. (1952). *Lectures on Cauchy's Problem in Linear Partial Differential Equations*, Dover, New York.

Hafez, M., South, J., and Murman, E. (1979). Artificial Compressibility Methods for Numerical Solutions of Transonic Full Potential Equation, *AIAA J.*, vol. 17, pp. 838–844.

Hall, M. G. (1981). Computational Fluid Dynamics–A Revolutionary Force in Aerodynamics, AIAA Paper 81-1014, Palo Alto, California.

Hanjalić, K. and Launder, B. E. (1972). A Reynolds Stress Model of Turbulence and Its Application to Asymmetric Shear Flows, *J. Fluid Mech.*, vol. 52, pp. 609–638.

Hansen, A. G. (1964). *Similarity Analyses of Boundary Value Problems in Engineering*, Prentice-Hall, Englewood Cliffs, New Jersey.

Harlow, F. H. and Fromm, J. E. (1965). Computer Experiments in Fluid Dynamics, *Sci. Am.*, vol. 212, pp. 104–110.

Harlow, F. H. and Nakayama, P. I. (1968). Transport of Turbulence Energy Decay Rate, Los Alamos Scientific Laboratory Report LA-3854, Los Alamos, New Mexico.

Harlow, F. H. and Welch, J. E. (1965). Numerical Calculation of Time-Dependent Viscous Incompressible Flow of Fluid with Free Surface, *Phys. Fluids*, vol. 8, pp. 2182–2189.

Harris, J. E. (1971). Numerical Solution of the Equations for Compressible Laminar, Transitional, and Turbulent Boundary Layers and Comparisons with Experimental Data, NASA TR-R 368.

Harten, A. (1978). The Artificial Compression Method for Computation of Shocks and Contact Discontinuities: III. Self-Adjusting Hybrid Schemes, *Math. Comput.*, vol. 32, pp. 363–389.

Hartnett, J. P. and Eckert, E. R. G. (1957). Mass-Transfer Cooling in a Laminar Boundary Layer with Constant Fluid Properties, *Trans. ASME*, vol. 79, pp. 247–254.

Hayes, W. D. (1966). La Seconde Approximation Pour les Écoulements Transsoniques Non Visqueux, *J. Méc.*, vol. 5, pp. 163–206.

Hayes, W. D. and Probstein, R. F. (1966). *Hypersonic Flow Theory*, 2d ed., Academic, New York.

Healzer, J. M., Moffat, R. J., and Kays, W. M. (1974). The Turbulent Boundary Layer on a Rough Porous Plate: Experimental Heat Transfer with Uniform Blowing, Thermosciences Division, Report No. HMT-18, Department of Mechanical Engineering, Stanford University, California.

Helliwell, W S., Dickinson, R. P., and Lubard, S. C. (1980). Viscous Flow over Arbitrary Geometries at High Angle of Attack, AIAA Paper 80-0064, Pasadena, California.

Hellwig, G. (1977). *Partial Differential Equations: An Introduction*, B. G. Teubner, Stuttgart.

Herring, H. J. and Mellor, G. L. (1968). A Method of Calculating Compressible Turbulent Boundary Layers, NASA CR-1144.

Hess, J. L. and Smith, A. M. O. (1967). Calculation of Potential Flow about Arbitrary Bodies, *Progress in Aeronautical Sciences*, vol. 8, Pergamon, New York, pp. 1–138.

Hildebrand, F. B. (1956). *Introduction to Numerical Analysis*, McGraw-Hill, New York.

Hindman, R. G. (1981). Geometrically Induced Errors and Their Relationship to the Form of the Governing Equations and the Treatment of Generalized Mappings, AIAA Paper 81-1008, Palo Alto, California.

Hindman, R. G. and Spencer, J. (1983). A New Approach to Truly Adaptive Grid Generation, AIAA Paper 83-0450, Reno, Nevada.

Hindman, R. G., Kutler, P., and Anderson, D. A. (1979). A Two-dimensional Unsteady Euler Equation Solver for Flow Regions with Arbitrary Boundaries, AIAA Paper 79-1465, Williamsburg, Virginia.

Hindman, R. G., Kutler, P., and Anderson, D. A. (1981). Two-dimensional Unsteady Euler-Equation Solver for Arbitrarily Shaped Flow Regions, *AIAA J.*, vol. 19, pp. 424–431.

Hinze, J. O. (1975). *Turbulence*, 2d ed., McGraw-Hill, New York.

Hirschfelder, J. O., Curtiss, C. F., and Bird, R. B. (1954). *Molecular Theory of Gases and Liquids*, Wiley, New York.

Hirsh, R. S. and Rudy, D. H. (1974). The Role of Diagonal Dominance and Cell Reynolds Number in Implicit Methods for Fluid Mechanics Problems, *J. Comp. Phys.*, vol. 16, pp. 304–310.

Hirt, C. W. (1968). Heuristic Stability Theory for Finite-Difference Equations, *J. Comp. Phys.*, vol. 2, pp. 339–355.

Hockney, R. W. (1965). A Fast Direct Solution of Poisson's Equation using Fourier Analysis, *J. Assoc. Comput. Mach.*, vol. 12, pp. 95–113.

Hockney, R. W. (1970). The Potential Calculation and Some Applications, *Methods in Computational Physics*, vol. 9, Academic, New York, pp. 135–211.

Holst, T. L. (1979). Implicit Algorithm for the Conservative Transonic Full-Potential Equation using an Arbitrary Mesh, *AIAA J.*, vol. 17, pp. 1038--1045.

Holst, T. L. (1980). Fast, Conservative Algorithm for Solving the Transonic Full-Potential Equation, *AIAA J.*, vol. 18, pp. 1431–1439.

Holst, T. L. and Ballhaus, W. F. (1979). Fast, Conservative Schemes for the Full Potential Equation Applied to Transonic Flows, *AIAA J.*, vol. 17, pp. 145–152.

Hong, S. W. (1974). Laminar Flow Heat Transfer in Ordinary and Augmented Tubes, Ph.D. dissertation, Iowa State University, Ames.

Hornbeck, R. W. (1963). Laminar Flow in the Entrance Region of a Pipe, *Appl. Sci. Res., Sec. A*, vol. 13, pp. 224–232.

Hornbeck, R. W. (1973). *Numerical Marching Techniques for Fluid Flows with Heat Transfer*, NASA SP-297.

Horstman, C. C. (1977). Turbulence Model for Non-Equilibrium Adverse Pressure Gradient Flows, *AIAA J.*, vol. 15, pp. 131–132.

Hosny, W. M., Davis, R. T., and Werle, M. J. (1978). Improvements to the Solution of the Viscous Shock Layer Equations, Department of Aerospace Engineering and Applied Mechanics, Report AFL 78-11-45, University of Cincinnati, Ohio.

Howarth, L. (1938). On the Solution of the Laminar Boundary Layer Equations, *Proc. Roy. Soc. London, Ser. A*, vol. 164, pp. 547–579.

Howarth, L. (1951). The Boundary Layer in Three-dimensional Flow. Part I: Derivation of the Equations for Flow along a General Curved Surface, *Philos. Mag.*, vol. 42, pp. 239–243.

Hwang, S. S. and Pletcher, R. H. (1978). Prediction of Turbulent Jets and Plumes in Flowing Ambients, Engineering Research Institute Technical Report 79003/HTL-15, Iowa State University, Ames.

James, M. L., Smith, G. M., and Wolford, J. C. (1967). *Applied Numerical Methods for Digital Computation with FORTRAN*, International Textbook Company, Scranton, Pennsylvania.

Jameson, A. (1974). Interative Solution of Transonic Flows over Airfoils and Wings Including Flows at Mach 1, *Comm. Pure Appl. Math.*, vol. 27, pp. 283–309.

Jameson, A. (1975). Transonic Potential Flow Calculations using Conservation Form, *Proc. AIAA 2nd Computational Fluid Dynamics Conference*, Hartford, Connecticut, pp. 148–161.

Jeffrey, A. and Taniuti, T. (1964). *Nonlinear Wave Propagation with Applications to Physics and Magneto-Hydrodynamics*, Academic, New York.

Jensen, V. G. (1959). Viscous Flow Round a Sphere at Low Reynolds Numbers ($\leqslant 40$), *Proc. Roy. Soc. London, Ser. A*, vol. 249, pp. 346–366.

Jobe, C. E. (1974). The Numerical Solution of the Asymptotic Equations of Trailing Edge Flow, Technical Report AFFDL-TR-74-46, Air Force Flight Dynamics Laboratory, Dayton, Ohio.

Jobe, C. E. and Burggraf, O. R. (1974). The Numerical Solution of the Asymptotic Equations of Trailing Edge Flow, *Proc. Roy. Soc. London, Ser. A*, vol. 340, pp. 91–111.

Johnson, F. and Rubbert, P. (1975). Advanced Panel-Type Influence Coefficient Methods Applied to Subsonic Flows, AIAA Paper 75-50, Pasadena, California.

Johnson, G. M. (1980). An Alternative Approach to the Numerical Simulation of Steady Inviscid Flow, Proc. Seventh Int. Conf. Num. Methods Fluid Dyn., *Lecture Notes in Physics*, vol. 141, Springer-Verlag, New York, pp. 236–241.

Jones, W. P. and Launder, B. E. (1972). The Prediction of Laminarization with a Two-Equation Model of Turbulence, *Int. J. Heat Mass Transfer*, vol. 15, pp. 301–314.

Karamcheti, K. (1966). *Principles of Ideal-Fluid Aerodynamics*, Wiley, New York.

Kays, W. M. (1972). Heat Transfer to the Transpired Turbulent Boundary Layer, *Int. J. Heat Mass Transfer*, vol. 15, pp. 1023–1044.

Kays, W. M. and Crawford, M. E. (1980). *Convective Heat and Mass Transfer*, 2d ed., McGraw-Hill, New York.

Kays, W. M. and Moffat, R. J. (1975). The Behavior of Transpired Turbulent Boundary Layers, *Studies in Convection: Theory, Measurement, and Applications*, vol. 1, Academic, New York, pp. 223–319.

Keller, H. B. (1970). A New Difference Scheme for Parabolic Problems, *Numerical Solutions of Partial Differential Equations*, vol. 2, (J. Bramble, ed.), Academic, New York.

Keller, H. B. and Cebeci, T. (1972). Accurate Numerical Methods for Boundary-Layer Flows. II: Two-dimensional Turbulent Flows, *AIAA J.*, vol. 10, pp. 1193–1199.

Kentzer, C. P. (1970). Discretization of Boundary Conditions on Moving Discontinuities, Proc. Second Int. Conf. Num. Methods Fluid Dyn., *Lecture Notes in Physics*, vol. 8, Springer-Verlag, New York, pp. 108–113.

Kitchens Jr., C. W., Sedney, R., and Gerber, N. (1975). The Role of the Zone of Dependence Concept in Three-dimensional Boundary-Layer Calculations, *Proc. AIAA 2nd Computational Fluid Dynamics Conference*, Hartford, Connecticut, pp. 102–112.

Klineberg, J. M. and Steger, J. L. (1974). On Laminar Boundary Layer Separation, AIAA Paper 74-94, Washington, D.C.

Klinksiek, W. F. and Pierce, F. J. (1973). A Finite-Difference Solution of the Two- and Three-dimensional Incompressible Turbulent Boundary Layer Equations, *J. Fluids Eng.*, vol. 95, pp. 445–458.

Klopfer, G. H. and McRae, D. S. (1981a). The Nonlinear Modified Equation Approach to Analyzing Finite-Difference Schemes, AIAA Paper 81-1029, Palo Alto, California.

Klopfer, G. H. and McRae, D. S. (1981b). Nonlinear Analysis of the Truncation Errors in Finite-Difference Schemes for the Full System of Euler Equations, AIAA Paper 81-0193, St. Louis, Missouri.

Klunker, E. (1971). Contributions to Methods for Calculating the Flow about Thin Lifting Wings at Transonic Speeds–Analytical Expression for the Far Field, NASA TN D-6530.

Knechtel, E. D. (1959). Experimental Investigation at Transonic Speeds of Pressure Distributions over Wedge and Circular Arc Sections and Evaluation of Perforated-Wall Interference, NASA TN D-15.

Korkegi, R. H. (1956). Transition Studies and Skin-Friction Measurements on an Insulated Flat Plate at a Mach Number of 5.8, *J. Aero. Sci.*, vol. 25, pp. 97–192.

Kowalski, E. J. (1980). Boundary-Fitted Coordinate Systems for Arbitrary Computational Regions, *Numerical Grid Generation Techniques*, NASA Conference Publication 2166, pp. 331–353.

Krause, E. (1969). Comment on "Solution of a Three-dimensional Boundary-Layer Flow with Separation," *AIAA J.*, vol. 7, pp. 575–576.

Kreskovsky, J. P., Shamroth, S. J., and McDonald, H. (1974). Parametric Study of Relaminarization of Turbulent Boundary Layers on Nozzle Walls, NASA CR-2370.

Kutler, P. and Lomax, H. (1971). The Computation of Supersonic Flow Fields about Wing-Body Combinations by "Shock-Capturing" Finite Difference Techniques, Proc. Second Int. Conf. Num. Methods Fluid Dyn., *Lecture Notes in Physics*, vol. 8, Springer-Verlag, Berlin, pp. 24–29.

Kutler, P., Warming, R. F., and Lomax, H. (1973). Computation of Space Shuttle Flowfields using Noncentered Finite-Difference Schemes, *AIAA J.*, vol. 11, pp. 196–204.

Kwon, O. K. and Pletcher, R. H. (1979). Prediction of Incompressible Separated Boundary Layers Including Viscous-Inviscid Interaction, *J. Fluids Eng.*, vol. 101, pp. 466–472.

Kwon, O. K. and Pletcher, R. H. (1981). Prediction of the Incompressible Flow over a Rearward-Facing Step, Engineering Research Institute Technical Report 82019/HTL-26, Iowa State University, Ames.

Laasonen, P. (1949). Über eine Methode zur Lösung der Wärmeleitungsgleichung, *Acta Math.*, vol. 81, pp. 309–317.

Larkin, B. K. (1964). Some Stable Explicit Difference Approximations to the Diffusion Equation, *Math. Comput.*, vol. 18, pp. 196–202.

Launder, B. E. (1979). Stress-Transport Closures: Into the Third Generation, *Proc. First Symposium on Turbulent Shear Flows*, Springer-Verlag, New York.

Launder, B. E. and Spalding, D. B. (1972). *Mathematical Models of Turbulence*, Academic, New York.

Launder, B. E. and Spalding, D. B. (1974). The Numerical Computation of Turbulent Flows, *Comput. Methods Appl. Mech. Eng.*, vol. 3, pp. 269–289.

Lax, P. D. (1954). Weak Solutions of Nonlinear Hyperbolic Equations and their Numerical Computation, *Comm. Pure Appl. Math.*, vol. 7, pp. 159–193.

Lax, P. D. and Wendroff, B. (1960). Systems of Conservation Laws, *Comm. Pure Appl. Math.*, vol. 13, pp. 217–237.

Leonard, B. P. (1979a). A Stable and Accurate Convective Modelling Procedure Based on Quadratic Upstream Interpolation, *Comput. Methods Appl. Mech. Eng.*, vol. 19, pp. 59–98.

Leonard, B. P. (1979b). A Survey of Finite Differences of Opinion on Numerical Muddling of the Incomprehensible Defective Confusion Equation, *Finite Element Methods for Convective Dominated Flows*, AMD, vol. 34, The American Society of Mechanical Engineers.

Levine, R. D. (1982). Supercomputers, *Sci. Am.*, vol. 246, pp. 118–135.

LeBail, R. C. (1972). Use of Fast Fourier Transforms for Solving Partial Differential Equations in Physics, *J. Comp. Phys.*, vol. 9, pp. 440–465.

Li, C. P. (1973). Numerical Solution of Viscous Reacting Blunt Body Flows of a Multicomponent Mixture, AIAA Paper 73-202, Washington, D.C.

Li, C. P. (1977). A Numerical Study of Separated Flows Induced by Shock-Wave/Boundary-Layer Interaction, AIAA Paper 77-168, Los Angeles, California.

Li, C. P. (1981). Application of an Implicit Technique to the Shock-Layer Flow Around General Bodies, AIAA Paper 81-0191, St. Louis, Missouri. See *AIAA J.*, vol. 20, 1982, pp. 175–183.

Liebmann, L. (1918). Die Angenäherte Ermittelung Harmonischer Funktionen und Konformer Abbildungen, *Sitzungsber., Math. Phys. Kl. Bayer. Akad. Wiss.*, vol. 3, p. 385.

Liepmann, H. W. and Roshko, A. (1957). *Elements of Gasdynamics*, Wiley, New York.

Lighthill, M. J. (1953). On Boundary Layers and Upstream Influence. II. Supersonic Flows without Separation, *Proc. Roy. Soc. London, Ser. A*, vol. 217, pp. 478–507.

Lighthill, M. J. (1958). On Displacement Thickness, *J. Fluid Mech.*, vol. 4, pp. 383–392.

Lin, T. C. and Rubin, S. G. (1973a). Viscous Flow over Spinning Cones at Angle of Attack, *AIAA J.*, vol. 12, pp. 975–985.

Lin, T. C. and Rubin, S. G. (1973b). Viscous Flow over a Cone at Moderate Incidence: I. Hypersonic Tip Region, *Comput. Fluids*, vol. 1, pp. 37–57.

Lin, T. C. and Rubin, S. G. (1979). A Numerical Model for Supersonic Viscous Flow over a Slender Reentry Vehicle, AIAA Paper 79-0205, New Orleans, Louisiana.

Lin, T. C., Rubin, S. G., and Widhopf, G. F. (1981). A Two-Layer Model for Coupled Three Dimensional Viscous and Inviscid Flow Calculations, AIAA Paper 81-0118, St. Louis, Missouri.

Lindemuth, I. and Killeen, J. (1973). Alternating Direction Implicit Techniques for Two Dimensional Magnetohydrodynamics Calculations, *J. Comp. Phys.*, vol. 13, pp. 181–208.

Lock, R. C. (1970). Test Cases for Numerical Methods in Two-dimensional Transonic Flows, AGARD Report 575.

Lubard, S. C. and Helliwell, W. S. (1973). Calculation of the Flow on a Cone at High Angle of Attack, R&D Associates Technical Report, RDA-TR-150, Santa Monica, California.

Lubard, S. C. and Helliwell, W. S. (1974). Calculation of the Flow on a Cone at High Angle of Attack, *AIAA J.*, vol. 12, pp. 965–974.

Ludford, G. (1951). The Behavior at Infinity of the Potential Function of a Two-dimensional Subsonic Compressible Flow, *J. Math. Phys.*, vol. 30, pp. 131–159.

Lugt, H. J. and Ohring, S. (1974). Efficiency of Numerical Methods in Solving the Time-Dependent, Two-dimensional Navier-Stokes Equations, *Numerical Methods in Fluid Dynamics*, Peutech, London.

Macagno, E. O. (1965). Some New Aspects of Similarity in Hydraulics, *La Houille Blanche*, vol. 20, pp. 751–759.

MacCormack, R. W. (1969). The Effect of Viscosity in Hypervelocity Impact Cratering, AIAA Paper 69-354, Cincinnati, Ohio.

MacCormack, R. W. (1971). Numerical Solution of the Interaction of a Shock Wave with a Laminar Boundary Layer, Proc. Second Int. Conf. Num. Methods Fluid Dyn., *Lecture Notes in Physics*, vol. 8, Springer-Verlag, New York, pp. 151–163.

MacCormack, R. W. (1976). An Efficient Numerical Method for Solving the Time-Dependent Compressible Navier-Stokes Equations at High Reynolds Number, NASA TM X-73,-129.

MacCormack, R. W. (1981). A Numerical Method for Solving the Equations of Compressible Viscous Flow, AIAA Paper 81-0110, St. Louis, Missouri.

MacCormack, R. W. and Baldwin, B. S. (1975). A Numerical Method for Solving the Navier-Stokes Equations with Application to Shock-Boundary Layer Interactions, AIAA Paper 75-1, Pasadena, California.

MacCormack, R. W. and Paullay, A. J. (1972). Computational Efficiency Achieved by Time Splitting of Finite Difference Operators, AIAA Paper 72-154, San Diego, California.

Madavan, N. K. and Pletcher, R. H. (1982). Prediction of Incompressible Laminar Separated Flows Using the Partially Parabolized Navier-Stokes Equations, Engineering Research Institute Technical Report 82127/HTL-27, Iowa State University, Ames.

Madni, I. K. and Pletcher, R. H. (1975a). A Finite-Difference Analysis of Turbulent, Axisymmetric, Buoyant Jets and Plumes, Engineering Research Institute Technical Report 76096/HTL-10, Iowa State University, Ames.

Madni, I. K. and Pletcher, R. H. (1975b). Prediction of Turbulent Jets in Coflowing and Quiescent Ambients, *J. Fluids Eng.*, vol. 97, pp. 558–567.

Madni, I. K. and Pletcher, R. H. (1977a). Prediction of Turbulent Forced Plumes Issuing Vertically into Stratified or Uniform Ambients, *J. Heat Transfer*, vol. 99, pp. 99–104.

Madni, I. K. and Pletcher, R. H. (1977b). Buoyant Jets Discharging Nonvertically into a Uniform, Quiescent Ambient–A Finite-Difference Analysis and Turbulence Modeling, *J. Heat Transfer*, vol. 99, pp. 641–647.

Malik, M. R. and Pletcher, R. H. (1978). Computation of Annular Turbulent Flows with Heat Transfer and Property Variations, *Heat Transfer 1978, Proc. Sixth Int. Heat Transfer Conference*, vol. 2, Hemisphere, Washington, D.C., pp. 537–542.

Malik, M. R. and Pletcher, R. H. (1981). A Study of Some Turbulence Models for Flow and Heat Transfer in Ducts of Annular Cross-Section, *J. Heat Transfer*, vol. 103, pp. 146–152.

Marconi, F. (1980). Supersonic, Inviscid, Conical Corner Flowfields, *AIAA J.*, vol. 18, pp. 78–84.

Martin, E. D. and Lomax, H. (1975). Rapid Finite-Difference Computation of Subsonic and Slightly Supercritical Aerodynamic Flows, *AIAA J.*, vol. 13, pp. 579–586.

McDonald, H. (1970). Mixing Length and Kinematic Eddy Viscosity in a Low Reynolds Number Boundary Layer, United Aircraft Research Laboratory Report J2 14453-1, East Hartford, Connecticut.

McDonald, H. (1978). Prediction of Boundary Layers in Aircraft Gas Turbines, *The Aerothermodynamics of Aircraft Gas Turbine Engines*, Air Force Aero Propulsion Laboratory Report AFAPL-TR-78-52, Wright-Patterson Air Force Base, Ohio.

McDonald, H. and Briley, W. R. (1975). Three-dimensional Supersonic Flow of a Viscous or Inviscid Gas, *J. Comp. Phys.*, vol. 19, pp. 150–178.

McDonald, H. and Camerata, F. J. (1968). An Extended Mixing Length Approach for Computing the Turbulent Boundary Layer Development, *Proc. Computation of Turbulent Boundary Layers-1968 AFOSR-IFP-Stanford Conference*, vol. 1, Stanford University, California, pp. 83–98.

McDonald, H. and Fish, R. W. (1973). Practical Calculations of Transitional Boundary Layers, *Int. J. Heat Mass Transfer*, vol. 16, pp. 1729–1744.

McDonald, H. and Kreskovsky, J. P. (1974). Effect of Free Stream Turbulence on the Turbulent Boundary Layer, *Int. J. Heat Mass Transfer*, vol. 17, pp. 705–716.

McDonald, J. W., Denny, V. E., and Mills, A. F. (1972). Numerical Solutions of the Navier-Stokes Equations in Inlet Regions, *J. Appl. Mech.*, vol. 39, pp. 873–878.

McEligot, D. M., Smith, S. B., and Bankston, C. A. (1970). Quasi-Developed Turbulent Pipe Flow with Heat Transfer, *J. Heat Transfer*, vol. 92, pp. 641–650.

McGuirk, J. J. and Rodi, W. (1977). The Calculation of Three-dimensional Turbulent Free Jets, *Proc. Symposium on Turbulent Shear Flows*, Pennsylvania State University, University Park.

McLean, J. D. and Randall, J. L. (1979). Computer Program to Calculate Three-dimensional Boundary Layer Flows over Wings with Wall Mass Transfer, NASA CR-3123.

McRae, D. S. (1976). A Numerical Study of Supersonic Cone Flow at High Angle of Attack, AIAA Paper 76-97, Washington, D.C.

McRae, D. S. and Hussaini, M. Y. (1978). Numerical Simulation of Supersonic Cone Flow at High Angle of Attack, ICASE Report 78-21.

Middlecoff, J. F. and Thomas, P. D. (1979). Direct Control of the Grid Point Distribution in Meshes Generated by Elliptic Equations, AIAA Paper 79-1462, Williamsburg, Virginia.

Mills, R. D. (1965). Numerical Solutions of the Viscous Flow Equations for a Class of Closed Flows, *J. R. Aeronaut. Soc.*, vol. 69, pp. 714–718.

Minaie, B. N. and Pletcher, R. H. (1982). A Study of Turbulence Models for Predicting Round and Plane Heated Jets, *Heat Transfer 1982, Proc. Seventh Int. Heat Transfer Conference*, vol. 3, Hemisphere, Washington, D.C., pp. 383–388.

Miner, E. W. and Lewis, C. H. (1975). Hypersonic Ionizing Air Viscous Shock-Layer Flows over Sphere Cones, *AIAA J.*, vol. 13, pp. 80–88.

Mitchell, A. R. and Griffiths, D. F. (1980). *The Finite Difference Method in Partial Differential Equations*, Wiley, Chichester.

Miyakoda, K. (1962). Contribution to the Numerical Weather Prediction–Computation with Finite Difference, *Jap. J. Geophys.*, vol. 3, pp. 75–190.

Moore, J. and Moore, J. G. (1979). A Calculation Procedure for Three-dimensional Viscous, Compressible Duct Flow, Parts I and II, *J. Fluids Eng.*, vol. 101, pp. 415-428.

Moretti, G. (1969). Importance of Boundary Conditions in the Numerical Treatment of Hyperbolic Equations, *Phys. Fluids*, Supplement II, vol. 12, pp. 13–20.

Moretti, G. (1971). Complicated One-dimensional Flows, Polytechnic Institute of New York, PIBAL Report No. 71-25.

Moretti, G. (1974). On the Matter of Shock Fitting, Proc. Fourth Int. Conf. Num. Methods Fluid Dyn., Boulder, Colorado, *Lecture Notes in Physics*, vol. 35, Springer-Verlag, New York, pp. 287–292.

Moretti, G. (1975). A Circumspect Exploration of a Difficult Feature of Multidimensional Imbedded Shocks, *Proc. AIAA 2nd Computational Fluid Dynamics Conference*, Hartford, Connecticut, pp. 10–16.

Moretti, G. (1978). An Old-Integration Scheme for Compressible Flows Revisited, Refurbished and Put to Work, Polytechnic Institute of New York, M/AE Report No. 78-22.

Moretti, G. (1979). Conformal Mappings for the Computation of Steady Three-dimensional Supersonic Flows, *Numerical/Laboratory Computer Methods in Fluid Mechanics*, (A. A. Pouring and V. I. Shah, eds.), ASME, New York, pp. 13-28.

Moretti, G. and Abbett, M. (1966). A Time-Dependent Computational Method for Blunt Body Flows, *AIAA J.*, vol. 4, pp. 2136–2141.

Moretti, G. and Bleich, G. (1968). Three-dimensional Inviscid Flow about Supersonic Blunt Cones at Angle of Attack, Sandia Laboratories Report SC-RR-68-3728, Albuquerque, New Mexico.

Moretti, P. M. and Kays, W. M. (1965). Heat Transfer to a Turbulent Boundary Layer with Varying Free-Stream Velocity and Varying Surface Temperature–An Experimental Study, *Int. J. Heat Mass Transfer*, vol. 8, pp. 1187-1202.

Murman, E. M. and Cole, J. D. (1971). Calculation of Plane Steady Transonic Flows, *AIAA J.*, vol. 9, pp. 114–121.

Murphy, J. D. and Prenter, P. M. (1981). A Hybrid Computing Scheme for Unsteady Turbulent Boundary Layers, *Proc. Third Symposium on Turbulent Shear Flows*, University of California, Davis, pp. 8.26–8.34.

Murray, A. L. and Lewis, C. H. (1978). Hypersonic Three-dimensional Viscous Shock-Layer Flows over Blunt Bodies, *AIAA J.*, vol. 16, pp. 1279–1286.

Napolitano, M., Werle, M. J., and Davis, R. T. (1978). A Numerical Technique for the Triple-Deck Problem, AIAA Paper 78-1133, Seattle, Washington.

NASA (1972). Free Turbulent Shear Flows, vol. 1, in Proceedings of the Langley Working Conference on Free Turbulent Shear Flows, NASA SP-321.

Nardo, C. T. and Cresci, R. J. (1971). An Alternating Directional Implicit Scheme for Three-dimensional Hypersonic Flows, *J. Comp. Phys.*, vol. 8, pp. 268-284.

Nelson, R. M. and Pletcher, R. H. (1974). An Explicit Scheme for the Calculation of Confined Turbulent Flows with Heat Transfer, *Proc. 1974 Heat Transfer and Fluid Mechanics Institute*, Stanford University Press, Stanford, California, pp. 154–170.

Ng, K. H. and Spalding, D. B. (1972). Turbulence Model for Boundary Layers Near Walls, *Phys. Fluids*, vol. 15, pp. 20–30.

O'Brien, G. G., Hyman, M. A., and Kaplan, S. (1950). A Study of the Numerical Solution of Partial Differential Equations, *J. Math. Phys.*, vol. 29, pp. 223-251.

Orszag, S. A. and Israeli, M. (1974). Numerical Simulation of Viscous Incompressible Flows, *Annual Review of Fluid Mechanics*, vol. 6, Annual Reviews, Inc., Palo Alto, California, pp. 281-318.

Owczarek, J. A. (1964). *Fundamentals of Gas Dynamics*, International Textbook Company, Scranton, Pennsylvania.

Palumbo, D. J. and Rubin, E. L. (1972). Solution of the Two-dimensional, Unsteady Compressible Navier-Stokes Equations using a Second-Order Accurate Numerical Scheme, *J. Comp. Phys.*, vol. 9, pp. 466–495.

Pan, F. and Acrivos, A. (1967). Steady Flows in Rectangular Cavities, *J. Fluid Mech.*, vol. 28, pp. 643–655.

Patankar, S. V. (1975). Numerical Prediction of Three-dimensional Flows, *Studies in Convection: Theory, Measurement, and Applications*, (B. E. Launder, ed.), vol. 1, Academic, New York, pp. 1–78.

Patankar, S. V. (1980). *Numerical Heat Transfer and Fluid Flow*, Hemisphere, Washington, D.C.

Patankar, S. V. (1981). A Calculation Procedure for Two-dimensional Elliptic Situations, *Numer. Heat Transfer*, vol. 4, pp. 409–425.

Patankar, S. V. and Spalding, D. B. (1970). *Heat and Mass Transfer in Boundary Layers*, 2d ed., Intertext Books, London.

Patankar, S. V. and Spalding, D. B. (1972). A Calculation Procedure for Heat, Mass and Momentum Transfer in Three-dimensional Parabolic Flows, *Int. J. Heat Mass Transfer*, vol. 15, pp. 1787–1806.

Patankar, S. V., Pratap, V. S., and Spalding, D. B. (1974). Prediction of Laminar Flow and Heat Transfer in Helically Coiled Pipes, *J. Fluid Mech.*, vol. 62, pp. 539–551.

Patankar, S. V., Basu, D. K., and Alpay, S. A. (1977). Prediction of the Three-dimensional Velocity Field of a Deflected Turbulent Jet, *J. Fluids Eng.*, vol. 99, pp. 758–762.

Patankar, S. V., Ivanović, M., and Sparrow, E. M. (1979). Analysis of Turbulent Flow and Heat Transfer in Internally Finned Tubes and Annuli, *J. Heat Transfer*, vol. 101, pp. 29-37.

Patel, V. C. and Choi, D. H. (1979). Calculation of Three-dimensional Laminar and Turbulent Boundary Layers on Bodies of Revolution at Incidence, *Proc. Second Symposium on Turbulent Shear Flows*, Imperial College, London, pp. 15.14–15.24.

Peaceman, D. W. and Rachford, H. H. (1955). The Numerical Solution of Parabolic and Elliptic Differential Equations, *J. Soc. Ind. Appl. Math.*, vol. 3, pp. 28–41.

Peyret, R. and Viviand, H. (1975). Computation of Viscous Compressible Flows Based on the Navier-Stokes Equations, AGARD-AG-212.

Phillips, J. H. and Ackerberg, R. C. (1973). A Numerical Method for Integrating the Unsteady Boundary-Layer Equations When There are Regions of Backflow, *J. Fluid Mech.*, vol. 58, pp. 561–579.

Pletcher, R. H. (1969). On a Finite-Difference Solution for the Constant-Property Turbulent Boundary Layer, *AIAA J.*, vol. 7, pp. 305–311.

Pletcher, R. H. (1970). On a Solution for Turbulent Boundary Layer Flows with Heat Transfer, Pressure Gradient, and Wall Blowing or Suction, *Heat Transfer 1970, Proc. Fourth Int. Heat Transfer Conference*, vol. 1, Elsevier, Amsterdam.

Pletcher, R. H. (1971). On a Calculation Method for Compressible Boundary Layers with Heat Transfer, AIAA Paper 71-165, New York.

Pletcher, R. H. (1974). Prediction of Transpired Turbulent Boundary Layers, *J. Heat Transfer*, vol. 96, pp. 89–94.

Pletcher, R. H. (1976). Prediction of Turbulent Boundary Layers at Low Reynolds Numbers, *AIAA J.*, vol. 14, pp. 696–698.

Pletcher, R. H. (1978). Prediction of Incompressible Turbulent Separating Flow, *J. Fluids Eng.*, vol. 100, pp. 427–433.

Pletcher, R. H. and Dancey, C. L. (1976). A Direct Method of Calculating Through Separated Regions in Boundary Layer Flow, *J. Fluids Eng.*, vol. 98, pp. 568–572.

Polezhaev, V. I. (1967). Numerical Solution of the System of Two-dimensional Unsteady Navier-Stokes Equations for a Compressible Gas in a Closed Region, *Fluid Dyn.*, vol. 2, pp. 70–74.

Prandtl, L. (1926). Ueber die ausgebildete Turbulenz, *Proceedings of the 2nd International Congress for Applied Mechanics*, Zürich, pp. 62–74.

Pratap, V. S. and Spalding, D. B. (1976). Fluid Flow and Heat Transfer in Three-dimensional Duct Flows, *Int. J. Heat Mass Transfer*, vol. 19, pp. 1183–1188.

Pulliam, T. H. and Steger, J. L. (1978). On Implicit Finite-Difference Simulations of Three Dimensional Flow, AIAA Paper 78-10, Huntsville, Alabama.

Raetz, G. S. (1957). A Method of Calculating Three-dimensional Laminar Boundary Layers of Steady Compressible Flow, Report No. NAI-58-73 (BLC-114), Northrop Corporation.

Rai, M. M. (1982). A Philosophy for Construction of Solution Adaptive Grids, Ph.D. dissertation, Iowa State University, Ames.

Rai, M. M. and Anderson, D. A. (1980). Grid Evolution in Time Asymptotic Problems, *Numerical Grid Generation Techniques*, NASA Conference Publication 2166, pp. 409-430.

Rai, M. M. and Anderson, D. A. (1981). The Use of Adaptive Grids in Conjunction with Shock Capturing Methods, AIAA Paper 81-1012, St. Louis, Missouri.

Rai, M. M. and Anderson, D. A. (1982). Application of Adaptive Grids to Fluid Flow Problems with Asymptotic Solutions, *AIAA J.*, vol. 20, pp. 496-502.

Raithby, G. D. (1976). Skew Upstream Differencing Schemes for Problems Involving Fluid Flow, *Comput. Methods Appl. Mech. Eng.*, vol. 9, pp. 153-164.

Raithby, G. D. and Schneider, G. E. (1979). Numerical Solution of Problems in Incompressible Fluid Flow: Treatment of the Velocity-Pressure Coupling, *Numer. Heat Transfer*, vol. 2, pp. 417-440.

Raithby, G. D. and Schneider, G. E. (1980). The Prediction of Surface Discharge Jets by a Three-dimensional Finite-Difference Model, *J. Heat Transfer*, vol. 102, pp. 138-145.

Raithby, G. D. and Torrance, K. E. (1974). Upstream-Weighted Differencing Schemes and Their Application to Elliptic Problems Involving Fluid Flow, *Comput. Fluids*, vol. 2, pp. 191-206.

Rakich, J. V. (1978). Computational Fluid Mechanics–Course Notes, Department of Mechanical and Aerospace Engineering, North Carolina State University, Raleigh.

Ralston, A. (1965). *A First Course in Numerical Analysis*, McGraw-Hill, New York.

Reyhner, T. A. (1968). Finite-Difference Solution of the Compressible Turbulent Boundary Layer Equations, *Proc. Computation of Turbulent Boundary Layers–1968 AFOSR-IFP-Stanford Conference*, vol. 1, Stanford University, California, pp. 375-383.

Reyhner, T. A. and Flügge-Lotz, I. (1968). The Interaction of a Shock Wave with a Laminar Boundary Layer, *Int. J. Non-Linear Mech.*, vol. 3, pp. 173-199.

Reynolds, A. J. (1975). The Prediction of Turbulent Prandtl and Schmidt Numbers, *Int. J. Heat Mass Transfer*, vol. 18, pp. 1055-1069.

Richardson, L. F. (1910). The Approximate Arithmetical Solution by Finite Differences of Physical Problems Involving Differential Equations, with an Application to the Stresses in a Masonry Dam, *Philos. Trans. R. Soc. London, Ser. A*, vol. 210, pp. 307-357.

Richtmyer, R. D. (1957). *Difference Methods for Initial-Value Problems*, Interscience Publishers, New York.

Richtmyer, R. D. and Morton, K. W. (1967). *Difference Methods for Initial-Value Problems*, 2d ed., Interscience Publishers, Wiley, New York.

Rizzi, A. and Eriksson, L. E. (1981). Transfinite Mesh Generation and Damped Euler Equation Algorithm for Transonic Flow around Wing-Body Configurations, *Proc. AIAA 5th Computational Fluid Dynamics Conference*, Palo Alto, California, pp. 43-69.

Roache, P. J. (1972). *Computational Fluid Dynamics*, Hermosa, Albuquerque, New Mexico.

Roberts, G. O. (1971). Computational Meshes for Boundary Layer Problems, Proc. Second Int. Conf. Num. Methods Fluid Dyn., *Lecture Notes in Physics*, vol. 8, Springer-Verlag, New York, pp. 171-177.

Rodi, W. (1975). A Review of Experimental Data of Uniform Density Free Turbulent Boundary Layers, *Studies in Convection: Theory, Measurement, and Applications*, vol. 1, (B. E. Launder, ed.), Academic, New York.

Rotta, J. (1951). Statistische Theorie nichthomogener Turbulenz, *Z. Phys.*, vol. 129, pp. 547-572.

Rouleau, W. T. and Osterle, J. F. (1955). The Application of Finite Difference Methods to Boundary-Layer Type Flows, *J. Aero. Sci.*, vol. 22, pp. 249-254.

Rubbert, P. E. and Saaris, G. R. (1972). Review and Evaluation of a Three-dimensional Lifting Potential Flow Analysis Method for Arbitrary Configurations, AIAA Paper 72-188, San Diego, California.

Rubesin, M. W. (1976). A One-Equation Model of Turbulence for Use with the Compressible Navier-Stokes Equations, NASA TM X-73-128.

Rubesin, M. W. (1977). Numerical Turbulence Modeling, *AGARD Lecture Series No. 86 on Computational Fluid Dynamics*, pp. 3-1 to 3-37.

Rubin, S. G. (1981). A Review of Marching Procedures for Parabolized Navier-Stokes Equations, *Proceedings of Symposium on Numerical and Physical Aspects of Aerodynamic Flows*, Springer-Verlag, New York, pp. 171–186.

Rubin, S. G. and Harris, J. E. (1975). Numerical Studies of Incompressible Viscous Flow in a Driven Cavity, NASA SP-378.

Rubin, S. G. and Khosla, P. K. (1981). Navier-Stokes Calculations with a Coupled Strongly Implicit Method–I. Finite-Difference Solutions, *Comput. Fluids*, vol. 9, pp. 163–180.

Rubin, S. G. and Lin, T. C. (1971). Numerical Methods for Two- and Three-dimensional Viscous Flow Problems: Application to Hypersonic Leading Edge Equations, Polytechnic Institute of Brooklyn, PIBAL Report 71-8, Farmingdale, New York.

Rubin, S. G. and Lin, T. C. (1972). A Numerical Method for Three-dimensional Viscous Flow: Application to the Hypersonic Leading Edge, *J. Comp. Phys.*, vol. 9, pp. 339–364.

Rubin, S. G. and Lin, A. (1980). Marching with the PNS Equations, *Proceedings of 22nd Israel Annual Conference on Aviation and Astronautics*, Tel Aviv, Israel, pp. 60–61. See *Isr. J. Technol.*, vol. 18, 1980.

Rudman, S. and Rubin, S. G. (1968). Hypersonic Viscous Flow over Slender Bodies with Sharp Leading Edges, *AIAA J.*, vol. 6, pp. 1883–1889.

Rusanov, V. V. (1970). On Difference Schemes of Third Order Accuracy for Nonlinear Hyperbolic Systems, *J. Comp. Phys.*, vol. 5, pp. 507–516.

Saffman, P. G. and Wilcox, D. C. (1974). Turbulence Model Predictions for Turbulent Boundary Layers, *AIAA J.*, vol. 12, pp. 541–546.

Salas, M. D. (1975). The Anatomy of Floating Shock Fitting, *Proc. AIAA 2nd Computational Fluid Dynamics Conference*, Hartford, Connecticut, pp. 47–54.

Salas, M. D. (1979). Flow Properties for a Spherical Body at Low Supersonic Speeds, presented at the Symposium on Computers in Aerodynamics, Twenty-Fifth Anniversary of the Aerodynamics Laboratories, Polytechnic Institute of New York.

Saul'yev, V. K. (1957). On a Method of Numerical Integration of a Diffusion Equation, *Dokl. Akad. Nauk SSSR*, vol. 115, pp. 1077–1079. (In Russian)

Schiff, L. B. and Steger, J. L. (1979). Numerical Simulation of Steady Supersonic Viscous Flow, AIAA Paper 79-0130, New Orleans, Louisiana.

Schlichting, H. (1968). *Boundary-Layer Theory*, 6th ed., translated by J. Kestin, McGraw-Hill, New York.

Schlichting, H. (1979). *Boundary-Layer Theory*, 7th ed., translated by J. Kestin, McGraw-Hill, New York.

Schneider, G. E. and Zedan, M. (1981). A Modified Strongly Implicit Procedure for the Numerical Solution of Field Problems, *Numer. Heat Transfer*, vol. 4, pp. 1–19.

Schubauer, G. B. and Tchen, C. M. (1959). Section B of *Turbulent Flows and Heat Transfer*, vol. 5, in High Speed Aerodynamics and Jet Propulsion, Princeton University Press, New Jersey.

Schumann, U. (1980). Fast Elliptic Solvers and their Application in Fluid Dynamics, *Computational Fluid Dynamics*, Hemisphere, Washington, D.C., pp. 402–430.

Schwartztrauber, P. N. and Sweet, R. A. (1977). The Direct Solution of the Discrete Poisson Equation on a Disc, *SIAM J. Numer. Anal.*, vol. 5, pp. 900–907.

Shah, R. K. and London, A. L. (1978). *Laminar Flow Forced Convection in Ducts*, Academic, New York.

Shang, J. S. (1977). An Implicit-Explicit Method for Solving the Navier-Stokes Equations, AIAA Paper 77-646, Albuquerque, New Mexico.

Shang, J. S. and Hankey Jr., W. L. (1975). Supersonic Turbulent Separated Flows Utilizing the Navier-Stokes Equations, *Flow Separation*, AGARD-CCP-168.

Shankar, V. (1981). Treatment of Conical and Nonconical Supersonic Flows by an Implicit Marching Scheme Applied to the Full Potential Equation, *Proc. ASME/AIAA Conference on Computers in Flow Predictions and Fluid Dynamics Experiments*, Washington, D.C., pp. 163–170.

Shankar, V. and Chakravarthy, S. (1981). An Implicit Marching Procedure for the Treatment of Supersonic Flow Fields using the Conservative Full Potential Equation, AIAA Paper 81-1004, Palo Alto, California.

Shankar, V. and Osher, S. (1982). An Efficient Full Potential Implicit Method Based on Characteristics for Analysis of Supersonic Flows, AIAA Paper 82-0974, St. Louis, Missouri.

Shapiro, A. H. (1953). *The Dynamics and Thermodynamics of Compressible Fluid Flow*, vol. I, Ronald Press, New York.

Sichel, M. (1963). Structure of Weak Non-Hugoniot Shocks, *Phys. Fluids*, vol. 6, pp. 653–663.

Smith, R. E. and Weigel, B. L. (1980). Analytic and Approximate Boundary Fitted Coordinate Systems for Fluid Flow Simulation, AIAA Paper 80-0192, Pasadena, California.

Southwell, R. V. (1940). *Relaxation Methods in Engineering Science*, Oxford University Press, London.

Spalding, D. B. (1972). A Novel Finite-Difference Formulation for Differential Expressions Involving Both First and Second Derivatives, *Int. J. Numer. Methods Eng.*, vol. 4, pp. 551–559.

Srivastava, B. N., Werle, M. J., and Davis, R. T. (1978). Viscous Shock-Layer Solutions for Hypersonic Sphere Cones, *AIAA J.*, vol. 16, pp. 137–144.

Srivastava, B. N., Werle, M. J., and Davis, R. T. (1979). Numerical Solutions of Hypersonic Viscous Shock-Layer Equations, *AIAA J.*, vol. 17, pp. 107–110.

Steger, J. L. (1977). Implicit Finite-Difference Simulation of Flow about Arbitrary Geometries with Application to Airfoils, AIAA Paper 77-665, Albuquerque, New Mexico.

Steger, J. L. (1978). Coefficient Matrices for Implicit Finite-Difference Solution of the Inviscid Fluid Conservation Law Equations, *Comput. Methods Appl. Mech. Eng.*, vol. 13, pp. 175–188.

Steger, J. L. (1981). A Preliminary Study of Relaxation Methods for the Inviscid Conservative Gasdynamics Equations using Flux Splitting, NASA Contractor Report 3415.

Steger, J. L. and Kutler, P. (1976). Implicit Finite-Difference Procedures for the Computation of Vortex Wakes, AIAA Paper 76-385, San Diego, California.

Steger, J. L. and Sorenson, R. L. (1980). Use of Hyperbolic Partial Differential Equations to Generate Body Fitted Coordinates, *Numerical Grid Generation Techniques*, NASA Conference Publication 2166, pp. 463–478.

Steger, J. L. and Warming, R. F. (1979). Flux Vector Splitting of the Inviscid Gasdynamic Equations with Application to Finite-Difference Methods, NASA TM D-78605.

Steinhoff, J. and Jameson, A. (1981). Multiple Solutions of the Transonic Potential Flow Equation, *Proc. AIAA 5th Computational Fluid Dynamics Conference*, Palo Alto, California, pp. 347–353.

Stephenson, P. L. (1976). A Theoretical Study of Heat Transfer in Two-dimensional Turbulent Flow in a Circular Pipe and Between Parallel and Diverging Plates, *Int. J. Heat Mass Transfer*, vol. 19, pp. 413–423.

Stewartson, K. (1974). Multistructured Boundary Layers on Flat Plates and Related Bodies, *Advances in Applied Mechanics*, vol. 14, Academic, New York, pp. 145–239.

Stone, H. L. (1968). Iterative Solution of Implicit Approximations of Multidimensional Partial Equations, *SIAM J. Numer. Anal.*, vol. 5, pp. 530–558.

Swaminathan, S., Kim, M. D., and Lewis, C. H. (1983). Three-dimensional Nonequilibrium Viscous Shock-Layer Flows over Complex Geometries, AIAA Paper 83-0212, Reno, Nevada.

Szema, K. Y. and Lewis, C. H. (1980). Three-dimensional Hypersonic Laminar, Transitional and/or Turbulent Shock-Layer Flows, AIAA Paper 80-1457, Snowmass, Colorado.

Tannehill, J. C. and Anderson, D. A. (1980). Computation of Three-dimensional Supersonic Viscous Flows in Internal Corners, Technical Report AFWAL-TR-80-3017.

Tannehill, J. C., Vigneron, Y. C., and Rakich, J. V. (1978). Numerical Solution of Two-dimensional Turbulent Blunt Body Flows with an Impinging Shock, AIAA Paper 78-1209, Seattle, Washington.

Tannehill, J. C., Holst, T. L., and Rakich, J. V. (1975). Numerical Computation of Two-dimensional Viscous Blunt Body Flows with an Impinging Shock, AIAA Paper 75-154, Pasadena, California.

Tannehill, J. C., Venkatapathy, E., and Rakich, J. V. (1982). Numerical Solution of Supersonic Viscous Flow over Blunt Delta Wings, *AIAA J.*, vol. 20, pp. 203–210.

Taylor, A. E. (1955). *Advanced Calculus*, Ginn and Company, Boston.

Telionis, D. P. and Tsahalis, D. Th. (1976). Unsteady Turbulent Boundary Layers and Separation, *AIAA J.*, vol. 14, pp. 468–474.

Telionis, D. P., Tsahalis, D. Th., and Werle, M. J. (1973). Numerical Investigation of Unsteady Boundary-Layer Separation, *Phys. Fluids*, vol. 16, pp. 968–973.

Thareja, R., Szema, K. Y., and Lewis, C. H. (1982). Effects of Chemical Equilibrium on Three-dimensional Viscous Shock-Layer Analysis of Hypersonic Laminar or Turbulent Flows, AIAA Paper 82-0305, Orlando, Florida.

Thom, A. and Apelt, C. J. (1961). *Field Computations in Engineering and Physics*, C. Van Nostrand, Princeton, New Jersey.

Thomas, L. H. (1949). Elliptic Problems in Linear Difference Equations over a Network, *Watson Sci. Comput. Lab. Rept.*, Columbia University, New York.

Thomas, P. D. and Lombard, C. K. (1978). The Geometric Conservation Law–A Link between Finite-Difference and Finite-Volume Methods of Flow Computation on Moving Grids, AIAA Paper 78-1208, Seattle, Washington.

Thomas, P. D., Vinokur, M., Bastianon, R. A., and Conti, R. J. (1972). Numerical Solution for Three-dimensional Inviscid Supersonic Flow, *AIAA J.*, vol. 10, pp. 887–894.

Thommen, H. U. (1966). Numerical Integration of the Navier-Stokes Equations, *Z. Angew. Math. Phys.*, vol. 17, pp. 369–384.

Thompson, J. F. (1980). Numerical Solution of Flow Problems using Body Fitted Coordinate Systems, *Computational Fluid Dynamics*, vol. 1, (W. Kollmann, ed.), Hemisphere, Washington, D.C.

Thompson, J. F., Thames, F. C., and Mastin, C. W. (1974). Automatic Numerical Generation of Body-Fitted Curvilinear Coordinate System for Field Containing any Number of Arbitrary Two-dimensional Bodies, *J. Comp. Phys.*, vol. 15, pp. 299–319.

Thompson, J. F., Thames, F. C., Mastin, C. W., and Shanks, S. P. (1975). Use of Numerically Generated Body-Fitted Coordinate Systems for Solution of the Navier-Stokes Equations, *Proc. AIAA 2nd Computational Fluid Dynamics Conference*, Hartford, Connecticut, pp. 68–80.

Ting, L. (1965). On the Initial Conditions for Boundary Layer Equations, *J. Math. Phys.*, vol. 44, pp. 353–367.

Tsahalis, D. Th. and Telionis, D. P. (1974). Oscillating Laminar Boundary Layers and Unsteady Separation, *AIAA J.*, vol. 12, pp. 1469–1475.

van Driest, E. R. (1951). Turbulent Boundary Layer in Compressible Fluids, *J. Aero. Sci.*, vol. 18, pp. 145–160.

van Driest, E. R. (1952). Investigation of Laminar Boundary Layer in Compressible Fluids using the Crocco Method, NACA TN-2597.

van Driest, E. R. (1956). On Turbulent Flow Near a Wall, *J. Aero. Sci.*, vol. 23, pp. 1007–1011.

Van Dyke, M. (1969). Higher-Order Boundary-Layer Theory, *Annual Review of Fluid Mechanics*, vol. 1, Annual Reviews, Inc., Palo Alto, California, pp. 265–292.

Varga, R. S. (1962). *Matrix Iterative Numerical Analysis*, Wiley, New York.

Vigneron, Y. C., Rakich, J. V., and Tannehill, J. C. (1978a). Calculation of Supersonic Viscous Flow over Delta Wings with Sharp Subsonic Leading Edges, AIAA Paper 78-1137, Seattle, Washington.

Vigneron, Y. C., Rakich, J. V., and Tannehill, J. C. (1978b). Calculation of Supersonic Viscous Flow over Delta Wings with Sharp Subsonic Leading Edges, NASA TM-78500.

Vinokur, M. (1974). Conservation Equations of Gas-Dynamics in Curvilinear Coordinate Systems, *J. Comp. Phys.*, vol. 14, pp. 105–125.

Viviand, H. (1974). Conservative Forms of Gas Dynamic Equations, *La Recherche Aérospatiale*, No. 1974-1, pp. 65–68.

von Neumann, J. and Richtmyer, R. D. (1950). A Method for the Numerical Calculation of Hydrodynamic Shocks, *J. Appl. Phys.*, vol. 21, pp. 232–237.

Wachspress, E. L. (1966). *Iterative Solution of Elliptic Systems*, Prentice-Hall, Englewood Cliffs, New Jersey.

Wang, K. C. (1971). On the Determination of the Zones of Influence and Dependence for Three-dimensional Boundary-Layer Equations, *J. Fluid Mech.*, vol. 48, pt. 2, pp. 397-404.

Wang, K. C. (1972). Separation Patterns of Boundary Layer over an Inclined Body of Revolution, *AIAA J.*, vol. 10, pp. 1044-1050.

Wang, K. C. (1973). Three-dimensional Laminar Boundary Layer over a Body of Revolution at Incidence, Part VI: General Methods and Results of the Case at High Incidence, MML TR 73-02c, Martin Marietta Laboratories, Baltimore, Maryland.

Wang, K. C. (1974). Boundary Layer over a Blunt Body at High Incidence with an Open-Type of Separation, *Proc. Roy. Soc. London, Ser. A*, vol. 340, pp. 33-55.

Wang, K. C. (1975). Boundary Layer over a Blunt Body at Low Incidence with Circumferential Reversed Flow, *J. Fluid Mech.*, vol. 72, pt. 1, pp. 49-65.

Warming, R. F. and Beam, R. M. (1975). Upwind Second-Order Difference Schemes and Applications in Unsteady Aerodynamic Flows, *Proc. AIAA 2nd Computational Fluid Dynamics Conference*, Hartford, Connecticut, pp. 17-28.

Warming, R. F. and Beam, R. M. (1977). On the Construction and Application of Implicit Factored Schemes for Conservation Laws, Symposium on Computational Fluid Dynamics, New York. See *SIAM-AMS Proceedings*, vol. 11, 1978, pp. 85-129.

Warming, R. F. and Hyett, B. J. (1974). The Modified Equation Approach to the Stability and Accuracy Analysis of Finite-Difference Methods, *J. Comp. Phys.*, vol. 14, pp. 159-179.

Warming, R. F., Kutler, P., and Lomax, H. (1973). Second- and Third-Order Noncentered Difference Schemes for Nonlinear Hyperbolic Equations, *AIAA J.*, vol. 11, pp. 189-196.

Waskiewicz, J. D., Murray, A. L., and Lewis, C. H. (1978). Hypersonic Viscous Shock-Layer Flow over a Highly Cooled Sphere, *AIAA J.*, vol. 16, pp. 189-192.

Weinstock, R. (1952). *Calculus of Variations with Applications to Physics and Engineering*, McGraw-Hill, New York.

Welch, J. E., Harlow, F. H., Shannon, J. P., and Daly, B. J. (1966). The MAC Method, Los Alamos Scientific Laboratory Report LA-3425, Los Alamos, New Mexico.

Werle, M. J. and Bertke, S. D. (1972). A Finite-Difference Method for Boundary Layers with Reverse Flow, *AIAA J.*, vol. 10, pp. 1250-1252.

Werle, M. J. and Dwoyer, D. L. (1972). Laminar Hypersonic Interacting Boundary Layers: Subcritical Branching in the Strong Interaction Regime, ARL 72-0011, Wright-Patterson Air Force Base, Dayton, Ohio.

Werle, M. J. and Vatsa, V. N. (1974). A New Method for Supersonic Boundary Layer Separations, *AIAA J.*, vol. 12, pp. 1491-1497.

Werle, M. J. and Verdon, J. M. (1979). Solutions for Supersonic Trailing Edges Including Separation, AIAA Paper 79-1544, Williamsburg, Virginia.

Werle, M. J., Polak, A., and Bertke, S. D. (1973). Supersonic Boundary-Layer Separation and Reattachment–Finite Difference Solutions, Report No. AFL 72-12-1, Department of Aerospace Engineering, University of Cincinnati, Ohio.

White, Jr., A. B. (1982). On the Numerical Solution of Initial/Boundary-Value Problems in One Space Dimension, *SIAM J. Numer. Anal.*, vol. 19, pp. 683-697.

White, F. M. (1974). *Viscous Fluid Flow*, McGraw-Hill, New York.

Whitham, G. B. (1974). *Linear and Nonlinear Waves*, Wiley, New York.

Whittaker, E. T. and Watson, G. N. (1927). *A Course in Modern Analysis*, 4th ed. (reprinted 1962, Cambridge University Press).

Wieghardt, K. and Tillman, W. (1951). On the Turbulent Friction Layer for Rising Pressure, NACA TM-1314.

Wigton, L. B. and Holt, M. (1981). Viscous-Inviscid Interaction in Transonic Flow, AIAA Paper 81-1003, Palo Alto, California.

Wilcox, D. C. and Traci, R. M. (1976). A Complete Model of Turbulence, AIAA Paper 76-351, San Diego, California.

Williams, J. C. (1977). Incompressible Boundary Layer Separation, *Annual Review of Fluid Mechanics*, vol. 9, Annual Reviews, Inc., Palo Alto, California, pp. 113–144.

Winslow, A. (1966). Numerical Solution of the Quasi-linear Poisson Equation, *J. Comp. Phys.*, vol. 1, pp. 149–172.

Wolfstein, M. (1969). The Velocity and Temperature Distribution in One-dimensional Flow with Turbulence Augmentation and Pressure Gradient, *Int. J. Heat Mass Transfer*, vol. 12, pp. 301–318.

Wornom, S. F. (1977). A Critical Study of Higher-Order Numerical Methods for Solving the Boundary-Layer Equations, *Proc. AIAA 3rd Computational Fluid Dynamics Conference*, Albuquerque, New Mexico.

Wu, J. C. (1961). On the Finite-Difference Solution of Laminar Boundary Layer Problems, *Proc. 1961 Heat Transfer and Fluid Mechanics Institute*, Stanford University Press, Stanford, California.

Wylie Jr., C. R. (1951). *Advanced Engineering Mathematics*, McGraw-Hill, New York.

Yanenko, N. N. (1971). *The Method of Fractional Steps: The Solution of Problems of Mathematical Physics in Several Variables*, (M. Holt, ed.), Springer-Verlag, New York.

Yanenko, N. N., Kovenya, V. M., Tarnavsky, G. A., and Cherny, S. G. (1980). Economical Methods for Solving the Problems of Gas Dynamics, Proc. Seventh Int. Conf. Num. Methods Fluid Dyn., *Lecture Notes in Physics*, vol. 141, Springer-Verlag, New York, pp. 448–453.

Young, D. (1954). Iterative Methods for Solving Partial Difference Equations of Elliptic Type, *Trans. Amer. Math. Soc.*, vol. 76, pp. 92–111.

Zachmanoglou, E. C. and Thoe, D. W. (1976). *Introduction to Partial Differential Equations with Applications*, Williams & Wilkins, Baltimore.

INDEX